Edward Spon, Oliver Byrne

Spons' Dictionary of Engineering
Volume I

AF181040

Salzwasser

Edward Spon, Oliver Byrne

Spons' Dictionary of Engineering
Volume I

1. Auflage | ISBN: 978-3-84605-254-9

Erscheinungsort: Frankfurt, Deutschland

Erscheinungsjahr: 2020

Salzwasser Verlag GmbH

Reprint of the original, first published in 1869.

SPONS'

DICTIONARY OF ENGINEERING.

VOLUME I.

SPONS'

DICTIONARY OF ENGINEERING,

Civil, Mechanical, Military, and Naval;

WITH TECHNICAL TERMS

DICTIONARY OF ENGINEERING.

ABACUS, *and Instruments for Calculating.*

ABACUS. FR., *Tableau servant à calculer—Abaque;* GER., *Rechenbrett, Säulenplatte;* ITAL., *Abaco;* SPAN., *Abaco.*

A variety of more or less simple mechanical contrivances have been invented, almost from time immemorial, to simplify and facilitate the ordinary calculations of daily life; most of these contrivances, besides having little real utility, are so well known, and have been so frequently described, that a detailed description of them is here unnecessary; it is sufficient to say that the Chinese and other inhabitants of Central Asia still use these simple mechanical aids in performing calculations.

Napier devised a sort of abacus, or instrument for calculating, based upon the principle of rendering movable the columns of the ordinary multiplication table. The rods or bones of which this abacus is composed are termed Napier's rods or bones.

It is necessary to observe that each rod is divided into nine squares, and each square into two triangles, by a diagonal line drawn from the left lower angle. Fig. 1 represents one of these rods with the figure 3 in the right-hand triangle of the first square, zero and single figures, 1, 2, 4, &c., to 9, being always placed in a similar position on the other rods.

In Fig. 1 the second square from the top contains 6, or twice 3; the third 9, or three times 3; the fourth 12, or four times 3; and so on, to nine times 3, or 27; whence the figure at the top, and multiples expressed by single digits, are placed in the right-hand triangles, and the tens in the left-hand triangles.

It is clear that the faces which bear zero on the top must necessarily bear zero on all the triangles. For example, take three of the bones bearing on the top of their faces the figures 1, 9, and 2 respectively, and place them together, as shown in Fig. 2: the first line will read 192, the second 384 = 192 × 2, with this proviso, that *the figures placed between the same diagonal lines*, as 1, 2, *have to be added;* three times 192 reads 576 when the 3 and 2 between the diagonal lines are added. and so on to 9 times 192, which reads 1728 when the figures 9, 8, and 1, 1, between the diagonal lines are added. The apparent difficulty of this arrangement is by no means such as to deter any one from using Napier's rods, as a short practice suffices to render the use of those rods easy.

1.

2.

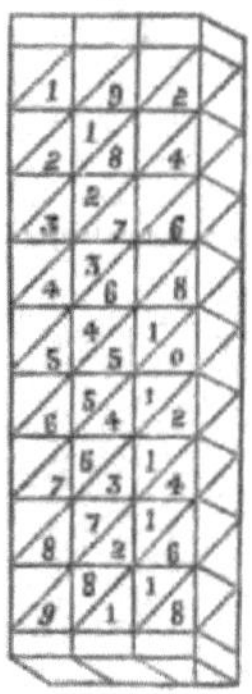

3.

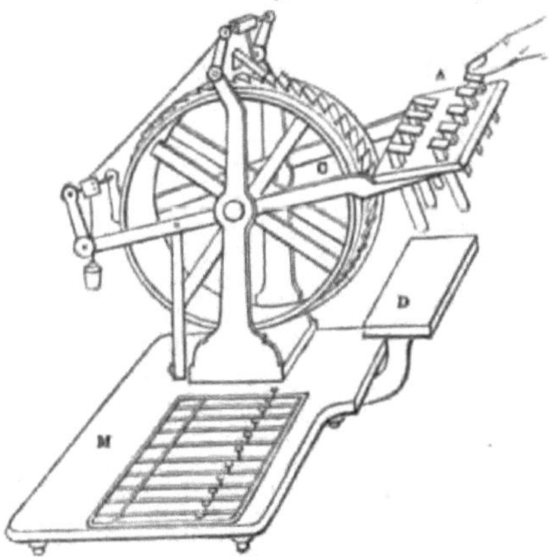

Augustus Barre's calculating instrument, Fig. 3, has a finger-board, A, furnished with ten keys of unequal length, each of which, in consecutive order, is marked with one of the figures 1, 2, 3, 4, &c., to 10. When the key which bears the number that an operator desires to add to another number, is pressed by the finger until it comes in contact with the stationary table D, the range of the angular motion given to the wheel C, which indicates the number to be added, is determined by the length of the key operated upon. The effect of pressing a finger upon the key, and lever C with which it is connected, is therefore to impart an angular motion to both the wheel and the lever proportional to the digit which the key bears, that is, this motion will be smaller for the lower and greater for the higher digits. The moving of the key starts a click or catch, connected with a small weight, in consequence of which the wheel begins to turn; the motion is stopped as soon as the key pressed upon by the finger resumes its primitive position of rest.

Calculating Instruments of Dubois and of Dunlop.—The calculating instruments of Dubois assist in performing the elementary operations of arithmetic. In adding and subtracting, Dubois applies a series of small movable rules, upon the surface of which is painted or engraved the nine digits. In performing multiplication and division, Dubois makes use of an arrangement previously applied by Petit, in 1671. This arrangement consisted of engraving the figures upon rectilinear rods, similar to Napier's rods. There may be a real merit in applying old well-known methods in a useful manner, but this remark does not apply to Dubois' adaptation of Petit's arrangement; for Dubois' instrument, on account of its size, is far less commodious in use than many of the antique instruments which have been employed to effect a similar purpose.

Dunlop has introduced two calculators; his calculator No. 1, to perform multiplication and division, consists of a series of numbers, arranged in a tabular form made up of movable parts, so adjusted as to make it quite easy to find the simple multiples of any given set of figures; in reality this contrivance is an extension of Napier's method.

Form I., page 3, represents, on the left side, the inner rear side of the first page of Dunlop's tabulated form; the right-hand side represents the movable slips, which are partly folded up and covered by each other, while on the right ends they are open and a digit is to be seen thereon. The use of the said slips is sufficiently indicated by the printed inscriptions which they bear. It must be noted that the figures printed between brackets are in Dunlop's table distinguished by a red colour.

Ex.—Suppose one desires to know the price of 32½ yards of silk at 13*s.* 2*d.* a-yard; to perform this calculation one operates in the following manner: uncover the slip of the tens, No. 3, the slip 2 of the units, and slip ½ of parts of units, the end of which is visible in A, Form I.; place the sheets which cover those you are in want of for your purpose upon the left-hand sheet, and Dunlop's calculator will then exhibit itself to you as seen in Form II. At B C you read the multiplicand 32½, and in order to find the product by 13*s.* 2*d.*, look first upon the column marked 13*s.* on the lowest line; add together the two figures 0*l.* 6*s.* 6*d.*, 1*l.* 6*s.* 0*d.*, 19*l.* 10*s.* 0*d.*, total 21*l.* 2*s.* 6*d.*; this is the price of 32½ yards at 13*s.* Look next to the column marked 2*d.*, which gives a result of 5*s.* 5*d.*; add this to the former result, and the sum total will be found to be 21*l.* 7*s.* 11*d.*

The Dunlop calculator, No. 2, based upon the same principle, is designed for the calculation of weights.

The calculator, No. 3, is an instrument designed to facilitate the addition of partial products in cases when the multiplier is composed of several numbers; it consists of a box containing small slate rules movable in horizontal grooves; the partial products are written down upon these small slates, and since these are made to slide, it is easy to place the products in the order they respectively must occupy.

Counters for Public Carriages.—There exist two kinds of these instruments: namely, *graphical counters* and *purely mechanical ones*, all of which require the use of clockwork. The so-called graphical counters are so arranged as to note down the information, which it is of interest to the parties concerned to know, upon a piece of paper moved by clockwork; they are in fact self-registering instruments; but it is quite evident that such an instrument, since it requires the daily changing of the sheet of paper, would be a very inconvenient instrument to be applied by the proprietor of a large number of public carriages, as it is clear that it would require a considerable number of clerks to take down every night the papers put up in each respective carriage in the morning, and to note down the particulars registered automatically. The mechanical counters, on the contrary, are so constructed that it is possible to read off at a glance, by means of a mechanical contrivance of more or less complicate structure, the work performed during the day and the money received and to be accounted for. There exist contrivances of this kind wherein the graphical and mechanical arrangements are combined.

4.

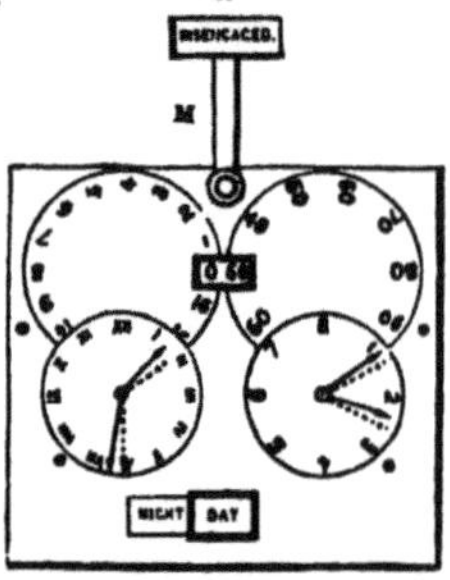

Various instruments both graphical and mechanical have been invented and used as automatic counters in carriages. We describe that of Bertrand and Addenet. The counter contrived by them is represented in Figs. 4, 5; the instrument is fixed on the carriage, behind the coachman's box, while towards the passengers, inside the carriage, are exhibited—1, a dial-plate of a clock or timepiece, indicating the time; 2, a dial-plate, provided with hands, indicating the number of kilometres, or distance run over, the hands being mechanically connected with the carriage wheels; 3, on the top of the said dial-plates a rectangular opening, showing the amount of the fare to be paid to the coachman; and 4, at the bottom of the instrument another opening is exhibited, bearing the words:—

or

Night, | Day,

since this refers to a different tariff of fares.

The figures which indicate the amount of the fare, are engraved on two discs, one of which, in the Paris carriages, marks the centimes, the other the francs. Motion is imparted to these discs by means of clockwork, from the timepiece already alluded to. At the time of starting, the counter indicating the amount of fare, reads, 0 francs 50 centimes, and the amount of fare due, runs

TABULATED FORM.—I.

A

										Parts of units, from 1/32nd to 7/8ths.								1/32	1/16	1/8	1/4	3/8	1/2	5/8	3/4	7/8
										Units, from one to nine.								1	2	3	4	5	6	7	8	9
										Tens, from ten to ninety.								1	2	3	4	5	6	7	8	9
										Hundreds, from 100 to 900.								1	2	3	4	5	6	7	8	9
										Thousands, from 1000 to 9000.								1	2	3	4	5	6	7	8	9
										Tens of thousands, from 10,000 to 90,000.								1	2	3	4	5	6	7	8	9
Rates	1/8 ..	(1/32 d.) 2d.	(1/16 d.) 3d.	(1/8 d.) 4d.	(1/4 d.) 5d.	(3/8 d.) 6d.	(1/2 d.) 7d.	(5/8 d.) 8d.	(3/4 d.) 9d.	(7/8 d.) 10d.	(1d.) 11d.	(2s.) 8s.	(3s.) 9s.	(4s.) 10s.	(5s.) 11s.	(6s.) 12s.	(7s.) 13s.									

C

TABULATED FORM.—II.

		£ s. d.	£ s. d.	£ s. d.	£ s. d.	£ s. d.	£ s. d.	£ s. d.	£ s. d.	£ s. d.	£ s. d.	£ s. d.	£ s. d.	£ s. d.	£ s. d.	£ s. d.	£ s. d.									
	1/2	0 0 1	0 0 1½	0 0 2	0 0 2½	0 0 3	0 0 3½	0 0 4	0 0 4½	0 0 5	0 0 5½	0 4 0	0 4 6	0 5 0	0 5 6	0 6 0	0 6 6						1/2	5/8	3/4	7/8
	2	0 0 4	0 0 6	0 0 8	0 0 10	0 1 0	0 1 2	0 1 4	0 1 6	0 1 8	0 1 10	0 16 0	0 18 0	1 0 0	1 2 0	1 4 0	1 6 0		2	3	4	5	6	7	8	9
	30	0 5 0	0 7 6	0 10 0	0 12 6	0 15 0	0 17 6	1 0 0	1 2 6	1 5 0	1 7 6	12 0 0	13 10 9	15 0 0	16 10 0	18 0 0	19 10 0			3	4	5	6	7	8	9
	B									Hundreds, from 100 to 900.								1	2	3	4	5	6	7	8	9
										Thousands, from 1000 to 9000.								1	2	3	4	5	6	7	8	9
										Tens of thousands, from 10,000 to 90,000.								1	2	3	4	5	6	7	8	9
Rates	..	(1/32 d.) 2d.	(1/16 d.) 3d.	(1/8 d.) 4d.	(1/4 d.) 5d.	(3/8 d.) 6d.	(1/2 d.) 7d.	(5/8 d.) 8d.	(3/4 d.) 9d.	(7/8 d.) 10d.	(1d.) 11d.	(2s.) 8s.	(3s.) 9s.	(4s.) 10s.	(5s.) 11s.	(6s.) 12s.	(7s.) 13s.									

up 10 centimes at a time. It is possible by this arrangement to use the instrument even for short distances. When a carriage is disengaged, there is exhibited outside, on the top, a small flag, M, *m*, bearing the word *disengaged*. The putting up of this flag by the coachman, has the effect, by means of proper mechanical contrivances, to disconnect the toothed wheelwork of the discs, and to bring the figures connected therewith, back to the first reading. If, with a view to fraud, the coachman should neglect this operation, the instrument is so arranged that it guards against his fraudulent intention, and compels him to account for all money received, on his return home.

5.

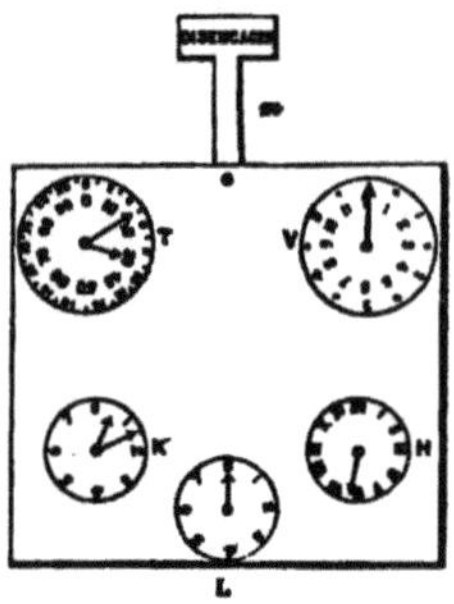

Fig. 5 represents an instrument employed in Paris omnibuses, and is a modification of that shown in Fig. 4.

T is a dial that shows the sum total; the outside figures on this dial represent francs, and the inner figures show the centimes.

L, a dial indicating how often the carriage has passed the barriers, and consequently shows the number of return tolls paid back to the driver, on his returning into the city.

V, a dial indicating the number of passengers; on the outside of this dial, figures show the total amount in francs, to which the constant 0.50 is added.

H, a dial indicating the time during which the carriage has been at rest.

K, a dial indicating the number of kilometres, or distance run over, while the carriage was unoccupied.

It is stated that the movements of the hands, of these different dials, are extremely simple, and may always be relied upon.

The mode of transmitting the motion from the carriage wheels to the instrument is new, and may be described thus:—

Suppose a wire placed, so as to be quite free to move, but in the most limited space possible, inside a flexible sheath; the wire is, in this particular case, spiral, of very tightly woven steel; the inside wire may then be taken to be a single central wire, which preserves its primitive length irrespective of the curve which the sheath makes, and also, irrespective of the variations of the ends of the sheath. The sheath, fixed to the counting instrument on the one hand, and to the axle of the wheels of the carriage on the other, has more or less play, according to the play of the springs which support the carriage; but the central wire always preserves the same length, and it is this wire which transmits to the kilometrical, or distance counter, the motion of the carriage-wheels. One of the front wheels of the carriage bears upon its stock an eccentric, which, at every revolution of the wheel, transmits an alternate motion to a bolt or beam placed on the axle-tree; upon this beam is fixed a piece which catches the wire contained in the sheath, and by this means a rack-wheel belonging to the distance counter is moved. Upon the evidence of several, who have been for many years connected with the Public Carriage Department in Paris, this arrangement, and every contrivance connected therewith, is pronounced to answer the purpose for which it was designed, in every respect. See COUNTER, *steam-engine*. PLANIMETER. SLIDE RULE.

ABATTIS. FR., *Abattis;* GER., *Verhau, Verhack;* ITAL., *Abbattuta, Tagliata;* SPAN., *Abatis.*

An abattis is generally constructed with large branches of trees, sharpened and laid with the points outward, in front of a fortification or any other position, to obstruct the approach of assailants. Abattis should be so placed as not to be exposed to the fire of artillery. In *redoubts* or entrenchments they are usually fixed in an upright position against the *counterscarp*, or at the foot of the *glacis*, the plane of which last is broken so as to permit of their being laid out of the enemy's sight, and so as not to interfere with the musketry fire from the *parapet* in their rear. See Fig. 6. Abattis is an excellent mode of blocking up a road; and when the branches are well

6.

and properly placed, and interwoven one with the other, the disengagement of them is extremely difficult, and to form an opening sufficient for the passage of artillery, or even of cavalry, requires a long time. An abattis can easily be made by a few men, with half-a-dozen felling-axes and a cross-cut saw, and in a short space of time, if trees of sufficient size are near, or on the spot. It is more easily formed and gives more effective defence than *palisades*.

An abattis should not be planted out of musketry-range; for this and all other obstacles are to break up the order of the enemy's advance, to impede and keep him under musketry fire. The application of the abattis should be considered as purely local and not one of the common resources for securing entrenchments, such as *palisades*, *chevaux-de-frise*, and *fougasses*, the materials for the construction of these last being capable of conveyance from a distance. Hence localities may enable the engineer to obstruct a road, by dragging trees from the hedge-side and connecting the defences of a position, by levelling groups of trees with their branches towards the enemy.

Shrubby trees are not adapted to form a good abattis; they are easily forced and drawn out by the hand. Heavy trees with the trunk cut half-through form insurmountable obstacles; this last is called an *Entanglement*. See Fig. 7.

7.

ABATTOIR. Fr., *Abattoir*; Ger., *Schlachthaus*; Ital., *Macello*; Span., *Matadero*.

A public slaughter-house in a city is termed an abattoir.

ABSTRACTING DIMENSIONS. Fr., *Epitomé pour servir de guide à bien prendre les dimensions*; Ger., *Einen Gegenstand aufmessen, Masse Zusammenstellen*; Ital., *Elenco delle misure*; Span., *Resúmen de dimensiones*. See Labour, *trained*.

ABUTMENT. Fr., *Culée, Butée*; Ger., *Widerlager*; Ital., *Coscia*; Span., *Estribo, Botarel*. See Arch.

ABUTTING JOINT. Fr., *Joint plat*; Ger., *Stumpfe Fuge*; Ital., *Commettitura piana*: Span., *Juntura plana*.

In carpentry, an abutting or a *butt joint* is a joint in which the plane of the joint is at right angles to the fibres, and the fibres of both pieces in the same straight line.

ACCELERATION. Fr., *Accélération*; Ger., *Beschleunigung*; Ital., *Accelerazione*; Span., *Aceleracion*.

Acceleration is the increase of velocity in a moving body, caused by the continued addition of motive force. When bodies in motion pass through equal spaces in equal times, that is, when the velocity of the body is the same during the period that the body is in motion, it is termed uniform motion, of which we have a familiar instance in the motion of the hands of a clock over its face; but a more correct illustration is the revolution of the earth on its axis. In the case of a body moving through unequal spaces in equal times, or with a varying velocity, if the velocity increase with the duration of the motion, it is termed accelerated motion; but if it decrease with the duration of the motion, it is termed *retarded motion*. A stone thrown up in the air affords an illustration of each of these cases, the motion during the ascent being retarded by the force of gravity, and accelerated by the same during the descent of the stone. All bodies have a tendency to preserve their state, either of rest or of motion; so that if a body were set in motion, and this moving force were withdrawn, the body, if unopposed by any force, would continue to move with the same velocity it had acquired at the instant the moving force was withdrawn. And if a body in motion be acted upon by a constant force, as the force of gravity, the motion becomes accelerated, the velocity increasing as the times, and the whole spaces passed through increasing as the squares of the times; whilst the proportional spaces passed through during equal portions of time will be as the odd numbers 1, 3, 5, 7, &c.; and the spaces passed over in any portion of time, taken as a unit, will be equal to half the velocity acquired at the end of such time. Thus, at the end of one second, the velocity of a body falling freely near the surface of the earth is said to be $32\frac{1}{6}$ ft.; at the end of 2 seconds, 2 times $32\frac{1}{6}$ ft.; at the end of 3 seconds, 3 times $32\frac{1}{6}$ ft.; at the end of 4 seconds, 4 times $32\frac{1}{6}$ ft., and so on; or generally, the velocity acquired by a falling body is equal to the product of the time of the body's fall in seconds by $32\frac{1}{6}$ feet, which may be expressed by the simple equation—

$$(\text{Velocity in feet}) = (\text{Time in seconds}) \times 32\tfrac{1}{6} \text{ or } v = t \times 32\tfrac{1}{6}.$$

The space described by a body in one second will be half of $32\frac{1}{6}$ feet $= 16\frac{1}{12}$ feet; because the velocity of the body in the middle of the time will be the mean velocity with which it moves during that time. In like manner, the space described by the body in 4 seconds, will be 4 times $2 \times 32\frac{1}{6}$ ft.; because $4 \times 32\frac{1}{6}$ ft. is the velocity at the end of 4 seconds, and therefore $2 \times 32\frac{1}{6}$ will be the mean velocity, or the velocity in the middle of the time. But 4 times $2 \times 32\frac{1}{6} = 16 \times 16\frac{1}{12} = 4^2 \times 16\frac{1}{12}$, that is, the space described by a falling body in 4 seconds is equal to the square of the time multiplied by the space described in 1 second. In the same manner, the relation of the space, s, in feet, and the time t in seconds, is expressed generally thus, $s = t^2 \times 16\frac{1}{12}$.

Morin's Apparatus for Demonstrating the Laws of falling Bodies, by means of Uninterrupted Indications.—The apparatus constructed by Morin, according to the instructions given him by Poncelet, to effect this object, consists of a cylinder A A, Figs. 8, 9, moved by means of a vertical axis, set in motion by means of clockwork, regulated by the pendulum D. The surface of this cylinder is covered with a sheet of paper, a conical leaden weight, d, is made to move, guided and kept in its proper position, by guide-rods, at a small distance from the cylinder; to this leaden weight is attached a small hair pencil with a fine point, and this pencil being previously dipped into colouring-matter, the point of it touches and marks the paper. This cylindro-conical weight, and the hair pencil, are represented on a large scale in Fig. 10. When an experiment is desired to be made with this apparatus, the weight d is kept at the upper part of the apparatus by a set of tweezers E; after the cylinder has been set in motion, and this motion has become uniform, the string F is pulled, by which the tweezer is unfastened, and consequently the weight d falls gliding down along its guide-rods, while the hair pencil marks simultaneously, on the surface of the paper placed upon the cylinder, a curved line, from which may be adduced the laws of the moving body.

When at the end of this experiment the paper is withdrawn from the cylinder, it will be observed that it contains two lines, one G Q, Fig. 11, a straight line, perpendicular to the axis of the cylinder (this line was marked out before the weight was allowed to fall); the other, a curve line G M H, to which G Q is a tangent. When to different points of this curved line, as for instance, M m, tangents are drawn, and when through the points T and t, where these tangents meet the straight line G Q, perpendicular lines are traced, it will be observed, that all these perpendiculars pass through one and the same point F. This is a property of the curve, known as the *parabola*, and the point F, wherein T F and t F meet, is the focus of the curved line. When from F, a straight line is drawn perpendicular to G Q, a perpendicular, G G′, is found, constituting the axis of the parabola,

8. 9. 10. 11.

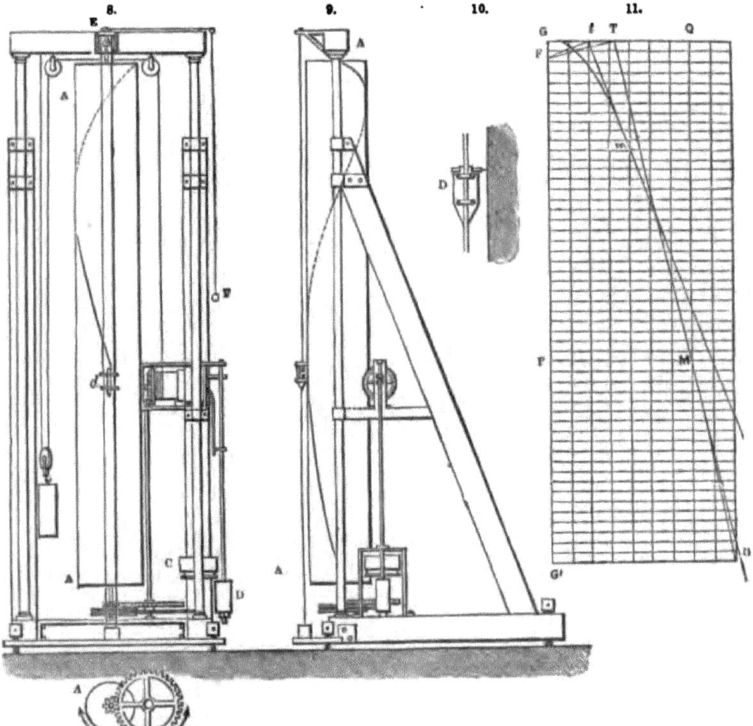

of which G is the top, or summit. That point is the starting-point of the moving weight, which point could not be very readily perceived without this construction, since the vertex or summit of the curve is only exhibited by the contact of the curved line m M and the straight line G Q. Let us now examine any point M of the curve, and draw the rectangular co-ordinates M Q and M P with respect to the axes G G′ and G Q. The vertical line M Q represents the space, e, travelled over by the moving weight in a given time, t. The horizontal line M P represents the arc of the circle described in the same lapse of time by any point of the surface of the cylinder; let r be the radius of the cylinder, ω the velocity, which is taken for granted to be constant; the arc in question, therefore, has for its measure $\omega r t$. But since the curve described is a parabola, there exists between the co-ordinates of the point M, the relation

$$(M P)^2 = 2 p (M Q); \qquad [1]$$

calling p the semiparameter. By substituting for M P and M Q their respective values, we have

$$\omega^2 r^2 t^2 = 2 p e,$$

whence

$$e = \frac{1}{2} \frac{\omega^2 r^2}{p} t^2. \qquad [2]$$

Equation [2] shows *that the space described by d is proportional to the square of the time.* The quantity p, is the double of the distance G F from the summit, or vertex, to the focus; by designating this distance by h, we have

$$e = \frac{1}{2}\,\frac{\omega^2 r^2}{2h}\,t^2,$$

while the acceleration due to the gravity is

$$g = \frac{\omega^2 r^2}{2h}. \qquad [3]$$

One might calculate $g = 32 \cdot 2$, from [3], but h cannot be measured to a sufficient degree of accuracy. But we deduce [4] from [3] by putting u for ωr,

$$h = \frac{u^2}{2g}; \qquad [4]$$

that is to say, *the distance G F from the top of the parabola to the focus, is the height due to the velocity of any given point of the cylinder.* The parameter of the parabola becomes greater when the cylinder rotates more rapidly.

The law of the velocities may be deduced from equation [2], by taking the variable e with respect to the time; whence

$$v = \frac{\omega^2 r^2}{p}\,t; \qquad [5]$$

it hence follows that the velocities are proportional to the time. Geometrical considerations establish the same law. For the curve G M H described by the moving-weight is the representative curve of the motion. By taking G P as the axis of x, and G G′ as the axis of y, the equation of the curve becomes

$$y = \frac{x^2}{2p}.$$

And the angular coefficient of the tangent, or the differential of y, with respect to x, is

$$\frac{dy}{dx} = y' = \frac{x}{p}. \qquad [6]$$

This angular coefficient is proportional to the velocity; now x is proportional to the time, and therefore *the velocity is proportional to the time.* But equation [6] would not give exactly the value of the velocity; since, for $x = \omega r t$, we should have

$$y' = \frac{\omega r}{p}\,t,$$

a value differing from expression [5]. This is because the units of time and space are not represented by the same length, which condition ought to exist in order that the angular coefficient of the tangent to the curve of space, be equal to the velocity of the moving weight.

The units of work conserved in a body weighing W lbs., moving in any direction $a\,b$, or $c\,d$, straight or curved, with a motion being either retarded or accelerated, may be readily found when the velocity v, in feet a second, is known at P, any point of the path described by the body W. Fig. 12. The units of work accumulated in a moving body is equal to the square of the velocity in feet a second, multiplied by the weight of the body in lbs., and divided by $2 \times 32 \cdot 2$. The *mass* (m) of a body is a constant quantity at all heights and in all latitudes, while the weight W and the value of g are variable; but $m = \frac{W}{g}$ under all circumstances. There is much uncertainty and error involved in the methods employed by philosophers to find the value of g in different places. In this work, for the want of knowing better, g is put $= 32 \cdot 2$ feet. That is, a body falling from a state of rest is supposed to be moving at the end of the first second with a velocity of $32 \cdot 2$ feet a second. When we say absolutely and without other explanation that the quantity g, which expresses the acceleration produced by gravity, is the measure of this force, we give an incorrect idea, since g is in reality only the velocity imparted to or taken from a body by gravity during each second of its action, and the velocity which is expressed in feet cannot measure a force which should be compared with pounds. The product of the mass m, and the velocity $v = \frac{w}{g}v$, has received the name *momentum*; it is a conventional phrase, to which we attach no other signification than that of the product of the *mass* into the velocity imparted to or taken from it. If the weight $w = 193$ lbs. where $g = 32\frac{1}{6}$ ft., then the mass will be $= \frac{193}{32\frac{1}{6}} = 6$. In Paris g is said to be $= 32\ 1817$ ft., in which place w would be $= 193 \cdot 0902$ lbs.; but the mass remains unaltered, for $\frac{193 \cdot 0902}{32\ 1817} = 6$ also. A body W at the point P, weighing 230 lbs. moving in any direction with a velocity of 14 feet a second has accumulated in it 700 units of work, for $\frac{14^2 \times 230}{2g} = 700$.

12.

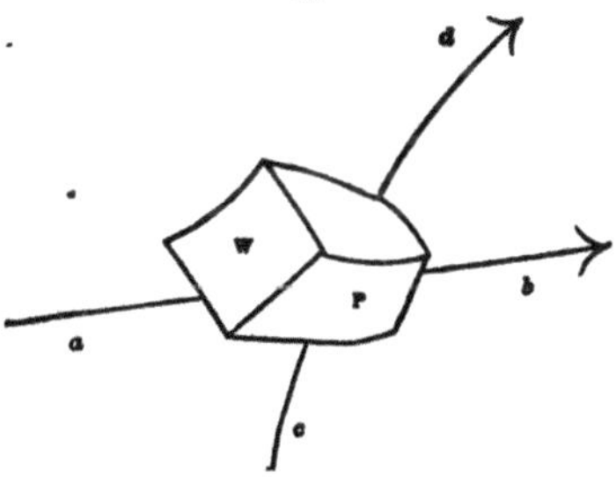

Suppose two weights, F and E, Fig. 13, weighing $4 \cdot 9$ lbs. and 7 lbs. respectively, to be connected by a cord, I C D, that goes over a fixed pulley C, as in *Atwood's machine*; the space through which E must descend to acquire a given velocity, say 2·3 feet a second, may be found on the *principle of work* without direct reference to the acceleration of the bodies in motion. Thus, the units of work in $F = \frac{(2\cdot3)^2 \times 4\cdot9}{2 \times (32\cdot2)} = \cdot4025$; the units of work in $E = \frac{(2\cdot3)^2 \times 7}{2 \times (32\cdot2)} = \cdot575$; therefore, the total accumulated work in the bodies E and F at the required position $= \cdot9775$. Now if we suppose x to be the space passed over by each of the weights, then the work of gravity on $F = 4\cdot9 \times x$; and the work of gravity on $E = 7 \times x$; as the work performed on F has been produced by the work of E, the work existing in the bodies is also represented by the difference of $7 \times x$ and $4\cdot9 \times x = 2\cdot1 \times x$. Therefore $2\cdot1 \times x = \cdot9775$, whence $x = \frac{\cdot9775}{2\cdot1}$ feet $= 5\frac{41}{70}$ inches.

13.

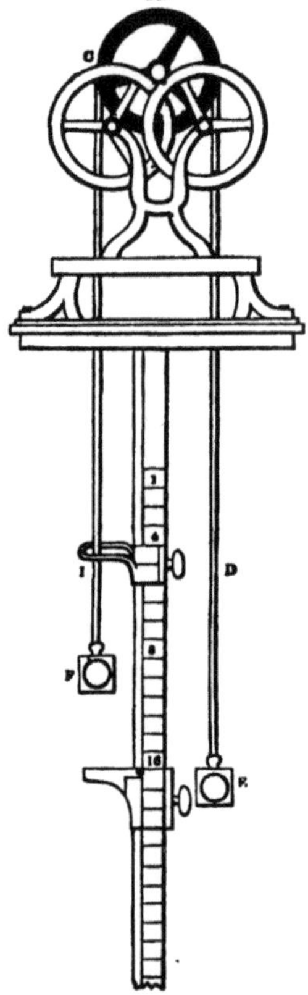

Again, suppose a weight of 9 lbs. to act upon a weight of 7 lbs. over a pulley C, Fig. 13; the time taken for the greater weight to descend a given number of feet (100), and the common velocity of both bodies, may be determined on the *principle of work*, without direct reference to acceleration. For 9 lbs. − 7 lbs. = 2 lbs., and 100 × 2 = 200, the units of work in both weights. Then if v be put for the velocity, the units of work in both bodies will also be expressed by $\frac{v^2 \times (7 + 9)}{64\ 4}$; whence $\frac{4\ v^2}{16\cdot1} = 200$, and $v = 28\cdot3725$ feet, the velocity at the end of 100 feet. Then the mean velocity $= 14\cdot1862$, and $\frac{100\cdot}{14\cdot1862} = 7\cdot05$ seconds, the time of descent.

ACHROMATIC LENS. Fr., *Lentille achromatique;* Ger., *Achromatische Linse;* Ital., *Lente acromatica;* Span., *Lente acromático.*

Those optical instruments and lenses which suffer the rays of light to pass through them, without decomposition, are called *achromatic*, which signifies without colour. See Optical Instruments.

ACRE. Fr., *Acre* (4840 *yards carrés*) ·404671 *hectare;* Ger., *Acher.—Morgen Landes* = 3291·2 *sq. yards;* Ital., *Jugero;* Span., *Acre* = 4046·87 *metros cuadrados.*

A measure of land containing 4 square roods, or 160 square perches, is termed an acre; the English acre of land contains 4840 square yards.

ADDRESSING MACHINE. Fr., *Machine pour faciliter l'impression d'un grand nombre d'adresses de lettres;* Ger., *Eine Vorrichtung damit man sehr schnell Briefen-adressen schreiben kann;* Ital., *Macchina da indirizzi;* Span., *Máquina para imprimir sobres, &a.*

14

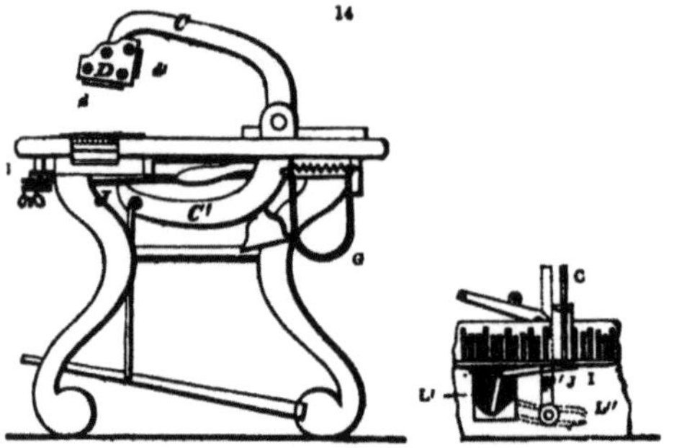

An addressing machine is a machine for inserting the addresses of letters and other similar articles.

The addressing machine of N. E. & G. W. Warren, Fig. 14, consists of a curved arm, C, operating on a *platen* and worked by a treadle.

The curved levers or arms, C, C′, are operated by the bent spring G, in combination with the adjustable head D, and the faces *d*, *d′*. The *pall* F′, *rock-shaft* L′, slotted arm L″, and adjustable rod J, are worked by the *quad ratchet* I.

ADHESION. Fr., *Adhésion;* Ger., *Anziehungskraft;* Ital., *Aderenza;* Span., *Adhesion.*

Adhesion is the union of the surfaces of bodies when brought together, and is measured by the force which is requisite to separate them. Adhesion may be either natural or artificial. It is not to be confounded with *Cohesion*, with *Gravitation*, nor yet with the pressure of the atmosphere upon an external surface when the air is removed from beneath it. The power or degree of strength with which bodies unite is called their force of adhesion. Bevan found that a nail driven into Christiana deal required 170 lbs. to extract it; in green sycamore, it required 312 lbs.; in dry oak, 507 lbs.; in dry beech, 667 lbs. A screw holds three times as strongly as a nail of similar length, and in most light timbers a nail driven across the grain holds with twice the force of one driven with the grain. In oak and elm there is not so much difference. Well-glued surfaces of dry ash hold with a force of 715 lbs. to the square inch if the glue be new; Scotch fir with an adhesive force of 562 lbs. to the square inch.

The adhesive force on railroads may be estimated approximately from the simple expression $c \times t$, usually written ct, in which c is the co-efficient of adhesion for the driving-wheels of the locomotive, and t the weight of the locomotive in tons, which rest on the driving-wheels. The adhesive force of the driving-wheels, $c \times t$, must always be greater than the retractive force, $22 \cdot 4 \times t \times h$ nearly, in which h is put for the vertical rise in feet for each 100 feet of road. Approximate results may be readily obtained by putting $c = 670 \cdot$ when the rails are dry; $= 560 \cdot$ when the rails are very dry; $= 450 \cdot$ under ordinary circumstances; $= 314 \cdot$ in wet weather; and $= 225 \cdot$ in snow and frost. On horse-railroads, or tramways in large towns, c varies from $300 \cdot$ to $400 \cdot$ in snow and frost. The influence of the resistances operating on railway trains in motion will be generally discussed when we treat of the experiments of MM. Krillemin, Guebhard, and Dieudonné. See DYNAMOMETER, *Railway Car.*

The force of adhesion will be better understood from its practical relation to friction, and to tractive and retractive forces.

Suppose the area (A) of one of the two cylinders of a locomotive = 400 square inches, Fig. 15; stroke (S) of piston = 1·5 feet; mean pressure (P) on the square inch = 96 lbs., and the diameter (D)

15.

16.

of the driving-wheels = 5 feet; then the tractive force $= \frac{\text{A S P}}{\text{D}} = \frac{400 \times 1 \cdot 5 \times 96}{5} = 11520$ lbs., n being very nearly $= \frac{28\text{ M}}{\text{D}}$ when M is the miles an hour and n the revolutions of the driving-wheels a minute. Then the actual horse-power (H) of the locomotive, Fig. 15, $= \frac{n\text{ A S P}}{11000} = \frac{\text{A S P M}}{376\text{ D}}$. About 25 per cent. is generally allowed for the friction of the locomotive machinery and the power required to work the pumps.

Suppose a locomotive, Fig. 16, weighing 18 tons (t), to be placed on an incline rising 8 feet (h) in 100; the length (S) of the stroke of the piston = 2 feet; area (A) of piston = 320 square inches; the pressure (P) = 75 lbs. on the square inch; (c) the coefficient of adhesion $= 560 \cdot$, and the diameter (D) of the driving-wheels = 4·5 feet. Required the *tractive force, retractive force,* and the force of *adhesion.*

$$\text{Tractive force} = \frac{\text{A S P}}{\text{D}} - 22 \cdot 4 \times t\,h = \frac{320 \times 2 \times 75}{4 \cdot 5} - 22 \cdot 4 \times 18 \times 8 = 7441 \text{ lbs.}$$

The retractive force $22 \cdot 4 \times t \times h$ being $= 3225 \cdot 6$ lbs.

The base (b) for the rise 8 in 100 = 99·68, whence the *force of adhesion* $= \frac{c \times t \times b}{100} = \frac{560 \times 18 \times 99 \cdot 68}{100} = 10047 \cdot 744$ lbs.

But since 10047·744 is greater than 3225·6 lbs., the locomotive can ascend the incline with a tractive force of 10047·744 − 3225 6 = 6822·144 lbs., and without the driving wheels slipping.

Putting T for the weight in tons moved on wheels, and suppose T to include the weight of carriages on common roads, and the weight of carriages, locomotive, and tender on railroads, then on railroads the tractive coefficient (k) in lbs. to the ton in T varies from 4 to 8 lbs. On railroads in good condition, with axles well lubricated, $k = 4$ lbs. to the ton in T; on railroads and tramways under ordinary circumstances, $k = 7$; for roads not in very good condition, $k = 8$.

In ordinary traffic—

	k
On very smooth stone pavement	= 12
On ordinary street pavements in good condition	= 20
On some pavements and turnpike roads	= 30
On turnpike roads newly laid with coarse gravel and broken stones ..	= 50

On common roads in bad condition, $k = 150$, and k becomes as high as 560 on natural loose ground or on sand.

While comparing the lbs. in (k) and the tons in (T), it must not be forgotten that t has been put for merely the weight of the locomotive in tons, which rests on the driving wheels. To illustrate this matter, let it be required to find the retractive force of a train (T) = 150 tons, Fig. 17, moving with a speed (M) = 25 miles an hour on a horizontal line of railroad in the best condition, or when $k = 4$.

The retractive force is nearly $= \text{T}\,(k + \sqrt{\text{M}}) = 150\,(4 + \sqrt{25}) = 1350$ lbs.: this force must be less than $c \times t$, the *adhesive force.* The actual horse power (H) of the locomotive is nearly equal to $\frac{\text{M T}}{375}\,(k \times \sqrt{\text{M}})$. Let it be required to find the horse-power (H) necessary to draw a train (T), Fig. 18, = 137 tons, up an incline of (h) = 9 feet in 100, with a speed (M) of 25 miles, when k = 6.

$$H = \frac{M\,T}{375}\,(22\ 4\ h + k + \sqrt{M}) = \frac{25 \times 137}{375}\,(22{\cdot}4 \times 9 + 4 + \sqrt{25}) = 1923\ 5.$$

17.

18.

19.

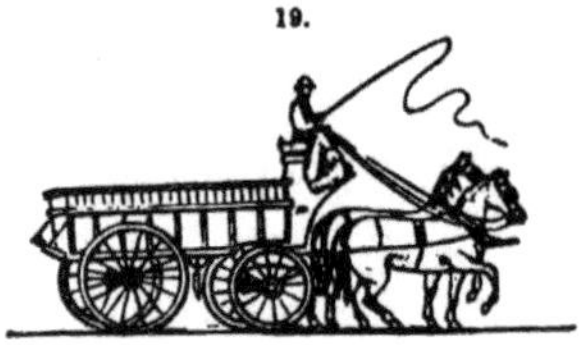

20.

The adhesive force $\frac{c\ t\ h}{100}$ must be greater than T $(22{\cdot}4\,h + k + \sqrt{M})$. If ($d$) be put for the number of consecutive working hours of a horse, (v) the velocity in feet a second, and (t') the weight of a horse in lbs., then, Fig. 19 we have the approximate formulæ—

$$F = T\,(k + \sqrt{M});\ v = 1{\cdot}466\ M;\ \text{and}$$

$$F = \frac{550}{v\sqrt{d}} = \frac{375}{M\sqrt{d}} = \text{the ability of a horse.}$$

Whence the tractive ability (F) of a horse running five miles an hour in four (d) consecutive hours = $\frac{375}{5\sqrt{4}}$ = 37·5 lbs. Lastly, let it be required to find the tractive force F of a load T = 10 tons, to be drawn M = $2\frac{1}{4}$ miles an hour, up a turnpike road, Fig. 20; h = 8 feet in 100, k = 50, the road being newly laid with coarse gravel. The following formulæ will approximately apply:—

$$F = T\,(22{\cdot}4\,h + k + \sqrt{M});\ M = {\cdot}682\ v;$$

$$\text{and}\ \ F = \frac{550}{v\sqrt{d}} - \frac{t'\,h}{100}.$$

$$F = 10\,(22{\cdot}4 \times 8 + 50 + \sqrt{2{\cdot}25}) = 2307\ \text{lbs.}$$

Suppose a horse to weigh t' = 1000 lbs. and to work continually, d = 1 hour, up this turnpike road; the tractive ability of this horse will be $\frac{375}{2\frac{1}{4}\sqrt{1}} - \frac{1000 \times 8}{100} = 86\frac{2}{3}$ lbs. Hence the number of horses required = $\frac{2307}{86\frac{2}{3}}$ = 27 nearly.

ADIT. Fr., *Passage, Galerie (d'écoulement d'eau dans les mines)*; Ger., *Zugang, Stollen*; Ital., *Adito*; Span., *Galeria de una mina.*

The horizontal opening by which a mine is entered, or by which water and ores are carried away is termed an adit. The woodcut represents an exaggerated section of part of the underground workings of a mine; b is the shaft, a and c the adits, and l the lode; c is called the shallow adit and a the deep adit.

20*.

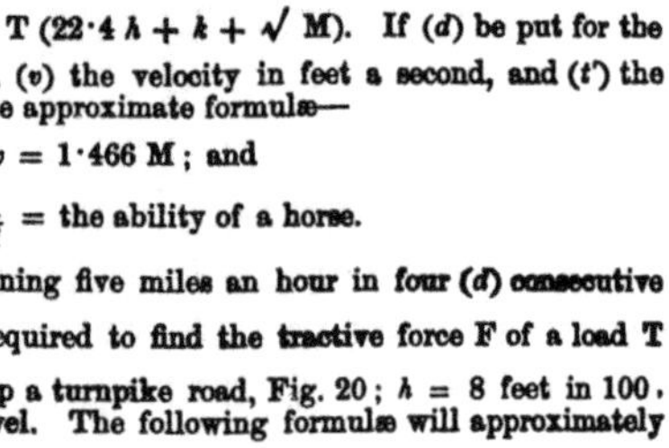

ADZE. Fr., *Herminette*; Ger., *Krummaxt, Hohleisen*; Ital., *Ascia*; Span., *Azuela.*

An adze is a tool for chipping, formed with a thin arching blade, and its edge at right angles to the handle. The edge is only bevelled on the inside. See Hand Tools.

AFTER-DAMP. Fr., *Mofettes*; Ger., *Böses, oder tödtendes Wetter*; Ital., *Mefite*; Span., *Mofeta.*

Choke-damp is often termed *after-damp* · it is the carbonic acid gas which accumulates in mines and wells; this gas is called choke-damp because it often destroys life by preventing the respiration of air. See ANEMOMETER.

AGRICULTURAL IMPLEMENTS. FR., *Ustensiles employés à l'agriculture*; GER., *Landwirthschaftliche Geräthe*; ITAL., *Macchine ed utensili agricoli*; SPAN., *Utiles agrícolas.*

Many of the agricultural implements introduced in this article are not only well-arranged to effect the purposes for which they are designed, but, at the same time, they will be found, as regards construction, to interest civil engineers generally, either in suggesting the application of some peculiar mechanical principle, or in pointing out combinations of machinery which may be found useful beyond the limits of the field or farmyard.

The annexed woodcut, Fig. 21, represents a portable steam-engine and windlass combined, as constructed by C. Burrell; the windlass has a single *sheave* of 5 ft. diameter, round which the rope

21.

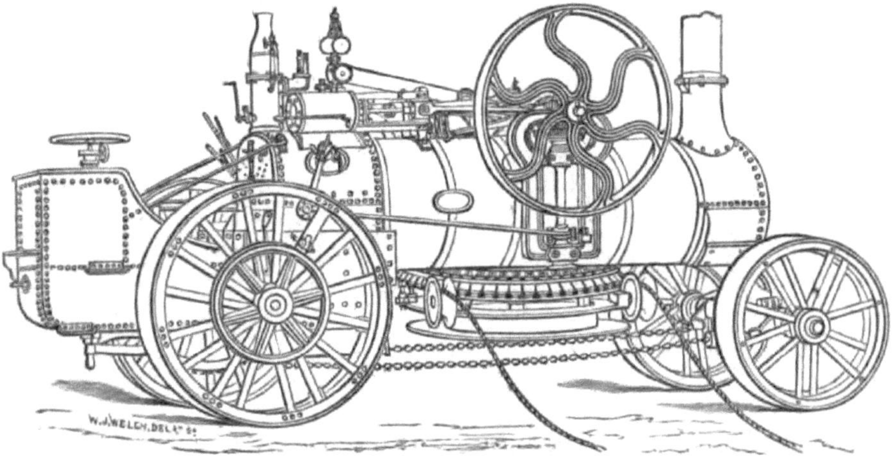

passes, and it is formed of a double series of small leaves, which on the least pressure clasp and hold the rope until it takes the straight line on the other side, when the clips freely open and liberate the rope. By this simple appliance all crushing and short bends, which are so detrimental to the profitable use of wire-rope, are entirely avoided; this, coupled with the fact that on each passage of the implement the rope is only twice bent, and then only round large diameters, will at once show this system of using wire-rope to be most advantageous. The small leaves are made of chilled cast-iron, which is not liable to much wear, but the leaves, when worn, can be replaced at a trifling cost. The power is conveyed to the windlass by an upright shaft from the crank shaft.

Fig. 22 represents the rope porters to be used along with the engine for agricultural purposes just described: these porters are placed along the fields at intervals of 40 yards, thereby

22.

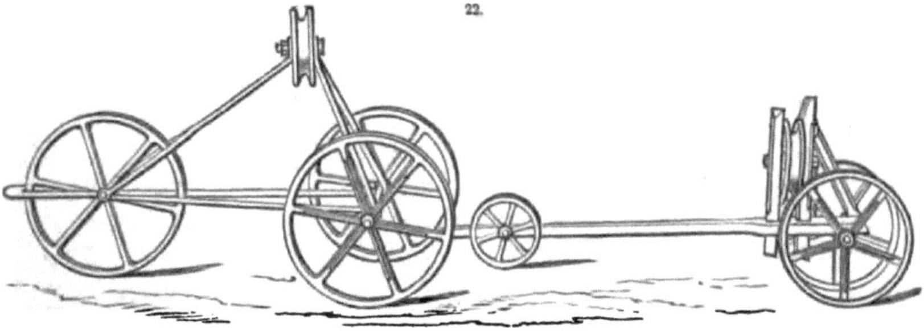

keeping the rope entirely off the ground. The outside ones are mounted on three wheels, so as to allow them to be moved by the rope.

Fig. 23 represents what is termed the anchor, which is shown attached to the working apparatus. This anchor is made to resist the side strain of the implement worked, by the cutting of the disc wheels into the ground. The anchor is moved along the headland by the motion of a 5-feet *sheave*, which is turned by the ploughing rope, and as the plough goes away from the anchor, the sheave winds up a rope stretched along the headland and keeps the anchor opposite its work. The frame is made entirely of wrought iron. As the disc can be steered in any direction, the anchor may be moved along a crooked headland. The box at the back is intended as a counterpoise to prevent the anchor being pulled over when heavy work is being done. This machine is managed by a boy, who also attends to the shifting of the rope porters.

23.

Fig. 24 represents what, for agricultural purposes, is termed a liquid manure distributor, designed by W. Crosskill. A pump and hose being fixed to this cart it may, when yoked, be often found useful as a watercart, either for the transport of water, from a distant river, well, or canal, or it may be applied to water roads and streets.

24.

Fig. 25 represents the portable farm railway of W. Crosskill, which may sometimes be of use to contractors, engineers, or builders.

Fig. 26 represents an improved horse gear or horse-power for driving machinery; it has a strong cast-iron *bed-plate* supporting the bearings of the horizontal ground shaft, and the step for the vertical shaft. To prevent accident, and as a protection from dust and dirt, the whole of the gearing, and working parts, are encased by a cast-iron dome cover, secured to the bed-plate, by screw bolts. The main top bearing is adjustable by set screws, so as to ensure uniformity of wear, and steadiness of motion.

Fig. 27 represents the bone-rasping and grinding mill of Picksley, Sims, and Co. It is simple in construction, strong in its working parts, and produces at the first operation, 25 per cent. more dust than the ordinary bone mill at present in use.

The working process is as follows:

Unbroken bones are thrown into the hopper, fall upon the cutting bed, and are pressed by feed rams against the teeth of revolving cylinders in rapid motion.

The reduced bones fall into an oscillating or revolving *riddle*, attached to the mill, in order to separate them into the two usual qualities, namely, dust and half-inch bones.

25.

26.

27

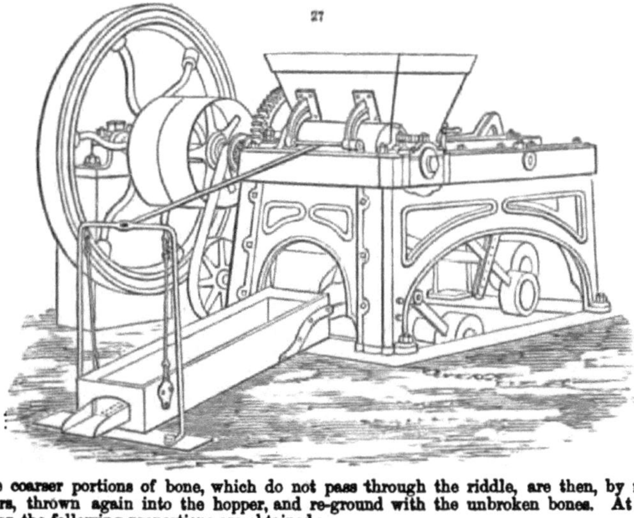

The coarser portions of bone, which do not pass through the riddle, are then, by means of elevators, thrown again into the hopper, and re-ground with the unbroken bones. At the first operation the following proportions are obtained :—

Dust. . . . . . . 45 per cent. of the entire quantity ground.
½-inch bone . . 30 " " "
Coarser matter (to be re-ground) 25 per cent.

This mill is adapted for grinding every description of bones, irrespective of size and quality.

The feeding of the mill is regulated alternately by the driving shaft, and by a counter balance-weight placed beneath the mill; by this contrivance, the bones are pressed against the cutters without any undue *strain* being thrown on the working parts, and the possibility of breakage is diminished.

The mill has an apparatus attached, for passing small pieces of iron which may accidentally get into the hopper with the bones.

Unusual facilities are given for keeping the cutters sharp and in working order, as they can be readily disengaged, sharpened on a grindstone, and replaced in working order by any intelligent labourer.

Directions for Proper Use.—The mill should be set level upon a solid foundation of stone, and secured by means of screw bolts.

The caps on the bearings should be firmly screwed down, merely leaving sufficient play for the shafts to revolve without unnecessary friction.

The driving strap should be placed tightly upon the *driving pulley*, and the mills driven at the following speeds:—

2-horse power mill,	250 revolutions a min.			8-horse power mill,	150 revolutions a min.		
4-horse do.	225	"	"	10-horse do.	150	"	"
6-horse do.	200	"	"	12-horse do.	125	"	"

The oil boxes on all the shafts should be kept well supplied with oil.

When the knives require grinding they may be readily removed from the cylinder, by using a key of $\frac{5}{8}$ steel as a drift; this operation is performed by holding one end of the drift against the small end of key which keeps the knives in place, and striking the other end with a hammer until the key is backed sufficiently to be withdrawn; by this means the spiral segment which keeps the knives in place can be disengaged, and when the segment is removed, the knives are liberated, and may be taken out.

In replacing the knives, care should be taken that the keys are so driven in, that they may clear the frame at the head and point, and the knife edge should pass the cutter bar without touching.

The knives should be ground daily, as upon their sharpness depends the satisfactory working of the mill, both as to quantity and fineness of the dust produced.

As the knives wear, they should be kept up in the slots by strips of wood being placed underneath them.

Before starting it is desirable that the mill be inspected, to see that all bolts and nuts, are secure, and the knives firmly fixed in their places, and that the cylinder has sustained no injury in a previous operation.

The chief aims of the application of mechanical power, as a substitute for manual labour, are to effect improvements in the results of labour, and to render them less expensive. The use of the hand-flail to separate and detach corn from its ears is now pretty generally superseded by the thrashing machine, which, in its main features, may be called a contrivance devised to supersede by mechanical means the use of the hand-flail, and thus to economize at the same time both time and labour, and secure a less wasteful mode of separating the corn and chaff from each other. A thrashing machine essentially consists of a rapidly revolving cylinder, with raised edges or beaters parallel to its axis and standing out from its surface. The cylinder or drum is covered by a concave surface at some two or three inches distant from the surface described by the edges of these revolving beaters. A feeding board extends radially and horizontally outwards from the cylinder, and near its termination are placed two feeding rollers, which, in revolving towards one another, not only rapidly draw the straw forward, but also hold it from going too fast, which, under the action of the beaters, would be liable to happen. The beaten straw, with the chaff and grain lying loose among it, is delivered on the floor behind the cylinder, and the operations of separation by fork, riddle, and fanner may be afterwards performed by hand; but in the more improved modern machines these operations also are effectually done by mechanical contrivances, usually, so connected with the thrashing machine as to operate with it at the same time and by

28.

the same motive power. The annexed woodcut, Fig. 28, represents a portable, combined, single blast thrashing, straw-shaking, riddling, and winnowing machine, constructed by Picksley, Sims, and Co.

Many minor improvements in agricultural implements have been recently made. We insert brief descriptions, illustrated by woodcuts, of some of the most useful.

An improved Ox-yoke.—The mortise through the bow of an ox-yoke greatly weakens the bow, and the key sometimes gets misplaced, and even lost, although attached to the yoke by a leather thong; the thong may break, and just when the key is most needed it becomes of no practical use. To remedy this is the design of the improvement shown in Fig 29. Two hinged plates are secured to the top of the yoke, as shown in Fig. 29, the free ends engaging with notches cut in the bow, and holding them securely in place until they are forcibly raised by hand.

29.

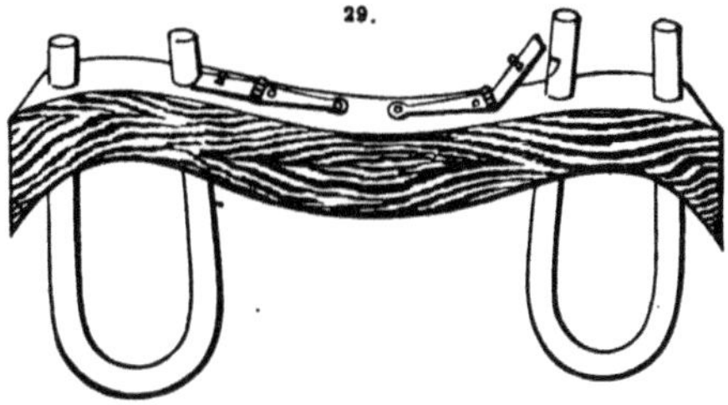

Sheep Shear.—Fig. 30 represents *an improved sheep shear*, the movable cutter A pivoted to the face of the stationary cutter B, which is divided into fingers, or bars, each one presenting a cutting edge to the action of the movable blade. A slot in the free end of the spring handle, and a screw in the end of the vibrating cutter, with a stop on C, on the opposite side of the plate B, governs the throw of the blade. The forks of the plate readily enter the matted fleece, thus facilitating the operation of shearing, and the action of the blade ensures a drawing cut, requiring less power and producing a cleaner cut than ordinary shears. The form of the cutter and its throw can be regulated to suit any hand; this implement may serve also for clipping horses.

30.

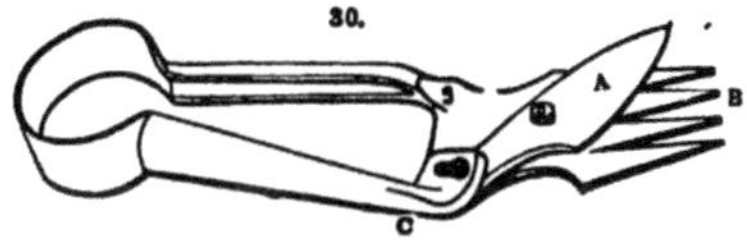

A wagon, the contents of which could be readily emptied or discharged by its attendant, has long been a requirement, not only upon a farm, but more especially in the grading of streets, railways, and the like. Dumping, as hitherto performed, required a considerable exercise of muscular force when the discharge of the materials was from a four-wheeled wagon; the operation

31.

32.

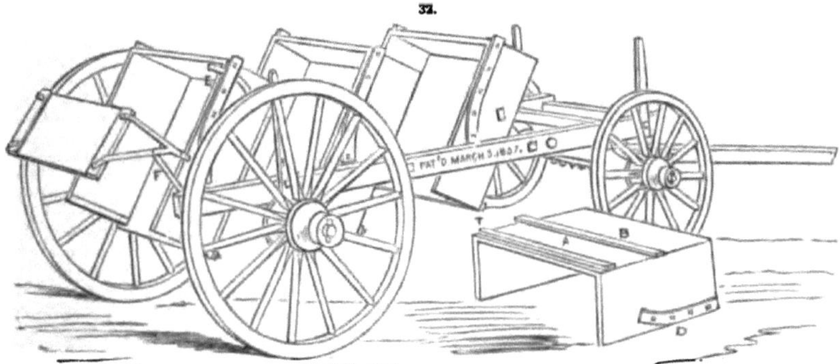

also, comparatively speaking, involved great loss of time. The wagon, represented in Figs. 31, 32, appears to be effectually secured in dumping. The essential features of the invention will be readily understood by a reference to the cuts. This wagon consists of a box, or body, composed of separate sections arranged in line with each other between the longitudinal sides, or bed-pieces, of the wagon frame, each section being pivoted or suspended upon these bed-pieces by suitable laterally projecting trunnions or pivots, so that it may be placed in a horizontal position to hold the materials, or it may be tilted with its open or rear end downwards to discharge the materials therefrom. When the sections are all in a horizontal position, as shown in Fig. 31, the sections are connected by suitable latch-pieces, or catches, at their sides, in such a way as to be firmly held in place. These sections may be filled by shovelling, or other means, in the same manner as an ordinary wagon or cart box. When it is desired to dump or discharge the load placed upon the wagon, the several sections composing the box are disconnected, and the sections are tilted as before mentioned, and shown in Fig. 32, whereupon the materials drop from the sections by their own gravity, and the sections are consequently emptied with great speed and facility.

33.

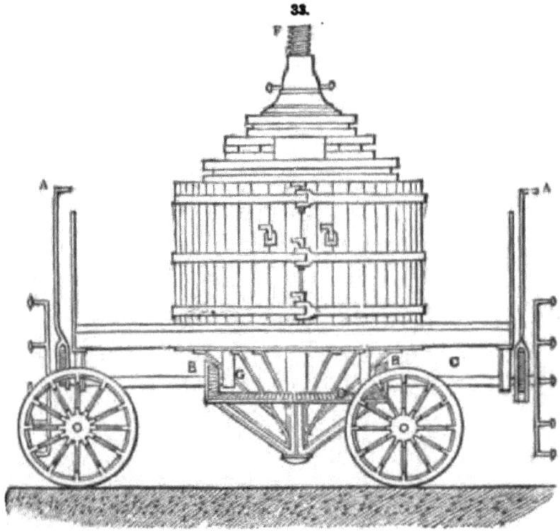

The Châtillonais.—For the convenience of such agriculturists as are in the habit of making wine, cider, or perry, we give a cut, Fig. 33, and short description of a very useful press known as the *Châtillonais*, and highly esteemed in France. The mechanism is placed on wheels, and the machinery for pressing is below the trough. The pressing is so performed that free passage is given to the screw; the ratchet brace A, placed upon the handle of the axis C, is provided with 14 crank-handles, and by means of C moves B, which has 90 teeth; the axis of the latter carries a conical cog-wheel with 10 teeth, and this wheel grips into the large wheel D provided with 102 teeth; the multiplication is therefore $6 \cdot 43 \times 10 \cdot 2 = 65 \cdot 586$, that is to say, that the screw makes one revolution for 65·586 strokes or revolutions of the handle. The average diameter of the screw is 0·1015, and its thread is 25; this gives for the inclination of the thread $\alpha = 4^\circ\ 30'$.

The radius of the handle is 0^m 40 at most, that is to say, $L = 0 \cdot 4$, $r = 0 \cdot 05075$, $n = 65 \cdot 586$, $\alpha = 4^\circ\ 30'$, and $\nu = 5^\circ\ 42' \cdot 40$. If these figures be substituted in the general formula we have the following result:—

$$P = F \times \frac{0 \cdot 4}{0 \cdot 05075} \times 65 \cdot 586 \times \frac{1}{tg\ 10^\circ\ 12'\ 40'' + 1 \cdot 33\ tg\ 5^\circ\ 42'\ 40''} = 1651 \cdot 5\ F.$$

Place for F 15 kilogrammes for one man, or in all 30 kilogrammes, P = 49545, and, neglecting friction, we have at least 40,000 kilogrammes of useful work.

$$\frac{49 \cdot 545 \text{ kil.} \times 0 \cdot 025}{65 \cdot 586 \times 2\pi \times 0 \cdot 4 \times 30 \text{ kil.}} = 0 \cdot 25.$$

When the friction is taken into consideration, we have only 0·207, or about 21 per cent.

The improved Reaper of Messrs. Howard.—In this reaper, Fig. 34, double concentric cams are employed, the one for directing the motion of the gatherers, and the other for guiding the rakes; the former being caused to drop down into the grain, in order to bring it up to the cutters, and then to rise again, so as to clear the cut grain on the platform, which is removed by the rakes governed by the second annular or concentric cam. The platform is hinged to the centre, by a kind of *drag bar*, and the delivery is effected by a central shaft, which is driven by a *pitch chain*, thus enabling light gearing to be used for operating the cutter bar. The mower of Messrs. Howard, also, possesses several improvements, amongst which we may point out a simple mode of lifting and varying the angles of the cutter. The "Clipper Mower," which we illustrate, has many peculiarities. The pole is independent of the draught, as the tractive strain is exerted through a sliding attachment on the under side of the pole. By this system there is a tendency not only to draw the machine directly forward, but at the same time to lift the shoe off the ground. The inside shoe and its attachments are so arranged that the fingers and knives can be changed from a level cut to an angle of thirty degrees, while the machine is in motion.

Colvin's Cow-milking Machine.—The engraving, Fig. 35, represents three cow-milking machines, operated by power, and attended by one man; two of these machines are shown, each milking a cow, and one exhibits the milking completed, and the cow turned back out of the way, so that the cow that has been milked may pass out to make way for another to come into the stall to be milked, so as not to stop the power while changing the cows. The stanchion is the same as any ordinary stanchion, with the exception that it opens out, to let the cows pass through, and facilitates the changing of them; in this manner, cows can be very quickly brought to the machine. The operation occupies less time than it would take to go to the cow in the yard, or stable; the cows soon learn to come to the machine if fed a few times while being milked, or by being enticed through giving them some salt. The milk is conducted by suitable tubing into large cans partially sunk in the floor; three machines are sufficient to milk sixty cows in the time it would take six men to milk them by hand. The moving power is imparted to the machines by hand, by a dog running in suitable gear, or other prime mover. The milkers are worked by pumps, the pistons of which are driven by power; they are attached by a jointed iron pipe to allow of the movement of the cow forward, backward, or sideways, always adapting itself to her motions; the teat-cups are made of corrugated india-rubber closely enveloping the teats, and will fit any cow. The pumps oscillate in such manner as to give the natural motion of a calf sucking, or to impart the motion of the human hand while milking; the space between the elastic diaphragm in the milker and the pump being filled with water, which, in working the pumps, oscillates in the tube, and produces a vacuum at each alternate stroke. By the working of this machine it is clear that no dust nor any dirt can fall into the milk.

34.

35.

Howard's Double-action Haymaker, for a single horse, is shown in Fig. 36. The axle in this machine is made of solid steel; this axle being strong is not liable to bend. The lines are well formed, and the forward action of this implement or machine effects a complete separation of the grass, while its back-action leaves the crop light and loose. The fork barrels are so arranged as to render clogging almost impossible; the forks are mounted in sets of three, and placed in a *zig-zag* position; this arrangement equalizes the work, while it separates and distributes the

36.

crop. The machine shown in Fig. 36 is fitted with a wire screen to prevent the grass from lodging on the front. The usual method of reversing the motion of such machines has hitherto been, either by loose sliding pinions operated by means of clutches on the fork barrels, or by sliding the fork barrels themselves; this last plan having the disadvantage of altering the relative positions of the forks, and rendering the machine liable to clog. In Howard's haymaker, the gear-work is strong and simple, and the motion can be changed in an instant to the backward or forward action by a simple eccentric movement of the main axle, and thus the disadvantages above pointed out are obviated. A similar eccentric movement is employed to raise or lower the fork barrels, so

37.

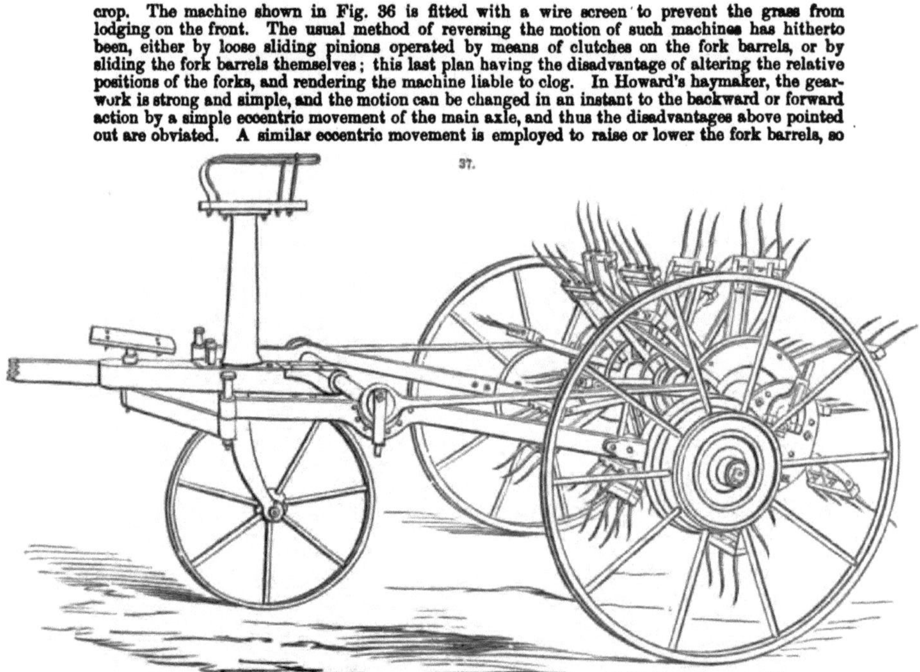

as to adapt the machine to the nature of the crop. When the forks are set for the forward action, no change is required when the backward action of the machine has to be brought into play.

Fig. 37 represents *Howard's Double-action Haymaker, designed for two horses:* the wire screen to prevent the grass from lodging in front may be applied to this machine in a similar manner to that shown in Fig. 36.

To prepare the machine for work, take off the travelling wheels, grease the axles, see that the gearing is clean, and supply a little of the *best* machine oil to the two oil holes in each fork barrel and in each side plate. When the machine is in work, the axles must be greased and the gearing cleaned once a-day, and the fork barrels and side plates oiled two or three times a-day.

For the first tedding or breaking the swarthe, the forward action should be used. To put the machine into gear, move the lever opposite to the letters "F A" on the side plate. It is generally better to work the machine *across* the swarthe, as it spreads the grass more evenly.

The backward action is to be used when the grass is partially dried, to lighten it up, and thoroughly expose it to the action of the sun and air. The backward action may also be used with great advantage for opening windrows.

The machine should be raised from and lowered to the ground to suit the state of the crop; the heavier the crop is, the higher the fork barrels should be. To alter the height of the machine, move the lever fixed to the end of the shaft bar. When working with the backward action only, set the machine near to the ground.

When the single haymaker is operated with, the best method of raising or lowering the fork barrels is as follows:—Close the fork-heads, raise the shafts gently till the heads rest on the ground, and then slacken the handle-nuts until the bolts can be raised or lowered into the required notch.

Should any of the parts of the machinery shown in Figs. 38 to 49 be accidentally broken, or require to be removed, they can be supplied separately and detached.

Fig. 38 shows the off side eccentric; Fig. 39, the covering plate; Fig. 40, centre star and barrel; Fig. 41, side star with pinion; Fig. 42, wheel; Fig. 43, fork head casting for spring; Fig. 44, fork head casting; Fig. 45, loose pinion; Fig. 46, near side eccentric (outside); Fig. 47, near side eccentric (inside); Fig. 48, side star; Fig. 49, wheel box.

38.

39.

42.

43.

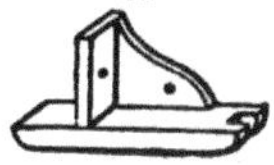

40.

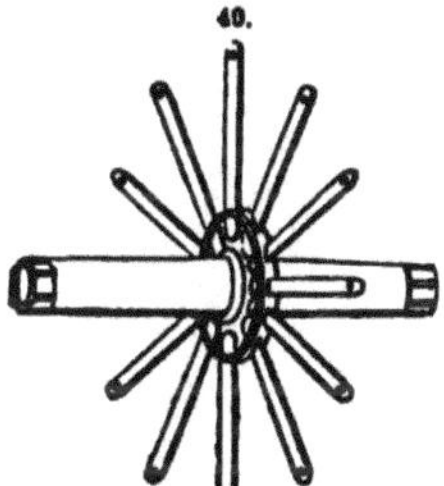

44.

45.

46.

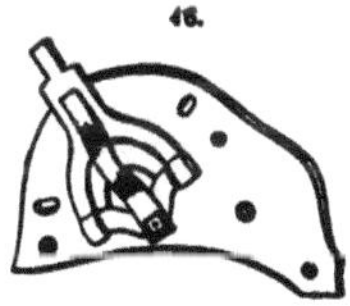

41.

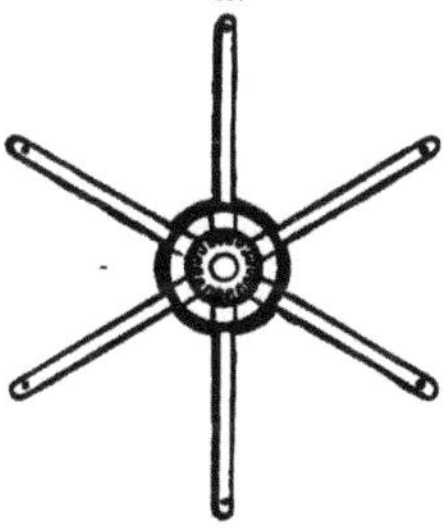

49.

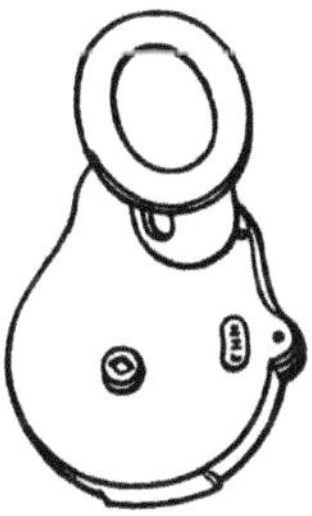

48.

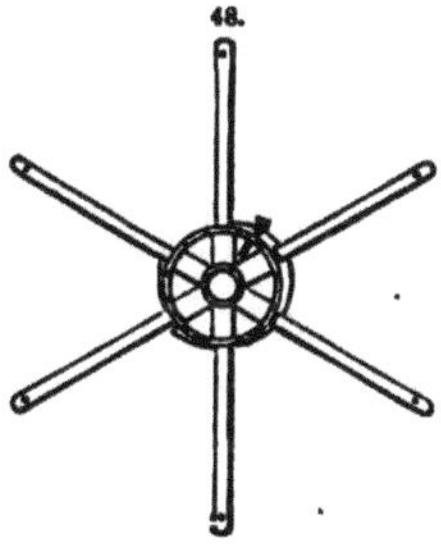

47.

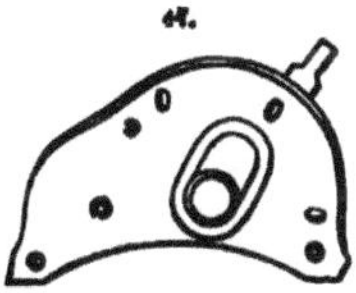

One of *Howard's Horse-rakes* is shown in Fig. 50, and is intended for raking heavy meadow crops, and for windrowing. Although this rake is of a large size, it is within the power of one man, and may therefore be used for general purposes. It can be fitted with a pole instead of shafts, and it has been found to leave the hay and corn in a looser or less compressed state than rakes of smaller size. These rakes have from 24 to 28 steel teeth each, the wheels are 42 inches high, the extreme width between the wheels from 7½ to 8½ ft.; the heaviest of these rakes does not weigh more than 5 cwt.

Fig. 51 shows a horse-rake on the same principle, but made so that the shafts can be readily removed to the end of the rake, by which means the implement can be drawn endwise through gateways or along narrow roads. This form of hay rake is well suited to mountainous districts where roads are narrow.

50.

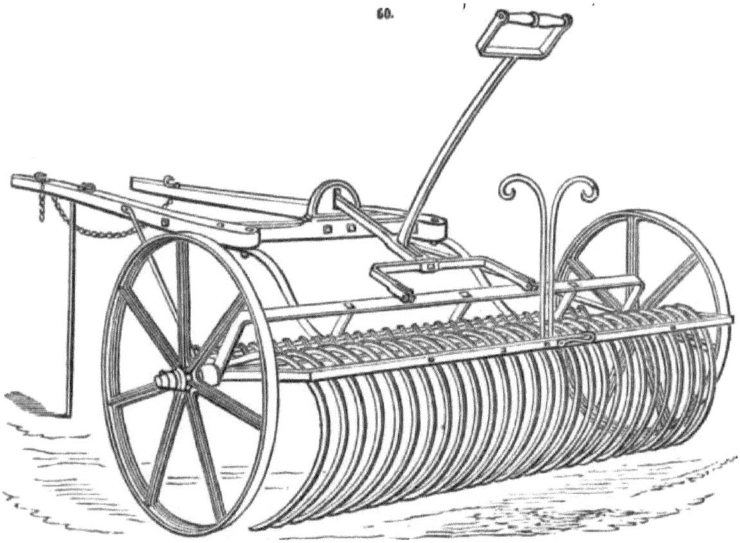

51.

On the Application of Steam-Power to Cultivation.—This article is taken from a paper published in the 'Proceedings of the Institute of Mechanical Engineers, 1865-6,' the joint production of John Fowler and David Greig, of Leeds. In considering the mechanical problem to be solved in the application of steam-power to agriculture, it is requisite before referring to the design of any particular machine to examine the general principles on which the application of mechanical power to cultivation can be best effected. To do this effectually, it is necessary to ascertain the nature and extent of the difficulties to be overcome, and these may be stated to be the following:—

I. The irregularities of level in the surface to be acted upon.

II. The varying positions of the machinery upon the ground rendered necessary as the work proceeds.

III. The difficulty of getting heavy engines of sufficient strength moved about where no roads exist.

IV. The production of a rope of sufficient strength, hardness, and elasticity, to stand the work.

V. The changes in the state of the soil from effects of the weather.

I. The first idea which naturally occurs in applying steam-power is that of attaching the motive power direct to the implement, as is done in the case of horses. But experience has proved that the power required to move a steam-engine over land, of sufficient power and weight for traction purposes, is quite impracticable, from the fact that such an engine would weigh at least 12 tons, and would in many cases absorb as much as 30-horse power in the mere act of moving itself at the rate of 2½ miles an hour. Moreover, when the land gets at all wet and greasy on the top, it becomes quite impossible to make such an engine travel over the soil; while, moreover, the compression caused by its travelling over the land would in most cases neutralize the good otherwise effected by the cultivating implement. Under these circumstances it becomes absolutely necessary to convey the power over the surface of the land by means of a rope, allowing the prime mover.

The use of wire rope for this purpose met at first with great difficulty, and was, from various causes, attended with great drawbacks, but these having been gradually overcome, it is now pretty generally applied. The first system of using rope was by placing the engine in a stationary position at the side or corner of the field to be cultivated, as shown in the diagram, Fig. 52,

52.

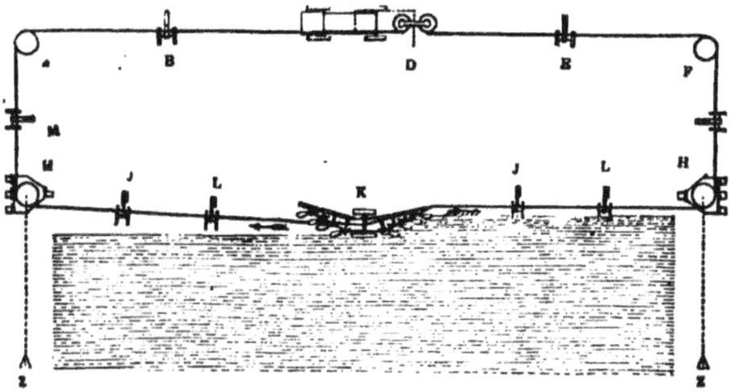

leading the rope all round the margin of the field; the two ends of the rope were attached to two winding drums at the engine, giving out and taking in the rope alternately, and the plough or cultivating implement being attached to the middle of the rope, was hauled backwards and forwards across the field. This rectangular arrangement involved a great deal of fixing machinery in the field before commencing operations, including fixing the engine and windlass, fixing a pulley or snatchblock at each of the two corners of the field nearest to the engine, and a large number of rope porters, or carrying pulleys; it also entailed two movable anchors, one at each end of the line of traverse of the implement, which had to be shifted by some means each time that the traverse was reversed, so as to lead the implement into a fresh line. In Fig. 52, D is the engine, B E J L M are rope porters, C the windlass, A and F pulleys, H H movable anchors, K the plough, and Z Z stationary points. The general construction of the rope porters is shown in Figs. 53, 54, 55, 56, 57, 58. Figs. 53, 54, 55, show the larger kind used for the permanent lines

53.

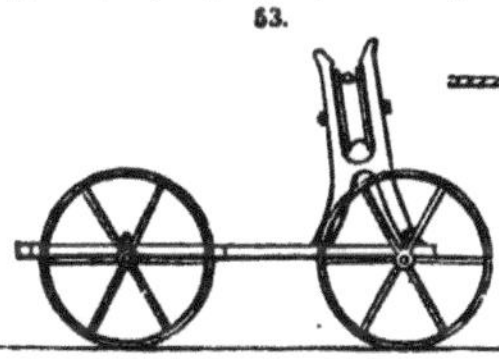

54.

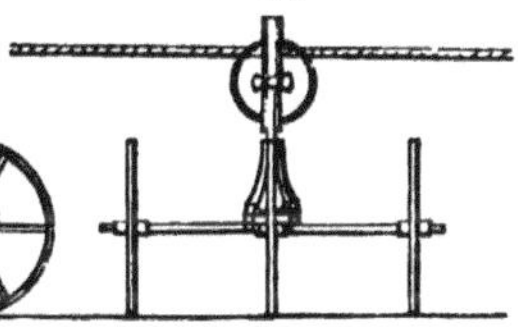

55.

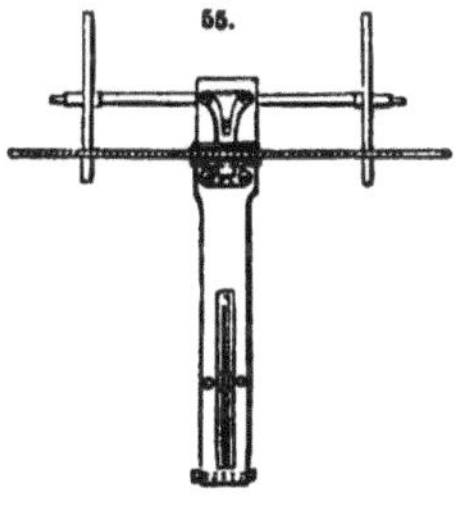

56.

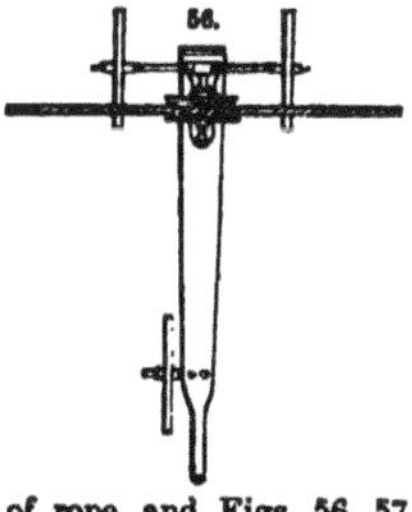

57.

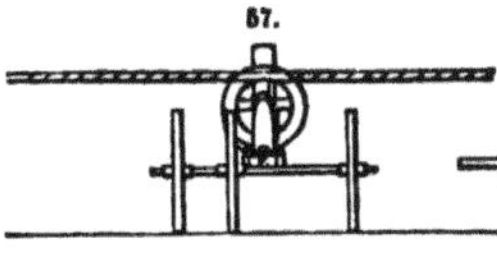

58.

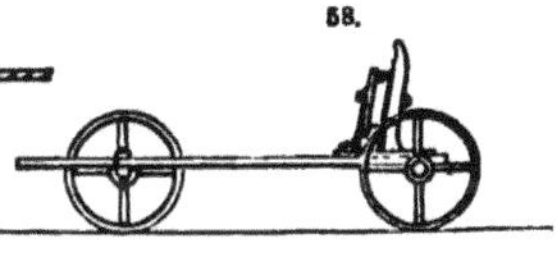

of rope, and Figs. 56, 57, 58, show the small porters for the rope attached to the implement, which are withdrawn and placed again by boys as the implement passes across the field. In employing such an arrangement of tackle, the consideration of the complication of the parts, the numerous pulleys and frequent bending of the rope over the pulleys, which were of necessity small in diameter, and the great time required for fixing the apparatus, early led to the conclusion that such plan of applying power could not prove permanently successful, and so it is now superseded by more direct and simple arrangements.

The second mode of using wire-rope, shown in Fig. 59, was merely a modification of the first, and consisted in placing the stationary engine and windlass in the centre of one side of the field, and leading the ropes away diagonally across the field to two movable anchors placed at each end of the line of traverse of the implement. A pair of horizontal leading pulleys attached to the windlass allowed the rope to pass off at the varying angles which the progress of the work required until both the movable anchors came in a straight line with the windlass. By this triangular plan the two fixed pulleys in the corner of the field, in Fig. 52, were dispensed with, and fully one-fourth of the rope with its requisite porters was saved. This arrangement was a great improvement on the former, and the encouragement that it elicited led to a further step,

59

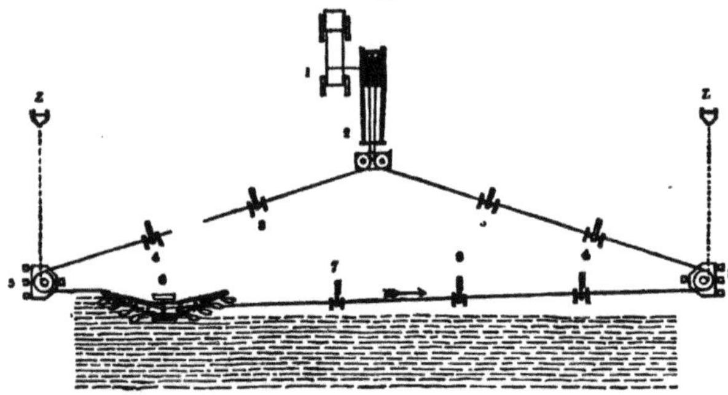

which suggested the important principle on which all subsequent machines have been constructed, namely, that of direct pull.

In Fig. 59, 1 is the engine; 2 the windlass; 3, 4, 7, 8, rope porters; 5, 10 movable anchors; 6 the plough, and Z, Z, stationary points.

The third plan of working with rope, with direct pull upon the implement, is shown in Fig. 60, and consisted in placing two horizontal winding drums under a travelling engine which moved

60.

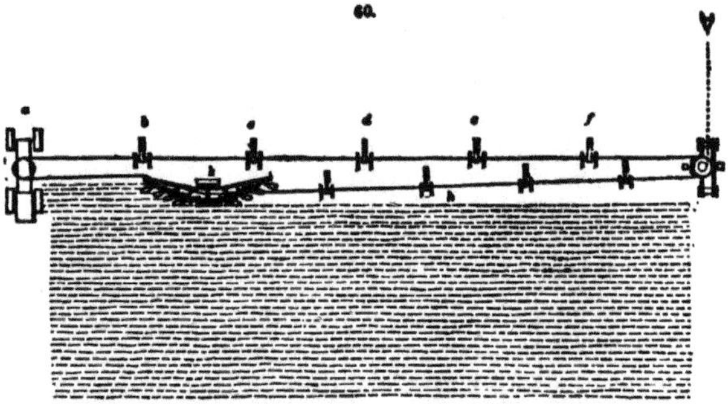

slowly along the headland of the field, keeping always in line with the work. The travelling motion was obtained by means of a pinion gearing into a large internal toothed wheel fixed upon one of the carrying wheels of the engine, and connected to it by a friction clip to prevent any risk of injury from overstrain. The rope was stretched from one winding drum of the engine across the field to a movable anchor on the opposite headland, and then back to the implement to which it was attached, and another rope from the other drum was also attached to the implement. The work was performed by the engine winding up one drum as it gave off rope from the other, the implement being thereby pulled backwards and forwards across the field.

In Fig. 60, *a* is the engine, *b c d e f h* rope porters, *g* the movable anchor, *k* the plough, and Z stationary point.

The movable anchor is shown in Figs. 61, 62, and consists of a carriage with a horizontal pulley, A, mounted on it, round which the hauling rope, B, of the plough worked while the sharp-edged carrying wheels, C, entered the ground and resisted the side pull of the rope. The anchor carriage was moved forward each time of changing the direction of the implement by means of a stationary rope, D, stretched along the headland and made fast at the end, as shown in Fig. 60. This rope was attached to a small drum, E, on the anchor carriage, and a slow motion was communicated to the drum from the pulley, A, by the two pair of wheels and pinions, F, being thrown into gear, the anchor carriage thus pulled itself along the headland a sufficient distance each time, so as always to keep in line with the implement and engine. The box, G, on the carriage was weighted sufficiently to serve as a counterpoise to the pull.

The experience gained in this plan of working showed that the principle of direct pull of the engine upon the implement was the correct one, but the cumbersome arrangement of the two winding drums and the difficulty of coiling the whole length of rope required for reaching across the field, together with the crushing of the rope arising from the soft material of which it was then made, and the small diameter of the drums necessarily employed, indicated the need for a still further modification in the apparatus. The next step was the employment of an endless rope stretched across the field, as in the preceding case, with this difference, that the power was now communicated to the rope by friction instead of by winding on and off a drum, as in the plan last described. In order to secure such an amount of hold on the rope as to give sufficient pull, it was found necessary to employ two driving drums with four grooves each, as shown at A A in

61.

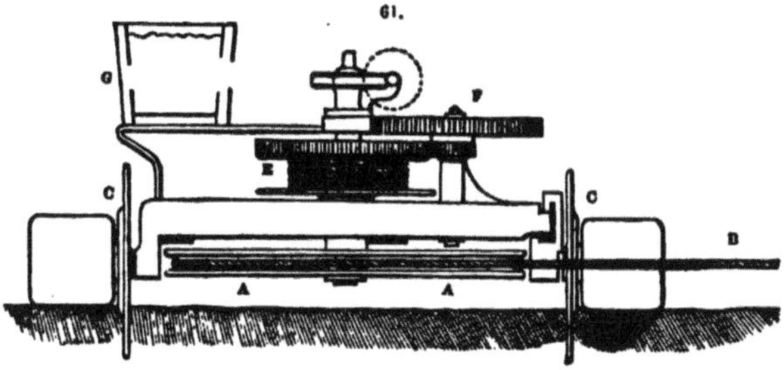

62.

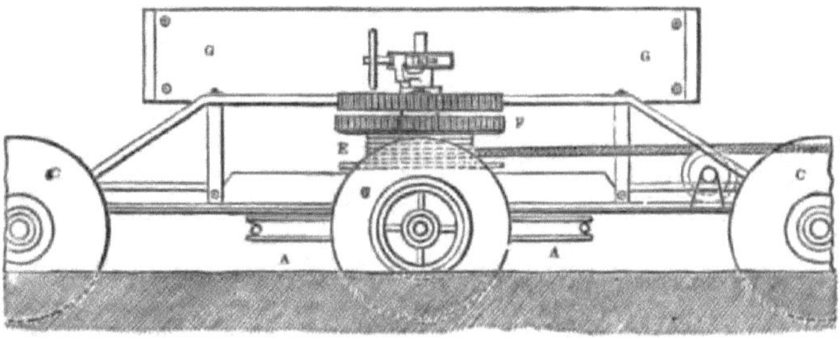

Figs. 63, 64; and the rope was led four times half round both drums, as in Fig. 64, the two drums being geared together by the pinion B. By this means sufficient hold was obtained to

63.

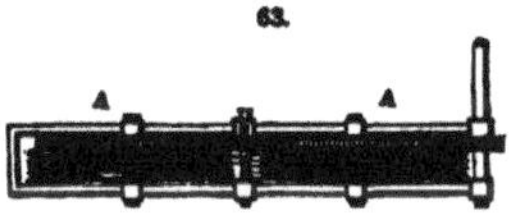

64.

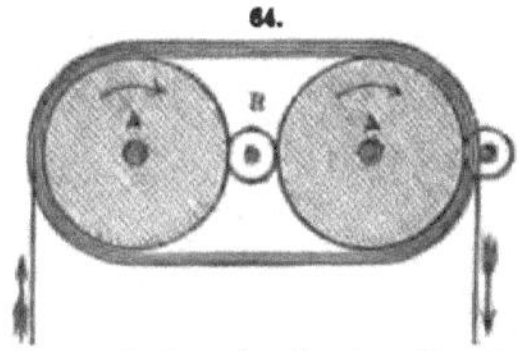

overcome the resistance of the work. In order to meet the variations in the length of the rope occasioned by the irregularities in the boundary of the field, two light barrels worked by hand were mounted on the cultivating implement to which both ends of the rope were attached, and by these barrels a portion of rope was let out or taken up by hand as required to keep it at the proper degree of tightness. When new, this apparatus worked very well, but the wear and tear of rope from its numerous bends, and more especially from another cause, which required some time to develop itself, rendered it necessary to abandon this plan. This great difficulty was the impossibility of keeping the eight grooves of the driving drum all of equal diameter. The two leading grooves were found to be always wearing at double the rate of the others, and all the grooves having to revolve at the same rate, a constant surging of the rope was occasioned by the difference in speed of the circumference of the different grooves. This involved destructive wear of the rope and loss from friction, and every revolution of the drums caused a further grinding away, thus increasing the errors in the diameters of the grooves. As an instance of the deterioration thus occasioned, it may be mentioned that the apparatus got into so bad a condition that the engine could not perform one-half the work that was done by it when new.

These evils led to a modification of this plan of driving, by the employment of a single driving drum with two V-grooves, as shown at C, Figs. 65, 66, round which the rope was made to take two three-quarter turns, one in each groove. This was effected by using two guide pulleys, D D, one on each side of the driving drum C, which transferred the rope from one groove to the other of the driving drum. In this case, as there was only one driving drum with two grooves in it instead of two drums with four grooves in each drum, the wear and tear were greatly diminished; and this plan of apparatus, although retaining to some extent the evils of the former, is still working successfully in several places. The objections still remaining, however, are the number of bends to which the rope is subjected; and from the grip on the rope being obtained by its forcing itself into the V-groove by the tension put upon it, serious wear and tear result. In this case there is a compound surging, for from the point where the rope first touches the drum, the pressure, forcing the rope into the groove, increases as the rope passes round the drum, causing the rope to lie deeper in the groove, whereby it virtually lessens the diameter of the drum, in consequence of which the rope must keep on surging endways at the same time that it sinks deeper into the groove. Although these movements are so small as to be imperceptible to the eye, they are actually taking place continually, and the result is serious wear and tear from the continuous grinding motion over the whole rope in succession.

The clip drum is shown in Figs. 67, 68, 69, 70, 71, 72, 73, 74.

66.

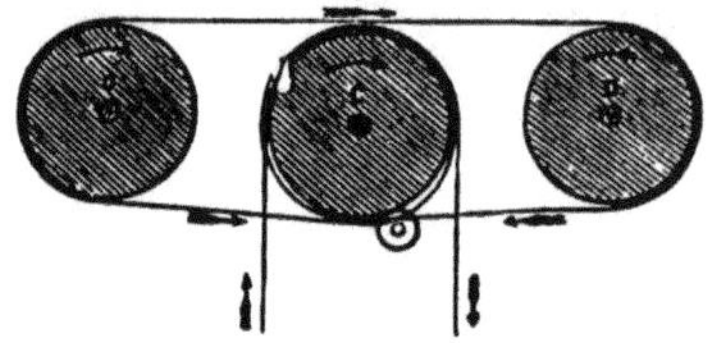

65.

67.

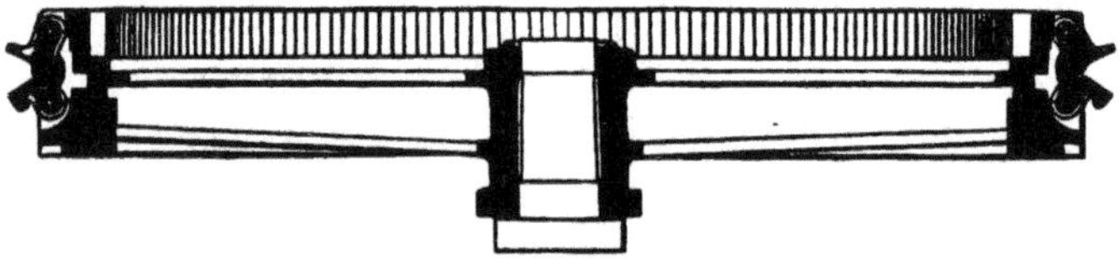

68.

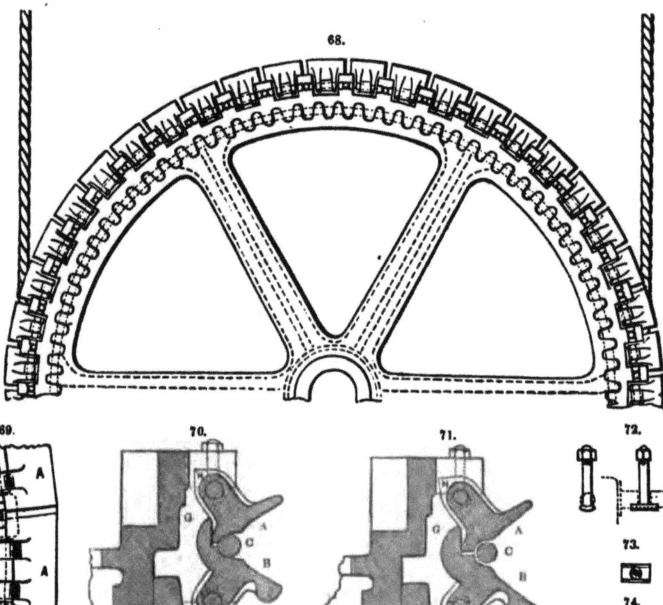

69.

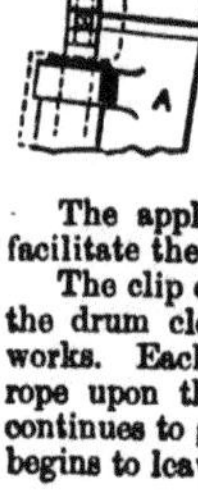

70.

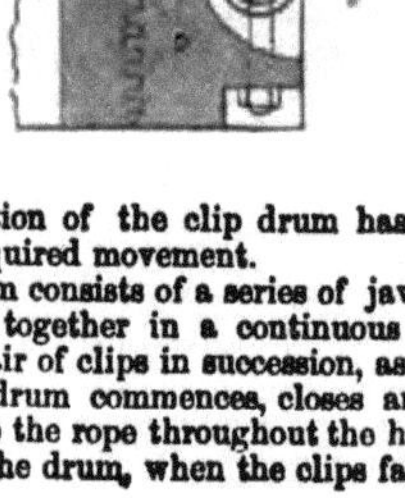

71.

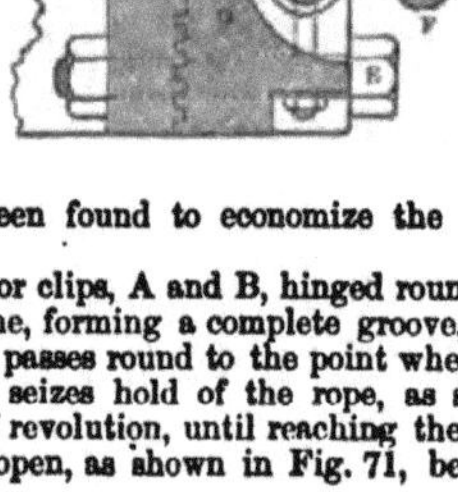

72.

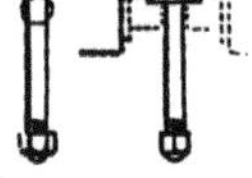

73.

74.

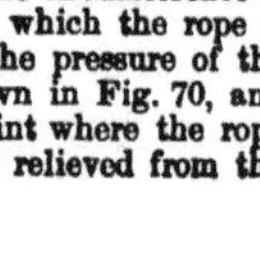

The application of the clip drum has been found to economize the motive power, and to facilitate the required movement.

The clip drum consists of a series of jaws or clips, A and B, hinged round the circumference of the drum close together in a continuous line, forming a complete groove, in which the rope C works. Each pair of clips in succession, as it passes round to the point where the pressure of the rope upon the drum commences, closes and seizes hold of the rope, as shown in Fig. 70, and continues to grip the rope throughout the half revolution, until reaching the point where the rope begins to leave the drum, when the clips fall open, as shown in Fig. 71, being relieved from the pressure of the rope. The amount of grip is in all cases proportionate to the pull upon the rope, and such as effectually to prevent any slipping.

The only provision requisite to suit the clip drum for working with any size of rope is to adjust the width of opening of the clips to the particular diameter of rope to be driven, by widening or contracting the distance between the centres of motion of each row of clips. This adjustment is effected in a very simple and complete manner, by having the lower row of clips, B, centred upon a ring D, Figs. 70, 71, which forms the circumference of one-half the depth of the drum, and this ring is screwed upon the body of the drum by a thread chased round its entire circumference, so that by turning the ring round in either direction the distance between the centres of the upper and lower clips is simultaneously increased or diminished in every pair to exactly the same extent, all of them being kept in perfectly parallel positions. The ring D is held in the desired position by the bolt E, Fig. 71, which prevents it from turning.

The lower clip, B, of each pair having a heavy overhanging lip, F, on the outside, is enabled to lift the upper clip, A, by means of a small finger, G, projecting from its inner end, and pressing upon the tail of the upper clip, so that the clips always remain open until receiving the pressure of the rope, and they fall open again, and release the rope the moment the pressure is withdrawn. The stop H on the upper clip, coming in contact with the body of the drum, prevents the clips from falling open too far. Figs. 72, 73, 74, show the bolts that serve as keeps for holding the ends of the pins on which the clips are centred.

The action of the clips is similar to the closing of a hand upon the rope, laying hold at once so firmly that the rope cannot slip, and retaining this hold uniformly until the rope is released altogether by the opening of the clips, so that all friction or surging from an imperfect hold is avoided, as well as any shifting of the rope at the beginning and end of its contact with the drum, such as is inevitably the case in round or V-shaped grooves; at the same time, by means of the ring D, on which the lower row of clips are centred, the hold upon the rope can be adjusted to any desired amount, according to the power required to be transmitted, and it can be absolutely depended upon when once adjusted to continue working uniformly with the same amount of hold.

An important practical advantage found to result from the working of this clip drum is that the rope is subjected to a continual pressure upon its sides whilst passing round the driving drum, thus avoiding all tendency to the rope of being flattened by the pull, as in an ordinary round bottomed groove, where the pressure of the rope is upon the bottom of the groove only. Also the groove in the clips being so curved as to fit the rope closely round a considerable portion of its circumference, the pressure preserves the form of the rope, and serves to consolidate it by continually closing down all protruding wires, and preventing the deterioration of the rope by such parts being caught in passing the subsequent guide pulleys. In the working of this apparatus, it will be seen, from Fig. 60, that one-half of the total length of rope is never in contact with the driving drum, the other half alone being passed round it backwards and forwards successively, and, in many cases, the actual result has been that the portion of the rope which passes round the drum, and has all the work to do of transmitting the hauling power, has lasted longer than the other portion which has no such work to do, but is simply exposed to the bend round the pulley of the movable anchor on the opposite headland, the friction from the guide pulleys being exactly the same in both cases. Another important advantage is that no tension is required upon the rope leaving the drum; all that is requisite is that the rope be taken away, and not allowed to kink.

It may be remarked that these advantages of the clip drum render it specially adapted for use in other positions where a rope is the medium of conveying power, and the saving that it has effected in the wear and tear of the rope employed in cultivation has been fully corroborated by the result obtained in its use for other purposes.

The second point of difficulty for consideration is the continually varying positions of the machinery upon the ground rendered necessary as the work proceeds, in consequence of which it is necessary for some means to be provided whereby the ropes will admit of the two extreme points being moved nearer together or farther apart, as the varying boundary of the fields may require. With a pair of winding drums this is easily effected, by not allowing the unwinding drum to begin by giving off rope until the rope becomes tight in each case. For this purpose a heavy break has to be applied to the paying-out drum to save the rope from trailing on the ground, for if the rope is not kept from touching the ground a serious loss of power is the result, as the difference in draught required to pull a rope lying on the ground and one properly carried is as much as ten to one. Hence it becomes a very important point that the rope should be efficiently carried off the ground.

Figs. 75, 76, show the construction of the compensating break that is employed when a pair of winding drums are used, so as to compensate for changes in the length of rope that is required as the work proceeds. The winding drum A is driven by the pinion B, which is coupled to the driving shaft C by the clutch D, but the driving shaft is geared to the paying-out drum E by means of the second shaft, F, having corresponding pinions at each end, so as to allow the two drums to run in opposite directions. The two pinion shafts, C and F, are however made to revolve at slightly different speeds by the two outside pinions, G and H, that gear together, being of different size, the pinion H being one-ninth smaller than the other, and consequently the paying-out drum E is compelled to revolve one-ninth slower than the winding-drum A. This causes the slack in the rope to be all taken up by a few revolutions of the drums, and further strain on the rope prevented by the pinion G being connected to its shaft by a friction-break, I, so that it is allowed to slip on the shaft. The rope is thus kept constantly stretched tight by the friction of the break T.

When the endless rope and clip drum are employed for working the implement instead of two winding drums, a very ingenious and efficient plan is adopted, whereby the rope is kept tight without any loss of power, the slack of rope is taken up or more rope is given off, as the field may require, and the implement cannot be started until the rope be tight. This is effected by means of what is termed the slack gear, which is shown in Figs. 77, 78, 79, 80. It consists of two small barrels, A and B, mounted on the plough and connected by gearing with a relative speed

75.

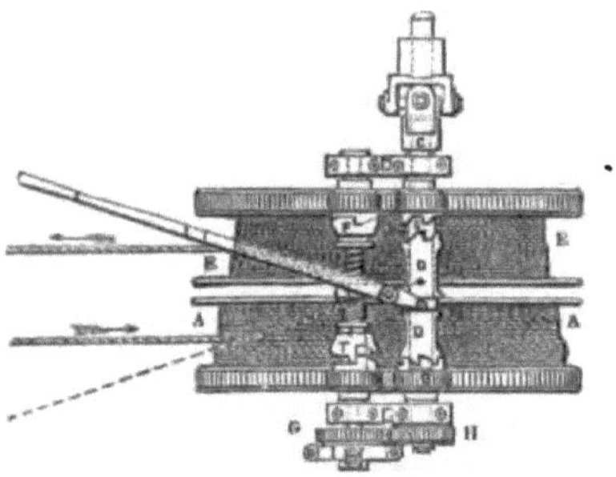

76.

of five to one, so that the pulling rope C, in drawing off one foot length of rope from the barrel A, winds up 5 ft. of the slack rope D on the other barrel B, until all the slack is taken up. The implement then starts at once, when the rope becomes tight, and on its arrival at the other end of the field the act of the man taking his seat at the other end of the implement reverses the action of the barrels, so that what was the slack rope barrel B, becomes the pulling one, and *vice versâ*. The driving of the barrels A and B is effected by a pitch-chain E, which passes over a wheel F,

77.

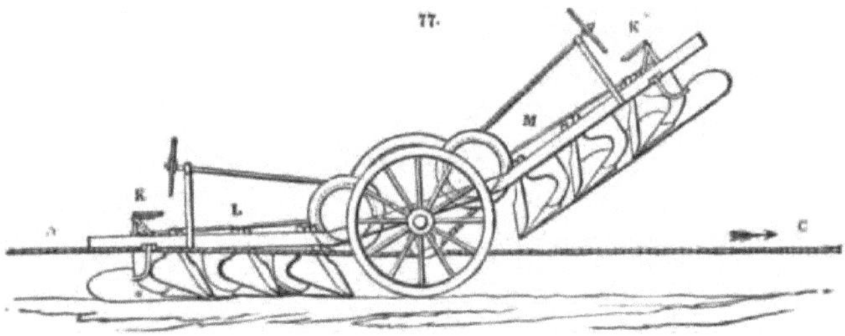

78.

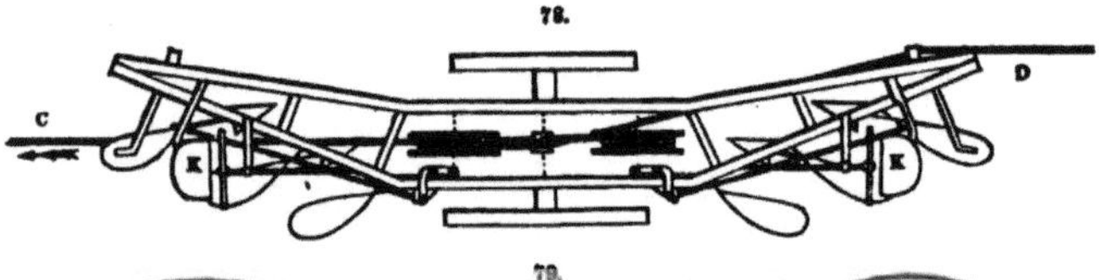

79.

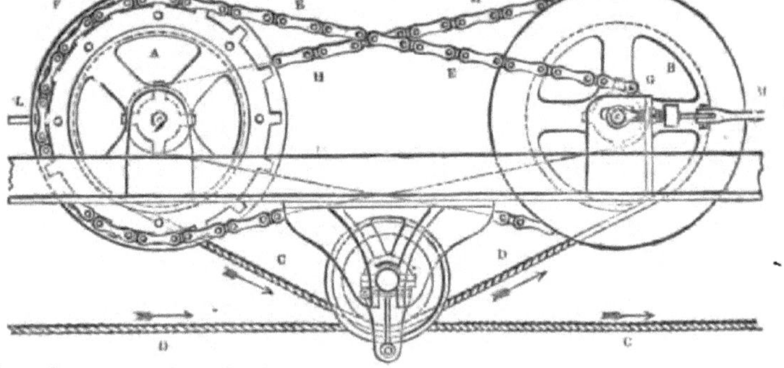

of large diameter, on the pulling-barrel A, and over another, G, of one-fifth the diameter on the slack barrel B, a second chain H being placed on the opposite side of the barrels, working over a pair of wheels of corresponding sizes to the former, but reversed in their relative positions. The self-acting levers and clutches, L and M, are both kept in gear by springs, but are thrown out of gear alternately by the act of the man taking his seat upon the seat K, Figs. 77, 78, first at one end and then at the opposite end of the plough, and by that means pulling the rod L or M. The same action is thus obtained in whichever direction the implement is travelling, one of the chains E or H with its pair of driving wheels being always kept in gear with the rope barrels whilst the other is out of gear. An advantage arising from the use of the slack gear is the elasticity thereby afforded to the rope, should the progress of the implement be obstructed by its coming in contact with stones or roots in the ground: in this case, the rope not being absolutely tight, has a margin for taking up further slack, which acts as a spring easing the strain caused by stopping the implement suddenly. The third difficulty to be considered is that of getting heavy engines of sufficient strength to be moved about over the ground where no roads exist.

80.

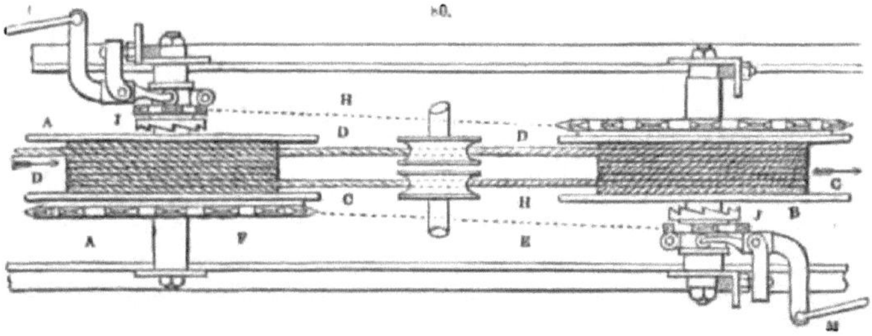

This has been a serious drawback to the introduction of steam cultivation, and one which has led to more breakage of tackle and machinery than all the action of the machinery in performing its work of cultivation. Two causes have contributed to this result, namely, a mistaken idea at first prevailing, that lightness was an essential point, which led to paring down the metal in all parts of the machinery, instead of making the machinery so strong that it could not be broken by the full steam-power, and then increasing the width of the carrying-wheels to such an extent as to ensure carrying the engine over the heaviest and wettest fields. The other mistake was that the speed of working on the road-wheel was not reduced sufficiently so as to allow the engine sufficient leverage to get out of any difficulty it might happen to get into; and the want of judgment on the part of the men using these machines, often led to their being put in places of unnecessary difficulty. The first of these mistakes has been met by making the machinery so strong that the steam when full on is the weakest part of the whole machine. This has naturally led to great weight, but that is no real obstacle, provided the carrying power of the wheels is increased in proportion to the increase of the weight to be carried. In fact, the weight is an advantage in steadiness for working, so long as the machine can be kept from sinking too much into the ground.

Carrying-wheels are now being made for special purposes as much as 30 inches wide on the rim, as shown in Figs. 81, 82, where the dotted lines A A show the portion that is added to

81.

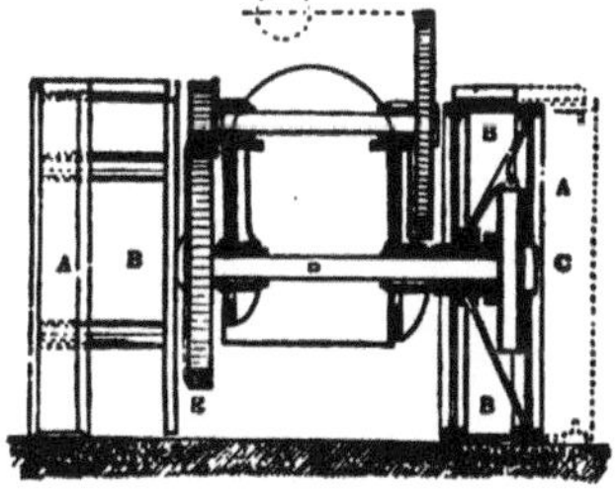

82.

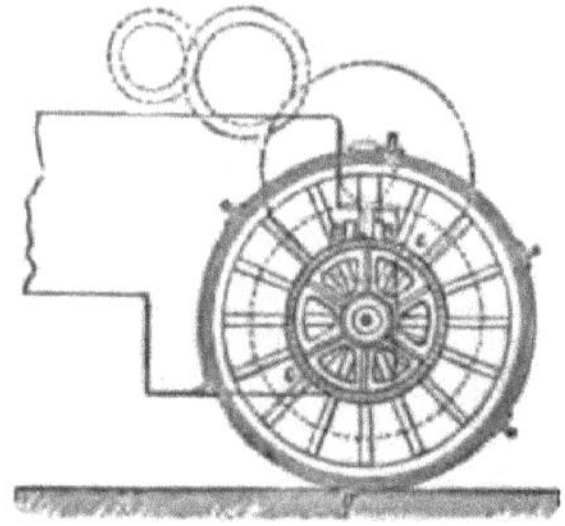

the ordinary 20-inch wheels B B, and these wheels have been proved to carry a 12-ton engine over any land in a fit state of cultivation. The wheels are driven, each separately, by means of a friction-clip C, which prevents any risk of breakage from excessive strain of driving, the shaft D being driven by the pinion and spur-wheel E.

The next point was to reduce the speed on the driving-wheel, so as to give the engine sufficient leverage to get out of any difficulty; and for this purpose two different driving speeds are provided, one giving 2½ miles per hour and the other only 1 mile per hour for travelling, when the engine is working at its full speed of 140 revolutions per minute. In order to obtain sufficient adhesion under specially difficult circumstances for the exertion of the full tractive power of the engine, the additional provision has been made of temporarily fixing transverse T irons by means of bolts upon the rims of the wheels, as shown in Figs. 81 and 82. With regard to the men, time and experience, combined with the extra work caused to them by getting into difficulties, are the means of gradually reducing the difficulty arising from want of judgment on their part

Another plan for meeting the difficulty of getting such heavy machines moved about has been adopted with the most satisfactory results. This consists in combining the power of two small-sized engines, as shown in Fig. 83, the second engine being worked in place of the movable anchor

83.

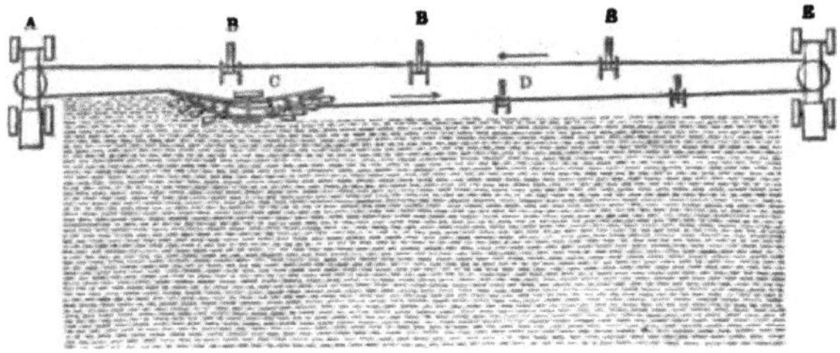

in the previous plan of working shown in Fig. 60. Each engine is provided with a clip-drum, which is essential to carrying out this system of cultivation; and the rope is worked as an endless rope between the two engines by having both its ends attached to the cultivating implement. As the power of both engines is applied at the same time to the rope in each direction, the heaviest class of operations can be performed by them; and the loss of power in working the rope is very much lessened by the fact that both lines of rope are always in effective tension, and are thereby well carried with half the number of rope porters. Another advantage derived from the adoption of this plan is that the engines are better adapted for the other work of the farm, as the farmer has then two engines of 7 or 8 horse-power instead of one engine of 10 or 14 horse-power; and by having two of them a regular system of cartage on the farm can be carried on, the engines being specially arranged for traction purposes.

In Fig. 83, A, E, are the engines; B, B, B, D, rope porters; and C, the plough.

The fourth difficulty to be surmounted was the production of a rope of sufficient strength and hardness, combined with elasticity, to stand the required work; and this was a very serious point. as the inability to accomplish it nearly upset at one time the profitable employment of steam cultivation.

The first rope used was made of iron wire; but it was worn out so quickly, not doing so much as 200 acres, that it soon became evident such material would not stand the strain and friction attending the work; whilst by increasing the strength of the rope its weight was so much increased as to consume nearly the whole engine-power in overcoming its friction. These difficulties became so serious, that great exertions were made to get a rope of steel sufficiently hard to stand the wear of trailing on the ground and also the friction caused by coming in contact with the numerous pulleys of the machinery then employed; and in 1857 two steel ropes were applied which answered the purpose admirably, and performed with the then imperfect machinery upwards of three times the amount of work that was done by the first iron rope. From this point it was established undoubtedly that all risk of the difficulty with the rope causing a check to the application of steam to cultivation was now safely overcome, the introduction of the steel rope having effectually accomplished the object in view. The machinery for working the rope, however, required great improvement and alteration before getting to the point of thorough efficiency with a minimum of wear: the chief objects in these improvements being to have as few bends as possible, and those bends over large pulleys. A great saving in the wear of rope has also been effected by the improved means of keeping the rope tight, preventing it from dragging upon the ground. From time to time, as the various improvements in the machinery have been effected, the increased quantity of work done by the rope before being worn out has been very marked; so that the cultivation of from 2000 to 4000 acres can now be accomplished with one steel rope, the amount varying with the nature of the soil and the width of the implement used.

Although much of this increase of duty depends upon the construction of the machinery, still a great part of the success is to be attributed to the superior manufacture of the steel wire. At first the steel ropes, although much superior to those of iron wire, were very irregular in their quality and durability, often varying as much as one-half in these respects; and up to the present day steel ropes made of the common qualities of steel wire vary in their quality to the same extent. After a series of careful experiments, combined with accurate testing, a quality of wire has now been produced for the purpose, which can be obtained of complete uniformity in tensile strength, and possessing a high degree of hardness, combined with the requisite flexibility and toughness for working. To this great advance in the manufacture of steel wire rope is to be attributed in a great measure the present success of steam cultivation. The tensile strength of this wire has been increased from 1500 lbs. to in some cases 2400 lbs. for No. 14 wire gauge. Steel wire of the common sort has indeed been made to attain nearly the same tensile strength; but this is always accompanied by the defect of brittleness, which is a fatal defect in the working of a wire rope. If the quality of steel rope should continue to improve at the same rate as during the last three years, the cost of wire rope will be reduced to an unimportant item by the acre.

At the commencement of steam cultivation the iron wire rope ran a mileage of not over 750 miles before being worn out, costing 1*s.* 7*d.* the mile of running. The first steel rope ran 1800 miles, costing 1*s.* a-mile; and the present steel ropes are running on an average 9000 miles, costing only about 2½*d.* a-mile, running with a tension upon them of about 25 cwt., and this notwithstanding that the price of rope has been increased from 60*l.* to 84*l.* for the ordinary length of rope of 800 yards. The steel rope at present used in steam cultivation is $\frac{11}{16}$ths inch diameter, and weighs about 2 lbs. a-yard, making a total of about 14 cwt. for the length of 800 yards.

The fifth class of difficulties are those arising from variations in the state of the soil caused by the effects of the weather.

These difficulties have been principally felt in wet weather, in moving the engine, and also from the stickiness of some land when in a half-wet state, which is too often the condition of the land whilst being cultivated. In such cases all the tackle would become literally covered with clay, and the power required to move the rope and the machine would be very great. This difficulty should not indeed exist, as no land ought to be touched when in such a state; but clay land has hitherto been very often worked when wet, from want of sufficient force to get all the work done before the wet sets in, and also from the inability of horses to perform the work while the land is in a dry state. As an illustration may be taken a clay-land field ploughed by horses while very wet, after which, if the next year be dry, it will be literally impossible to work the same ground with horses until some rain comes to soften it, as the horses' shoulders and the implement would not be able to stand such jarring work.

With steam-power, however, there is no difficulty in working the land in the driest condition, which is the proper time for such work; and if this is strictly attended to, it will never get into an extremely hard state. Supposing the clay land is ploughed wet by steam-power, more power will be expended in pulling the dirty rope and the sinking plough than even if the land be so dry that the soil breaks up into large pieces of as much as 1 cwt. each, though the latter could not be the case but for the wet-kneading that the land received before by being ploughed wet by horses. If the farmer were only to keep his machine off the land in wet weather, and work it night and day in dry weather, he would see the great advantage that would accrue from working at the proper time. In fact, the principle of the old maxim, "Make hay while the sun shines," applies to cultivation of the land as well as to the making of hay.

Another system of steam cultivation, shown in Fig. 83, has been adopted to meet special circumstances, by the use of two large engines, each of which is supplied with a winding drum, instead of the clip drum and endless rope employed with the light engines in the plan last described. The two large engines are placed at opposite ends of the field, the same as in Fig. 83; but they act alternately instead of in combination, one pulling the plough in one direction, while the other moves forward into position for the return bout, and *vice versâ*.

84.

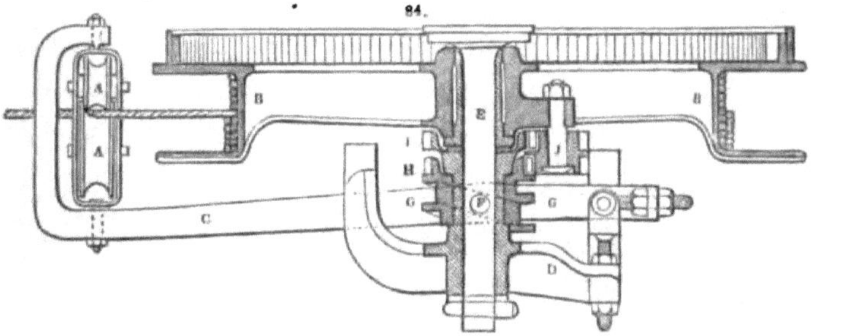

85.

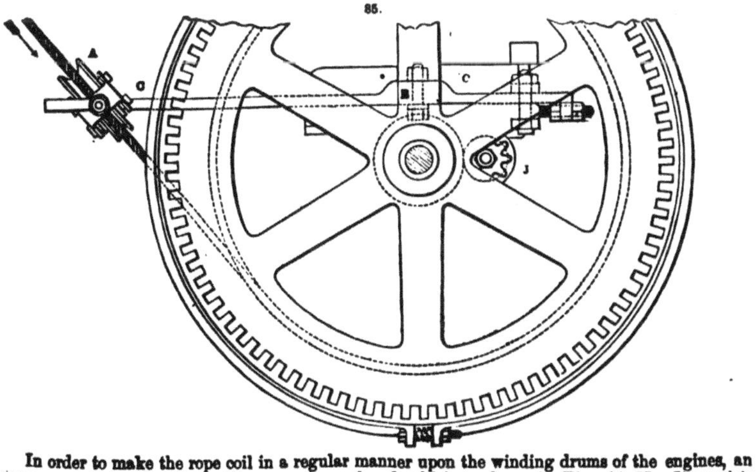

In order to make the rope coil in a regular manner upon the winding drums of the engines, an arrangement of self-acting coiling gear is employed, which is shown in Figs. 84, 85. It consists of a pair of guide rollers A A, between which the rope passes when coiling on or off the large winding drum B. These guide rollers are carried at the end of an arm C, which is centred at the other end upon a bracket D carried loosely upon the centre spindle E, round which the drum revolves. The arm C has a stud fixed in it at F, working in a spiral groove G, formed on the lower part of the spur wheel H, which also turns loosely round the centre spindle E. A second spur wheel I is carried close above the wheel H, and is fixed upon the upper end of a cylindrical casing carried up from the bracket D, and passing through the wheel H. A pinion J carried by a stud fixed in the winding drum B gears into both the wheels H and I, but the upper wheel I has one more tooth than the lower one H; and consequently in each revolution of the winding drum the pinion J being also carried round with the drum causes the lower wheel H to be advanced one tooth, the upper wheel I being held stationary by the arm C, which is held at the outer end by the tight-stretched rope passing through the guide rollers A. The spiral groove G is thus gradually turned round, and acting upon the stud F in the arm C, causes this arm with its guide rollers A to be gradually raised and lowered, thereby guiding the rope from top to bottom of the drum in regular coils whilst it is being wound on or off the drum.

The purpose for which this system of working with two large engines was arranged was for travelling about and doing work by hire, so as to meet the requirements of those who have not sufficient land or capital to purchase machinery for their own use. The plan has the advantage of requiring no fixing, and the machines are ready to start work the moment they get into the field; and as soon as the implement stops, the ropes are in their places and the machines ready for removal. As fields of all shapes are met with, it is of importance that the machines should be of such a character that no loss of time should be occasioned by the management of the rope. So far as the working of these machines goes, it is entirely satisfactory, but the first drawback to the adoption of this plan is the price. Secondly, there is the difficulty of taking two large engines about; and from the fact that a heavy break has to be put on the paying-out drum in order to keep the rope tight, considerable power is lost. But still the time saved in doing small irregular fields more than counterbalances those disadvantages.

The traction part of the machinery having now been considered, the most mechanical means of performing steam cultivation has to be referred to.

The implements hitherto used for steam cultivation have been something similar to those employed with horse-power; but recently a system has been arranged for throwing up the land in the roughest possible way, and leaving it in such a state as to expose the largest amount of surface to be acted on by the air, which is the only truly practical way of dealing with heavy land. The development of different classes of implements will always be going on, to meet different varieties of land, and the various operations which will ultimately be required. It is proposed here only to refer to the best principle of loosening the land for the purposes of cultivation, looking at the question entirely from a mechanical point of view.

Cultivation by rotary implements has been much advocated, and may appear at the first glance the right means of applying steam-power; but when the nature of the substance to be dealt with is considered, this plan is mechanically wrong in the way of operating on the soil, from the fact that rotary implements must necessarily strike on the top of the hard land, thus absorbing a quantity of power in entering the hard substance. As an illustration of this, reference may be made to the method adopted in breaking up a macadamized road: the pick is used so as to lever the material upwards, and by entering it underneath the hard substance the latter is easily broken up. A rotary digger must however be used in the contrary way, or else it will be acting against the onward motion of the machine, thereby increasing the power required for traction. The difficulties that interfere with getting such a machine over the surface of the land have also to be considered, and the damage done to the land by the transit of such a heavy machine over the soil to be cultivated; and these objections, with the serious error in the mode of applying the power, must prevent such a system from proving practically successful.

The cultivation of the land consists merely in loosening a certain quantity of soil, and what has to be considered is how to loosen the greatest quantity with the smallest amount of power. In all the experiments tried by Fowler and Greig it has been found that this is never so economically done as by wedging the soil off to a loose side, and entering the wedge underneath, where the soil is softer. Much objection has been raised to the old mode of working with the plough; but it is not the implement that is at fault in that case, but the power that is defective; and by the aid of steam that implement can now be driven at such a pace as to throw the land sideways in a manner quite equal to the effect of any digging by hand. The great point requiring attention is that the tools should be so arranged that each follows its neighbour, taking its own cut and wedging off the soil to a loose side. If this is done, the speed of 2½ miles per hour at which the implement is driven will throw the loosened material at least 2 feet clear from its previous position, and by the rapid motion it will be left in the state most desirable to the farmer, and in the best possible condition to receive the action of the atmosphere, and this will be effected with the least amount of power. Few implements embody this principle; but without it there is a great loss of power, which is accounted for by the fact that a tyne or cutter, in making its way through the solid ground, always takes twice the power to draw it that would be required if it were taking its cut close to where another cut had been taken before. This is a point of great importance, and should never be lost sight of in the construction of implements for heavy work. In the implements for light operations it is by no means so necessary; but still the principle holds good to a certain extent, and should be attended to as far as practicable.

In order that steam cultivation may be brought to the greatest perfection, it is of the utmost importance that the use of horses in cultivation should be altogether abandoned. For this purpose a number of implements are required, adapted to get over a large breadth of land in a day, so as to do the very light operations of the farm and exclude horses entirely from such work, as their use inevitably increases the expense of after-operations, besides being detrimental to the land. If the horses on a farm are done away with altogether, or as far as practicable, the cartage becomes the next question to be dealt with; and before steam cultivation takes its proper place, the heavy part of the cartage must be done by the ploughing-engines. Considering, however, that to be good ploughing-engines they must be good traction-engines, there is nothing to stand in the way of cartage by steam-power but the want of good roads about the farm, which are an essential point in a highly cultivated farm, whether steam cultivation be employed or not. The experience lately gained in the use of the traction-engine is, that loads can be conveyed over moderately good roads at an expense of *2d.* a-ton the mile; and there is therefore no doubt that the farm cartage can also be done economically. This operation will require some time to develop itself, as the vehicles for conveying the materials will have to be bought, and proper roads made in every direction through the farm, before steam cartage can be carried on conveniently; but at present there can be no reason why all corn should not be taken to market, and coals and so forth brought back, by steam-power. At the time this was written Fowler and Greig had no doubt that before ten years' time two-thirds of the cartage of the farm would be generally done by steam, and also that the railways would be fed by traction-engines, and that these would become quite common: although, through mistaken ideas, serious attempts had been made to stop their use on the public roads. This prediction is being carried out, for it has been proved that a 10-horse engine will convey a load of 20 tons independent of itself over a road with gradients not exceeding 1 in 15; and the wear and tear is very slight indeed in properly constructed traction-engines.

In order properly to understand the advantages of steam cultivation, it is necessary to draw a contrast between steam-power and horse-power as applied to cultivation.

In the case of horses, the utmost available force which can possibly be brought to bear on an implement is 9 cwt., and this is obtained by employing six horses, or two more than can work profitably together on the land. The practical limit of draught is therefore 6 cwt., as horses cannot give off the same amount of draught in the fields as on hard roads. Hence with a team of four horses giving off 6 cwt. total draught on an implement which acts on a width of land of from 10 to 12 inches, the utmost total power for an inch in width of the soil acted upon will be only 70 lbs. At the same time the resistance will be very much increased by the pressure of the horses' feet in doing the work. For if a horse be taken when the land is in a rather plastic state, and walked across the track of the steam-plough, and made to travel to and fro transversely on every 10 inches width until a breadth of 6 yards is trodden over, it is then found that if the steam cultivator has just sufficient steam to perform its work properly before it arrives at the ground so trodden down, it will be completely stopped before it gets through the 6 yards; and considering the momentum of the flywheel, this experiment shows plainly that the power required is something very material, and experience shows one-third additional draught to be required on land that has been trodden down to the same extent as in cultivation by horse-power. It is clear therefore that a considerable part of the 6 cwt. draught of the horses is expended in undoing the compression caused by their own weight; and as the 4 tons' weight of the horses themselves must be lifted up and down all the inequalities of the ground, there is only a very small portion of the animal force left to be usefully exerted upon the implement, and this only at the slow speed of 1¼ mile an hour, at which the horses travel.

But with steam the case is very different: a draught of 35 cwt. is available upon the implement, giving the farmer the means of employing a force of from 70 lbs. to 280 lbs. to each inch in width of the soil moved. And considering that only a 1½-ton load is passing over the land instead of 4 tons, much less force is employed to move the same measurement of soil.

The comparison therefore stands thus:—With horses there is a total force of 6 cwts., with the drawback of having to convey 4 tons of useless load over the land; while with steam-power there is a total force of 35 cwt. conveying only 1½ ton of useless load. The result of experience is, the less weight carried over the land the better; and when the great weight of horses, compared with the force they exert, is considered, and also the number of footprints left by them on an acre, it cannot but excite surprise that such an unmechanical means of cultivation should have existed so long. The number of footprints left by four horses in ploughing a 12-inch furrow is above 300,000 per acre; whereas the steam-plough, which has a width of from 3 to 4 feet, is carried on two wheels 6 inches in width.

The facts that have been stated afford a good reason why horses should be kept off the ground altogether. If the necessary precautions in this respect are but strictly followed, a complete revolution in agriculture would soon be witnessed, as no mechanical means of pulverizing the land is required, and less than one-half the number of operations at present necessary would be found sufficient.

Land, like metal in a furnace, requires the greatest attention in order to perform the different operations at the same time, when it is in the right state for their being effected thoroughly. The present system of management is entirely inadequate to effect this object, from the want of sufficient force at a given time. In some years there are barely two months suitable for all the cultivation of the season: that is, provided the greatest judgment is exercised, and the land never touched except when in the proper state for the purpose of cultivation. But with such management as is here recommended, the result will be always an adequate crop.

The advantages of steam cultivation having now been described, and its gradual development during the past few years having been traced, together with the various systems employed, it is proposed in conclusion to enumerate what may be considered as the principal points which are essential to the production of good steam-cultivating machinery, so far as regards its mechanical arrangements.

First, a sufficiently powerful engine with a wide bearing surface, and plenty of leverage to move itself out of difficult situations, and of simple construction in all its parts.

Second, a hauling apparatus with a drum of large diameter, and so arranged as to bend the rope as seldom as possible and with the drum placed horizontally on a vertical axis, so as to allow the rope to work in any direction without requiring guide pulleys.

Third, a direct pull upon the implement, with as short a length of rope as possible, and that of good quality, light, hard, tough, and flexible.

Fourth, an arrangement for keeping the rope tight, so as to carry it clear of the ground and avoid loss by friction.

Fifth, an implement in which the shares or tynes follow each other consecutively, wedging off the soil to a loose side.

Lastly, as small an amount of manual labour as practicable.

AGRICULTURAL ENGINES. Fr., *Machines à vapeur locomobiles appliquées à l'agriculture*; Ger., *Dampfmaschinen verwendet zu landwirthschaftlichen Zwecken*; Ital., *Macchine a vapore agricole*; Span., *Maquinaria agrícola*.

The portable agricultural steam engine of Holmes and Sons, of Norwich, is shown in Fig. 86. The construction of this engine is simple; the working parts being all outside, the whole can be

86.

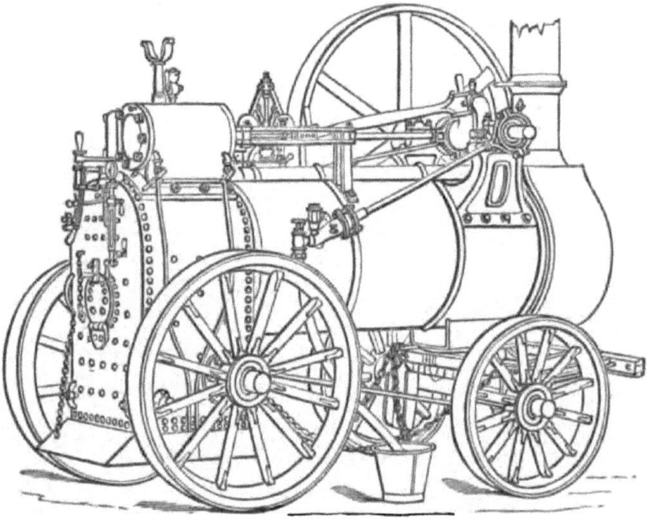

adjusted readily. It is fitted with a *cranked shaft*, so that a wheel or driving drum can be fixed at either side. It has a *governor*, of simple form, having few parts, which works with precision. Every part of the engine is of great strength, wrought iron being used where practicable. The *piston-rod*, pins, and small parts are made of steel. All the brass bearings are wide and easy to adjust. The *feed-pump* is fitted so as to prevent any liability of accident in frosty weather, and in such a manner that it cannot easily get out of order. The *boiler* is of sufficient capacity, and capable of doing heavy work without priming. The *cylinder* is large, with 14-inch stroke.

The engine of Ransomes and Sims, Fig. 87, is specially designed and constructed to economize fuel and to regulate its consumption according to the power required. It has a *feed water-heater*, or reservoir placed in the *smoke-box* of the engine, in which the water is heated by the exhaust steam and hot air passing from the *fire-box* to the *chimney*. A double feed-pump is in connection with this heater by means of pipes shown on the side of the engine. One pump draws the cold water from the supply tank and discharges it into the heater. The other is supplied with hot water from the heater, and forces it into the boiler. The *slide-valve* is on the gridiron principle, and the pressure of the steam is removed from the back of the *valve* by means of metallic equilibrium relief-rings kept up to the faces of the *slide-case cover* by means of spiral springs; these valves are found to be as easily moved when the steam is on as when it is off. The *cut-off*, or expansion-valve, is worked by a movable *eccentric*. The adjustment of this eccentric is very simple; it is held in its place by a nut screwed to a bolt fixed to the fast eccentric, under which is placed a pointer, which slides on a plate graduated with the various grades of expansion, and fixed to the other eccentric, and the engine-driver has only to adjust the loose eccentric until the pointer reads the mark in the grade which corresponds with the point at which the steam is cut off in the stroke of the piston. The general construction of the engine is very strong. The average consumption of fuel is about 3·5 lbs. of ordinary coal an hour.

Clayton, Shuttleworth, and Co.'s Portable Steam Engine.—This engine, Fig. 88, in its construction presents a method of heating the exterior surfaces of the cylinder and *steam chest*. The cylinder is placed in the smoke-box surrounded by a *jacket* forming an annular space, which, being filled with steam, heats the cylinder, and, at the same time, protects it from the injury that would be caused by direct contact with the heated gas. By this arrangement condensation and radiation are prevented.

There has been a great objection to placing the cylinder and steam chest in the interior of the boiler, as it rendered the engine more difficult to examine and repair; but in the engine under notice this has been obviated by leaving both ends of the cylinder and steam chest exposed; the

87.

covers can be removed, and the piston and slide-valve cleaned or adjusted as expeditiously as in an outside cylinder engine. This engine is also made to reverse. See ENGINES, *varieties of.*

Works appertaining to this subject:—

'The Implements of Agriculture,' by J. A. Ransome, royal 8vo, 1843.

'The Farm Engineer, a Treatise on Barn Machinery,' by R. Ritchie, royal 8vo, 1849.

'Agricultural Engineering,' by G. H. Andrews, 12mo, 1852–3.

'On Steam Cultivation,' by John Fowler, jun. Proceedings Inst. Mechanical Engineers, 1857.

'The Book of Farm Implements,' by James Slight and R. S. Burn, royal 8vo, 1858.

'The Official Illustrated Catalogue of the Exhibition of 1862,' British Section, 2 vols., imperial 8vo, 1862.

'On the Application of Steam Power to Agriculture,' by John Fowler and D. Greig. Proceedings Inst. Mechanical Engineers, 1865.

'Steam Cultivation,' Engineering, vol. iv., 1867.

'Études sur l'Exposition,' 1867. Lacroix, Paris, 1867–8.

'On Steam Cultivation,' by Baldwin Latham. Transactions of the Society of Engineers, 1868.

'Officieller Ausstellungs-Bericht herausgegeben durch das K.K. Österreichische Central - Comité.' Vienna, 1868.

88.

AICH'S METAL. FR., *Métal d'Aich;* GER., *Aich's Metall;* ITAL., *Lega d'Aich;* SPAN., *Metal de Aich.* See ALLOYS.

AIR-BRICK. FR., *Brique creuse;* GER., *Hohlziegel;* ITAL., *Grata dello spiraglio, dello sfiatatoio;* SPAN., *Ladrillo perforado.*

An air-brick is a brick of the ordinary size, made of earthenware, built into the walls of a building, but perforated, Fig. 89, to admit air under the floors or into the rooms. It is sometimes made of cast iron, with a slide worked by a small knob, Fig. 90, to enable the openings to be closed if required.

89.

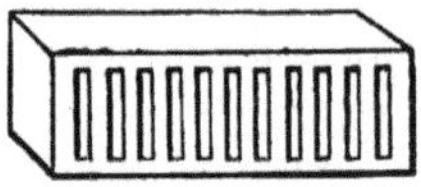

90.

AIR-GRATING is similar to an air-brick, but of larger size and of less thickness in proportion to its other dimensions. It is used as in the case of the air-brick, to admit air to the interior of buildings.

AIR-CHAMBER. FR., *Chopinette, trachée, réservoir d'air*; GER., *Windkessel, Windraum*; ITAL., *Serbatoio d'aria*; SPAN., *Cámara de aire.*

A cavity containing air to act as a spring for equalizing the flow of a liquid in pumps and other hydraulic machines.

Fig. 91 is a section of a locomotive feed-pump; the water is drawn in by the action of the plunger in the barrel A, through the feed-pipe B, and valves E, which rest on their seats F, and held in place by the cages G; the water entering the air-chamber D, in the top of which the air is compressed, forcing the water out of the delivery-pipe C beyond its middle position when the piston is at the end of its stroke. The feed-pumps of American locomotives are supplied with air-chambers *d*, *d*, on the suction side, as well as D, D, for the delivery. This pump was invented by Walter McQueen, an American engineer. The elastic force of the cushions D, *d*, of condensed air in the air-chambers, relieves the pipes, valves, and joints from sudden shocks; besides, the action of the air by its alternate compression and expansion secures a steady supply of water to the boiler. The barrel of the pump is generally of brass, the plunger, working in the pipe A, a solid bar of iron. The valves are of the cup form, Fig. 91. The rise of the valves is seldom more than $\frac{1}{4}$th of an inch; and sometimes, while the rise of the inlet-valve is restricted to $\frac{1}{4}$th of an inch, the delivery-valve rises only $\frac{3}{16}$ in. and the check-valve $\frac{1}{4}$.

91.

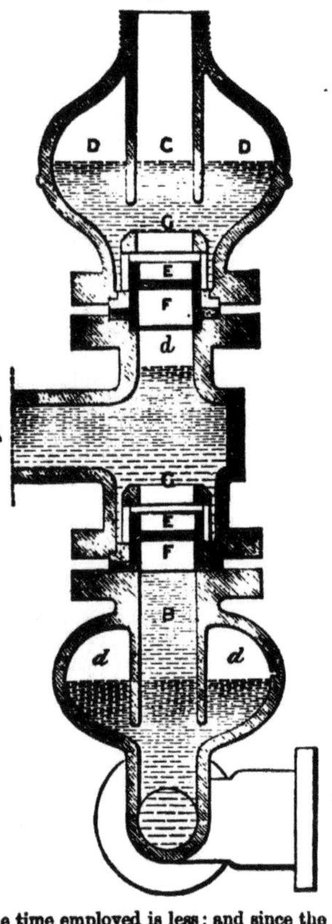

If a plunger P, Fig. 92, working in a pipe P A B, $5\frac{1}{2}$ inches in diameter, forces the water *a b g h*, 9 inches from *a* to *g*, into a smaller pipe *h f*, in a second, the same quantity of water passes through the smaller pipe in the same time, but with an increased velocity. A head of water H, giving the same pressure as the plunger P, would have the same effect. Let E F = 3 inches at the narrowest part of the small pipe, then the velocity of the water at E F will be 30·25 inches a second, while at A B in the larger pipe the velocity is only 9 inches a second; for $\frac{(5 \cdot 5)^2 \times 9}{3^2} = 30 \cdot 25$. Now let W be the weight of a column of water which produces the pressure P, which may be supposed constant, and put t for the time of its operation, long or short, and $m = \frac{W}{g} = \frac{W}{32 \cdot 2}$; whence $P\,t = m\,v$, v being the velocity, therefore $P = \frac{m}{t}\,v$. This expression shows that the effort required to impart or destroy a quantity of motion $m\,v$ is so much the greater as the time employed is less: and since the reciprocal action of bodies is more rapid compared to the spaces described, their compressions, flexures, and penetrations are less for the same quantity of motion destroyed.

92.

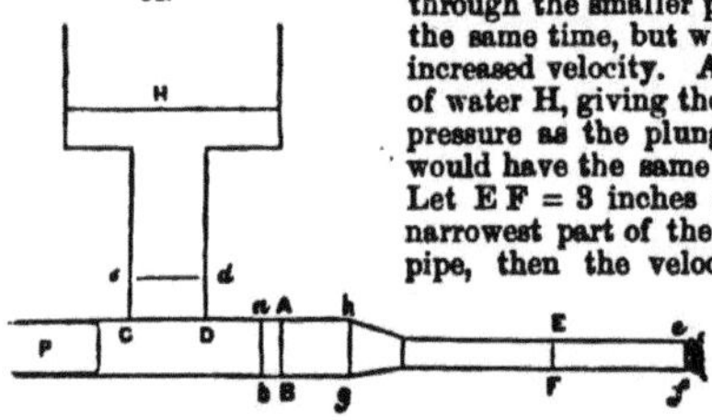

We have here explained why the shock of hard bodies, the transmission or destruction by bodies slightly flexible, compressible, or extensible, occasion such great efforts and such ruptures and accidents; and how it is, on the other hand, by the interposition of soft and compressible bodies, that the intensity of efforts and their consequences are so much diminished. We may see by the expression $P = \frac{m}{t}\,v$ that a finite velocity could never impart in an infinitely small time to a mass m. except by an infinite effort, which shows the error in the hypothesis of the instantaneous transmission of motion by forces, to which we are then compelled to give a special name, and thus suppose a special nature in calling them *forces of percussion*. This error is often too explicitly admitted in the teachings of rational mechanics. Nothing like an instantaneous operation really occurs in nature; quantities of motion are imparted and destroyed in greater or less periods of time, sometimes, indeed, imperceptible to our senses and means of observation, but never instantaneous.

Suppose the force acting on the plunger to be 48·3 lbs. on the square inch, and that it is required to find the pressure acting on a valve at $e\,f$, which valve opens $\frac{1}{8}$ths of an inch against the action of a spring; then $\frac{48\cdot 3}{32\cdot 2} = \frac{3}{2} = m$; we have before shown that the velocity of the water in the small pipe E f = 30·25 inches a second, $= \frac{30\cdot 25}{12}$ feet a second; whence the force acting on each square inch of the valve will be—

$$\frac{3}{2} \times \frac{242}{7} = \frac{30\cdot 25}{12} = 130\cdot 7 \text{ lbs.}$$

for the time the water in the small pipe is moving over $\frac{7}{8}$ in. $= \frac{7}{96}$ ft. $= \frac{7}{242}$ seconds. When the spring at $e\,f$, or the pressure of the water and steam in a boiler, becomes too great to be overcome by the action of the plunger P, the air in the air-chamber D, Fig. 91, by being condensed, conservates the balance of the force of the water passing through the small pipe, and delivers such concentrated force when the plunger injects more water. Care should be taken in making use of what is here termed the quantity of motion or *momentum*, for when we know the product of the mass m, of a body, and the velocity imparted to, or taken from it, we have the measure of effort produced by the force during the period of action; but we see that this measure cannot be taken as a term of comparison except for analogous cases, where the velocities are really imparted or destroyed by force; and it does not follow that the product P t, of the force, by its period of action (equal, when there is a change of motion, to the quantity of motion imparted or destroyed), should always serve as a measure of the effort of forces, as is sometimes admitted for certain instruments and certain kinds of work. It is often seen that an effort may continue a long time without producing a mechanical effect. Thus, horses pulling upon a mired waggon, without starting it, develop considerable efforts, which multiplied by the period of their action would give an enormous product without any useful effort resulting in any mechanical work, and nothing but fatigue and exhaustion of the horses. Take, for example, the draught of a plough, which in strong earth requires a mean total force of 794 lbs. Suppose the furrow to be 400 feet long, the horses in one take 100″, and in the other 200″ to plough it. We shall have in the first case, P t = 794 × 100″ = 79,400; and in the second, P t = 794 × 200″ = 158,800; and yet in both cases they have accomplished the same work. An instrument giving the product of efforts by the times or periods of duration would by no means lead to an exact appreciation of the mechanical effects produced. The true measure of these effects is the product of the effort exerted by the path described in its duration, which is usually estimated in *units of work*.

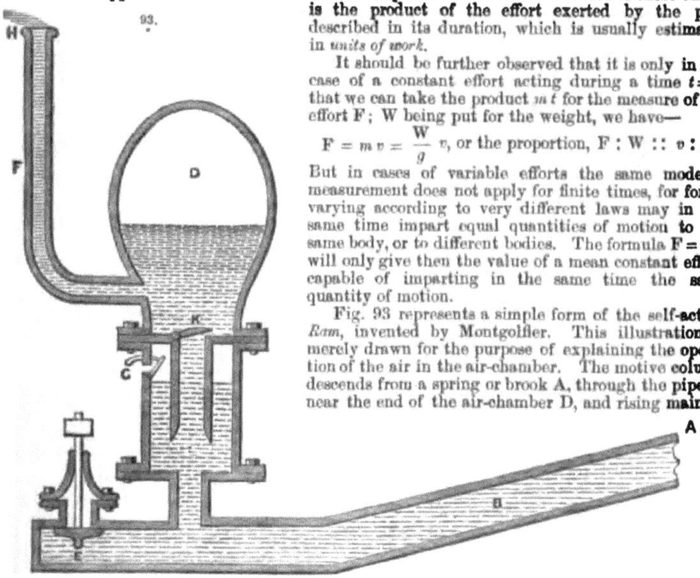

93.

It should be further observed that it is only in the case of a constant effort acting during a time $t = 1''$ that we can take the product $m\,t$ for the measure of the effort F; W being put for the weight, we have—

$$F = m\,v = \frac{W}{g}\,v, \text{ or the proportion, } F : W :: v : g.$$

But in cases of variable efforts the same mode of measurement does not apply for finite times, for forces varying according to very different laws may in the same time impart equal quantities of motion to the same body, or to different bodies. The formula $F = m\,v$ will only give then the value of a mean constant effort, capable of imparting in the same time the same quantity of motion.

Fig. 93 represents a simple form of the self-acting *Ram*, invented by Montgolfier. This illustration is merely drawn for the purpose of explaining the operation of the air in the air-chamber. The motive column descends from a spring or brook A, through the pipe B, near the end of the air-chamber D, and rising main F, which are attached, as shown in the figure. At the extreme end of B, the orifice is opened and closed by a valve E. This valve opens downwards, and may be either a spherical one or a common spindle valve, as shown in the figure. It is the play of this valve that renders the machine self-acting. To accomplish this, the valve is made of, or loaded with, such a weight as just to open when the water B is at rest, that is, it must be so heavy as to overcome the pressure against its under-side when closed, as represented in Fig. 93. Now supposing this valve open, the water flowing through B soon acquires an additional force that carries up the valve against its seat; then a portion of the water will enter the air-chamber D and rise in F, the valve of the air-chamber preventing its return. When this has taken place, the water in B has been brought to rest, and as in that state its pressure is not quite sufficient to sustain the weight of the valve, E opens, descend . the water in B is again put in motion, when its whole pressure begins to act, and again closes E, as before, when another portion is driven into the air-vessel D, and pipe F; and thus the operation is continued, as long as the spring affords a sufficient supply and the apparatus remains in order. The pressure which closes the valve E is that due to the short range of the valve and the increased velocity acquired by the water in passing from a large to a smaller pipe. The surface of the water in the spring or source should always be kept at the same elevation, so that its pressure against the valve E may always be uniform, otherwise the weight of E would have to be altered as the surface of the spring rose and fell.

This ingenious machine may be adapted to numerous localities in every country; but when the perpendicular fall from the source to the valve E is but a few feet, and the water is required to be raised to a considerable height through F, then the *length* of the ram or pipe B must be increased, and to such an extent that the water in it is not forced back into the spring when E closes, which will always be the case if B is not of sufficient length. If a ram of large dimensions, and made like Fig. 93, be used to raise water to a great elevation, it would be subject to an inconvenience that would soon destroy the beneficial effect of the air-chamber D. For if air be subjected to great pressure in contact with water, it in time becomes incorporated with or absorbed by the latter. This sometimes occurs in water-rams like this, for when used they are incessantly at work both day and night. To remedy this, Montgolfier ingeniously adapted a very small valve, opening inwards to the pipe beneath the air-chamber, and which was opened and shut by the ordinary action of the machine. Thus, when the flow of the water through B is suddenly stopped by the valve E, a partial vacuum is produced immediately below the air-chamber D by the recoil of the water, at which instant the small valve opens, and a portion of air enters and supplies that which the water absorbs. Sometimes this *Snifting-valve*, as it has been named, is adapted to another chamber immediately below that which forms the air-chamber, as at G. In small rams a sufficient supply of air is found to enter at the valve E. The compressed air in D not only conservates the work expended in raising the water to H, but also the work required to overcome the friction of the water in the pipe F H; and consequently the air in the air-chamber has the power to close the valve K when the water at H begins to return.

94.

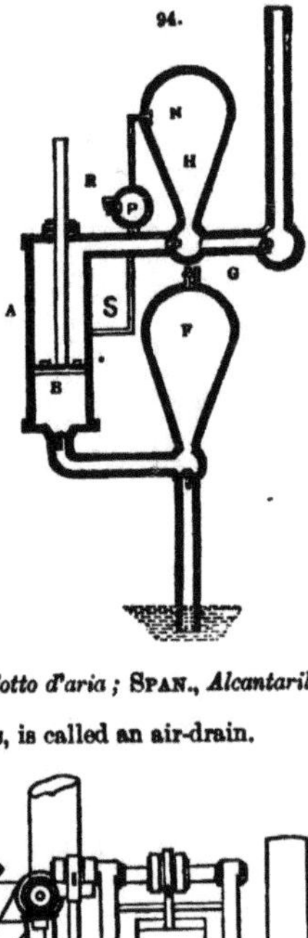

P. H. Vander Weyde has invented a second air-chamber, Fig. 94, to be attached to the induction-pipe, which has a small opening by which air is admitted from without. By the operation of the pumps, air is gradually withdrawn from this chamber and conducted to the principal air-chamber attached to the delivery-pipe.

H and F are air-chambers, P and N valves, and R and G stop-cocks, all arranged so as to supply the constant loss of air taking place in air-chambers. See HYDRAULICS. PUMPS.

AIR-DRAIN. FR., *Conduit d'air;* GER., *Luftkanal;* ITAL., *Condotto d'aria;* SPAN., *Alcantarilla para la conduccion de aire.*

A cavity in the external walls of a building, to prevent dampness, is called an air-drain.

AIR-ENGINE. HEATED-AIR ENGINE.

AIR-ENGINE. FR., *Machine à air chaud;* GER., *Atmosphärische Maschine;* ITAL., *Macchina ad aria calda;* SPAN., *Máquina de aire.*

95.

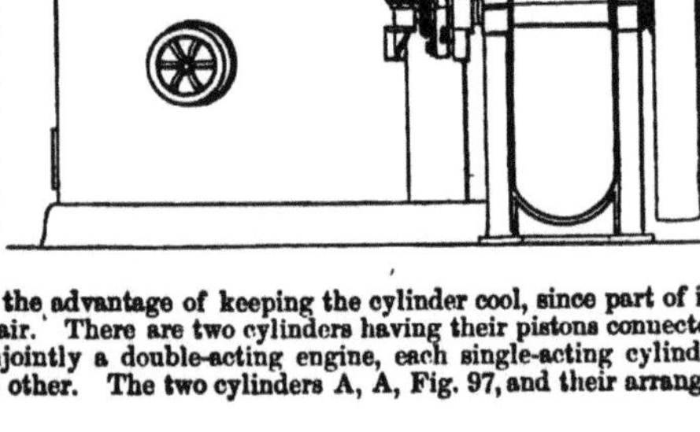

The engine, Figs. 95, 96, 97, 98, invented by Philander Shaw, works with heated air in a close combustion chamber, where it comes in direct contact with the fuel, and is mixed with the gaseous products of combustion. The air and gases pass through the engine, and effect its movement, part of the power being made use of to pump fresh cold air into the combustion chamber. There is no special air-pump provided for this purpose, as the cylinder is made single-acting, and the piston has a large trunk which forms an annular space in the cylinder, and the latter is made use of to act as the air-pump. This arrangement has the advantage of keeping the cylinder cool, since part of its surface is always in contact with cold air. There are two cylinders having their pistons connected together by a beam, so as to form conjointly a double-acting engine, each single-acting cylinder acting during the return stroke of the other. The two cylinders A, A, Fig. 97, and their arrangement, are shown in section in Fig. 96. Fig. 97 shows a section of the engine on a horizontal plane. The combustion chamber is a vessel constructed of boiler plates, and lined with two rows of fire-bricks, all round between which a passage is left for the circulation of air. The air, passing in cold, takes up the heat from the inner row of bricks, and prevents the outer row becoming very hot and wasting heat by radiation. The piston B, Fig. 96, carries the trunk B′, which leaves the annular space D to act as an air-pump, the valves E and F regulating the inlet of the cold air in the down-stroke, and its outlet into the combustion chamber during the up-stroke. There is a regenerator or air-heater interposed between the air-pump and the combustion chamber; this is shown at K, Fig. 98; it consists of a series of vertical pipes, through which the exhaust air from the engine passes up a chimney. The waste heat of this air is utilized in the regenerator by being partly taken up by the cold air passing outside the tubes and arriving at the combustion chamber at a higher temperature. The fire-door is at the top of the chamber, which is provided with a dome *a*, Fig. 96. The grate is slightly inclined towards the ash-door *f*. All these doors are carefully closed during the working of the engine, as the pressure within would be reduced by leakage. The valves are worked by cams and levers, and no special arrangements are required, in working this engine, to keep the cylinder cool.

96.

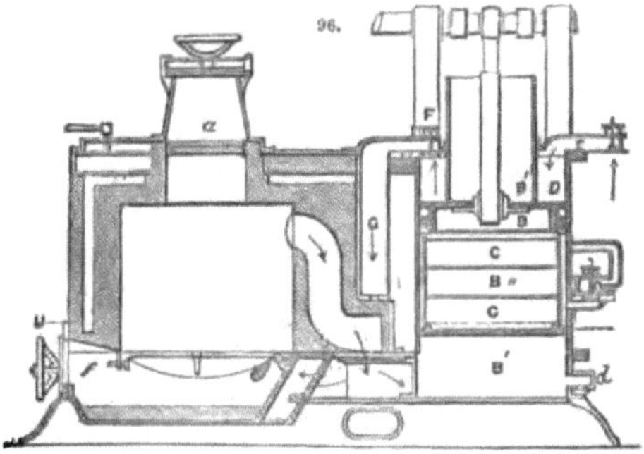

97.

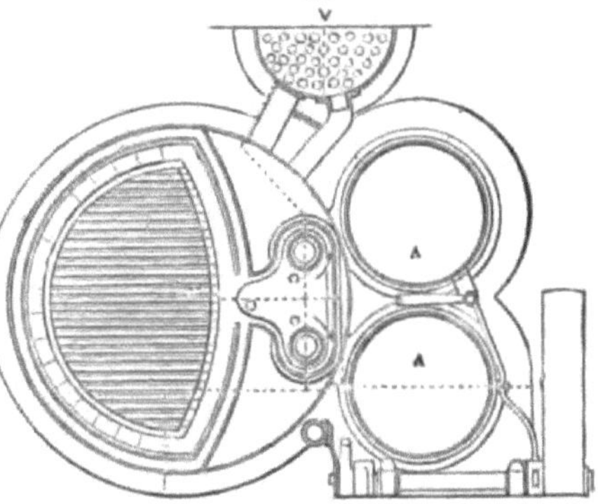

98.

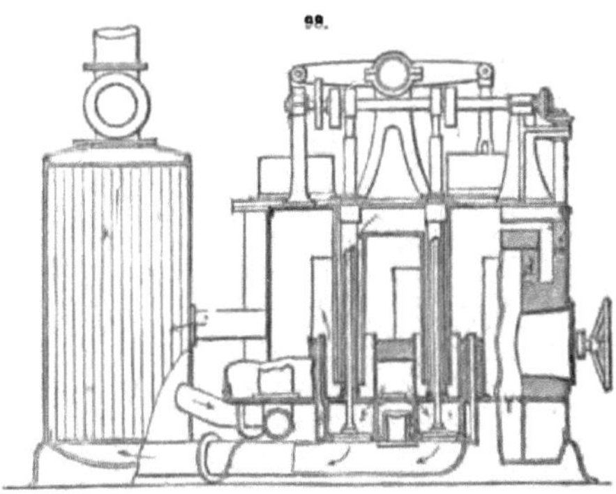

By another arrangement, which we illustrate in the annexed engravings, Figs. 99, 100, 101, the upper end of the cylinder *a*, Figs. 99, 100, is made of such capacity that when the piston *b* reaches the top of the stroke at *c*, a space *d* may be left, in which the portion of air not required may be compressed, and, consequently, will not enter the furnace, and, by its subsequent expansion, will assist the down-stroke of the piston. This space may be formed between the piston and cylinder lid, or the air may be compressed in a side passage or reservoir. In these engines Wenham prefers carrying the crank shaft *e*, in bearings bolted to the top of the flange of the cylinder, and working the crank by a return connecting rod *f*, with the guides *g*, from the piston-rods above, similar in form to that known as a *steeple engine*; in other respects, as in the furnace and valve arrangements, the parts are substantially the same as in most hot-air engines, constructed by others as well as Wenham. In the enigne, Figs. 99, 100, however, Wenham places a disc of fire-clay above the furnace, as shown at *h*, Fig. 99; this is supported on the rim of the fire-pan, and has a central hole, equal in diameter to the bore of the fuel hopper *l*, which touches the disc, and the bottom edge is thus protected from the direct heat of the fire. Around the disc are a series of perforations, J, shown in plan, Fig. 101, leading to the annular space of the stove; a jet of pure flame rises through each of these at every stroke of the engine, and at the same time the fire is prevented from rising above its proper level.

AIR-ESCAPE. Fr., *Appareil pour l'échappement de l'air;* Ger., *Vorrichtung zum entweichen der Luft;* Ital., *Sfiatatoio;* Span., *Venteador.*

An air-escape is a contrivance for letting off the air from water-pipes. It consists of a hollow ball F, Fig. 102, attached to the upper part of the pipe, in which a ball-cock C is placed, adjusted in such a way that when any air collects in the pipes A and E, it will ascend in the vessel, and, by

99.

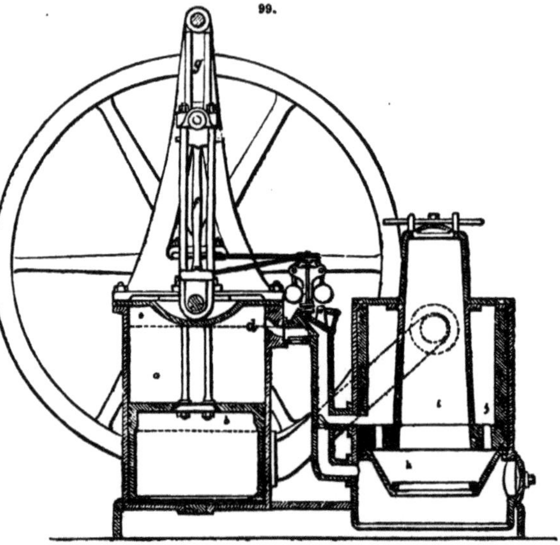

100.

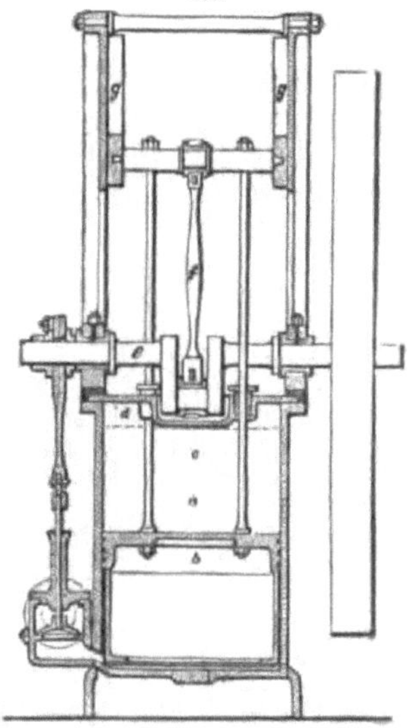

101.

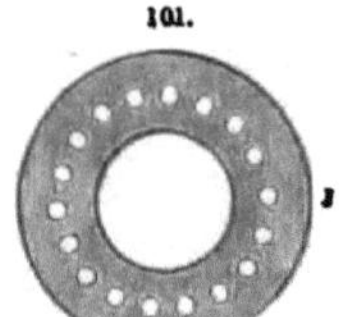

102.

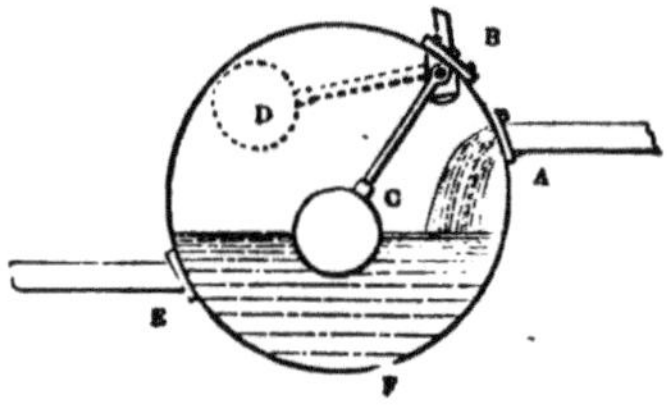

displacing the water, cause the ball to descend, and thus open the cock B, and allow the air to escape. No water, however, can follow it, for when the fluid rises to a certain height the ball rises to D, and shuts the cock.

AIR-GUN. Fr., *Fusil à vent ;* Ger., *Windbüchse ;* Ital., *Schioppo ad aria ;* Span., *Fusil de viento.*

An air-gun is an instrument resembling a musket, made to discharge projectiles by the force of compressed air.

AIR-HOLES. Fr., *Soufflures ;* Ger., *Luftblase ;* Ital., *Bolle d'aria ;* Span., *Respiraderos.*

Holes or cavities in a casting, produced by bubbles of air in the liquid metal, are termed air-holes.

AIR-PIPES. Fr., *Ventilateurs, Conduit ou carreau à air*, Ger., *Luftröhren ;* Ital., *Tubi dell'aria ; Tubi da Ventilazione ;* Span., *Ventiladores.*

Air-pipes are pipes used to draw foul air from a ship's hold, mines, and other close places

AIR-PUMP. Fr., *Pompe à air, machine pneumatique*; Ger., *Luftpumpe*; Ital., *Tromba, Macchina pneumatica*, Span., *Bomba de aire*.

Any pump or machine employed to exhaust the air from a closed vessel is termed an air-pump.

In the air-pump designed by E. S. Ritchie, Fig. 103, the lower valve is conical, held in place by a triangular stem fitting the tube; it is raised by the valve-rod passing up through a stuffing-box in the piston. An enlarged section, Fig. 104, shows the manner in which the attachment is made which allows a motion of the rod sidewise, so that any slight change of form of the packing of the piston, or stuffing of the rod, cannot prevent the valve from shutting properly. The cone of the valve is ground to a perfect fit to its seat, but the valve is also furnished with a disc of oiled silk, which projects just beyond its outer edge, and touches the flat surface of the valve-seat; the valve-rod extends up, and its upper end is secured in a hole drilled in the upper plate, of depth sufficient to allow motion vertically to open the valve. The piston is of thick brass, made in two parts; the upper piece has a hole drilled larger than the piston-rod, the lower part of a conical form, ground to fit a cone on the piston-rod; this forms the piston-valve. The lower piece of the piston covers the end of the piston-rod, but allows it enough motion to open the valve; a series of small holes through the plate gives a free passage for the air to the valve. A third valve is placed outside the cylinder, made of oiled silk in the usual way. In the upper plate of the cylinder is inserted a steel lever, one end of which covers the valve-rod, the other end, when the lower valve is closed, is flush with the plate, but when the valve is raised it projects into the cylinder.

103.

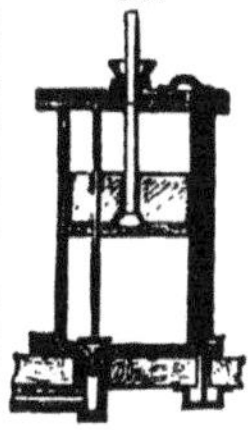

104.

In action the first motion, upward, of the piston-rod closes the piston-valve; the first motion of the piston opens the lower valve; as the piston ascends, the air above it is forced out through the upper valve, and air from the receiver flows unobstructedly into the cylinder The piston strikes the tail of the lever, and at the instant of arriving at the top closes the lower valve. The first downward motion of the piston-rod opens the piston-valve, the air remaining in the interstices above the piston distributes itself equally throughout the cylinder, but none can pass the lower valve back into the receiver. When the piston again reaches the bottom of the cylinder, the interstices below are filled with air as rarefied as a pump with ordinary valves can exhaust.

The air-pump of Robert Gill, Fig. 105, consists of a cylinder *a*, to the lower flange of which the bell-mouthed vessel *b*, projecting upwards into the cylinder, is secured. Between the outer side of the vessel and the inner side of the cylinder, the tubular bell-shaped piston *c* is situated. The hollow piston-rod *d* is screwed into the upper part of the piston, and the leather flap-valve *e* is held firmly between these two parts by the gripping action of the screw. The lower part of the piston-rod *d* is tubular, and a small leather-packed piston *f* is fitted to the rod, carrying at its lower end the valve *h*. A part of the rod above the valve *h* is encircled by a helical spring. The upper cover of the cylinder is provided with an air-escape valve *i*, which is immersed in a reservoir of oil to ensure its tightness, and the air extracted from the receiver escapes through the small hole *o*. The cup and tube *k* are used for supplying the liquid to the cylinder.

105

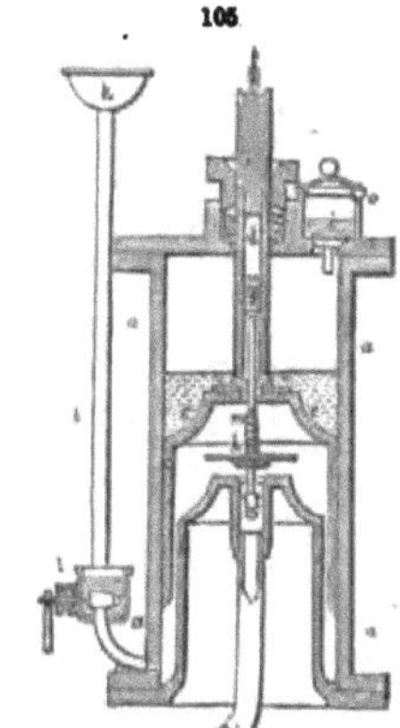

To prepare the apparatus for work, the upper cover of the cylinder is taken off, and the cock *l* opened. The fluid is then poured down the tube *k* into the cylinder until the piston is completely covered, the cover being replaced and the cock shut, the apparatus is then ready for action. On moving the piston upwards, the valve *h* would, if free, be also drawn up by the friction of the small piston *f* in the tubular piston-rod. The spring *m*, however, is of greater length than the distance between the under side of the piston *c* and the valve *h* when the piston is at its lowest position, consequently the spring *m* is compressed until the piston *c* is raised through a distance corresponding to the length of the spring. Being in this state, the elasticity of the spring overcomes the tendency to raise the valve *h*, produced by the friction of the small piston *f*; the valve *h* is therefore maintained close against its seat until the elasticity of the spring ceases to act, or, in other words, until the piston *c* is elevated through a distance equal to the spring length, on arriving at which point the valve *h* begins to rise. The foregoing arrangement of the spring and valve is necessary for the following reason.—If the valve were free to rise, the apparatus being charged with oil, some of it would flow through the valve opening, and enter the tube leading to the receiver, into which it might pass. As the piston, however, is raised higher, its tubular part being gradually emerged from the annular space in the lower parts of the cylinder, the level of the oil under the piston is lowered, the space for holding the liquid being gradually enlarged, so that the valve is left uncovered by it The apparatus is shown in the figure in the position of inaction of the spring, and with the valve open.

Before raising the piston the space below it is completely full of oil, and the joints through which air might intrude are all covered with this liquid; consequently, the space left by the motion of the piston must remain perfectly empty, at least as regards air. The vacuous space below the piston being now in communication with the receiver through the tube N, becomes filled with air more or less rarefied; at the same time the air contained in the upper part of the cylinder being compressed by the ascent of the piston, raises the valve *i*, and escapes into the atmosphere through the aperture *e*. As the pressure above the piston is greater than that below it, a small leakage of oil takes place downwards through the interstice round the little piston-rod, and dropping upon the shield attached to the valve *h*, falls into the annular space, and there accumulates during the ascent of the piston. The interstice for leakage allows only a small quantity of oil to pass, otherwise the valve *h* might be overflowed before the piston begins to descend, and consequently before the valve was closed. It will be seen that in every part of the piston's motion its lower edge is immersed in the oil, which prevents any lodgment of air between it and the sides of the cylinder.

At the commencement of the downward motion of the piston, the valve *h* is closed immediately by its piston *f*, and the valve *i* is closed by its own weight and the atmospheric pressure; the space below the piston becomes smaller and smaller as the piston descends; at the beginning of the upward stroke all the space below the piston was full of oil, and during the upstroke more oil has passed down into the annular space; the consequence is, that when the piston reaches the bottom of its stroke, the space below being completely full, it is evident that the air extracted from the receiver must be completely expelled through the apertures *e e*, together with that small quantity of oil which passed downwards through the interstice around the rod, during the ascent of the piston. The space below the piston is thus infinitesimally reduced, and consequently the air is completely expelled, however rarefied it may have been on entering from the receiver.

The most perfect air-pump is that invented by Hermann Sprengel, Fig. 106, which consists of a glass tube *c d*, longer than a barometer, open at both ends, and in which mercury is allowed to fall down, supplied by the funnel A, with which the tube is connected at *c*. The lower end *d*, of this tube dips into a small glass bulb B, into which it is fixed by means of a cork. This glass bulb has a spout at its side, situated a few millimètres higher than the lower end of the tube *c d*. The first portions of mercury which run down will consequently close the tube, and form a safeguard

107.

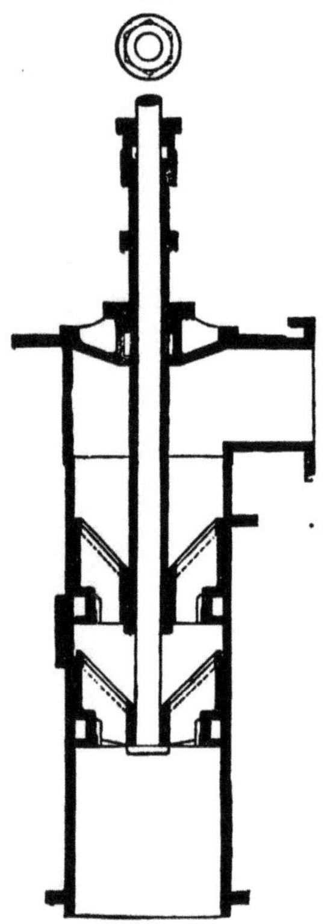

106.

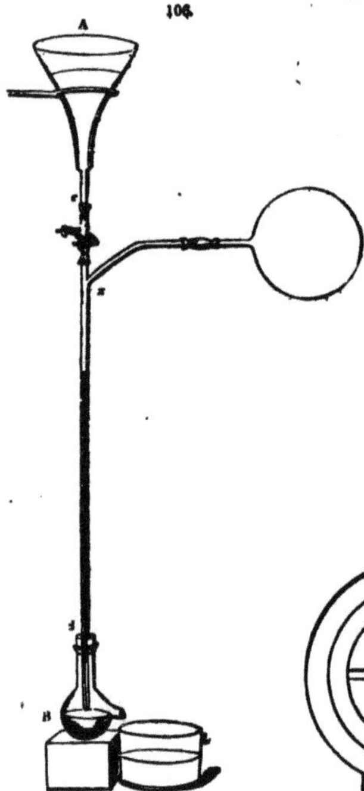

108.

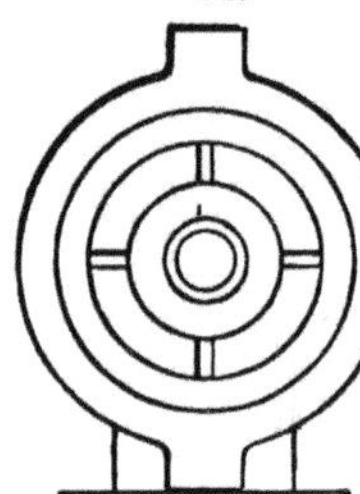

against the air which might enter from below, if the equilibrium should be disturbed. The upper part of *c d* branches off at *x* into a lateral tube, to which the receiver R is affixed. As soon as the stop-cock at *c* is opened, and the mercury allowed to run down, the exhaustion begins, and the whole length of the tube, from *x* to *d*, is seen to be filled with cylinders of mercury and air, having a downward motion. Air and mercury escape through the spout of the bulb B, which is above the basin H, where the mercury is collected. This has to be poured back from time to time into the funnel A, to pass through the tube again and again, until the exhaustion is completed. As the exhaustion is progressing, it will be noticed that the enclosed air between the mercury cylinders becomes less and less, until the lower part of *c d* presents the aspect of a continuous column of mercury, about 30 inches high. Towards this stage of the operation a considerable noise begins to be heard, similar to that of a shaken water-hammer, and common to all liquids shaken in a vacuum. The operation may be considered completed when the column of mercury does not enclose any air, and when a drop of mercury falls upon the top of this column without enclosing the slightest air-bubble. The height of this column now corresponds exactly with the height of the column of mercury in the barometer; or, what is the same, it represents a barometer, whose Toricellian vacuum is the receiver R.

The pump, Figs. 107, 108, of the pumping engine at Hartford, U.S., is noticed here on account of the peculiar action and arrangement of its valves and pistons. The engine is a double-acting condensing crank and beam engine, with a single cylinder 32·375 inches diameter, five feet stroke, with an adjustable cut-off. The injection water is taken from the well. There is a heavy fly-wheel of 22 feet diameter on the crank-shaft, and power is communicated by a pinion of twenty-seven teeth on the end of the shaft, gearing into a spur-wheel on either side, of eighty teeth, and on each spur-wheel shaft are two cams, each giving motion to a set of pumps by means of bell-cranks. Each set of pumps consists of two pistons or boxes in one chamber or cylinder, one above the other, see Figs. 107, 108; the piston-rod of the upper one being a tube, and the piston-rod of the lower box passing through it, the valves are butterfly-valves hinged in the middle; and each piston commences its stroke slowly, and increasing in a short space to its uniform velocity, and at the end decreasing for like distance till it stops. Thus while the lower box is rising, the upper is descending, the water passing up through the valves of the upper box; but just before the lower box has completed its up-stroke, the upper box has completed its down-stroke, and commences to rise, the lower decreasing in velocity and the upper increasing: and, *vice versâ*, during the rise of the upper box the lower one descends, and commences to rise in time to relieve the upper box at the end of its stroke. Thus the stroke of one box laps on to that of the other, and the absolute movement or stroke of 17·875 inches of each box may be considered 16·125 inches effective.

AIR-SHAFT. Fr., *Puits d'aérage, Buse d'aérage*; Ger., *Luftschacht, Wetterschacht*; Ital., *Pozzo di mina*; Span., *Pozo de ventilacion.*

An air-shaft is a passage for air into a mine, usually opened in a perpendicular direction, and meeting the *adits* to cause a free circulation of fresh air through the mine.

AIR-STOVE. Fr., *Calorifère à air chaud*; Ger., *Ofen zum heitzen mit warmer Luft*; Ital., *Calorifero* Span., *Calorífero de aire caliente.*

An air-stove is an enclosed fire-place so constructed as to admit a stream of air to pass round it or through it, and this impinging upon a heated surface is rarefied, carried upwards, and warms the apartment. See Ventilating and Warming.

AIR-TRAP. Fr., *Arrangement pour prévenir l'air impur des égouts et des rigoles de se disperser dans les bâtiments*; Ger., *Eine Vorrichtung zum zurückhalten von fauler Luft in Cloaken*; Ital., *Valvola ad aria*; Span., *Aparato para impedir el escape de aire impuro de sumideros.*

An air-trap is a trap immersed in various ways in water, to prevent foul air rising from sewers or drains.

AIR-VALVE. Fr., *Soupape à air*; Ger., *Luftventil, Luftklappe*; Ital., *Valvola di sicurezza per il vuoto*; Span., *Válvula de aire.*

An air-valve is a safety-valve fixed at the top of a steam-boiler, and opening inwards to prevent rupture from the pressure of the atmosphere upon the sides of the boiler, should a vacuum occur within from the steam becoming condensed, or partially so.

AIR-WAY. Fr., *Passage pour l'air*; Ger., *Luftweg*; Ital., *Corso dell'aria*; Span., *Conducto de aire.*

A tubular passage for air flowing in pipes is termed an air-way.

AJUTAGE. Fr., *Ajutage, Ajutoir*; Ger., *Aufsatz (für Springbrunnen)*; Ital., *Tubo d'aggiunta*; Span., *Tubo para regularizar la salida de aqua.*

A tube through which water is discharged is called an *ajutage*, as the *ajutage* of a fountain.

ALGEBRAIC SIGNS. Fr., *Signes algébriques*; Ger., *Algebraische Zeichen*; Ital., *Segni algebraici*; Span., *Signos Algebráicos.*

The sign +, termed *plus*, or more, is, like each of the other algebraic signs, a mere conventional mark; it indicates addition; thus, $18 + 19 = 37$, that is, 18 added to 19 equal 37.

= being the sign put for, equal to; equal; or equals. $a + b = c$, or, in words, a added to b makes a sum equal to c.

+ is sometimes used to indicate that figures have been omitted from the end of a number, or that the number is approximately exact, as the square root of 5 is 2·2360679 +.

This little cross mark, when written diagonally like ×, stands for *multiplied by*; as, $15 \times 8 = 120$.

× also stands for *times* and *into*, as, $a \times b = ab$. Multiplication is also often indicated by placing a dot between the factors, or by writing the latter, when not numerals, one after another, without any sign; as, $a \times b \times c \times d = a \cdot b \cdot c \cdot d = abcd$; $1 \times 2 \times 3 \times 4 = 24 = 1\ 2\ 3 \cdot 4$.

$a + a + a + a + a = 5$ times a, written $5a$, must not be confounded with $a \times a \times a \times a \times a = a$ in the fifth power, written a^5. If $a = 2$, then in the first case $5a = 10$; and in the second $a^5 = 32$.

− Minus, less; which indicates subtraction; as, $7 - 3 = 4$; that is, 7 less or diminished by 3 is equal to 4: if p be put for 7, q for 3, and r for 4, then $p - q = r$.

÷ or : stands for divided by; as, $x \div y$; that is, x divided by y. Now, to take a particular case, if $x = 24$, and $y = 3$, then $x \div y = 24 \div 3 = 8$. However, division is more generally indicated by writing the divisor under the dividend, with a line between them; as, $\frac{x}{y}$; that is, x divided by y. Supposing x to be put for 20, and y for 3, then $\frac{x}{y} = \frac{20}{3} = 6\frac{2}{3}$.

Small figures, 1, 2, 3, 4, &c., termed *indices*, are placed above and at the right hand of quantities to denote that they are to be raised to powers whose degree is indicated by the figure, as, a or a^1, that is, the first power of a; a^2, the square or second power of a; a^3, the cube or third power of a. a^4, the fourth power of a; and the like. $4^3 = 4 \times 4 \times 4 = 64$. Small figures, in indicating powers, are often preceded by the negative sign (−), and thus show or indicate the reciprocal of the corresponding power, as, a^{-1}, a^{-2}, a^{-3}, a^{-4}, &c., are respectively equivalent to $\frac{1}{a^1}$, $\frac{1}{a^2}$, $\frac{1}{a^3}$, $\frac{1}{a^4}$, &c.

$\surd$, or $\surd$ signifies root;—indicating, when used without a figure placed above it, the square root; as, $\sqrt{9} = 3$; $\sqrt{289} = 17$; $\sqrt{16a^2} = 4a$; and so on. This symbol is called the radical sign. To denote any other than the square root, a figure (called the *index*), expressing the degree of the required root, is placed above the sign; as, $\sqrt[3]{a}$, $\sqrt[4]{a}$, $\sqrt[9]{a}$, &c.; that is, the cube root, fourth root, ninth root, &c., of a. $\sqrt{\ }$ is merely a modification of the letter r, which was used as an abbreviation of the Latin word *radix*, root. The root of a quantity is also denoted by a fractional index at the right-hand side of the quantity and above it, the denominator of the index expressing the degree of the root; as, $a^{\frac{1}{2}}$, $a^{\frac{1}{3}}$, $a^{\frac{1}{5}}$; that is, the square, cube, and fifth roots of a, respectively.

———— Vinculum,
( ) Parenthesis,
[ ], or { }, Brackets,

These signs indicate that the quantities to which they are appended, or which are enclosed by them, are to be taken together; as, $3(8 + 11) = 57$; that is, 8 is first added to 11, and then the parenthesis shows that the sum 19 has to be multiplied by 3. $3(8 \times 2 - 11) = 15$; that is, from twice 8 take 11, and multiply the remainder by 3. But $3(8 \times 2) - \overline{11} = 37$, must not be taken for $3(8 \times 2 - 11)$. If $a = 2$ and $b = 3$, then, $(a + b)^2 = (2 + 3)^2 = 25 = \overline{a + b}^2$.

$$a \times [b + a^2(3b - a^2)] = 2[3 + 4(9 - 4)] = 46.$$

$$a \surd \{(a + b) + a^2[3b - a^2]\} = 2 \surd \{(2 + 3) + 4[9 - 4]\} = 10.$$

This last expression must not be confounded with the form

$$a \surd \{(a + b + a^2)[3b - a^2]\} = 2 \surd \{9 \times 5\} = 6\sqrt{5} = 13 \cdot 4164078.$$

Is to, or the ratio of, is expressed by two dots, one placed over the other, as, (:) between a and b, $(a : b)$, signifies a is to b, or the ratio of a to b. : : signifies as, or equals. This arrangement of dots is employed to indicate proportion; as, $a : b :: c : d$; that is, a is to b as c is to d; or, the ratio of a to b equals the ratio of c to d. In numbers, for example, if $a = 7$, $b = 3$, $c = 35$, then $d = 15$, for $7 : 3 :: 35 : 15$; as, $a \times d = b \times c$; or $\frac{a}{b} = \frac{c}{d}$.

☞ To understand the proper application of the symbols $+$; $-$; $\times$; $\div$; $\surd$; $\sqrt{\ }$; ( ); and : :: :; is of much importance to the practical mechanic and engineer, especially if he has neglected the study of mathematics; for any person who understands common arithmetic, and how these symbols are applied, may determine the numerical value of any general formula in particular cases.

For example, the slip (n) of screw propellers, or paddle-wheels, is given by the formula

$$n = \frac{\sqrt{g^3}}{\sqrt{g^3} + \sqrt{d^3}}$$

in which g = the area of resistance, and d = the acting area. Now, if n be required when $g = 49$ square feet and $d = 225$ square feet, we have

$$n = \frac{\sqrt{49^3}}{\sqrt{49^3} + \sqrt{225^3}} = \frac{\sqrt{117649}}{\sqrt{117649} + \sqrt{11390625}} = \frac{343}{343 + 3375} = \frac{343}{3718} = \cdot 09225.$$

To illustrate this method of substituting numbers for their representative letters in established formulæ, we select the following problems, respecting the crank and fly-wheel, from Byrne's 'Essential Elements of Practical Mechanics.'

Problem.—To find the position of the crank corresponding to its maximum and minimum velocity in a single-acting engine:—Let O B and O D be the required positions on the crank, and let P represent the constant pressure of the connecting-rod, supposed to act in a vertical line. Put Q = the constant resistance, acting at 1 foot from the axis of the fly-wheel, equivalent to the work of the engine.

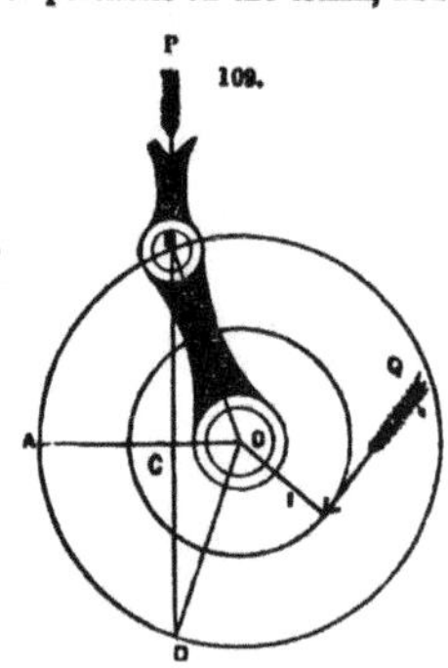

109.

The motion will be accelerated from B to D. This acceleration will commence when the moving pressure is equal to the resisting pressure, and it will cease under the same condition. The former will correspond to the position of minimum, the latter to that of maximum velocity. Hence, at these two points the moment of P must be equal to the moment of Q, and the point D will be as much below the horizontal line A O as the point B is above it.

$$\therefore \ P \times CO = Q \times 1. \qquad [1]$$

Again, by the equality of work, putting

$$r = OB,$$

Units of work by P in 1 revolution $= 2rP$.

" " Q " $= Q \times 2 \times 3 \cdot 1416$.

$$\therefore \ 2rP = Q \times 2 \times 3 \cdot 1416. \qquad [2]$$

Dividing equation [1] by equation [2],

$$\frac{CO}{r} = \frac{1}{3 \cdot 1416} = \cdot 31831$$

∴ From the table of natural sines $\cdot 31831$ = cosine of $71^\circ 27'$; for $\frac{CO}{r}$ is the cosine of the angle B O C.

Problem.—To find the dimensions of the fly-wheel, such that its angular velocity may at no point differ from the mean velocity beyond a certain limit:—Let d and p be the maximum and minimum velocities of the wheel at the distance of 1 foot from the axis, W the weight of the wheel, and k the distance of the centre of gyration from the axis.

Work of P from B to D = P × B D = P × 2 r × sin 71° 27′ = 2 r P × ·948.

Work of the constant pressure Q from D to B = $\frac{Q \times 2 \times 3{\cdot}1416 \times 142^{\circ}54'}{360^{\circ}}$ = 2 r P × ·3968, by putting for Q × 2 × 3·1416 its value 2 r P, found in the last problem.

Now the difference of these will give the *work* that goes to increase the speed of the wheel between the points B and D, that is, *work* going into the wheel between B and D =

$$2\,r\,P \times {\cdot}948 - 2\,r\,P \times {\cdot}3968 = r\,P \times 1{\cdot}1022.$$

$$\text{Accumulated work at B} = \frac{p^2\,k^2\,W}{2g}.$$

$$\text{Accumulated work at D} = \frac{d^2\,k^2\,W}{2g}.$$

Hence the accumulated work gained from B to D = $\frac{k^2\,W}{2g}(d^2 - p^2)$, but this must be equal to the work before found;

$$\therefore\ \frac{k^2\,W}{2g}(d^2 - p^2) = r\,P \times 1{\cdot}1022 \qquad [3]$$

Let V be the mean velocity of the wheel at one foot from the axis, and let the extreme velocities d and p differ from this mean velocity by the nth part; then

$$d = V + \frac{V}{n} \text{ and } p = V - \frac{V}{n};$$

$$\therefore\ d^2 - p^2 = \frac{4\,V^2}{n} \qquad [4]$$

Let N be the number of double strokes performed a minute, then

$$V = \frac{2 \times 3{\cdot}1416 \times N}{60} = {\cdot}10472 \times N; \qquad [5]$$

Let U be the work of the engine, then

$$U = 2\,r\,P\,N \quad \therefore\ r\,P = \frac{U}{2\,N} \qquad [6]$$

Substituting the values given in the equation [4], [5], [6], in equation [3], and reducing, $2g = 32\frac{1}{6}$

$$W = \frac{n\,U}{k^2\,N^3} \times \frac{32\frac{1}{6} \times 1{\cdot}1022}{4 \times ({\cdot}10472)^2}$$

$$\therefore\ W = \frac{n\,U}{k^2\,N^3} \times 808{\cdot}2 \qquad [7]$$

which is the expression for the weight of the fly-wheel in pounds.

If H be put for the horse-power of the engine, then U = 33000 H; substituting this in equation [7], and reducing to tons,

$$W = \frac{n\,H}{k^2\,N^3} \times 11907 \qquad [8]$$

which is the expression in units of tons. Let R = the mean radius of the fly-wheel, e = depth of the rim, then from a well-known property of the centre of gyration

$$k^2 = R^2 + \frac{e^2}{4};$$

substituting this in equation [8], then

$$W = \frac{n\,H}{\left(R^2 + \frac{e^2}{4}\right)N^3} \times 11907.$$

Neglecting $\frac{e^2}{4}$ as being comparatively small, then

$$W = \frac{n\,H}{R^2\,N^3} \times 11907. \qquad [9]$$

It may be observed that the weight of the wheel varies inversely as the cube of the number of strokes performed by the engine per minute.

If a = the area of the section of the rim in square feet, and 450 lbs. be taken as the weight of a cubic foot of the metal, then W = 2 π R $a\,\frac{450}{2240}$ tons nearly. Substituting in equation [9], and solving the resulting equation for R,

$$\text{then } R = \left(\frac{n\,H}{N^3 a} \times \frac{11907 \times 2240}{2 \times 3{\cdot}1416 \times 450}\right)^{\frac{1}{3}}$$

$$\therefore\ \frac{21}{N}\sqrt[3]{\frac{n\,H}{a}} = R,$$

which is an expression for the mean radius of a cast-iron fly-wheel of a single-acting engine, when there are given the number of strokes of the piston, the horse-power, the area of the mean section of the rim, and the proportional variation from a mean velocity. Proceeding in the same way, for the double-acting engine, cosine B O A = 2 × ·31831,

$$\text{and } R = \frac{12}{N}\left(\frac{n\,H}{a}\right)^{\frac{1}{3}}$$

In the fly-wheel B A W, Fig. 110, C A = R; C B = r, the outer and inner radius of the ring. W = weight of the ring in pounds; w the weight of the arms, breadth of each, D E = b. If y be the centre of gyration of the ring and arms, then, putting y = C y,

110.

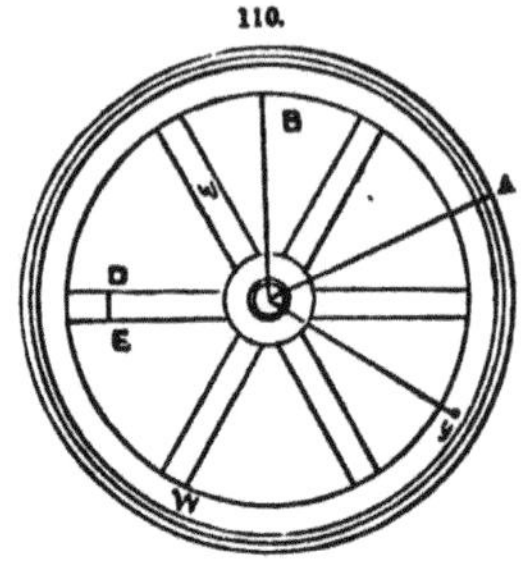

$$y = \sqrt{\frac{6\,W\,(R^2 \times r^2) + w\,(4\,r^2 \times b^2)}{12\,(W+w)}}.$$

In practice, the centre of gyration, including the ring and arms, may be assumed at $y = r$ the length of the inner radius from the centre, C. Putting a for the angular velocity or number of revolutions a minute, at the end of the time t in which the fly-wheel would concentrate the same power as the steam engine, t may be taken = 128 seconds; but when the work is irregular, t may be taken as high as 170 seconds. Taking these average quantities, the weight of a fly-wheel for a given horse-power H will be

$$\text{Weight} = \frac{(11)^3 \times (10)^2 \times 128\,H}{a^2\,r^2}.$$

Question.—Required the weight of a fly-wheel when the engine is of 56 horse-power; the inner radius of the ring = 10 feet making 42 revolutions a minute?

$$\text{Weight} = \frac{133100 \times 128 \times 56}{42^2 \times 10^2} = 5409 \text{ lbs. nearly.}$$

We append other algebraic symbols, which have been employed by writers in treating of quantities that could be numerically expressed. Many of the following signs or symbols are obsolete, rare, or of far less importance to the mechanic or engineer than those which we have explained and illustrated.

$\perp$ Divided by; as, $a \perp b$; that is, a divided by b; $6 \perp 3 = 6 \div 3 = 2$. [*Rare.*]

$>$ Is greater than; as, $a > b$; that is, a is greater than b; $6 > 5$.

$\sqsubset$ Is greater than; — the same as $>$. [*Rare.*]

$<$ Is less than; as, $a < b$; that is, a is less than b; $3 < 4$.

$\sqsupset$ Is less than; — the same as $<$. [*Rare.*]

$\nless$ Is not less than;— the contradictory of $<$; as, $a \nless b$; that is, a is not less than b; or, a may be equal to, or greater than, b, but cannot be less than it.

$\ngtr$ Is not greater than; — the contradictory of $>$; as, $a \ngtr b$; that is, a is not greater than b; or, a may be equal to, or less than, b, but cannot be greater than it.

$\Leftrightarrow$ Is equivalent to; — applied to magnitudes or quantities which are equal in area or volume, but are not of the same form, or capable of superposition; as, $a^2 \Leftrightarrow bc$; that is, the square whose side is a is equal to the rectangle whose sides are a and b. [*Rare.*]

$\leftrightarrows$ Of the form of; as $a \leftrightarrows (2\,n + 1)$; that is, the term a is of the form $2\,n + 1$; $17 \leftrightarrows (2 \times 8 + 1)$; that is, the odd number 17 is of the form $2 \times 8 + 1$. [*Rare.*]

$\rightleftarrows$ Is divisible by; as, $a \rightleftarrows b$; that is, b is an exact factor of a; $8 \rightleftarrows 2$. [*Rare.*]

$\backsim$ The difference between; — used to indicate the difference between two quantities without designating which is the greater; as, $a \backsim b$; that is, the difference between a and b.

— : The difference between;—the same as $\backsim$. [*Rare.*]

$\propto$ Varies as; is proportional to; as, $a \propto b$; that is, a varies as b, or is dependent for its value upon b.

$\because$ Geometrical proportion; as, $\because a : b : : c : d$; that is, the geometrical proportion $a : b : : c : d$ [*Rare.*]

.. Minus; the arithmetical ratio of, } used to indicate arithmetical proportion; as, $a \cdot\cdot b$
:: Equals; is equal to; } $: : c \cdot\cdot d$; that is, $a - b = c - d$. [*Rare.*]

∞ Indefinitely great; infinite; infinity;—used to denote a quantity greater than any finite or assignable quantity.

0 Indefinitely small; infinitesimal;—used to denote a quantity less than any assignable quantity; also, as a numeral, naught; nothing; zero.

$\angle$ Angle; the angle; as, $\angle$ A B C = $\angle$ D E F; that is, the angle A B C is equal to the angle D E F;—less frequently written $>$ or $\searrow$.

$\wedge$ or $\wedge$ The angle between; as, $a\hat{}b$, or A $\wedge$ B; that is, the angle between the lines a and b, or A and B, respectively.

By some geometers, the angle between two lines, as a and b, is also indicated by placing one of the letters denoting the enclosing lines over the other: as, $\frac{a}{b}$; that is, the angle between the lines a and b; sin. $\frac{a}{b}$; that is, the sine of the angle between the lines a and b.

∟ Right angle; the right angle; as, ∟ A B C; that is, the right angle A B C.

$\perp$ The perpendicular; perpendicular to; is perpendicular to: as, draw A B $\perp$ C D: that is, draw A B perpendicular to C D.

$\parallel$ Parallel; parallel to; is parallel to, as, A B $\parallel$ C D; that is, A B is parallel to C D.

$\veebar$ Equiangular; is equiangular to; as, A B C D $\veebar$ E F G H; that is, the figure A B C D is equiangular to the figure E F G H. [*Rare.*]

$\stackrel{1}{=}$ Equilateral; is equilateral to; as, $\stackrel{1}{=}$ D E F; that is, the figure A B C is equilateral to the figure D E F. [*Rare.*]

$\bigcirc$ Circle; circumference; 360°.

$\frown$ Arc of a circle; arc.

△ Triangle; the triangle; as, △ A B C = △ D E F; that is, the triangle A B C is equal to the triangle D E F.

□ Square; the square; as, □ A B C D; that is, the square A B C D.

▭ Rectangle; the rectangle; as ▭ A B C D = ▭ E F G H; that is, the rectangle A B C D equals the rectangle E F G H.

f or F; Function; function of; as $y=f(x)$; that is, y is, or equals, a function of x.

Various other letters or signs are frequently used by mathematicians to indicate functions; as, f, ϕ, ϕ', ψ, π, and the like. Some of these are used also without the parenthesis; as, $\phi\, x$, function of x.

d Differential; as, dx; that is, the differential of x.

δ Variation; as, $\delta\, x$; that is, the variation of x.

Δ Finite difference.

D Differential co-efficient; derivative; — sometimes written also d_c. The variable, with respect to which the differential co-efficient is taken, is indicated by writing the letter designating it at the right hand below; as, $D_t\, \phi$; that is, the differential co-efficient of ϕ with respect to t.

The letters d, δ, Δ, D, and sometimes others, are variously employed by different mathematicians, prefixed to quantities to denote that the differentials, variations, finite differences, or differential co-efficients of these quantities are to be taken; but the ordinary significations are those given above. An index is often placed at the right hand of d, to indicate the result of one or more repetitions of the process denoted by that sign; as, $d^2 x$, $d^3 x$, &c.; that is, the second, third, &c., differential of x, or the result of differentiating x two, three, &c., times. This common device is clumsy and inadequate.

· Fluxion; differential; as $\dot{x}$; that is, in modern notation, dx. [*Obs.*]

$\int$ Integral; integral of; — indicating that the expression before which it is placed is to be integrated; as, $\int 2\, xdx = x^2$; that is, the integral of $2\, xdx$ is x^2. This sign is merely a modified form of S, which is itself the abbreviation of the Latin word *summa*, sum, the integral being the sum of the differentials. It is repeated to indicate that the operation of integration is to be performed twice, or three, or more times, as $\int\int$, $\int\int\int$, &c. For a number of times greater than three, an index is commonly written at the right hand above; as, $\int^m$, xdx^m; that is, the mth integral, or the result of m integrations of xdx^m. The variable, with respect to which the integral is taken, is sometimes indicated by writing the letter designating it at the right hand below; as, $\int_x \phi$; that is, the integral of ϕ with respect to x.

$\int_b^a$ denotes that the integral is to be taken between the value b of the variable and its value a. $\int^a$ denotes that the integral ends at the value a of the variable, and $\int_b$ that it begins at the value b. These forms must not be confounded with the similar one indicating repeated integration, or with that indicating the integral with respect to a particular variable.

Σ Sum; algebraic sum;—commonly used to indicate the sum or summation of finite differences, and in nearly the same manner as the symbol $\int$.

Residual.

(⌶) A symbol used in abbreviations of quantics whose terms have the same numerical co-efficients as a corresponding expression formed by involution; as $(a, b, c, d ⌶ x, y)^3$, which denotes the quantic $ax^3 + 3\, bx^2y + 3\, cxy^2 + dy^3$, the numerical co-efficients of which are the same as those obtained by expanding $(x+y)^3$.

(⌶) A symbol for a quantic which has no numerical co-efficients; as, $(a, b, c, d ⌶ x, y)^3$, which denotes the quantic $ax^3 + bx^2y + cxy^2 + dy^3$.

π The number $3 \cdot 14159265 +$; the ratio of the circumference of a circle to its diameter, of a semicircle to its radius, and of the area of a circle to the square of its radius. In a circle whose radius is unity, it is equal to the semi-circumference, and hence is used to designate an arc of 180°. In a circle whose diameter is unity, it is equal to the circumference, or an arc of 360°.

The ratio of the circumference of a circle to the diameter;—the same as π; a graphic modification of the letter C, for circumference.

The base of the hyperbolic system of logarithms; the same as ϵ; a graphic modification of the letter B, for base.

M The modulus of a system of logarithms;—used especially for the modulus of the common system of logarithms, the base of which is 10. In this system it is equal to $0 \cdot 4342944819 +$.

g The force of gravity. Its value for any latitude is expressed by the formula $g = 32 \cdot 17076\, (1 - 0 \cdot 00259 \cos.\, 2\, \lambda)$, in which λ is the latitude given, and $32 \cdot 17076$ (that is, $32 \cdot 17076$ feet per second) the value of g at the latitude of 45°.

° Degrees: as 60°; that is, sixty degrees.

′ Minutes of arc; as, 30′; that is, thirty minutes.

″ Seconds of arc; as, 20″; that is, twenty seconds.

R° Radius of a circle in degrees of arc, equal to 57°·29578.

R′ Radius in minutes of arc, equal to 3437′·7468.

R″ Radius in seconds of arc, equal to 206264″·8.

′, ″, ‴, &c. Accents used to mark quantities of the same kind which are to be distinguished; as, a', a'', a''', &c., which are usually read a prime, a second, a third, &c.; $a\, b'\, c'' + a'\, b''\, c + a''\, b\, c'$.

When the number of the accents would be greater than three, the corresponding Roman numerals are used instead of them; as, a', a'', a''', a^{iv}, a^{v}, a^{vi}, &c. The accents are often written below also; as, $a_{,}$, $a_{,,}$, $a_{,,,}$, a_{iv}, a_{v}, &c. Figures, and also letters, are sometimes used for the same purpose; as, a^1, a^2, a^m, a^n, a_0, a_2, a_3, a_m, a_x, and the like.

Indices or small figures are also often used to indicate the repetition of an operation; as, d^2x, d^3x, d^4x, &c., indicating that the operation of differentiation has been performed upon x two, three, four, &c., times.

sin. x. The sine of x; that is, of the arc represented by x. In the same manner cos. x, tan. x, cot. *x*, sec. *x*, cosec. *x*, versin. *x*, and co-vers. *x*, denote respectively the cosine, tangent, cotangent, secant, cosecant, versed sine, and co-versed sine of the arc represented by *x*.

$\sin.^{-1}x$. The arc whose sine is *x*. In the same manner $\cos.^{-1}x$, $\tan.^{-1}x$, $\cot.^{-1}x$, $\sec.^{-1}x$, $\operatorname{cosec.}^{-1}x$, $\operatorname{versin.}^{-1}x$, and $\operatorname{co-vers.}^{-1}x$, are used to denote respectively the arc whose cosine, tangent, cotangent, secant, cosecant, versed sine, or co-versed sine is *x*.

This sign must not be confounded with the negative index designating the reciprocal of a quantity, which would be applied to a parenthesis inclosing one of these expressions; as, $(\sin. x)^{-1}$, which is equivalent to $\frac{1}{\sin. x}$.

Oliver Byrne, the inventor of the art and science of Dual Arithmetic, in developing which he employs in his works two new signs, an arrow (↓) and a comma (,); these signs (↓, ↑') do not interfere with those we have previously enumerated, as dual arithmetic essentially differs from arts previously known. Besides, those who examine dual arithmetic in all its bearings will find that a branch of greater importance has not been contributed to mathematical science; hence, how these new signs are applied may interest many.

↓ 6, 5, 3, 7, 9, 2, 3, 3, is a dual number of the ascending branch, composed of eight consecutive digits;

2 ↑ '0 '6 '7 '4 '3 '8 '1 '2 is a dual number of the descending branch, composed of the same number of dual digits; for the 2, to the left of the arrow, which in this case points up, is a coefficient;

↓ 6, 5, 3, 8, 0, '7 '6 '6 is a mixed dual number, the first four digits being of the ascending branch, and the last three of the descending. 2 ↓ '1 4, '2 2' 2, '3 '4 '5 is a dual number in the lowest terms, because the first digit does not exceed 3, and none of the other digits exceeds 5; all natural or common numbers may be readily reduced to dual numbers of the last form.

The natural number 1·869 = ↓ 6, 5, 3, 7, 9, 2, 3, 3,; 1·869 = 2 ↑ '0 '6 '7 '4 '3 '8 '1 '2; = ↓ 6, 5, 3, 8, 0, '7 '6 '6; = 2 ↓ '1 4, '2 '2 2, '3 '4 '5; and = a vast number of other dual numbers. The dual logarithm of *a* is written ↓, (*a*), the dual logarithm of 2 or ↓, (2) = 69314718, which is a whole number.

↓, (1·869) = 62540354, and each dual number which represents 1·869 may be instantly reduced to the same dual logarithm, namely, 62540354. These observations apply to all numbers generally. Without the use of tables, in a variety of ways, and under different circumstances, Byrne in his works has shown by easy, independent, and direct processes, how any two of the three corresponding numbers

(NATURAL NUMBER); (DUAL NUMBER); (Dual Logarithm);

may be found, the remaining one being given.

ALLOYS. FR., *Alliages*; GER., *Legirungen*; ITAL., *Leghe*; SPAN., *Aleacion*.

ALLOYS, METALLIC; *employed in the mechanical and useful arts.*

Every metallic alloy, as regards utility, may be considered a new species of metal, because the qualities of the constituents are in most cases not recognized again in the compound; the compound in such cases shows properties which do not belong to the simple metals, and which cannot be determined by theoretical speculations. By changing the proportions of tin to copper, we obtain bronze of different qualities, varying extremely in colour, hardness, and sound. A few per cent. of tin causes copper to be hard and more tenacious. The addition of a little lead causes brass to be more ductile, while a large addition makes it brittle. We shall not here enter fully into the peculiarities of alloys, as these peculiarities are given in our articles on particular metals employed in the mechanical and useful arts. Metallic elements do not at first sight appear to combine in certain ratios and form definite compounds; still it cannot be denied that some metals do; and we are justified, by the general law of affinity, in assuming that all metals combine chemically. We succeed always in melting various metals together, but we do not very often succeed in separating the excess of any one or more metals in the alloy, or the refractory nature of another. As a general rule, we may state that all the metals which form alkalies have a particular tendency to unite with those which form acids. *Potassium* combines readily with *antimony* and *arsenic*, more so than other metals. In considering the nature of protoxides in their chemical relations, we may successfully form a series in which the ability of metals to combine is represented.

This accounts for the peculiarities of the alloys of selenium, arsenic, antimony, and tellurium; which resemble very much the combinations of metal and sulphur, or phosphorus, or chlorine. All these substances form acids in their most simple combination with oxygen. Alloys and compounds of this kind are peculiarly inclined to be brittle and fusible. When two metals are near in the series of affinities for oxygen, they do not combine very readily; and they may often be separated by crystallization only, when their degree of fusibility is sufficiently distinct. This happens when both metals absorb the same, or nearly the same, quantity of oxygen in forming oxide. All chemical combinations liberate heat; silver and platinum, when melted together, produce a high temperature; so do zinc and copper. In most cases, we obtain a mere mechanical mixture of metals in an alloy; this is always characterized by forming distinct crystals with one metal, between which the other is visible. When an alloy is formed which contains equivalents, no such disconnected crystals are observed. An irregularly composed alloy is a mere mechanical mixture, like wax and fat, and never forms a uniform body of metal; it is of either a granulated or crystallized texture, the latter of which is not compact. Between the crystals of such an alloy, one of the metals is always found in a nearly pure condition. The alloy of iron and silver, in which the silver is mechanically enclosed between the crystals of iron, is an instance of these compounds.

Lead and tin combine in certain proportions, and whatever excess there may be of either metal, it is enclosed between the crystals of the alloy. The same is the case with zinc and tin, bismuth and tin—and, in fact, with all other metals. The number of definite compounds appears to be very large, and in all cases a metal is never obtained pure whenever another is present. In cooling a melted alloy, that composition which is most refractory crystallizes first; and that which is most fluid is compelled to occupy the spaces between the crystals of the most refractory. Thus, copper and tin are very fusible; but in cooling, copper-tin crystallizes first, and tin-copper last—which latter occupies the spaces between the first. Iron and arsenic are very fusible, but in cooling, iron-arsenic crystallizes first, and arsenic-iron last; the surface of the cooled mass shows a perfect network of bright lines, in regular forms. In all these compounds the first crystallized metal, or alloy, contains some of the latter, and the latter some of the first.

When a bar of cold lead is held in mercury, the first absorbs some of the latter throughout its body; the pores of the lead are filled with mercury—but the mercury in this case contains lead, as well as the lead mercury. When iron is gently heated imbedded in carbon, as is the case in making blistered steel, the carbon penetrates the very heart of the iron-rods; but no iron is imparted to the carbon, because its particles are not movable. By these and similar means the pores of a refractory, solid metal, may be filled with another metal, provided the atoms of the latter are movable. Lead does not readily combine with iron, but if we heat wrought-iron turnings in melted lead, the solid iron will absorb lead. Gray cast-iron will not absorb lead, because its pores are filled with carbon. Borings of cast-iron absorb arsenic readily, when imbedded in arsenious acid and heated; but wrought-iron does not absorb arsenic by these means until carbon is brought into contact with these substances. In properly applying these rules, we may form alloys of the most heterogeneous metals, and in regular proportions.

It is extremely difficult to form a definite compound of zinc and copper, or tin and copper; but if we take either the oxides of all these metals, or the oxide of any one, and the metal in small particles, and imbed the whole in carbon—heating it so as not to melt it, although bringing it near its melting point—and then apply the heat gradually and slowly, we are enabled to form definite compounds, which may be melted by a heat which does not evaporate either the one or the other metal. We shall point out the formation of alloys by metals in other articles, but we wish to draw the attention to these facts, because it is often very difficult to form an alloy of a definite composition. Yet as these alloys are the most valuable, we refer to the above mode of forming them.

Fusibility of Alloys.—Alloys are more fusible than the mean temperature at which the metals melt singly would indicate. This is a very important law in our investigations, and affords, when properly applied, the most valuable results. When tin melts at 500°, and pure copper at 2500°, equal parts of copper and tin do not melt at $\frac{2500 + 500}{2} = 1500°$, but at a lower heat. Platinum, and also chromium, are infusible in our furnaces, but may be made very fusible by the addition of arsenic. Pure iron is extremely refractory, but when combined with arsenic and phosphorus, it may be melted in a cast-iron pot, without adhering to it. Thus, when an alloy of two metals is fusible at a lower heat than the mean of the two, a composition of three metals is still more fusible than their various degrees of melting indicate; and by multiplying the number of component parts, a low degree of fusibility of any metal may be obtained. If the composition is according to the laws of chemical affinity, the melting point is lowest. In all cases of practice we must reflect on the application of the metal before we form an alloy, the object of which is to facilitate the smelting operation. If phosphorus causes iron to be very fusible, and coal, flux, and labour, may be saved in smelting bog ores which contain phosphorus, the first inquiry must be whether the cold-short iron thus obtained is of any value.

Arsenic and zinc cause lead and silver ores to be easily melted, but the loss in silver by this operation is so great as to make it desirable to remove these volatile metals before any smelting is entered upon. We may flux copper by arsenic, but not by lead, because we can remove the first in refining, but not the latter sufficiently to form a good quality of copper for sheets. Arsenic, lead, or zinc may be in iron which is destined for wrought-iron, for these metals are easily removed in the refining process. It is always more safe in smelting gold ores to have lead, copper, or silver present, and if possible all of them; for gold is very volatile and extremely divisible, hence much of it may be lost when no other metals which act by their quantity are present. Lead fuses at 600°, tin at 500°, and bismuth at 400°; but a composition of the three may be made which melts at 212°, a heat far below the most fusible of the ingredients. This shows how great the advantages are which may be obtained by forming alloys in the smelting furnaces. It is the degree of fusibility of the slags and that of the metals which determines the expenses of smelting. The number of alloys is infinite. We cannot establish tables of fusibility *a priori;* these must be determined by practice. In all smelting operations it is necessary to find the most profitable conditions by experiment.

If an alloy is more fusible than a single metal, it follows that, when one or the other constituent is removed, the fusibilty of the metal is impaired. An alloy of lead and arsenic is more fusible than pure lead; but when the alloy is exposed to heat for a long time, arsenic will evaporate; and as it was the cause of fluidity, the lead will not be so fusible after losing its flux. Iron is made fusible by the presence of carbon or other substances, but when these are removed it is very refractory, and can hardly be melted. Upon this principle the formation of wrought-iron is founded; and in applying it to other metals there is no doubt but similar results are obtained. Tin is refined by oxidizing or evaporating sulphur, arsenic, and other matter; this causes the tin to be less fusible, but more tenacious. When zinc is melted in an iron pot, and exposed to the air, it forms dross on its surface, like other metals; its fluidity is consequently diminished, but its malleability is improved. A layer of carbon, or, what is still better, a layer of common salt, or bone ashes, prevents such phenomena with zinc and similar metals.

Specific Gravity.—When metals are melted together and form an alloy, there is produced a remarkable change in their specific gravity; which is sometimes greater and at other times less than the mean. A condensation of volume is manifested in melting together zinc and gold; the specific gravity of the compound is greater than the mean of the constituents; gold also and silver condense, or gold with lead or bismuth; silver combined with copper, lead, tin, bismuth, and zinc or antimony shrinks; also copper with tin, zinc, or antimony; lead with zinc, bismuth, or antimony;

mercury with tin or lead, bismuth or antimony. An expansion takes place when gold is melted with copper, iron, or tin; also platinum with copper; iron with antimony, bismuth, or tin; also copper and lead; tin and zinc, lead or antimony; zinc and antimony, and mercury with bismuth. In consequence of this contraction or expansion we are justified in assuming a similar change in the aggregate form.

The hardness of alloys is generally greater than may be inferred from the nature of the constituents; still there are exceptions to this rule. Silver or arsenic renders iron hard; but these metals are soft by themselves: copper and tin, two very soft metals, may be made extremely hard by melting them together in certain proportions. Hard zinc and copper make soft brass, and a compound of lead and iron is extremely soft. Antimony causes all metals to become hard; iron, with a little antimony, cuts glass very readily, but is very brittle.

The ductility of alloys is in some cases greater than the elements indicate; that of lead and zinc is very tenacious. In most instances the alloy is more brittle than the original metals. Lead and antimony is very brittle. Two or more brittle metals melted together are always brittle. A compound of lead and gold is very brittle. There is no precise limit at which we know when brittle or malleable alloys are formed; it is necessary to find this by practice; but there are strong indications that the alloy of an equivalent composition is always harder than a mere mechanical mixture of metals. Any alloy, when slowly heated and gradually cooled—annealed—is softer than a compound which is suddenly chilled. In annealing, the various elements which are in combination endeavour to separate. Heat here, as everywhere, weakens affinity. This causes a finely crystallized, or a granulated fracture; the component particles endeavour to assume a round form. When a hot alloy is suddenly chilled, the particles of which it is composed contract suddenly, and form a close, compact body; at least the ultimate crystals are condensed into the smallest space, which causes them to be hard. This is most strikingly exemplified in some kinds of iron.

Any kind of iron in market may be considered an alloy; there is no such thing as pure iron. Some wrought-iron contains much foreign matter, and still is soft as lead, and fibrous. If such iron is heated to a certain degree, that is, beyond the degree at which it has been manufactured, it becomes cast-iron, is brittle, and when suddenly chilled becomes hard, similar to steel. Such iron may be annealed and softened like steel, or any other alloy. The hardness of an alloy is therefore less dependent upon its composition than upon the arrangement and form of the ultimate particles.

Oxidation.—Alloys oxidize more rapidly than single metals; to this rule there are, however, remarkable exceptions. Hardened or crystallized metal oxidizes more rapidly than tempered or annealed metal, which is most strikingly exemplified in iron and steel. It appears that metals expand on being suddenly chilled. The spaces thus formed between the crystals are fissures into which oxygen may penetrate with facility; and as the crystallized form of the particles affords much surface for combination, an oxidation is readily effected. An alloy of antimony and iron, two metals which are remarkable for crystallization, oxidizes very readily; such an alloy may be ignited by mere filing, particularly when it is a little heated previously to that operation. Chrome and lead act in a similar manner. Antimony and potassium burn spontaneously, and if a little warm, with explosion. An alloy of tin and lead—soft solder—burns with a vividness equal to carbon, and if some zinc is present only a low heat is required to ignite the mixture.

An amalgam of potassium and mercury is so sensitive to oxygen that the mercury is oxidized to a high degree. In many cases, and under peculiar arrangements, the oxidation is carried to the highest degree: we have seen that in roasting sulphurets, phosphurets, and other compounds, the metal is always oxidized to the highest degree. The same cause acts in both cases. When two or more metals are melted together, there is a perfect penetration of the one by the other. In heating the mass, the tendency to separation is augmented, and as the ultimate particles of one metal cannot congregate and form large particles, they are exposed to the action of oxygen in their minutest form, and consequently will combine with the largest quantity of oxygen. If one kind of metal is more oxidizable than the other, or if the mass is porous, the oxidation goes on, not only faster, but must proceed to the highest degree, because the atoms of metal may be attacked by oxygen on all sides. While some alloys are easily oxidized, others resist that influence more than pure metals.

Pure iron is very easily oxidized, and often throughout, when in heavy masses. It has been found that a lot of bloom four inches thick, of puddled iron, was oxidized in a *reheating furnace* by giving merely a drawing heat to such a degree, that when drawn out between rollers into bars, these were found to be converted partly into magnetic and partly into protoxide of iron. This iron was puddled under the influence of a cinder highly charged with chlorides. Metals are porous, and easily penetrated by other matter. Thus, water enters successfully into the body of cast or wrought iron; and if a metal is present, or any other substance which decomposes water, the iron is easily oxidized. We find, when carbon is present, that oxygen has little effect on iron; but when any of those substances which form acids, such as arsenic, antimony, or sulphur, are combined with the metal, it is an easy prey to oxygen. An alloy of chromium and iron resists oxygen as well as gold; and phosphuret of copper is not so easily affected as pure copper. Bronze will resist the influence of oxygen much longer than pure copper.

It appears that the degree of affinity has less influence on the oxidation of metals than the aggregate form of their particles and the strength of their union. Hardened steel oxidizes more readily than annealed steel, and chilled cast-iron more so than gray cast-iron. Hammered or rolled lead is less penetrable by mercury than cast lead. German silver, which is composed of oxidizable metals, is little affected by oxygen, and less so when rolled than when cast. Copper has a strong affinity for sulphur, and is more readily attacked by it than iron, when finely divided; yet, when in a compact body, it will resist that substance far more successfully than iron. Some kinds of brass are remarkably liable to oxidation, while others resist it more effectually. A surface of metal, which is highly polished, and particularly when polished by rubbing it with a hard substance, is far less subject to oxidation than a rough surface. If it is desirable to resist oxidation, or in fact the influence of any other matter upon metal, those alloys must be formed which have naturally little affinity for that particular substance, and which, in the mean time, form the most intimate union, so that the penetration of foreign matter into the body of the metal is prevented. It is not the compactness of zinc or lead which prevents their oxidation in the atmosphere; it is the cover of oxide, which forms a close body, and prevents the further penetration of oxygen. We may assert that the density of gold and silver has as much influence in preventing their oxidation as their want of affinity for oxygen. Affinity between the metals of an alloy has, in consequence of an intimate union, a large share in preventing oxidation.

Iron is easily oxidized, but it is less subject to that influence when combined with phosphorus than when alloyed with silver or gold, particularly the former: this is chiefly because silver has but little affinity for it, and is thus excluded from its crystals, and forms a layer between them. There is a separation; oxygen finds access, and a rapid action of it is the consequence. Carbon protects iron successfully, not in consequence of its greater or less affinity for oxygen or iron, but chiefly on account of its form. Carbon is elastic, and will fill the spaces between the particles of metal. When gray or white cast-iron contains five or six per cent. of carbon, the latter will form a body, when liberated, which cannot be condensed into the same space again by any mechanical means; and even in the form graphite it occupies nearly the space of the iron. Still, cast-iron is porous. All substances foreign to iron, which are contained in the finest kind of cast steel, cannot, when liberated, be condensed into the same space which they occupied in the steel; and such steel, when glass hard, is very porous, there is not even cohesion between its particles; it is brittle.

Component Elements of Metallic Alloys in General Use.

Yellow brass, 2 parts copper, 1 zinc. Rolled brass, 32 parts copper, 10 zinc 1·5 tin.
Brass—casting, common, 20 parts copper, 1·25 zinc, 2·5 tin.
Hard brass, for casting, 25 parts copper, 2 zinc, 4·5 tin.
Gun-metal, 8 parts copper, 1 tin.
Copper flanges for pipes, 9 parts copper, 1 zinc, 0·26 tin.
Brass that bears soldering well, 2 parts copper, 0·75 zinc.
Muntz's metal, which can be rolled and worked at red heat, 6 parts copper, 4 zinc.
Statuary metal, 91·4 parts copper, 5·53 zinc, 1·7 tin, 1·37 lead.
German silver, 20 parts copper, 15·8 nickel, 12·7 zinc, 1·3 iron.
Frick's German silver, 53·39 parts copper, 17·4 nickel, 13 zinc.
Metal for medals, 100 parts copper, 8 zinc. Pinchback, 5 parts copper, 1 zinc.
Chinese silver, 65·2 parts copper, 19·5 zinc, 13 nickel, 2·5 silver, and 12 cobalt of iron
Britannia metal, 1 part zinc, 1 antimony } 1 zinc, 2 antimony, 1 bismuth.
When fused add 1 antimony, 1 bismuth }
Aich's metal, an alloy of iron, copper, and zinc, called also sterro metal.
Babbit's anti-atrition metal, 25 parts tin, 2 antimony, 0·5 copper.
Bell-metal for large bells, 3 parts copper, 1 tin; for small bells, 4 parts copper, 1 tin.
Newton's fusible alloy, 8 parts bismuth, 5 lead, 3 tin; melts at 212°.
Rose's fusible alloy, 2 parts bismuth, 1 lead, 1 tin; melts at 201°.
Very fusible alloy, 5 parts bismuth, 3 lead, 2 tin; melts at 199°.

Solders.—Tin solder, coarse, 1 part tin, 3 lead; ordinary tin solder, 2 tin, 1 lead; melts at 500° and 360° respectively. Soft spelter-solder, for common brass-work, 1 part copper, 1 zinc. Hard spelter-solder for iron, 2 parts copper, 1 zinc. Solder for steel, 19 parts silver, 3 copper, 1 zinc. Solder for fine brass-work, 1 part silver, 8 copper, 8 zinc. Pewterers' soft solder, 2 parts bismuth, 4 lead, 3 tin. Pewterers' common solder, 1 part bismuth, 1 lead, 2 tin. Gold solder, 24 parts gold, 2 silver, 1 copper. Hard silver-solder, 4 parts silver, 1 copper. Soft silver-solder, 2 parts silver, 1 brass-wire.

Fusibility of Alloys.

Parts Tin.	Parts Lead.	Parts Bismuth.	Melts at	Parts Tin.	Parts Lead.	Parts Bismuth.	Melts at
2	3	5	199°	4	5	4	266°
1	1	2	200·75	4	6	4	270
3	5	8	202	1	..	1	286
3	3	8	202	45·3	..	54·6	325
3	6	8	208	3	2	..	333
3	8	8	226	2	1	..	360
2	4	4	243	8	..	1	392
6	1	5	245	1	3	..	500
1	1	1	254				

Tin melts at 442°. Bismuth at 480°. Lead at 600°.

Specific Heat of Various Substances.

Water	1·0000	Glass	·1770	Silver	·0557
Hydrogen	3·2936	Iron	·1098	Tin	·0514
Steam	·8470	Copper	·0949	Platinum	·0355
Nitrogen	·2754	Phosphorus	·1880	Mercury	·0330
Air	·2669	Zinc	·0927	Lead	·0310
Oxygen	·2361	Alcohol	·0700	Bismuth	·0290
Carbonic Acid	·2210	Ether	·0600	Carbon	·2570

Ductility and Malleability of Metals.

Ductile and Malleable Metals in alphabetical order.	Brittle Metals in alphabetical order.	Metals in order of their Wire-drawing Ductility.	Metals in order of their Laminable Ductility.
Cadmium.	Antimony.	Gold.	Gold.
Copper.	Arsenic.	Silver.	Silver.
Gold.	Bismuth.	Platinum.	Copper.
Iron.	Cerium.	Iron.	Tin.
Iridium.	Chromium.	Copper.	Platinum.
Lead.	Cobalt.	Zinc.	Lead.
Magnesium.	Columbium.	Tin.	Zinc.
Mercury.	Iridium.	Lead.	Iron.
Nickel.	Manganese.	Nickel.	Nickel.
Osmium.	Molybdenum.	Palladium.	Palladium.
Palladium.	Osmium.	Cadmium.	Cadmium.
Platinum.	Rhodium.		
Potassium.	Tellurium.		
Silver.	Titanium.		
Sodium.	Tungsten.		
Tin.	Uranium.		
Zinc.			

Works that treat of Metallic Alloys:—Hervé (A.), 'Nouveau Manuel complet des Alliages Métalliques,' 18mo, Paris, 1839. 'The Useful Metals and their Alloys,' by Scoffern, Truran, &c., crown 8vo, 1856. Calvert (C.) and Johnson (R.), 'On the Relative Powers of Metals and Alloys to Conduct Heat,' Philosophical Transactions, London, 1858. Calvert (C.), 'On the Specific Gravities of Alloys,' Philosophical Magazine, London, 1859. Percy's (Dr.) 'Metallurgy,' 2 vols., 8vo, 1861–64. Ziurek (O. A.), 'Technologische Tabellen und Notizen,' 8vo, Braunschweig, 1863. Guettier (A.), 'Guide pratique des Alliages Métalliques,' fcap. 8vo, Paris, 1865. Overman (F.), 'A Treatise on Metallurgy,' royal 8vo, New York, 1866. Larkin (J.), 'The Practical Brass and Iron Founders' Guide,' crown 8vo, Philadelphia, 1867. Graham (W.), 'The Brassfounders' Manual,' 12mo, 1868.

ALLUVIAL DEPOSITS. Fr., *Débris d'alluvion;* Ger., *Anschwemmungen;* Ital., *Strati d'alluvione;* Span., *Depósitos de aluvion.*

Alluvial deposits are those deposits of earth, sand, gravel, and other transported matter, made by rivers, floods, and other causes upon land, not permanently submerged beneath the waters of lakes or seas.

ALUMINUM. Fr., *Aluminium;* Ger., *Aluminium;* Ital., *Aluminio;* Span., *Aluminio.*

Aluminum is the *metallic base* of alumina. This metal is white, but with a bluish tinge, and is remarkable for its resistance to oxidation, and for its lightness, having a *specific gravity* of only about 2·6. The electrical indifference of this metal is most striking, and its effects upon other metals have been extensively investigated. Aluminum is produced from alumina by a peculiar smelting process. When pure alumina, the oxide of this metal, is mixed with finely pulverized carbon, and exposed in a porcelain tube to a red heat, and, in the same time, chlorine is conducted over it, a dry chloride of aluminum is formed, accompanied by a vivid combustion. When this substance is placed in a porcelain crucible, upon whose bottom some pieces of pure potassium are deposited, and the crucible is well covered and luted, and then gently heated over a spirit-lamp, a reduction of the alumina is performed by the potassium, with the production of a high heat at the moment when these two metals decompose each other. Here is a reduction of one metal by the other, as we have seen it performed in reducing sulphurets.

This operation is, therefore, not confined to sulphur, oxygen, phosphorus, and similar substances; it applies to all metals and their combinations, and it requires nothing but a proper selection of the decomposing substance, and the conditions under which it may be performed. The aluminum thus obtained is similar to the alkaline metals; it is very refractory, and does not melt at the heat of melted cast-iron; it is hard, tenacious, and not oxidized at common temperatures, but requires a high red-heat for oxidation. This metal has been observed, alloyed with iron, in Indian steel, and it has been said that the excellent qualities of that steel are owing to its presence. Experiments which have been made with this view have shown that iron combined with aluminum is remarkably strong. In endeavouring to combine aluminum with other metal, we are in the same predicament as with the alkaline metals; silex is reduced before alumina is affected by carbon; and if any advantages are to be derived from an alloy of this kind, silex ought not to be present at its formation. The process used at the Talyndre Works for the manufacture of alumina has been described in the *Revue Universelle.*

At the Talyndre Works they are working a very valuable ore, furnishing pure alumina by two very simple operations, which now renders the preparations of aluminum an actual metallurgical operation in the Ollionelles, near Toulon. Its average composition is—alumina, 60 per cent.; oxide of iron, 25; silica, 3, and water 12 per cent. = 100. After being pulverized under an edge-runner, it is mixed with soda, and heated in a reverberatory furnace. The mass, although not even agglutinating, becomes changed into an aluminate of soda, and a double silicate of soda and alumina is obtained, mixed with oxide of iron, silica, and a little of the alumina which has not reacted. The aluminate of soda is dissolved out with water (the impurities remaining undissolved), and thrown in fine streams through a current of carbonic acid, by which means alumina is thrown down, and carbonate of soda remains. The precipitated alumina is separated by decantation, and washed with warm water to remove the last traces of soda. In practice no soda is lost, except a small portion converted into silicates, the remainder being recovered by evaporation. The alumina is completely dried, and is ready for final treatment. The manufacture of the sodium has been but little modified. The final reaction which yields the aluminum is effected in a reverberatory furnace. To the double chloride of aluminum and sodium is added about 5 per cent. of sodium; and lastly, cryolite as a flux. By this means the metallic aluminum is economically and speedily produced.

To prepare the Chloride of Aluminum.—A mixture is made of alumina and charcoal, and the whole made into paste with some oil, so that it may be shaped into small lozenges. This mixture is placed in a well-made fire-clay retort, Fig. 111; as soon as the retort is brought to a dark red heat, a current of dry chlorine gas is passed into the retort, whereby at first water is expelled from the apparatus, and soon after vapours of chloride of aluminum are produced in abundance. The vapours of the chloride of aluminum are collected in the bell-jar F, which is so narrow that it soon gets sufficiently hot to cause the fusion of the chloride of aluminum, which is hence obtained in a concrete mass. The bell-jar, used as a condenser, is fixed to the neck of the retort A by means of a funnel E, luted on with fire-clay mixed with chopped cow-hair; the tube at the other end of the condenser serves to carry off the carbonic oxide gas, which must be ignited after the air has been entirely expelled from the apparatus. In order to prevent the neck of the retort A from chocking up, it should not project out of the furnace more than five or six inches. With the apparatus just described, 12 lbs. of chloride of aluminum may be prepared daily. The aluminum is obtained from the salt known as ammonia alum, a double salt of alumina and ammonia as bases, both combined with sulphuric acid; it so happens that, at higher temperatures, both the ammonia and sulphuric acid are volatile, and hence alumina is left in a state nearly pure, if the salt previously used be pure.

111.

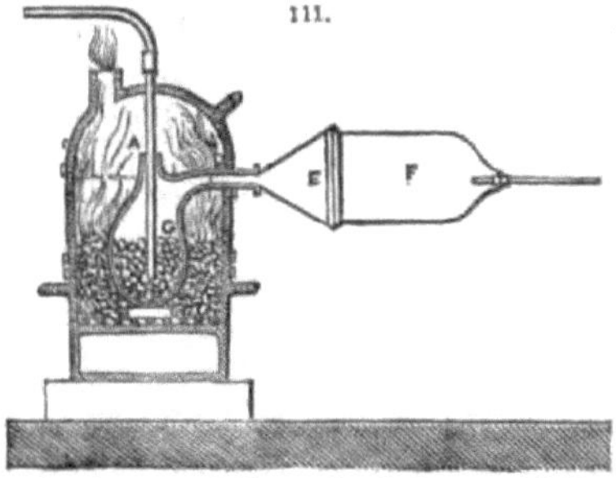

ALUMINUM BRONZE. FR., *Bronze d'Aluminium;* GER., *Aluminium bronze;* ITAL., *Bronzo d'alluminio;* SPAN., *Bronce de aluminio.*

Aluminum bronze is an alloy of copper and aluminum mixed in different proportions. This alloy has a golden hue, and is extensively used in the manufacture of cheap jewellery. One hundred parts copper and ten parts aluminum, measured by weighing, when combined forms a very durable alloy, which may be forged and worked in the same manner as copper. This alloy, of 100 c. to 10 al., takes a fine polish by being burnished, and is the same colour as pale gold; 100 c. and 5 al. forms an alloy, which takes a fine polish by burnishing: it resembles pure gold in colour, but is less durable than the former alloy of 100 c. and 10 al. Farmer's Aluminum Bronze is a mixture in which the copper may vary from 65 to 80 per cent. of the whole, and the zinc, or other light-coloured metals, from 35 to 20 per cent. of the whole; the aluminum being from $\frac{1}{10}$th to $\frac{1}{15}$th of the whole quantity of the light-coloured metals used.

AMALGAM. FR., *Amalgame;* GER., *Amalgam;* ITAL., *Amalgama;* SPAN., *Amalgama.*

An amalgam is a compound of mercury with some other metal, or a native compound of mercury and silver. See AMALGAMATING MACHINE. GOLD and SILVER.

AMALGAMATING MACHINE, *in silver mining.*

AMALGAMATING MACHINE. FR., *Machine à amalgamer;* GER., *Amalgamir Maschine;* ITAL., *Macchina per amalgamare;* SPAN., *Máquina para amalgamar.*

Varney's amalgamating pan is shown in Figs. 112, 113, 114, 115, 116, 117, 118, 119, 120. Fig. 112 is a vertical section of this amalgamator; Fig. 113, a plan of the parts beneath pan; Fig. 114, elevation of the amalgamator complete; Fig. 115, view of interior of amalgamator; Fig. 116, view of one-half the lower dies with wood in slots; Fig. 117, view of under-side of one-half of muller with shoes attached; Figs. 118 and 119 stand for gear on vertical shaft; and Fig. 120, pillow-block for the driving-shaft.

The body of the amalgamator consists of a pan or tub A, Figs. 112 and 114, with cover B, through which is an opening for the introduction of the pulp to be ground and amalgamated. The pan is supported on suitable framework, shown in Fig. 113. From the centre of the pan, and extending from its bottom, to which it is cast, some distance above the cover stands the vertical tube D, through the interior of which is a hole passing vertically through the pan, in order that the shaft C may work through it. On the bottom of the pan, and secured to it by bolts *e*, is fixed the lower muller *a*, consisting of a circular iron-plate having a round hole *d* in its centre, considerably larger than the base of the tube D. This die may, if desired, be made in sections.

That portion of the hole through the muller not occupied by the tube D, is so filled with wood, as to present a plain surface from the tube to the circumference of the muller. The diameter of this muller is somewhat less than that of the interior of the pan, by which means a space *a'* is left to be filled with quicksilver. Above the lower muller is the upper one *b*, of like general form and size, having twelve shoes *c*, the form and relative positions of which will be understood by supposing a plate of the diameter and thickness of the lower muller attached to the under-side of the upper one, and sawn into twelve equal parts on lines drawn from the circumference of the plate to the outside of the tube D. The saw must also be supposed to be held inclined at an angle of about forty-five degrees, thus forming radial grooves from the inner to the outer opening.

Each shoe is fastened to the muller by a bolt, or a wrought-iron rivet, cast into the shoe and riveted into a counter-sink on the upper side of the muller, as shown at *f*, Fig. 112; the bosses and recesses *j*, keep the die in its place. In the lower muller are radial slots, similar to those in the upper one. These slots may be either inclined laterally or be made vertical. The slots in the lower muller are filled with wood, so as to grind on its end, in order that it may be kept slightly worn, in advance of the wear of the die; thus furnishing a cavity for the admission of pulp between the surfaces, by which the grinding capacity of the machine is greatly increased.

112.

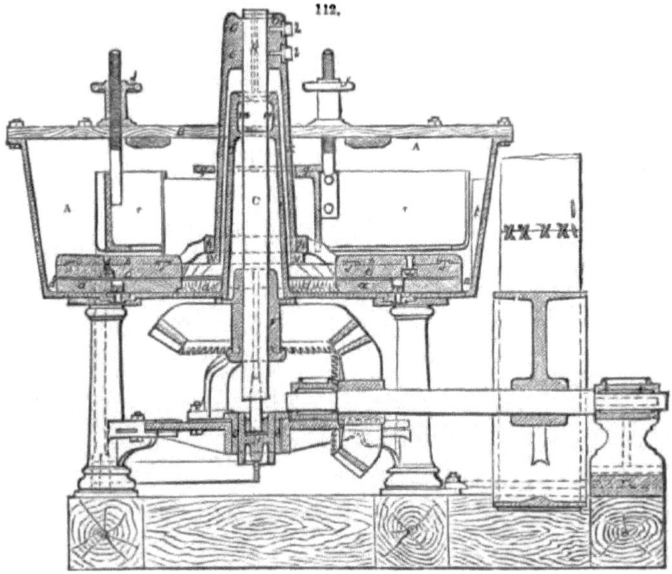

113.

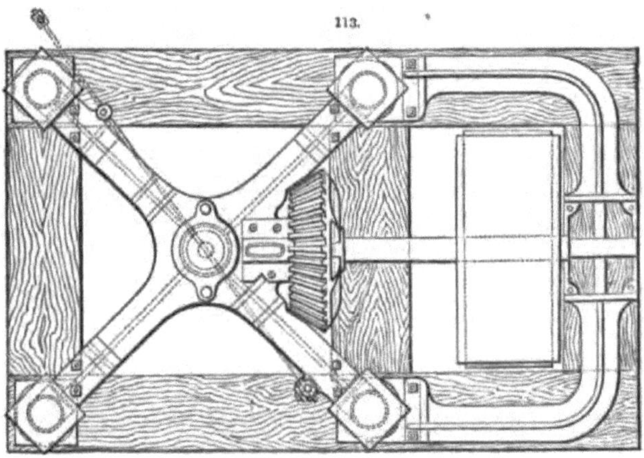

Over and around the tube D, but not in contact with it, is placed the larger tube E, exactly perpendicular to the lower face of the upper muller, and having around its lower extremity the flange V, upon which rests the ring *h*, which is cast with, and forms a part of, the upper muller. This is connected with the muller by means of six curved arms *i*, two pairs of which are much nearer together than the others, and the space between them is filled by a projection from the periphery of the flange V, for the purpose of carrying with it the upper muller when the flange makes a revolution. To the shaft C is fastened the large tube E by the feather *k*, and set-screws *l* in the hub G. The shaft C passes through a *Babbit's metal bearing* at *m*, and through the boss F of the driving-wheel, in which is a feather sliding vertically in the shaft. The shaft is stepped, by the ordinary method, into the vertical sliding-box H, which is itself held in the laterally adjustable box O.

The step-box rests upon an iron bar, one end of which is supported by a screw bolt *w*, Fig. 115, and the other is held by a bolt and hand-wheel *x*, Figs. 114 and 115, by which it can be either raised or lowered; raising or lowering the upper muller at the same time.

114.

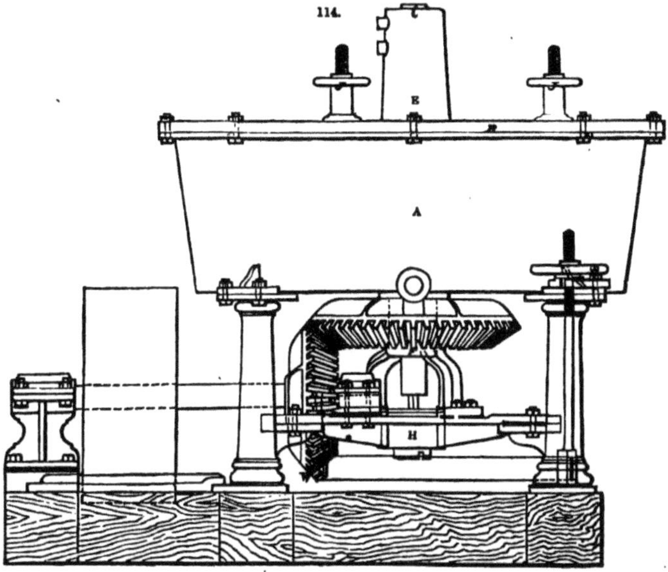

115

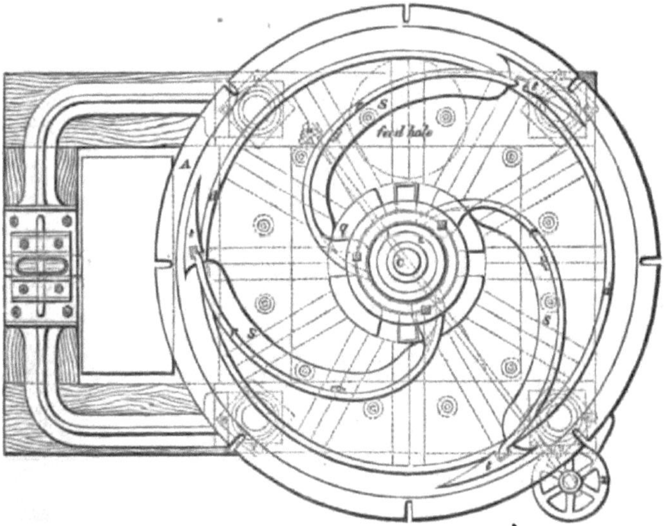

Within the body of the pan are suspended three curved plates *r*, Figs. 112 and 115, extending from near the surface of the upper muller upwards, and stretching in length from the inner side of the pan around to a point near the outside of the large boss, opposite that from which they started.

The lower edges of the curved plates are bent inwards, as shown at *s*, Fig. 115, forming flanges. The inner ends of the curved plates are secured rigidly to the ring *q*, of sufficient diameter to surround and clear the tube E; the whole being suspended by a rod attached to each plate, passing through the cover and hand-wheels J, by which it may be adjusted. The outer ends of the curved plates slide vertically in grooves in the projections *t*, cast upon the inner side of the pan. The operation of this apparatus is as follows:—The space *a'* about the periphery of the lower muller is filled with quicksilver, and the pan nearly filled with pulp, of the proper consistency to flow easily; the shaft C is now made to revolve at a proper speed, from sixty to eighty revolutions per minute, by which the upper muller is rotated. The pulp between the mullers, by means of the centrifugal force developed, is made to pass out through the radial channels between the dies, as well as between the grinding surfaces of the upper and lower mullers; also into and over the quicksilver, thereby causing *amalgamation*.

116.

117.

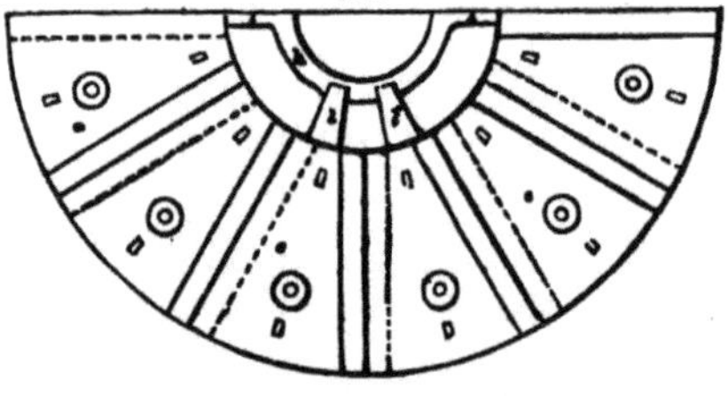

118.

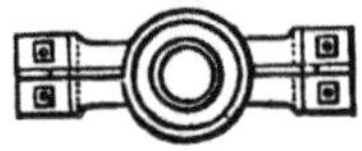

119.

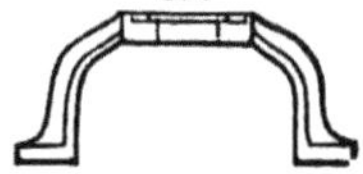

120.

The outward motion of the pulp has the effect of keeping the quicksilver entirely away from the grinding surface, thereby obviating what has often proved a very serious difficulty, the grinding of the mercury.

The rotation of the upper muller causes the pulp in the pan to revolve with it. This current is met by the cuneiform projections and curved plates, and thereby turned towards the central opening in the upper muller. The radial slots between the shoes, running from the central opening to the outward one, allow currents of considerable size to pass with great velocity; and the pump filling these slots, being continually thrown outwardly, tends to produce a vacuum. By this the pulp in the body of the pan is set in motion, causing a rapid and abundant flow downwards at the centre, and upwards along the inner surface of the pan. The pulp is thus made to circulate, until complete pulverization of the quartz and amalgamation of the metals have taken place.

Reference.—A, pan or tub; B, cover; C, vertical shaft; D, central tube; E, exterior tubes; F, boss of driving-wheel; G, hub of outer tube; H, sliding bearing; J, hand-wheels; O, adjustable box; V, flange of outer tube; *a*, lower muller; *a'*, space for mercury; *b*, upper muller; *c*, shoes of muller; *d*, hole in centre of lower muller; *e*, bolts securing lower muller; *f*, bolts securing shoe to upper muller; *h*, ring on upper muller; *i*, curved arms of muller; *k*, sliding key; *j*, lugs to keep shoes in place; *l*, set-screws; *m*, Babbit's metal bearing; *q*, rings supporting curved plates; *r*, curved plates; *s*, flanges of curved plates; *t*, brackets supporting plates; *w*, fulcrum of lever; *x*, hand-wheel for lifting muller.

AMALGAMATION PAN. FR., *Cuvette à amalgamer;* GER., *Amalgamamir-pfanne;* ITAL., *Bacino per amalgamare;* SPAN., *Máquina para amalgamar.*

This pan, shown in Fig. 121, differs from ordinary pans which are employed for the same purpose, especially in the shape of the bottom. The bottom of this pan is inclined towards the centre, and is shaped like the inverted frustum of a cone, to which the shoes are bolted, and the corresponding dies are fastened to the bottom of this frustum. When the pulp is thrown in and the *Mullers* set in motion, that portion of the pulp which finds its way between the grinding surfaces is thrown towards the circumference, from whence it again descends to the centre by gravitation, and passes between the mullers. A constant circulation is thus established without the aid of the curves or wings employed in the Wheeler Pan, which have sometimes been found an impediment in starting the machine after the sand had become packed from stopping. The charge for this pan is about 1400 lbs., and the time required for working it is from two to four hours. The time varies in accordance with the fineness, state of division, and other characteristics of the pulp. When the ore has been sufficiently reduced and amalgamated, the pulp is, after dilution, discharged into the *separator*, and the amalgamating pan immediately recharged without stopping the machine. After the pulp has been run off into the separator, it is further thinned down with water to such a consistency as will allow the mercury and amalgam to settle, whilst it still retains sufficient plasticity to hold the coarser particles of ore in suspension in water. If the compound be in a proper state of dilution, the mercury and amalgam will gradually precipitate, and at the same time no perceptible difference will be felt in the consistency of the pulp situated near the bottom and that at the top of the vessel. When, however, too large a portion of water has been added, the coarser particles will be felt to distinctly separate from the slime, and to strike against the hand when placed near the bottom of the separator. This pan usually makes between fifty and sixty revolutions a minute.

The Hepburn and Peterson Pan is much employed in the reduction establishments of the Pacific Coast, and, in addition to being an excellent amalgamator, it is also a good grinder; but it has the disadvantage of requiring the expenditure of from four to five horsepower for efficient working. The charge of the Wheeler Pan is not only less than that of the Hepburn and Peterson Pan, but the grinding power of the Wheeler Pan is also less considerable. See SILVER.

121.

AMALGAMATING-PAN. FR., *Machine à amalgamer;* GER., *Amalgamir-pfanne;* ITAL., *Apparecchio per amalgamare;* SPAN., *Máquina para amalgamar.*

AMALGAMATOR, *Attwood's.* In gold mining.

Attwood's amalgamator is shown in Figs. 122, 123, 124, 125: it is designed to save the gold as it issues from the stamping mill. Fig. 122 is a sectional elevation; Fig. 123, plan; Fig. 124, lower end of tyers; and Fig. 125, end of steam-chest.

In this arrangement Attwood does not make use of blankets, but the ground ore, issuing from the battery-screens, flows directly on to the amalgamator, where it is gently stirred by the action of the cylinders A, turning in the direction indicated by the arrows, and then passes on to a riffle-board B, covered by amalgamated copper-plates, where a great portion of the amalgam, escaping from the cast-iron mercury boxes *a*, will

122.

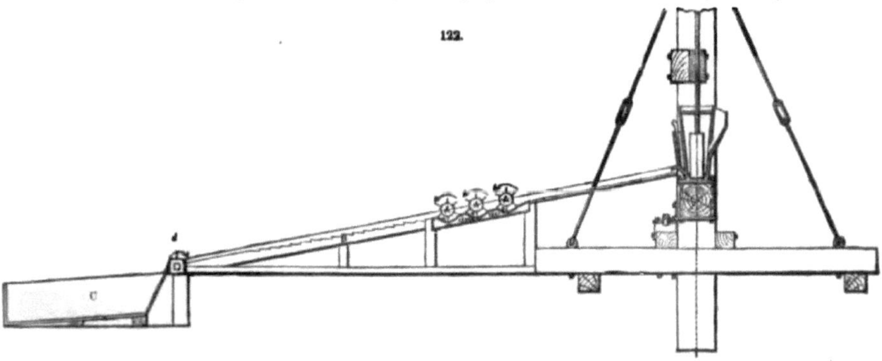

be collected. In order that the mercury in the boxes under the rollers may not become too cold, and its affinity for gold be thus rendered sluggish, they are cast with a double bottom, through which a current of steam can be made to pass, and which is easily regulated by an ordinary tap.

From the riffles B, the ground material passes into the tye C, of which the bottom is inclined at a considerable angle, and which is provided at the lower end with a slot *c*, for regulating the depth of water within it. This is done by means of the stops *c'*. In order to catch any globules of soft amalgam or mercury, which may become detached from the surface of the amalgamated plates, a small cistern D, running the whole width of the riffle-board, is provided; in this is an agitator *d*, turning in the direction indicated by the arrow, and which constantly keeps the box, to the depth of its arms, free from accumulations, so as to form a depression in which the mercury and amalgam may become deposited.

To use this apparatus, one of the stops *c'* is placed in the slot *c*, and the mill started in the usual way; the sand which has passed through the amalgamator soon reaches the tye, and the heavier portions begin to accumulate behind the stop, whilst the lighter particles are carried off by the current. The removal of the light sand is facilitated by gently sweeping the surface of the deposit upwards against the stream with a light broom, a boy being stationed there for that purpose; and when the pyrites which is deposited accumulates to the height of the top of the first stop, another is inserted, and the operation carried on continuously. When one of the tyes has been filled in this way, the tongue E is so turned as to direct the sand and water into the other, which is thus filled whilst the first is being emptied.

123.

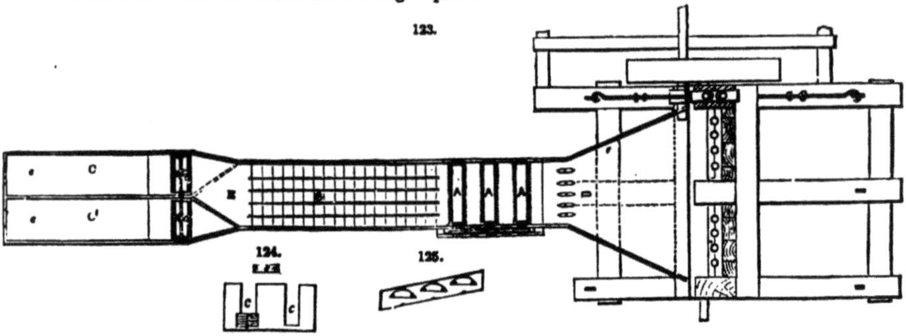

124.

125.

It is evident that by this means the pyrites will be collected in the tyes in a very concentrated form, and that the amount of labour required is but small; we have, however, never seen this apparatus in operation, and are without any precise data showing its efficiency, as compared with the blankets and riffles now in general use.

Reference.—A, agitating rollers with iron blades; *a*, mercury troughs; B, copper riffles; C C′, tyes; *c c*, apertures at bottom of tyes; *c′ c′*, stops for bottom of tyes; D, trough for collecting mercury and amalgam; *d*, agitator in trough; E, tongue for directing course of tailings into tyes C and C′.

AMBULANCE. Fr., *Ambulance;* Ger., *Feldhospital;* Ital., *Ambulanza, Ospitale ambulante;* Span., *Ambulancia militar.*

A variety of contrivances have been invented for the conveyance of soldiers wounded in battle; the main object of these contrivances, termed ambulances, is to remove the men with the least possible suffering, and to place them, consequently, in an easy position, consistent with the limited space allotted for this purpose, or until they can be removed to an infirmary or military hospital. Fig. 126 represents a carriage designed to carry five wounded men; such carriages are employed

126.

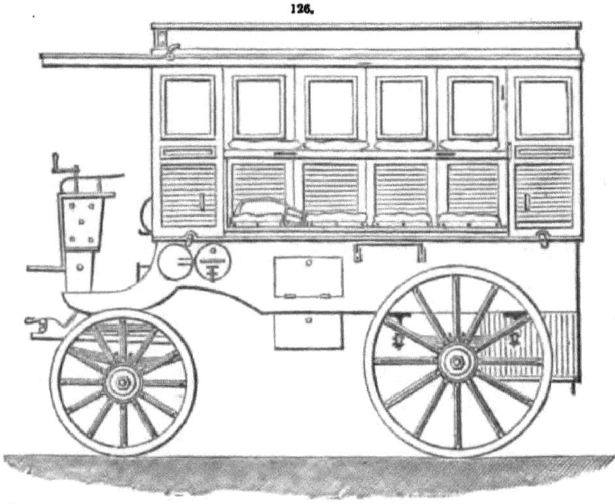

in the Italian army, and constructed according to directions given by Dr. Bertani. The inside of the body of the carriage, Fig. 126, contains five beds, namely, two on the side-seats, which are placed lengthwise in the body of the carriage, and two above these supported by beams which are fixed to iron pillars or supports; the fifth bed is placed at the bottom of the body of the carriage between the seats. The carriage is made to open sideways, so as to give facility to place the wounded men on the beds, which rest on flat steel springs that act with ease, in consequence of their nicely adjusted power of elasticity. Tanks, made of zinc, are placed under the supports of the beds; these tanks are fitted with suitable gutta-percha tubing, which carries off the fluids that escape, to funnels which communicate with openings under the carriage where the fluids are discharged. The upper beds, by means of a suitable mechanical contrivance can be lowered so as to obviate any difficulty that might arise in placing the men upon them with ease and rapidity; the beds, with the wounded men, are replaced by the same mechanical contrivance. The frame-work of the vehicle shown in Fig. 126 is solidly and well built; the body is supported upon six springs of the best make; ventilation is amply provided; and great care has been taken to exclude strong light, which might interfere with the comfort of the temporary occupants.

Fig. 127 is an end view of the carriage shown in Fig. 126; the door at the back is shown in Fig. 127; it is strongly made, and on its inside there are pockets which hold bottles of medicine and other necessaries. The driver's seat is constructed to carry three persons—two and the driver. This convey-

127.

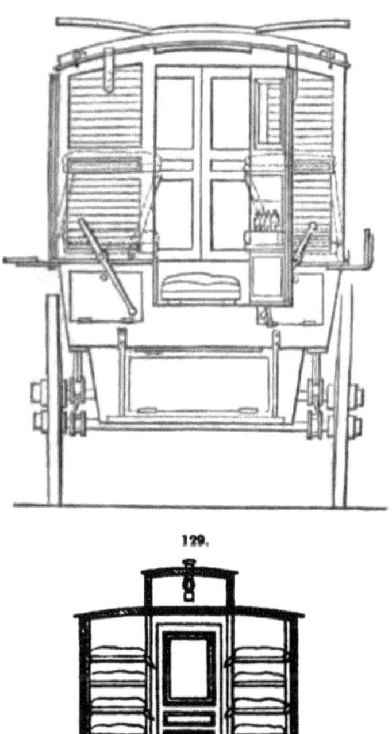

129.

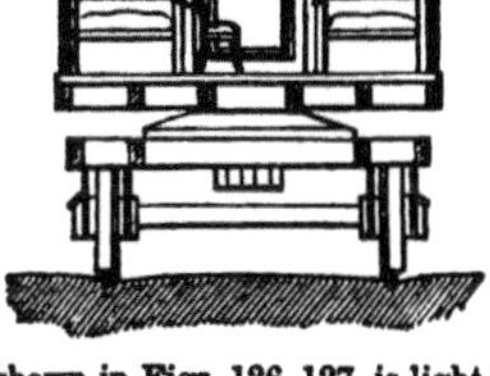

128.

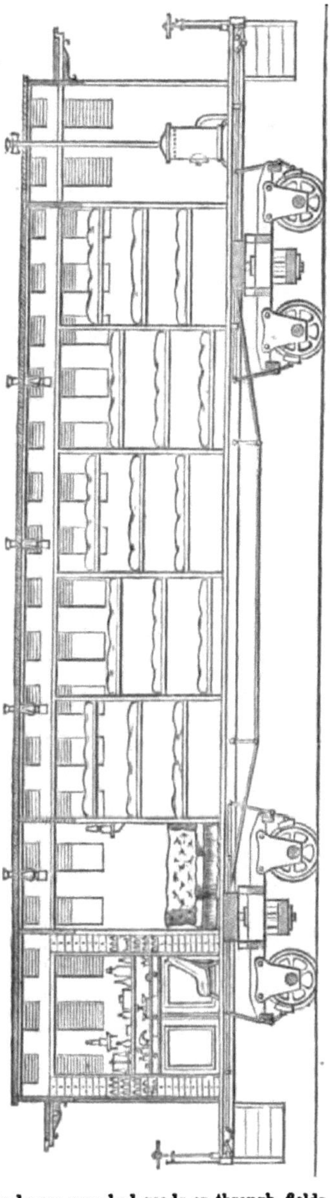

ance, shown in Figs. 126, 127, is light, and may be easily drawn over bad roads or through fields by two horses.

Fig. 128 represents an American hospital railway-waggon, fitted up for thirty men. This waggon or carriage is about sixty-five feet long; it is divided into two parts, one of which contains a chemist's shop and a surgery, together with a small compartment for attendants. At the opposite end there is a small compartment for the guard and breaksman. The beds are supported by strong straps of vulcanized india-rubber. This ambulance, Fig. 128, is properly lighted and ventilated, and, when necessary, it can be heated by a stove. Fig. 129 is an end view of the carriage shown in Fig. 128. Strong breaks, similar to those used on ordinary American railway-carriages, are applied to the wheels of this ambulance both fore and aft.

AMMUNITION. Fr., *Munition de guerre*; Ger., *Kriegs-ammunition*; Ital., *Munizione*; Span., *Municiones de guerra*.

See Ordnance. Small-arms.

ANCHOR. Fr., *Ancre*; Ger., *Anker*; Ital., *Ancora*; Span., *Ancla*.

An anchor is an instrument employed for obtaining a temporary hold of the ground, either under water or on land, but principally in the former position. By its use a strain may be resisted, as in the case of the ship at single anchor, or moored; or a fixed point may be obtained on which to exert power, as in the case of a vessel aground, the anchors of which, being laid out at a distance, enable her to be *hove off* by power exerted on board.

The shapes of anchors are various, to fulfil special objects. The weights range from the 112-cwt. anchors of the 'Great Eastern,' down to those of the smallest yacht; but the proportion to tonnage now employed in the Royal Navy is given in the following Table:—

Anchors.	Line of Battle Ships.				Frigates.				Corvettes.						Despatch Vessels.			Gun-Boats.	
	Above 3500 Tons.	Above 2700, under 3500 Tons.	Above 2000, under 2700 Tons.	Under 2000 Tons.	Above 3500 Tons.	Above 2700, under 3500 Tons.	Above 2000, under 2700 Tons.	Under 2000 Tons.	Above 1600 Tons.	Above 1100, under 1600 Tons.	Above 960, under 1100 Tons.	Above 700, under 960 Tons.	Above 500, under 700 Tons.	Under 500 Tons.	Above 800 Tons.	Above 600, under 800 Tons.	Under 600 Tons.	Above 250 Tons.	Under 250 Tons.
	Cwts.	Cwts.	Cwts.	Cwts.	Cwts.	Cwts.	Cwts.	Cwts.	Cwts.	Cwts.	Cwts.	Cwts.	Cwts.	Cwts.	Cwts.	Cwts.	Cwts.	Cwts.	Cwts.
Bower ..	100	95	90	75	85	75	70	60	50	45	40	30	25	20	25	20	14	9	7
Stream ..	30	25	25	20	25	20	20	18	14	12	10	9	7	6	7	6	6	5	3
Kedge ..	9	8	7	5	7	5	5	4	4	4	3	3	3	2	..	..	..	..	..

The average appears to be about $\frac{1}{350}$th part of the tonnage for bower anchors of the larger classes; one-third of this for the stream, and $\frac{1}{11}$th for the kedge.

Among solid anchors the best form is that of the British Admiralty, shown in Fig. 130; it consists of a *shank* A, with two hooked arms E, E, termed *flukes*, and a *stock* B; the shank A has a ring or *shackle* C, at the end. That end of the shank which is next the stock is called the small round; the point F, where the arms and shank unite, is termed the *crown*; and the rounded angle at its junction with the arms, the *throat*. The arms, for about half their length, are made either round or polygonal; the remaining half consists of three parts, namely, the blade, the palm, and the bill. The blade or wrist is the continuation of the arm towards the palm or fluke D, which is a broad, flat, triangular plate of iron fixed on the inside of the blade. The *bill* or *pea* G, is the extremity of the arm. I, I, are projections intended to enable a wooden stock to be applied in case of necessity. The *forelock* and its chain are shown at H, Fig 130.

Among hinged anchors the best forms are those of the Porter or Trotman, shown in Fig. 131, and the Martin, Fig. 132, which has the peculiarity of holding with both flukes.

130.

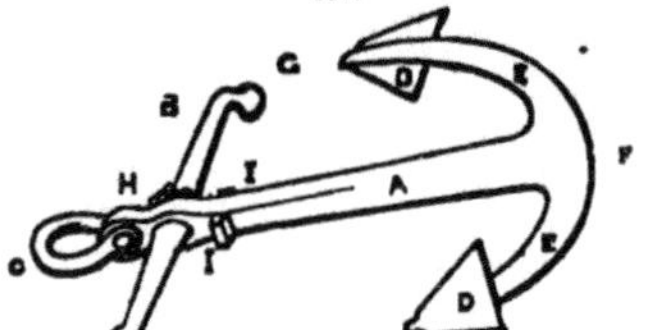

131.

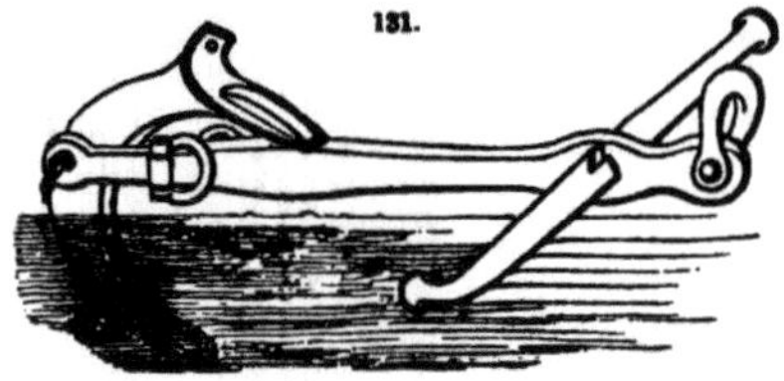

The chief requisites of a good anchor are solidity, holding-power, and non-liability to "fouling." Also the anchor must be easy to "cat" and "fish," and certain to "bite" when a strain is applied.

And here a comparison may be useful between the old form of anchor, as the Admiralty (*vide sup.*) and hinged anchors. While one fluke of the old form must always be standing up out of the ground ready to go through the bottom of an iron ship, or to inflict serious damage on a wooden one, and is always likely to catch the "bight" of the chain cable as it is dragged round when a gale arises and the ship swings to her anchor, and thus cause a foul anchor,—the hinged anchor has no such inconvenience, and in the case of Martin's has a far better hold of the ground. Against this, in the Admiralty anchor are to be put the solid advantages of superior strength and simplicity.

A rude but efficient form of four-fluked anchor is made of wood and stone, Fig. 133, and is still in use in eastern countries. The anchor of most civilized nations is made of wrought iron, but for this, as in chain cables, steel might with advantage be substituted. Since the introduction of chain cables, a great part of the weight formerly considered necessary may be dispensed with, if the tensile strength remains the same. This is on account of the fact that the weight of chain lying on the mud or ooze at the bottom contributes so much to hold the ship, that she rarely "looks

132.

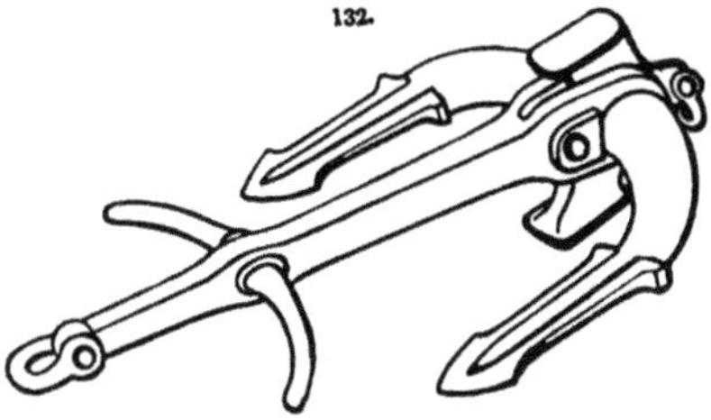

133.

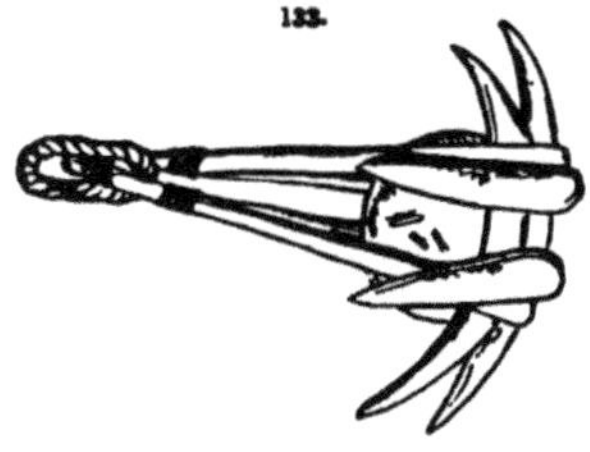

at her anchor" unless in a heavy gale. A useful form of land-anchor is that where, instead of an anchor of the ordinary type, posts of wood or bars of iron are made to do the same duty, Fig. 134.

The laying out of large anchors in case of disaster at sea is best accomplished by the use of two boats of similar size, between which the anchor is slung, Fig. 135. A, A, are spars lashed fore and aft in the boats; B, B, are larger and shorter spars, which rest on A, A, and support the anchor hung below; C, C, stock of anchor; D, the anchor.

Every man-of-war has two "bower" anchors—the best and the small; a sheet anchor, same size as the best bower; a spare anchor, ditto; and a stream anchor; with two or more "kedge" anchors of smaller size. The two bower anchors, as their name implies, are always at the bows,—best bower starboard side, small bower port side, secured to the "cat-heads" and "fish-bollard;" sheet anchor starboard side, abaft fore-chains; spare ditto ditto port side; the others as most convenient, but all generally outside the ship's gunwale, where they may easily be got at in case of necessity. The mushroom anchor for moorings is shown in Fig. 136.

134.

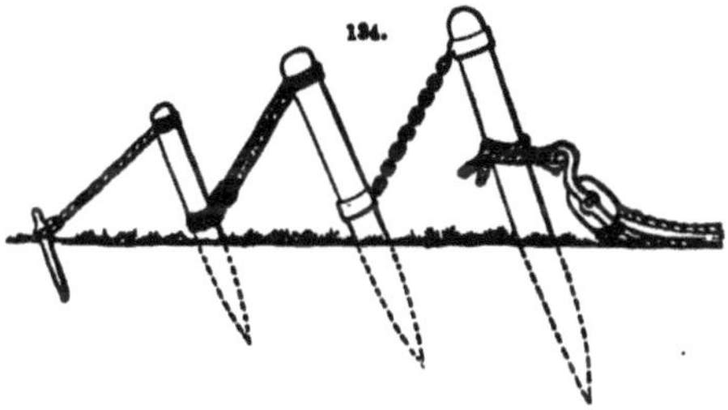

137.

135.

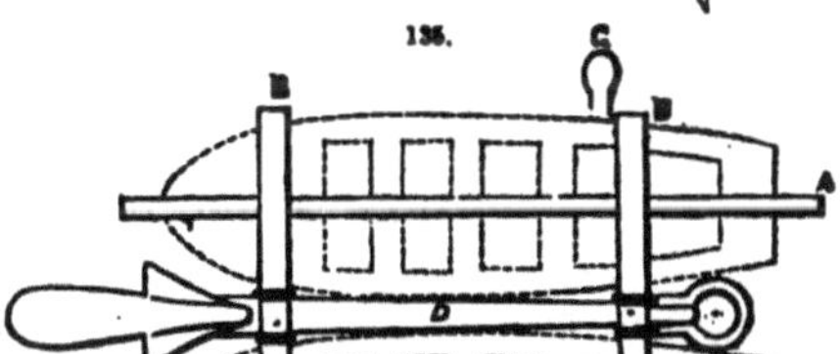

136.

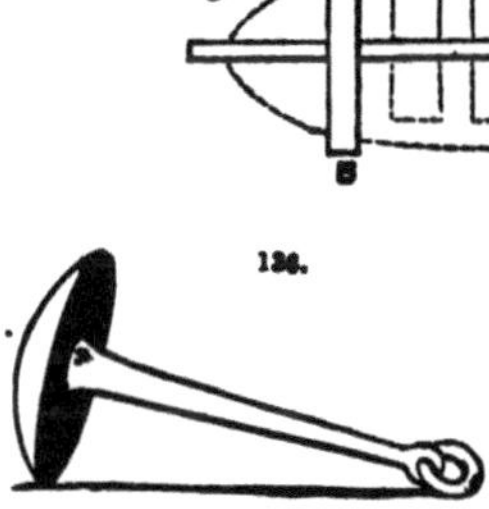

138.

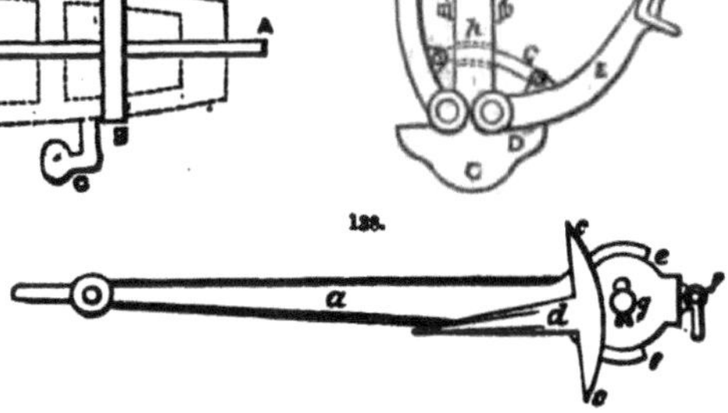

Fig 137 represents the anchor of E. R. C. Morgan; it consists in the arrangement of two flukes, E, hinged to the anchor-shank B, by separate bolts, D, independently of each other; the two flukes are connected by means of a segment G, passing through a slot *h*, in the shank B.

The anchor of E. Snell, Fig. 138, has four horns, *c*, *c*, fixed to the movable arms *d*, which take double grip when the anchor is moved by the cable along the holding-ground. A strong key-bolt *g*, acts as a hinge at the turning point; the range of the motion is limited by the *cams e, e*. A shackle *f*, is fixed at the crown. The horns *c, c*, cause the anchor to lie on the ground in the right position, and compel the palms to penetrate the ground and take hold at once. The use of a stock on the shank *a*, in this anchor is dispensed with. It has great holding power, lightness, facility of stowage, non-liability to foul, and facility of withdrawal from foul ground or obstructions.

The anchor of F. J. Latham is shown in Fig. 139. It consists of a shank A, with two flukes, D, pivoted to a stock B, which vibrate on either side of the latter to an extent determined by the contact of a crown-piece C, with the stock.

139.

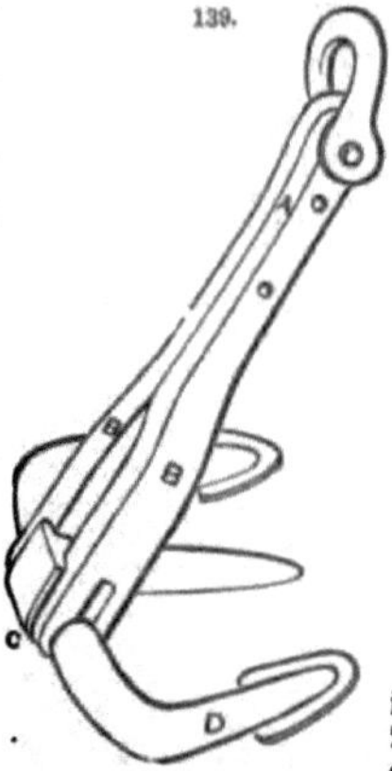

140.

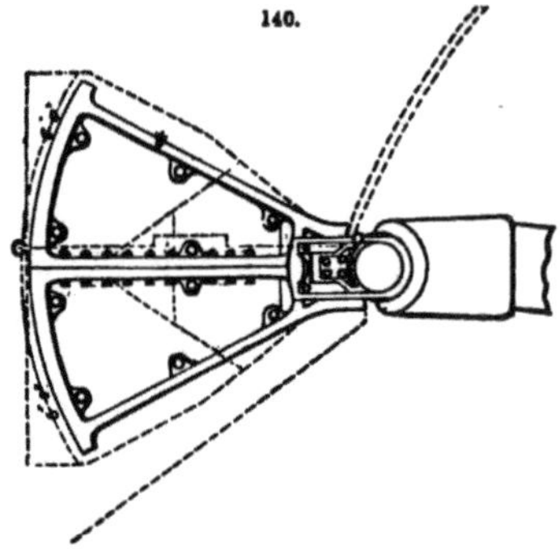

The anchor of the Victoria Docks, London, is a heavy iron casting, resembling a sextant in form, Fig. 140. The length of the curved part of the back is 12 feet, the arc being struck from the centre of the gate pivot with a radius C, D, of 11 feet. In consequence of its great size it was cast in two pieces, which were bolted together through the middle rib; it is held firmly in place by ten vertical bolts, 2 inches in diameter, taking hold of a mass of solid brickwork, 10 feet thick, by means of bed-plates nearly 70 feet in area. There are in addition two long raking bolts, passing through strong plates, bedded further back in the brickwork of the wall.

The anchor strap, which is of wrought iron, is 7 inches deep and 2 inches in thickness, increased to 5 inches in the middle of the length. The ring or upper axis of the gate which it surrounds is 18 inches in diameter. It is formed of a piece of forged iron firmly riveted to a wrought-iron plate ⅜ of an inch thick, on the top of the gate, which is stiffened by additional gusset-pieces and angle-irons. The strap is adjusted by means of keys; and provision is made for examining them and the strap with facility.

Anchors, Chain-cables, and Hawsers of Merchant-ships,

According to Lloyd's Rules.

Ship's Tonnage.	Anchors. Number. Bower.	Stream.	Kedge	Anchors. Weight. Bowers (excluding stock).	Bowers (Admrlty Test.)	Stream (including stock).	Kedge (including stock).	2nd Kedge (including stock).	Stud-chain Cables. Minimum Size.	Proved to Admiralty Test.	Length.	Hawsers and Warps. Stream. Chain.	Rope.	Hawser.	Warp.	Length.
Tons.				Cwts.	Tons.	Cwts.	Cwts.	Cwts.	In. 16ths.	Tons.	Faths.	In. 16ths.	In.	In.	In.	Faths.
50	2	1	1	2·25	4·7	1·00	0·50	..	0 11	8·5	120	0 7	5	3	..	90
75	2	1	1	2'75	5·2	1·50	0·75	..	0 12	10·1	120	0 7	5	3	..	90
100	2	1	1	4·00	6·4	1·75	1·00	..	0 13	11·9	150	0 8	5·5	3	..	90
125	2	1	1	5·25	7·6	2·00	1·00	..	0 14	13·75	180	0 8	5·5	3·5	..	90
150	2	1	1	6·00	8·2	2·50	1·25	..	0 15	15·75	180	0 9	6	4	..	90
175	2	1	1	7·25	9·5	2·75	1·25	..	1 0	18·0	180	0 9	6	4	..	90
200	3	1	1	8·25	10·4	3·00	1·50	..	1 1	20·3	180	0 10	6·5	4	..	90
250	3	1	2	10·00	12·0	4·75	2·25	1·00	1 2	22·75	210	0 10	7	5	..	90
300	3	1	2	12·00	13·9	5·00	2·50	1·25	1 3	25·5	210	0 11	7·5	5·5	..	90
350	3	1	2	13·50	15·2	6·00	3·00	1·50	1 4	28·1	240	0 11	7·5	5·5	..	90
400	3	1	2	15·25	16·7	6·50	3·25	1·75	1 5	31·0	240	0 12	8	6	..	90
450	3	1	2	16·75	18·0	7·00	3·50	1·75	1 6	34·0	270	0 12	8·5	6·5	..	90
500	3	1	2	18·00	19·0	8·00	4·00	2·00	1 7	37·2	270	0 13	9	7	..	90
600	3	1	2	21·00	21·6	9·00	4·50	2·25	1 8	40·5	270	0 13	9·5	7	4	90
700	3	1	2	23·50	23·5	10·00	5·00	2·50	1 9	44·0	300	0 14	10	8	5	90
800	3	1	2	25·50	25·2	10·50	5·25	2·75	1 10	300	300	0 14	10	8	5	90
900	3	1	2	27·75	26·9	11·00	5·50	2·75	1 11	51·2	300	0 15	10	9	5·5	90
1000	3	1	2	30·00	28·6	12·00	6·00	3·00	1 12	55·1	300	0 15	10	9	5·5	90
1200	3	1	2	32·00	30·1	13·00	6·50	3·25	1 13	59·1	300	1 0	10	9·5	6	90
1400	3	1	2	34·00	31·6	13·50	6·75	3·25	1 14	63·25	300	1 0	10	10	6	90
1600	3	1	2	36·50	33·4	14·00	7·00	3·50	1 15	67·5	300	1 1	11	10·5	6·5	90
1800	3	1	2	38·00	34·5	14·50	7·25	3·50	2 0	72·0	300	1 1	11	11	7	90
2000	4	1	2	40·00	35·7	15·00	7·50	3·75	2 1	76·5	300	1 2	11	11	7	90
2500	4	1	2	42·00	37·1	17·00	8·50	4·25	2 2	81·3	330	1 2	12	12	8	90
3000	4	1	2	45·00	39·2	19·00	9·50	4·75	2 4	91·1	360	1 3	12	12	8	90

Two of the bower anchors must not be less than the weight set forth above, but in the third a reduction of 15 per cent. is allowed. All anchor-stocks must be of acknowledged and approved description.

Unstudded close-link chains of 1 inch in diameter and under, are admitted as cables, if proved to two-thirds the test required for stud-chains. But in all such cases a short length, not less than 12 links, must be tested up to the full strain for stud-link chains.

In cases where parties are desirous of using or supplying chains of smaller size than is set forth above, a reduction is allowed not exceeding $\frac{1}{16}$th of an inch in chains of 1 inch to $1\frac{1}{2}$ inch diameter, and $\frac{1}{8}$th of an inch in chains above $1\frac{1}{2}$ inch diameter, provided they be subjected to the Admiralty strain for the size for which they are to be substituted; and further, that a few links, not less than twelve, to be selected by the tester, shall be proved to the breaking strain, and show a margin of at least 10 per cent. beyond the Admiralty proof for a chain of the full size required by the Table.

For steamers the anchors and cables will not be required to exceed, in weight and length, those of a sailing vessel of two-thirds their total tonnage.

ANCHOR-TRIPPER, or ANCHOR-STOPPER. FR., *Arrangement mécanique pour vitement détacher l'ancre et le faire tomber dans l'eau*; GER., *Eine Vorrichtung zum schnellen Werfen der Anker*; ITAL., *Apparecchio per gettar l'ancora*; SPAN., *Aparejo para soltar el ancla.*

An anchor-tripper is a device for the purpose of relieving or forcing the anchor from the davit or cat-head. In the anchor-tripper of W. Stacy, Fig. 141, the tripping line E, turns the hook D, which holds the ring, thus permitting the ring to slide off its suspending hooks. A is a block; B, the davit or cat-head; and C a rope passing through the block and over the davit; F is a guide-block over which the rope passes, and G is a belaying-pin. Fig. 142 illustrates the manner in which the anchor-tripper of B. H. Heitmann is applied: A is a hawse-pipe, and B the anchor. To throw the anchor off from the rail with this tripper, it is only necessary to raise a lever *b*, which rests upon the rail; when this lever is raised, both the shank-painter and the ring-stopper are instantly discharged by the motion given to the rotating bar or keeper D, with its troughs *f*, *f*, chain *d*, and latches *c*, *c*.

The anchor-tripper of G. Gibson is shown in Fig 143. This invention consists in so forming or arranging upon the upper edge and inner side of the bulwarks of a vessel, and at or near the bow, a

141.

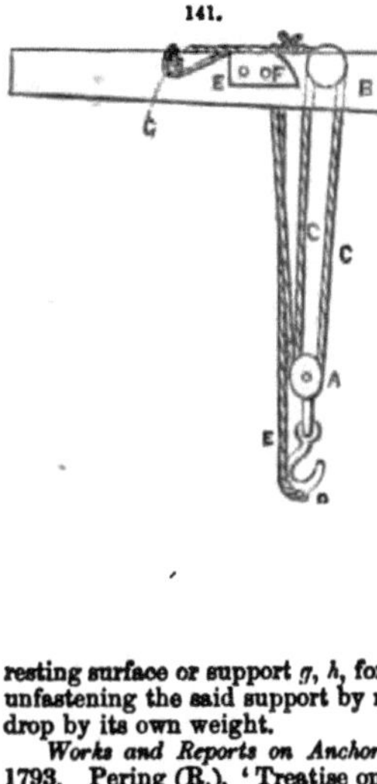

142.

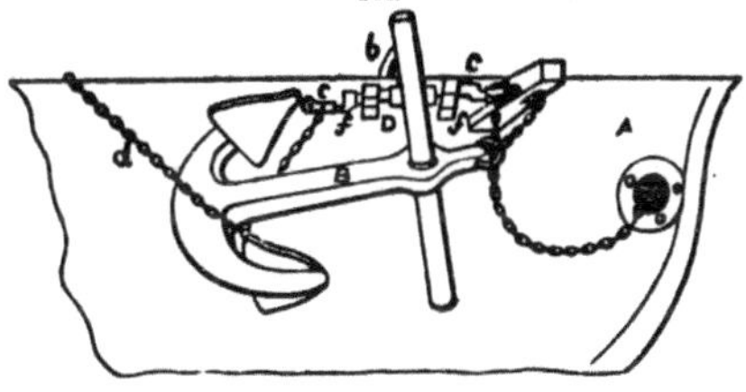

143.

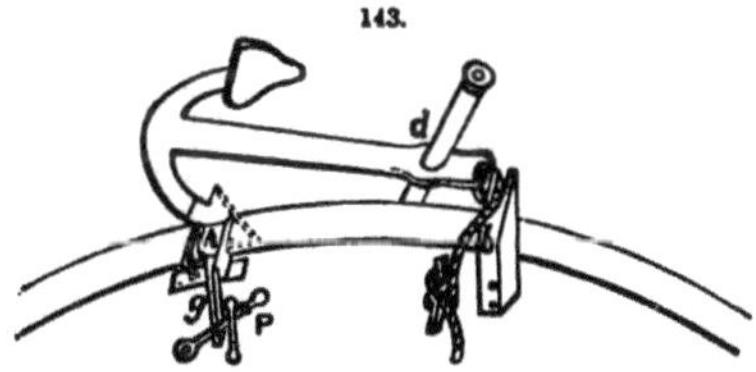

resting surface or support *g*, *h*, for the fluke of the anchor, that when desired, by simply releasing or unfastening the said support by means of the detaining lever P, the anchor *d*, will readily fall and drop by its own weight.

Works and Reports on Anchors:—'Anchor-making' in Steel's 'Rigging and Seamanship,' 4to, 1793. Pering (R.), 'Treatise on the Anchor, with some Observations on Chain-cables,' royal 8vo, 1819. Rodger (Lieut. W.), 'Improvements in Anchors,' 8vo, 1830. Jamieson (A.), 'On the Mechanical Properties of Porter's Patent Anchor,' 8vo, 1843. Cotsell (G.), 'Treatise on Ships' Anchors,' 12mo, 1856. 'Parliamentary Reports on Anchors,' 1860, 1864. 'Rapports de la Commission du Ministère de la Marine sur l'Exposition de 1867,' Paris, 2 vols., 1868.

ANEMOMETERS EMPLOYED IN MINING OPERATIONS.

ANEMOMETER. FR., *Anémomètre*; GER., *Windmesser*; ITAL., *Anemometro*; SPAN., *Anemómetro.*

The anemometers of which we treat are instruments designed to measure the force and velocity of currents of air in mines through chimney-shafts and in other places. Meteorologists employ machines, termed anemometers, which they employ to measure the force and velocity of the wind; of these machines we take no particular notice here.

Means employed to Measure the Velocity of Air.—The difficulty of measuring the velocity of the air with precision has been, and still is, the chief obstacle against the attainment of conclusive experiments which shall indicate the laws of the effort exerted.

The most general mode adopted by experimenters consists in throwing to the wind light bodies, such as feathers, thistledown, the smoke of powders, or the essence of turpentine, and in observing the distances described, with the corresponding times, in the movement of translation. But this simple method affords but little precision, on account of the small distances in which they can be observed; of which we shall speak presently.

Anemometers, composed of a small light fan-wheel, whose motion is transmitted to a counter which registers the number of turns, are most certain, and convenient for use, though they must previously be tested, or the relation existing between the velocity of the wind and the number of turns of the wings must be accurately determined; this determination presents great difficulties.

Most generally, we accomplish this test by placing the instrument upon the horizontal arm of a species of horse-gin (Fig. 153) with a vertical axis, which is made to turn as uniformly as possible. We then observe simultaneously the number of turns of the wings and the velocity of translation of the instrument, and then suppose that the effect produced by this movement of the apparatus in the air the same as that which would be due to the action of the wind, impressed with the velocity of transport of the anemometer, upon the wings of the instrument at rest.

In examining the different systems of measuring the velocity of air in mines, we shall describe the construction and practical application of the anemometers now generally employed: the first we shall describe is a very light anemometer which M. Combes, Inspector-General of Mines, constructed to measure the small velocities of air, principally in the ventilation of mining works.

Anemometer of M. Combes.—This instrument is similar to Woltiman's mill for gauging streams of a considerable section. It is composed of a very delicate axle A (turning in agate caps), upon which are mounted four plane wings, equally inclined as to a plane perpendicular to the axis. In the middle of the axle (Fig. 144) is cut an endless screw, which drives a small wheel, R, with a hundred teeth, so that the latter advances one tooth for each revolution of the axle bearing the wings. The axle of the first wheel carries a small cam, which acts upon the teeth of a second wheel, R′. The last is held fast by a claw or very flexible steel spring, which is attached to the horizontal plate upon which the instrument is mounted. At each revolution of the first wheel with a hundred teeth, driven by the endless screw, the cam starts one tooth of the second wheel with fifty teeth. The two wheels are marked at intervals of ten teeth; the first from one up to ten, the second from one to five. The index-pointers, fixed upon light uprights, which bear the axle of the wings, serve to mark the number of teeth which each wheel has advanced, and thus to indicate the number of revolutions of the axle of the wings. By means of a detent and two cords, which move it, we may, at a distance, arrest the rotation of the wings, or allow them to turn, under the impulse of the current of air which strikes them.

144.

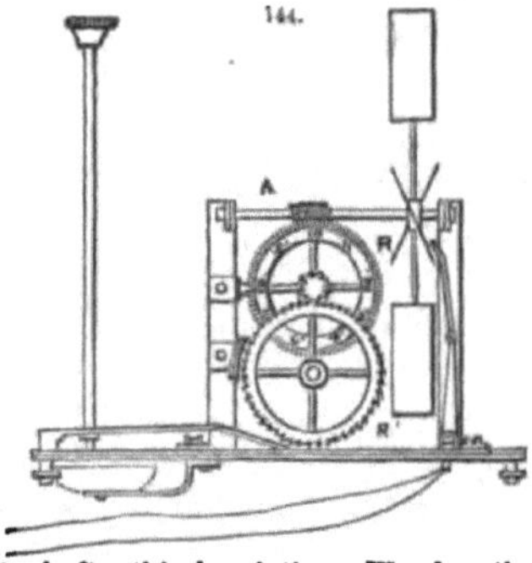

The manner of using this instrument is easily understood after this description. We place the limbs at zero, and the instrument in the axis of the air tubes, keeping the limbs immovable by means of a catch, which is loosened at the moment of commencing the observation, and made fast at the end of the same.

It is well to prolong the observation as long a time as possible, and for two or three minutes at least, if it can be done. The division of the limbs does not admit of counting over 5000 turns, which, for a velocity of air 9·84 feet per second, would only correspond with a duration of about 2·8 minutes.

The test or error of these instruments may differ very much from each other, though their dimensions may seem identical in all points. It should then be made for each one in particular, and repeated, as far as possible, whenever we wish to use it after an interruption.

Thus the anemometer, whose trial was reported by M. Combes, gave

$$v = 0\cdot8458 \text{ foot} + 0\cdot3005\, n,$$

v being the velocity of the air in feet per second,
and n the number of turns of the wings in 1″.

Another anemometer of the same model gave the relation

$$v = 0\cdot4921 \text{ foot} + 0\cdot3821\, n.$$

Remarks upon the Use of the Instrument.—This little instrument is handy for the measurement of small velocities, since we see that it can appreciate those from 0·492 to 0·82 foot per second. In this case it works long enough to give sufficiently exact indications in practice, still with this condition, that the current shall be continuous and tolerably regular, such as is the case with mines whose ventilation is produced by permanent causes, slightly varying from one instant to another.

The Resistance of the Air.—The phenomena produced by bodies moving in air are similar to those presented by liquids, and the resistance which it opposes to the motion of these bodies is of the same kind. Still, it is proper to distinguish between what occurs in uniform motion, and that which takes place in variable motion.

In the first case, the velocity remaining the same, the fluid molecules, successively driven aside by the body, experience the same displacements, receive the same velocities, and in different instants of its motion the body meets the same resistance. But in variable motion, accelerated, for example, the fluid molecules receive greater and greater degrees of velocity; and as they belong to an elastic fluid, the fluid prow formed in front of the body acquires a density and mass continually increasing, whence it follows that the mass displaced increases in the same time as the velocity imparted to it. We conceive then, *a priori*, that the greater the acceleration of motion $\frac{v}{t}$, so will be the resistance; and so we may foresee that, in accelerated motion, the expression of the resistance of the air must comprise, besides other terms, one peculiarly due to the acceleration of motion itself. It was reserved, however, for the experiments of Morin at Metz for the first proving of this matter, to which we shall allude presently.

Results of Experiments.—The celebrated Borda made, in 1763, experiments upon the laws of the resistance of air, by means of a kind of fan-wheel, with a vertical axle and horizontal arms, a little over 7·15 feet in length. He placed at the end of this arm the surfaces and different formed bodies on which he wished to operate, and he observed the uniform velocities of the fly-wheel under the action of different weights. He thought the influence of the friction of this apparatus might be overlooked, which has occasioned some uncertainty in his results; for it is difficult to admit that, in dealing with so small a resistance, the portion of the motive weight engaged in overcoming the friction should not be comparable to that surmounting the resistance of the air.

Borda placed in succession, at the ends of the arms of his apparatus, square surfaces of 9·56, 6·38, and 4·25 inches at the sides, and set them in motion with weights of 8·8, 4·4, 2·2, 1·1, and 0·5 lbs., and consequently with different velocities. From the dimensions and data relative to this apparatus, the author has calculated the resistances of the air corresponding with the different velocities, and the results expressed in *yards* are given in the following Table:—

Results of Borda's Experiments upon the Resistance of Air.

Surface of 9·591 inches each side, or of ·07099 square yard.			Surface of 6·394 inches each side, or of ·03155 square yard.			Surface of 4·263 inches each side or of ·01402 square yard.		
Resistance of Air.	Velocities.	Squares of Velocities.	Resistance of Air.	Velocities.	Squares of Velocities.	Resistance of Air.	Velocities.	Squares of Velocities.
lb.	yards.		lb.	yards.		lb.	yards.	
0·16695	3·787	14·34	0·16713	5·938	35·26	0·1592	9·053	81·94
0·07895	2·690	7·237	0·08356	4·199	17·64	0·0796	6·397	40·93
0·04168	1·891	3·579	0·0416	2·978	8·86	0·0399	4·505	20·30
0·02084	1·334	1·780	0·02083	2·091	4·28	0·01986	3·84	10·14
....	..	..	0·01036	1·382	1·91	0·00944	2·252	5·07

If we represent these results graphically, in taking the resistances for abscissæ, and the squares of velocities for ordinates, we find all the points relative to the same surface are situated in a straight line, thus indicating that the resistance increases as the square of the velocity. The small extent of surfaces used by Morin could not manifest with certainty the existence of a constant term in the expression of resistance.

Comparing these results with the formula $R = K_1 A V^2$ (expressed in yards and square yards), we have for K_1 the following values:—

Square of 9·585 in. or 0·26575 yd., side, $K_1 = 0·1618$.
Square of 6·390 in. or 0·17716 yd., side, $K_1 = 0·1472$.
Square of 4·263 in. or 0·1181 yd., side, $K_1 = 0·1382$.

The coefficient K_1 is for A in square yards, and V in yards.

It is to be observed that Borda having neglected the influence of friction, which increases with the resistance and the motive weights employed, the apparent diminution of the resistance for the smaller surfaces may be attributed to this cause.

Experiments by M. Thibault upon Bodies in Motion in the Air.—We are indebted to Thibault for numerous and very well executed experiments, published at Brest in 1832. He used for his experiments a fly-wheel with two wings, turning on a horizontal axle, and moved by a weight, which the resistance of the air itself soon rendered uniform. This very light wheel was composed of a steel axle 2·13 ft. in length by 0·016 ft. square, terminated by journals with a diameter of 0·0082 ft. The arms of the fly were each formed of an iron rod 8·97 ft. long by 0·045 ft. wide in the direction of movement near the axle, and 0·016 ft. near the ends, with a constant thickness of 0·019 ft. in a direction parallel to the axle. The side of the arm striking the air was beveled.

The wings were mounted upon the arms of the flyer, and at first directed in planes passing through the axis; then by means of suitable arrangements they were inclined, 1st, in turning them around the radius; 2nd, in turning them round parallel to the axis, so that their direction left the axis either in the front or in rear. The inclinations thus obtained were varied at intervals of five degrees, and were carefully measured. The motion of the fan-wheel was produced in all cases by the same motive weight of 8·82 lbs., and the duration of twenty turns made with uniform motion was observed.

Morin calculated the results of the experiments of M. Thibault, in applying the formula

$$R = K_1' A + K_1 A V^2,$$

which represents, as we shall see hereafter, all the results of the experiments made at Metz. In giving to the coefficient K_1' relative to the constant resistance, independent of the velocity, the value $K_1 = 0·08002$ (units of yards), derived from these experiments upon a fan-wheel, we are enabled to deduce the value of the coefficient K_1 dependent upon the velocity. The inclination of the surface of the wings towards the direction of the motion was also taken, by introducing in the second term of the formula, in place of the area $A = 0·12323$ square yard, its projection upon a plane perpendicular to the direction of the motion.

Experiments of M. Thibault upon the Resistance of Air.

Inclination of Surfaces.	Time of 20 revolutions of Wheel.	Velocity of centre of the Wings.	Total resistance of each Wing.	Part of the resistance independent of the Velocity.	Resistance proportional to the square of the Velocity.	Ratio of the resistance to the square of the Velocity.	Projection of surface upon plane perpendicular to Motion.	Resistance per square yard projected, and per yard of Velocity.
°	seconds.	yards.	lb.	lb.	lb.	lb.	sq. yard.	lb.
90	68·40	2·752	0·1660	..	0·1563	0·02063	0·12323	0·16737
85	68·07	2·765	0·1658	..	0·1561	0·02041	0·12278	0·16619
80	67·90	2·772	0·1658	..	0·1561	0·02031	0·12139	0·16729
75	67·70	2·781	0·1658	0·0097	0·1561	0·02018	0·11904	0·16953
70	65·56	2·828	0·1655	..	0·1558	0·01948	0·11581	0·1682
65	64·76	2·906	0·1655	..	0·1558	0·01845	0·11169	0·1651
60	62·47	3·014	0·1651	..	0·1554	0·01710	0·10673	0·1602
55	61·15	3·078	0·1648	..	0·1551	0·01637	0·10095	0·1621
50	60·25	3·124	0·1648	..	0·1551	0·01588	0·09440	0·1682
45	56·75	3·318	0·1642	..	0·1545	0·01403	0·08714	0·1610
40	52·83	3·563	0·1635	..	0·1538	0·01211	0·07921	0·1529
35	48·50	3·882	0·1622	..	0·1525	0·01011	0·07084	0·1428
30	43·00	4·378	0·1602	..	0·1505	0·00785	0·06161	0·1274
25	36·75	5·122	0·1569	..	0·1472	0·00561	0·05208	0·1077
20	30·50	6·173	0·1549	..	0·1452	0·00381	0·04214	0·0904
15	24·50	7·683	0·1411	..	0·1314	0·00222	0·03189	0·0697
10	19·	9·910	0·1220	..	0·1123	0·00114	0·02137	0·0535

This Table contains the data of the experiments of M. Thibault, and the results of his calculations. The figures entered in the Table show that the resistance to the square yard of surface projected perpendicularly to the direction of the motion, and per yard of velocity, where the value of the coefficient K_1 of the formula, $R = K_1 A V^2$, does not decrease so long as the angle of inclination is not below from 50° to 60°.

Remarks upon Wing-regulators and Windmills.—It follows in the case of fan fly-wheels used as regulators of motion, where the wings are inclined and turn round the radius of the fly-wheel, that when the motive power is too feeble we do not have a diminution of resistance until the wings have passed the inclination of from 50° to 60°; and as these regulators should also serve to prevent the acceleration of motion when the motive power increases, and consequently then afford the greatest resistance, it would be well, in the normal state, to place them at an angle of about 35° with the plane perpendicular to the direction of the motion.

It seems that something analogous to this is produced in windmills, the sails of which are, by some special mechanism, made to incline when the wind has acquired too much intensity.

Experience shows, in fact, that this disposition, the aim of which is to check the velocity from being too greatly *accelerated* by the effect of squalls, does not fully attain its object, and that the mill, whose normal velocity is from five to six turns in one minute, by a good breeze from 16 to 19 feet of velocity per second, reaches that of from twenty-nine to thirty turns, and more, with greater winds.

Experiments upon different formed Surfaces.—M. Thibault has successively repeated the same experiments with concave cylindrical surfaces; he arrived at the same consequences, and has established the fact that, with an equal projection of surface, upon a plane perpendicular to the direction of the motion, the resistance increases gently with the curvature.

As for hollow surfaces, with double curvature, such as frame surfaces, the resistance increases with the curvature, and more rapidly than in the preceding case.

A comparison was made of the resistance offered by bent sails, with that experienced by plane sails with the same surface as that of the sails developed; the two surfaces of folded sails were each 0·1302 square yard, and the lower side was brought near the upper, as is usual with sails under the action of wind; and Thibault found that the resistance of the bent surface was the same as that of the plane surface, notwithstanding the diminution of the projection of the first surface upon the direction of the motion. A comparison is thus made between the increase of the resistance due to the curvature, and the diminution due to the narrowing of the projected surface.

This consequence is important, inasmuch as it facilitates the applications relative to the action of wind upon the sails of vessels.

Influence of the Inclination of the Wings.—It was found that when the vanes are inclined so that the axis of rotation is found in front of their plane, in regard to the direction of motion, Fig. 145, the resistance diminishes rapidly as the inclination increases, and that at the inclination of 55° it is not more than 0·5715 of the perpendicular resistance; while when the axis of rotation is found behind the plane of the wings, the resistance goes on increasing even up to the angle of 55°, Fig. 146, for which it is equal to 1·2293 times the perpendicular resistance.

145.

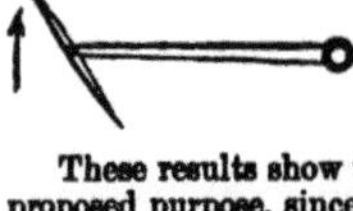

146.

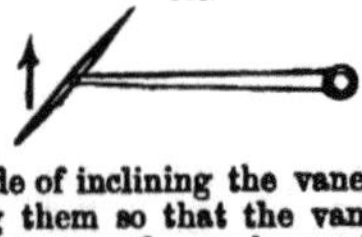

147.

These results show that this mode of inclining the vanes of fly-regulators answers readily the proposed purpose, since in disposing them so that the vanes may be inclined at will in either direction, Fig. 147, the resistance experienced may be rendered greater or less, according to the necessities of the case.

The same experiments, repeated upon curved surfaces, with different degrees of inclination, have led to similar consequences, while indicating a still greater intensity of resistance than is experienced by plane surfaces. This explains the advantage which navigation derives from the movements of rotation impressed upon sails parallel to the axis of the masts.

Influence of the Approximation of the Surfaces which are exposed to the Resistance of the Air.—M. Thibault has also made some experiments to ascertain whether two equal surfaces (placed one behind the other, a very small distance apart) experience a less total resistance than when isolated. For this purpose he mounted upon his fly-wheel four wings, placed in pairs, the one behind the other, at a distance which he has not given, and he found for the case upon which he operated that the resistance of the posterior was not over $\frac{1}{4}$ of that of the anterior surface. This result, which can be applied to railroad trains, is important, and it was desirable that more complete experiments should be made upon this subject. See DYNAMOMETER, *Railway Car.*

Influence of the Form of Surfaces.—The same experimenter having placed at the extremities of his fly-wheel various surfaces of the same area, but of which two were square, two circular, and two in the form of a right-angled triangle, so that the centre of their figure was in all cases at the same distance from the axis, has observed that under the action of the same motive weight the fly-wheel took, in all cases, the same velocity, which shows that the resistance is independent of the form of the plane surfaces experimented upon.

Resistance of Air to the Motion of Spherical Bodies.—This particular case, which is of special interest in the study of the motion of projectiles in the air, has for a long time occupied the attention of philosophers and geometricians. Newton was the first to experiment upon this subject, in observing the fall of spherical bodies. Hutton and other observers have studied this resistance in the case of small velocities, by means of a rotating apparatus; and more lately the latter, in comparing the velocities of projectiles at different distances from the piece of ordnance, has extended his researches to great velocities.

Here, however, we limit ourselves to indicating the results more especially applicable to industrial questions.

From a summary of Newton's experiments upon the fall of glass globes in air, with velocities comprised between zero and 29·528 feet per second, at a mean temperature of 53·6°, and at a pressure of 2·46 feet, the value of the coefficient K_1 was about 0·0007137, so that the resistance experienced by spheres moved in the air, at velocities comprised within these limits, would be

$$R = 0{\cdot}0007137\ AV^2 = 0{\cdot}0007137\ \frac{D^2}{1{\cdot}273}\ V^2, \text{ for units of feet;}$$

$$\text{or } R = 0{\cdot}05781\ AV^2 = 0{\cdot}05781\ \frac{D^2}{1{\cdot}273}\ V^2, \text{ for units of yards;}$$

but in great velocities the coefficient of the resistance increases with the velocity; and after a discussion of Hutton's experiments, and those of the Commission at Metz, General Piobert has proposed, for a representation of the law of the *resistance of the air to the motion of projectiles*, the formula

$$R = 0{\cdot}03546\ AV^2\ (1 + {\cdot}002103\ V), \text{ units of yards,}$$
$$R = 0{\cdot}00043778\ AV^2\ (1 + 0{\cdot}0070102\ V), \text{ units of feet;}$$

which would indicate that, with these velocities, the expression of the resistance must contain a term proportional to the cube of the velocity, and that the constant term is without a sensible influence.

Experiments at Metz upon Bodies moving in Air.—Numerous experiments, with the joint labour of MM. Piobert, Didion, and Morin, were made at Metz in 1835 and 1837, which were more particularly made by M. Didion, in which they made use of chronometric apparatus to observe the law of the descent in air of different formed bodies, and of different dimensions. These experiments were made where the experimenters could avail themselves of a vertical fall of 46·916 feet.

The bodies employed were suspended upon a silk cord, wound round a pulley, which in its motion bore a style whose trace upon the plate of the chronometric apparatus, impressed with a known uniform motion, and observed at every experiment, furnished the law of motion of the descent of the body.

Special experiments were made to determine the passive resistances of the apparatus, to keep an account of them in the calculations.

Without going into a detailed discussion of the results, and the tests applied to them, we simply indicate the method adopted for the calculations.

MM. Morin, Piobert, and Didion, from their experiments upon the resistance of water, concluded that in the expression for the resistance of fluids there existed a constant term, and that of a term proportional to the square of the velocity. This conclusion was confirmed by the experiments which they made upon the resistance of air, obtained from uniform motion.

A first series of experiments, made upon a thin plate 1·267 sq. yd., gave for the expression of the resistance of the air,

$$R = 0{\cdot}0663\text{ lb. } A + 0{\cdot}1372\ AV^2; \text{ in units of yards;}$$

but as the fall of 46·9 feet was not sufficient to give at the end of it a strictly uniform motion, and as we shall presently see that the resistance in variable motion must comprise a third term dependent upon the acceleration $\frac{v}{t}$ of motion, it follows that the term $0{\cdot}1372\ AV^2$, which comprises implicitly this third term, is a little too great, and should be diminished.

The existence of a constant term in the expression of the resistance was manifested in the experiments made upon a wheel with wings 1·09 yd. internal diameter, bearing square wings 0·2187 yd. by 0·2187 yd., twenty in number, presenting thus a total surface of 0·9568 sq. yd.

The results of these experiments were very exactly represented in the case of uniform motion by the formula

$$R = 0{\cdot}008892 \text{ lb. } A + 0{\cdot}001907\ AV^2 \cdot \text{ in units of feet,}$$

and

$$R = 0{\cdot}08002 \text{ lb. } A + 0{\cdot}1548\ AV^2; \text{ in units of yards,}$$

as may be seen in the following Table, in which the values found, at different uniform velocities, for the coefficient of the term proportional to the square of the velocity are very nearly constant.

EXPERIMENTS UPON THE RESISTANCE OF AIR TO THE MOTION OF A WHEEL WITH PLANE PLATES.

Uniform velocity of the centre of resistance of wings, in yards per second	yds. 2·89	yds. 4·11	yds. 5·17	yds. 5·89	yds. 6·69	yds. 7·20	yds 7·83	yds. 8·28
Resistance of wings reduced to the mean density of the air	lbs. 1·338	lbs. 2·602	lbs. 3·941	lbs. 5·183	lbs. 6·502	lbs. 7·867	lbs. 9·166	lbs. 0·458
Coefficient K_1 of the square of the velocity	·15818	·15618	·15077	·15355	·14986	·15494	15494	·15818
Mean K_1 .. .. = ·1548								
Velocity answering to the formula	yds. 2·918	yds. 4·129	yds. 5·108	yds. 5·87	yds. 6·58	yds. 7·25	yds. 7·83	yds. 8·37

A review of the coefficient gives slight variations from those recorded by Morin, the mean of which would be $K_1 = 0{\cdot}1004$ instead of $0{\cdot}1002$.

This comparison of the results of experiments with those of the above formula show within what limits of exactness the latter represents the real effects.

Method of Reckoning the Effects of Acceleration.—It has been already shown that in elastic fluids the resistance must depend upon the acceleration of motion; and if these considerations are admitted, it follows that the resistance of the air in variable motion must be represented by a formula of the form of

$$R = K_1'A' + K_1\,AV^2 + K_2\,A\,\frac{v}{t}.$$

The experiments upon uniform motion having already furnished the approximate values of K_1' and K_1, it remains to find that of K_2, or rather the term $K_2\,A\,\frac{v}{t}$.

Without going into the details of the calculations, we limit ourselves to pointing out the method followed, since it shows a remarkable example of the advantages to be derived from a graphic representation of the law of motion.

In the actual case, this law being represented by a continuous curve, whose abscissæ indicate the number of turns, or the spaces described, and whose ordinates express the times, it is clear that for one of these tangents, M P, for example, Fig. 148, the ratio of N P to M N, in the triangle M N P, will be the same as that of e to t, representing by e the infinitely small increase of the abscissa in passing from the point M to the infinitely near point M′, and by t the corresponding increase of time or of the ordinate: this ratio $\frac{e}{t}$ of the elementary path to the element of time in which it was described is precisely what is termed the velocity, which we express by the relation $V = \frac{e}{t}$; and we see that we may by means of the graphic trace of Fig. 148, form a table of the simultaneous values of the times and velocities, and so construct a new curve, whose abscissæ shall be the times T, and whose corresponding ordinates shall be the velocities V.

This new curve, Fig. 149, yields to analogous considerations; the tangents, at the different points, give us the ratio $\frac{AB}{BC}$, which is equal to the acceleration $\frac{v}{t}$, v being the elementary increase of the ordinate or of the velocity V, and t being always the elementary increase of the time.

148.

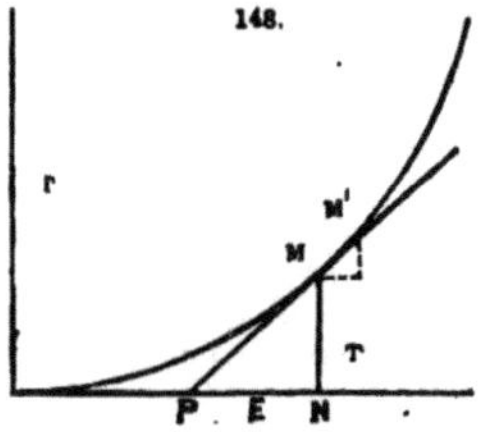

149.

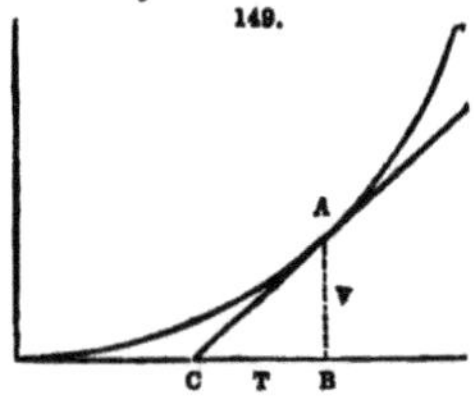

Consequently, knowing at each instant the total resistance R, or the portion of the motive effort employed in overcoming the resistance of the air, as well as the coefficients K_1' and K_1, we may calculate the term $K_2\,A\,\frac{v}{t}$ and so deduce the value K_2.

This process may be abridged by operating upon that part of the curve relating to the end of the fall, since the variations of inclination of the tangents of the first curve are so small, that instead of tracing them, we may determine them by the value of the quotient $\frac{E-E'}{T-T'}$ of the difference of two consecutive spaces divided by that of the corresponding times.

This ingenious mode of discussion led M. Didion to assign to the coefficients of the formula, which represents the law of the resistance of air to the accelerated motion of descent of a plate 1·196 sq. yd. of surface, the following values:

$$R = 0\cdot06633 \text{ lb.} + 0\cdot1295\, V^2 + 0\cdot27652\, \frac{v}{t},$$

which is reduced in case of uniform motion to

$$R = 0\cdot06633 \text{ lb.} + 0\cdot1295\, V^2,$$

for one square yard of surface, V being in yards.

Proof of the Exactness of this Formula.—To show, *a posteriori*, that this formula, composed of three terms, represents the law of the resistance in accelerated motion more exactly than those which only contain a term proportional to the square of the velocity, or two terms, the one constant, and the other proportional to the square of the velocity, M. Didion has first sought for the values of the constant coefficients which it was proper to admit for each of these formulæ, so as to render them as exact as possible, and, after having found them, he calculated, by a very simple analytical method, the values of the times corresponding to the regularly increasing spaces described by the bodies, such as would be furnished by these formulæ, and he has compared them with the real times furnished by the curve of the law of motion. From the results of this comparison, which for one particular case are entered in the following Table, we see that the formula with three terms of resistance, represents, quite truly, the law of accelerated motion of the descent of a body in air, while the suppression of the term depending upon the acceleration $\frac{v}{t}$ does not admit of so exact a representation of this law, even in determining the coefficients so as to reproduce the calculated duration for one of the spaces, and that is also the case when we suppress the constant term.

The only results inserted in the Table are those of an experiment, during which the temperature was at 62°·24′ (Fah.) and the barometric pressure at 2·465 feet of mercury.

Comparison of Times and Velocities of the Fall of a Plate One Mètre Square = 1·196 Sq. Yd., Observed and Calculated.

Spaces described.	Observed Durations.	Observed Velocities.	Durations calculated by the Formulæ (1) $R = 0\cdot66 + 0\cdot129\,V^2 + 0\cdot276\,\frac{v}{t}$	(2) $R = 0\cdot066 + 0\cdot166\,V^2$	(3) lb. $R = 0\cdot2076\,V^2$	Velocities calculated by formula (1).
yds.	seconds.	yds.	seconds.	seconds.	seconds.	yds.
0·0999	0·176	..	0·178	0·160	0·160	..
0·1993	0·254	..	0·253	0·227	0·226	..
0·2998	0·306	..	0·310	0·278	0·277	..
0·3994	0·359	..	0·358	0·322	0·321	..
0·4809	0·400	..	0·400	0·360	0·358	..
0·5997	0·428	..	0·428	0·394	0·393	..
0·6996	0·474	..	0·473	0·419	0·417	..
0·7996	0·508	..	0·506	0·460	0·457	..
0·8995	0·537	..	0·536	0·488	0·487	..
0·9995	0·566	..	0·566	0·518	0·515	..
1·2018	0·619	..	0·622	0·570	0·567	..
1·3998	0·679	..	0·679	0·619	0·617	..
1·5988	0·725	..	0·723	0·665	0·663	..
1·7990	0·771	..	0·771	0·710	0·707	..
1·9991	0·815	..	0·820	0·748	0·746	3·46
2·9987	1·013	..	1·013	0·947	0·943	5·46
3·9919	1·187	6·07	1·186	1·123	1·120	6·05
4·8098	1·346	6·50	1·346	1·289	1·287	6·49
5·9974	1·493	6·91	1·497	1·452	1·451	6·86
6·9970	1·636	7·25	1·639	1·607	1·606	7·14
7·9966	1·771	7·50	1·776	1·760	1·760	7·37
8·9962	1·910	7·60	1·912	1·912	1·912	7·56
9·8133	2·034	7·62	2·042	2·062	2·064	7·69

Influence of the Extent of Surfaces.—To establish this influence, M. Didion used a square plate, each side of which was 0·5468 yd., and so having an area of 0·299 sq. yd., or equal to a quarter of that of the first plate. In calculating the time of the fall by the same method as for the plate of 1·196 sq. yd., and by means of the same formula

$$R = \left\{0\cdot06638 \text{ lb.} + 0\cdot1295\, V^2 + 0\cdot276\, \frac{v}{t}\right\} A \text{ yd.},$$

he found between the results of observation and those of calculation a coincidence quite sufficient to permit him to conclude that, between the extended limits in which he had operated, the resistance of the air is proportional to the extent of the surfaces. The temperature and barometric pressure were sensibly the same as in the experiments above referred to.

COMPARISON OF TIMES AND SPACES DESCRIBED IN THE FALL OF A PLATE OF 0·299 SQ. YD. SURFACE, FROM OBSERVATION AND CALCULATION.

Spaces described.	Duration.		Spaces described.	Duration.		Spaces described.	Duration.	
	Observed.	Calculated.		Observed.	Calculated.		Observed.	Calculated.
yds.	seconds.	seconds.	yds.	seconds.	seconds.	yds.	seconds.	seconds.
0·0995	0·174	0·173	0·8322	0·490	0·485	4·809	1·240	1·215
0·2001	0·246	0·242	0·9000	0·519	0·515	5·997	1·361	1·330
0·2996	0·301	0·297	0·9995	0·547	0·543	6·997	1·476	1·412
0·4002	0·356	0·343	1·993	0·775	0·767	7·996	1·586	1·527
0·4811	0·387	0·384	2·998	0·951	0·939	8·996	1·693	1·646
0·5993	0·425	0·420	3·998	1·102	1·085	9·546	1·799	1·738
0·6999	0·460	0·454						

Consequence of these Results.—We see by this Table that the calculated times of the falls are sensibly the same, though a trifle less than the observed times, which shows that if the coefficient of resistance varies with the extent of surface, it tends to diminish with the diminution of surface, rather than to increase, as some authors have concluded from experiments made by observation of the motion of rotation.

In recapitulating, we may, without fear of notable error, admit in practice that the resistance of the air is proportional to the extent of the surfaces.

Experiments upon Parachutes.—One of the most useful questions among our researches upon the resistance of air which our means of observation enabled us to resolve, was an exact determination of the resistance experienced by parachutes. Their concave form causing, with the same surface, a marked increase of resistance, it was easy, in this case, to obtain a uniform motion of descent, which was indicated by the curve representing the law of motion, which in this case degenerated into a straight line, whose inclination furnished the value of the uniform velocity.

The parachute employed was composed of a frame of whalebones, disposed into four equidistant meridian planes, fastened upon a common rod, and strengthened by stays. This frame was covered with taffeta, strongly stretched, and it was suspended upon a rod, at the lower part of which was attached the additional weights.

The exterior diameter of the parachute was 1·461 yd. measured perpendicularly from the sides of the polygon, and 1·312 yd. measured between the nearest points of the arcs formed by the rim. Its perpendicular projection to the direction of motion varied from 1·433 sq. yd. to 1·444 sq. yd. of surface. The versed sine of curvature of this parachute was 1·41 foot to the plane of the ends of the whalebones.

A discussion of the experiments in which the velocity was uniform has shown that the resistance of the air to the motion of this parachute could also be represented by an expression composed of two terms, and that it was equal to 1·936 times that of a plane of the same surface, that is to say, nearly double.

It follows, from this, that it may be expressed by the formula

$$R = 1{\cdot}936 \text{ A sq. yd.} \left[0{\cdot}06638 \text{ lb.} + 0{\cdot}1295 V^2 \text{ yd.}\right] = \text{A sq. yd.} \left(0{\cdot}1285 \text{ lb.} + 0{\cdot}2507 V^2\right),$$

for units of yards of surface and velocity, at the ordinary density and temperature of the air.

Case where the Parachute presents its Convexity to the Air.—In reversing the parachute, and causing it to descend with its convex surface downwards, a much less resistance was found, and equal 0·768 of that of the plane surface with the same area. So that in this case the resistance is represented by the formula

$$R = 0{\cdot}768 \text{ A sq. yd.} \left(0{\cdot}06638 \text{ lb.} + 0{\cdot}1295 V^2\right) = \text{A} \left(0{\cdot}0509 \text{ lb.} + 0{\cdot}0994 V^2\right).$$

We see by this that the resistance of the same body varies in the ratio of 1·936 to 0·768, or from 2·5 to 1, according as it presents to the air its concavity or convexity.

Case where the Motion of the Parachute was Accelerated.—In this expression of resistance we also admit the necessity of introducing a term dependent upon the acceleration of motion $\frac{v}{t}$, and this expression for the parachute employed is

$$R = A \left(0{\cdot}1290 \text{ lb.} + 0{\cdot}2513 V^2 + 0{\cdot}2394 \frac{v}{t}\right),$$

in units of yards for area and velocity.

A comparison of the observed times of the fall with those deduced from this formula has shown that it represents the circumstances of motion with all desirable accuracy.

Resistance to the Motion of Inclined Planes in Air.—These experiments were made by means analogous to those above described, by causing to descend two jointed planes, 1·0963 yd. long by 0·5486 yd. wide, whose angles were varied, at intervals of 5°, from 5° up to 180°, where they form a single plane. The results regularly observed from 180° to 130° have shown that the resistance decreases proportionally with the angles, so that, calling a the angle of one of the planes with the direction of motion, the resistance was expressed for uniform motion by the formula

$$R = \frac{a}{90} A\ (0 \cdot 06638 \text{ lb.} + 0 \cdot 1295\ V^2), \text{ in units of yards.}$$

A comparison of the observed resistance with those calculated by this formula show a satisfactory agreement.

COMPARISON BETWEEN THE OBSERVED AND CALCULATED RESISTANCES, FOR DIFFERENTLY INCLINED PLANES.

Angles formed by each of the Planes with the Direction of Motion.	Resistances in the ratio to those of a Plane perpendicular to the Direction of Motion.		Angles formed by each of the Planes with the Direction of Motion.	Resistances in the ratio to those of a Plane perpendicular to the Direction of Motion.	
	Observed.	Calculated.		Observed.	Calculated.
90°·	1·0000	1·000	77°·5	0·846	0·861
87·5	0·996	0·972	70·	0·773	0 778
82·5	0·865	0·917	67·5	0·737	0·750
80·	0·856	0·889	65·	0·728	0·722

It should be remarked that these results relate to the case of two equal and jointed planes, moved in the air, with the edge of intersection in front, and are by no means applicable to the case of isolated planes.

The law of the variation of the resistance proportionally to the angles is also that which these experimental philosophers found for water, in operating upon cones of different acuteness.

General Conclusions from the Experiments at Metz.—In conclusion, the reported experiments which have been made with chronometric mechanism, giving the times, to nearly some thousandths of seconds, and the velocities acquired at any instant nearly to a hundredth, in observing the law of descent in air of different sized plates, of two plates inclined towards each other, and that of a wheel with wings, for which the velocities have not exceeded from 29 to 33 feet a-second, have conducted us to the following conclusions:—

1st. In the uniform motion of a body in air, the resistance experienced is proportional to the extent of its surface, and to another factor composed of two terms, the one constant and the other proportional to the square of the velocity.

As it was easily foreseen that the number of molecules of the air shocked by the displacement of the body must increase in the same ratio with its density, the general expression of the resistance should contain a factor relative to this density; so that calling d the density of the air at the temperature and pressure observed, and d_1 its density at 50° (Fah.) and at 76 centigrades (or 29·92 inches) of barometric pressure, and preserving the preceding notations, this resistance is represented by the following formulæ:

Thin plates perpendicular to the direction of motion $R = A \frac{d}{d_1} \left\{ 0 \cdot 066 \text{ lb.} + 0 \cdot 199\ V^2 \right\}$

Parachutes .. .. $R = A \frac{d}{d_1} \left\{ 0 \cdot 129 \text{ lb.} + 0 \cdot 251\ V^2 \right\}$

Parachutes reversed .. .. $R = A \frac{d}{d_1} \left\{ 0 \cdot 051 \text{ lb.} + 0 \cdot 994\ V^2 \right\}$

Two jointed plates, inclined towards each other .. $R = A \frac{d}{d_1} \frac{a}{90} \left\{ 0 \cdot 066 \text{ lb.} + 0 \cdot 129\ V^2 \right\}$

The wings of a fan-wheel .. .. $R = A \frac{d}{d_1} \left\{ 0 \cdot 08002 \text{ lb.} + 0 \cdot 1545\ V^2 \right\}$

It may be observed that this last formula accords in a satisfactory manner with the results of M. Thibault's experiments.

2nd. In accelerated motion we must add to the preceding expression a term proportional to the acceleration of motion, and the resistance is then represented by the following formulæ:

Thin plates perpendicular to the direction of motion $R = A \frac{d}{d_1} \left\{ 0 \cdot 066 \text{ lb.} + 0 \cdot 129\ V^2 + 0 \cdot 276 \frac{v}{t} \right\}$

Parachutes .. .. $R = A \frac{d}{d_1} \left\{ 0 \cdot 128 \text{ lb.} + 0 \cdot 251\ V^2 + 0 \cdot 2394 \frac{v}{t} \right\}$

With respect to the various modes of ascertaining the velocities of currents of air in mines, in order to determine the quantities circulating in a given time, we extract the following, on the nature and use of anemometers, from a paper by John J. Atkinson and John Daglish, published in the 'Transactions of the North of England Institute of Mining Engineers:'—

In the ventilation of mines, great advantages are well known to arise from dividing the air into a series of currents or splits; and its proper distribution amongst these is an object of very great importance, because upon it depends, to a certain extent, not only the actual amount of the gross quantity that will be put into circulation, in a given time, by any ventilating power that may be employed, but also the relative degrees of safety and salubrity that will prevail in the different districts into which the mine is divided.

In order to effect this distribution of the air in such a manner as to obtain the most efficient general ventilation, and, at the same time, to allot to each district or split its proper share or proportion of the whole, it is essential to have some satisfactory mode of ascertaining the velocities of the currents, and the quantities of air circulating in each of the splits in the unit of time.

The various methods that have been employed for this purpose may be divided into three groups.

First.—By travelling at the same velocity as the current, and noting the distance passed over in a unit of time.

Second.—Determining from observation the rate at which small floating particles are carried along by the current, and assuming their velocities to be identical with that of the air-current itself. Smoke from exploded gunpowder, burning turpentine or amadou, small pieces of down, and small balloons filled with hydrogen, have been all more or less employed for this purpose.

Third.—By using anemometers, or apparatus of various forms; and these may be divided into three classes:—(*a*) Anemometers having vanes or wands, made to *revolve* by the current of air impinging upon them, the rate at which they revolve being indicated by pointers on dials forming a part of the instrument—the pointers being made to revolve by means of wheels connecting them with the axis of the vanes or wands. The anemometers of Combes, Biram, Whewel, Osler, and Robinson, are instances of this class of instruments now in use in this country, all of which require a correction for friction. (*b*) Instruments which are affected by the *force* or impulse of the wind, without being subjected to any continuous revolving motion, such as Dr. Lind's, Henaut's, Bougier's, and Dickinson's anemometers. (*c*) Anemometers of a more complex character, such as Leslie's.

First Group.—Perhaps the primitive mode of ascertaining the velocity of currents of air in mines was that of choosing a part of the gallery forming the air-way having as uniform sectional dimensions as could be found, and after measuring off a distance of 100 to 150 yds. in length, taking a lighted candle and walking in the direction of the current, holding the flame in such a position as to be fully exposed to the influence of the current, but taking care to walk at the particular rate required, to cause the flame to burn in an upright position, without being deflected from the vertical, either by the current or by the progress of the person carrying it. The time required to traverse the distance measured off, being noted by a seconds' watch, enabled the average rate of walking to be determined; and the average rate so found, from three or four trials, was assumed to be the velocity of the air-current; and this, multiplied by the average sectional area of the part of the air-way selected for the experiment, was taken to represent the quantity of air passing in the unit of time. Formerly, when this mode of measuring the air in mines was in use, it would afford a close approximation to the truth; but, with the ventilation now existing in many of our large mines, it would not be practicable to walk as quickly as the currents travel in the principal splits; and running is not a sufficiently steady pace. One of the objections to this, as well as to all other methods that require a considerable distance to be traversed, over which to observe the velocity, is the difficulty of obtaining a gallery of equal area throughout, over a sufficient distance; but in cases where this is attainable, this method admits of great accuracy for velocities up to 400 feet per minute; Atkinson and Daglish state that they have been able to obtain as accurate results by this method as by any other, as can be seen by referring to Table I. and Fig. 149. In Fig. 149, and in the other figures employed for the like purpose, the bent or crooked lines are obtained by taking the *actual* velocities of the air-currents (ascertained as hereafter described), and the revolutions of the anemometer, or other *indicated* velocities, as co-ordinates. If the indicated velocity were the same as the actual velocity, a line drawn through the points where they would intersect each other in the diagrams, would exactly coincide with the simple straight diagonal; but as the one exceeds the other, so the crooked lines drawn through their points of intersection depart more or less from the diagonal.

The close approximation of this line to the diagonal shows that great accuracy can be attained by walking with a lighted candle. It ought, however, to be mentioned that the place where the experiments were made was in all respects suitable, and specially adapted for the purpose, being perfectly level, and of an accurately uniform sectional area throughout the whole distance of 200 feet.

Second Group.—One of the principal of the second group of modes employed for the measurement of air consists in observing the velocity of the smoke from an exploded charge of gunpowder in a part of the gallery of nearly uniform sectional area; and this, until recently, was the means most generally adopted in the coal mines of this country for ascertaining the velocity of air-currents; and although it has of late been largely superseded by the use of Benjamin Biram's anemometer, the practice is still in considerable use, and, so far as regards shaft velocities, remains the only method. It is, therefore, desirable to ascertain how far the results obtained by this and similar methods of measuring air-currents can be relied upon for accuracy, and to investigate the various sources of error connected with them, with a view of either avoiding or making proper allowances for their effects, so far as may be practicable.

The sudden explosion of gunpowder in the confined passages of mines produces several effects, which tend to cause inaccuracies in the results obtained by noting the passage of the smoke as an index of the velocity of the current.

Experiments prove (as indeed might have been anticipated, considering the small quantities of gunpowder used) that in general neither the increase of bulk due to the introduction into the current of the *products of combustion*, nor that due to the *elevation of temperature*, have any appreciable effect on its velocity. But other experiments show that the *force of the explosion*, when a considerable quantity of gunpowder is used in a feeble current, gives an impulse to the current, and creates a velocity in excess of the normal one. A revolving anemometer was placed in an air-passage traversed by a feeble current, so regulated as to be just strong enough to produce thirty revolutions of the instrument a-minute. The explosion of a cubic inch of gunpowder at a distance of 70 feet did not in any way affect the instrument; but when the charge of gunpowder was increased to 20 cubic inches, the explosion caused a sudden and violent increase of its rate of revolving, acting as a temporary impulse, the revolutions very quickly decreasing to the original number again. The same effect is also clearly shown in the second series of experiments, page 77. The amount of error arising from this source, and which tends to *increase* the apparent velocity, depends on the quantity of gunpowder used, the sectional area of the air-way, and the velocity of the current,—increasing with the quantity of gunpowder employed, but decreasing as the sectional area of the air-way and the velocity of the current are increased, so that the explosion of a large quantity of gunpowder in a feeble current of air passing over a short distance in a gallery of small sectional area will be attended with the greatest errors; but as, under the ordinary conditions of the currents and air-ways of mines, 1 cubic inch of gunpowder does not give rise to any sensible error from the cause alluded to, and as it affords sufficient smoke to be readily observable at a distance of 200 feet, that quantity has been adopted as a standard, and used in the experiments made by Atkinson and Daglish.

It appears to be very desirable that a standard quantity of gunpowder should be employed in all cases, whether in the ordinary measurement of air or in conducting experiments, to enable comparisons to be made, as any variation in this respect will give rise to discrepant results.

If a charge of gunpowder be exploded in an air-current, and the velocity of its smoke be timed over a series of consecutive and equal distances in an uniform air-way, it will be found to be apparently most rapid near the point of ignition, and to decrease gradually as it flies to a greater distance from that point. This is a most serious source of error, and may be regarded as fatal to the accuracy of this method of determining the velocity of a current of air.

The following experiments, selected from many others giving similar results, establish what has been just stated.

The charge in each of these experiments was 1 cubic inch of gunpowder, which was exploded 10 feet to the windward of the commencement of the first space, or interval of 25 feet, and the time was noted when the smoke reached the commencement and also the end of each of the two intervals of 25 feet, into which the total distance of 50 feet was divided.

Equal Quantities of Gunpowder at Different Velocities.

Time in passing over First Interval of 25 feet.	Time in passing over Second Interval of 25 feet.	Total Time.	Average Velocity of the Air-current, as indicated by the Smoke.
15″	21″	36″	83 feet a-minute.
12½″	15½″	28″	107 „
6½″	9″	15½″	193 „
5″	5½″	10½″	285 „

In the above experiment it will be noticed that in all cases the time occupied in passing over the second interval is greater than that occupied in passing over the first; and it is further observable that this difference decreases as the velocity of the air increases. At the low velocity of 83 feet per minute the times are 15″ and 21″, being a difference of 40 per cent. of the lesser time; whilst at the higher velocity of 285 feet a-minute, the difference between 5″ and 5½″ only amounts to 10 per cent.

The charge of gunpowder, in the two following series of experiments, was varied in quantity, and exploded 20 feet to the windward of the first interval; the time being noted, as before, when the smoke reached the commencement and also the end of each of the two intervals of 50 feet.

Different Quantities of Gunpowder and Equal Velocities

First Series.

Quantity of Gunpowder. Cubic Inches.	Time in passing over First Interval of 50 feet.	Time in passing over Second Interval of 50 feet.	Total Time.	Average Velocity of Air-current, as indicated by the Smoke, in feet a-minute (presumed average).
1	37″	45″	82″	75
4	27″	35″	62″	„
20	15″	24″	39″	„

Second Series.

1	5″	5″	10″	600
4	5″	5″	10″	„
20	4″	5″	9½″	„

It will be observed in these experiments:—1st. That in a slow current (75 feet a-minute), with 20 cubic inches of gunpowder, the time occupied by the smoke in passing over the second interval of 50 feet was 60 per cent. more than it occupied in passing over the first interval of 50 feet, showing that the *apparent* velocity gradually decreased. These experiments further show, that this gradual loss of velocity is greatest where the charge of powder employed is greatest. 2nd. That the apparent velocity, and, therefore, the apparent quantity of air, is more than doubled (being in the proportion of eighty-two to thirty-nine) with the low velocity, by using 20 cubic inches of gunpowder instead of 1 inch. 3rd. That these discrepancies are by far the greatest at low velocities, and are hardly apparent at high velocities, as will be seen by the *second series*.

The smoke resulting from the explosion of gunpowder is not of the same density as the air-current. This has been observed by previous experimentists, and has been confirmed by Atkinson and Daglish, by substituting turpentine smoke, which can be observed at a distance of 50 feet, but which dissipates, and cannot be accurately observed at 100 feet.

One Cubic Inch Gunpowder.							Turpentine.				
First Interval of 25 feet.		Second Interval of 25 feet.		Total.		First Interval of 25 feet.		Second Interval of 25 feet.		Total.
17	+	21	=	38	.. ..	11	+	14	=	25
13½	+	16½	=	30	.. ..	10	+	12	=	22
7	+	9	=	16	.. ..	8	+	8	=	16

It will be noticed that, whilst gunpowder smoke required 38″, 30″, and 16″ respectively to travel a distance of 50 feet, in a current having the same velocity, turpentine smoke required only 25″, 22″, and 16″. But it may be observed that experiments made with turpentine smoke are very unsatisfactory. The turpentine cannot, like gunpowder, be ignited in a large quantity simultaneously, but resembles more the ignition of a train of gunpowder; added to this, the resulting smoke is very difficult to discern, and is soon dissipated.

Experimentists who have written on this subject have also noticed another source of error in all currents, especially in the more feeble, in the eddies and streams of varying velocity which almost always exist; and when any small particles or light bodies are introduced into the currents, a part of them get into the axis of greatest velocity, and give a result higher than the average; other portions fly too slow; and even on the average of the first and last particles traversing the distance, the results are too low.

Similar remarks are applicable to the use of smoke; at least Mons. Jochams ('Annals des Travaux Publics de Belgique,' vol. ix.) came to these conclusions on comparing the results of his experiments by these modes with the corresponding and simultaneous observations made with Combes' anemometer. The results of his experiments, indicating the distances traversed per second by the different agents, are given in the following tabulated form:

	Velocities deduced from Mons. Combes' Anemometer.	Velocities of the Current of Air observed with				
		Powder Smoke.		Amadou Smoke.		Down.
		First of Smoke.	Average of Smoke.	First of Smoke.	Average of Smoke.	
Mètres	4·84	..	..	..	..	4·31
"	1·53	1·67	1·25	1·56	1·19	
"	2·34	2·78	2·00	2·50	2·00	

In reference to the down especially, if it got out of the axis of the air-way where the most rapid current prevailed, it adhered to the damp walls of the gallery, and was, consequently, greatly retarded.

The various sources of error connected with the use of gunpowder smoke are given in the following tabular form:—

Causes of Error in Experimenting on the Velocity of Air-Currents in Mines, by means of Gunpowder Smoke.

Cause.	*Effect.*
1.—The expansion of the whole column of air, by the addition to it of the results of the combustion of gunpowder, and by the heat developed (*of slight magnitude*).	Tending to *increase* apparent velocity owing to two causes, *viz.*:— 1.—The conversion of a small portion of solid gunpowder into gas. 2.—The further expansion of this, owing to the high temperature of ignition.
2.—The explosive force of gunpowder (*of considerable magnitude*).	Tending to *increase* the apparent velocity, and can be avoided with care.
3.—Diffusion or deposition of the smoke.	Tending to *decrease* very considerably the apparent velocity.
4.—Eddies and currents. 5.—The density of the smoke.	Giving rise to serious irregularities, materially affecting the accuracy of the results.

Precautions to be used in Experimenting with Gunpowder Smoke.—By the use of fixed quantities and distances, and the avoidance of extreme velocities, an *approximation* to accuracy in the measurement of air-currents by gunpowder smoke may be attained; and the numerous experiments made by Atkinson and Daglish suggest the following precautions as being necessary:—

1st. Always to use 1 cubic inch of gunpowder as a standard.

2nd. The *velocity* of the current never to be less than 100 feet a-minute, nor to exceed 500 feet a-minute. In order to attain this, a gallery of such area must be selected as will afford this velocity of current.

3rd. The *time* not to be less than twelve seconds, nor to exceed thirty seconds.

4th. To explode the gunpowder 10 feet to the windward of the first mark.

Therefore, in slow currents of from 100 to 250 feet per minute velocity, the distance to be taken over which the smoke passes will be 50 feet; and for the higher velocities of from 250 to 500 feet, the distance will be increased to 100 feet.

The following Table of Experiments, made by timing gunpowder smoke, by walking so as to keep the flame of an exposed light in a vertical position, and by the use of a Biram's anemometer respectively, is given with a view of showing the comparative degrees of accuracy of these different modes of measuring currents of air; and the results are graphically exhibited in Fig. 149.

TABLE I.—EXPERIMENTS MADE WITH GUNPOWDER SMOKE, WALKING, AND AN ANEMOMETER (BIRAM'S 6-INCH) IN THE SAME CURRENT OF AIR.

	Gunpowder Smoke.			Walking.	Anemometer.	
	Equal Distances.		Equal Times.	200 feet	Recorded Revolutions. = R.	True Velocities calculated. $v = \cdot 97\ R + 40$, nearly.
	50 feet.	200 feet.	20 seconds.			
	A B.	C D.	E F.	G H.		
	Feet a-minute.	Feet a-minute.	Feet a-minute.	Feet a-minute.	Revolutions.	Feet a-minute.
1	125	100	..	88	44	83
2	158	162	..	..	90	127
3	166	..	171	162	103	140
4	250	214	228	..	188	222
5	285	..	..	..	220	253
6	300	255	263	255	233	266
7	353	307	307	307	293	324
8	461	413	428	413	407	435
9	545	500	502	500	482	507

Explanation of Table I. and Fig. 149.—A B column. In those experiments the time during which the powder smoke travelled a distance of 50 feet was observed, and from it the velocity per minute calculated; thus, in the first experiment the time so occupied was 24", indicating a velocity of 125 feet per minute; this is the ordinary mode of finding the velocity of air by powder smoke.

149.

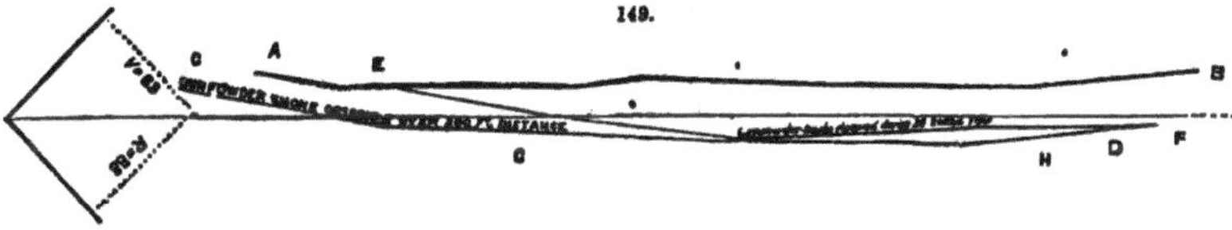

It will be observed that, owing to the distance travelled over by the smoke being short, the velocities of the smoke are in excess of the true velocities of the air.

In column C D, the distance over which the time of the powder smoke is noted is 200 feet instead of 50 feet; and it will be observed that the apparent velocities here are less than the true ones, excepting in the two first experiments, where the contrary result is due probably to the explosive force of the gunpowder in a feeble current.

Column E F.—Experiments made by using "equal times" instead of "equal distances," and varying the distance, so that in each experiment the *time of observation* was as nearly as possible the same.

Column G H.—Experiments made by walking 200 feet with a lighted candle, and noting the time, and calculating velocity per minute therefrom.

Anemometer columns give the velocity as recorded on the dial of a Biram's anemometer. These readings require correction by the formula, for this instrument nearly, $V = \cdot 97\ R + 40$,

where V = the velocity of the air-current in feet a-minute;

and R = the revolutions of the anemometer, as shown by the index on its dial, in the same time.

These experiments were conducted with the greatest care, in a gallery of a mine specially adapted for the purpose, by being made perfectly level, and of uniform sectional area; during the experiments the velocity was kept as constant as practicable throughout, by keeping the water-gauge of that part of the mine at a uniform height. The observations were taken with a large seconds' watch, specially adapted to this purpose, and all experiments were repeated until a correct average could be obtained; but even under these circumstances, which in general will not prevail for ordinary measurements in mines, great discrepancies are observable between the results obtained by timing smoke over equal distances, and those obtained either by timing smoke during equal times, or over equal distances of different lengths. Doubtless such an empirical rule could be found for so regulating the distance to be traversed, the amount of powder to be exploded, and the duration of the experiments, as that with great care nearly accurate results could be obtained; still the difficulty of obtaining galleries fully adapted for the purpose, by their uniformity of sectional area, &c., and the numerous chances of error in observation are so great, that it is most desirable that there should be some more ready and accurate mode of ascertaining the velocities of currents of air; anemometers of various constructions have of late been more or less employed for the purpose, and it is therefore important to ascertain how far the indications of such instruments can be relied upon for accuracy.

Revolving Anemometers.—The anemometer most generally used in the coal mines of England is that invented by Benjamin Biram, shown in Figs. 150, 151, 152. It consists of a series of vanes,

150.

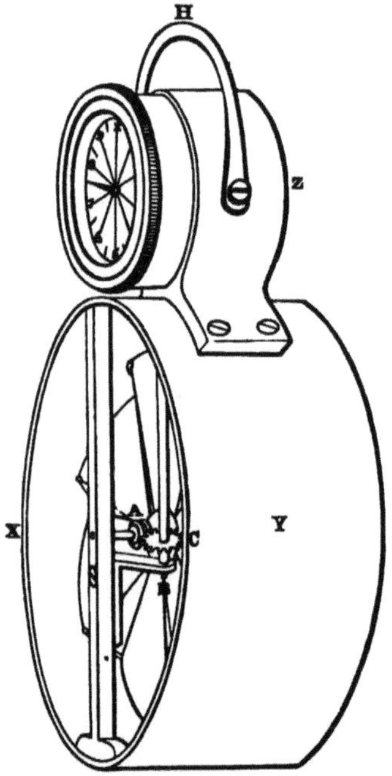

151.

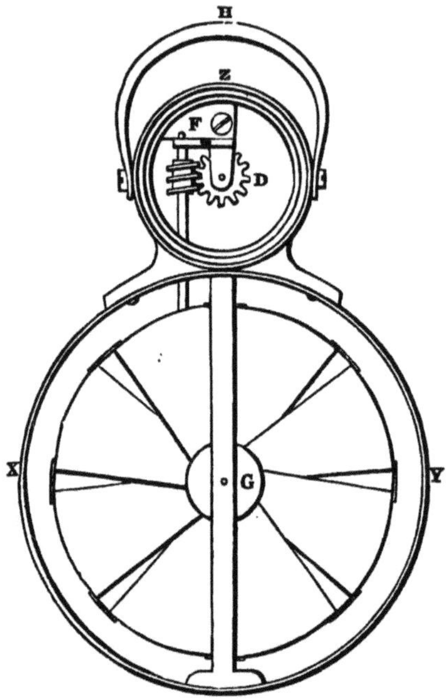

D, E, Fig. 152, which resolve with the action of the air-current—the number of revolutions, or rather numbers proportional to the revolutions, being registered by pointers, P, on the face of a dial forming a part of the instrument itself. It is made of three sizes, 4, 6, and 12 inches; is very portable; and is not, with proper care, liable to get out of order, especially the smaller size. A certain force of current is required to overcome the friction, and put the instrument into motion. The plate spur-wheel C, Fig. 150, as it moves in a horizontal plane, relieves the step B, Fig. 150, from undue pressure, and thus tends to lessen the amount of friction. Some of these instruments will continue to revolve in a current as low as 30 feet a-minute; but with most of them a velocity of about 50 feet is required.

152

Every one who has occasion to use this anemometer should be aware that it does not register the actual velocity of the air, especially in feeble air-currents, nor yet the number of revolutions of the wands, but only a number proportional to the latter; and although it is of great value, as indicating an increase or decrease in the velocity from time to time, such as the periodical variations in any particular current, it is of comparatively little value, *as generally used*, for ascertaining real velocities, such, for instance, as occur in changing or splitting air-currents, when it is of great importance to know the actual quantities. To obtain with this instrument accurate results, available for all purposes, it is necessary, as with Combes' anemometer, to apply a formula to its recorded revolutions, or rather to the number indicated by the index, in order to ascertain the actual velocity of any current; each particular instrument requiring special experiments to be made with it, in order to determine the value of the constants required to be employed in the formula. These constants remain the same for the same instrument, so long as it remains in the same condition, and are independent of the velocities of the currents of air in which it is employed. However, it is necessary to state that these adjustments are carefully made by the principal manufacturer, John Davis, optician, Derby, who is a man of considerable mechanical skill.

In anemometers made like that shown in Fig. 152, the mechanism, dial, and pointers are placed in the centre. The arrangement does not essentially differ from that shown in Figs. 150, 151. X Y is the cylindrical case in which the fan-wheel revolves, supported by the upright bars G, S; H, handle by which the instrument is held.

In Figs. 150, 151, part of the mechanism is contained in a small box, Z, over the fan-wheel, which is thus allowed more play. A, F, are small, delicate axles, upon which are cut endless screws, A, F, which drive small wheels, C, D.

The registering apparatus is in front of the wheel, and consists of six small circles, marked respectively X, C, M, X M, C M, and M, the divisions on which denote units of the denominations of the respective circles; in other words, the X index in one revolution passes over its ten divisions, and registers (10 × 10), or 100 feet; the C index, in he same way, 1000 feet · and so on up to 10 million feet; so that an observer has only to record the position of the several indices, at the first observation (by writing the lowest of the two figures on the respective circles between which the index points, in their proper order), and deduct the amount from their position at the second observation, to ascertain the velocity of the air which has passed during the interval. This, multiplied by the area in feet of the passage where the instrument is placed, will show the number of cubic feet which has passed during the same period.

To ascertain the rate at which air is moving, proceed thus:—suppose 100 revolutions = 200 feet per minute.

88] 200 [2·27 miles an hour.

To obtain the constants of this formula, as applicable to any particular instrument, it is absolutely necessary, in making the experiments, to know correctly the true velocity, as a standard of comparison. As before explained, none of the ordinary modes employed for ascertaining the real velocities are reliable; Atkinson and Daglish, therefore, had a Whirling Machine constructed, the wand of which, in revolving, described a circle of 25 feet in circumference; the number of its revolutions being indicated by a pointer on a dial.

In the first instance, this Whirling Machine was turned by the hand, but as this did not give a sufficiently uniform velocity, a small drum, and a rope with a descending weight attached to it, was employed, to give motion to the machine; and worked thus it gave extremely accurate results; so far, at least, as regards the uniformity of its own velocity. By fixing the anemometer on the end of the wand, the velocity with which it passes through the air can be ascertained and compared with the revolutions of the anemometer, as indicated on its dial. Fig. 153 represents this machine.

153.

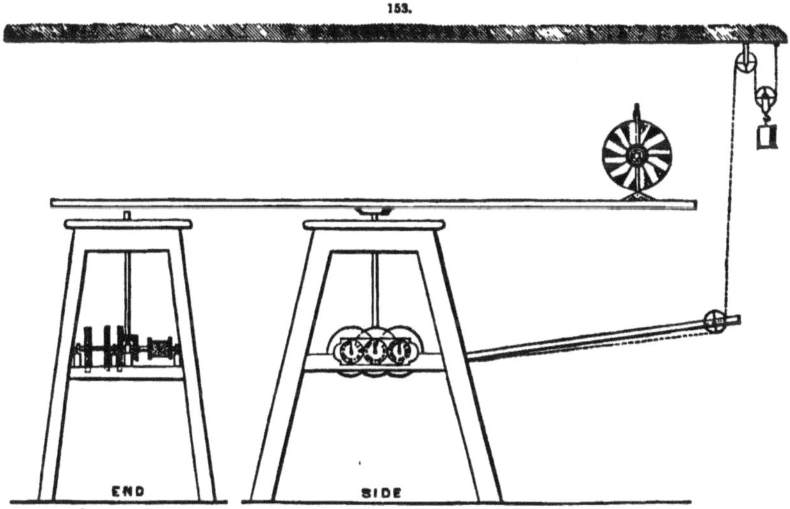

It has been stated by some writers that there is a difference between the force or impulse of air moving upon a body at rest, and the resistance which a body moving through a still atmosphere meets with in its passage, supposing the velocity to be the same in each case; and besides this, the effect of a body moving in a circle, in a still atmosphere, may not be the same as when moving in a straight line. The experiments of Hutton and others appear, however, to indicate that the force of impact of a wind against a stationary body is always proportional to the resistance which a solid, moved through a still atmosphere, meets with at the same velocity.

Comparison of Governors.—The following experiments were made with the Whirling Machine, and the results are given both in a diagram, Fig. 154, and in a tabulated form, Table II.

154.

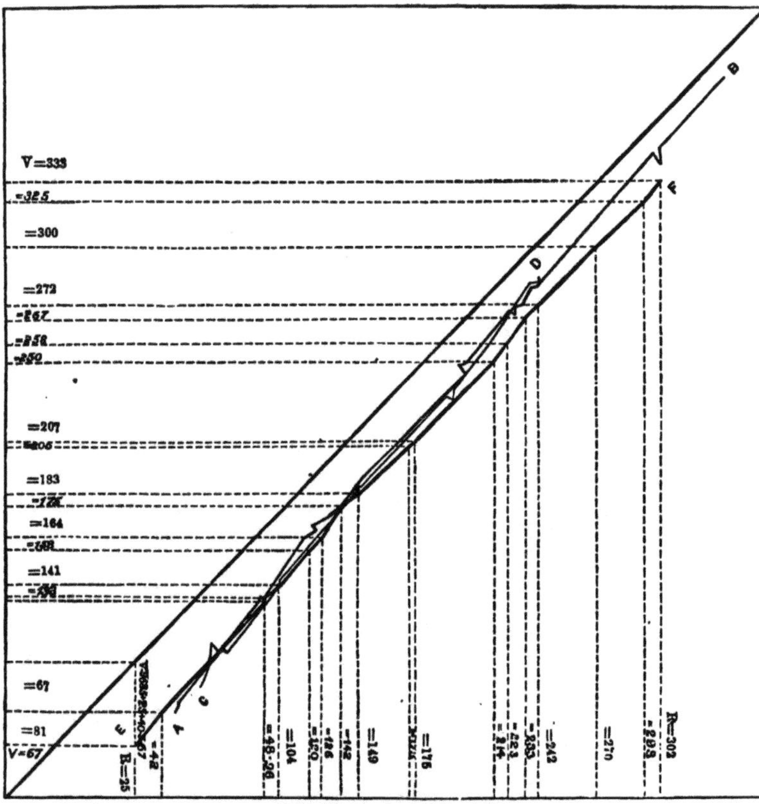

Scale 40 feet a-minute, half an inch.

In this diagram the different bent or crooked lines are drawn through the points found by taking the actual velocities of the anemometer through a still atmosphere, and the numbers indicated by the pointers of the anemometer as being passed over in the unit of time, as the co-ordinates of a line or curve in each series of experiments made at different velocities under the same conditions, and with the same anemometer. A sufficient number of the co-ordinates are transferred from Table II. to Fig. 154, to connect the Table with the Figure.

The anemometers at the commencement of the experiments were made to revolve by the Whirling Machine in a circle only 10 feet in circumference; but as the rate of revolution appeared to be somewhat irregular, a flat board, intended to regulate the motion and render it uniform, was fixed at the end of the revolving wand opposite to that at which the anemometer was fixed; and the experiments shown by the lines A B, C D, were made, the former with the regulating-board projecting downwards, and the latter with it projecting upwards, from the revolving wand.

The discrepancies between these lines are so slight, that the mere position of the regulating-board does not appear to have any sensible effect; the changes in the state of lubrication of the anemometer, and errors of observation, being sufficient to account for the slight differences that exist.

The line E F, in the same diagram, exhibits the results of certain experiments made with the same anemometer in the same sized circle, but without any flat board or governor on the wand; and as these experiments give results differing considerably from the two lines obtained with the governor on the wand, we infer that the governor really produces some effect.

As, however, all the three lines are sensibly straight, the inference is that the effect does not arise from any greater uniformity in the rate of motion with, than exists without, the board; because so long as the lines are straight, it follows that an average speed, however irregular, would, on the whole, give the same general results as a uniform speed equal to such average; and hence we infer that the governor only produces an effect that is disadvantageous, arising probably from disturbing the stillness of the atmosphere in which the experiments are conducted.

On these considerations we are led to reject the experiments made with the governors, and, so far, adopt those shown by the line E F, in which no governor was used.

This view was partly confirmed on using a longer wand, and causing the anemometers to revolve in a circle 25 feet in circumference, in lieu of one of only 10 feet in circumference; as, under these circumstances, the irregularity of the rate of revolution is much less apparent.

The results of the experiments give a straight line in the diagram, and lead to a formula of the form of $V = m R + a$, the same as arrived at by M. Combes, where V = velocity of air R = revolutions of anemometer, or rather the numbers indicated by its index, in the unit of time; and m and a, constants, suited to the friction of the anemometer, the form of the vanes, and the density of the air.

Table II.—Experiments made with a 6-inch Biram's Anemometer to ascertain the effect of placing a Board on the end of the Wand of the Whirling Machine, to act as a Governor, the Circle described by the Wand being 25 feet in circumference.

No of Experiment.	Without the Governor or Board. Line E F.					Board Upwards. Line A B.					Board Downwards. Line C D.				
	Velocity.			Error.		Velocity.			Error.		Velocity.			Error.	
	Number indicated by Index of Anemometer = R.	Actual Velocity = V.	Velocity calculated by formula V = ·9685 R + 40·5.	More.	Less.	Number indicated by Index of Anemometer = R.	Actual Velocity = V.	Velocity calculated by Index of Anemometer V = ·9197 R + 40·5.	More.	Less.	Number indicated by Index of Anemometer = R.	Actual Velocity = V.	Velocity calculated by formula V = ·9 R + 45·3.	More.	Less.
1	25	67	..	..	..	43	85	..	..	..	56	99	..	..	..
2	42	81	81	..	..	49	89	..	..	..	65	104	..	..	..
3	96	133	..	..	..	50	89	..	..	..	71	108	109	1	..
4	97	133	..	..	..	53	92	..	..	..	72	113	..	..	..
5	98	133	..	..	..	53	92	..	..	..	72	112	..	..	..
6	104	141	..	..	..	95	129	..	..	..	75	106	..	..	..
7	120	158	157	..	1	97	131	..	..	..	75	106	..	..	..
8	126	164	..	..	..	97	131	130	..	1	107	142	..	..	..
9	142	175	..	..	..	98	131	..	..	..	107	142	..	..	..
10	149	183	..	..	..	98	133	..	..	..	107	142	..	..	..
11	173	205	208	3	..	127	156	..	..	..	108	142	142	..	..
12	175	207	..	..	..	128	156	..	..	..	150	180	..	..	..
13	214	250	..	..	..	131	158	160	2	..	152	181	..	..	..
14	223	258	..	..	..	132	158	..	..	..	153	183	..	..	..
15	233	267	..	..	..	137	168	..	..	..	153	183	..	..	..
16	242	272	..	..	..	137	168	167	..	1	195	225	..	..	..
17	270	300	..	..	..	137	168	..	..	..	196	228	..	..	..
18	293	325	..	..	..	140	171	..	..	..	198	225	..	..	..
19	302	333	333	..	..	140	171	..	..	..	198	225	225	..	..
20	..	..	..	..	..	141	172	..	..	..	199	225	..	..	..
21	..	..	..	..	..	158	183	..	..	..	223	248	..	..	..
22	..	..	..	..	..	208	233	..	..	..	223	248	..	..	..
23	..	..	..	..	..	210	233	..	..	..	252	267	..	..	..
24	..	..	..	..	..	211	235	..	..	..	253	273	..	..	..
25	..	..	..	..	..	211	235	..	..	..	256	273	275	2	..
26	..	..	..	..	..	236	260	..	..	..	..	..	..	..	..
27	..	..	..	..	..	237	258	..	..	..	..	..	..	..	..
28	..	..	..	..	..	238	260	..	..	..	..	..	..	..	..
29	..	..	..	..	..	238	260	..	..	..	..	..	..	..	..
30	..	..	..	..	..	239	260	..	..	..	..	..	..	..	..
31	..	..	..	..	..	240	260	..	..	..	..	..	..	..	..
32	..	..	..	..	..	242	264	..	..	..	..	..	..	..	..
33	..	..	..	..	..	249	267	..	..	..	..	..	..	..	..
34	..	..	..	..	..	250	269	..	..	..	..	..	..	..	..
35	..	..	..	..	..	252	271	..	..	..	..	..	..	..	..
36	..	..	..	..	..	256	275	..	..	..	..	..	..	..	..
37	..	..	..	..	..	270	287	..	..	..	..	..	..	..	..
38	..	..	..	..	..	313	333	..	..	..	..	..	..	..	..
39	..	..	..	..	..	317	333	..	..	..	..	..	..	..	..
40	..	..	..	..	..	317	329	..	..	..	..	..	..	..	..
41	..	..	..	..	..	319	333	..	..	..	..	..	..	..	..
42	..	..	..	..	..	320	333	..	..	..	..	..	..	..	..
43	..	..	..	..	..	345	358	..	..	..	..	..	..	..	..
44	..	..	..	..	..	355	367	..	..	..	..	..	..	..	..
45	..	..	..	..	..	355	367	367	..	..	..	..	..	..	..

The experiments represented by the line E F in the diagram are fairly represented by the formula V = ·9685 R + 40·5.

The rejected line A B, in this diagram represents experiments made with an anemometer revolving in a horizontal circle 25 feet in circumference, through a still atmosphere, with a governor-board 2 feet in area fixed upwards, at the contrary end of the wand to that on which the anemometer was fixed, and give rise to a similar formula, namely, V = ·9197 R + 40·5.

While the rejected line C D, represents experiments made under similar conditions with the same anemometer, excepting that the governor in this case was turned downward, and give V = ·9 R + 45·3; so that, on the whole, the use of the governor appears to increase the numbers indicated by the index of the anemometer, excepting, perhaps, at very low velocities.

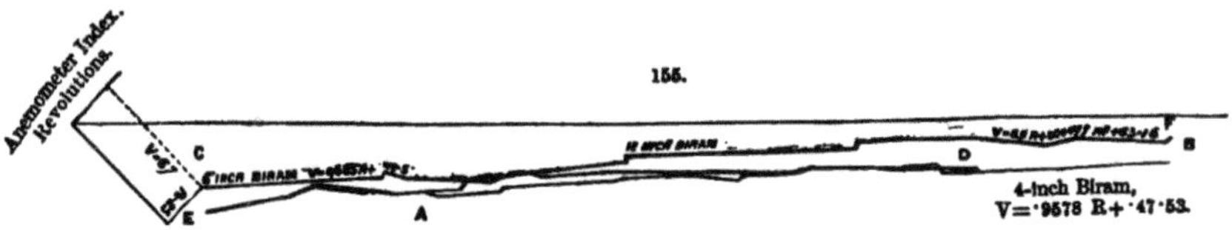

155.

The experiments from which this diagram was constructed are as follows:—

TABLE IV.—EXPERIMENTS MADE WITH THE THREE SIZES OF BIRAM'S ANEMOMETERS TO ASCERTAIN THE CONSTANTS m AND a IN THE FORMULA $V = mR + a$ FOR EACH INSTRUMENT, THE TRUE VELOCITIES BEING ASCERTAINED BY PASSING THE ANEMOMETERS THROUGH A STILL ATMOSPHERE AT KNOWN VELOCITIES BY THE WHIRLING MACHINE, IN A CIRCLE OF 25 FEET IN CIRCUMFERENCE.

	LINE A B, 4-IN. ANEMOMETER.					LINE C D, 6-IN. ANEMOMETER (same as E F, Fig. 154).					LINE E F, 12-IN. ANEMOMETER (same as C D, 156, and A B, 158).				
	VELOCITY.			ERROR.		VELOCITY.			ERROR.		VELOCITY.			ERROR.	
No. of Experiment.	Number indicated by Index of Anemometer = R.	Actual Velocity = V.	Velocity calculated by formula V = ·9578 R + 47·53.	More.	Less.	Number indicated by Index of Anemometer = R.	Actual Velocity = V.	Velocity calculated by formula V = ·9685 R + 40·5.	More.	Less.	Number indicated by Index of Anemometer = R.	Actual Velocity = V.	Velocity calculated by formula V = ·89 R + ·000477 R² + 63·16.	More.	Less.
1	107	150	150	..	..	..	..	..	..	..	19	79	78	..	1
2	110	156	..	..	..	..	..	..	..	..	46	96	..	..	..
3	132	175	..	..	..	..	..	..	..	..	55	102	..	..	..
4	132	173	..	..	..	..	..	..	..	..	67	108	..	..	..
5	142	183	..	..	..	..	..	..	..	..	93	137	..	..	..
6	152	192	193	1	..	..	..	..	..	..	117	158	..	..	..
7	167	206	..	..	..	..	..	..	..	..	119	158	..	..	..
8	173	212	..	..	..	..	..	..	..	..	151	183	..	..	..
9	180	217	..	..	..	..	..	..	..	..	186	208	..	..	..
10	200	234	..	..	..	..	..	..	..	..	185	208	..	..	..
11	203	242	..	..	..	..	..	..	..	..	215	236	..	..	..
12	205	242	244	2	..	..	..	..	..	..	257	276	..	..	..
13	209	247	..	..	..	..	..	..	..	..	267	283	..	..	..
14	249	281	..	..	..	..	..	..	..	..	268	283	..	..	..
15	250	283	..	..	..	..	..	..	..	..	288	300	..	..	..
16	210	247	..	..	..	..	..	..	..	..	305	317	318	1	..
17	275	306	..	..	..	..	..	..	..	..	313	325	..	..	..
18	273	306	..	..	..	..	..	..	..	..	313	325	..	..	..
19	287	317	..	..	..	..	..	..	..	..	327	342	..	..	..
20	286	321	..	..	..	..	..	..	..	..	333	350	..	..	..
21	337	371	..	..	..	..	..	..	..	..	345	358	..	..	..
22	368	400	401	1	..	..	..	..	..	..	378	392	392	..	..
23	..	..	..	..	..	..	..	..	..	..	382	392	..	..	..

Comparison of Anemometers.—Fig. 155 exhibits graphically the results of experiments made with three Biram's anemometers of the different sizes, on the Whirling Machine, in a circle of 25 feet in circumference, without any governor. We observe that, for the 4-inch and 6-inch anemometers, the lines are sensibly straight, and the formula consequently simple, and of the form previously indicated. With the 12-inch anemometer, however, the line is curved, and the formula complicated, and troublesome in consequence; and, although the friction of this large anemometer is somewhat greater than that of the smaller ones, the difference in this respect is trifling; even the large instrument would apparently be kept in motion by a velocity of 64 feet per minute, or little more than 1 foot per second; and as friction must be allowed for in all cases, its amount is of no great importance, within moderate limits. It is not very easy to account for the curvature of the line given by the large instrument, when not observed in the smaller ones; may, however, arise from one of two causes, or possibly partly from both. The large instrument is moved sensibly quicker in the circle of revolution at its outer than at its inner extremity, while this difference of velocity in the two sides of the anemometer is less palpable in the smaller ones; and this difference may tend to cause all the instruments to depart from the straight line, but, from being less in amount with small instruments, may not be observable within the velocities attained in the experiments: or, apart from this cause, all such anemometers may give curved lines when plotted in this way, but perhaps quicker curves in large than small instruments: considerations as to strength, portability, comparative freedom from liability to derangement, and original cost, all appear to be favourable to the use of the 6-inch or 4-inch Biram anemometer, in preference to the 12-inch ones.

The constant multiplier, *m*, in these formulæ—which depends to some extent, for its amount, upon the spur-gear, and the arbitrary marks and numbers on the dial—is greatest on the 6-inch, and least in the 12-inch anemometer; while in the smallest, or 4-inch, it is intermediate; proving that it does not necessarily follow the size of the instrument, but really depends partly upon the conditions just mentioned.

156.

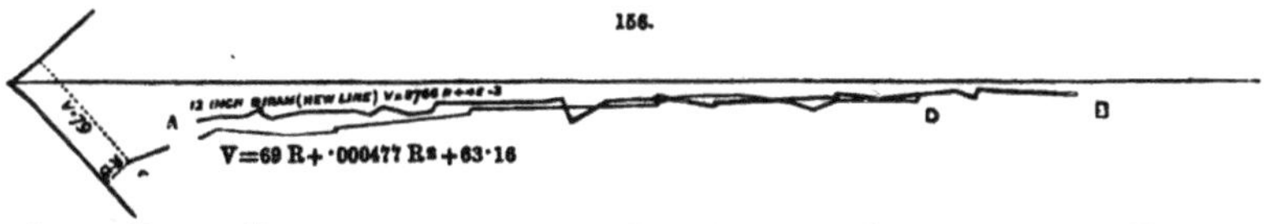

SECOND SET OF EXPERIMENTS, MADE WITH A BIRAM'S 12-INCH ANEMOMETER, WITH WHIRLING MACHINE, 25-FEET CIRCLE, FIG. 156, MADE TO CORROBORATE EXPERIMENTS, FIG. 155.

	A B (New Series).					C D.				
	Velocity.			Error.		Velocity.			Error.	
No. of Experiment.	Number indicated by Index of Anemometer = R.	Actual Velocity = V.	Velocity calculated by formula V = ·8766 R + ·0001428 R² + 35·65.	More.	Less.	Number indicated by Index of Anemometer = R.	Actual Velocity = V.	Velocity calculated by formula V = ·69 R + ·000477 R² + 63·16.	More.	Less.
1	64	96	..	..	..	19	79	78	..	1
2	77	104	..	..	..	46	96	..	..	..
3	88	114	114	..	..	55	102	..	..	..
4	93	118	..	..	..	67	108	..	..	..
5	94	121	..	..	..	93	137	..	..	..
6	95	123	..	..	..	117	158	..	..	..
7	110	133	..	..	..	119	158	..	..	..
8	134	156	..	..	..	151	183	..	..	..
9	136	158	..	..	..	186	208	..	..	..
10	143	167	..	..	..	185	208	..	..	..
11	143	167	..	..	..	215	236	..	..	..
12	150	167	170	3	..	257	276	..	..	..
13	161	185	..	..	..	267	283	..	..	..
14	170	192	..	..	..	268	283	..	..	..
15	173	192	..	..	..	288	300	..	..	..
16	175	192	..	..	..	305	317	318	1	..
17	216	233	..	..	..	313	325	..	..	..
18	227	242	..	..	..	313	325	..	..	..
19	244	258	..	..	..	327	342	..	..	..
20	262	275	..	..	..	333	350	..	..	..
21	269	283	..	..	..	345	358	..	..	..
22	272	283	..	..	..	378	392	392	..	..
23	293	308	..	..	..	382	392	..	..	..
24	295	308	..	..	..	..	..	..	..	..
25	348	358	..	..	..	..	..	..	..	..
26	357	367	..	..	..	..	..	..	..	..
27	399	408	..	..	..	..	..	..	..	..
28	403	417	..	..	..	..	..	..	..	..
29	405	417	..	..	..	..	..	..	..	..
30	410	417	417	..	..	..	..	..	..	..
31	424	433	..	..	..	..	..	..	..	..
32	448	458	..	..	..	..	..	..	..	..
33	224	250	..	..	..	..	..	..	..	..
34	346	358	..	..	..	..	..	..	..	..
35	362	375	..	..	..	..	..	..	..	..

Comparison of Anemometers.—Fig. 156. In consequence of the sensible difference in the formula required for the 12-inch, and for the 6-inch and 4-inch anemometers, as exhibited in Fig. 155, another set of experiments were made to prove whether this was owing to the construction of the larger instrument, or to some error in observation in the previous experiments. The results are given in Fig. 156.

In this diagram two sets of experiments, made with the same 12-inch anemometer, at different times, by the Whirling Machine, with a circle 25 feet in circumference, are compared with each other.

The line C D is the same as the line E F of Fig. 155, and is a curve, as has been already stated; the line A B is also a curve, but approaches rather more nearly to a straight line, the friction being at the same time much smaller than in the same instrument tried in the experiments shown by Fig. 155, so that the friction does not appear to depend so much upon the size as upon the condition of the instrument. This result agrees with those obtained in other experiments with the 4-inch anemometer (page 78), which latter clearly prove that the friction does not vary much if the instrument itself remains unaltered; but from the construction of the larger instruments, and from the nature of the material of which their vanes are formed, they are extremely liable to become disarranged, and altered in form; in fact, it is most difficult to avoid putting them out of form when using them, and it is extremely probable that the alteration in the additive constant is due to this cause.

This view is corroborated by Fig. 162, in which it will be observed that the constant for one of the 4-inch anemometers is altered in consequence of its falling off the wand and becoming deranged, although no alteration in the instrument was observable from the accident. Fig. 163 also elucidates this.

Both the lines deduced from the experiments made with the 12-inch or large anemometers are, it will be seen, curves, and require complicated formulas, while it is somewhat doubtful as to whether the departure from a straight line is due to the circular motion caused by the revolving wand of the Whirling Machine, or is inherent in the nature of the instruments themselves, in which latter case only would it arise on employing the instrument for measuring currents of air in the ordinary way.

157.

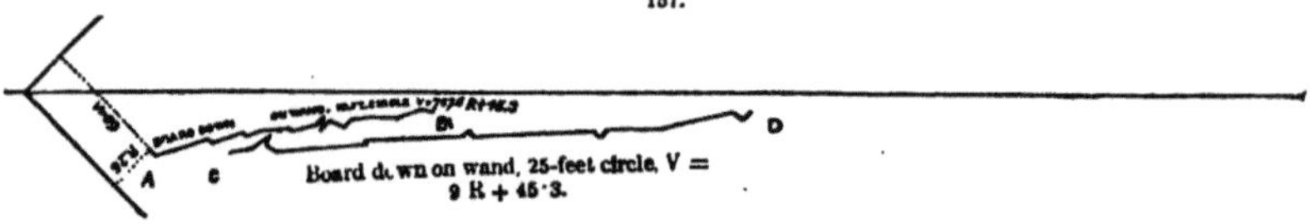

EXPERIMENTS MADE WITH A 6-INCH BIRAM'S ANEMOMETER TO ASCERTAIN THE EFFECT OF REVOLVING IN CIRCLES OF DIFFERENT DIAMETERS, WITH A GOVERNOR OR BOARD FIXED DOWNWARD ON THE WAND OF THE WHIRLING MACHINE, FIG. 157.

No. of Experiment.	LINE C D, 25-FEET CIRCLE. (C D, 154.)					LINE A B, 10-FEET CIRCLE.				
	Number indicated by Index of Anemometer = R.	Actual Velocities = V.	Velocity calculated by formula V = ·9 R + 45·3.	ERROR. More.	ERROR. Less.	Number indicated by Index of Anemometer = R.	Actual Velocities = V.	Velocity calculated by formula V = ·7676 R + 35·3.	ERROR. More.	ERROR. Less.
1	56	99	..	..	..	25	69	..	..	..
2	65	104	104	..	..	26	70	..	..	..
3	75	106	..	..	..	27	68	66	..	2
4	71	108	..	..	..	53	86	..	..	..
5	72	112	..	..	..	54	88	..	..	..
6	72	113	..	..	..	67	97	..	..	..
7	107	142	142	..	..	68	95	..	..	..
8	107	142	..	..	..	75	102	..	..	..
9	107	142	..	..	..	89	115	..	..	..
10	108	142	..	..	..	87	112	112	..	..
11	150	180	..	..	..	98	122	..	..	..
12	152	181	..	..	..	99	120	..	..	..
13	153	183	..	..	..	103	121	..	..	..
14	153	183	..	..	..	103	122	..	..	..
15	195	225	..	..	..	107	129	..	..	..
16	196	223	..	..	..	113	130	131	1	..
17	198	225	224	..	1	115	132	..	..	..
18	198	225	..	..	..	128	145	..	..	..
19	199	225	..	..	..	128	145	..	..	..
20	223	248	..	..	..	139	152	..	..	..
21	223	248	..	..	..	141	155	154	..	1
22	252	267	268	1	..	..	..	..	..	..
23	253	273	..	..	..	..	..	..	..	..
24	256	273	..	..	..	..	..	..	..	..

Comparison of Circles.—Fig. 157 shows the effect of causing an anemometer to revolve in a 10-feet as compared with a 25-feet circle, on the Whirling Machine, with a governor fixed downwards in both cases; the line C D is the same as the line C D in Fig. 154, and is for the 25-feet circle. On taking the circle at 10 feet, the constant additive for friction remains unaltered, but the instrument itself revolves quicker than in the larger circle for equal velocities, so that the constant multiplier is reduced.

The anemometer was the same 6-inch Biram in each case.

Comparison of Circles, Fig. 158, *to corroborate Experiments, Fig.* 157.—This series of experiments, like those of Fig. 157, give a comparison of the effect of causing anemometers to revolve in circles

158.

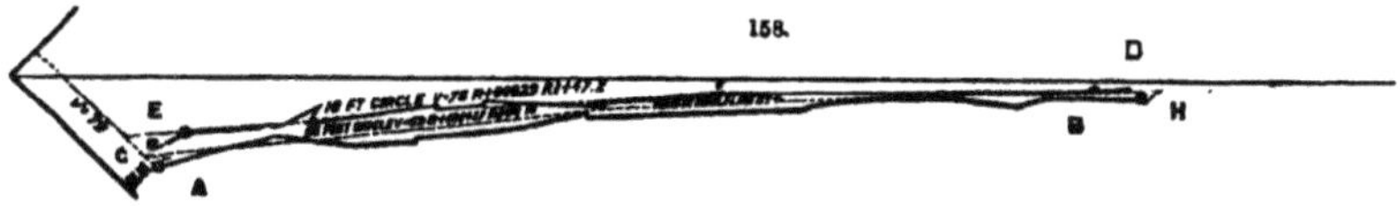

of different sizes, with a 12-inch in lieu of a 6-inch anemometer, and without governors. Here, again, the instrument revolves quicker, for a given velocity through the air, in a 10-feet than in a 25-feet circle, showing that the result is not accidental, but has a cause, depending for its amount upon the size of the circle; and this fact renders it probable, that if the circles were indefinitely large, or, what is equivalent, if the anemometers were moved in a straight line through a still atmosphere, the numbers indicated by the index of the instrument would be somewhat less than those obtained in a 25-feet circle; the difference, however, would in all probability be small.

Experiments, Fig. 158, made with a 12-inch Biram's Anemometer, by the Whirling Machine, with a 25-feet and a 10-feet Circle, without Governor-boards.

No. of Experiment.	Line C F D, 10-feet Circle.					Line A B.				
	Number indicated by Index of Anemometer = R.	Actual Velocities = V.	Velocity calculated by formula $V = \cdot 75 R + \cdot 00039 R^2 + 47 \cdot 2$.	Error. More.	Error. Less.	Number indicated by Index of Anemometer = R.	Actual Velocity = V.	Velocity calculated by formula $V = \cdot 69 R + \cdot 000477 R^2 + 63 \cdot 16$.	Error. More.	Error. Less.
1	31	74	..	..	..	19	79	78	..	1
2	39	77	..	..	..	46	96	..	..	..
3	64	100	..	..	..	55	102	..	..	..
4	69	103	..	..	..	67	108	..	..	..
5	72	107	..	..	..	93	137	..	..	..
6	73	109	..	..	..	117	158	..	..	..
7	83	113	..	..	..	119	158	..	..	..
8	86	117	..	..	..	151	183	..	..	..
9	93	116	..	..	..	186	208	..	..	..
10	97	126	..	..	..	185	208	..	..	..
11	132	155	..	..	..	215	236	..	..	..
12	133	154	..	..	..	257	276	..	..	..
13	151	170	..	..	..	267	283	..	..	..
14	152	170	..	..	..	268	283	..	..	..
15	191	210	..	..	..	288	300	..	..	..
16	192	210	..	..	..	305	317	318	1	..
17	234	245	..	..	..	313	325	..	..	..
18	235	245	..	..	..	313	325	..	..	..
19	265	275	..	..	..	327	342	..	..	..
20	270	280	..	..	..	333	350	..	..	..
21	311	320	..	..	..	345	358	..	..	..
22	316	325	..	..	..	378	392	392	..	..
23	367	375	..	..	..	382	392	..	..	..
24	374	380	..	..	..	..	..	..	..	..

Comparison of Anemometers.—Lines formed by Experiments made with Combes' Anemometer, Fig. 159.—In the various experiments that have been made with these instruments on the Continent, it has been found that the excess of the actual wind velocities, over the particular wind velocities which are required to overcome their fractional resistance, is simply proportional to the number of revolutions performed by them in the unit of time; or, what is the same, to the numbers indicated by the pointers, giving rise to an expression of the form of $V = m R + a$, by means of which the velocity of the wind V can be found when we know the number of revolutions, R, of the instrument in the unit of time, a and m being constants for the same instrument, whatever be the velocity V, or revolutions R.

The above expression is the equation to a straight line. The constants a and m can be determined for any instrument by means of two experimental trials of the number of revolutions corresponding to different ascertained wind velocities; thus, if V and R are the velocity and corresponding number of revolutions or indications given by the pointers in the unit of time in one of such trials, and V' and R' the same respectively on the other trial, then $V = m R + a$, and $V' = m R' + a$, from whence $m = \frac{V - V'}{R - R'}$ and $a = \frac{V' R - V R'}{R - R'}$.

The irregularities in the line made by this instrument are, doubtless, caused by errors in observation, and derangement of the pointers, as it is much more difficult to read off than Biram's instrument, under the peculiar circumstances of revolving on a wand in a circle; but these irregularities would probably not exist when using the instrument in the ordinary way for measuring a current of air. The line for the Biram's anemometer is more uniform, because it can be read off with more ease and accuracy in this kind of experiment.

TABLE OF EXPERIMENTS WITH COMBES AND ROBINSON'S ANEMOMETERS, COMPARED WITH BIRAM'S ANEMOMETER, ON WHIRLING MACHINE, 25-FEET CIRCLE, FIG. 159.

	COMBES' LINE, C D.					ROBINSON'S ANEMOMETER.					LINE E F. BIRAM'S 6-IN. (same as E E, 160).					
	VELOCITY.			ERROR.		VELOCITY.			ERROR.		VELOCITY.				ERROR.	
No. of Experiment.	Number indicated by Index of Anemometer = R.	Actual Velocities = V.	Velocity calculated by formula V = ·738 R + 78.	More.	Less.	Number indicated by Index of Anemometer = R.	Actual Velocities = V.	Velocities calculated by formula V =	More.	Less.	Number indicated by Index of Anemometer = R.	Actual Velocities = V.	Velocities calculated by formula V = ·9568 R + 60·92.	Velocities calculated by formula V = ·8851 R + ·0000842 R² + 65·32	More.	Less.
..	201	225	226	..	..	67	149	..	..	..	61	125	..	..	..	..
..	213	233	..	..	..	327	368	..	..	..	67	125	124	124	..	1
..	317	308	..	..	..	542	587	..	..	..	140	187	..	..	..	..
..	324	317	..	..	..	777	798	..	..	..	145	200	..	..	..	..
..	325	325	..	..	..	..	..	..	..	..	187	225	..	..	..	..
..	332	325	..	..	..	..	..	..	..	..	190	237	..	..	..	..
..	336	358	..	..	..	..	..	..	..	..	197	237	..	..	..	..
..	433	408	..	..	..	..	..	..	..	..	200	250	..	..	..	..
..	359	342	..	..	..	..	..	..	..	..	200	250	..	..	..	..
..	369	358	..	..	..	..	..	..	..	..	215	250	..	..	..	..
..	433	392	..	..	..	..	..	..	..	..	235	275	..	..	..	..
..	448	408	..	..	..	..	..	..	..	..	287	337	..	..	..	..
..	480	442	..	..	..	..	..	..	..	..	292	337	..	..	..	..
..	491	433	..	..	..	..	..	..	..	..	367	412	..	..	..	..
..	529	458	460	2	..	..	..	..	..	..	375	412	..	..	..	..
..	529	467	..	..	..	..	..	..	..	..	418	450	..	..	..	..
..	552	475	..	..	..	..	..	..	..	..	422	450	..	..	..	..
..	613	533	..	..	..	..	..	..	..	..	442	487	..	..	..	..
..	616	533	..	..	..	..	..	..	..	..	452	487	..	..	..	..
..	618	542	..	..	..	..	..	..	..	..	477	512	..	..	..	..
..	618	533	..	..	..	..	..	..	..	..	482	512	..	..	..	..
..	641	550	..	..	..	..	..	..	..	..	485	512	..	..	..	..
..	690	583	..	..	..	..	..	..	..	..	577	600	..	..	..	..
..	722	608	..	..	..	..	..	..	..	..	592	612	..	..	..	..
..	752	625	..	..	..	..	..	..	..	..	605	625	..	..	..	..
..	756	633	..	..	..	..	..	..	..	..	617	637	..	..	..	..
..	740	617	..	..	..	..	..	..	..	..	685	712	..	..	..	..
..	797	650	..	..	..	..	..	..	..	..	690	712	..	..	..	..
..	811	675	678	3	..	..	..	..	..	..	750	775	..	..	..	..
..	..	..	..	..	..	..	..	..	..	..	775	800	..	..	..	..
..	..	..	..	..	..	..	..	..	..	..	785	812	..	..	..	..
..	..	..	..	..	..	..	..	..	..	..	785	812	811	811	..	1

Moving Anemometers in a Straight Line through a Still Atmosphere, Fig. 160.—This figure exhibits the results of different sets of experiments made with a 6-inch and 4-inch Biram's anemometer, by a person walking in the still atmosphere of a large granary, and carrying the instrument in his hand; and another set (for the purpose of comparison with those made as above) by the Whirling Machine, in a 25-feet circle, without governors on the revolving-wand. The results obtained in these experiments appear to be both remarkable and important, inasmuch as they indicate that the formulas deduced from the walking experiments do not agree, even within moderate limits, with those obtained by the use of the Whirling Machine; and it becomes a question as to whether the results of either mode are reliable, and if either, as to which mode is so.

All the lines in this diagram are more or less curved. The line E, E, relating to the revolving experiments is, however, less curved than the other lines relating to the experiments made by walking in a straight line, and carrying the instrument through a still atmosphere. The line E, E, indeed, may be regarded as being nearly straight. Of the more recent experiments, the line D, D, may be taken as approaching the straight line: also the line C, C, up to *a*, where it rapidly falls off. This is probably owing to errors at the high velocities.

The line A, A, obtained by walking in a straight line, is much more curved than the line, E, E, of the 25-feet circle,—the multiplier, *m*, of the numbers indicated by the index of the anemometer being much smaller, and the multiplier of the square of the numbers indicated by the index of the

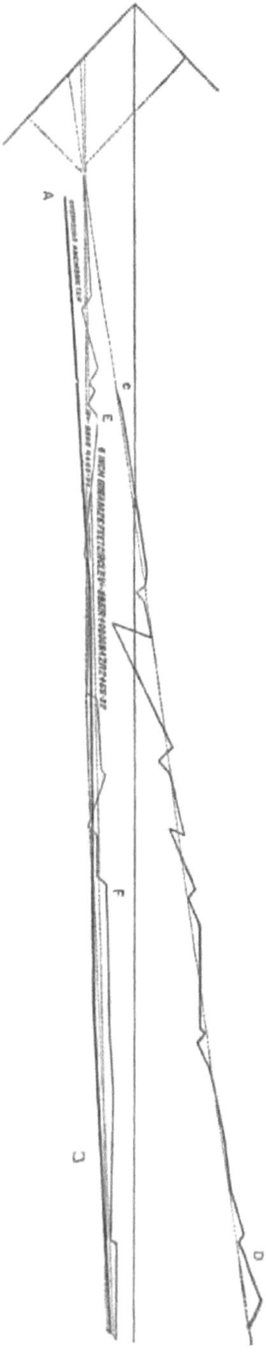

160.

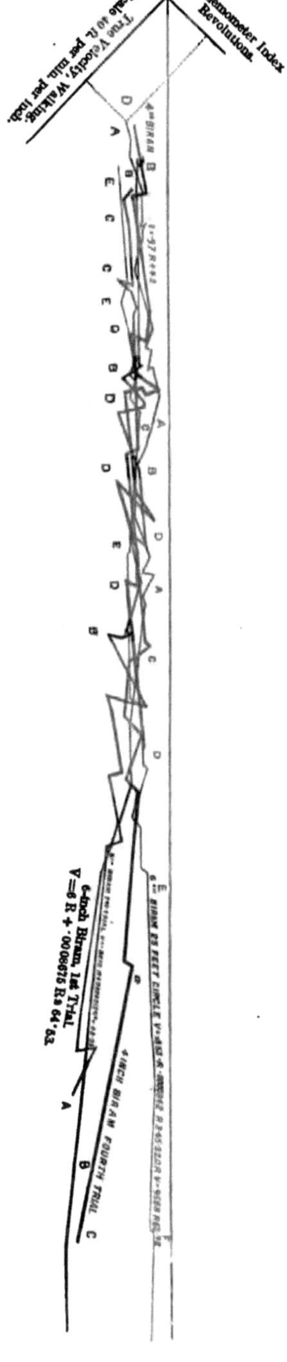
Anemometer Index Revolutions.
True Velocity, Walking.
Scale 40 ft. per min. per inch.
6-inch Biram, 1st Trial.

EXPERIMENTS, FIG. 160, MADE BY WALKING 500 FEET IN A GRANARY IN A STILL ATMOSPHERE, WITH A 6-INCH AND 4-INCH BIRAM'S ANEMOMETER IN THE HAND, COMPARED WITH EXPERIMENTS MADE WITH THE SAME ANEMOMETER ON THE WHIRLING MACHINE.

WALKING.

No. of Experiment.	LINE A A, 1ST TRIAL. Number indicated by Index of Anemometer = R.	Actual Velocities = V.	Velocity calculated by formula V = ·6 R + ·0006675 R^2 + 64·83.	ERROR. More.	ERROR. Less.	LINE B B, 2ND TRIAL. Number indicated by Index of Anemometer = R.	Actual Velocity = V.	Velocity calculated by formula V = ·8612 R + ·0004212 R^2 + 43·35.	ERROR. More.	ERROR. Less.
1	45	96	..	..	..	78	108	..	..	..
2	59	103	103	..	..	82	116	115	..	1
3	68	111	..	..	..	83	112	..	..	..
4	95	125	..	..	..	93	133	..	..	..
5	95	125	..	..	..	103	144	..	..	..
6	97	129	130	1	..	110	149	..	..	..
7	98	129	..	..	..	184	219	..	..	..
8	110	141	..	..	..	187	242	..	..	..
9	111	144	..	..	..	192	227	..	..	..
10	112	139	..	..	..	194	227	..	..	..
11	116	147	145	..	2	203	262	..	..	..
12	121	151	..	..	..	208	242	..	..	..
13	134	174	..	..	..	208	256	..	..	..
14	177	197	..	..	..	212	242	..	..	..
15	186	211	..	..	..	215	246	..	..	..
16	191	208	..	..	..	250	285	282	..	3
17	194	214	..	..	..	285	326	..	..	..
18	198	225	..	..	..	289	333	..	..	..
19	200	222	..	..	..	296	325	..	..	..
20	220	234	..	..	..	314	349	..	..	..
21	243	270	..	..	..	325	395	..	..	..
22	243	283	..	..	..	335	389	..	..	..
23	272	312	..	..	..	334	384	..	..	..
24	308	329	..	..	..	335	389	..	..	..
25	310	344	..	..	..	340	384	..	..	..
26	317	337	..	..	..	359	434	..	..	..
27	324	353	..	..	..	375	441	..	..	..
28	330	357	..	..	..	389	416	..	..	..
29	388	435	..	..	..	404	476	..	..	..
30	422	468	..	..	..	410	468	..	..	..
31	455	535	..	..	..	444	517	..	..	..
32	520	625	..	..	..	447	508	..	..	..
33	541	652	..	..	..	646	769	..	..	..
34	549	638	..	..	..	687	811	..	..	..
35	549	638	..	..	..	707	769	..	..	..
36	566	682	682	..	..	710	789	..	..	..

No. of Experiment.	LINE D D, 3RD TRIAL. Number indicated by Index of Anemometer = R.	Actual Velocity = V.	Velocity calculated by formula V = ·97 R + 42.	ERROR. More.	ERROR. Less.	LINE C C, 4TH TRIAL (taken to "a"). Number indicated by Index of Anemometer = R.	Actual Velocity = V.	Velocity calculated by formula V = R + 46.	ERROR. More.	ERROR. Less.
1	51	91	91	..	..	90	140	..	..	..
2	100	125	..	..	..	91	132	..	..	..
3	93	134	..	..	..	103	149	149	..	..
4	168	190	..	..	..	131	190	..	..	..
5	184	208	..	..	..	144	186	..	..	..
6	171	215	..	..	..	145	190	..	..	..
7	197	232	..	..	..	157	198	..	..	..
8	218	232	..	..	..	160	203	..	..	..
9	212	238	237	..	1	237	285	..	..	..
10	207	238	..	..	..	240	276	..	..	..
11	197	250	..	..	..	307	338	..	..	..
12	261	300	..	..	..	355	380	..	..	..
13	276	303	..	..	..	352	380	..	..	..
14	245	306	..	..	..	355	399	401	2	..
15	287	306	..	..	..	380	434	..	..	..
16	259	313	..	..	..	400	434	..	..	..
17	296	333	..	..	..	418	467	..	..	..
18	317	353	..	..	..	431	467	..	..	..
19	307	357	..	..	..	441	480	..	..	..
20	410	440	441	..	..	517	570	..	..	..
21	..	..	..	..	..	525	570	571	1	..
22	..	..	..	..	..	650	760	..	..	..
23	..	..	..	..	..	..	..	..	..	..
24	..	..	..	..	..	..	..	..	..	..
25	..	..	..	..	..	..	..	..	..	..
26	..	..	..	..	..	..	..	..	..	..
27	..	..	..	..	..	..	..	..	..	..
28	..	..	..	..	..	..	..	..	..	..
29	..	..	..	..	..	..	..	..	..	..
30	..	..	..	..	..	..	..	..	..	..
31	..	..	..	..	..	..	..	..	..	..
32	..	..	..	..	..	..	..	..	..	..
33	..	..	..	..	..	..	..	..	..	..
34	..	..	..	..	..	..	..	..	..	..
35	..	..	..	..	..	..	..	..	..	..
36	..	..	..	..	..	..	..	..	..	..

MACHINE, 25-FEET CIRCLE.

No. of Experiment.	LINE E E (same as E E, 160). Number indicated by Index of Anemometer = R.	Actual Velocity = V.	Velocity calculated by formula V = ·9568 R + 60·92.	Velocity calculated by formula V = ·8851 R + ·0000842 R^2 + 65·32.	ERROR. More.	ERROR. Less.
1	61	125	..	..	..	..
2	67	125	124	124	..	1
3	140	187	..	..	..	..
4	145	200	..	..	..	..
5	187	225	..	..	..	..
6	190	237	..	..	..	..
7	197	237	..	..	..	..
8	200	250	..	..	..	..
9	200	250	..	..	..	..
10	215	250	..	..	..	..
11	235	275	..	..	..	..
12	287	337	..	..	..	..
13	292	337	..	..	..	..
14	367	412	..	..	..	..
15	375	412	..	..	..	..
16	418	450	..	..	..	..
17	422	450	..	..	..	..
18	442	487	..	..	..	..
19	452	487	..	..	..	..
20	477	512	..	..	..	..
21	482	512	..	..	..	..
22	485	512	..	..	..	..
23	577	600	..	..	..	..
24	592	612	..	..	..	..
25	605	625	..	..	..	..
26	617	637	..	..	..	..
27	685	712	..	..	..	..
28	690	712	..	..	..	..
29	750	775	..	..	..	..
30	775	800	..	..	..	..
31	785	812	..	..	..	..
32	785	812	811	811	..	1
33	..	..	..	..	..	..
34	..	..	..	..	..	..
35	..	..	..	..	..	..
36	..	..	..	..	..	..

anemometer being very much larger in the experiments made by walking in a straight line than in the revolving experiments. It appears to be anomalous that the lines relating to experiments in a 10-feet circle should be more curved than those relating to experiments in a 25-feet circle, on the one hand; while the line relating to experiments made in a straight line should, on the other hand, be more curved than either.

Future experiments may possibly throw some light upon this apparent anomaly.

It is stated in Weisbach's 'Mechanics,' that from the experiments of Du Buat and those of Thibault, it would appear that the forces of still air against a flat surface, moved at different velocities through it in a straight line, are proportional to the forces of impact of winds of the same velocities, but less than them in the ratio of 1·4 to 1·85; so that if V_r be the velocity of the wind, and V_m that of the body, in the two cases, then when the velocities are such as to make the resistance in the one case equal to the force of impact in the other, we have $1 \cdot 85\, V_r^2 = 1 \cdot 4\, V_m^2$, and hence $V_r = \sqrt{\frac{1 \cdot 4}{1 \cdot 85}}\, V_m = 0 \cdot 87\, V_m$, from which we see that the velocity of wind only appears to require to be 0·87 of that of a body moved through still air in a straight line, to give rise to the same force against the body; and hence it may be that the velocities calculated from our formulas, deduced from the walking or Whirling Machine experiments, would require, when the anemometer is employed in the ordinary way for measuring a current of air, to be multiplied by 0·87, to give the true velocity of the wind; but on this point Atkinson and Daglish entertain very grave doubts, as such results do not appear, *a priori*, to be probable, and the subject is one in which mistakes are very liable to occur.

In like manner, when a flat surface is moved against a still atmosphere in a circle, the experiments of Hutton, Borda, and Thibault appear to indicate that at a given velocity the force or resistance is only about 1·5, compared with a force of 1·85 for the same velocity when the flat surface is at rest, and the air moves, as a wind, against it; so that if V_o be the velocity of a flat surface moving in a circle, and V_r the velocity of a wind giving rise to the same force, then, in order that the force in the one case may be equal to the resistance in the other, we have the velocities such that $1 \cdot 85\, V^2 = 1 \cdot 5\, V_o^2$, from whence $V_r = \sqrt{\frac{1 \cdot 5}{1 \cdot 85}}\, V_o = \cdot 9\, V_o$, from which it would appear that the velocities of wind, as deduced from the formulas found by the experiments made with the Whirling Machine, should be reduced by multiplying them by ·9 in order to get the actual velocities of the wind. There is, however, said to be more grave doubts still as to the correctness of this conclusion, and for the present it is preferred to leave it out of the formulas given.

The discrepancy between the line E, E, in Fig. 160, representing V_o, and the lines B, B, A, A, representing V_m, are much greater than the differences just alluded to would appear to indicate, and hence there is reason to suppose that the data are not reliable; indeed, it may be seen that Weisbach, in adopting 1·85, 1·4, and 1·5 as multipliers of the motive column to give the force or resistance due to a wind, to moving a flat surface in a straight line through still air, and then in a circle, respectively, makes no distinction between a large and a small circle; while a glance at Fig. 158 appears to show that the size of the circle has considerable influence upon the amount of the resistance the body meets with in its revolutions through still air at any velocity.

It should, however, also be stated that the discrepancies between the line E, E, on the one hand, and the lines B, B, A, A, and other lines on the other hand, in Fig. 160, chiefly prevail at the higher velocities, and probably arise from the errors of observation, owing to the experimenter having to run at considerable speeds; the jolting motion arising from this, and the disturbance of the air by his body, must be considerable. The formula for the lines D, D, C, C, is obtained only up to a velocity of 500 feet per minute, up to which point the diagram exhibits a straight line, conforming very closely with that given by the Whirling Machine.

The height of a column of a fluid, h, in feet, required to generate a velocity of V_1 in feet per second, apart from friction, is expressed by $h_1 = \frac{V_1^2}{2\,g}$, when $g = 32\frac{1}{6}$, giving $h_1 = \frac{V_1^2}{64\frac{1}{3}}$; or, where h_2 is the height of the column due to V_2, the velocity in feet per minute, this becomes $h_2 = \frac{V_2^2}{231,600}$; and from the experiments of the authors just mentioned, it would appear that the impact of a fluid in motion would support a vertical column of the height f_r in feet, expressed by $f_r = \frac{1\ 85\, V_2^2}{231,600} = \frac{1 \cdot 85\, V_1^2}{64\frac{1}{3}}$; or $f_r = \frac{V_2^2}{125,189} = \frac{V_1^2}{34 \cdot 77}$. (See Acceleration.) While if the fluid is at rest, and the body (having in each case a flat surface) is moved in a straight line, the resistance f_m in height in feet of a column of the fluid would only amount to $f_m = \frac{1 \cdot 4\, V_2^2}{231,600} = \frac{1 \cdot 4\, V_1^2}{64\frac{1}{3}}$ or $f_m = \frac{V_2^2}{165,429} = \frac{V_1^2}{45\ 952}$, or if the body were moved in a circle, the resistance would support a column of the fluid f_o feet in height found by $f_o = \frac{1 \cdot 5\, V_2^2}{231,600} = \frac{1 \cdot 5\, V_1^2}{64\frac{1}{3}}$ or $f_o = \frac{V_2^2}{154,400} = \frac{V_1^2}{42 \cdot 889}$.

The experiments shown in Fig. 161 were made for the purpose of observing whether the action of Biram's anemometer varied much with the condition of the instrument, that is, whether the same formula and constants were required for the same instrument when properly cleaned and oiled, and again, after being much used and in a dirty condition; and it is certainly satisfactory to find, as will be observed in Fig. 161, that the action of the same instrument is very little altered through these varied conditions.

The line A, A, on this diagram shows the velocity found by correcting the readings of the anemometer by the formula (for this particular instrument), $V = 1{\cdot}017\,R + 30$; and it will be remarked how very close the line approximates to the true velocity of the instrument passing through the air, as found by the Whirling Machine.

EXPERIMENTS, FIG. 161, MADE WITH A 6-INCH BIRAM'S ANEMOMETER, IN CLEAN AND DIRTY CONDITIONS RESPECTIVELY, ON THE WHIRLING MACHINE, IN A 25-FEET CIRCLE.

No. of Experiment.	LINE B B.—INSTRUMENT CLEAN.					LINE C C.—INSTRUMENT DIRTY.				
	Number indicated by Index of Anemometer = R.	Actual Velocity = V.	Velocity calculated by formula $V = 1{\cdot}017\,R + 30$.	ERROR.		Number indicated by Index of Anemometer = R.	Actual Velocity = V.	Velocity calculated by formula $V = R + 39$.	ERROR.	
				More.	Less.				More.	Less.
1	93	125	125	..	..	62	100	..	..	..
2	120	150	152	2	..	80	119	..	..	..
3	133	162	165	3	..	100	137	..	..	..
4	412	450	449	..	1	365	400	..	..	..
5	412	450	449	..	1	395	427	..	..	..
6	422	450	459	9	..	425	462	..	..	..
7	427	462	464	2	..	585	625	..	..	..
8	582	625	622	..	3	585	625	..	..	..
9	585	625	625	..	..	..	..	..	..	..
10	592	637	632	..	5	..	..	..	..	..

EXPERIMENTS, FIG. 162, MADE WITH SEVERAL ANEMOMETERS (BIRAM'S) ON WHIRLING MACHINE, IN 25-FEET CIRCLE, TO FIND CONSTANTS m AND a.

No. of Experiment.	LINE E E F.—4-IN. BIRAM.					LINE C D.—4-IN. BIRAM.					LINE J I.—4-IN. BIRAM.				
	Number indicated by Index of Anemometer = R.	Actual Velocity = V.	Velocity calculated by formula V =	ERROR.		Number indicated by Index of Anemometer = R.	Actual Velocity = V.	Velocity calculated by formula V =	ERROR.		Number indicated by Index of Anemometer = R.	Actual Velocity = V.	Velocity calculated by formula V =	ERROR.	
				More.	Less.				More.	Less.				More.	Less.
1	205	250	..	..	..	187	237	..	..	..	77	131	..	..	..
2	210	262	..	..	..	200	250	..	..	..	82	137	..	..	..
3	215	262	..	..	..	255	300	..	..	..	182	225	..	..	..
4	340	387	..	..	..	337	387	..	..	..	200	237	..	..	..
5	345	387	..	..	..	340	387	..	..	..	230	262	..	..	..
6	360	400	..	..	..	350	400	..	..	..	242	275	..	..	..
7	405	462	..	..	..	477	537	..	..	..	312	350	..	..	..
8	412	475	..	..	..	480	537	..	..	..	320	362	..	..	..
9	437	500	..	..	..	490	550	..	..	..	375	412	..	..	..
10	440	500	..	..	..	..	..	..	..	..	460	500	..	..	..
11	..	..	..	..	..	..	..	..	..	..	487	525	..	..	..
12	..	..	..	..	..	..	..	..	..	..	510	550	..	..	..
	LINE D D D, 160.—4-IN. BIRAM.					LINE K K L, 162.—4-IN. BIRAM.					LINE A B, 159.—6-IN. BIRAM.				
1	77	118	..	..	..	76	125	..	..	..	87	125	..	..	..
2	112	150	..	..	..	107	150	..	..	..	102	137	..	..	..
3	152	187	..	..	..	110	156	..	..	..	112	150	..	..	..
4	342	387	..	..	..	335	375	..	..	..	295	337	..	..	..
5	357	400	..	..	..	372	412	..	..	..	337	375	..	..	..
6	377	412	..	..	..	385	425	..	..	..	342	385	..	..	..
7	377	412	..	..	..	..	..	..	..	..	440	487	..	..	..
8	460	500	..	..	..	492	537	..	..	..	450	500	..	..	..
9	500	537	..	..	..	497	537	..	..	..	452	500	..	..	..
10	520	562	..	..	..	507	550	..	..	..	557	612	..	..	..
11	..	..	..	..	..	..	..	..	..	..	640	700	..	..	..

Fig. 162 exhibits a series of lines formed by various 4 and 6 inch Biram's anemometers, with which experiments have been tried. It will be observed that, although the general tendency of the lines formed by the different anemometers is in the same direction, thus giving one general form of formula applicable to all of them; still each anemometer has a decided line of its own, requiring that special constants, found by direct experiment, should be applied to each instrument.

161.

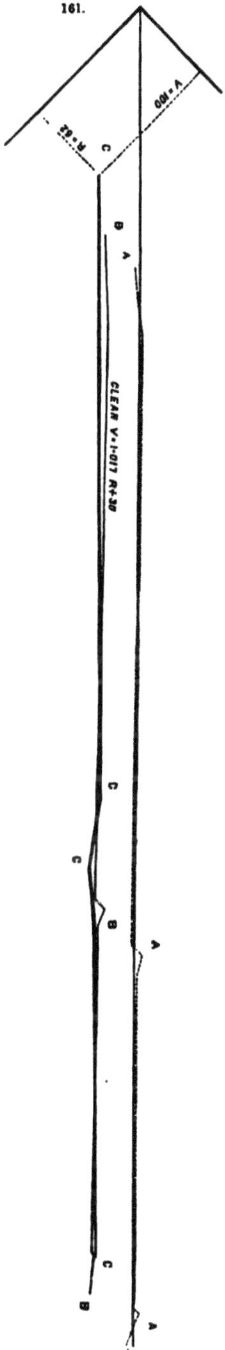

162.

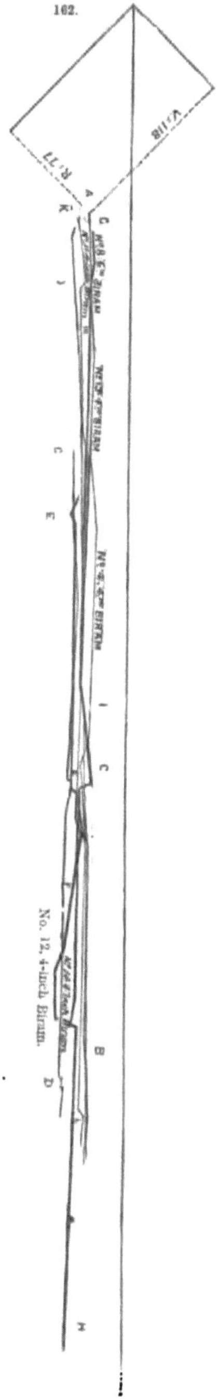

163.

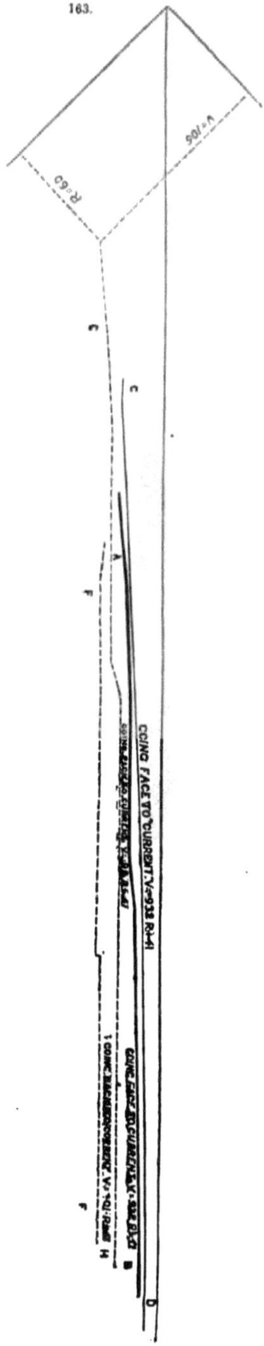

In making one of the experiments, the anemometer fell off the wand of the Whirling Machine (the line E F), and although it sustained no apparent damage, still the line made by this instrument was altered at and from this point. The instrument was apparently not damaged, but the constants required for the correction of its readings were altered, owing possibly to some alteration in the form of the vanes; showing that although, as previously explained, Fig. 161, the fact of the instrument being in a clean or dirty condition does not seriously affect its readings, still any sudden shock, or other violent treatment tending to vary the form of the vanes, or otherwise alter the mechanical condition of the instrument, would do so.

But even in this case the discrepancy caused by the fall seems to be much less in amount than the inaccuracy of the original readings; and the application of the formula, as first found, would still render the readings much more correct than if they were used in their uncorrected state.

Atkinson and Daglish made experiments for the purpose of observing the comparative rates of revolution of the same Biram's anemometer, when passed through a still atmosphere, by the Whirling Machine, in the first instance with its back, and again with its face towards the direction of its motion; the former being the ordinary mode of operating.

As there were considerable differences in the results given by these experiments, evidently depending upon whether the front or the back of the instrument received the impulse of the air, so that when plotted to form a diagram the lines appeared considerably apart from each other, showing that the same formula could not apply to the instrument under the variation of conditions involved in the experiments, another series of similar experiments was made, further to test the truth of the conclusion indicated by the former series. The results of these experiments only served to confirm the former ones.

The two lines F, F, A B, Fig. 163, are formed from experiments registered in the following Table; and the two lines G H, C D, from these experiments; the line G H, in each case being that given when the back of the anemometer received the impulse of the current, as is correct in practice.

EXPERIMENTS, FIG. 163, MADE WITH A 4-INCH BIRAM'S ANEMOMETER ON THE WHIRLING MACHINE, IN A 25-FEET CIRCLE, TO ASCERTAIN THE DIFFERENCE BETWEEN THE ANEMOMETER GOING THE BACK TO CURRENT AND FACE TO CURRENT.

No. of Experiment.	LINE F F.—ANEMOMETER GOING BACK TO CURRENT.					LINE A B.—ANEMOMETER GOING FACE TO CURRENT.				
	Number indicated by Index of Anemometer = R.	Actual Velocity = V.	Velocity calculated by formula V = 1·01 R + 41.	Error.		Number indicated by Index of Anemometer = R.	Actual Velocity = V.	Velocity calculated by formula V = ·932 R + 41.	Error.	
				More.	Less.				More.	Less.
1	172	212	214	..	..	157	187	187	..	..
2	182	225	..	..	..	192	218	..	..	..
3	315	362	..	..	..	240	262	..	..	..
4	317	362	361	..	1	290	312	..	..	..
5	405	450	..	..	..	307	325	326	1	..
6	417	462	462	..	..	332	350	..	..	..
	..	..	..	..	..	425	437	..	..	..
	..	..	..	..	..	450	462	..	..	..
	..	..	..	..	..	450	462	460	..	2

No. of Experiment.	LINE G H.—ANEMOMETER GOING BACK TO CURRENT.					LINE C D.—ANEMOMETER GOING FACE TO CURRENT.				
	Number indicated by Index of Anemometer = R.	Actual Velocity = V.	Velocity calculated by formula V = ·98 R + 41.	Error.		Number indicated by Index of Anemometer = R.	Actual Velocity = V.	Velocity calculated by formula V = ·932 R + 41.	Error.	
				More.	Less.				More.	Less.
1	60	106	..	..	..	112	144	145	1	..
2	65	109	..	..	..	117	150	..	..	..
3	122	162	161	..	..	256	275	..	..	..
4	135	172	..	..	..	297	312	..	..	..
5	213	250	..	..	..	390	400	..	..	..
6	230	262	..	..	..	440	450	..	..	..
7	317	350	..	..	..	467	475	475	..	..
8	367	400	..	..	..	..	..	..	..	..
9	442	475	474	..	..	..	..	..	..	..

These results prove that the *additive* constant remains the same in the same instrument when used either side foremost, but that, on the other hand, the amount of the necessary *multiplicative* constant depends upon the form of the wands receiving the impulse of the wind, as had been anticipated and mentioned in the remarks upon Figs. 161, 162.

In general principles the anemometers of Whewell and Osler are similar to those of Combes and Biram, but by additional apparatus the two former are made self-registering, both in reference to the force and direction of the wind; this, of course, adds to the friction of the instrument, and they are not, therefore, adapted for ascertaining the velocities of feeble currents.

The anemometer of Dr. Robinson is constructed on the assumption that the force of impact of the air against hollow hemispherical cups is twice as great on the concave as on the convex side of the cups, and that the vanes revolve at the rate of one-third of the velocity of the current, except in so far as the velocity of revolution is modified by friction.

The mechanism of this instrument is very strong, and allows of the revolutions being recorded throughout a whole day; it would, therefore, be a very suitable anemometer to have near a furnace, or in the principal intake or return from a mine.

Pressure Anemometers.—Perhaps amongst the best known of the pressure anemometers are M. Bongui's, Dr. Lind's, that of Henaut, described by Ponson, and Dickinson's, one of Her Majesty's Inspectors of Mines. The anemometer of Bongui consists of an apparatus like a spring-balance, furnished with a float-board or plain surface of given area, and the pressure or impulse is indicated by marks on the sliding-rod of the spring; it is figured and described in the 'Edinburgh Encyclopædia.'

The anemometer of Dr. Lind resembles the photometer of Pitôt; it determines the velocity of the wind by its action on a small quantity of water in a U-shaped tube. As the same instrument is much used in coal mines as a water-gauge for indicating the difference of pressure between the downcast and upcast air-column, it will not be at all necessary to give a detailed description of it. From numerous experiments, Dr. Lind considered that the pressure of the wind in direct impulse is nearly proportional to the square of its *velocity*. The following Table is calculated from this, but considerably enlarged by other experiments.

TABLE OF THE FORCE AND VELOCITY OF DIFFERENT WINDS FOR THE GRADUATION OF ANEMOMETERS.
('Edinburgh Encyclopædia.')

Height of the column of water in Dr. Lind's Anemometer.	Force on a sq. foot, in pounds avoirdupois.	Force on a sq. foot, in pounds, ounces, and drams avoirdupois.	Feet in one second.	Miles in one hour.	Feet in one second.	Miles in one hour.	Character of the Winds.
			Computed from Rouse's Experiments.		Computed from Dr. Hutton's Experiments.		
		lbs. oz. dr.					
0·0009515	0·005	0 0 1·280	1·43	1	1·63	1·11	Hardly perceptible .. .. Rouse.
0·0038060	0·020	0 0 5·120	2·93	2	3·26	2·22	Just perceptible .. .. Rouse.
0·0083732	0·044	0 0 11·264	4·40	3	4·84	3·30	
0·0133210	0·079	0 1 4·224	5·87	4	6·52	4·44	Gentle winds .. .. .. Rouse.
0·023	0·123	0 1 15·488	7·33	5	.8·09	5·51	
0·025	0·130	0 2 1·280	7·55	5·14	8·33	5·67	A gentle wind .. .. .. Lind.
0·050	0·260	0 4 2·560	10·67	7·27	11·77	8·00	Pleasant wind .. .. .. Lind.
0·092	0·492	0 7 13·952	14·67	10·00	16·16	11·01	Pleasant brisk gale .. .. Rouse.
0·10	0·521	0 8 5·378	15·19	10·35	16·66	11·35	Fresh breeze .. .. .. Lind.
0·11	1·107	1 1 11·392	22·00	15·00	24·30	16·57	Brisk gale .. .. Rouse.
0·368	1·968	1 15 7·808	29·34	20·00	32·39	22·00	Very brisk .. .. .. Rouse.
0·5	2·604	2 9 10·624	33·74	23·00	37·26	25·40	Brisk gale .. .. Lind.
0·585	3 075	3 1 3·200	36·67	25·	40·51	27·62	Very brisk .. .. .. Rouse.
0·84	4·429	4 6 13·824	44·01	30·	48·60	33·13	High wind .. .. .. Rouse.
1·0	5·208	5 3 5·248	47·73	32·54	52·70	35·93	High wind .. .. .. Lind.
1·146	6·027	6 0 6 912	51·34	35·	56·69	38·65	
1 5	7·873	7 13 10·688	58·68	40·	64·79	44·00	Very high .. .. Rouse.
1·9	9·963	9 15 6·528	66·01	45·	72·89	49·69	Great storm .. .. .. Denham
2·0	10·417	10 6 10·496	67·50	46·02	74·53	50·81	Very high .. .. Lind.
2·68	12 300	12 4 12 800	73 35	50·	81 02	55·24	Storm or tempest .. .. Rouse.
3·	15·625	15 10 00·000	82 67	56·37	91·28	62·23	Storm .. .. Lind.
3·37	17·715	17 11 7·040	88·02	60·	97·20	66·27	Great storm .. .. .. Rouse.
4·	20·833	20 13 5·248	95 46	65·08	105·40	71·86	Great storm .. .. .. Lind.
4 08	21·435	21 6 15 360	96·82	66·	106·92	74·79	Great storm .. La Condamina.
5·	26·041	26 0 10·496	106 72	72·76	117·84	80·10	Very great storm .. .. Lind.
6·	31·490	31 7 13 440	117·36	80·	129·59	88·54	Hurricane .. .. .. Rouse.
6·	31·250	31 4 00 000	116·91	79·71	129·09	88·01	Hurricane .. .. .. Lind.
7·	36·548	36 8 12·288	126·43	86·20	139·65	95·21	Great hurricane .. .. Lind.
8·	41·667	41 10 10·752	135·00	92·04	149·07	101·63	Very great hurricane .. Lind.
9·	46·875	46 14 00·000	143·11	97·57	168·11	107 80	Most violent hurricane .. Lind.
9·36	49·200	49 3 3·200	146 70	100·	162·04	110·48	Hurricane that tears up
10·	52·083	52 1 5·248	150·93	102·90	166·66	113·63	trees and throws down
11·	57·293	57 4 11·008	158 29	107·92	171·72	117·08	buildings .. .. .. Rouse.
11·12	58·450	58 7 3·200	160 00	109·	176·55	120·37	Observed by Rochon.
12·	62·500	62 8 0·000	165·34	112·73	182·57	124·47	
1	2	3	4	5	6	7	8

Borda, however, found that the force of the wind was greater by $\frac{1}{10}$ part than Rouse's Table gives. Hutton also showed that the forces at very great velocities increased in a somewhat higher ratio than the squares of the velocity.

Henaut's anemometer, Figs. 164, 165, is similar in its principle and action to that of Dickinson; in the latter the impulse is received on a plain surface A, of oiled skin about 3 inches square, suspended from the top *p*, the variations of which, from the perpendicular *p b d g*, are noted on a scale *d d n*, which is marked off by direct experiments. This instrument is extremely portable, and not easily put out of order; but whilst it possesses the great value, with other instruments of this class, of not requiring any watch or other means of noting the time, it is, in common with them, subject to the great disadvantage of vibrating continually, especially in a rapid current, and of not recording the variation of the velocity within limits of 20 feet per minute; it is, however, very useful in steady currents of from 200 to 700 feet per minute. The supports *g g*, are secured to a base *c t*, which is levelled by screws *r v*.

164.

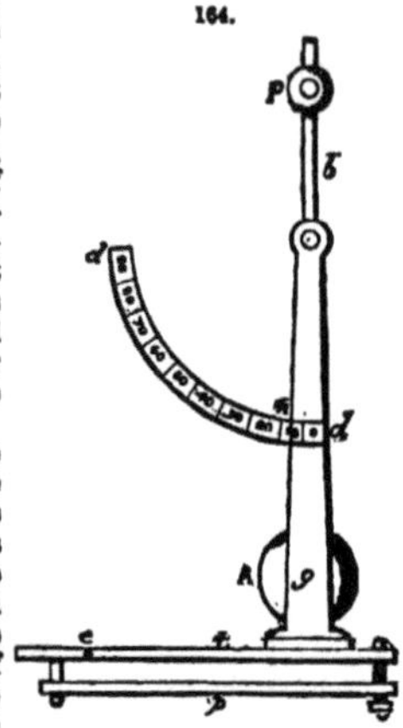

165.

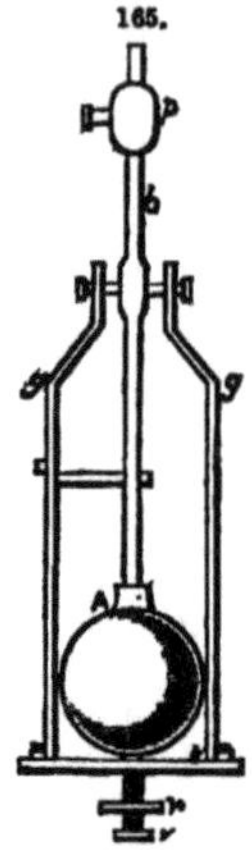

Complex Anemometers.—The principle of the interesting method proposed by Professor Leslie for finding the velocity of an air-current by its cooling action, can be studied at length in his 'Treatise on Heat.' From his experiments he deduced the following:—"A thermometer is held in the open, still atmosphere, and the temperature marked; it is then warmed by the application of the hand, and the time noted which it takes to sink back to the normal point; this is termed the fundamental measure of cooling. The same observation is made on exposing the bulb to the impression of the wind, and the time required for the bisection of the interval of temperatures is termed the occasional measure of cooling; then divide the fundamental by the occasional measure of cooling, and the increase of the quotient above unity, being multiplied by 4½, will express the velocity of the wind in miles per hour."

ANGLE-BEAD. Fr., *Corniche de cornière, Chapelet angulaire;* Ger., *Winkel oder Eck Karniess;* Span., *Guarda-vivo.*

Angle-bead, sometimes termed *Staff-bead*, shown in Fig. 166, is a small round moulding, often cut into short embossments, like pearls in a necklace. In plastering, angle-beads are made flush with the finished surface on each return to assist in floating the plaster; they are nailed to plugs or to wood-bricks built into the wall. Angle-beads are in some cases made to show double on each face of a corner, thus forming a triple bead; however, in superior apartments a triple bead is not employed, but the plaster is well gauged and brought to an *arris*, a thin copper angle-bar being in most cases fitted in to preserve the corner from accidental fracture. Wooden angle-beads are fixed to the jambs of arched recesses, the bead round the head of the arch is formed with plaster, and in good work it is necessary to conceal the joint between the plaster and wooden beads by an *impost*.

166.

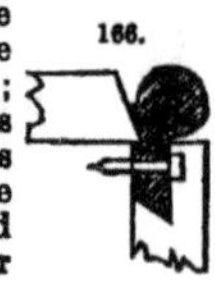

ANGLE-BRACE. Fr., *Attache angulaire;* Ger., *Winkel Strebe;* Ital., *Calastrello d'angolo, Traversa;* Span., *Tirante.*

Any framing when situated on the inner side of an angle, for the purpose of tying the work together, is termed an angle-brace. In Fig. 167, representing the framing of the external angle of a building, A is the angle-brace, B dragon-piece, and C C wall-plates.

The term angle-brace is also applied to a tool for boring in corners and other confined localities where there is not space to use the cranked handle of a common brace. It is composed of metal, and works by means of two bevel-pinions, and a winch-handle which turns at right angles to the centre of the hole to be pierced. See Brace.

167.

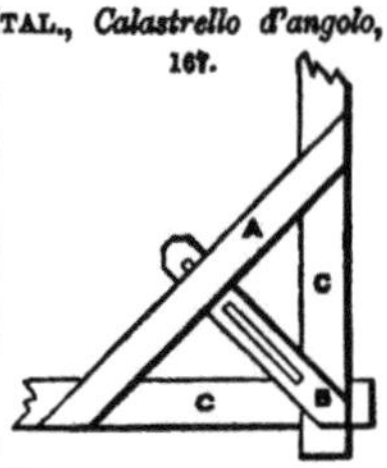

Angle-Bracket.

ANGLE-BRACKETS. Fr., *Tasseaux angulaires, Support angulaire;* Ger., *Winkellager;* Ital., *Gattello, Measola d'angolo;* Span., *Modillones angulares.*

A bracket projecting from the angle of a wall or building, instead of at right angles, as from the face of a wall, is termed an angle-bracket.

Let D B A, Fig. 168, be the elevation of the bracket of a cove, to find the angle-bracket.

First, when it is a mitre-bracket in an interior angle, the angle being 45°, divide the curve A D into any number of equal parts 5 4 3 2 1, and draw through the divisions of the lines 5 *b*, 4 *c*, 3 *d*, 2 *e*, 1 *f*, perpendicular to B A, and cutting it in *b c d e f*, and produce them to meet the line C O, representing the centre of the seat of the angle-bracket; and from the points of intersection *w i j k l* draw lines *w* 4, *i* 3, *j* 1, *k* 5, at right angles to C O, and make them equal—*w* 4 to *b* 5, *i* 3 to *c* 4, *j* 1 to *d* 3, and so on; and through G 4 3 1 5 0 draw the curve of the edge of the bracket. The dotted lines on each side of C O on the plan show the thickness of the bracket, and the dotted lines *u s*, *n t*, *v r*, show the manner of finding the bevel of the base.

The same figure shows the manner of finding the bracket for an obtuse exterior angle. Let C J F be the exterior angle, bisect it by the line J E, which will represent the seat of the centre of the bracket. The lines J I, *s* 4, *o* 3, *p* 2, *q* 1, *r*, are drawn perpendicular to J E, and their lengths are found as in the former case.

To find the angle-bracket of a cornice for interior and exterior angles. Let K A B, Fig. 169, be the elevation of the cornice-bracket, F A the seat of the mitre-bracket of the interior angle, and L R that of the mitre-bracket of the exterior angle. From the points K *b c d t* B, or wherever a change

168.

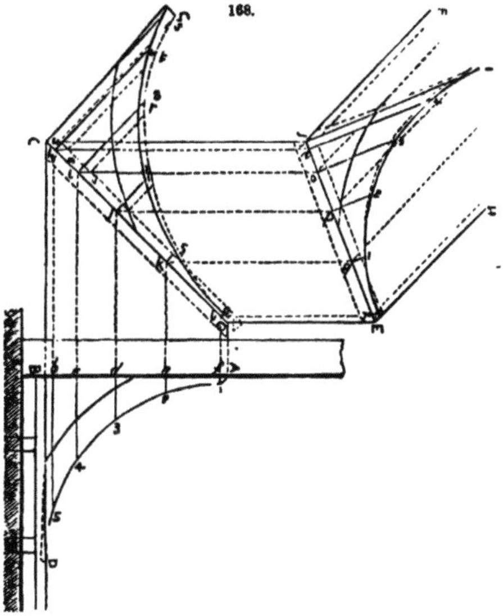

169.

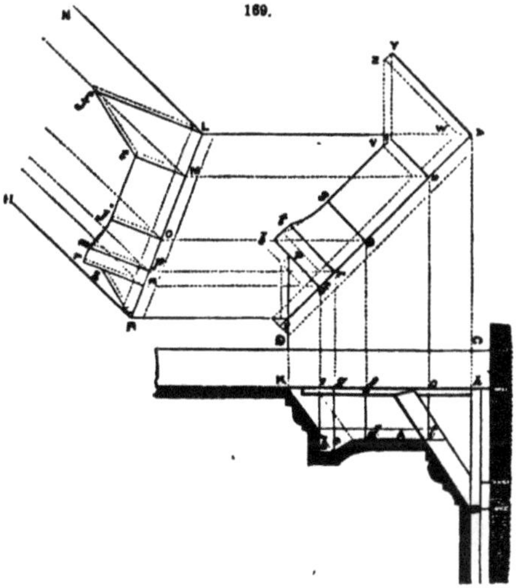

in the form of the contour of the bracket occurs, draw lines perpendicular to K *h* or Q C, cutting K *h* in *e g f c h*, and cutting the line F A in F *f r* Q *u* A. Draw the lines F R, R H, and A L, L N, representing the plan of the bracketing, and the parallel lines from the intersections *n p o* M, as shown dotted in the engraving, then make A Y and L *f* each equal to *h* B, *u o* and M *t* to *c t*, Q S and *o j* equal to *f d*, *r t* and *p m* equal to *g c*, *f b* and *n r* equal to *e b*, *f p* and *n s* equal to *e b*, and join the points so found to give the contour of the brackets required. The bevels of the face are found as shown by the dotted lines W Z V *b*.

170

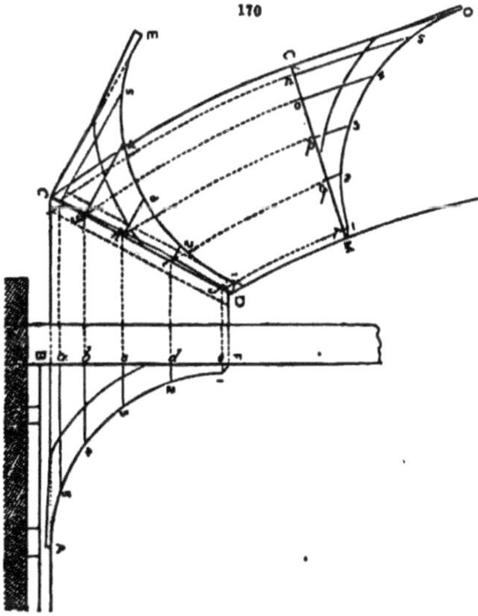

171.

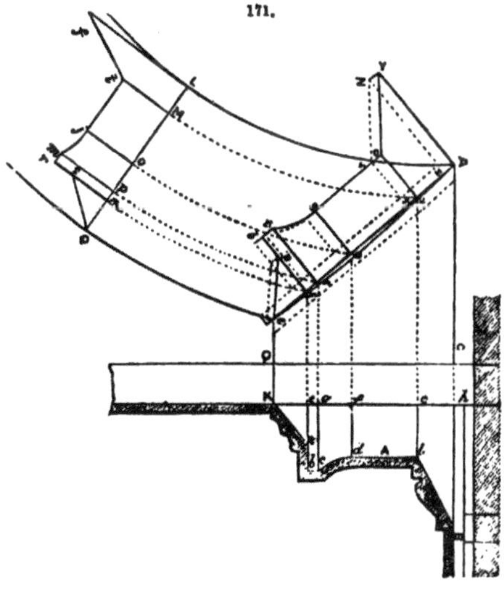

To find the angle-bracket at the meeting of a concave curved wall with a straight wall

Let B C D F, Fig. 170, be the plan of the bracketing on the straight wall, and C C, M D the plan on the circular wall. A B F the elevation on the straight wall and M C O on the circular wall.

Divide the curves A F, M O into the same number of equal parts; through the divisions of A F draw the lines A C, 5 *a f*, 4 *b g*, and so on, perpendicular to B F, and through those of M O draw the parallel lines, part straight and part curved, 5 *n f*, 4 *o g*, 3 *p h*, and so on. Then through the intersections *f g h i* of the straight and curved lines draw the curve C D, which will give the line from which to measure the ordinates *f* 5, *g* 4, *h* 3, *i* 2, *j* 1.

172.

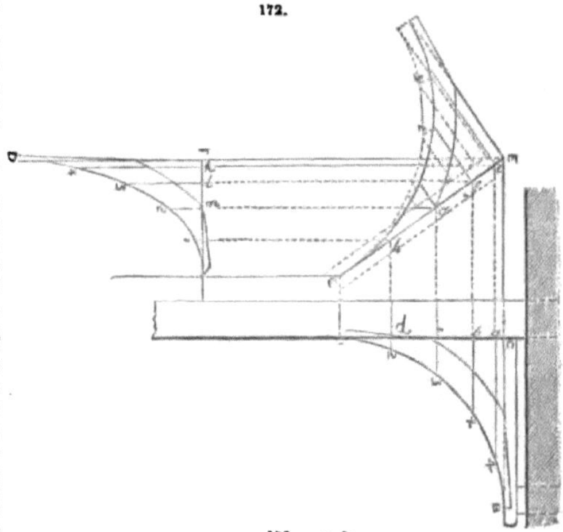

To find the angle-bracket when the wall is a convex curve.

Let A *e* Q C, Fig. 171, be the plan of the bracketing on the straight wall, and A *e* Q L the plan on the curved wall. From the points K *t b c d b* B of the bracket K A B, where its contour changes, draw perpendiculars as before. Draw L Q a radius to the curve of the wall L A, and set on it the divisions M *o p n*, equal and corresponding to *c f g e* of the elevation K A B; and draw L *f*, M *t*, *o j*, *p m*, *n r*, *n s*, perpendicular to Q L, and make them equal to *h* B, *c l*, *f d*, *g c*, *e b*, *e t*, of the elevation; then join the points by the lines *f t*, *t j*, *j m*, *m r*, *r s*, *s* Q, to obtain the contour of the bracket equal and corresponding K A B. Through the points M *o p n* draw concentric curves, meeting the perpendiculars from the corresponding points of K A B; from the intersections of the straight and curved lines, *u o r f*, draw the lines A Y, *u o*, *o s*, *r t*, *f b*, perpendicular to *e* A, and make them equal to the corresponding lines of the elevation, as before; then join the points Y *o s t b p e* to obtain the contour of the angle-bracket.

173.

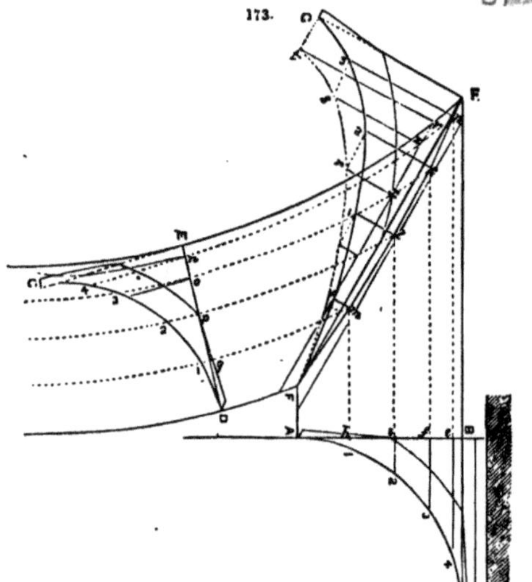

The examples shown in Figs. 172, 173, are projected in a similar manner.

ANGLE OF FRICTION.

Fr., *Angle de frottement*; Ger., *Reibungswinkel*; Ital., *Angolo d'attrito*; Span., *Angulo de rozamiento*.

The angle of friction is also called the *angle* of *repose*. In an Arch, that angle at which the arch-stones cease to have any tendency to slip from their own weight or to exert any thrust on the pier. Rondelet found by repeated experiments that even where the surfaces were wrought in the best manner the angle was seldom less than 28°, and it was in some cases as much as 36°. The angle varies with the weight as well as with the roughness of the stone. With soft stones the tendency to slide is greater than with hard stones, where the amount of roughness is the same, particularly where the weight is great, as the inequalities are broken down when the stone is soft, and the dust facilitates sliding.

On Roads and Railways it is that inclination or gradient at which the carriages are on the point of moving by their own gravity; it varies with the roughness or smoothness of the road or rails, and with the diameter and friction of the wheels upon their axles. (See Morin, 'Nouvelles Expériences sur le Frottement.')

The following, from Poncelet and others, show the relative angle of repose on various kinds of roads and railways:—

Wheels with iron tires on a—	Ratio of Slope.	Wheels with iron tires on—	Ratio of Slope.
Road of sand and gravel	1 in 16	Oak planks, not planed	1 in 98
„ of broken stone, in ordinary condition	1 „ 25	Stone trackway, well laid	1 „ 179
„ of broken stone, in perfect condition	1 „ 67	Railway	1 „ 280
Well-made pavement	1 „ 71	An iron-shod sledge on hardened snow	1 „ 30

For further information on this subject, see Friction.

ANGLE-IRON. Fr., *Cornière, ou Fer de cornière;* Ger., *Winkel, oder Eckeisen;* Ital., *Ferro d'angolo;* Span., *Hierro angular.*

Any rolled bar of iron, Figs. 174, 175, 176, 177, of an angular shape, is usually termed *angle-iron*; it is employed for forming the edges of iron-safes, bridges, and ships; or to be riveted to the corners of boilers, tanks, &c., to connect the side-plates. See Boilers; Bridges; Iron-Ship Building.

When we treat of the different structures in which angle-iron is employed, we will discuss the strength, suitable form, dimensions, and other mechanical properties of *angle-iron*; here we merely give cross-sections, dimensions, and weights of some of the principal forms manufactured in the Earl of Dudley's extensive works at Round Oak, near Dudley. For *Tee* and angle iron, Kirkaldy lately found that the specific gravity varied from 7·7310 to 7·5297, and he took 7·6006 as the mean result of ten experiments. Whence if the area, in square inches, of the cross-section of any specimen of angle or of T iron be multiplied by 3·29887 (3·3 nearly), the product will give the weight of a lineal foot of such specimen in pounds. For example, Fig. 179 is the cross-section of one of a series of angle-irons of the Round Oak Works; length of each side = 6 in. mean readth = 1 in.; area = 11 sq. in.; therefore 11 × 3·29887 = 36·28757 lbs. the lineal foot, or 36¼ lbs, nearly. The mean breadth of this series varies from ¾ to 1 in. However, it is necessary to state that a lineal foot of this iron, Fig. 179, weighed 38 lbs.

174.

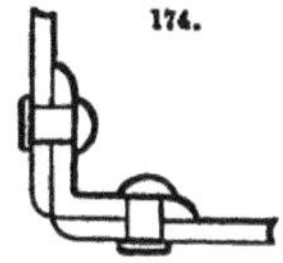

175.

Length of sides, 2½ and 2½; breadth, $\frac{1}{8}$ in. Weight the lineal foot, 2 lbs. 2 oz.

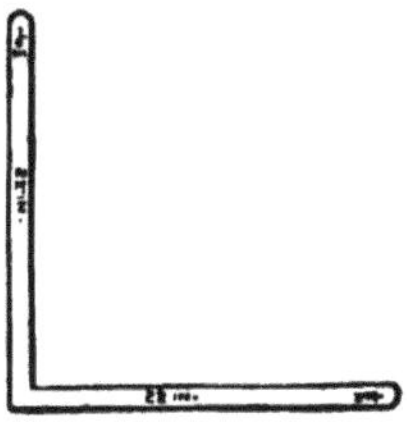

176.

2½ and 8; breadth, $\frac{9}{16}$ in. Weight the linea foot, 17·5 lbs.

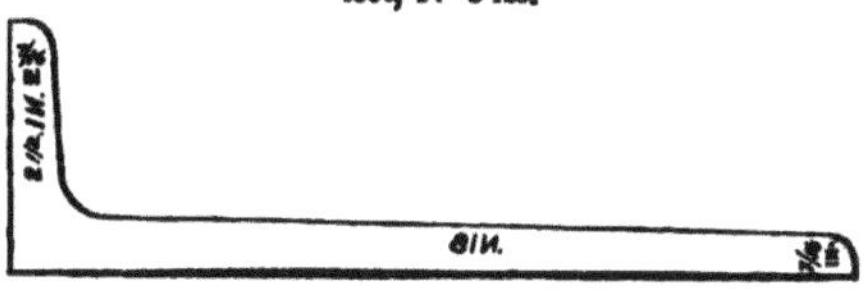

177.

3 and 3; breadth, $\frac{5}{8}$ in. Weight 12 lbs. the lineal foot.

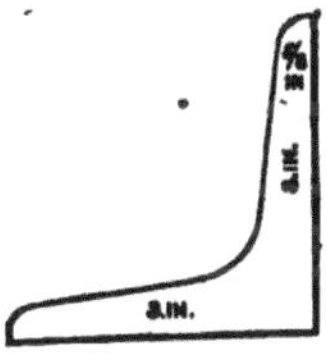

178.

2 and 2; mean breadth, $\frac{5}{16}$ in. Weight 5 lbs. 4 oz. the lineal foot.

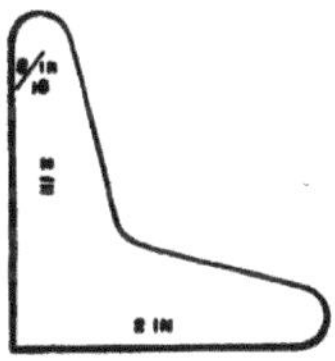

179.

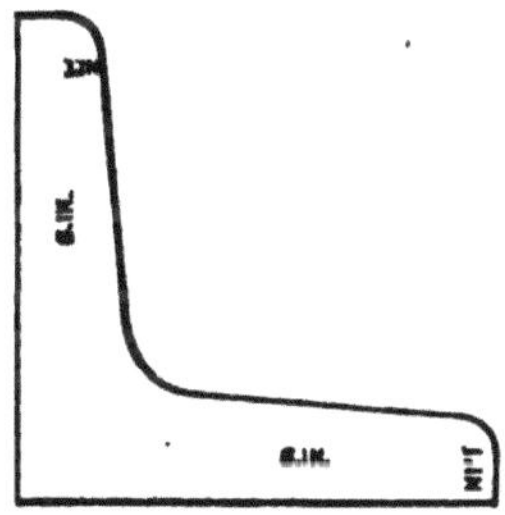

180.

Length of sides, 5 and 5; breadth, 1 in. Weight the lineal foot, 31·25 lbs.

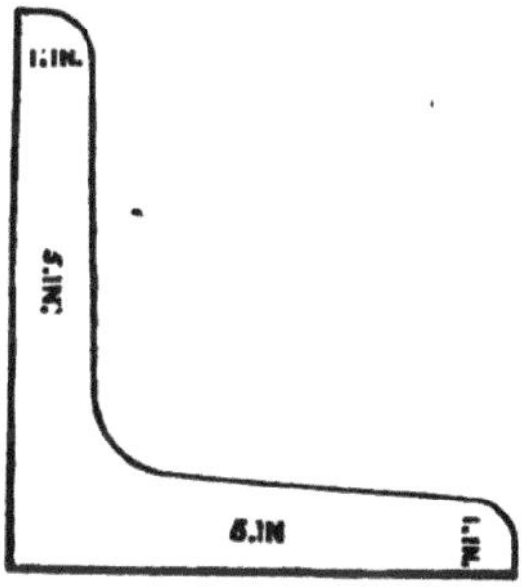

181.

Length of sides, 3 and 3; breadth, $\frac{3}{8}$ in. Weight the lineal foot, 7·5 lbs.

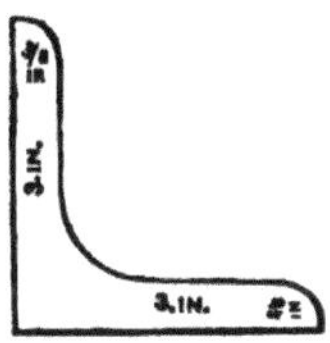

182.

Length of sides, 2½ and 2½; breadth, $\frac{5}{16}$. Weight the lineal foot, 5 lbs. 12 oz.

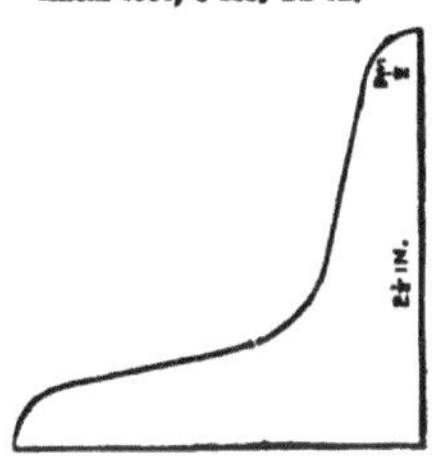

183.

Length of sides, 4 and 2, breadth, $\frac{1}{2}$ in. Weight of lineal foot, 9·5 lbs.

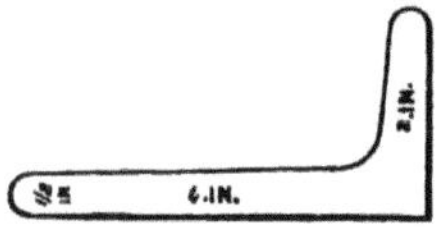

184.

Length of sides, $1\frac{5}{8}$ and $1\frac{5}{8}$; breadth, $\frac{3}{16}$ in. Weight of lineal foot, 2 lbs. $6\frac{1}{2}$ oz.

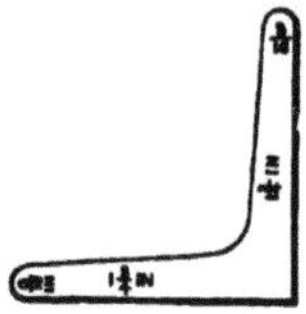

185.

Length of sides, $3\frac{1}{2}$ and 2; breadths, $\frac{3}{8}$ in. and $\frac{1}{2}$ in. respectively. Weight the lineal foot, 7 lbs.

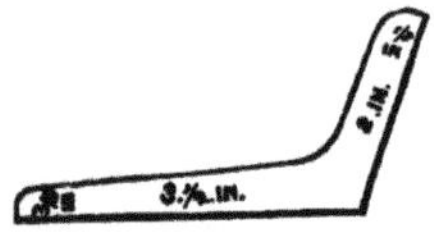

186. T-iron.

Length: web, 3 in.; flange, $2\frac{1}{2}$ in.; thickness, $\frac{7}{16}$. Weight the lineal foot, 7·25 lbs.

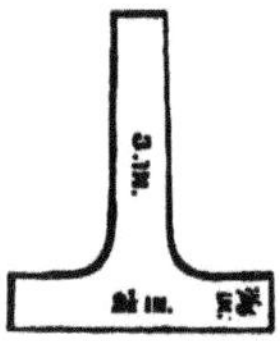

187. Section.

Weight the lineal foot, 3 lbs.

188. T-iron.'

Length: web, $2\frac{1}{2}$; flange, 4; breadth, $\frac{3}{8}$ in. Weight the lineal foot, 8·5 lbs.

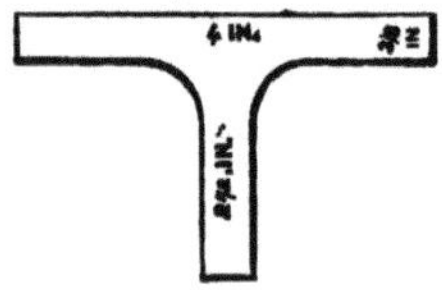

189. T-iron.

Length: web, 3; flange, 6; breadth, $\frac{1}{2}$ in. Weight the lineal foot, 15 lbs.

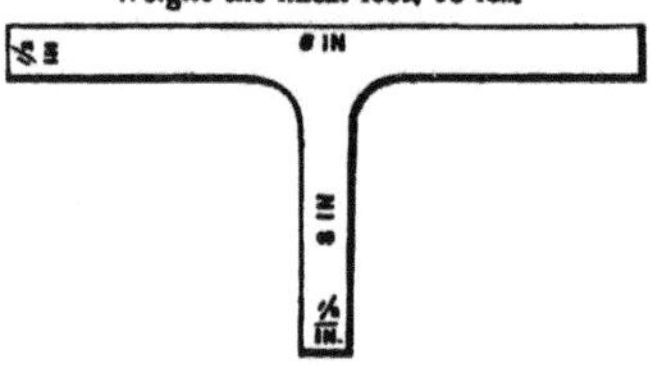

190. T.

Length: web, 4; flange, 5; breadth, $\frac{1}{2}$ in. Weight the lineal foot, 14·5 lbs.

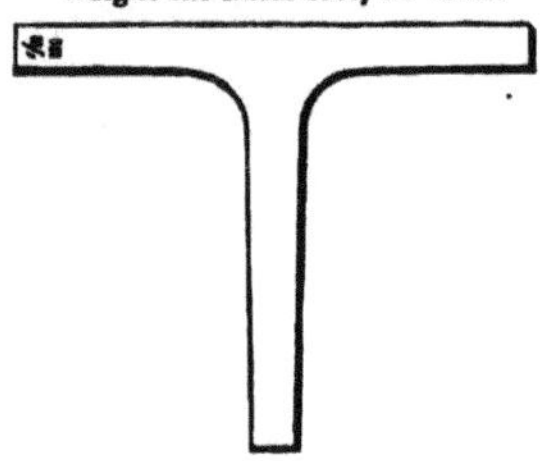

191. T. Section.

Length: web, 5; flange, 4; breadth, $\frac{1}{2}$ in. Weight the lineal foot, 16 lbs.

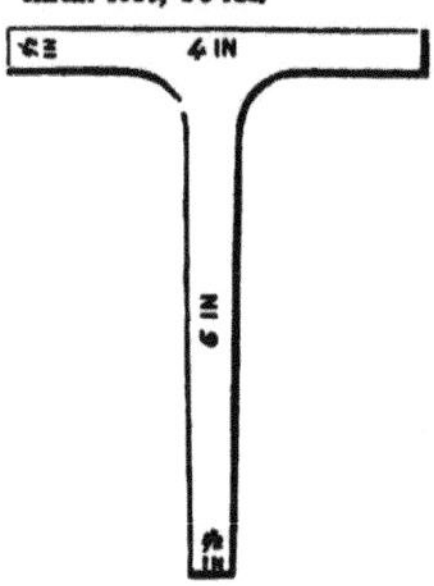

192. Section.

Length of each arm, 2 in., web, $2\frac{1}{4}$; breadth, $\frac{7}{16}$ in. Weight the lineal foot, 8 lbs. 2 oz.

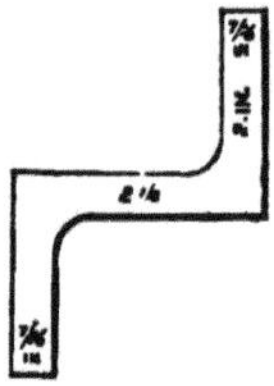

193. Section.

Length of each arm, 2 in.; breadth varying uniformly from ·4 in. to ·25 in. Weight the lineal foot, 5·75 lbs.

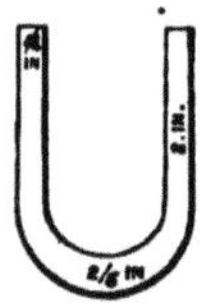

194. Section.

Length of each arm, ⅞ in.; web, 2·75 in.; breadth varying from ⅛ in. to ⅜ in. Weight the lineal foot, 6 lbs. 14 oz.

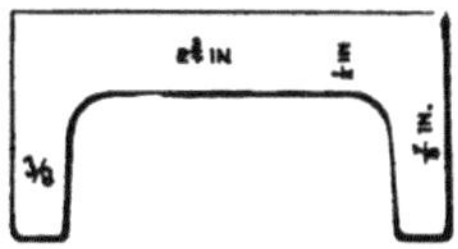

195. Section.

Length of each arm, ½ in.; least breadth, ¼ in.; web, 1½ in.; ⅛ in. thick. Weight the lineal foot, 1 lb. 6 oz.

196. Section.

Length of each arm, ⅜ in.; web, 1 in.; breadth, ⅛ in. Weight the lineal foot, 11 oz.

197. Section.

Length, 1⅜ in.; breadth, ⅜ in. Weight the lineal foot, 3 lbs. 15⅜ oz.

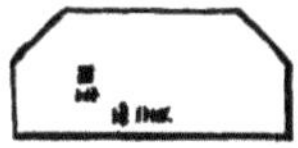

198. Section.

⅛ in. thick; the other dimensions are given on the section. Weight the lineal foot, 1 lb. 5 oz.

199. Section.

$\frac{1}{16}$ in. thick; the other dimensions are given on the section. Weight the lineal foot, 5 lbs. 1½ oz.

200. Section.

¾ in. thick. Weight the lineal foot, 2 lbs.

201. Section.

Length, 4 in.; breadth varying from ⅛ in. to ⅜ in. Weight, 9 lbs.

202. Section.

⅜ in. thick; chord, 2⅞ in., as shown on the section. Weight the lineal foot, 3 lbs. 4 oz.

ANGLE-RIDGE. FR., *Arêtier*; GER., *Gradsparre.*

Angle-rafter, more commonly called "hip-rafter," in hipped roofs is the piece of timber which runs from the angle of the building to the ridge of the roof into which it is framed, and of which it forms the continuation to the eaves, where it butts on the *dragon-piece*. It is usually from 1½ inch to 2 inches thick, and from 6 inches to 8 inches deep, according to the length; the ends of the jack-rafters are nailed to it in the same manner as the common rafters are nailed to the ridge.

ANGLE OF REPOSE. FR., *Angle de repos*; GER., *Ruhewinkel*; ITAL., *Angolo limite d'attrito.* SPAN., *Angulo de reposo.*

See ANGLE OF FRICTION.

ANGLE-STAFF. FR., *Corniche de cornière*; GER., *Winkel Karniess*; SPAN., *Guarda-arista.*

The strips of wood employed in the inside of buildings, upon the exterior vertical angles, to protect the plastering, are called angle-staffs.

Angle-staffs are of two kinds, namely, square staffs and round staffs, also called *angle-beads*, the former being mostly employed when the walls are papered over, and the latter when the angles are seen.

ANGLE-TIE. FR., *Attache angulaire*; GER., *Winkel Strebe*; ITAL., *Calastrello d'angolo, Traversa*; SPAN., *Tirante.*

See ANGLE-BRACE.

ANGLE OF TRACTION. FR., *Angle de traction*; GER., *Zug-Winkel*; ITAL., *Angolo di trazione*; SPAN., *Angulo de tràccion.*

The angle formed by the inclination of the traces with the surface of the roadway is termed the angle of traction.

ANGULAR MOTION, OR VELOCITY. FR., *Mouvement angulaire*; GER., *Winkelbewegung oder Winkelgeschwindigkeit*: SPAN., *Velocidad angular.*

The velocity of a point moving in a circle, whose radius is taken as a unit, is measured by the length of the arc that may be described by the point in a given time; but this arc measures an angle, and that angular measure is termed the angular velocity. As it is a part of our design to generalize and follow principle wherever it lead us, when such generalization tend to important practical results; hence, we purpose to explain angular velocity in general terms.

In calculating the motions of geared machinery, the angular velocity of a body has often to be determined when the number of rotations (n) in a minute is given. The distance traversed by the point situated 1 foot from the axis is 2π for each revolution, consequently $2n\pi$ for n revolutions, that is, $2n\pi$ in 1 minute, or 60 seconds. The distance traversed by this point in 1 second is therefore $\frac{2n\pi}{60}$ or $\frac{n\pi}{30}$, which is the angular velocity required. If we call this angular velocity ω, $\omega = \frac{n\pi}{30}$; that is, *to obtain the angular velocity, multiply the number of revolutions a-minute by the ratio of the circumference to the diameter, and divide the product by* 30. If, for instance, a wheel makes 45 revolutions a-minute, its angular velocity is $\omega = \frac{45 \times 3 \cdot 1416}{30} = 4 \cdot 7124$. From this formula is deduced $n = \frac{30\,\omega}{\pi}$; that is, *to find the number of revolutions a-minute, multiply the angular velocity by* 30, *and divide by the ratio of the circumference to the diameter.* For instance, if $\omega = 5$ $n = \frac{5 \times 30}{3 \cdot 1416} = 47 \cdot 7$, or about 48 revolutions a-minute.

The motion of rotation signifies the movement which all points of a single body make, in describing arcs or circles round the same axis, when the planes are perpendicular to that axis, and retain the same distance from each other.

Such is the movement of a cog-wheel or fly-wheel, and of all revolving parts of machines.

The first property peculiar to this species of motion is, that *all points of a body describe in the same time arcs of the same number of degrees.* Let us suppose the axis of rotation to be perpendicular to the plane of Fig. 203, and let O be the point where it meets this plane. Let us take any point; for instance, A; let a be the projection, upon the plane of the figure, of the initial position of the point A, and let a' be the projection of its position at the end of the time t; the arc described by the point A will be equal to the arc $a\,a'$, having O for its centre. Let B be another point of the body; b, the projection of its first position; b', the projection of its final position; and $b\,b'$, the projection of the arc, which it has described. Unite Oa, Oa', Ob, Ob'. Oa' will be the position of A with regard to the axis after the space of time t and Oa before the first instant. From the point B to the axis also draw a line. Ob will be its position before the first instant, and Ob' after the lapse of the time t. Therefore the points A and B will maintain their relative positions; the angle formed by the two planes drawn from A and B remains unchanged. But this angle measures at the first instant $a\,\mathrm{O}\,b$, and at the final instant $a'\,\mathrm{O}\,b'$; therefore these two angles are equal. If we cancel the common part $a'\,\mathrm{O}\,b$, there remains $a\,\mathrm{O}\,a' = b\,\mathrm{O}\,b'$. Therefore the arcs $a\,a'$ and $b\,b'$ are the arcs corresponding to the equal angles at the centre, which have the same number of degrees. The same may be said of all points of the body, as A and B are taken indiscriminately.

203.

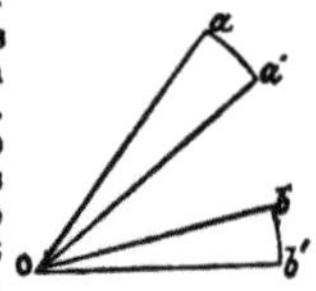

It therefore follows that *the velocities of different points of the body, at the same instant, are proportional to their distances from the axis of rotation.* Suppose, for instance, that the arcs $a\,a'$ and $b\,b'$ have been described in a very short time $\Delta\,t$. The similitude of these arcs gives the proportion $a\,a' : b\,b' = \mathrm{O}\,a : \mathrm{O}\,b$, from which is derived $\frac{a\,a'}{\Delta\,t} : \frac{b\,b'}{\Delta\,t} = \mathrm{O}\,a : \mathrm{O}\,b$. $\frac{a\,a'}{\Delta\,t}$ and $\frac{b\,b'}{\Delta\,t}$ show respectively the velocity of the points A and B. By calling the distances Oa and Ob, r and r', and the velocities v and v', the result will be, $v : v' = r : r'$.

II. To determine completely the motion of a body turning round an axis, it is sufficient to know the motion of one of its points. The motions of all points of a body are generally compared to that of one particular point, 1 foot, or 1 mètre, distant from the axis. The motion of this point is generally called, as previously stated, the *angular velocity*, and is usually expressed by the letter ω. On comparing the velocity of any point A to that of the point situated at the unity of distance from the axis, we get, in accordance with the rule explained above, $v : \omega = r : 1$ or $v = \omega\,r$; that is, *the velocity of any point of a body is equal to the angular velocity multiplied by the distance of that point from the axis of rotation.* If, for instance, the *angular velocity* be 3, the velocity of a point, situate at 0·40 ft. from the axis, would be $v = 0 \cdot 40$ ft. $\times 3 = 1 \cdot 20$ ft., from which is drawn the formula $\omega = \frac{v}{r}$; that is, *the angular velocity is obtained by dividing the velocity of any point by the distance of that point from the axis.* For instance, let us seek the angular velocity of the earth. R is the radius of the equator; the velocity of a point situated on this circle is therefore $\frac{2\pi\,\mathrm{R}}{86400}$, there being 86400 seconds in 24 hours, the time employed by the point in one revolution. Divide this velocity by the distance R of the point under consideration from the axis, and the result will be the angular velocity $\omega = \frac{2\pi}{86400} = \frac{3 \cdot 1415926}{43200} = 0 \cdot 000072722$; consequently $r = \frac{v}{\omega}$; that is, *the distance of any point of a body from the axis is the quotient of the velocity of the point by the angular velocity.* If, for instance, $v = 1 \cdot 50$ ft. and $\omega = 0 \cdot 8$ ft., then $r = \frac{1 \cdot 50 \text{ ft.}}{0 \cdot 8} = 1 \cdot 875$ ft.

When the angular velocity is constant, the rotary motion is uniform.

If the arc described by a point situate 1 foot, or 1 mètre, from the axis be a at the beginning of the first instant, and t the time employed in traversing that arc, $a = \omega\,t$.

If the angular velocity be variable, the limit between the *angular velocity* and the increase of the time is called *angular acceleration.* This *angular velocity* is generally represented by $\gamma = \frac{d\,\omega}{d\,t}$; and as ω is itself derived from the arc a, the acceleration is the second derivative of that arc;

$d\,\omega$ being put for an extremely small quantity termed the differential of ω, and $d\,t$ for differential of t, which is an extremely short time.

When the angular acceleration is constant, $d\,\omega = \gamma\,d\,t$.

If ω_0 be the primitive angular velocity, we obtain $\omega = \omega_0 + \gamma\,t$, whence $d\,a = \omega_0\,d\,t + \gamma\,t\,d\,t$.

Supposing the arc a to begin at the position occupied by the point at starting, by integrating, we find $a = \omega_0\,t + \frac{1}{2}\,\gamma\,t^2$. In these two proportions, the formulas of *uniformly variating motion* are expressed.

When several rotatory movements are to be considered, a geometrical sign is used to represent the *angular velocities*. On the axis of rotation an arbitrary point O, Fig. 204, is taken, and from this point is projected a length O P proportional to the angular velocity ω. O P is placed in such a direction, that a spectator, placed at P and looking in the direction P O, sees the body turning in the direction of the hand of a watch; that is, if O P were horizontal, the movement above would be from left to right. O P′ would represent the angular velocity if the movement were in the opposite direction, namely, from left to right when passing underneath.

204.

It is therefore said that the angular velocity, or the rotation, *is represented in quantity and direction* by the line O P in the first instance, and O P′ in the second.

When a solid body revolves round an axis, through the action of any force, the reactions upon that axis can be determined. Let O Z, Fig. 205, be the axis of rotation connected with the points A and A, and with its extremity B, which rests against a perpendicular, fixed plane. From the centre of gravity G of the body pass a plane perpendicular to the axis, and cutting it at the point O; take the point O for the origin of the axes of the co-ordinates of the different points, the axis O Z for the axis of z, and two axes O X and O Y perpendicular to the first. Let M be any point of the body, m its mass. From M draw upon the axis the perpendicular M C = r, and let a be the angle it makes at the end of the time t, with a parallel C V to the axis x; and at the end of the same time let M P = x, C P = y, and P Q = z, be the co-ordinates of the point M. This point, rotating round O Z, describes a circle having C for centre. It may therefore be considered as being acted upon by a tangental force T and by a normal force N, the powers of which are respectively $T = m\,r\,\frac{d\,\omega}{d\,t}$ and $N = m\,\omega^2\,r$, if the angular velocity be ω at the end of the time t. As the same may be said of all other points of the system, it follows that the body may be regarded as entirely subject to forces analogous to T and to forces analogous to N. The body is therefore moved in reality by a system of forces F, F′, F″, &c., the reactions of supports and the mutual action exercised by the material points of the body. The system of forces F and of the mutual actions is therefore equivalent (see *Equivalent* FORCES) to the system of forces analogous to T and N. There must therefore be equality between the sum of the projections of these two systems of forces on each of the three axes, and between the sum of their momentums with regard to these axes.

205.

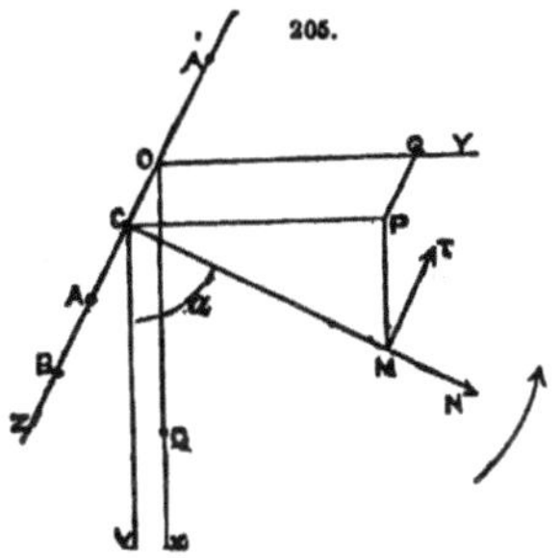

The projections of the force T on the axes of x, y, and z, are respectively

$$-\,T \sin. a,\ +\,T \cos. a,\ \text{zero};\quad \text{or},\ -\,m\,y\,\frac{d\,\omega}{d\,t},\ \ +\,m\,x\,\frac{d\,\omega}{d\,t},\ \ \text{zero}.$$

The relation of the momentums to the same axes is, adopting the conventional signs with regard to momentum,

$$-\,T \cos. a.z,\ \ -\,T \sin. a\,z,\ \ +\,T\,r;\quad \text{or},\ -\,m\,x\,z\,\frac{d\,\omega}{d\,t},\ \ -\,m\,y\,z\,\frac{d\,\omega}{d\,t},\ \ +\,m\,r^2\,\frac{d\,\omega}{d\,t}.$$

The projections of the force N are,

$$+\,N \cos. a,\ \ +\,N \sin. a,\ \ \text{zero};\quad \text{or},\ +\,m\,\omega^2 a + m\,\omega^2 y,\ \ \text{zero}.$$

The momentums of the same force with regard to the same axes are expressed,

$$-\,N \sin. a.z,\ \ +\,N \cos. a.z,\ \ \text{zero};\quad \text{or},\ -\,m\,\omega^2\,y\,z,\ \ +\,m\,\omega^2\,x\,z,\ \ \text{zero}.$$

In passing from M to another point of the body, x, y, z, r, and a, will change; but ω and $\frac{d\,\omega}{d\,t}$ will remain the same. On the other hand, the mutual forces will disappear when the sum of their projections on to any axis is nought, as they are equal and opposed to each other; the sum of their momentums, with regard to an axis, will also be nought, for the same reason.

Let R be the reaction exercised upon the point A perpendicular to the axis, omitting friction, and R′ the same reaction upon A′; and S the reaction exercised by a point B in the direction of the axis. O A = h and O A′ = h'; we can then describe the six conditions of equivalence; putting Σ for sum, $\Sigma\,m\,x$ signifies the sum of the products $m\,x$, then,

$$\Sigma\,F_x + R_x + R'_x = -\,\frac{d\,\omega}{d\,t}\ \Sigma\,m\,y + \omega^2\,\Sigma\,m\,x, \qquad [1]$$

$$\Sigma\,F_y + R_y + R'_y = +\,\frac{d\,\omega}{d\,t}\ \Sigma\,m\,x + \omega^2\,\Sigma\,m\,y, \qquad [2]$$

$$\Sigma\,F_z + S = 0, \qquad [3]$$

$$\Sigma\,\mathfrak{M}_x\,F - R_y\,.\,h + R'_y\,.\,h' = -\,\frac{d\,\omega}{d\,t}\ \Sigma\,m\,x\,z - \omega^2\,\Sigma\,m\,y\,z, \qquad [4]$$

$$\Sigma \mathfrak{M}_y F + R_x h - R'_x . h' = - \frac{d\omega}{dt} \Sigma myz + \omega^2 \Sigma mxz, \qquad [5]$$

$$\Sigma \mathfrak{M}_z F = \frac{d\omega}{dt} \Sigma mr^2. \qquad [6]$$

The last equation gives $\frac{d\omega}{dt} = \frac{\Sigma \mathfrak{M}_z F}{\Sigma mr^2}$; that is, *the angular acceleration is expressed by the sum of the momentums of the forces with regard to the axis of rotation, divided by the vis inertia of the body with regard to the same axis*, which expression can also be arrived at in a shorter manner. The first five equations make known the five unknown quantities, R_x, R_y, R'_x, R'_y, and S; from the equations [1] and [5] R_x and R'_x may be found, and from [2] and [4] R_y and R'_y are deduced. Equation [3] will give the value of S.

If the movement is reckoned only for the first instant (it can also be another space of time), the axis of x can be passed through the centre of gravity. If we call M the total mass of the body and a the distance of the centre of gravity from the axis O Z, $\Sigma mx = Ma$ and $\Sigma my = 0$.

If, again, we suppose the body symmetrical with regard to the plane X O Z, we get $\Sigma mxz = 0$ and $\Sigma myz = 0$; and the six equations of the problem become

$$\Sigma F_x + R_x + R'_x = M\omega^2 a,$$

$$\Sigma F_y + R_y + R'_y = Ma\frac{d\omega}{dt}, \quad \Sigma_z + S = 0,$$

$$\Sigma \mathfrak{M}_x F - R_y h + R'_y h = 0,$$

$$\Sigma \mathfrak{M}_y F + R_x h - R'_x h' = 0,$$

$$\Sigma \mathfrak{M}_z F = \frac{d\omega}{dt} \Sigma mr^2.$$

The quantities $Ma\frac{d\omega}{dt}$, $M\omega^2 a$ express the tangental force and the normal force of the centre of gravity, considered as a material point where all the mass M is concentrated.

Two particular cases deserve consideration:—1. Where the axis of rotation is horizontal, the body has its centre of gravity in that axis, and is subject to no external force except that of gravitation. 2. When the axis of rotation is vertical, and the body, having its centre of gravity in the axis, is only subject to the force of gravitation and to forces acting horizontally.

1. In the first case we get

$$a = 0, \Sigma F_x = P, \Sigma F_y = 0, \text{ and } \Sigma F_z = 0,$$

$$\Sigma \mathfrak{M}_x F = 0, \Sigma \mathfrak{M}_y F = 0, \Sigma \mathfrak{M}_z F = 0,$$

$$R_x + R'_x = -P, R_y + R'_y = 0, S = 0.$$

The sixth equation gives $\frac{d\omega}{dt} = 0$, from which it follows that the movement is uniform; the equations [4] and [5] can be reduced to

$$- R_y . h + R'_y h' = - \omega^2 \Sigma myz,$$

$$R_x . h - R_x . h = + \omega^2 \Sigma mxz;$$

or, if the plan X O Z were symmetrical,

$$R_y . h = R'_y h' \text{ and } R_x . h = R'_x . h'.$$

The first of these, compared with the second equation, gives $R_y = 0$ and $R'_y = 0$; that is, the reactions at A and A' are then vertical.

2. In the second case suppose the axis O Z, Fig. 206, to be vertical, and the centre of gravity to be situated in the axis O X at the first instant, P to be the weight of the body, and that we have as above $\Sigma my = 0$ and $\Sigma mx = Ma$, we find

$$R_x + R'_x = \omega^2 Ma,$$

$$R_y + R'_y = Ma\frac{d\omega}{dt},$$

$$- P + S = 0,$$

$$- R_y . h + R'_y . h' = - \frac{d\omega}{dt} \Sigma mxz - \omega^2, \Sigma myz,$$

$$Pa + R_x . h - R'_x . h = - \frac{d\omega}{dt} \Sigma myz + \omega^2, \Sigma mxz,$$

$$\Sigma \mathfrak{M}_z F = \frac{d\omega}{dt} \Sigma mr^2.$$

206.

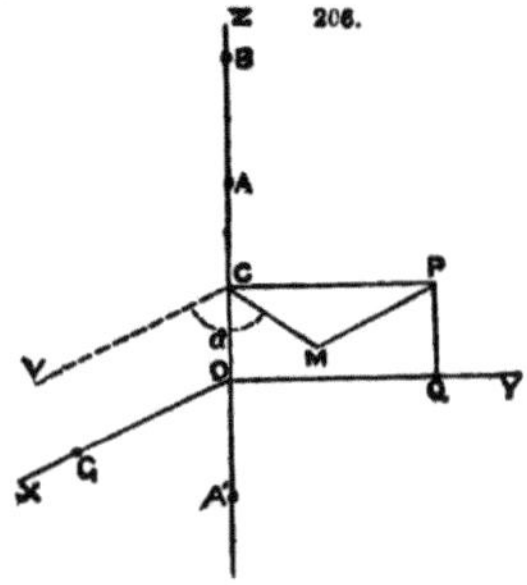

If we wish the reactions of R and R' to be nought, we must according to the two first equations, make $a = 0$, that is, the centre of gravity must be in the axis of rotation. The equation $- P + S = 0$ gives $S = P$; that is, the charge of the pivot which sustains the axis of rotation is equal to the weight of the body. The two following equations can be reduced to

$$\frac{d\omega}{dt} \Sigma mxz + \omega^2, \Sigma myx = 0,$$

$$\text{and } \frac{d\omega}{dt} \Sigma myz - \omega^2, \Sigma mxz = 0;$$

from which follows $\Sigma m x z = 0$ and $\Sigma m y z = 0$; that is, the axis of rotation is one of the principal axes of the body.

In millstones these two conditions are fulfilled by means of lead introduced into vertical holes made for that purpose. If this were not done, it might happen that, although remaining horizontal when at rest, they may, when in motion, exercise lateral reactions upon this axis, and cease to be horizontal.

General Conditions of the Uniformity of Motion or of Equilibrium of a Solid Body, Free in Space, and subjected to any Forces.—It is evident that a solid body, entirely free, can receive and take but one of the three following motions:—A motion of translation without rotation, a motion of rotation without translation, and a simultaneous motion of translation and rotation.

Every motion of translation may be resolved into three other motions similar in relation to any three rectangular axes drawn in space; and it is evident that if each of these component motions is separately zero, the resultant motion of translation will be so likewise. This condition is, moreover, necessary and sufficient.

Now, in order that these three motions shall be zero for each of these axes, the sums of the components parallel to the axes should separately be zero. Then, if we call X, Y, and Z, the sums of the components of the exterior forces applied to an invariable solid, these forces cannot impart a motion of translation if we have at the same time $X = 0$, $Y = 0$, $Z = 0$, and the motion will remain uniform or the body be in equilibrium as to translation.

So also every motion of rotation of a body, or of material points composing it, may be resolved into three motions of rotation around three rectangular axes drawn through any point. In order that the body shall receive no motion of rotation, it is only requisite that the rotations around each of the three axes shall be separately zero, which requires the sums of the moments (not momentums) of forces in relation to each of the three axes, that is, the sums of the masses or weights multiplied by their perpendicular distances, to be separately zero, so that if we call L, M, and N, these three sums, we must have at the same time $L = 0$, $M = 0$, $N = 0$. When these conditions are satisfied, the work developed in imparting a motion of rotation will be zero, and it will continue to move uniformly or will rest in equilibrium.

In order that the body receive no motion of translation, nor of rotation, or that its motion be in no wise altered, all that is requisite is: 1st. That the sum of all the components of the forces soliciting the body, in relation to any three rectangular axes, shall be separately zero. This is expressed by the relations

$$X = 0,\ Y = 0,\ Z = 0,$$
$$L = 0,\ M = 0,\ N = 0,$$

which we call the six equations of uniform motion, or the equilibrium of an invariable body, free and solicited by any forces.

Centrifugal Force.—It is well known that, if we tie a stone or other heavy body to a cord, impress it with a circular motion of which the hand is the centre, the cord will experience a tension, the greater as the motion is more rapid. From observation of this fact came the use of the sling as an implement of war among the ancients, and which is now but a boy's play. Similar effects are seen in waggons running swiftly in short curves, in circuses, when the horses and riders are naturally induced to lean towards the centre of the curves they describe to prevent being overthrown. The reader may readily find other effects from the same cause: all of them prove that in curvilinear motion the bodies are subjected to a peculiar force tending to drive them from the centre, which force is called the *centrifugal force.*

Measure of the Centrifugal Force.—To understand what takes place when a material point is submitted to the action of the centrifugal force, let us examine, first, how this force is developed in circular motions. When a material point or an elementary mass m passes from one element of a curve which it describes to another, it tends by virtue of its inertia to continue its motion in the direction of the prolongation of this element, or of the tangent $b\,d$ of the curve, and is what is termed *flying off at a tangent*, as is the case with the sling at the moment one suddenly lets go his hold upon the cord. If the mass m takes the direction of the next element, it is then retained upon the curve, either by the resistance of the curve itself, upon which it then exerts a pressure, or by the tension which it develops in the cord. This pressure or tension is itself the measure of the centrifugal force, in contradistinction to which it is sometimes called the centripetal force.

This force is in the direction of the radius of the curve or of the corresponding circle, and if we call V the velocity with which the mass m is impressed in the direction of $a\,b$, and take the length $b\,d$ to represent it, it is clear that the velocity destroyed by the resistance of the cord or the centripetal force, will be represented by the side $d\,c$ of the parallelogram $b\,c\,d\,f$, whose side $d\,c$ is parallel to the radius $o\,b$, in the direction of which this force is exerted. Now, an inspection of the figure shows that the angles $a\,b\,O$ and $b\,d\,c$ are equal as internal and external, and the angles $d\,c\,b$ and $c\,b\,O$ as alternate and internal; and as moreover the angles $c\,b\,O$ and $a\,b\,O$ being formed on both sides of the radius by two equal and consecutive elements of the circle or of the polygon whose infinite number of sides replace it, it follows that the angles $b\,d\,c$ and $d\,c\,b$ are equal, and the triangle $b\,d\,c$ is isosceles. Then the velocity $b\,a$ with which the mass m is moved in the direction of the following elements $b\,t$, is the same as that it had in the direction of the preceding element. Thus, *in circular motion, the centrifugal force does not alter the velocity of rotation;* which is conformable with the *principle of work*, since this force, in the direction of the radius, or normal to the path described, produces no work in the direction of motion, so long as there is no path described in its own direction and by its action. This being settled, the velocity destroyed in the element of time t by the centripetal force has, according to the figure,

207.

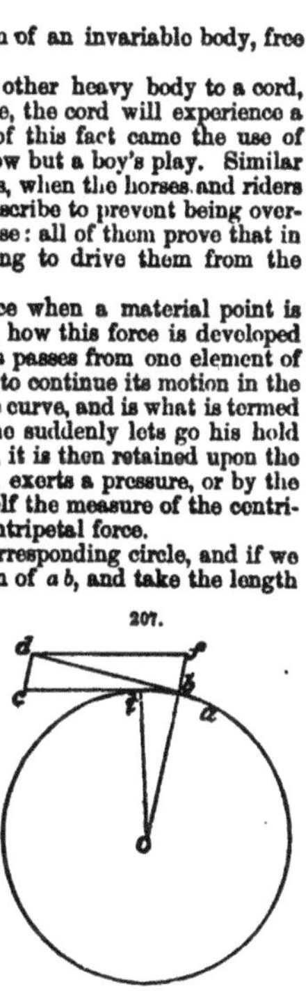

dc for its measure, and the centripetal and centrifugal forces, which are equal and directly opposite, have for a common measure $F = m \cdot \frac{c}{t}$.

Now, the triangle bdc and Obt having equal angles, are similar; we have then $bO : bt :: bd : dc$, whence $dc = \frac{bd \times bt}{bO} = \frac{Vs}{R}$. In calling R the radius of the circle described, and s the elementary arc run over in the element of time t; and as we have $V = \frac{s}{t}$ or $s = Vt$, it follows that $dc = \frac{V \times Vt}{R} = \frac{V^2 t}{R}$; and finally, that the centrifugal force has for its measure $F = \frac{mV^2 t}{R \cdot t} = m\frac{V^2}{R}$; if, moreover, we call V_1 the angular velocity, or that at the unit of distance, we have $V = V_1 R$, and the expression for the centrifugal force becomes $F = m\frac{V_1^2 R^2}{R} = mV_1^2 R$.

What we have said of the centrifugal force applies to a material point describing any curved line, since in each of its positions an osculating circle may be substituted for the curve; the only difference being in the fact that the radius R of this circle varies for each position of the moving body, while that in the circle is constant.

Work Developed by the Centrifugal Force.—When instead of being retained by a circular curve or at a constant distance from the centre of rotation, the material point is removed farther from it, the centrifugal force will cause it to describe a certain path in the direction of the radius; it develops upon this body a work easily appreciated. In fact, if in an element of time the material point is displaced in the direction of the radius by a certain elementary quantity V, the corresponding work of the centrifugal force will be $Fr = mV_1^2 R r$, and the total work due to this force, when the material point shall have passed from R″ to R′ at a greater distance from the centre, will be given by the sum of all the analogous elementary works taken from R = R″ to R = R′. Now we see from *the principle of work* that this sum is equal to $\frac{1}{2} mV_1^2 (R'^2 - R''^2) = \frac{1}{2} m (V'^2 - V''^2)$ if we call $V' = V_1 R'$ and $V'' = V_1 R''$, the velocities of rotation of the point around the centre. We have then for the work of the centrifugal force $T = \frac{1}{2} mV_1^2 (R'^2 - R''^2) = \frac{1}{2} m (V'^2 - V''^2)$.

We remark that the second member of this relation is no other than the variation of the *vis viva* of rotation, experienced by the material point while partaking of this motion in its removal from the centre of rotation, whatever may be the curve or path described in this removal. This expression then could be directly deduced from the principle of *vis viva*.

In the case just considered, the centrifugal force tends to increase the absolute velocity of the body moved, and acts thus as a motive force which is developed in the motion of rotation. When, on the other hand, the body approaches the centre, the centrifugal force is opposed to it, and acts as a resistance in developing a work having indeed the same expression, but which is resistant, since the path described is in a direction contrary to the action of the force.

The preceding considerations will find their application in the study of the effects of certain hydraulic receivers.

Action of the Centrifugal Force upon Waggons.—When a coach with great speed turns upon a short curve, the effects of the centrifugal force is felt by the passengers, who are driven towards the outer curve with an intensity often dangerous for those placed on the outside, and which may even disturb the stability of the coach itself.

There is often a prejudice against the effects of this force upon railways, when it is proposed to use curves of small radius; but it is easily shown by figures, that in this regard the greatest velocities with the common radii of curves produce no danger.

208.

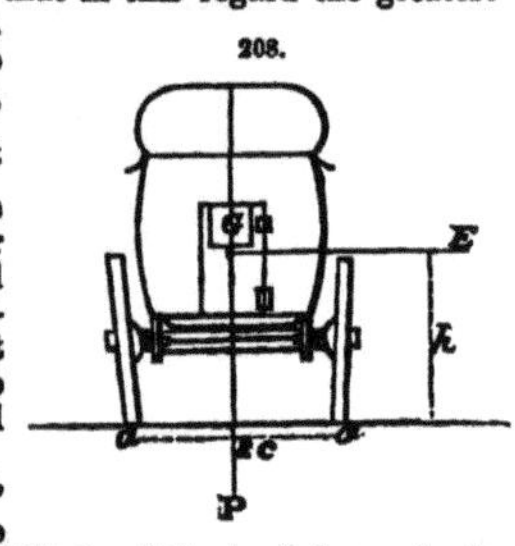

In fact, calling P the weight of the car or any carriage, h the height of its centre of gravity above the plane of the track, $F = \frac{P}{g} V_1^2 R$ the centrifugal force, $2c$ the width of the track. It is evident that when the car passes around the centre O of the curve, and is arrested by some obstacle, such as the falling or rising of the rail, it tends to upset outwards, in turning around the point a of instantaneous support. This motion is counterbalanced by the weight P of the carriage, and at the moment when the weight and centrifugal force are in equilibrium as to the point, we have between the moments of the two forces P and $F = \frac{P}{g} V_1^2 R$ the relation $Pc = \frac{P}{g} V_1^2 R h$, which shows that, with equal velocities and weights, the stability of the car will be so much the greater, and the equilibrium better secured, as the width $2c$ of the track is greater in its ratio with the height of the centre of gravity. The velocity of transit answering to this equilibrium upon common tracks, for which $2c = 4{\cdot}75$ ft. with cars whose centre of gravity when loaded is 3·28 ft. in height, and with curves 1312 ft. radius, will be given by the relation $V_1^2 R = \frac{gc}{h}$, whence $V_1 R = \sqrt{\frac{gc}{h} R} = 174{\cdot}9$ ft., a velocity beyond the greatest speed of railroads. This shows that in this regard the centrifugal force occasions no danger. But we should not forget that it brings the flanges of the outer wheels to bear against the rails, producing a cutting away which wears them out and greatly contributes to their running off the track.

Action of the Centrifugal Force in Fly-wheels.—For regulating the irregularities of machines we make use of rotating pieces of considerable weight and diameter, impressed with quite a great velocity, upon which the motion of rotation develops a centrifugal force of considerable intensity. Thus, for example, the fly-wheel of an iron rolling-mill, established at the iron-works of Fourchambault, weighs 13,232 lbs., its radius is 9·58 ft., the number of turns it makes is 60 in 1′, or 1 per second.

We have thus $V_1 = 6 \cdot 28$ ft. in 1″, and consequently, $V_1 R = 6 \cdot 28 \times 9 \cdot 58$.

If we consider a segment of the ring equal to $\frac{1}{6}$ of its circumference, corresponding to a single arm, its weight will be 2205 lbs.; and if its connection with the adjoining segment is broken, the arm will experience, in the direction of its length, a traction expressed by $\frac{2205}{32 \cdot 1817} \times 6 \cdot 28^2 \times 9 \cdot 58 = 25{,}887$ lbs., which shows that in fly-wheels the centrifugal force acquires a dangerous intensity, and that it is well to give great solidity to their connections. The velocity of rotation of these machines should be confined within certain limits. If, for example, we were to impart to the above fly a double velocity, or 120 turns in 1′, the centrifugal force of the segment just considered would be four-fold, or equal to 103,548 lbs.

Application to the Motion of Water contained in a Vase turning round a Vertical Axis.—In this case the liquid molecules are simultaneously subjected to the vertical action of their own weight, and to a centrifugal force developed horizontally; in order that they shall be in equilibrium under the action of these two forces, it is requisite that the resultant of these two forces should be normal to the surface assumed by the fluid mass, for if this resultant was inclined to the surface, the molecules would yield to its oblique action.

Let us consider a molecule m with the weight p and mass $\frac{p}{g}$, situated at the distance $mp = R$ from the axis of rotation A C. In a horizontal direction and perpendicular to the axis, it will be impressed with a centrifugal force expressed by $\frac{p}{g} V^2 R$. Let us take $m B = \frac{p}{g} V^2_1 R$, $m D = p$, and construct the parallelogram m B E D, whose diagonal normal to the surface assumed by the fluid intersects the axis at i. The similar triangles $m p i$ and m B E give us $m B$ or $\frac{p}{g} V_1^2 R : BE$ or $p :: mp$ or $R : pi$, whence $pi = \frac{g}{V_1^2}$.

209.

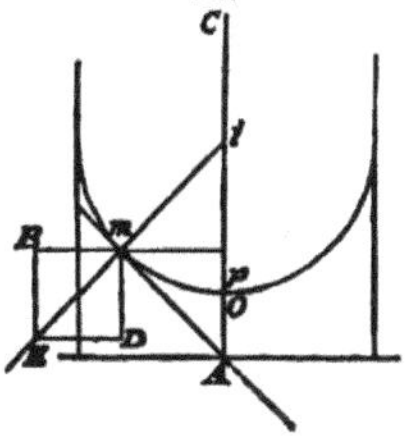

Thus the distance $p\,i$, which is called the subnormal, depends only upon the constant number g, and the angular velocity supposed also to be constant. Consequently this distance is constant, which, according to the known properties of the parabola, shows that the generating curve of the surface of the level is a parabola whose summit is at the point O, and whose axis is that of the rotation, and we readily see that its parameter is $\frac{2g}{V_1^2}$, so long as we have pp' or $2x : mp$ or $y :: mp$ or $y : pi$ or $\frac{g}{V_1^2}$, whence $y^2 = \frac{2g}{V_1^2} x$.

Surface of Water contained in a Bucket of a Hydraulic-wheel with a Horizontal Axle.—In following the reasoning of the preceding case, it is easy to see that, if we represent by ab the centrifugal force $\frac{p}{g} V_1^2 R$, and by ad the weight p of any molecule situated on the surface, we shall have the proportion ab or $\frac{p}{g} V_1^2 R : bc$ or $p :: R : OI$, whence $OI = \frac{g}{V_1^2}$; which shows that the distance O I is constant for all points of the surface of the liquid, and that consequently this surface is that of a cylinder, with a circular base of radius a I, whose axis is parallel to that of the wheel. This theorem, for which we are indebted to M. Poncelet, serves as the basis of the theory which this engineer has given upon the effects of water in bucket-wheels with great velocities.

210.

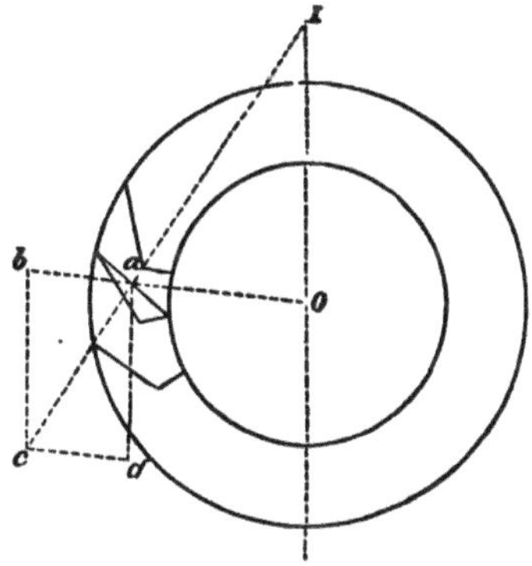

Regulators with Centrifugal Force.—The action of centrifugal force is utilised in the construction of an apparatus called a governor. It consists principally of a vertical spindle A H, Fig. 211, which receives from the machine to be regulated a motion of rotation. Upon this spindle are suspended two rods A P and A P′ jointed at A, and terminated by the equal weights or bobs P and P′. At the two joints B and B′ of the rods A P and A P′ are jointed two other equal rods B C and B′C, forming with the first a lozenge, and which at their ends C and C′ are also jointed with a collar traversed by the vertical spindle with which it turns, having at the same time a motion of translation in the direction of the length of this spindle. This collar has a yoke in which is fastened the fork of a lever D E, which acts upon the throttle-valves for steam, or upon any other piece.

The working of this contrivance is readily understood. By the effect of the rotatory motion of the vertical spindle, the balls of the regulator are thrown outwards from the axis, and so raise the collar a certain height. If the machine has attained and preserves its normal velocity, the balls and the collar are held in the same position, because there is established a state of equilibrium between the centrifugal force and the weights of the different parts of the apparatus. When the velocity increases, the centrifugal force increases, tending to spread outwards the balls and to raise the collar, and consequently the lever D E. Inversely, if the velocity diminishes, the balls approach the spindle; the collar and the end of the lever D E are lowered.

211.

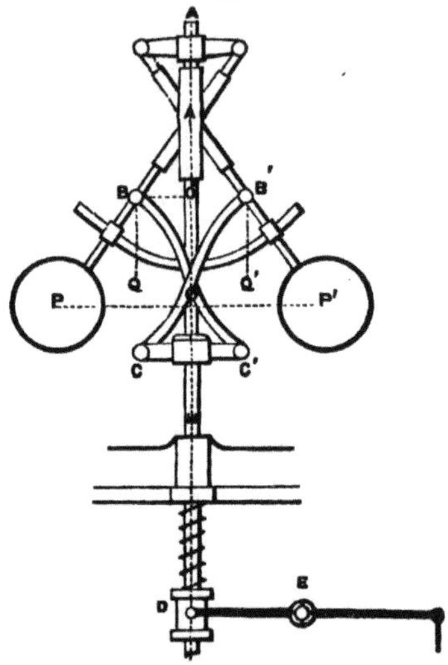

Let us examine the mechanical conditions of the action of this apparatus, and first suppose the collar C C', as well as the rods B C and B' C', to be in equilibrium with the lever D E, so that, neglecting friction, we may regard the rods A B and A B' as free to yield to the centrifugal force which tends to separate them, and to the weight of the balls which tends to bring them nearer to the spindle.

The centrifugal force of each ball is $\frac{P}{g} V_1^2 \times OP$, and its moment in relation to the axis of joints A is $\frac{P}{g} V_1^2 \times OP \times AO$. The moment of the weight P of each ball in respect to the same axis is $P \times OP$. Consequently the condition of equilibrium of each is $\frac{P}{g} V_1^2 \times AO = P$, whence $\frac{V_1^2}{g} = \frac{1}{AO}$; which shows that the distance of the balls' separation from the spindle depends not upon their weight, but solely upon the angular velocity of rotation, and enables us to so dispose of the weight of the balls as to satisfy other conditions. If we call T the time of the revolutions of the balls around the vertical spindle, we have $V_1 T = 2\pi = 6\cdot28$, whence $V_1 = \frac{2\pi}{T}$, and consequently $\frac{4\pi^2}{gT^2} = \frac{1}{AO}$, whence $T = 2\pi\sqrt{\frac{OA}{g}}$, which is double the duration of oscillations of a pendulum having for its height the height A O, at which the balls would be raised to the normal velocity.

The above formula enables us to determine approximately the height A O at which the balls are raised with a given velocity, and thus to establish their mean position. It gives, in fact, $AO = \frac{gT^2}{4\pi^2} = 0\cdot81517\,T^2$. Thus for $T = 1''$, $AO = 0\cdot81517$ ft.; $T = 2''$, $AO = 3\cdot2606$ ft.

In this calculation we have neglected the weight and the centrifugal force of the rods A B and A B'.

The preceding remarks are not sufficient to ensure the action of the pendulum as a regulating apparatus, since it is a requisite that it should be able to move the lever D E and the parts for the distribution of the steam or water upon which this lever operates, or in other terms, it should be able to overcome the resistances experienced in the motion of the collar, when the balls are separated or brought nearer to each other. These resistances can be estimated or measured when the apparatus is constructed, and if we call 2 Q the vertical force applied to the collar in the direction of the vertical spindle, $V'_1 = (1 + n')\,V_1$ another angular velocity, greater, for example, than the mean velocity V_1 by a fraction n' of the latter, it is easily seen that the force 2 Q can be resolved into two other forces parallel and equal to Q, applied at each of the joints B and B', and that then we shall have for the equilibrium corresponding to these new conditions, at the instant of its being broken, the relation $\frac{P}{g} V'^2_1 \times OP \times AO = P \times OP + Q \times BO'$. Calling a the distance $AB = AB'$ and b the length $AP = AP'$ of the rods to the centre of the balls, we remark that $b : a :: OP : BO'$, whence $BO' = \frac{a}{b}.OP$, and consequently $\frac{P V'^2_1}{g}.AO = P + Q.\frac{a}{b}$. We have previously found that the value of A O corresponding to the mean position of the balls was $AO = \frac{g}{V^2_1}$; the above relation becomes, then, $P\frac{V'^2_1}{V^2_1} = P + Q\frac{a}{b}$, whence we derive $\frac{P}{Q} = \frac{a}{b}\,\frac{V^2_1}{V'^2_1 - V^2_1} = \frac{a}{b\,(2n' + n'^2)} = \frac{a}{2\,b\,n'}$, as long as n'^2 is very small compared with n'.

We also see, then, from these considerations, due to M. Poncelet, that there exists a necessary relation between the ratio of the weight of the balls to the resistance and the degree of regularity of which the apparatus is susceptible.

We see, also, that for a degree of regularity desired or considered as necessary in the operation of the machine, the weight of the balls increases proportionally with the resistance which the collar opposes or experiences. Then, for example, if we have the proportions $a = 0\cdot666$, and if we have

$n' = \frac{1}{50} = 0 \cdot 02$, we find $\frac{P}{Q} = \frac{0 \cdot 66}{2 \times 0 \cdot 02} = 16 \cdot 5$, so that, if the resistance of the collar was only 22·05 lbs., the weight of each of the balls should be $P = 11 \cdot 03 \times 16 \cdot 5 = 181 \cdot 99$ lbs. This result shows that this apparatus cannot give a great degree of regularity to machines, without great dimensions and weights, if we would overcome, directly by the collar, considerable resistances.

It is from a disregard of these circumstances that many constructors have failed in the establishment of this kind of regulators, made for the purpose of raising sluice-gates, or in fixtures for the distribution of steam. This serious inconvenience may be avoided; and, with this simple and solid apparatus, we may obtain a proper regulation by arranging it in the manner which will be hereafter described. See GOVERNORS.

How to Estimate the Units of Work in a Rotating Body.—To take a simple case, let two balls, C and D, be connected by a rod, A D, and made to revolve round the centre, A; suppose C to weigh 50 lbs. and D 20 lbs.; the distance of C from A = 12 ft., and that of D = 27 ft. It is required to find the units of work in these balls when the point B, 1 ft. from the centre of motion A, moves at the rate of 19·3 ft. a-second.

212.

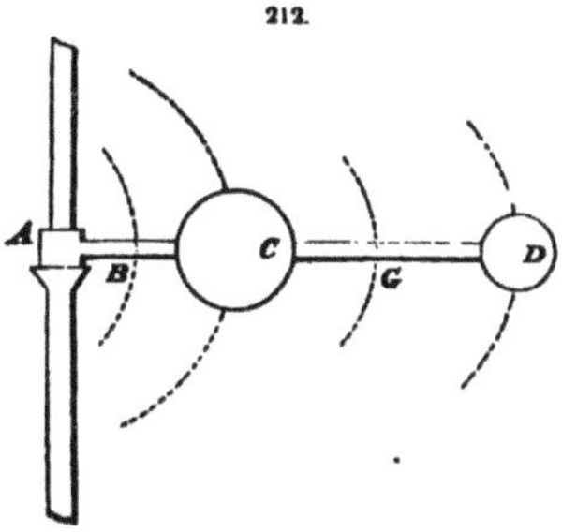

The *centre of gyration* is also required—that is, a point, G, in the rod where we may suppose the weight of the two balls collected—so that the amount of work may remain the same as when the bodies were apart.

Velocity of C = (19·3 × 12) feet a-second.

Velocity of D = (19·3 × 27) feet a-second.

Units of work in C $= \frac{(19 \cdot 3 \times 12)^2 \times 50}{64\frac{1}{3}} = 41688.$

Units of work in D $= \frac{(19 \cdot 3 \times 27)^2 \times 20}{64\frac{1}{3}} = 84418 \cdot 2.$

Total units of work = 126106·2.

Let x be the distance, A G, then the work in the two balls collected at G $= \frac{(19 \cdot 3\, x)^2 \times 70}{64\frac{1}{3}} = 126106 \cdot 2$. $\therefore x^2 = 311\frac{1}{4}$, and $x = 17 \cdot 639$ ft.

Now, if the two weights 20 + 50 = 70 lbs. be placed on the rod at the *centre of gyration* G, and move with a uniform velocity of (17·639 × 19·3) feet a-second, the amount of work in the bodies thus combined is the same as when posited at C and D.

From this simple case it is evident that when the *centre of gyration* of a rotating body is known, the accumulated work in that body is readily found. To find the *centre of gyration* in differently formed bodies requires the aid of a higher calculus, the introduction of which would be out of place in the present work. However, it is necessary to observe that the distance of this centre from the axis of rotation in a circular wheel of uniform thickness is equal to the radius of the wheel $\times \sqrt{\frac{1}{2}}$: in a rod revolving about its extremity it is equal to the length of the rod $\times \sqrt{\frac{1}{3}}$, and when the rod revolves about its centre it is equal to the length $\times \sqrt{\frac{1}{12}}$; and in a plane rim, like the rim of a fly-wheel, it is equal to the square root of one-half the sum of the squares of the radii forming the ring.

Question.—The weight of a fly-wheel = 8000 lbs., suppose the centre of gyration to be 10 ft. from the axis, the diameter of which = 14 inches; the wheel makes 27 revolutions a-minute; how many revolutions will it make before it stops, the friction of the axis being $\frac{1}{5}$ of the whole weight?

Velocity of centre of gyration a-second $= \frac{20 \times 3 \cdot 1416 \times 27}{60} = 28 \cdot 2744$ ft.

Work in the wheel $= \frac{(28 \cdot 2744)^2 \times 8000}{64\frac{1}{3}} = 99411 \cdot 77.$

Circumference of the axis in feet $= \frac{14}{12} \times 3 \cdot 1416 = 3 \cdot 6652.$

Work destroyed in x revolutions $= 3 \cdot 6652 \times x \times \frac{8000}{5} = 5864 \cdot 32\, x$. $\therefore x = \frac{99411 \cdot 77}{5864 \cdot 32} =$ 16·952 revolutions.

Question.—The weight of a fly-wheel is $1\frac{1}{2}$ ton, the distance of the *centre of gyration* from the axis = 8 ft., and the number of revolutions a-minute = 24; what number of strokes will this wheel give two forge-hammers, each weighing 250 lbs., each hammer having a lift of 3 ft., friction being neglected?

Velocity of the centre of gyration = 20·1 ft. a-second. Work in the wheel $= \frac{(20 \cdot 1)^2 \times 3360}{64\frac{1}{3}} =$ 22984·6.

Work of x lifts of the hammers $= 250 \times 2 \times 3 \times x = 1500\, x$. $\therefore 1500\, x = 22984 \cdot 6$. $\therefore x =$ 15·3 lifts to each hammer.

Question.—The diameter of a grindstone is 5·6 ft., and its weight = 386 lbs.; the circumference is made to revolve with a velocity of 6 ft. a-second; the circumference of the axis = 8 inches, the friction of it $= \frac{1}{4}$ of the weight: find the number of revolutions made by the stone when left to itself?

The *centre of gyration* from the axis of the grindstone $= 2 \cdot 8 \sqrt{\frac{1}{2}}$.

To find the velocity of the centre of gyration, $\frac{5 \cdot 6}{2} : 6 :: 2 \cdot 8 \sqrt{\frac{1}{2}} : 6 \sqrt{\frac{1}{2}}$. The square of

$6\sqrt{\frac{1}{4}} = 18$. Work in the stone $= \frac{18 \times 386}{64\frac{1}{4}} = 108$. Let x be the number of revolutions, then $\frac{8}{12} \times \frac{386}{7} \times x =$ the work destroyed in x revolutions. *Byrne's Essential Elements of Practical Mechanics.*

Animal Charcoal, Machinery for reburning.

ANIMAL-CHARCOAL MACHINE. Fr., *Machine à purifier le noir animal;* Ger., *Maschine zur Bereitung der Thierkohle;* Ital., *Macchina da carbone animale;* Span., *Máquina para purificar el carbon animal.*

The apparatus of J. F. Brinjes, of London, for reburning animal charcoal is shown in Figs. 213, 214, 215, 216.

213.

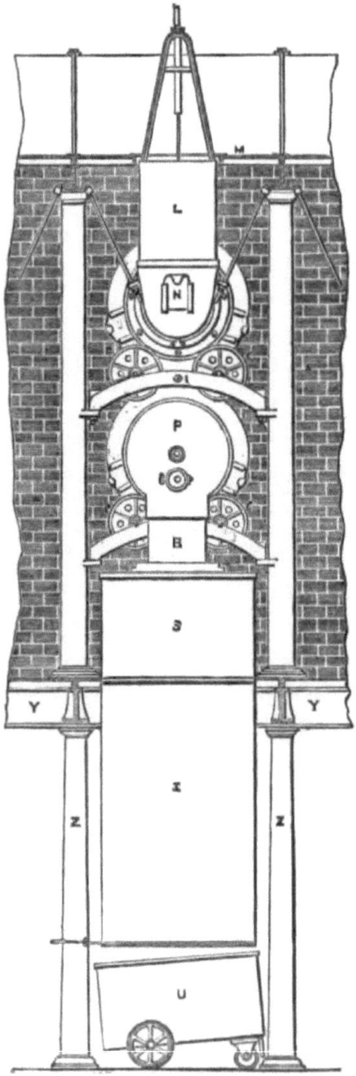

215.

Fig. 213 represents a front elevation of the apparatus. Fig. 214 is a sectional elevation of Fig. 213. Fig. 215 is a back elevation, and Fig. 216 is a section at the back of the apparatus. A is the brick-setting of the horizontal retorts, B, and C, each of which receives a circular reciprocating motion of nearly one entire revolution on its longitudinal axis. The upper retort, which acts as a drying-chamber for preparing the charcoal for the re-carbonization which takes place in the lower retort, is contained in a separate brick-chamber of its own, which is situated immediately above the roof of the furnace or fireplace D, the heat from which, after circulating round the lower retort, enters the upper chamber through openings left for that purpose in the roof of the furnace, and then acts upon the upper retort before passing off to the chimney. E E are passages provided with dampers, and leading to the main flue, F, below. The two retorts are provided with a series of internal flanges, a, a, at intervals of about 6 or 8 inches, and ledges are formed between the flanges for carrying up the charcoal as the retorts reciprocate. An opening is made through each flange, and all these openings are disposed in a line with each other.

In order to cause the charcoal to travel continuously along the retorts during the process of re-carbonizing, an angled projection, somewhat after the form of a three-sided pyramid, *b*, is cast

214.

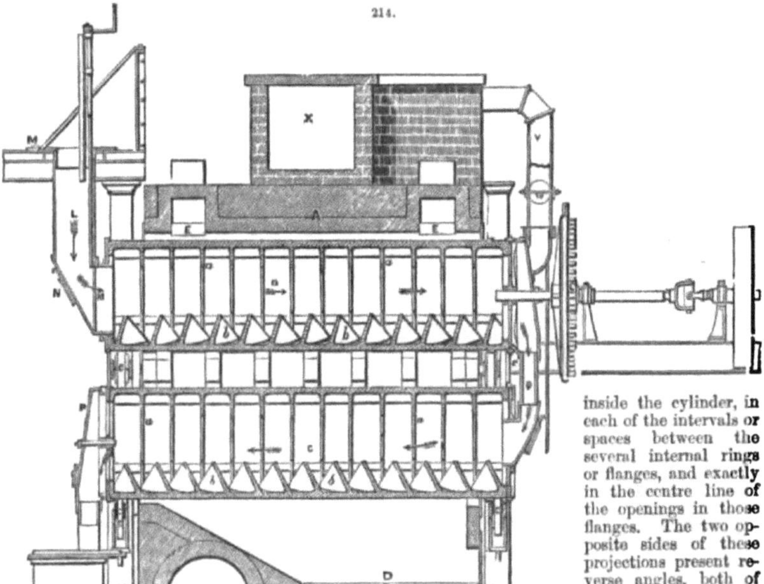

inside the cylinder, in each of the intervals or spaces between the several internal rings or flanges, and exactly in the centre line of the openings in those flanges. The two opposite sides of these projections present reverse angles, both of which direct the charcoal into the next interval or space on the partial rotation of the retort. The upper retort is driven direct by a mangle wheel and pinion arrangement G, and this motion is transmitted to the lower retort by means of the endless chain H, suspended from the rear end of the upper retort, and passing under the corresponding end of the lower retort. Both ends of the retorts are supported upon anti-friction pulleys, *c c*, carried in the transverse framing, I, bolted to the main supporting columns, K K. The feeding-hopper, L, opens into a floor, M, from which the charcoal is shovelled when being supplied to the retorts, the feed being adjusted by means of the sliding door, M, worked by a winch-handle and screw-spindle. N is a sliding door, covering an opening in the inclined side of the hopper, for the purpose of inspecting the interior of the retort; a spy-hole being also provided for the same purpose at O in the stationary front-cover, P, of the lower retort. The upper retort discharges its contents into the conduit Q, which conducts it to the lower retort; after traversing which it is discharged down the pipe R, into the closed box, or receiver, S. From this receiver it passes through the cooler, which consists of a number of long, narrow passages, T, placed side by side, and having intervening air-spaces between them, for the more effectual cooling of the contents. By the time the charcoal has traversed these coolers, it is sufficiently cool to be exposed to the action of the atmosphere, and is discharged into a small truck or waggon, V. The vapours which are evolved during the reburning of the charcoal are carried off by the pipe V, provided with a throttle-valve, W, into the chamber X, communicating with the chimney. The unpleasant consequences arising from the free escape of noxious effluvia are thus obviated. The entire arrangement is supported upon strong iron girders, Y, resting upon columns, Z, in the basement.

Each apparatus of two cylinders over one furnace is capable, in ordinary working, of reburning about 90 tons of animal charcoal a-week, with a consumption of about 10 tons of coal, or at the rate of only 1 ton of fuel to 9 tons of charcoal.

Animal charcoal is prepared by calcining bones. The bones are either placed in closed retorts, similar to those employed in the manufacture of gas for illuminating purposes, or, better still, in closed iron or earthenware pots, piled one above another in kilns, somewhat similar to pottery-kilns. Each pot contains about 50 lbs. weight of bones; and the time required for the complete calcination of a charge is from fourteen to eighteen hours. The pots, after they are withdrawn from the kiln, are kept closed for a time, in order to exclude the air; and when sufficiently cool, their contents are discharged into a magazine. The calcined bones, in the state of animal charcoal, are

216.

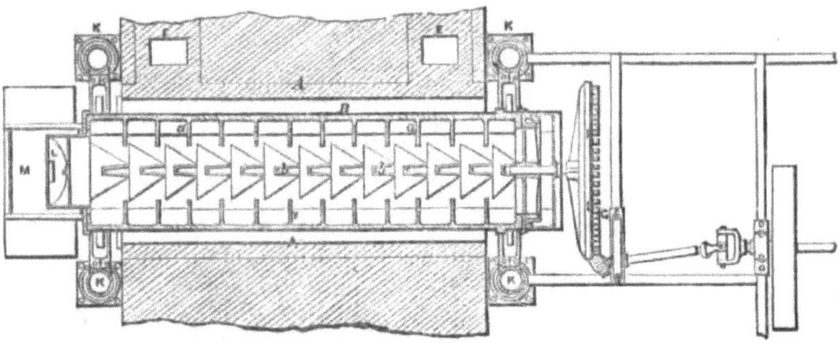

subsequently crushed by passing them through rollers, which are grooved in order to prevent the formation of dust or fine powder; the charcoal being required in a granulated state, and free from dust, for the purposes of the sugar-refiner. After being used some little time in the clarifying of sugar, the charcoal loses its decolourizing properties; and, in order to restore them, it is subjected to a revivifying process, which consists in first thoroughly washing it, for the purpose of removing the saccharine matter adhering to it, and then allowing it to dry. When dry, it is placed in close vessels, or retorts, and recalcined. On cooling, it is found to have almost entirely regained its former virtue. The immense consumption of animal charcoal in the purifying or decolourizing of sugar, about 70 tons of charcoal to 100 tons of sugar, renders the process of revivifying one of considerable commercial importance.

In France, the reburning or revivifying of animal charcoal has long been carried on, but until very recently in a crude and imperfect manner, compared with the mechanical appliances which have been for some years brought to bear upon this branch of industry in this country. The arrangement designed by Crespel-Delisse, of Arras, is shown in Fig. 217, which represents a transverse vertical section of the furnace, taken through two of a series of twenty retorts, placed side by side in pairs, and heated by one furnace. A is a brick-setting, and B the furnace, above and on either side of which are arranged the inclined retorts, C C, set in the combustion-chamber D. These retorts are of a rectangular section, and open at their upper ends on to the plate E, upon which the animal charcoal is spread for the purpose of drying it before it is shovelled into the retorts. Near the bottom of each retort is fitted a sliding door, F, which is kept shut whilst the reburning is going on, and opened to empty the retorts. The charcoal, as soon as the discharging-doors are opened, descends by its own gravity into closed receivers, G, where it is kept from contact with the atmosphere until sufficiently cool to be packed.

217.

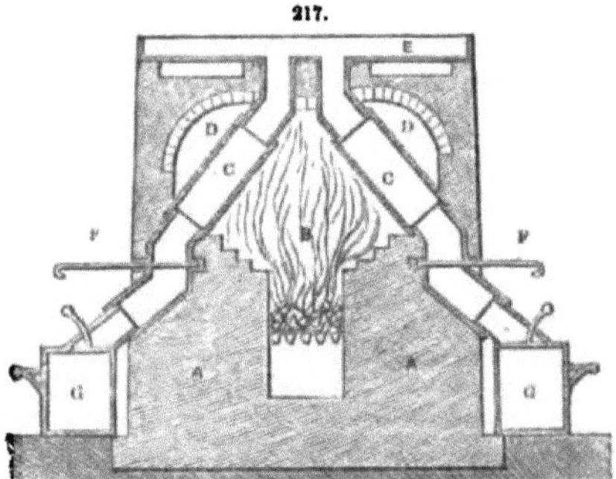

Each receiver is capable of containing one charge of the retort; and by the time the charge has sufficiently cooled, the succeeding one which has been introduced into the retort will be ready for discharging, consequently the process is almost continuous. From twenty-five to fifty minutes are required for the reburning of each full charge of the entire series of twenty retorts, which, working day and night, give on an average about one ton of animal charcoal in the twenty-four hours. The consumption of fuel is at the rate of about one ton of coal for five tons of revivified animal charcoal.

When crucibles or pots are used, the gas evolved through openings left for that purpose readily ignites, and so assists in the burning of the charcoal; the result being a probable saving of fuel, as compared with the consumption when ordinary closed retorts, similar to gas-retorts, are employed.

The first improvement of any note in the appliances for reburning animal charcoal was effected in 1846 by J. W. Bowman, who introduced revolving retorts in lieu of stationary apparatus, whereby the charcoal, by being constantly agitated and turned over during the process of reburning, is more readily and uniformly operated upon, thus effecting a saving of both time and fuel. Fig. 218 shows a longitudinal vertical section of Bowman's arrangement. A is a cylindrical horizontal retort, which revolves in bearings formed in the two fixed end-plates B, rotatory motion being given to the retort by means of endless chains, C, passing round large grooved pulleys, D, on each end of the retort, and over corresponding pulleys, E, on an overhead shaft, F, driven by steam or other power. Each extremity of the retort is provided with a door or cover, G, the front one of which is so fixed as to be readily removed, and when removed, suspended by the chain, H, which passes over overhead guide-pulleys, and has a counter-weight suspended thereto. I is a tube in the back-cover, through which any vapours driven off from the animal charcoal may pass away to a condenser. K is the fireplace or furnace for heating the retort; and to prevent it acting too violently on the retort, fire-bricks or lumps, L, with openings in them, are interposed between the fire and the retort. M M are dampers for regulating the fire. Inside the retort are fitted a number of ledges, N, which, during the revolution of the cylinder, cause the animal charcoal to be deflected towards the centre, thereby effectually burning and regulating the contents whilst subjected to the action of the fire. The flames and products of combustion have free play round the sides of the retort; and as the latter is constantly revolving, it becomes uniformly heated. In using this apparatus, a charge of animal charcoal is introduced through the front door, which is then closed, and the retort set in motion till the charge is properly reburnt, after which it is withdrawn and a fresh charge introduced, the process being intermittent.

218.

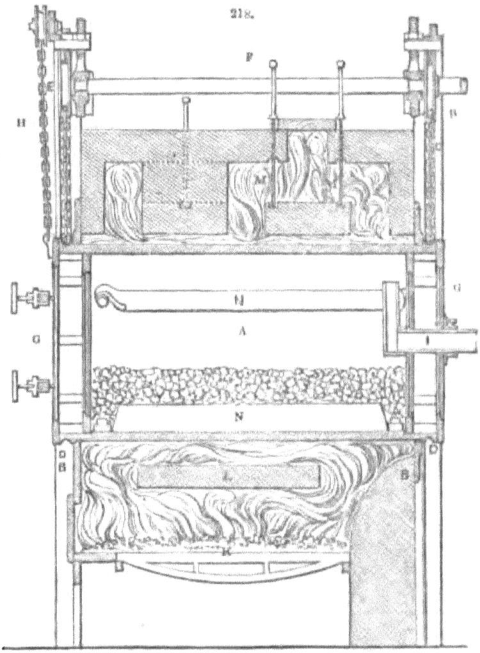

The improvements in the construction and setting of revolving retorts for reburning animal charcoal, introduced in 1832 by George Torr, are shown in Fig. 219. The sides and one end of the retort are contained within the combustion-chamber; thus the end of the retort is heated by the flames which are allowed to play round it.

Fig. 219 is a longitudinal vertical section of Torr's apparatus. The retort A is cylindrical, and revolves on a horizontal axis. That end of the retort situated inside the setting has a boss formed upon it, which passes out through the setting, and carries a large grooved pulley, B; a corresponding pulley being fitted on to the front end of the retort. From these two pulleys, endless chains, C, pass over corresponding pulleys, D, on the shaft, E, by means of which a continuous rotatory motion is transmitted to the retort. F is a pipe for carrying off the vapours, and G is a long plate fixed by arms, H, to the interior of the retort, for the purpose of turning over its contents. The addition of this plate alone makes a difference of nearly ten tons of charcoal in favour of Torr's arrangement as compared with Bowman's. I is the combustion-chamber, within which the retort revolves, and which is separated from the furnace, K, by the fire-clay lumps, L. Openings in the roof of the chamber afford a communication with the flue, M, leading to the chimney. The charging and emptying door is hinged to the retort at N. This process is also intermittent, as the retort is stopped whilst being filled and emptied.

219.

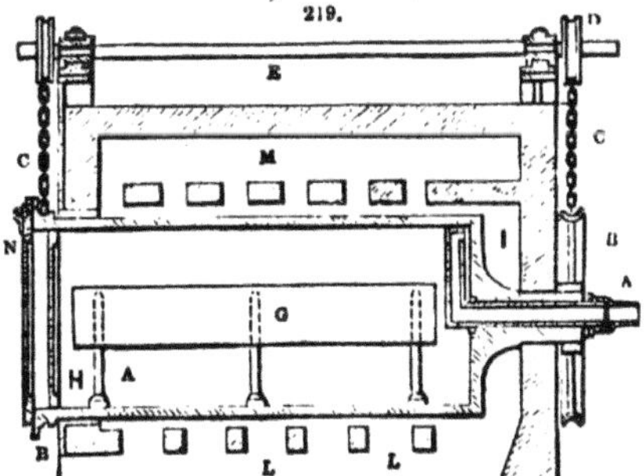

In 1856, James Bryant obtained a patent for the use of retorts having a reciprocating or alternating rotatory motion on their axes, in lieu of a continuous rotatory motion, as in Bowman and Torr's arrangements. Bryant's retorts were constructed and arranged in a similar manner to Torr's, with a space between their inner ends and the setting for the free circulation of the flame and gases from the furnace below. They were suspended at one or both ends by endless chains passing over pulleys on a driving-shaft above, which shaft received an alternating or reciprocating rotatory motion by means of the "mangle wheel" or pinion arrangement, and consequently the retorts were made to reciprocate on their axes. The interior of each of the retorts was provided with ledges similar to Bowman's, for the purpose of turning over and deflecting the charcoal towards the centre of the cylinder. According to Bryant's mode of setting, a number of retorts were ranged side by side, and their several actuating shafts were geared together, so as to work in concert, the first shaft of the series only being driven by the reversing arrangement.

The first arrangement rendering the process of reburning animal charcoal continuous was invented by Brinjes and Collins in 1858; the retorts, according to their system, being continuously charged at one end and emptied at the opposite end by the aid of an *Archimedian screw* inside the retort.

Fig. 220 represents a longitudinal vertical section of a portion of Brinjes and Collins' arrangement. The retort consists of a longitudinal cylinder, A, which does not revolve, but is set permanently in the brickwork, B, in such a manner as to leave a free space all round it for the circulation of the flames and gases from the furnace, C, beneath. Into one end of the retort opens the mouth of a feeding-hopper, B, whilst its opposite end opens direct into a chamber, E, which leads by the passage, F, into a number of narrow cooling-tubes, G, in passing down which the charcoal is sufficiently cooled to be ready for packing. These tubes open at their lower extremities into a box, H, shown also in side-elevation detached, fitted with two slides, in the form of gratings, which slide over other gratings fixed inside the box. These slides are moved by the double lever, I, on the spindle, K, which spindle receives a rocking motion from a crank-pin, L, through the intervention of the rod, M, and lever-arm, N. The crank-pin is carried on the end of the shaft, O, of an *Archimedian screw*, P, revolving slowly inside the retort A, in order to move the charcoal steadily along whilst being reburnt, and to discharge it into the chamber, E, and cooling-tubes, G. The screw derives its rotatory motion from a worm gearing into a worm-wheel, Q, on the screw-shaft; this shaft, which carries the worm, being driven by a belt and pulley. It will thus be seen that the apparatus is self-discharging; for at every revolution of the crank, L, the slides in the discharging-box will be alternately opened and closed, so as to allow a small quantity of the charcoal to fall through the lower grating into a receptacle. A pipe is connected with the chamber, E, for carrying off the vapours and gases evolved from the charcoal inside the retort, and conveying them to a condensing-worm or other condensing-apparatus.

220.

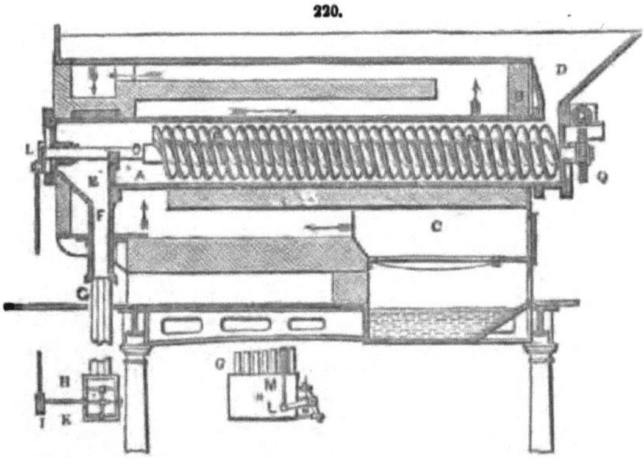

The idea of a continuous process of charging and emptying the retorts, first carried out by Brinjes and Collins, was subsequently followed up by Drummond, of Montreal, and a patent was obtained in this country for his arrangement, in the name of James Paterson, in the year 1862.

Drummond's plan, Fig. 221, consisted in placing two or more inclined revolving or reciprocating retorts at reverse angles one above another, so that the charcoal descends by its own gravity from the upper to the lower retort of the series. The figure represents a longitudinal vertical section of this arrangement. A and B are two cylindrical retorts, inclined in reverse directions, and placed one above the other. The flame from the furnace, C, has free play round the sides of both retorts, which are caused either to revolve continuously, or to have an alternating motion on their axes, by means of a worm, O, gearing simultaneously into two worm-wheels, E, fitted on to the end of the upper and lower retort respectively. The upper retort, A, is supplied with charcoal by the feeding-hopper, F, and as it revolves, its contents caused to descend gradually along the interior, until it reaches the lower end, where it is lifted up by a series of vanes, G, attached to the end-cover of the retort, and by them discharged into a pipe, H, leading to the higher end of the retort below. After

221.

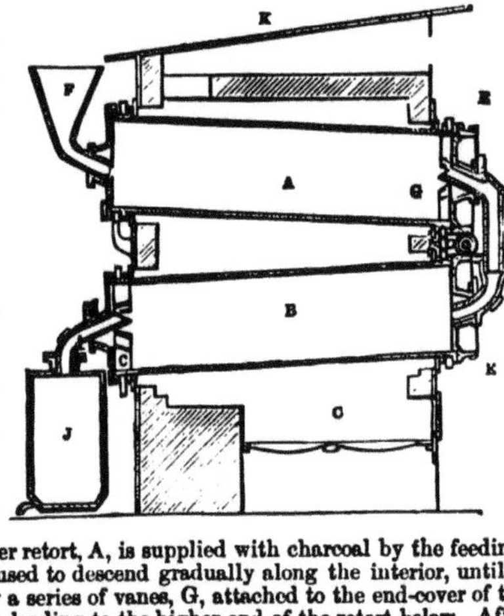

traversing the second retort, the charcoal is again lifted up by a set of revolving vanes, G, and discharged into the pipe, I, which conducts it to the closed vessel or receiver, J, where it remains until cool enough to be packed. By employing two or more retorts in connection with each other, and so arranging them that the last, or that in which the operation is completed, shall receive the greatest heat, whilst the first of the series, or that into which the charcoal is first supplied, receives the least amount of heat, a considerable saving of fuel is effected, as the otherwise waste heat, after having acted upon the lower retort, serves to heat the upper one, and thereby to gradually prepare the charcoal for the greater heat of the finishing retort. K is the upper surface or top plate of the *kiln* upon which the charcoal may be dried and then shovelled into the feeding-hopper, the plate being slightly inclined to facilitate this operation. Drummond, it appears, was the first to propose the use of two or more retorts placed one above another.

In 1862, Torr obtained another patent for apparatus for manufacturing and reburning animal charcoal, whereby the process is carried on continuously, instead of intermittently, as in his first arrangement.

Fig. 222 is a longitudinal vertical section of this subsequent arrangement. A is a revolving retort, placed horizontally, and provided with an *Archimedian screw*, B, in the interior thereof. In the interior of this cylinder, termed the "main cylinder," there is placed another and smaller cylinder, C, the axis of which coincides with the axis of the main cylinder, a space of about 1 inch being left between the exterior of the inner cylinder and the threads of the screw, B. This inner cylinder is open at both ends, and extends to within 6 or 8 inches of the back end of the main cylinder, and projects about 2 feet beyond the front of it. It is secured to the outer cylinder, and revolves with it.

222.

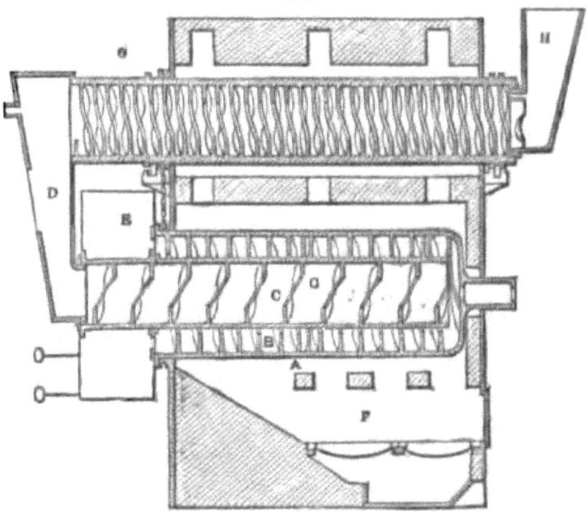

An *Archimedian screw* is also formed inside the inner cylinder, but in the reverse direction to that of the main cylinder; and its pitch and depth must be in accordance with the different diameters and pitch of the outer screw, so that the crushed bones or charcoal will travel with the same velocity and in a continuous stream through each cylinder. At the front end of the inner cylinder there is a stationary hopper, D, for supplying the bones or charcoal; and to the front ends of the cylinders, A and C, is secured a revolving cooling-box, E, consisting of a double drum of sheet iron, the inner drum having about the same diameter as the interior of the main cylinder. The outer side or face of this drum is closed; but the inner side, next to the main cylinder, is left open in the centre to receive the contents of the cylinder, A, after being operated upon. A slide is placed between the inner and outer drum, for the purpose of discharging the contents from the inner into the outer drum, where they are kept from contact with the atmosphere till sufficiently cooled to be discharged from the outer drum, by opening another sliding door. In order to economize fuel, the waste heat from the furnace, F, after passing round the outer or main cylinder, A, and before passing to the chimney, enters a brick chamber in which there is a revolving cylinder, G, by preference of the same diameter as the inner cylinder, C, and provided with an internal *Archimedian screw* attached to or cast on its inner surface. The crushed bones or charcoal are fed into the upper cylinder from the stationary hopper, H; and, after traversing the length of the cylinder, are discharged down the shoot and hopper, D, which direct them into the inner cylinder, C. After traversing this cylinder in one direction, the charcoal is discharged at the inner end of the cylinder into the main or outer cylinder, and returns in the contrary direction, being finally discharged into the revolving cooling-box.

In 1864, J. F. Brinjes contrived an arrangement of horizontal cylindrical retorts, having a circular reciprocating, instead of a continuous rotatory, motion on their axes. Fig. 223 represents a sectional elevation of this arrangement. A and B are the upper and lower retorts; the upper one receiving a circular reciprocating motion direct from a mangle wheel and pinion, or other convenient contrivance; and the lower one deriving a similar motion from the upper one by means of an endless chain passing over the end of the upper retort, and under the end of the lower one, suitable teeth or projections being furnished for taking into the links of the chain. These retorts are contained in separate chambers above the furnace or fireplace, C, openings being made in the roof of the lower chamber communicating with the upper one, so that a free circulation of the heat from the furnace is allowed to take place around both retorts before it escapes by the passages, D D, leading to the flue, E, and chimney. In the interior of each retort a number of internal flanges are fixed at regular intervals, and an opening is made in each flange of the series, such openings being in a line with each other, from end to end of the retorts—the arrangement so far being the same as that shown in Fig. 223. Along this line of openings a rocking cranked shaft passes, which carries a number of inclined vanes or plates, *a*, one in each of the intervals between the flanges. As the cylinders reciprocate round their axes, the cranked shafts, with their vanes, turn over, partly by their own gravity, so as to reverse the angle of the vanes; and consequently the charcoal, as it falls down the rising side of the retort, will come in contact with the vanes, and by that means be deflected into the adjoining interval. When the retort reverses its motion, the vanes turn over to the opposite angle, and the charcoal is again directed by the inclined surfaces into the next one of the intervals between the flanges, and so on till it has travelled from end to end of the retort. L is a transverse section of one of the retorts, showing the different positions of their vanes or deflectors. The charcoal is discharged from the end of the upper retort into a pipe, F, which conducts it into the end of the lower retort, through which it travels as above described, and is finally discharged into a double revolving cooling-box, G, which is kept cool by a water-jacket. The feeding-hopper is shown at H. It opens out to a floor or platform above the retorts, whereby the charcoal can be readily shovelled into it, and is provided with a sliding door for regulating the feed. I is a pipe communicating with the connecting-pipe, F, of the retorts, for the purpose of carrying off the vapour and effluvia evolved from the charcoal and conveying it into the chamber, K, leading to the chimney, a throttle-valve in this pipe serving to regulate the draught to the extent required.

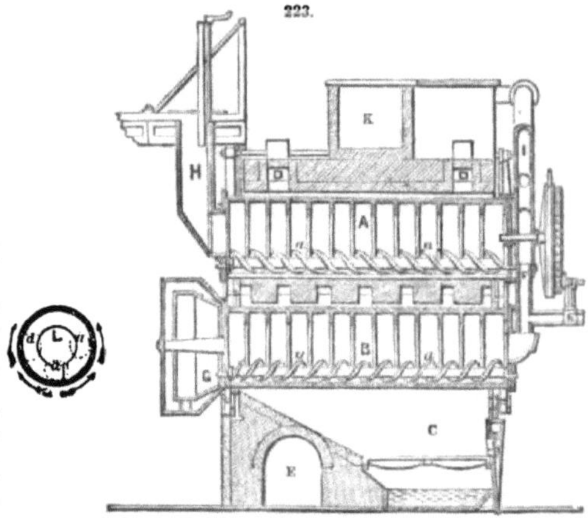
223.

ANNEALING-FURNACE. Fr., *Fourneau à récuire*; Ger., *Kühlofen, Auswännofen*; Ital., *Forno da ricuocere*; Span., *Horno para templar vidrio.*

See Furnaces.

ANNULAR PISTON. Fr., *Piston annulaire*; Ger., *Ringförmiger Kolben*; Ital., *Stantuffo anulare*; Span., *Embolo anular.*

See Details of Engines. Pumps.

ANTI-CORROSION. Fr., *Anti-corrosion*; Ger., *Gegen-Aetzung*; Ital., *Anticorrosivo*; Span., *Anti-corrosion.*

See Corrosion.

ANTI-FRICTION METAL. Fr., *Métal pour diminuer le frottement*; Ger., *Reibung vermindungs Metal*; Ital., *Lega di antifregamento*; Span., *Metal para disminuir el rozamiento.*

Babbit's metal is usually termed *Anti-friction metal*; it is composed of 50 parts tin, 5 antimony, and 1 copper. See Alloys. An alloy of tin and pewter is often used as an anti-friction metal for the bearings of engines.

Fenton's anti-friction metal is a mixture of tin, copper, and zinc; it is lighter than gun-metal, and of a soapy character, so that less oil or grease is required with it than with gun-metal.

The anti-friction metal in use on some of the Belgian railways is, in places exposed to much friction, composed of 20 parts copper, 4 tin, 0·5 antimony, 0·25 lead; and for parts subjected to great concussions, 20 copper, 6 zinc, 1 tin; for surfaces exposed to heat, 17 copper, 1 zinc, 0·5 tin, 0·25 lead. Mix the last-mentioned ingredients, and then add the copper.

For the bearings of axles and journals, a compound grease is often employed, and termed anti-friction grease. P. S. Develin oil, which is composed of hog's-lard and gutta-percha, when mixed with black-lead is termed anti-friction oil, and frequently used in the United States.

At Munich a composition is used consisting of 10·5 parts hog's-lard, melted with 2 of finely-powdered and sifted black-lead. The first of these ingredients is gently melted, and when liquid the black-lead is gradually added, the whole being stirred until the ingredients are thoroughly incorporated, and until the mixture is quite cool.

ANTIMONY. Fr., *Antimoine*; Ger., *Antimonium, Spiesszglanz*; Ital., *Antimonio*; Sp., *Antimonio.*

The properties of *antimony* are in many respects distinguished from those of other metals, particularly in its tendency to crystallize. When antimony is melted in a pot and suffered to cool on its surface, and the fluid part then cast off, a mass of beautiful crystals remains in the pot. Antimony is very brittle. It may be pulverized in a mortar. It is silver-white, and with a brilliant lustre. It fuses at about 800°, or at a dull red heat, and volatile at white heat. Its specific gravity is 6·7. The metal in its pure condition is not in use, but alloyed with other metals is much employed. The only useful ore of antimony is its sulphuret; no other kind is obtained in sufficient quantity to be smelted.

The sulphuret of antimony occurs in masses, consisting of crystalline needles, which are closely united. It is of metallic lustre, of a grey colour, and forms a grey powder. When gently heated, it turns black, or is iridescent. It is extremely fusible, and melts in the flame of a candle with the exhalation of a sulphureous smell. After being heated, the powder is very black. This ore consists of 72·86 metal and 27·14 sulphur. Its specific gravity is 4·1 to 4·6. Sulphuret of antimony occurs in and near the veins of sparry iron-ore, with heavy spar, blende, galena, quartz, and other minerals. Most of the metal in market is obtained from Germany.

Alloys.—All the antimony metal of commerce may be considered an alloy. It is never pure, but contains iron in all instances. Antimony and tin, melted together in equal parts, form a moderately hard, brittle, but very brilliant alloy, which is not soon tarnished, and is frequently employed for small speculums in telescopes. Of all the metals, antimony combines most readily with potassium or sodium. These alloys are obtained by smelting the carbonaceous compounds of these metals, or their oxides mixed with carbon. The presence of other metals, such as copper or silver, does not diminish the affinity of these metals for antimony. The alloy thus formed of the alkaline metals and antimony is not easily evaporated by a strong heat. Arsenic and antimony combine in all proportions, and form, unexpectedly, a tenacious alloy, which is very fusible, compact, and often of a granular texture. It has been remarked, in speaking of the alloys of iron, that this metal alloyed with iron causes the compound to be extremely hard. Eighty parts of lead and 20 of antimony form type-metal; to this commonly 5 or 6 parts of bismuth are added. Tin 80 parts, antimony 20, is music-metal; it is also composed of 62·8 tin, 8 antimony, 26 copper, and 3·2 iron. Plate-pewter also contains from 5 to 7 per cent. of antimony; 89 tin, 7 antimony, 2 copper, 2 iron, is one of these compositions. Britannia-metal contains frequently an equal amount of antimony. Queen's-metal is 75 tin, 8 antimony, 8 bismuth, and 9 lead.

Uses of Antimony.—Besides its employment in medicine, it is much used for forming alloys; of these, type-metal and anti-friction metal—which is type-metal with the addition of copper—are those most used. Crude antimony is employed for purifying gold.

Manufacture.—The smelting of this metal is very simple. It is easily revived from its ore, which, however, is attended with a heavy loss of metal. The crude ore is picked by hand; the pieces are broken to the size of an egg; and, by means of a hand-hammer, the gangue, such as quartz, barytas, or carbonate of lime, is removed. These pieces may be heated in an earthenware pot, in the bottom of which there is a small aperture. The sulphuret of this metal, melting at a very low heat, will flow out from the gangue, and may be gathered in another pot set below. The operation used to be performed in this manner; but, as it is expensive, the ore is at present melted in a reverberatory furnace, similar to that shown in Fig. 224, the hearth of which is very concave, and formed of sand. In the centre of the hearth, at its deepest part, there is a tap-hole which communicates with one of the long sides of the furnace. The ore, on being sorted, is spread over the hearth of the furnace, and is there melted. The tap-hole is stopped by dense coal-dust while the reduction is going on. About three-hundredweight of ore is charged at once, mixed with iron-ore or hammer-slag, and heated, with an occasional stirring. Eight or ten hours are sufficient to finish one heat, after which the metal is tapped, the scoria removed, and the furnace charged anew.

224.

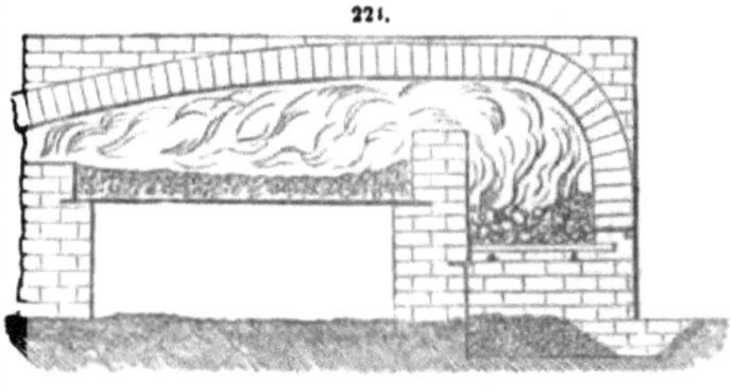

The metal thus obtained is not pure. It contains iron, sulphur, arsenic, lead, and copper; from most of these admixtures it may be freed to a certain extent, but not entirely. This metal is refined by remelting it in crucibles, arranged on the hearth of a reverberatory furnace similar to the one shown in Fig. 225. The pots contain about 30 lbs. of metal, which is covered with coal-dust. These are exposed to a low, uniform heat for some hours. Most of the metals are thus oxidized, and may be removed after emptying the crucibles.

225.

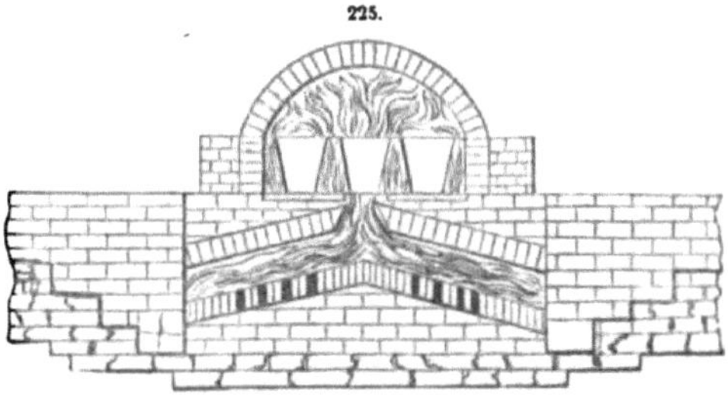

The smelting operation is in some instances divided into two manipulations; the one, or first, is a process of liquefaction, in which the crude antimony is melted in vertical pipes, and thus separated from the gangue, which remains in the retort while the former filtrates through the perforated bottom.

In this operation much of the antimony is lost. A part of it adheres to the gangue, which in pool amounts to 25 per cent., and is never less than 10 per cent. Part of the crude antimony also volatilizes, which increases the loss. This loss is, therefore, an important object where the ore is expensive; and it may be in most cases the best plan to stamp and wash it while crude, free from the rocky matter, and then subject it to reduction by direct smelting. The specific gravity of the ore is sufficiently great to remove the most of the gangue. Metallic sulphurets of other metals than antimony, of course, remain with it. The crude antimony, or the concentrated ore-sand, is smelted with metallic iron, or iron-ore; and since it is difficult to add just as much iron as is required to absorb all the sulphur, and as too much imparts iron to the metal, the practice is to add either carbonate or sulphate of potash, or soda, and also fine charcoal-powder, to the ore. One part of metallic iron to 2 or 2·5 parts of crude antimony, ought to absorb all the sulphur; but when no other flux is present, about 20 per cent. of antimony remains in the slags. By using 42 parts of iron to 100 of crude antimony, with 50 carbonate of soda and 5 charcoal, nearly the whole of the antimony is revived. Instead of metallic iron, any kind of pure iron-ore may be employed with more charcoal; but its metallic contents should come near the above-named quantity.

In refining the crude metal of antimony in crucibles, it is advantageous to soak the charcoal-powder with which the metal is covered, in a strong solution of carbonate of soda. When the metal is not sufficiently pure after the first refining, the operation is repeated. In all the operations with antimony, a high heat must be avoided, for the metal as well as the sulphuret is very volatile.

A fusible slag increases the yield of the ore.

ANVIL. Fr., *Enclume;* Ger., *Amboss;* Ital., *Tuendine;* Span., *Yunque, Bigornia.*

An anvil is an iron block, usually with a steel face, upon which metals are hammered and shaped. The ordinary smith's anvil, Figs. 226, 227, is one solid mass of metal,—iron in different

226.

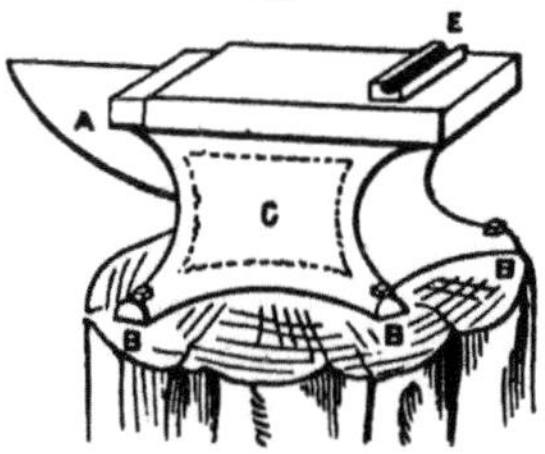

227.

states; C is the core or body; B, B, B, four corners for enlarging the base; D, Fig. 227, the projecting end; it contains one or two holes for the reception of set chisels in cutting pieces of iron, or for the reception of a shaper, as shown at E, Fig. 226. In punching flat pieces of metal, in forming the heads of nails or bolts, and in numerous other cases, these holes *a, a,* of ordinary anvils, are not only useful but indispensable. A is the beak-horn, which is used for turning pieces of iron into a circular or curved form, welding hoops, and for other similar operations. In the smithery, the anvil is generally seated on the root-end of a beech or oak tree; the anvil and wooden block must be firmly connected, to render the blows of the hammer effective; and if the block be not firmly connected to the earth, the blows of the hammer will not tell. The best anvils, anvil-stakes, and planishing-hammers, are faced with double shear-steel. The steel-facings are shaped and laid on the core at a welding heat, and the anvil completed by being reheated and hammered. When the steel-facing is first applied, it is less heated than the core. But the proper hardening of the face of the anvil requires great skill; the face must be raised to a full red-heat, and placed under a descending column of water, so that the surface of the face may continue in contact with the successive supply of the quenching fluid, which at the face retains the same temperature, as it is rapidly supplied. The rapidity of the flow of water may be increased by giving a sufficient height to its descending column; it is important that the cooling stream should fall perpendicularly to the face which is being hardened. Heat may escape parallel to the face, but not in the direction of the falling water. The operator, during this hardening process, is protected from spray and smoke by a suitable cover; and by confining the falling water to a tube which must contain the required volume. When an anvil is to be used for planishing metals, it is polished with emery and crocus powders. The skilful manner in which Henry Walker, of Red Lion Street, Clerkenwell, combines and carries out these apparently simple operations, in making anvils and planishing-hammers for silversmiths and metal-workers, gives him the reputation he so well deserves;—to describe a process is one thing, but the execution requires practice as well as skill. See Fullers. Shapers. Steam-hammer. Upsetting-block.

APERTURE. Fr., *Ouverture;* Ger., *Offnung;* Ital., *Apertura;* Span., *Abertura.*

In building, the term aperture is usually applied to doorways, windows, and other openings through the walls: the sides of a rectangular aperture are named "*jambs;*" the upper part, the "*head;*" and the bottom part, the "*sill.*"

APPROACHES. Fr., *Approches;* Ger., *Zugang, Laufgraben;* Ital., *Approcci;* Span., *Aproches, Ataques.*

Works thrown up by besiegers, to protect them in their advances towards a fortress, are termed *approaches.* See Fortification. This term is also applied to those parts of a road which are raised to suit the level of a bridge over a railway or canal.

APRON, in Shipbuilding. Fr., *Eperon, Contre étrave;* Ger., *Binnenvorsteven, Ueberlauf;* Span. *Albitana de proa, ó contraestrabe.*

The apron, also termed stomach-piece, is a strengthening compass-timber, which is bolted behind the lower part of the stem, and immediately above the foremost end of the keel.

APRON. Fr., *Radier;* Ger., *Platform, Schleussen bett;* Span., *Zampeado.*

In engineering structures an *apron* is a platform of wood, stone, or brick, placed at the base of the structure, to protect it from abrasion or heavy shocks. The platform which receives the water falling through the sluices of a lock-gate or embankment is called an apron; the planking or platform placed at the toe of a sea-wall, to protect its base from the scour occasioned by the returning wave, is also termed an apron.

An apron in carpentry is the horizontal piece of timber which takes the carriage-piece, or rough string, of a staircase; and also the ends of the joists which form the *half-space* or landings. It should be firmly wedged at both ends into the wall.

In plumbers' work, the apron is the lead sheeting or *flashing* dressed on to the slates in front of dormer-windows or skylights.

APRON-LINING. FR., *Contre-étrave*; GER., *Binnenvorsteven*; ITAL., *Fascia*; SPAN., *Travesera para sostener los largueros de una escalera.*

Apron-lining, as this term implies, is a lining placed outside the apron. It is applied by the joiner to the piece of wrought boarding which covers the rough apron-piece of the staircase.

APRON-PIECE. FR., *Contre-étrave*, GER., *Binnenvorsteven.*

See APRON, *in Building.*

A. P. FR., *Niveau d'Amsterdam*; GER., *Amsterdammer Pegel.*

228.

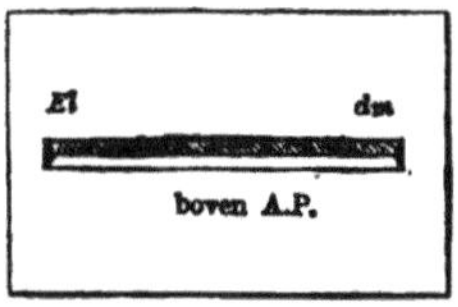

229.

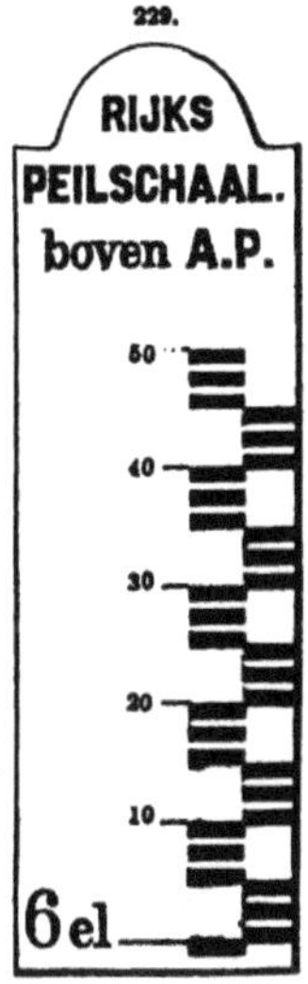

The letters A. P. are the initial letters of two Dutch words, Amsterdamsch Peil, meaning Amsterdam Level, and indicate the *Datum Line*, or mean level, from which all surveys are made in Holland, Belgium, and in the Northern parts of Germany.

The A. P. is an imaginary plane, drawn through the average high-water mark of the North and Zuiderzeen Seas on the Dutch coasts, and derives its name from the circumstance of its having been adopted at Amsterdam from the year 1670.

It was carefully revised during the present century, under the direction of the Dutch Government; and in consequence of this careful revision, is the most accurately fixed sea-level.

Figs. 228, 229, show the marks which are placed in various parts of Holland, to indicate the height above or below A. P. See DATUM LINE.

AQUEDUCT. FR., *Aqueduc*, GER., *Wasserleitung*; ITAL., *Acquedotto*; SPAN., *Acueducto.*

A conductor, conduit, or artificial channel for conveying water, is termed an aqueduct.

The Aqueduct of the Loch Katrine Waterworks.—The description of this aqueduct is taken from a paper by James M. Gale, published in the 'Proceedings of the Institution of Mechanical Engineers, 1864.' The length of this aqueduct is about 34 miles. The built and tunnelled part of the aqueduct is 22 miles long, and 8 ft. high by 8 ft. broad; sections of it are shown in Figs. 230, 231, 232. It has an inclination of 1 in 6336, as shown in longitudinal section, Fig. 233, and is capable of passing fifty million cubic gallons a-day. The valleys of Duchray, Endrick, and Blane, *e f h*, Fig. 233, which are crossed by the line of aqueduct, and

230.

231.

232.

prevent a uniform inclination being obtained throughout, make up an aggregate length of 3¾ miles, and are passed by cast-iron siphon-pipes 48 inches in diameter, with a mean fall of 1 in 1000 between their extremities. These pipes deliver about twenty million gallons a-day, and provision was made for laying two additional lines of pipes, at bridges and other places where masonry was required, in order to increase the supply of water to the city when necessary. The first work upon the line of aqueduct, upon leaving Loch Katrine, is a tunnel through the ridge which separates the

233.

Reference.—*a*, River Clyde. *b*, Glasgow. *c*, Mugdock Reservoir and Straining-well. *d*, Mugdock Tunnel. *e*, Blane Valley. *f*, Endrick Valley. *g*, Drummore Tunnel. *h*, Duchray Valley. *i*, Loch Katrine Tunnel. *k*, Loch Katrine.

valley of Loch Katrine from that of Loch Ard. The length of the tunnel is upwards of 1·25 mile, and is at a depth of more than 500 feet below the top of the hill. Twelve shafts were sunk on the line of tunnel, to facilitate the work, five of them being about 450 feet deep. The rock passed through by this tunnel, and by the greater part of the first ten miles of the aqueduct, which is principally a series of tunnels, is mica slate, of the hardest description. Along the margin of Loch Chon, the work, at some of the faces, did not progress more than three lineal yards in a month, although it was carried on night and day. The minor ravines in the first ten miles of the aqueduct are crossed by aqueduct-bridges of iron. Besides a number of smaller ones, there are five extensive aqueduct-bridges of this kind, one of which, near Culgarton, is shown in Figs. 234, 235, 236. It consists of a

234

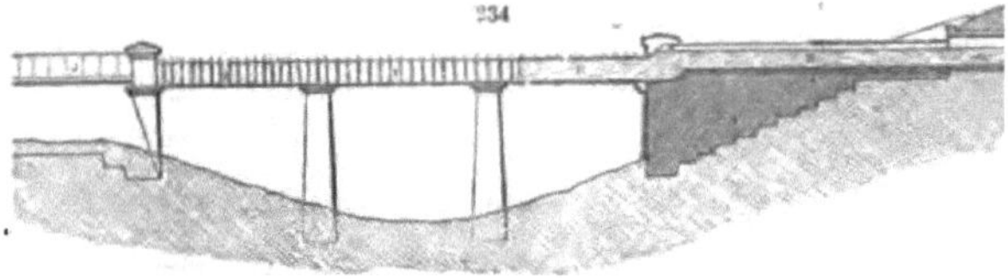

wrought-iron tube I, 8 ft. broad, and 6·5 ft. high inside, extending over the greater part of the ravines, supported at intervals of 50 ft. by stone piers; and a cast-iron trough, J, also 8 ft. broad, and 6·5 ft. high, supported on a dry stone-rubble embankment at either end of the wrought-iron tube I, extending over the remaining part of the valleys, where the ground is not much depressed. The bottom and sides of the wrought-iron tube I are $\frac{3}{8}$ in. thick, and the top $\frac{7}{16}$ in. thick, the whole being strengthened by angle and T iron. The plates of the cast-iron trough are $\frac{5}{8}$ in. thick, the dimensions of the largest being 4·5 ft. by 4 ft.; and they are connected and strengthened by flanges, with rust-joints. The level of the tube I is about 3 ft. lower than that of the cast-iron trough J J at each end, so as to ensure the tube being always completely filled with water, in order that the top of the tube may be kept at the same temperature as the sides, and that the tube may not be racked by the strain which would arise from the top plates becoming heated by the sun, if the water were not in contact with them. That the tube may be emptied at any time, for

235.

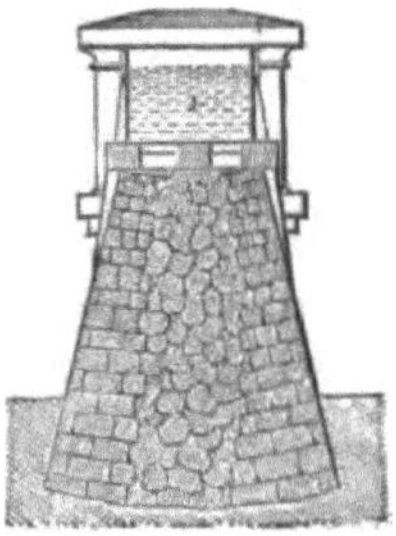

236.

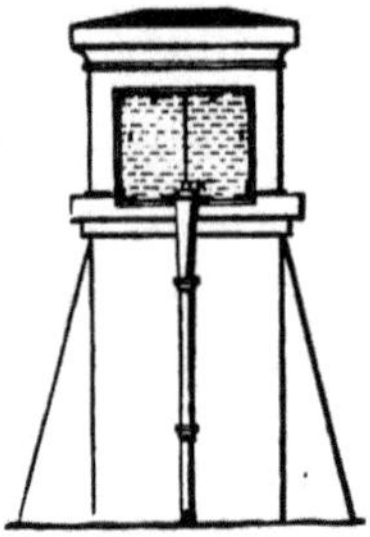

237.

painting or other purposes, a discharge-valve, K, Figs. 236, 238, is provided at one end of the tube, by which the water can be run off into the valley beneath. The junction between the wrought-iron tube and the cast-iron trough is shown in Figs. 238, 239, and in detail in Figs. 240, 241. It is made by bolting the cast-iron trough to a cast-iron bed-plate, L, Figs. 237, 238, and to upright

238.

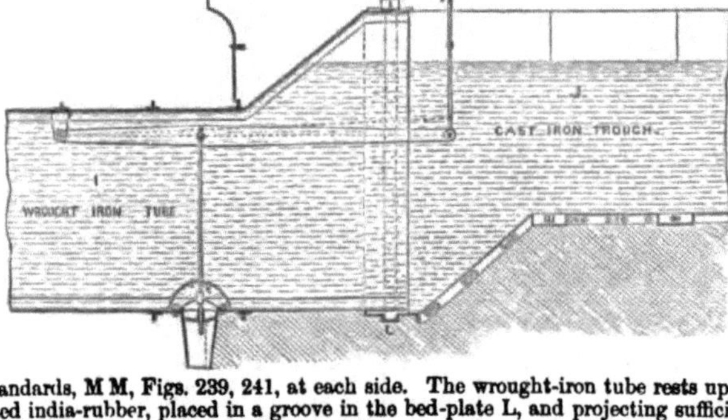

cast-iron standards, M M, Figs. 239, 241, at each side. The wrought-iron tube rests upon a bolster of vulcanized india-rubber, placed in a groove in the bed-plate L, and projecting sufficiently above the surface of the plate to allow for the requisite compression on the india-rubber for making a water-tight joint by the weight of the tube bearing on it, without allowing the tube to come down to a bearing upon the bed-plate L itself. A similar india-rubber bolster is carried up each side of the tube, and compressed against it by oak wedges, the bolster and wedges being contained in a recess in the upright standards M M, as shown in Figs. 239, 241. This arrangement leaves the wrought-iron tube free to contract and expand longitudinally under change of temperature, without risk of leakage. The heads of all the rivets are countersunk for a short distance on each side of the bearing parts of the tube. The india-rubber bolsters are, both at the bottom and sides of the tube, 2 in. in diameter. They are in separate pieces; the bolster under the bottom extending

239.

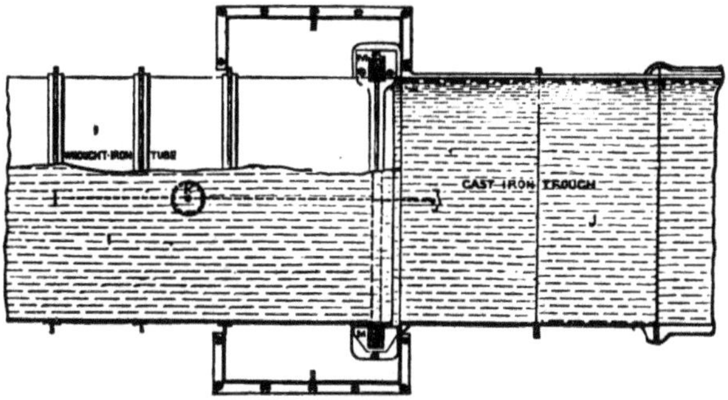

from the wedge-box M on one side, to the back of the wedge-box on the opposite side. The joints of the bolsters at the bottom corners are made by butting the bottom ends of the vertical bolsters upon the top of the transverse bottom bolster, the bottom ends of the vertical bolsters being slightly rounded out to fit the curvature of the bottom bolster. The side wedges are driven down tight on

240.

241.

the ends of the bottom bolster. There are three oak wedges in each wedge-box, M, Fig. 241, with an oak feather, or tongue, let in to break the joints between the wedges, and to guide the centre wedge while being driven down. A flat strip of india-rubber, I, is placed between the back of the wedge-box and the outermost wedge. The wedges were carefully fitted before the feather-grooves

242.

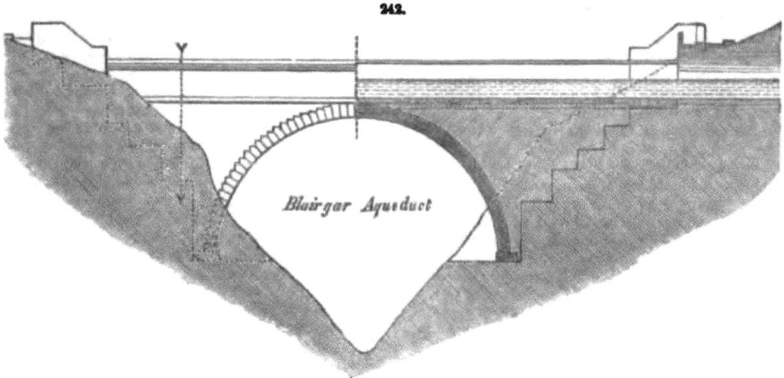

243.

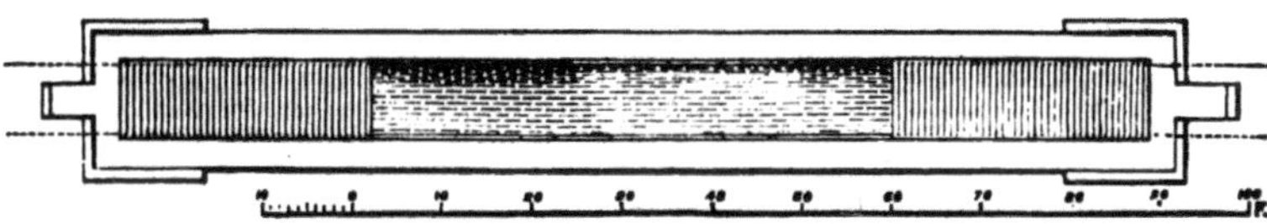

were made, and were put in with thick wet paint in the joints; the centre wedge was then driven down to tighten up the india-rubber bolster against the side of the tube, and the spaces on either side of the wedges in the standards M were filled with oakum and white-lead. The above construction of the iron aqueduct-bridges was the most applicable in the first portion of the aqueduct, as no good building-stone could be obtained within reasonable distance, and the roads were badly suited for the carriage of materials. From the eleventh mile to the reservoir at Mugdock, however, good building-stone was abundant, and all the aqueduct-bridges in that district are therefore of stone. One of these is shown in Fig. 242, in elevation, and longitudinal section in plan, Fig. 243; and Figs. 244, 245, show transverse sections of the same. There are, in all, twenty-five important

244.

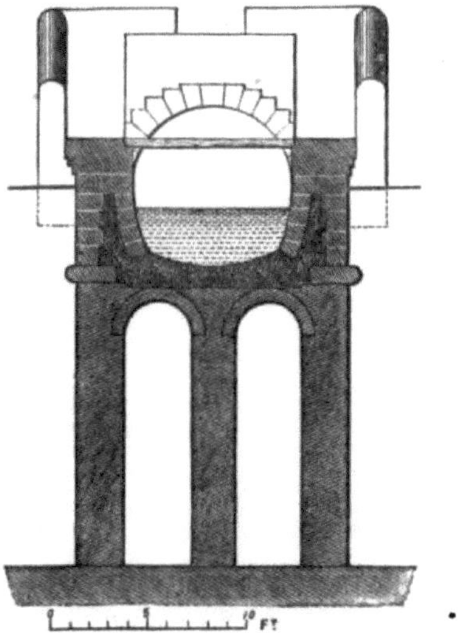

245.

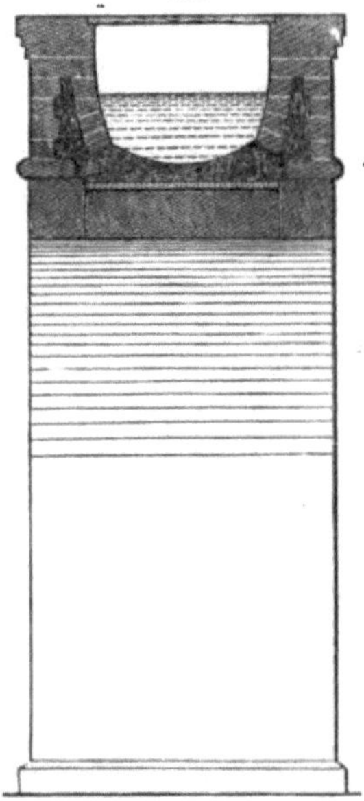

iron and stone bridges, some of them of considerable magnitude; and about eighty distinct tunnels, varying in length from 1½ mile downwards, and forming a total length of thirteen miles. Where the aqueduct was formed in open cutting, the ground was filled in and the surface restored, as shown in Fig. 232. At the cast-iron troughs of the iron aqueduct-bridges, and at the other bridges, the waterway is covered with planking, as shown in the sections of the Blairgar aqueduct-bridge, Figs. 244, 245, to prevent snow from choking the aqueduct. Grooves to receive stop-planks are cut in the masonry of the aqueduct at intervals, and most of the bridges are provided with overflow and discharge sluices. The latter are similar in construction to the outlet sluices of the locks, but of smaller dimensions. The three valleys of the Duchray, Endrick, and Blane, which are of great width and depth, the second being more than three miles wide, are passed by means of the 48-in. cast-iron siphon-pipes, carried down one side of the valley to the bottom, and up the opposite side. These pipes have the ordinary spigot and socket joints; a section of which is shown in Fig. 246, the joint being made with lead, N, and yarn, O. Some depressions on the line of these siphon-pipes are crossed by flanged pipes, supported upon stone piers, 18 ft. apart, as shown in Fig. 247, the joint being made by a ring of vulcanized india-rubber, P, as shown in section Fig 248. In the Endrick valley, two public roads, and the Forth and Clyde Railway, are crossed by these flanged pipes; and to support the pipes over these greater spans, cast-iron brackets are put in, abutting on the stone piers which are thickened to receive them. The pipes are further strengthened at these places by projecting webs cast on them, as shown by the enlarged transverse section of the pipe, Fig. 237. It was found that the expansion and contraction of these long lengths of flange-jointed pipes under changes of temperature injuriously affected the socket and spigot lead joints at each end; and to obviate this, a felt covering, about ⅜ in. thick, was laid on all round the pipes, and protected from the weather by a tarpaulin cover, laced tightly over the whole. This has the effect of almost entirely obviating the inconvenience that arose from the expansion and contraction.

The service reservoir of Mugdock, Fig. 233, has a water-surface of 60 acres, and is 50 feet deep, the top water-level being 312 feet above the level of the sea. It contains 548,000,000 gallons when full, equal to a supply for twenty-nine days at the present rate of consumption; and thus admits of repairs being made upon the line of aqueduct without interrupting the supply to the city. The reservoir is entirely artificial, being formed by two earthen embankments, 400 yards and 240 yards long respectively. The water is first received from the aqueduct into a basin at the upper end of the reservoir, from which it flows over four cast-iron gauge-plates, 10 feet long each, brought to a thin edge, into an upper pool or compartment of the reservoir having an area of about 2 acres. The depth of water passing over the gauge-plates is regularly gauged, the delivery from

246.

249.

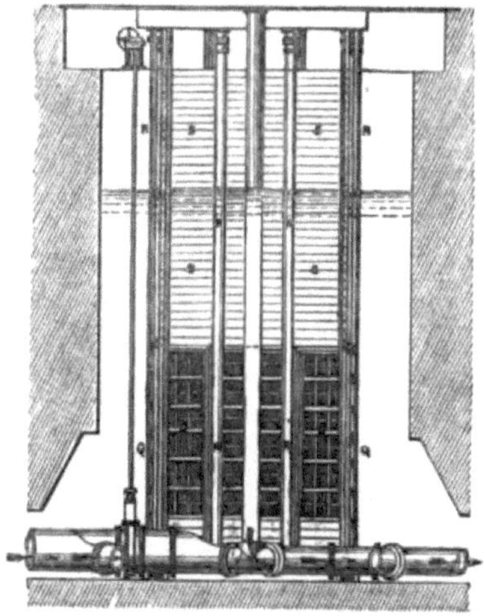

247.

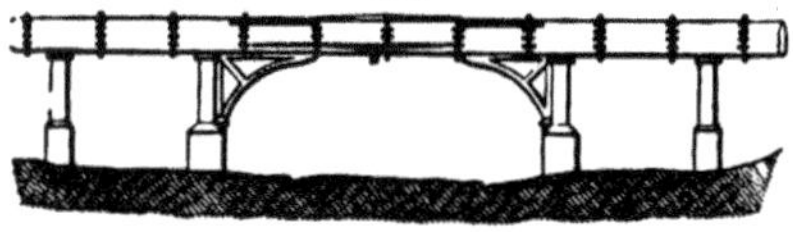

248.

250.

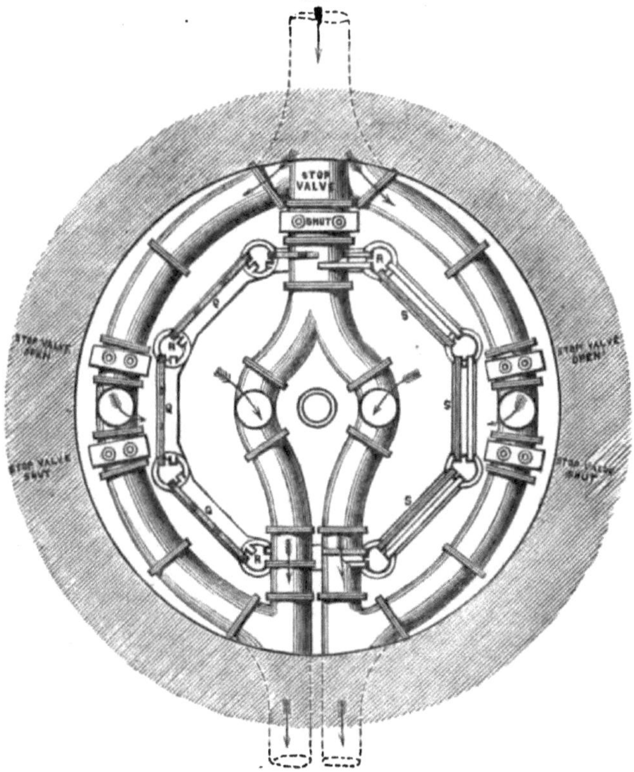

the aqueduct thereby computed, and the quantity of water passing every day into Glasgow is thus known. From the upper pool the water passes into the main reservoir over similar cast-iron gauge-plates. The water is drawn from the reservoir by pipes laid in a tunnel cut through the rock in the solid, at the end of the main embankment, no pipes being laid through the embankments themselves. At the end of the tunnel next the reservoir there is a stand-pipe with valves at different heights, which admit of water being drawn off at various levels. The water passes down the stand-pipe and along a 48-inch pipe in the tunnel for a distance of about 50 yards to a circular straining-well cut in the rock. Water can also be drawn direct from the aqueduct or from the upper compartment of the reservoir into the pipes leading to the city, without passing through the reservoir, by means of a line of 48-inch pipes laid through the bottom of the reservoir from the stand-pipe back to the upper end of the reservoir where the aqueduct enters.

251.

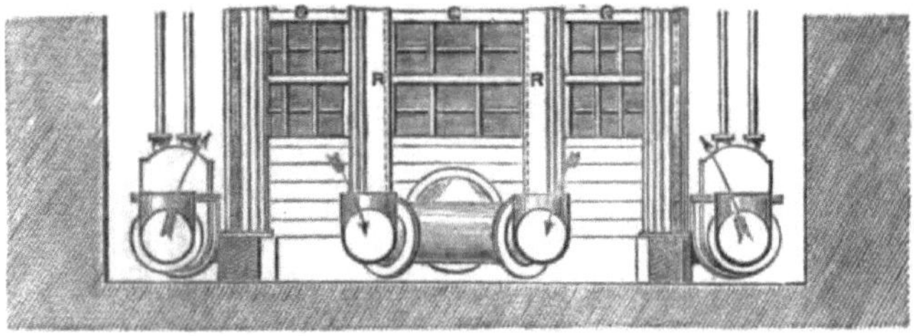

252.

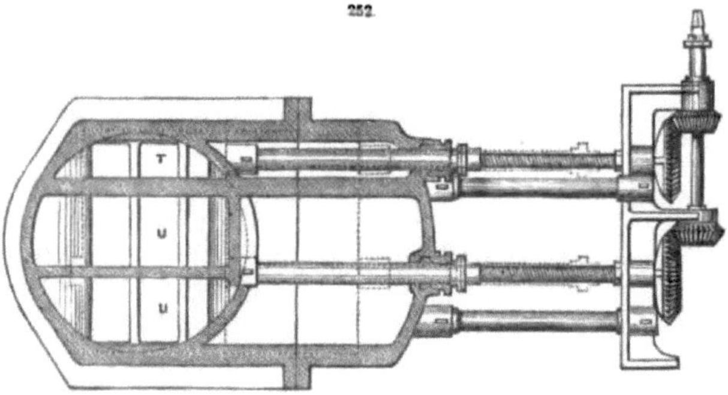

253.

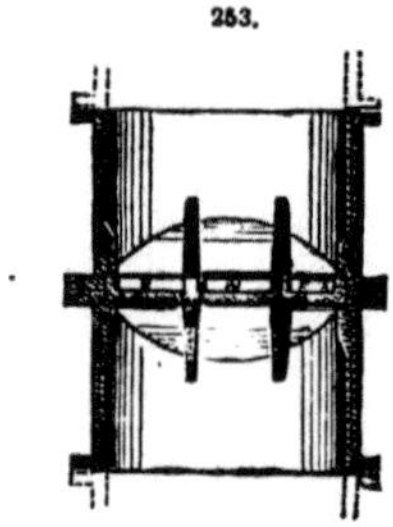

254.

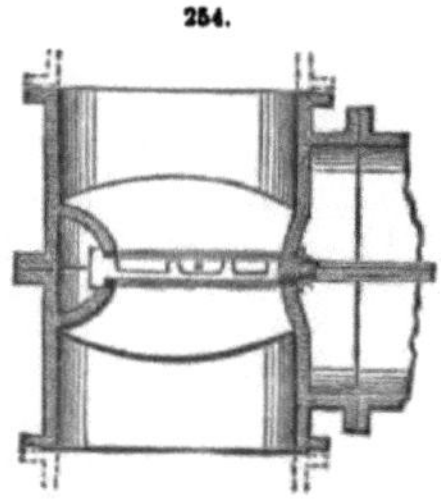

The straining-well is shown in vertical section in Figs. 249, 251; and Fig. 250 is a sectional plan. The well is 40 feet diameter and 63 feet deep, cut out of the solid rock. Within the straining-well, and forming an inner chamber of octagonal shape, 25 feet diameter, a series of oak frames Q Q, Fig. 249, is placed, covered with copper-wire cloth of 40 meshes to the inch; these are held in the light cast-iron pillars R, which have grooves cast in them to receive the frames. These wire-cloth strainers occupy only the lower part of the well, the space above being filled in with wood planking S S, up to the top water-level of the reservoir. The water passes from the outside through the wire-cloth strainers into the inner chamber, and is taken off thence to the city by two lines of cast-iron pipes 42 inches diameter, as shown by the arrows. The water undergoes no filtration, but in passing through these copper-wire strainers, any straws, or other floating matters, are separated from it.

The pipes in the bottom of the straining-well are provided with junctions and stop-valves, as seen in the plan, Fig. 250, so as to admit of the supply being drawn direct from the reservoir, while the strainers are being cleaned; which latter is done by emptying the well, and throwing a

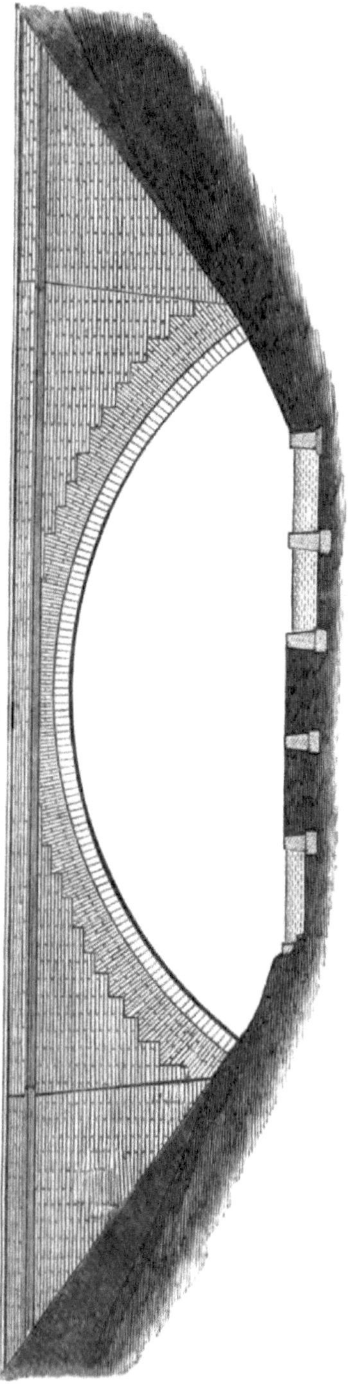

255.

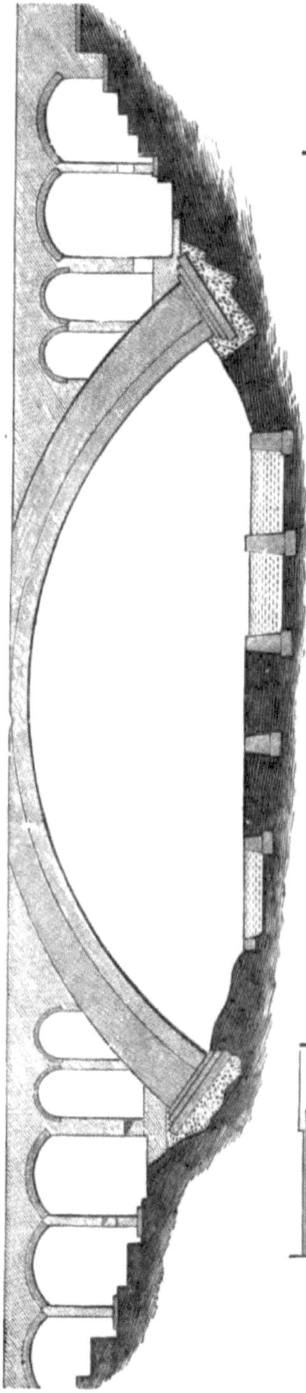

256.

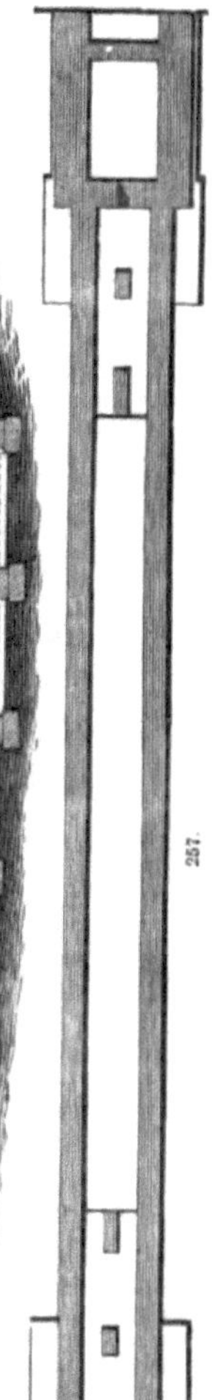

257.

jet of water upon the strainers from the inside outwards, by a leather hose with the head pressure of the reservoir, the foul water being carried off by a tunnel through the rock. The frames Q Q carrying the strainers can also be raised to the top of the well and taken out for repairs, by being drawn up through the grooves in the cast-iron pillars R, in which they are fitted. The top of the straining-well is roofed in and partly covered with glass, as a protection to the working gearing of the stop-valves. These valves are each divided into two halves, affording together a waterway of the full diameter of the 42-inch pipes. Each half of the valve is opened and shut by an iron rod, passing up through a cast-iron pipe, and terminating at a convenient height above the water-level in a long brass nut, into which works a stationary iron screw, turned by a crank and bevel-wheels. The two lines of 42-inch pipes laid in the tunnel, leading off from the straining-well, will deliver the whole 50,000,000 gallons a-day; but on emerging from the tunnel, which is 440 yards long, they are diminished to 36 inches diameter, and provision is made for additional pipes being laid when required. At the point where the pipes are reduced to 36 inches diameter, a self-acting throttle-valve is fixed on each line of pipes, the object of which is to shut off the water coming from the reservoir in the event of one of the pipes bursting, or any other accident occurring whereby the velocity of the water in the pipe is increased beyond that to which the valves are adjusted.

At intervals along the line of the mains to Glasgow and at several points in the city, stop-valves are fixed in the large pipes, one of which, for a 36-in. pipe, is shown in Figs. 252, 253, 254. To admit of these valves being easily closed or opened, the slide is divided into two compartments, T and U, one being considerably smaller than the other. The smaller slide, T, is the first opened, and the passage of the water through this opening so much reduces the pressure upon the larger slide U, that it can be opened with ease; the valve is thus easily worked by one man. To economize space, which is an object where large valves have to be placed in public streets, the total effective area of the valve has been reduced, in the case of these 36-inch valves, from 7 sq. ft., the area of the pipe, to 4¼ sq. ft.; the smaller slide, T, having an area of 1 sq. ft., and the larger, U, an area of 3¼ sq. ft. To pass this contraction with the velocity that the water in the pipes will have when the discharge is greatest, the loss of head will be from 4 to 6 inches; but this loss is more than compensated for by the economy of the valve and the reduction in the dimensions of all the parts.

The Washington aqueduct, constructed for the purpose of supplying to Georgetown and Washington, U.S., water from the river Potomac, at a point eleven miles above the last-mentioned city, consists for the greater part of its length of a masonry *conduit*, 9 feet in internal diameter. The fall of this conduit averages 9 in. per mile, its length being 11$\frac{1}{10}$ miles. The total length of the aqueduct is 16¼ miles, and it is capable of supplying to the town *reservoir* 100,000,000 gallons a-day. In carrying out this work, all unnecessary expenditure has been avoided by constructing its parts in as simple a manner as possible; thus the various water-gates and pipe-vaults have, in most cases, been arranged within the masonry embankments; and thus, whilst most of the fittings are out of the reach of frost, much of the expense, which would have been incurred by the erection of gate-houses, and so on, above ground, has been saved. The aqueduct-bridge over Cabin John Creek, on the line of the conduit, from the source of supply to the receiving reservoir, is shown in Figs. 255, 256, 257. It is a stone-arched bridge, its clear span is 220 ft. The arch is an arc of a circle of 134·2852 ft. radius, and its rise is 57·2624 ft. Fig. 255 is a side elevation; Fig. 256, longitudinal section; Fig. 257, sectional plan; Fig. 258, transverse section through the eastern *abutment*, taken through the springing of the arch; and Fig. 259, a transverse section through the crown of the arch. The radius of the *extrados* is 143·2695 ft., the depth of the *voussoirs* being 6 ft. 2 in. at the springings, and 4 ft. 2 in. at the crown. Outside the voussoirs is another series of arched stones, which make up the total thickness of arch at the springings to 20 ft., this thickness diminishing towards the crown. The width of the bridge on the face of the arch is 20 ft. 4 in. The abutments are formed by the solid rock on each side of the creek; the face of this rock being stepped down, as shown in Fig. 256, and the steps filled in with concrete, on which the footings of the arch bed. The channel through which the water is conveyed consists of a conduit of circular section, 9 ft. diameter inside and 9 in. thick, this conduit being imbedded in the masonry of the bridge. The haunches and abutments of the bridge are lightened by *relieving arches*, of which there are five on the western and four on the eastern side, extending through half the thickness of the bridge. The centering on which the arch was constructed was supported upon temporary piers, formed in the bed of the creek, as shown in Fig. 256; the vertical timbers bearing upon these piers, and the bracing connecting them, carrying a series of struts radiating to the lines of timbers beneath the lagging-boards. The key-stone was inserted in mid-winter, and the centre was not struck until some ten or twelve months later. The rise of temperature lifted at times the arch of the centering; but when the latter was removed, careful observations were made without any settlement being noticed.

259.

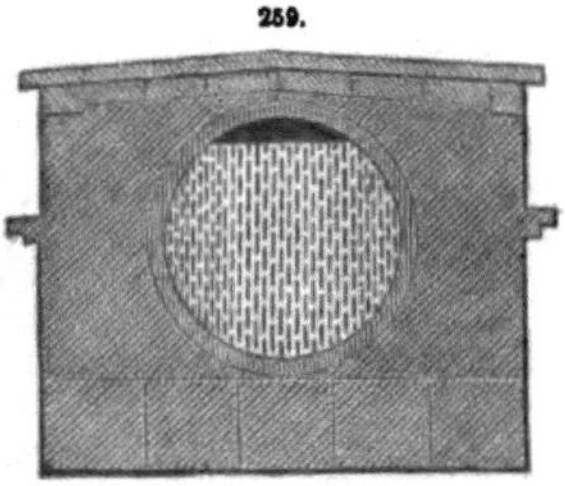

258.

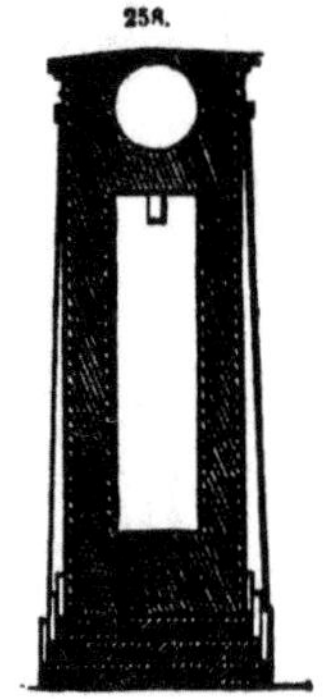

Works and Reports on Aqueducts:—Tower's (F. B.) 'Croton Aqueduct,' royal 4to, New York, 1843. Fontenay (T.), 'Construction des Viaducs, Ponts, &c., 2 vols. 8vo, and 'Atlas,' in 4to, 1852. Cautley (Sir P. T.), 'Report on the Ganges Canal Works,' 2 vols. 8vo, 1 vol. 4to, and Plates in folio, 1860. Becker, 'Der Wasserbau in seinem ganzen Umfange,' 1861. Birot (F.), 'Guide pratique du Conducteur des Ponts,' &c., 2 vols., 12mo, Paris, 1862. Hagen, 'Handbuch der Wasserbau Kunst,' 1863. 'On the Loch Katrine Waterworks,' by Jas. M. Gale: Proceedings Inst. M. E., 1864. Moncrieff (C. C. Scott), 'Irrigation in Southern Europe,' 8vo, 1868.

See also:—Belidor, 'Architecture Hydraulique.' Delaistre, 'Encyclopédée de l'Ingénieur. Minard, 'Cours de Construction.' 'Life of Telford,' 4to and fol. Sganzin, 'Cours de Construction.'

ARCH. Fr., *Arche, cintre;* Ger., *Bogen, Gewölbe;* Ital., *Arco;* Span., *Arco.*

An *arch* is a form of structure in which the vertical forces due to the weight of the materials of which it is composed are transmitted to the supports or *abutments* in a polygonal line, usually termed the "curve of equilibrium," from the fact that it becomes a curve when the arch-stones are infinitely numerous.

Arches are named from the curve or outline of the under-surface presented by a section taken at right angles to the axis. Thus, an arch whose outline is a semicircle is called a "semicircular arch," and one formed to an elliptical curve is called an "elliptical arch." Figs. 260 to 273 show the forms of arches usually constructed.

260.

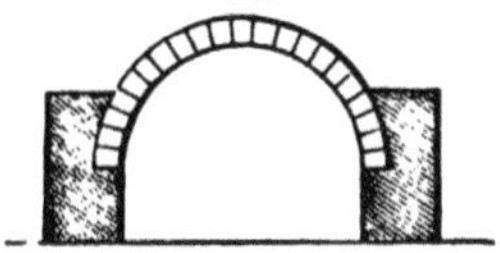

Semicircular Arch.

261.

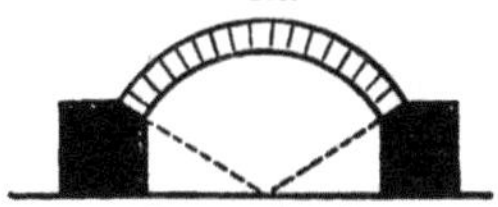

Segmental Arch.

262.

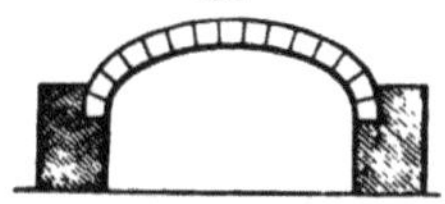

Elliptical Arch.

263.

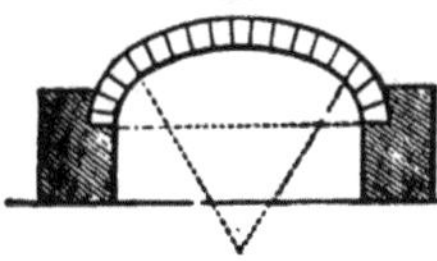

Three-Centre Arch.

264.

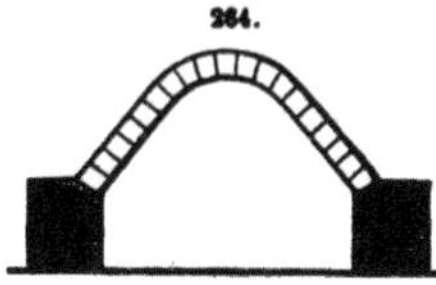

Parabolic Arch.

265.

Pointed Arch.

266.

Straight Arch.

267.

Cambered Arch.

268.

Groined Arch.

Where two arches intersect, they are called "groined arches."

269.

Fluing Arch.

270.

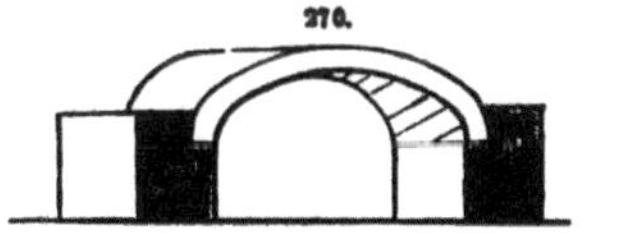

Skew Arch.

Where the opening at one end is less than at the other, as in Fig. 269, the arch is called a "fluing arch."

The terms "gauged" and "axed" are applied to brick arches when the bricks are gauged or axed to shape.

Where the axis of an arch is oblique to the face, it is called a "skew arch."

271.

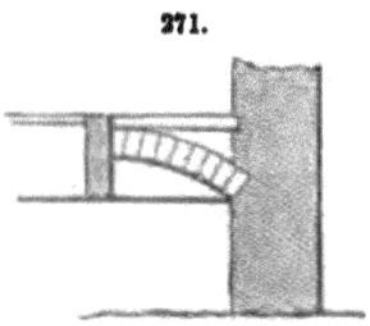

Trimmer Arch.

272.

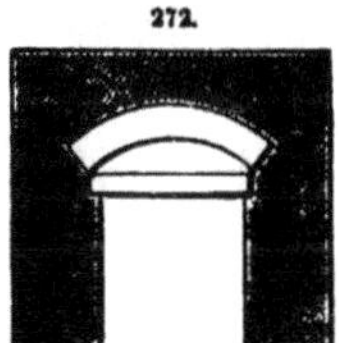

Relieving Arch.

273.

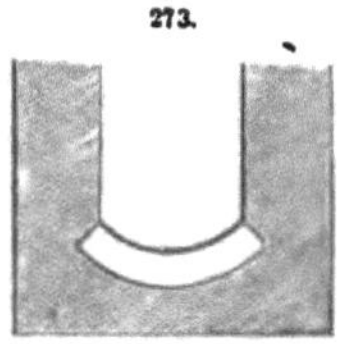

Inverted Arch.

Arches are also named from the use to which they are applied, as "trimmer arch," usually built under the hearths of fire-places, Fig. 271.

"Relieving arch," when placed over a lintel or bressummer to relieve it of the weight of the wall above, Fig. 272.

"Inverted arch," when placed under an opening or space to distribute the pressure of the walls over a more extended area, Fig. 273.

The parts of an arch are named as follows :—

274.

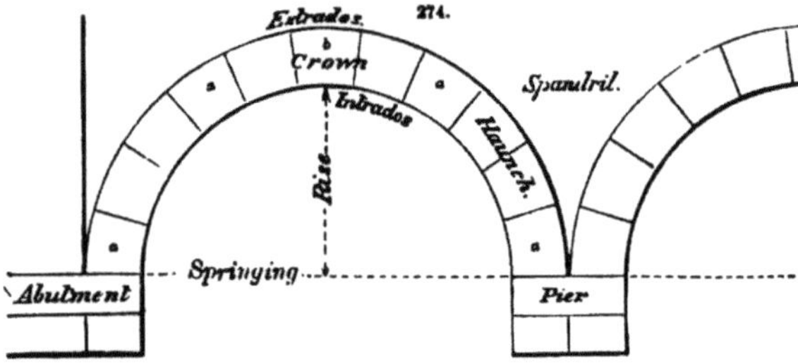

Extrados.—The outer surface or back.
Intrados.—The inner surface or "soffit."
Crown.—The upper part between the extrados and intrados.
Haunches.—The flanks or sides between the springing and crown.
Voussoirs.—The stones or blocks, *a a*, Fig. 274, of which the arch is formed.
Key-stone.—The highest or middle voussoir in the crown, as *b*, Fig. 274.
Sommering Lines.—The radiating lines corresponding to the direction of the bed-joints of the voussoirs.
Springing.—The level or point of the intrados at which the arch joins the piers.
Spandril.—The part between two arches springing from the same pier, or between the back of an arch and a perpendicular line erected from the point where the extrados commences at the springing.
Span.—The width between the piers at the springing line.
Rise.—The vertical distance between the crown and springing.
Skewback.—The part of the pier on which a segmental arch rests, or from which it springs, as *a*, in Fig. 275; the lower bed being horizontal, to correspond with the joints of the pier, and the upper bed inclined towards the centre of the arch to correspond with that of the voussoirs.

275.

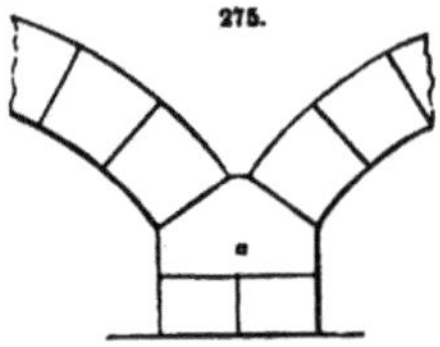

An arch derives its strength from the fact that, if the ends are prevented from spreading, its curve cannot be shortened except by the crushing of the materials. Hence it is obvious that there are *two* conditions at *least* which are essential to the stability of an arch—immobility of the supports, and sufficient strength in the materials to resist crushing. *Another* condition of equal importance is, that the arch should have sufficient thickness to contain its curve of equilibrium—when all these conditions are fulfilled, the arch is said to be *stable*.

No writer has propounded a theory by which the proper thickness of an arch at the crown can be obtained so as to be of any use in practice. The question involves the depth required for equilibrium, as well as to resist crushing. English engineers, for the most part, adopt an empirical formula, due to J. T. Hurst, 'Building News,' Feb. 27, 1857, and given in Rankine's work on 'Civil Engineering,' but in a slightly modified form.

If D = the depth or thickness of the arch at the crown, R = the radius of curvature, and C = a constant depending on the nature of the material, we have

$$D = C\sqrt{R}.$$

For Block-stone .. .. C = 0·3.
" Brickwork in mortar " = 0·4.
" Rubble-stone in mortar " = 0·45.

Where the brick or rubble work is built with Portland cement, a lower value of C may be taken, but to what extent there are no experiments at present to show.

For elliptical or other arches, with a varying curvature, R may be taken to represent the radius at the crown.

In a straight arch, D should equal $C\sqrt{S} + \frac{S}{12}$, S being the span of the arch.

According to theory, having obtained the *minimum* thickness D at the crown for an arch in equilibrium, the thickness at any other point, to prevent the arch from blowing up at that point, owing to the thrust, will equal $\frac{DV}{v^2}$, V and *v* being the rise of the arch at the crown, and at the point for which the thickness is required, respectively.

The force of an arch tending to spread, or to thrust out its abutments, is usually resolved, at the springing, into a horizontal direction, termed the "horizontal thrust," and is equal to that at the crown when the arch is in equilibrium.

Various theories have been advanced with the view of determining the value of this force, but most of them are useless to the engineer, from the difficulty involved in their application to practice

—some fail from the incorrectness of the data on which they are founded, and others because they apply only to a special form of structure instead of being adapted to the form required in practice.

When the thickness and loading of an arch have been properly determined, the horizontal thrust can generally be found with sufficient accuracy by assuming that the tendency of the arch is to separate at the crown and springing, whence

If G = distance from the springing to a vertical line drawn through the centre of gravity of the half-arch,
V = rise of the arch,
W = weight of half-arch,
P = the horizontal thrust in the same terms as W,

then $$P = \frac{W\,G}{V}.$$

As the centre of gravity is tedious to find, particularly in an arch extradossed to a horizontal line, the following method will sometimes be found convenient. Divide the half-arch into say four equal parts, and call their respective weights, W_1, W_2, W_3, and W_4, as in Fig. 276. Conceive the centre of gravity of each point to be in a line drawn vertically through the middle, then the moment of each part tending to overset the pier will be the weight of the part multiplied by the horizontal distance of its centre line from the springing; and if we take the width of each of the parts = $2a$, and the rise of the arch added to half its depth at the crown = r, the horizontal thrust will be nearly

276.

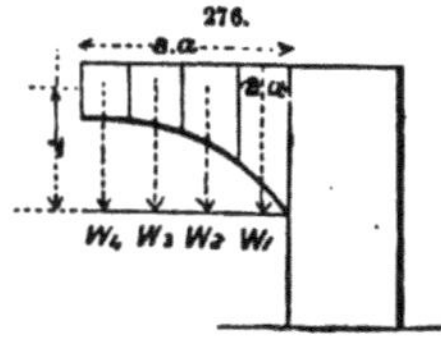

$$P = \frac{W_1 a + W_2 3a + W_3 5a + W_4 7a}{r}.$$

The *abutment*, Fig. 277, is the part which supports and takes the thrust of an arch. It is subject to three conditions:—1st. It should sustain the weight of the arch without crushing. 2nd. There must be sufficient adhesion between the courses of masonry to resist the tendency to slide. 3rd. It should be of sufficient thickness to resist the thrust of the arch without overturning.

277.

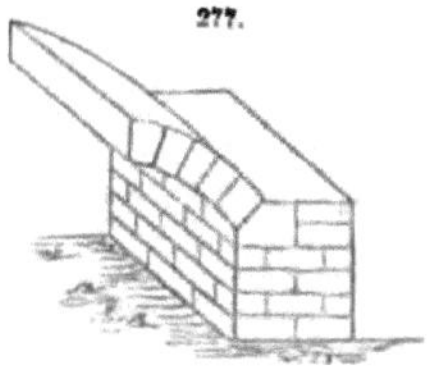

The use of granite, the harder limestones, or sandstones, fulfils the first condition in most cases. The second may be attained by the use of dowels, or a good hard-setting cement. When dowels are used, they are usually of copper, iron, or slate, from 1 to 2 inches square, and from 4 to 6 inches long. To fulfil the last condition we must resort to calculation as follows:—

Let H = height of abutment,
T = thickness of ditto,
P = horizontal thrust of half-arch,
W = weight of ditto,
C = weight of a cubic foot of the arch and abutment.

We have $$P\,H = \frac{C\,H\,T^2}{2} + W\,T;\ \text{or},\ T = \sqrt{\frac{2P}{C} + \left(\frac{W}{CH}\right)^2} - \frac{W}{CH}.$$

This formula gives the nett thickness of abutment to resist the thrust of the arch, on the supposition that the weight of the half-arch W acts on the inner edge of the abutment, instead of through the centre of gravity. In practice safety is attained by adding counterforts or wingwalls in addition to the thickness attained by the formula.

See Bridge, Construction, Dome, &c.

Works on Arches:—Atwood (G.), 'Dissertation on the Construction and Properties of Arches,' 2 Parts, 4to, 1801–1804. Gauthey (E. M.), 'Traité de la Construction des Ponts,' 3 vols., 4to, Paris, 1809–1816. Ware (S.), 'Tracts on Vaults and Bridges,' royal 8vo, 1822. Gwilt's (J.) 'Treatise on the Equilibrium of Arches,' 8vo, 1839. Scheffler, 'Theorie der Gewölbe und Futternzauern,' 1857. Woodbury's (Capt. D. P.) 'Treatise on the various Elements of Stability in the Well-proportioned Arch,' 8vo, New York, 1858. Cavalli (G.), 'Memoria sul delineamento equilibrato degli archi in murato,' 4to, Torino, 1859. Breymann, 'Bau Constructions,' Lehve, vol. i., 1860. Bland (W.), 'On the Principles of Construction in Arches, Piers, Buttresses,' &c., 12mo, 1862.

See also:—Rondelet, 'L'Art de Batir.' Sganzin, 'Cours de Construction.' Robison's 'Mechanical Philosophy.' Hann and Hosking, 'Theory, Practice, and Architecture of Bridges.'

Archimedian Screw, *for raising water*.

ARCHIMEDIAN SCREW.

Fr., *Vis d'Archimède*; Ger., *Archimedische Schraube*; Ital., *Vite d'Archimede*; Span., *Tornillo de Arquímedes*.

If, upon the surface of a cylinder, a *helix* of several spirals be traced, so that in a groove cut according to this curve are set small plates, all of the same height, and joining well upon each other, the combination will present, as it were, the thread of a screw, perpendicular to the surface of the cylinder, and of uniform thickness. A screw so formed, covered with a cylindrical envelope of staves, will constitute an *Archimedian screw*, for raising water. Its envelope with the barrel, the plates forming the thread of the screw, will be the steps, and the solid cylinder the *newel*, or core; the space comprised between the *newel*, the barrel, and the thread, will form a *helicoidal* canal. In the common screws, three equidistant blocks, or threads, are placed upon the same *newel*, and consequently three canals. The diameter of the screw, which is the interior diameter of the barrel, varies from 1 ft. to 2 ft.; that of the *newel* is a third of it, and the length of the screw is from twelve to eighteen times the diameter, as the strength may require. The angle made by the helix with the axis, or rather with a right line traced upon the *newel*, and consequently parallel to the axis, has undergone great variations. The Romans made it but 45°; at Toulouse, from plans btained from Holland, it is made about 54°; constructors at Paris generally make this angle 60°; and Eytelwein, in a small screw, carefully made, went as high as 78°. At the upper extremity of the axis there is a crank, and at the lower is a pivot, which is received in a socket, embedded in one of the small sides of the frame of the machine. If we place an Archimedian screw thus constructed in a body of water, giving it an inclination less than that of the helix upon the axis, which is usually from 30° to 45°, and impress upon it a motion of rotation, in an opposite direction to that of the helices, the inferior orifice of the canals passing in the water will draw up a certain quantity, which will rise from spiral to spiral, and will issue at the upper orifice. The screw is particularly adapted to the draining of water from places where we wish to lay, unobstructed by water, the foundations of any hydraulic structure, such as the pier of a bridge, or a lock.

Its simplicity, the small space it occupies, the facility of transporting and setting it up, as well as that of setting up many at the same place, causes its use to be very general in such drainings, and give it a preference even over other machines which have advantages in other respects.

For greater simplicity, let us take a screw, Fig. 278, formed by a tube, bent and wound round a cylinder. We first place it horizontally; if, through the orifice at the base, we introduce a bullet, in rolling, as upon an inclined plane, it will advance towards the other extremity of the tube, and it will stop upon the lowest point of the first spiral. By turning the machine, the point on which it rests will be raised; it will leave it, and, as if descending, it will pass to the following point, and in succession to the others, remaining always at the same level, but advancing towards the outlet of the tube, which it will finally attain, and pass through it. Now, incline the machine a little, and again introduce the bullet through the lower end; it will still settle itself upon the lowest point of the first spiral, when it will be raised by means of the motion of rotation, and will pass upon the following one, which will also be raised, but in a less quantity. In this manner, by a movement at once progressive and ascensional, it will gain the upper outlet: it will have risen by descending, the plane on which it rested rising more than itself. If the inclination of the screw had been such that the helix should present no point lower than that upon which the bullet is first placed, it would have continued to remain there. Finally, if the inclination had been still increased, the bullet could not have entered it; and if it had been introduced through the upper orifice of the tube, it would have descended in following all the windings, and have issued through the lower orifice.

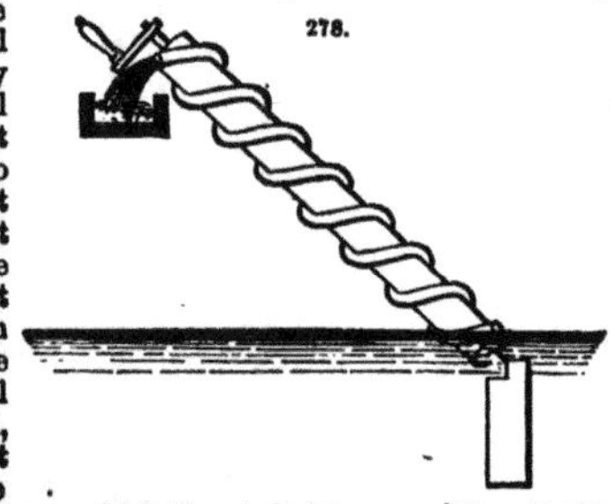

278.

What we have said of the bullet applies equally to the water which enters through the base into the spiral tube. It will flow to the lowest point of the spiral; it will then rise on both sides, in the two branches, to the level of the most elevated point of the branch of entry. The arc of the spiral, containing all the water it can then admit, is the *hydrophoric arc* of the screw. If, after the first spiral is filled, we make a revolution of the machine, the water it contains will advance, like the bullet, with a double motion, progressive and ascensional, and it will be found in the hydrophoric arc of the second spiral; it will be replaced in the first by a new and equal quantity of water. In the following revolutions, these two bodies of water, as well as those which follow after them, will ascend from spiral to spiral, even to the orifice of exit. Thus, at each revolution, the screw will evidently discharge a quantity of water equal to that contained by the hydrophoric arc.

But for this purpose, the base of the screw should be plunged in the well a certain quantity.

It should be at least so much submerged that the mouth of the helicoidal tube, after having traversed in its rotation the water of the well, on its arrival at the surface, shall be found at the summit of the hydrophoric arc of the first spiral; then this arc will be entirely filled; and it is evident that it could not be so if the level of the reservoir was below this point, whose position we shall soon determine. When the mouth, in pursuing its rotation, shall have passed this level, the atmospheric air will enter in the tube, will take the place vacated by the water, and at the end of the first revolution it will fill the upper part of the first spiral, that which is above the hydrophoric arc. It will be the same with the following spirals; the water and the air will be then disposed as indicated by the figure; each of the columns of the former fluid will be entirely supported by its spiral; it will not exert any pressure upon the inferior columns, and throughout the air will have the same density as that of the atmosphere.

It will not be so if the level of the well should be raised above the summit of the hydrophoric arc, even though the orifice of the tube may be found, in some portion of its revolution, outside the water. The air, it is true, will be introduced among the spirals, but the water will occupy more than the hydrophoric arc; it will rise, in the ascending branch, above the summit of this arc, that is to say, above the summit of the descending branch; it will bear upon the inferior column with all this excess, and will compress the air comprised between that and itself. Often this air, striving to regain its density, traverses the column which is above it. On the other hand, and by reason of the movements which take place, and of the irregularity with which the water and the air are reciprocally disposed, the last of these fluids may be found rarefied in certain parts; and we may see the atmospheric air introducing itself in the tube, passing briskly through the water of some spirals, and going to establish the equilibrium; these shocks and irregular movements diminish considerably the product of the machine.

Finally, when the base is plunged entirely in the well, the air cannot enter the screw; nothing but water can enter there. If the velocity of rotation be very great, the centrifugal force resulting from it may raise this water, and cause it to be discharged through the upper outlet (see TURBINE). But with a less velocity, the water will only reach a certain height in the tube; forming a continuous whole, it will press, with all the weight due to its vertical height, upon the orifice of entry, and will thus counteract the centrifugal force. In great machines, the air which is already in the helicoidal ducts, and that which arrives there through the upper opening, also produce irregularity in the motions, and the diminution of the product already alluded to. When, however, the canals are very large, and the machine is properly disposed and inclined, the exterior air arriving without commotion in all the spirals, these inconveniences no longer occur, and we obtain nearly the usual product.

Eytelwein, who made a particular study of the movements of water in different kinds of screws, published a series of experiments which show the bad effect of a too great or too little submersion of the base in the water to be drained; at least, for screws with small *ducts*. We give here some of the results obtained. He was provided with a model of a screw made with great care; it was 0·512 ft. in diameter and 3·608 ft. long: it had two helicoidal ducts, intersecting the axis at an angle of 78° 21′, and having, in the direction of the radius, a height of 0·138 ft. This screw was placed in a reservoir, in an angle of 50° to the horizon; and when it yielded the greatest product, the level was 0·042 ft. above the centre of the base. In the first column of the annexed Table, the height of the water above or below the centre of the base is indicated; and in the second, the volume of water raised at each revolution.

Height of Level.	Product per Revolution
ft.	cub. ft.
·400	0·008
·180	0·009
·082	0·009
·049	0·010
·041	0·012
·032	0 011
·016	0·011
·019	0·010

Though the Archimedian screw is simple in its character, still there is no theory to be found for the machine as it is now used. The essays of some learned mathematicians are far from enabling us to determine its effects exactly. That which Bernoulli and most authors have given, applies only to the case, now out of use, of a tube, with a very small diameter, rolled spirally round a cylinder. We make an elementary exposition of the principal features of it, both to guide our first impressions upon this subject, and to avoid leaving a gap in this work.

Let A M C N D, Fig. 279, be a vertical projection of the axis of the helicoidal tube, wound round the cylinder A B E D, and the circle $a\,n\,b\,m\,a$ a projection of the base of the cylinder, upon a plane perpendicular to its axis. Through the point F of the arc A M M′ C draw the tangent G H; it will make with the edge O I an angle I F H, which we designate by a; and through the extremity B of A B draw the horizontal B K. the angle E B K, or b, will measure the inclination of the screw.

279.

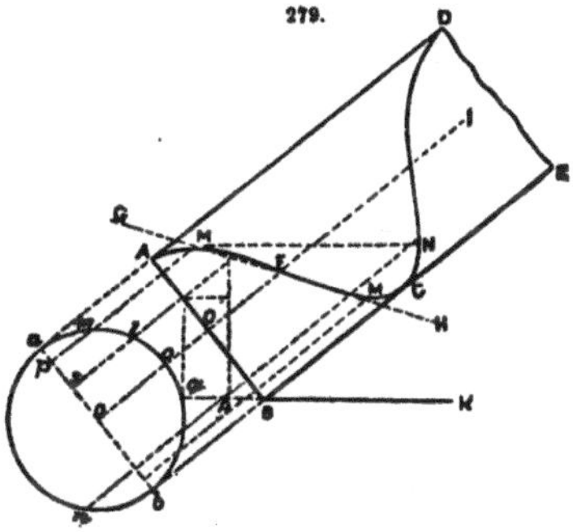

Let us determine the length of the hydrophoric arc M C N.

And first, the height L P of any point L of the helix, above the horizontal plane B K. Project L at l upon the circumference of the circle of the base, and draw the horizontal lr, we shall have L P = L r + r P. For greater simplicity make the radius $oa = 1$; designate by a the length of the arc A l ($= al$); the angle which the helix makes at A with the plane of the base, being the complement of a, we shall find $\text{L}r = \text{L}l \sin. b = \text{A}l \cot. a \sin. b = a \cot. a \sin. b$. We shall also have $r\text{P} = lq = l\text{B} \cos. b = sb \cos. b = (1 + \cos. a) \cos. b$. Then $\text{LP} = a \cot. a \sin. b + (1 + \cos. a) \cos. b$.

The summit or commencement of the hydrophoric arc of the spiral A C D will be at M, the most elevated point above B K. It corresponds consequently to the *maximum* value of L P. Differentiating the above expression, equalling the differential to zero, we have $\sin. a = \cot. a \tang. b$; which gives the value of the arc a, or dm, for the case of the *maximum*. Calling m this particular value at the point M, we have for the height of this point above B K, $m \cot. a \sin. b + (1 + \cos. m) \cos. b$.

If through M we imagine a horizontal plane, the point N, where it intersects the ascending branch of the spiral, will be the end of the hydrophoric arc; since the commencement and the end should have the same level. Project N upon the circumference of the base; it will fall upon the point n; call n the arc bn; the arc of the circle $a\,m\,b\,n$, corresponding to the arc of the helix A M C N, will be $\pi + n$; and for the elevation of N above the horizontal plane passing through B, we shall have $(\pi + n) \cot. a \sin. b + [1 + \cos. (\pi + n)] \cos. b$.

This elevation should be equal to that of M. Making the two expressions equal and reducing, we have $(\pi + n) \sin. m + \cos. (\pi + n) = m \sin. m + \cos. m$: an equation from which we may deduce the value of n, by means of successive substitutions.

This value being found, we shall know the arc mbn corresponding to the hydrophoric arc M C N. But an arc of the helix is equal to an arc of the corresponding circle, increased in the ratio of the radius of the tables to the cosine of the angle comprised between the two arcs, that is to say, divided by this cosine. Here the arc of the circle is $\pi - m + n$, the angle comprised between the two arcs is $90° - a$: the length of the hydrophoric arc will then be $\frac{\pi + n - m}{\sin. a}$; and, for a cylinder whose radius is r,

$$r\,\frac{\pi + n - m}{\sin. a}.$$

If s is the section of the helicoidal tube, the volume of water raised at each turn of the screw will be the above expression multiplied by s.

Calling N the number of turns made by the screw in a given time, L its length outside of the water, and observing that the height of the elevation is L sin. b, we shall have for the value of the useful effect, during this time,

$$\text{N L } s r (\pi + n - m) \frac{\sin. b}{\sin. a}.$$

The expression sin. m = cot. a tang. b, obtained by differentiating, and making equal to zero the general value of the elevation of any point of the first spiral, answers equally to the case of *maximum* and *minimum*; it gives the smallest as well as the greatest elevation. Moreover, the sin. m applies as well to the arc $a\,m'$ as to the arc $a\,m$, by taking $b\,m' = a\,m$. Consequently, if we project the point m' upon the hydrophoric arc, M'. which is its projection, will be the lowest part of the arc, as M is the highest point.

The expression cot. a tang. b, representing a sine, cannot exceed 1. When it is equal to it, the arcs $a\,m$ and $a\,m'$ will become $a\,o'$; the points M and M' will be merged in the point F; there will no longer be a hydrophoric arc, and no more water raised. But cot. a tang $b = 1$ gives tang. $b = \frac{1}{\cot.\,a}$ = tang. a or $b = a$; that is to say, that when the angle of inclination shall be equal to the angle made by the helix with the edge of the cylinder, the discharge will cease; it is necessary, then, in order that it may take place, that the first of these angles should be smaller than the second, as we have already remarked.

That of the values of b giving the greatest effect is impliedly embraced in the above expression of effect. For the same screw, moved with the same velocity, there will be no variable in this expression but sin. $b\,(\pi + n - m)$, and it will be necessary to determine the value of b which will render this quantity a *maximum*.

From what was said at the commencement of this article, in order that the hydrophoric arc should take all the water it can contain, the level of the fluid in the well should be as high as the point m, or as the point p, which is on the same horizontal line; and consequently should be raised above the centre of the base by the quantity $o\,p = r \cos.\, m = r \sqrt{1 - (\cot.\,a\ \text{tang.}\,b)^2}$. For the vertical elevation, we shall have

$$r \cos.\, b \sqrt{1 - (\cot.\,a\ \text{tang.}\,b)^2}.$$

In what has been said, we have supposed the hydrophoric arc had time to be filled with water, without any mention of the velocity of the water. It has, however, a great influence upon the amount of the product, especially when the bottom of the screw is entirely submerged. This influence is shown by the experiments of Eytelwein. They were made with the small screw already mentioned, with an inclination of 50°. In the first series, the end of the screw was entirely submerged; an unfavourable circumstance, the disadvantages of which are not sufficiently appreciated by workmen. The second was made under more favourable circumstances, with the base submerged only a suitable quantity. In practice, it will suffice to establish the screw in such a manner as that the end of the vertical diameter of the core may project a little above the surface.

Number of Revolutions in 1'.	Water raised per Revolution.
	cub. ft.
22	0·0099
41	0·0094
51	0·0088
74	0·0081
121	0 0068
56	0·0113
60	0·0118
73	0·0121
85	0·0123
98	0·0123
120	0·0118

Comparing the terms of the two series, when the velocity of the machine has been nearly the same, we see that when the inferior extremity was entirely submerged, the product was about one-third less.

We pass to the effect of which great screws are capable.

It is made known what this product would be, by giving, in the following Table, the results of experiments made with three pumps, of 1 ft., 1½ ft., and 2 ft. (French measure) in diameter, the latter limit never being exceeded. The length and velocity of each are given, as well as the angle of inclination at which it stopped delivering water; an angle which, according to theory, is equal to that made by the helix with the axis. The greatest effect was produced at an angle of 30°; Morin has taken it for the unit, and has compared with it those obtained under different angles; this comparison shows the great influence of the inclination.

Angle of Inclination.	Diameter .. .. = 0·066 ft. Length .. .. .. = 19·182 ft. Revolution in 1' .. = 90 Limit of Inclination = 60°			Diameter .. .. = 1·597 ft. Length .. .. .. = 27·69 ft. Revolution in 1' .. = 60 Limit of Inclination = 62°			Diameter .. .. .. = 2·10 ft. Length .. .. .. = 25·57 ft. Revolution in 1' .. = 41½ Limit of Inclination = 65°		
	Water raised in 1 hour.	Height of Elevation.	Series of Effects.	Water raised in 1 hour.	Height of Elevation.	Series of Effects.	Water raised in 1 hour.	Height of Elevation.	Series of Effects.
°	cub. ft.	ft.		cub. ft.	ft.		cub. ft.	ft.	
30	1486·8	8·98	1·00	4576	12·36	1·00	9149	10·66	1·00
35	1236·	10·10	0 93	3630	14·62	0·94	7164	13·12	0·97
40	872·3	11·25	0·74	2397	16·85	0·71	4841	14·90	0·74
45	443·8	12·36	0·50	1306	19·12	0·44	2613	16·49	0·44
50	307·2	13·48	0·31	508	20·23	0·18	893	..	..
55	91·8	14·62	0·10	180	21·35	0·07	367	17·84	0·07

Though the volumes of water indicated in the Table have been admitted, as the results of experiment, by a commission of engineers; still, as they are presented by a constructor of the

Archimedian screw, we may fear that there is some exaggeration; and in application, we should not reckon upon more than two-thirds of the product indicated.

It seems that the quantities of water raised by these machines, they having been reduced to the same number of turns in the same time, should be proportional to the capacity of the hydrophoric arc, and consequently to the cube of the diameters, if the screws were similar solids; yet Morin found that these quantities are very sensibly proportional to the $3\frac{1}{3}$ power of the diameter, or to $D^{3\frac{1}{3}}$. Consequently, by reducing one-third the quantities given in the preceding Table, the volumes of water raised in one hour, under different angles of inclination, by a screw of a given diameter D, would be such as are indicated in the adjoining Table.

Angle of Inclination.	Water raised in 1' by 40 Revolutions.
°	cub. ft.
30	364 $D^{3\frac{1}{3}}$
35	288 „
40	191 „
45	104 „

These screws are usually put in motion by men, who act indirectly upon the crank, through the intervention of beams or connecting-rods, upon which they impress a reciprocal motion which converts that of the crank into a rotatory. What is the number of men to be employed to produce a given effect?

A screw 1·607 ft. in diameter, and 19·19 ft. long, used for draining by M. Lamandé, engineer, moved by nine men (working in spells of two hours, and then relieved by a similar number of fresh hands), inclined about 35°, making forty turns per minute, raised in one hour 1589·2 cub. ft. of water 10·82 ft. For each of the nine workmen, this was 176·58 cub. ft. raised 10·82 ft., or 1910 cub. ft. raised 1 ft.; he did not work over five hours in the day; thus, the day's labour of each was only 9550 cub. ft. In another experiment, six workmen, working six hours, raised each 10,660 cub. ft., and consequently 1776 cub. ft. per hour.

According to these positive and authentic facts, we may admit that a workman, employed upon a well-arranged screw, can raise in one hour 1738 cub. ft. one foot in height, and that he may labour in this manner six hours per day. He might even work eight hours in the twenty-four, in a continuous draining, if the relays were properly established; so that the number of workmen to accomplish such a draining would be $\frac{Q'H'}{579}$, or, to prevent any mistake, $\frac{Q'H'}{463}$, Q' being the volume of water to be raised in one hour, and H' the height of the elevation.

We also employ for draining, screws without the envelope or barrel, consisting simply of a newel, upon which are placed the helicoidal threads. We place them in a canal or semi-cylindrical box enclosure, made of carpentry or masonry, and having a suitable slope: it is, as it were, a half-barrel, but immovable. But a very small interval is left between its sides and the edges of the threads. These machines, called *hydraulic screws* by the Germans, are much used in Holland, where they are frequently set in motion by windmills.

They have a great velocity imparted to them, lest a great quantity of water, raised at first, should fall back into the well, following the sides of the trough, before it has reached the point of discharge. They have the advantage of being independent, in their product, of the height of the water of the reservoir compared to their extremity, and, without shifting their place, they may drain a reservoir whose level is gradually reduced. But this advantage is more than counteracted by an inconvenience: very often the core or newel, at least if it is not large, bends, and the edges of the threads rub against the sides of the canal; which wears out the machine, and occasions a resistance, absorbing a portion of the motive force.

It may be necessary to make brief mention of a machine, which has some resemblance to the Archimedian screw, and which may be used for raising water to great heights: this is the *spiral pump*. It consists of a conical or cylindrical turning-shaft, upon which is wound, screw-fashion, a tube of lead or other material: one of its extremities takes up the water, and the other is enclosed exactly in the curved end of an upright tube, which conveys this water to the desired point.

This machine, invented and made, in 1746, by a tinman of Zurich, has been made the subject of a work by Daniel Bernoulli, who has given its theory, and proposed some improvements, which have been adopted in a construction made at Florence. Since then, Nicander and Eytelwein have devoted their attention to it: the latter reported that, in 1784, he had established such a pump near Moscow, with complete success; it conveys 4 09 cub. ft. in 1 a distance of 761 ft., and 75·46 ft. in vertical height. This author extols all the advantages of this machine, and recommends its use.

When the mouth takes up alternately water and air, these two fluids advance, from spiral to spiral, up to the upright pipe: they enter it; the air is discharged and escapes into the atmosphere, the water ascends gradually, and is discharged through the spout placed at the top of the pipe.

During the motion, the two fluids are disposed in the spirals as shown in the figure: the water on one side, the air on the other—the latter occupying less and less space. In the first spiral from the entrance-mouth the air is loaded, not only with the atmospheric weight, but that of the column of water of the second spiral; the air of the latter sustains also the weight of the third column; and so on; so that in the last spiral, that which is near the upright tube, it is as it were loaded with the weight of a column of water, whose height is the sum of the heights of this fluid in all the spirals. This same air supports, by the elastic force due to such pressure, the column of water in the upright tube; it can therefore support one whose height is equal to the sum of the heights of the water in the spirals. Thus the height to which we can raise water by means of a spiral pump, depends upon the length and the number of spirals of the helicoidal tube.

If the compressed air, on issuing from this machine, were properly received and directed, it would produce a blast, which might easily be made nearly continuous. An Archimedian screw, containing also in its spirals alternate masses of air and water, might yield an analogous effect, if it were disposed and moved in an order in some sort the inverse of that followed in draining. Blast-machines on this principle have been used in many instances with success. The Archimedian screw, in this latter case, is of great diameter compared with its core, and placed in a basin filled with water, with a certain inclination, so that the upper end of the axis shall be very near the liquid surface. When the screw turns, the upper mouth of the helical canal passing in the atmosphere during one-half of its revolution, there takes a certain quantity of air, which at first has its place above the first hydrophoric axis, and which then descends from spiral to spiral, issues through the lower mouth of the canal, and tends to rise in the water of the basin, with an elastic force measured by the height of the liquid surface above this mouth.

AREA.

The sunk space around the lower stories of a building is termed the area. See MENSURATION.

AREA-DRAIN. FR., *Isolement àiré.* GER., *Luftschicht.*

280.

An area-drain is a narrow open area, Fig. 280, generally less than 3 feet in width, constructed around the basement of a building, to prevent the approach of damp from the surrounding soil.

ARGENTAN. FR., *Argentan, alliage de cuivre, nickel et zinc*; GER., *Nickelsilber*; ITAL., *Packfong*; SPAN., *Argentan mezcla de cobre, niquel y zinc.*

Argentan is the name given to an alloy of nickel, copper and zinc; and is generally termed German silver. See NICKEL, *Alloys of.*

ARM-BAND. FR., *Support à fusil*; GER., *Gewehrhalter.*

An arm-band, Figs. 281, 282, 283, 284, is a piece of crooked iron, attached either to a wooden rail or stone block, fixed against the walls in barrack-rooms, to retain the soldiers' muskets when not in use, the butt-ends resting on the floor. Formerly, muskets were laid horizontally, one over another, in arm-racks, which were not so convenient to reach in cases of emergency.

281.

282.

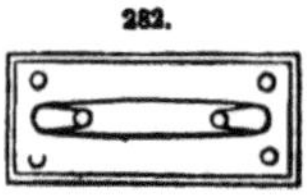

283.

284.

285

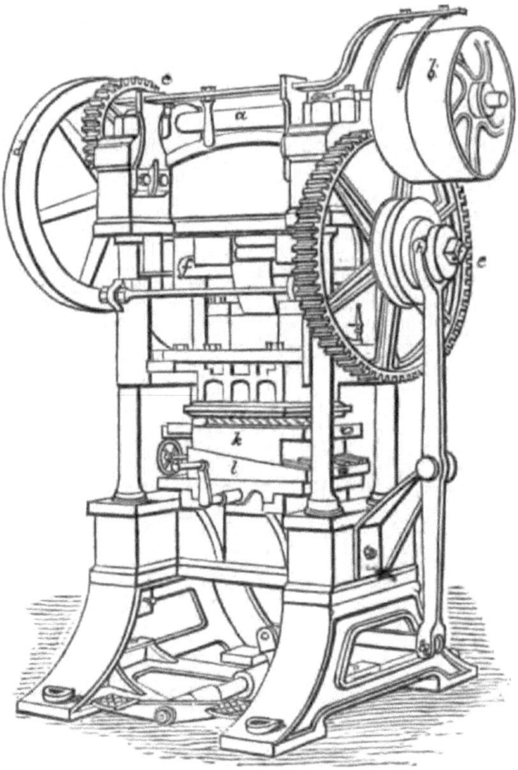

ARMING-PRESS. FR., *Presse de carton*; GER., *Deckelpresse*; ITAL., *Torchio da imbozzare*; SPAN., *Prensa para encuadernar.*

An arming-press is a machine used for embossing the back and sides of the cover of a book. Fig. 285 represents the rotary arming-press designed by John Gough, to be worked by steam-power. The shaft, *a*, carrying the strap-pulleys, *b*, is fixed upon the top of the press; and at one end of the shaft is placed a toothed-wheel, *c*, geared into the fly-wheel *d*. On the other end of this shaft is fixed a small pinion, geared into the large wheel, *e*, running loose upon the eccentric shaft *f*. This wheel is fitted on a clutch, *h*, sliding on the square end of the eccentric shaft.

This press, which is of large size, is fitted with two sets of gauges, for feeding front and back, so that, with two attendants, two blocks can be worked at the same time, rendering this press equal to two of a smaller size. The whole is supported by columns, which tie the head and top bearings, bed, and standards, in a compound and rigid manner. The support at each side of the eccentric prevents deflection of the eccentric shaft. The table, *k*, is supported by a sliding-wedge, *l*, the whole of its width, by which arrangement the whole power of the press is concentrated upon the work. The driving-gear and machinery, being placed on the head of the press, renders the usual dangerous gearing on the floor unnecessary.

ARMOUR. FR., *Armure, Cuirasse*; GER., *Panzer*; ITAL., *Corazzatura*; SPAN., *Armadura*.

ARMOUR-PLATES *for Ships of War.*

The first account we have of an armoured ship is in 1530 (see 'Istoria della Sacra Religione ed Illustrissima Milizia di San Giovanni Gierosolimitano,' di Giacomo Bosio, pub. 1594). The largest ship then known, one of the fleet of the Knights of St. John, was sheathed entirely with lead, and is said to have successfully resisted all the shot of that day. We have no account of any similar device until the Emperor Louis Napoleon initiated the idea of covering vessels intended to approach forts with iron, in consequence of the general acknowledgment of the impossibility of supporting the effects of shell-fire in ordinary wooden ships, or in thin iron ones when elongated shell are substituted for the spherical ones formerly used. In the United States, trial has been made of superposed plates of 1 and 2 inches thick, up to a total thickness of 8, 13, and 20 inches; but the resistance obtained in this way has never equalled the results where solid plates have been applied.

It has been roughly established that to keep out a hardened projectile fired from a modern rifled gun, the armour-plate must have a thickness at least as great as the diameter of the shot; and the best practice seems to require a solid backing of wood of from twice to four-and-a-half times the thickness of the iron. This will give great resistance when divided into a cellular form by iron edge-pieces or girders, as in the Chalmers' target. An iron lining on the inside is also necessary to diminish the risk of splinters, &c.

Armour-plates are rolled or forged up to the thickness of 20 inches by machinery constructed for the express purpose. They are then bent to the shape necessary by hydraulic-power; and having been planed on the edges and bored with conical holes for the bolts, are attached to the ship's side by long through-bolts, and screwed up with nuts from the inside.

As the artillery have within the last few years advanced with enormous strides in weight and power, it has been necessary to keep pace with these improvements, in the thickness and resisting-power of the armour; and each step forward in the science of artillery has hitherto been met by a corresponding increase in the defences opposed to it. But this cannot go on; and it seems that the limit of defensive armour the ship can carry will be sooner reached than the *ne plus ultra* of the gun or shot that can be brought against it.

It remains to be seen to how much of the ship must be given absolute immunity, and how much may be left to be pierced by shot without fatal consequences. Hitherto the practice has varied from protection over the whole hull with a moderate thickness of armour, to absolute protection confined to a small part of the hull. The Monitor system protects efficiently the hull and guns; but these are little raised above the surface, and the vessel is not intended to keep the sea. The cupola system is an advance on the Monitor, inasmuch as it is applied to vessels of a larger class, and which are seaworthy. But on this plan very few guns are carried in proportion to the tonnage, and the ship requires one special armour and the guns another. There is also the disadvantage that the rigging must be sacrificed if the guns are to be used over a large arc of training.

In 1865 a series of experiments were made by the British Government to determine the relative penetrating effects of two shot on an iron plate, provided they strike with the same *work* or *energy*, notwithstanding the one may be heavy with a low velocity, and the other light with a high velocity; and also to determine the relative resistances of a plate to penetration by two shot of similar form of head, and striking with work proportional to their respective diameters.

For these experiments the charges were determined with the aid of Navez's apparatus, by which the velocity of each projectile was observed at the distance of 100 yards from the muzzle. Cast-iron shot were supplied for this experiment, their weight being the same as the steel shot. The spherical projectiles were shells weighted up with lead.

The following Table shows the velocity which each projectile should have in order that the conditions might be fulfilled.

TABLE A.—SHOWING THE NECESSARY VELOCITIES AND CHARGES DETERMINED BY EXPERIMENT: 5·5-INCH PLATES. PROJECTILE, SOLID STEEL HEMISPHERICAL-HEADED SHOT.

Gun.	No. of Experiment.	Charge determined by Experiment.	Projectile. Mean Weight.	Projectile. Mean Diam.	Necessary Velocity at 100 Yards.	"Work" on Impact at 100 Yards.
		lbs.	lbs.	inches.	feet.	foot tons.
6·3-inch M. L. rifle-gun	1	15·848*	35·56	6·22	1917	906·2
" "	2	12·000*	71·12	"	1355	"
" "	3	11 219*	106·68	"	1107	"
" "	4	13·875*	35·56	"	1823	819·3
" "	5	10·500*	71·12	"	1271	796·2
" "	6	9·812*	106·68	"	994	731·5
7-inch M. L. rifle-gun	7	13·500	100·00	6·92	1130	885·8
" "	8	11 625	"	"	1022	724·8
100-pr. M. L. smooth-bore gun (9-in.) ..	9	15 437†	104·00	8·87	1254	1135·4
" " " ..	10	11·125†	"	"	1135	929·0

* Charge, powder nearly half the weight of shot. † Charge, powder one-seventh the weight of shot.

The charges having been thus ascertained, the guns used were placed in battery, at 100 yards from a row of 5½-in. iron plates, fixed by upright supports, but unbacked. The guns were fired directly at the plates; that is, the plane in which the shot moved was perpendicular to the face of the plates.

TABLE B.—ABSTRACT, SHOWING THE RESULTS OF THE EXPERIMENTS CARRIED OUT AT SHOEBURYNESS, 22/3/65, AGAINST 5·5-INCH UNBACKED PLATES.

1000 to 1010 show Effect of Equality of *vis viva* where V and W vary. In the last three rounds the charges were altered on the ground, and the effects are not comparable:—

35 lbs Shot, Spherical,	Length,	6·220 inches.	
63 " " Elongated,	"	8·500 "	Hemispherical-headed.
70 " " "	"	9·315 "	
106 " " "	"	13·458 "	

Photographic Number of Round.	Charge.	Weight of Projectile.	Velocity on Impact.	Approximate $\frac{Wv^2}{2g}$ in Foot Tons on Impact.	Effects with Steel Projectiles of 6·22-in. Diameter, fired from the 6·3-in. Gun of 140 cwt., R., Expl. No. 275.
	lbs.	lbs.	feet.		
1000	15·848	35·875	1920·0	917·0	Clean hole *through* plate, and 3 ft. 2 in. into earth.
1003	"	35·562	1925·9	913·8	" " and stopped by a balk of timber.
1001	12·000	63·687	1417·0	886·7	" " and 3 ft. 6 in. into earth.
1004	"	70·937	1345·0	889·8	Struck left edge of plate and broke in three pieces.
1006	"	71·250	1346·0	895·1	Clean hole *through* plate, and 5 ft. 6 in. into earth. "Earth loosened by former shot."
1002	11·219	106·625	1110·0	911·0	Clean hole *through* plate, and 3 ft. 3 in. into earth.
1007	"	106·812	1112·0	915·8	" " depth in earth not known.
1008	13·875	35·562	1829·0	824·9	" " about 1½ ft. into earth.
1009	10·500	70·875	1270·0	792·7	Stuck in plate; base projects 3¼ in. from face of plate; slight star in rear; outer lamina off plate.
1010	9·812	106·562	996·0	733·1	Stuck in plate; base projects 7 in. from face of plate; part of shot showing through in rear.

TABLE C.—TO DETERMINE THE RELATIVE RESISTANCES OF A PLATE TO PENETRATION BY TWO SHOT OF SIMILAR FORM OF HEAD, AND STRIKING WITH VIS VIVA PROPORTIONAL TO THEIR DIFFERENT DIAMETERS. SOLID STEEL SHOT, HEMISPHERICAL-HEADED.

Photographic Number of Round.	GUN.	CHARGE.	PROJECTILE.		Velocity on Impact.	Approximate $\frac{Wv^2}{2g}$ in Foot Tons on Impact.	$\frac{Wv^2}{2g}$ as proportional to Diameter.	EFFECTS.
			Weight.	Diam.				
		lbs.	lbs.	inches.	feet.		foot tons.	
1011	100-pr. smooth-bore.	15·437	104·125	8·87	1254·0	1135·4	1135·4	Clean hole *through* plate, and 3 ft. 3 in. into earth.
1009	6·3-inch M. L.	10·500	70·875	6·22	1270·0	792·7	796·2	Stuck in plate; base projects 3¼ in. from face of plate. Nose was nearly through.
1027	100-pr. smooth-bore.	11·125	104·000	8·87	1135·0	929·0	929·0	Shot rebounded, indent 3·25 in. deep; plate bulged and cracked in rear, over an area of 1 ft. 2 in. by 1 ft. Same plate as 1002, 1003, 1009.
1012	7-inch M. L., No. 217.	11·625	100·312	6·92	1004·0	701·1	724·8	Shot rebounded, indent 3·655 in. deep, 7·33 in. diameter; plate cracked in rear; shot much cracked.
1013	"	"	100·125		1012·0	711·0	724·8	Shot rebounded, indent 4 in. deep, 7·1 in. diameter; plate cracked in rear.
1011	100-pr. smooth-bore.	15·437	104·125	8·87	1254·0	1135·4	1135·4	Clean hole *through* plate, and 3 ft. 3 in. into earth.
1026	7-inch M. L., No. 217.	13·500	100·312	6·92	1131·0	889·7	885·8	Clean hole *through* plate; shot struck and broke support of wood 1 ft. square, and fell in rear. Same plate as 1000, 1004, 1010.
1008	7-inch B. L.	12·000	110·000	6·88	1090·0	906·2	906·2	Just penetrated the plate; 2 ft. into earth. Brown's steel.
765	6·3-inch M. L.	13·875	35·562	6·22	1829·0	824·9	819·3	Just *penetrated* the plate; 1½ ft. into loose earth.
1009	"	10·500	70·875	"	1270·0	792·7	819·3	See above. Did not penetrate; base 3¼ in. out.
908	68-pr. smooth-bore.	16·000	72·000	7·91	1365·0	930·2	930·2	Indent 2·8 in.; plate cracked behind. Firth steel. *Minute* 14,082.
1010	6·3-inch M. L.	9·812	106·562	6·22	996·0	733·0	731·5	Stuck in plate.

It appears from these Tables, round 1008, that a 6·22-in. projectile is *just* able to penetrate a 5½-in. plate, with a *work* on impact of about 825 foot tons; and assuming that the resistance of the plate varies as the square of its thickness, we shall have the following proportion to determine the work necessary to penetrate a 4½-in. plate with the same projectile; that is

$$5 \cdot 5^2 : 825 :: 4 \cdot 5^2 : x, \text{ and } x = 552 \text{ foot tons.}$$

Experiments against 4½-in. unbacked plates, under similar conditions to those detailed in the experiments against 5½-in. plates :—

TABLE D.—ABSTRACT, SHOWING THE RESULTS OF EXPERIMENTS CARRIED OUT AGAINST 4·5-INCH UNBACKED PLATES TO DETERMINE THE RELATIVE PENETRATING EFFECT OF PROJECTILES OF THE SAME DIAMETER AND FORM OF HEAD, BUT SO VARYING IN WEIGHT AND VELOCITY THAT THE VIS VIVA ON IMPACT WAS CONSTANT.

Date of Experiment, 13/3/66. *Brand of Powder*, Rifle L. G. 8/7/64. *Lot* 805.

Photographic Number of Round.	Number of Plate.	Charge.	Weight and Length of Projectile.	Observed Velocity at 220 Feet.	Calculated Velocity on Impact at 300 Feet.	Approximate $\frac{Wv^2}{2g}$ in Foot Tons on Impact.	Effects with Hemispherical-headed Steel Projectiles of 6·22-in. Diameter, fired from the Service 64-pr. M. L. Gun of 6·3-in. calibre.
		lbs.		feet.	feet.		
1047	..	6·69	63·87 lbs. 8·42 in.	1121·2	1112·2	547·8	Just penetrated. Shot rebounded about 6 yards; length of shot 8·05 in. Preliminary round.
1158	1	6·63	63·87 lbs. 8·42 in.	1128·3	1119·3	554·8	Just penetrated; broke plate behind in the usual manner; shot rebounded 4 ft.; length of shot 7·92 in.; diameters of hole 6 × 6·25 in.
1159	1	,,	70·94 lbs. 9·3 in.	1077·7	..	..	Miss. Struck support of plate, and glanced off into the earthwork.
1160	1	6·09	106·19 lbs. 13·39 in.	864·1	860·5	545·2	*Through* plate, breaking away rear in the usual manner. Shot fell 2 ft. in rear; length of shot 12·92 in.; diameters of hole 6·75 × 7·5 in.
1161	1	7·87	35·50 lbs. 6·22 in.	1483·6	1460·0	524·7	Stuck in plate, breaking it away behind; shot almost through.
1162	2	8·00	35·56 lbs. 6·22 in.	1506·7	1482·4	541·8	Just penetrated; broke plate behind in the usual manner; shot rebound d 4 ft.; diameter of shot 6·32 in.; diameters of hole 6·4 × 6·5 in.
1163	2	6·63	70·94 lbs. 9·3 in.	1069·2	..	..	Miss. Struck support, and glanced into earthwork.
1164	2	,, ?	63·81 lbs. 8·42 in.	1107·1	1098·2	533·6	Almost penetrated; broke away plate behind over an area of 1 ft. by 1 ft. Shot rebounded 5 ft. 9 in. Indent 4·35 in.; length of shot 7·88 in. Plates XIII., XIV.
1165	2	6·06	106·62 lbs. 13·39 in.	861·3	857·7	543·9	Stuck in plate, breaking it away at back something more than round 1164; shot almost through.
1166	3	,,	,,	863·1	859·5	546·2	*Through* plate. Shot turned over and entered earthwork to a depth of 1 ft.; length of shot 18·96 in.
1167	3	8·00	35·50 lbs. 6·22 in.	1509·2	1484·9	542·8	Made a hole clean through, but shot remained sticking in the plate, projecting as much in rear as in front.

On examining this Table it appears that all the projectiles but one struck with "work" slightly under that which was required, *viz.* 552 foot tons; and that 542 tons is only just capable of piercing a 4·5-in. plate. Thus in most instances the shot, *after penetration*, rebounded, and fell in front of the plate, showing that they had expended almost their entire force in the penetration. As 552 tons was calculated on data supplied by a shot, round 1008, which penetrated, *and had some little force left in it*, it is to be expected that a force of 542 tons should act as it did. It appears that a reduction of two ounces in the charge, and consequent diminution of "work" to 525 foot tons, was sufficient to prevent complete penetration, round 1161, although it apparently required but a small blow with a hammer to separate the piece of plate at the back of the point struck. As this effect was produced by the shot moving with the highest velocity, it is a convincing proof that, with steel shot, the penetration is not proportional to a higher power than the square of the velocity.

From these experiments the following *practical* conclusions may be drawn when the projectiles are fired *direct* :—An unbacked wrought-iron plate will be perforated with equal facility by solid steel shot, of similar form of head, and having the same diameter, provided they have the same *vis viva* on impact; and it is immaterial whether this *vis viva* be the result of a heavy shot and low velocity, or a light shot and a high velocity, within the usual limits of length, and so on, which occur in practice. An unbacked iron plate will be penetrated by solid steel shot, of the same form of head but different diameters, provided their striking *vis viva* varies as the diameter, nearly, that is, as the circumference of the shot. That the resistance of unbacked wrought-iron plates to absolute penetration by solid steel shot of similar form and equal diameter, varies as the square of their thickness, nearly.

These experiments have proved that, although in the case of cast iron a light projectile moving with a high velocity will indent iron plates to a greater depth than a heavier projectile with a low velocity, but equal "work," it is not as necessary that there should be a high velocity when the projectiles are of a hard material, such as *steel* and *chilled iron*; and this result will be much in favour of rifled guns, by enabling them to prove effective with comparatively moderate charges

Putting these results in an algebraic form, and taking the units as the pound and the foot,

$$\frac{W v^2}{2g} = 2 \pi R k b^2, \qquad [1]$$

where W = weight of shot in lbs., v = velocity on impact in feet, g = the force of gravity, 2 R = diameter of shot in feet, b = thickness of unbacked plate in feet, k = a coefficient depending on the nature of the wrought iron in the plate, and the nature and form of head of the shot.

The shot is supposed to be of the best quality of steel, and the plate of the best quality of wrought iron.

Solving equation [1] for b,

$$b = v\sqrt{\frac{W}{4\pi R g k}}, \qquad [2]$$

and for k,

$$k = \frac{W v^2}{4\pi R g b^2}. \qquad [3]$$

In order to determine k, a series of equations can be formed of the following conditions:—

$4\pi R g b^2 k - W_1 v_1^2 = 0, \quad 4\pi R_2 g b^2 k - W_2 v_2^2 = 0, \quad 4\pi R_3 g b^2 k - W_3 v_3^2 = 0,$ &c. &c. &c.

Substituting the experimental values of the different quantities, and eliminating k, it will be found that for hemispherical-headed shot $k = 5{,}357{,}200$.

Having thus determined the value of k, the work necessary to penetrate any unbacked plate of given thickness can be calculated.

Thus, to determine the *work* required to just penetrate a 5·5-in. plate, with a hemispherical-headed steel shot of 6·22 in. diameter,

$R = 3{\cdot}11$ in. $= 0{\cdot}25917$ ft. $\quad b = 5{\cdot}5$ in. $= 0{\cdot}45833, \quad k = 5{,}357{,}200$;

and substituting these values in equation [1],

$$\frac{W v^2}{2g} = 1{,}832{,}522 \text{ lbs.} = 818 \text{ foot tons.}$$

It will be seen that round 765, consisting of a spherical shot of 35·56 lbs. and 6·22 in. diameter, moving with a velocity of 1829 ft., and consequent *work* of 825 foot tons, just penetrated a 5·5-in. plate. This *work* is practically the same as that given in the above example, as a difference of 5 ounces in the weight of the shot would have reduced its *work* to 818 foot tons.

Oblique Fire.—Suppose the plate set at an angle, or the gun fired obliquely at an upright plate. The shot has then a tendency to glance off, and continue its motion in a new direction.

The force with which the shot, acting obliquely, will strike, is to that with which it would strike if acting directly as the sine of the angle of incidence is to unity. That is, the shot striking in a slanting direction may be supposed to have opposed to it a plate of a thickness equal to the diagonal formed by the line of direction.

TABLE E.—SHOWING THE RESULTS OF PRACTICE WITH STEEL PROJECTILES FIRED AT 4·5-INCH UNBACKED PLATES, PLACED AT AN ANGLE.

No of Rounds.	Target.	Gun.	Charge.		Projectile.			Striking Velocity.	$\frac{Wv^2}{2g}$ in Foot Tons on Impact.	Foot Tons per Inch of Shot's Circumference.	Observed Effects.
			Weight.	Brand of Powder.	Nature and Length.	Weight.	Diameter.				
			lbs.			lbs.	in.	feet.			
1	4·5-inch unbacked plate at an angle of 38°.	70-pr. M. L. Armstrong competitive gun.	14·0	Rifle L. G.	Cylindrical steel shot.	70·0	6·34	1470·0	1049	52·7	Struck centre of plate; made an indent 15 in. long and 7 in. broad, and a hole about 2 in. square through the plate. Two large pieces were torn off the back of the plate and driven to the rear. Shot broke up.
2	,,	,,	,,	,,	,,	,,	,,	,,	,,	,,	Made an indent 14 in. long, 4½ in. deep, and 6¾ in. broad; plate cracked through and opened out at back, from which a large piece was torn. Shot broke up.
3	,,	,,	,,	,,	,,	,,	,,	,,	,,	,,	Made an indent 14 in. long, 7 in. broad, and 5½ in. deep. Back bulged and cracked through. A large piece of the back torn off. Shot broke up.
			drams.								
2	Unbacked ⅜-inch plates placed at angles of 60°.	Wall piece	10	Rifle F. G.	Cylindrical steel shot with flat head.	0·344	0·87	1141	3·1051	1·136	Indent 0·53-in. Plate bulged behind.
2	45°	,,	,,	,,	,,	,,	,,	,,	,,	,,	Indent 0·68-in. Bulged as before.
2	30	,,	,,	,,	,,	,,	,,	,,	,,	,,	Indent 0·50-in. Slight bulge behind.
2	20	,,	,,	,,	,,	,,	,,	,,	,,	,,	Indent 0·54-in. No bulge behind.
2	10	,,	,,	,,	,,	,,	,,	,,	,,	,,	Small piece of plate scooped out. No bulge behind.
			lbs.								
630	2·5-inch plate unbacked (upright).	12-pr. B. L. Whitworth gun.	1·75	Rifle L. G.	Flat-headed steel shot.	12·1	{2·97 / 2·71}	1264	134·05	14·64	Through plate and into earth-butt behind.
..	Do. at 45°	,,	,,	,,	,,	12·1	,,	,,	,,	,,	Clean hole through plate, but not sufficient to admit shot, which rebounded.
..	,,	,,	,,	,,	,,	12·1	,,	,,	,,	,,	Struck near above round, clean hole through; shot fell inside at foot of plate. Not fair hit.
..	Do. upright	12-pr. M. L. Whitworth gun.	1·50	,,	,,	12·1	,,	1120	105·25	11·49	Penetrated plate.

Equation [1] will therefore become $$\frac{W v^2}{2 g} = \frac{2 \pi R k b^2}{\sin.^2 \theta}, \qquad [4]$$

and [2] $$b = v \sin. \theta \sqrt{\frac{W}{4 \pi R g k}}. \qquad [5]$$

It appears from this that the resistance of the plate increases as the value of θ diminishes.

It has already been shown that a 4·5-in. unbacked plate when fired at direct, requires a force represented by 28 foot tons per inch of shot's circumference to ensure penetration.

Suppose, however, that the plate is placed in such a position that it makes an angle of 38° with the ground. From equation [4] the force required to penetrate it in this position amounts to 1445 foot tons for a shot of 6·22 in. diameter, or 73·9 foot tons per inch of shot's circumference.

An experiment of this nature was actually tried by the Armstrong and Whitworth Committee. They caused 4·5-inch plates to be set up at an angle of 52° with the vertical, and fired at them from 200 yards' distance with the competitive Armstrong and Whitworth guns.

Table E gives the results of this experiment.

It appears that the projectiles were solid steel shot of 70 lbs. weight and 6·34 in. diameter; that they struck with a "work" of 1049 tons, or 52·7 tons per inch of shot's circumference, and that they *failed to pass through*, although the plate was cracked and opened at the back.

In all these experiments it is to be observed that the life of a smooth-bored gun firing charges of $\frac{1}{4}$ the spherical steel shot's weight will be more than equal to that of a rifled gun firing charges of $\frac{1}{8}$, if the guns are of equally good construction. Consequently, that the work done by the smooth-bore in these examples is not to be taken as absolute proof of what might be done with higher, yet safe charges.

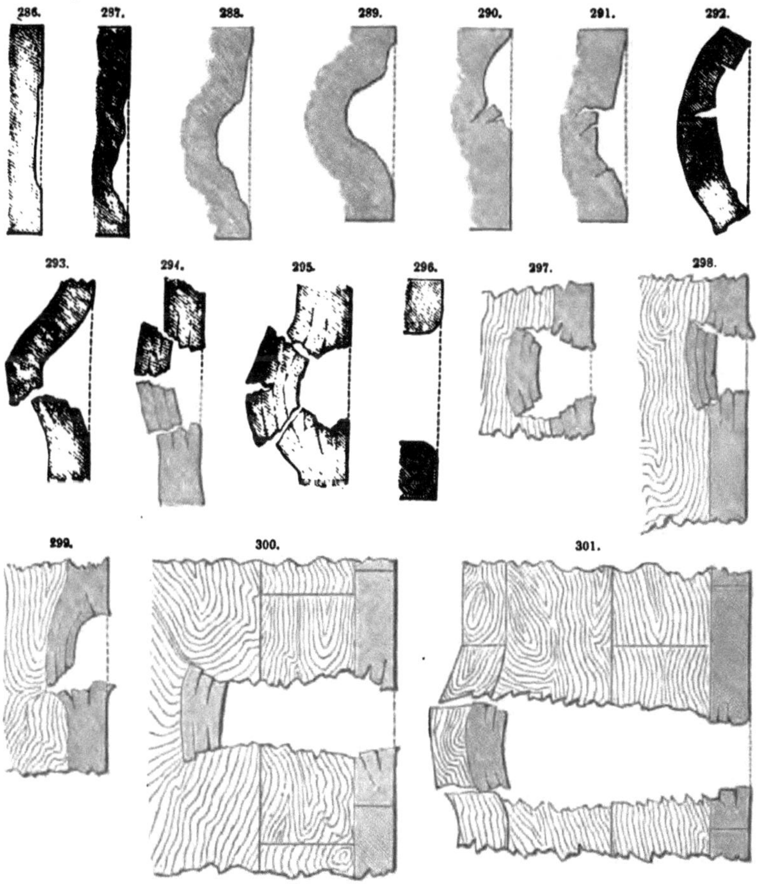

Figs. 286 to 301 exhibit the various effects of projectiles upon armour-plates, and upon armour-plates backed with wood.

Figs. 302 and 303 represent a cross-section and plan of the armour-plated British vessels 'Warrior,' 'Black Prince,' 'Defence,' 'Achilles,' 'Resistance,' 'Hector,' 'Valiant,' and 'Prince Albert.' The ships themselves are constructed of iron plates, ⅝-in. thick, and strengthened by iron ribs at intervals of 18 in. Outside these plates are two layers of teak-planking, making together a thickness of 18 in. Outside the planking, rolled iron plates, 4½ in. in thickness, are placed, and the whole structure is strengthened and held together by strong iron braces. From various experiments made by the British Government, it has been found that a 7-in. muzzle-loading gun, of 130 cwt., with a solid shot of 100 lbs., with a charge of 25 lbs., is capable of piercing the side of the 'Warrior' up to a range of 600 yards.

The 100-pounder smooth-bore gun, 9-in., of 125 cwt., with a solid spherical steel shot of 104 lbs., and 25 lbs. charge, is not capable of piercing the 'Warrior' at any distance over 100 yards.

The 9·22-in. rifled gun, of 12 tons, with a solid elongated steel shot of 221 lbs., and charge of 44 lbs., is capable of piercing the 'Warrior' up to 2000 yards. The 10·5-in. rifled gun, of 12 tons, with a solid elongated steel shot of 301 lbs., and charge of 45 lbs., is capable of piercing the 'Warrior' up to a range of 2000 yards. All these assertions of piercing at long ranges, for example at 2000 yards, are given from calculation, not from actual experiment, and ignore the angle at which the shot must strike, owing to its trajectory at these ranges. The American 15-in. gun, of 22 tons, with a spherical steel shot of 484 lbs., and a charge of 50 lbs., is capable of piercing the 'Warrior' up to a range of 500 yards. The American smooth bore 11-in. and 9-in. guns, fired with solid spherical steel shot and their maximum charges, are not capable of piercing the 'Warrior' at any range.

302.

303.

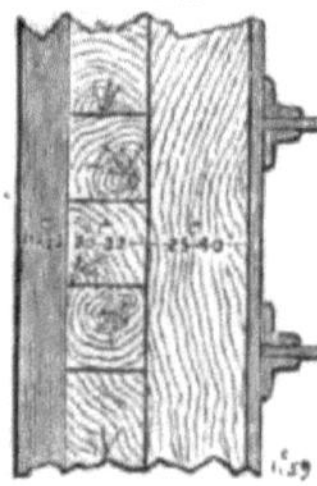

304.

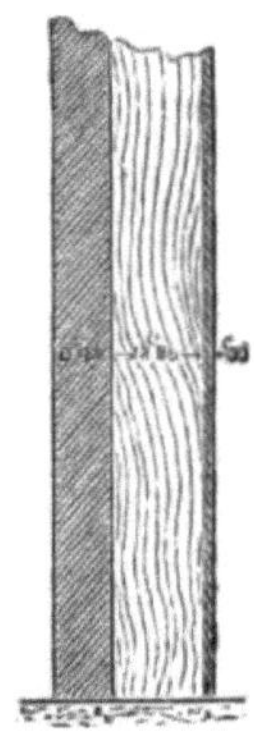

305.

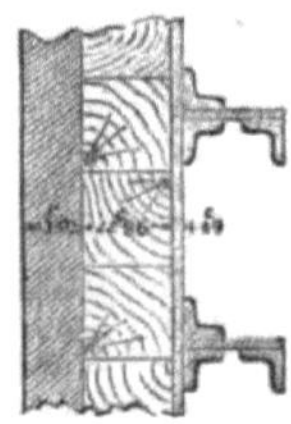

Figs. 304 and 305 are a cross-section and plan of the 'Minotaur,' 'Agincourt,' and 'Northumberland.' The inner skin of these vessels is the same as that of the 'Warrior.' The backing consists of 9 in. of teak, and is covered outside with plates 5½ in. thick. This armour-plating is fastened by three rows of heavy coned bolts, most of which pass through all the skins. A strip of iron 1·25 in. thick is fastened in rear at the junction of the plates. The rest of the construction is similar to that of the 'Warrior.'

From experiments made on a target of this construction, the results obtained were nearly the same as those obtained from the 'Warrior' target.

Figs. 306 and 307 show the construction of the 'Bellerophon' target. *a* is a plate 6 in. thick, forming the exterior covering, which is followed by a 10-in. backing of teak, worked longitudinally on the skin-plating, between the angle-iron stringers, and bolted with nut and screw bolts through the skin-plating. The latter is composed of two thicknesses of ¾-in. plating, with a layer of painted canvas between. The target is shown in the figure as it was erected for the purpose of experiment, supported by the Fairbairn target, so as to resemble the conditions of a ship's side as nearly as possible. This target was not subjected to a severe test: the most severe blow it was subjected to being from the 10·5-in. rifled gun, with a spherical steel shot of 165 lbs., and charge of 35 lbs.

This shot failed to penetrate the target; but there is no evidence to prove that the 10·5-in. gun would not have penetrated with a charge of 50 lbs.

Figs. 308 and 309 are a cross-section and plan of the British men-of-war 'Lord Warden' and 'Lord Clyde.' The frame-timbers of these ships are of English oak, 12½ in. thick, and are connected by iron diagonals 6 in. by 1¼ in. The inner planking is of the best English oak, 8 in. thick, and is covered with an iron skin of 1½ in. The outer teak-planking has a thickness of 10 in., and the armour-plate protecting it of 4½ in. The bolts sustaining the armour-plates are 2½ in. in diameter; their heads press against iron washers, which, in their turn, rest upon india-rubber washers let into the timber.

From the tests to which this target was subjected, it may be concluded that the 7-in. muzzle-loading rifled gun, fired with a solid elongated steel shot of 100 lbs., and charge of 25 lbs., is not capable of piercing the 'Lord Warden' at any range. The 9·22-in. rifled gun, of 12 tons, fired with elongated steel shot of 221 lbs., and 44 lbs. charge, is capable of piercing the 'Lord Warden' up to a range of 100 yds.

Figs. 310 and 311 show a target representing the construction of the French ironclads 'La Gloire' and 'La Flandre,' both of which are wooden ships protected by armour-plates placed in four rows. The dimensions of the plates forming the two upper rows are 5 ft. 9 in. by 2 ft. 7 in., thickness

4$\frac{3}{8}$ in., and of the two lower rows 5 ft. 9 in. by 2 ft. 5 in., thickness 5$\frac{5}{16}$ in.; inner planking 6 in., outside planking 10 in. This target was completely pierced by a 259 lbs. cylindrical chilled cast-iron projectile, shot from a 9·22 in. wrought-iron rifled gun weighing 12 tons, with a charge of 30 lbs.

306.

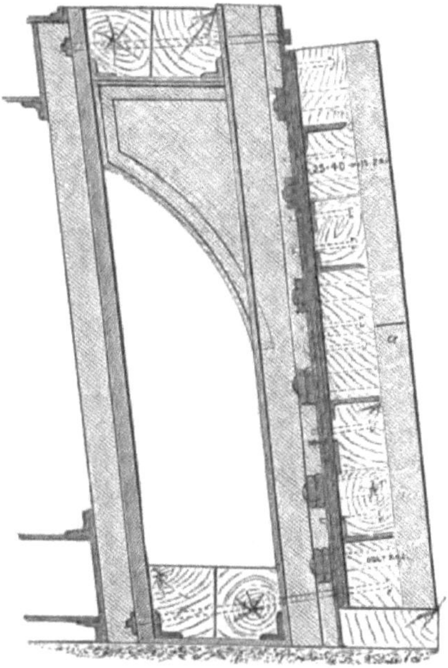

307

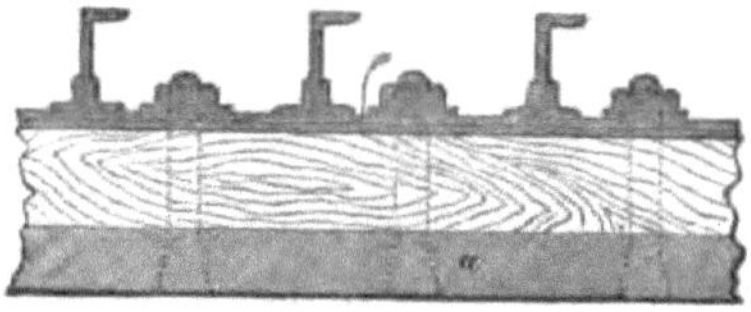

308.

309.

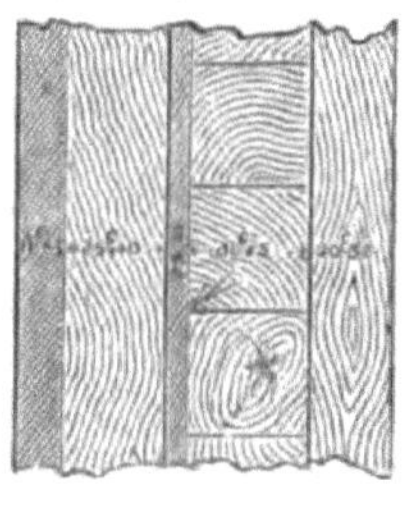

310.

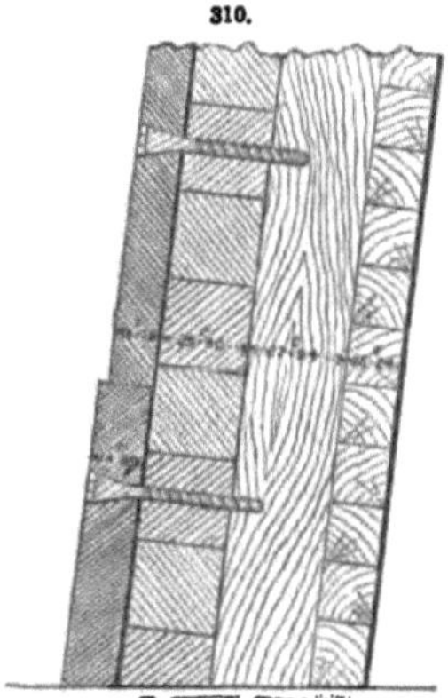

311.

A spherical steel shot weighing 73·80 lbs., fired from the 68-pounder smooth-bore with a charge of 16 lbs., penetrated the armour, driving the piece into the backing, and making an indent of 5·7 in.

Figs. 312 and 313 show a target representing the construction of the armour of the 'Hercules.' The upper half of this structure is faced with a wrought-iron plate 9 in. thick, and the lower half with a similar plate 8 in. thick. These plates are backed with 12 in. of teak, resting against a skin of two $\frac{3}{4}$-in. plates. The whole is secured to iron ribs, 10 in. deep, with vertical teak timber worked in between them. Behind these ribs are two linings of horizontal teak timber 18 in deep, confined by 7-in. iron ribs inside all, and an inside iron skin. The armour-plates are secured by 3-in. bolts. The total thickness of the target, exclusive of the 7-in. ribs, is at the top 51$\frac{1}{4}$ in., and at the bottom 47$\frac{1}{4}$ in. A 13-in. muzzle-loading rifled gun, of 23 tons, with a charge of 100 lbs., and an ogival-headed chilled shot weighing 577·5 lbs., struck the 8-in. plate, and passed through the target. Another shot, weighing 580·5 lbs., from the same gun, with a similar charge, struck the 8-in. plate, and penetrated the target to a depth of 22 in. No shot appears to have penetrated the 9-in. plate, the greatest effect being an indent of 8·27 in. in the plate.

312.

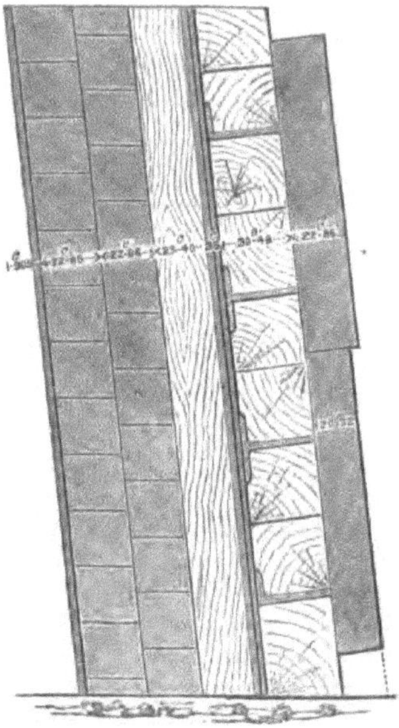

314.

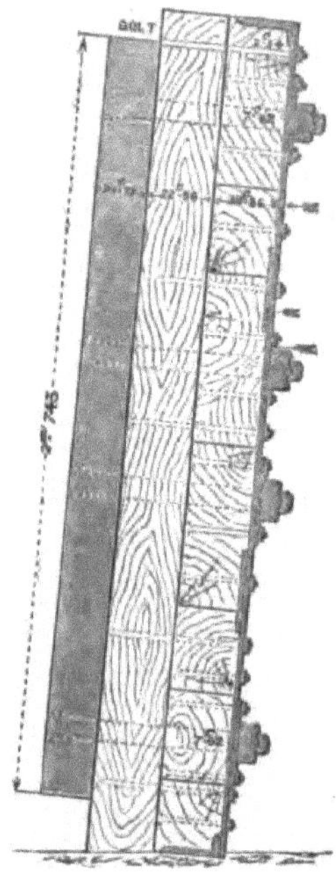

313.

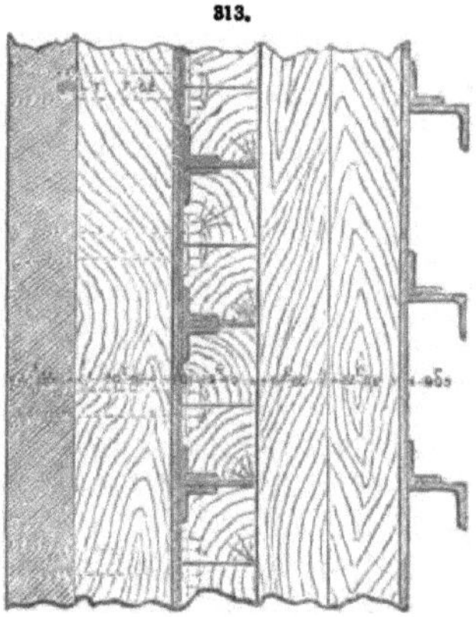

315.

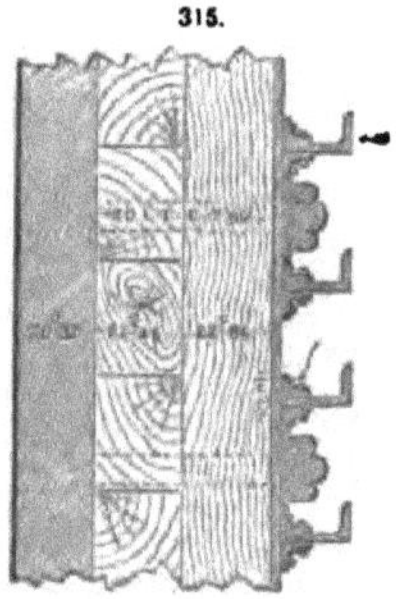

Figs. 314 and 315 show a section and plan of the so-called 8-in. shield, constructed to test the effect of different kinds of projectiles made of steel or of chilled cast-iron. This target was constructed of 8-in. armour-plate, backed by 18 in. of teak and a ¾-in. iron skin, secured to iron ribs 12 in. apart. The object of this experiment was to obtain a shield sufficiently strong to resist or keep out steel projectiles of 250 lbs., fired at 200 yds. from the 9-in. muzzle-loading rifled gun with 43 lbs. of powder, the head of the projectile being ogival, struck with a radius of one diameter.

The results of the experiments upon this target were, that it was proof against all projectiles when fired at obliquely; that it was not penetrated when fired at direct, except by the Palliser chilled projectile, which completely pierced the target. They proved that a pointed projectile 7·92 in. diameter can cut a hole in an 8-in. plate, provided it strike with the necessary work.

Laminated Armour.—Laminated armour consists of a number of thin plates bolted together, so as to form a shield of a certain total thickness, which depends on the number and individual thicknesses of the plates employed. This description of iron-plating has been extensively used in America, on account of its cheapness and facility of construction; it, however, offers much less resistance to shot than solid plating, at any rate while placed in close superposition

For experiments on this subject the targets were composed as follows:—

No. 1 Target consisted of seven $\frac{5}{8}$-in. wrought-iron plates, all breaking joint, faced with a 1·5 in. wrought-iron plate, the whole fastened together by $1\frac{1}{2}$-in. rivets and screws, 8 in. apart.

The target measured 12 ft. × 9 ft. × 6 in., and was fixed to an upright wooden frame.

No. 2 Target was composed of thirteen $\frac{5}{8}$-in. plates, faced with a 2-in. plate, and secured and supported in a similar manner to No. 1. The target measured 12 ft. × 9 ft. × $10\frac{1}{8}$ in.

TABLE F.—SHOWING THE EFFECTS OF FIRE AGAINST LAMINATED ARMOUR.

Photographic No. of Round.	Target.	Gun.	Charge.		Projectile.			Striking Velocity.	$\frac{Wv^2}{2g}$ in Foot Tons on Impact.	Foot Tons per Inch of Shot's Circumference.	Observed Effects.
			Weight.	Brand of Powder.	Nature and Length.	Weight.	Diameter.				
			lbs.			lbs.	in.	feet.			
Not taken	No. 1 target, laminated plates; total thickness 6 in.	40-pr. B. L. rifled.	5	Rifle L. G.	Elongated cast-iron.	41 1	4·7	1157	381·5	25·8	Indent 2·35 in.; plate in rear considerably bulged and cracked.
,,	,,	7-inch B. L. rifled.	14	,,	,,	110	6·9	1140	991	45·7	Clean through target, making a hole 13 × 13·5 in. in front plate, and sending a shower of splinters to the rear; back much torn.
,,	,,	,,	,,	,,	,,	,,	,,	,,	,,	,,	As before.
,,	,,	68-pr. S. B.	16	L. G.	Spherical cast-iron.	66·5	7 91	1380	878	35·3	As before.
,,	No. 2 target, laminated plates; total thickness $10\frac{1}{8}$ in.	40-pr. B. L. rifled.	5	Rifle L. G.	Elongated cast-iron.	41·1	4·7	1157	381 5	25·8	Indent 1·2 in.; slight bulge in rear.
,,	,,	7-inch B L. rifled.	14	,,	,,	110·0	6·9	1140	991	45·7	Indent 2·35 in.; rear of plates cracked and bulged.
,,	,,	68-pr. S. B.	16	L. G.	Spherical cast-iron.	66·5	7·91	1380	878	35·3	Indent 3·15 in.; as before.
,,	,,	,,	,,	,,	,,	,,	,,	,,	,,	,,	Indent 3 in.; as before.

From these results it appears that laminated armour is considerably weaker than solid armour. Thus a 4-in solid plate would have effectually stopped all the projectiles, whereas they easily penetrated 6 in. of laminated plates.

Fig. 316 represents Chalmers' system of armour-plating. A target upon this system, which had a $3\frac{3}{4}$-in. armour-plate, a compound backing, a second plate and a cushion, with stringers running at right angles to the ship's frames, between the second plate and the skin, the stringers being riveted to the latter, when fired at with steel and cast-iron projectiles from the following guns,—68-pounder smooth-bore, with cast-iron shot and shell, 16 lbs. charge; 100-pounder Armstrong, with cast-iron shot and shell, 12 and 14 lbs. charges; 300-pounder Armstrong, with cast-iron spherical shot, 50 lb. charge; and lastly, with solid steel shot, 301 lbs., from a 300-pounder Armstrong, with 45 lbs. charge,—proved that this system of backing affords great support to the armour-plates, and prevents their distortion from buckling. It is also of considerable advantage in adding strength and resisting power to the structure, and no other target designed for naval purposes has resisted so great a weight of shot with so little injury.

316.

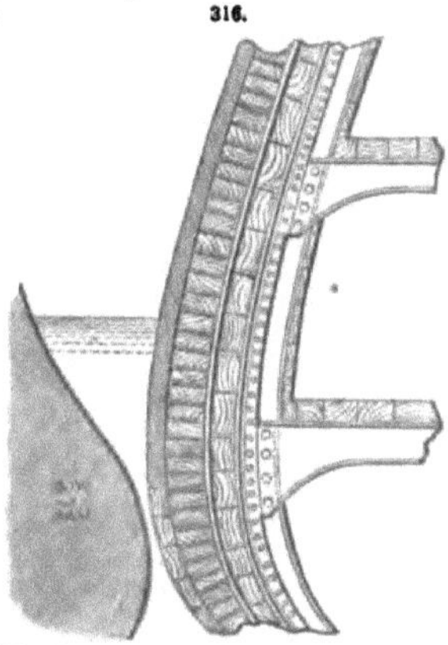

317.

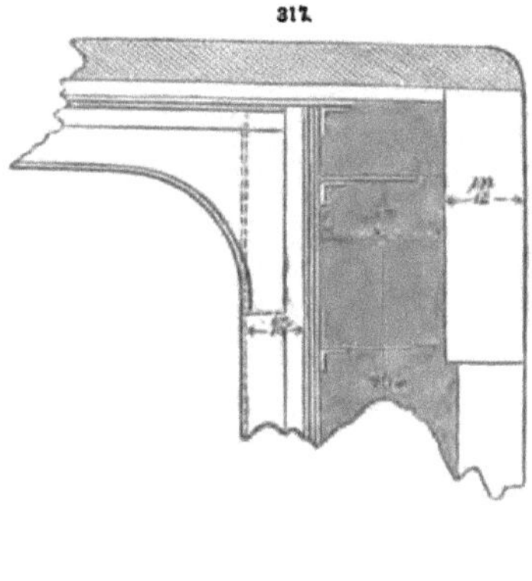

Fig. 317 is a sketch of the armour applied to the vessel 'Glatton.' It consists of 12-in. plates backed by 18 to 20 in. of teak upon an iron skin 2 in. thick.

The armour used in vessels of the large 'Monitor' type is shown in Fig. 318, and is formed of 15-in. plates, backed by 2 ft. 6 in. of teak upon two iron skins, shown at ×, each 1 in. in thickness.

318.

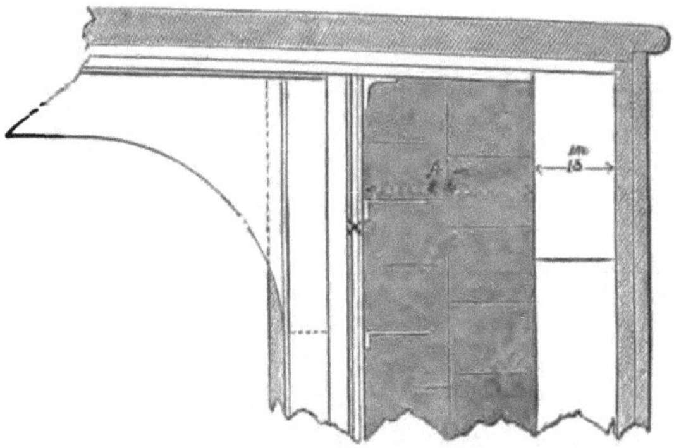

Cast-iron Projectiles as compared with Steel of the same Size and Form.—The difference between the effects of cast-iron and steel shot upon armour-plates is most marked. The latter material is the nearest approach to perfect hardness and cohesion at present in use, and the amount of *work* expended on the shot is less with steel than any other known material. With ordinary cast-iron, a large amount of *work* is expended in breaking up the projectile and hurling the fragments in all directions; but when steel shot are manufactured in the best manner, little *work* is expended on the projectile; and in one instance a 12-pounder Whitworth steel shot was of such perfect material, that after passing through 2½ in. of solid iron its temperature was apparently unaltered. Several experiments have been made with a view of ascertaining the amount of *work* lost by the breaking up of cast-iron, alteration of form in steel shot, and so on.

The following Table shows the absolute thickness of plate which can be penetrated by cast-iron shot fired from various guns with service-charges. The guns were at a distance of 100 yds. from the plates, with the exception of the 68-pounder, which was at 200 yds.

TABLE G.

Target.	Gun.	Charge.		Projectile.			Striking Velocity.	$\frac{Wv^2}{2g}$ in Foot Tons on Impact.	Foot Tons per Inch of Shot's Circumference.	Observed Effects.
		Weight.	Brand of Powder.	Nature and Length.	Weight.	Diameter.				
		lbs.			lbs.	in.	feet.			
Unbacked plates, 1·286	6-pr. B. L. rifled gun.	0·75	Rifle L. G.	Elongated cast-iron shot.	6·25	2·50	1010	44·21	5·63	Just penetrated through the plate.
1·803	12-pr. ,,	1·50	,,	,,	11·56	3·00	1140	104·1	11·05	As before.
2·350	20-pr. ,,	3·13	,,	,,	24·81	3·75	1120	215·8	18·32	As before.
2·820	40-pr. ,,	5·00	,,	,,	41·20	4·75	1150	377·8	25·32	As before.
3·85	68-pr. smooth-bore.	16	L. G.	Spherical cast-iron shot.	66·50	7·91	1380	878	35·3	As before.

If the results given by this Table are compared with the effect of steel projectiles, it will appear that the cast-iron shot requires about 2½ times the *work* of the steel shot to effect the same penetration, except when the velocity of the cast-iron shot is high.

Here observe that the distance of the 68-pounder was double that of the other guns, and that being a cast-iron 68-pounder, it was fired with a weak charge compared to what it would have stood had it been a modern smooth-bored wrought-iron gun, in which case the charge should have been 24 lbs., the velocity 1700 ft. per second at 100 yards' range.

TABLE H.—SHOWING THE DIFFERENCE BETWEEN THE EFFECTS PRODUCED BY CAST-IRON AND STEEL SHOT WHEN FIRED AT IRON PLATES.

Thickness of Plate.	Foot Tons per Inch of Shot's Circumference required for absolute Penetration.		Difference.	Proportional Excess for Cast-iron Shot.	Remarks.
	Cast-iron Shot.	Steel Shot.			
inches. 1·286	5·63	2·29	3·34	2·517	From results with— 6-pr. B. L. rifled gun.
1·803	11·05	4·49	6·56	2·461	12-pr. ,, ,,
2·350	18·32	7·64	10·68	2·398	20-pr. ,, ,,
2·820	25·32	11·00	14·32	2·302	40-pr. ,, ,,
3·850	35·30	20·50	14·80	1·722	68-pr. smooth-bore.

It will be seen that it is almost useless to fire cast-iron projectiles against iron defences, if *penetration* is required. As steel is a most expensive material for shot, and it has been proved that chilled iron is almost as good as steel, chilled iron will most probably be the generally used material for projectiles for battering purposes.

The proper form of front or head to be given to hardened projectiles has been a matter of much dispute. It has been found in practice, however, that the pointed form is the best. The flat-headed or round-headed shot punches out a piece of the armour-plate, Figs. 319, 320, and drives it into the backing; the shot has no means of ridding itself of this piece of armour-plate, *and has to push it in front of it through the backing*. Thus in targets penetrated by flat-headed or round-headed shot it has always been found that *the piece of armour-plate has passed through the target along with the shot.*

319.

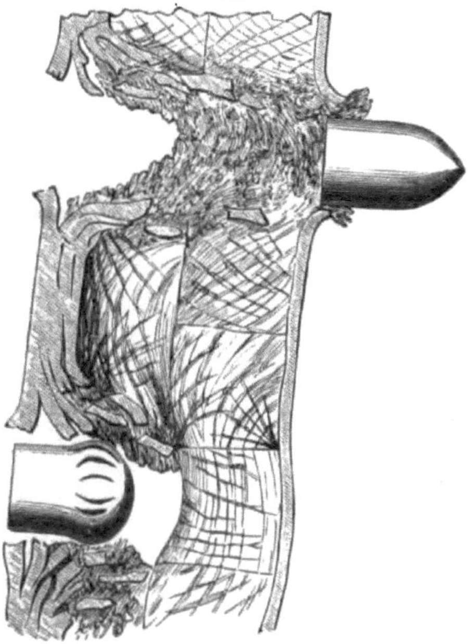

321.

322.

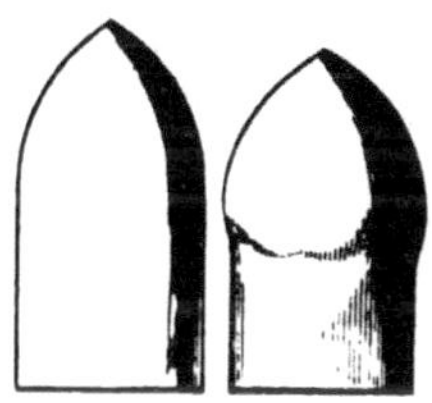

320.

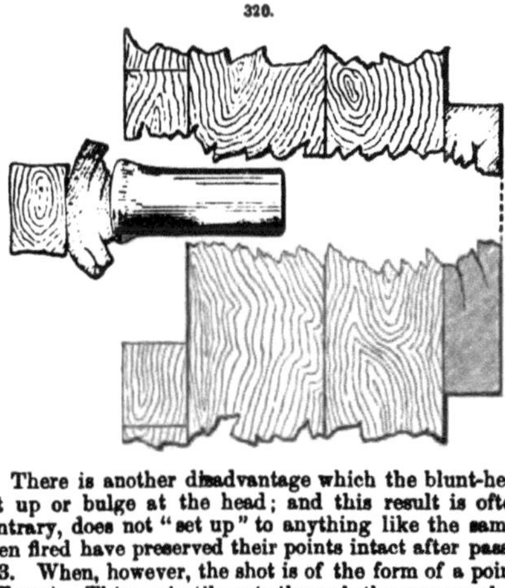

323.

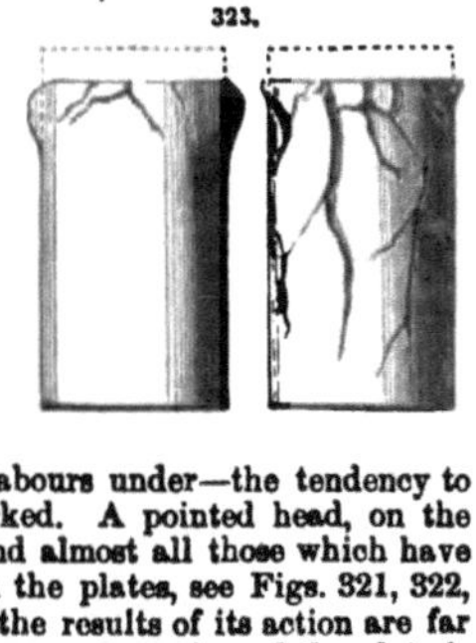

There is another disadvantage which the blunt-headed form labours under—the tendency to set up or bulge at the head; and this result is often very marked. A pointed head, on the contrary, does not "set up" to anything like the same extent; and almost all those which have been fired have preserved their points intact after passing through the plates, see Figs. 321, 322, 323. When, however, the shot is of the form of a pointed ogival, the results of its action are far different. This projectile cuts through the armour-plate, or rather tears through, and the plate is bent back, and forced into the backing round the edge of the hole; the shot thus passes through the backing without carrying any jagged armour in front of it. Fig. 319.

The following Table K, gives the results of some late experiments, which clearly show the great superiority of the pointed head.

In these experiments both steel shot and Palliser's chilled shot and shell were used. All the projectiles were fired from the same gun, under the same circumstances, the velocity of each round being observed. The targets consisted of a structure representing the side of an iron-clad vessel, protected by solid plates of 6 in. thickness, backed by 18 in. of teak, an iron skin of two ½-in. plates, the usual iron ribs, &c. &c. A second target, of unbacked 4·5-in. plates, inclined at an angle of 38° with the ground, was erected at the same distance.

The projectiles were of a mean weight of 115 lbs., and of the following forms of head:—

For Palliser's Chilled Shot.

1. Ogival head, struck with a radius of one diameter, and brought to a point.
2. Belgian form, head struck with a radius of 1·47 diameter, and pointed in the shape of a cone.
3. Elliptical, the height of the ellipse being equal to the diameter of the projectile.

For Steel Shot.

1. Hemispherical.
2. Ogival head, struck with a radius of one diameter, and brought to a point.

TABLE K.

Photographic No. of Round.	TARGET.	GUN.	CHARGE.		PROJECTILE.			Striking Velocity.	$\frac{W v^2}{2g}$ in Foot Tons on Impact.	Foot Tons per Inch of Shot's Circumference.	OBSERVED EFFECTS.
			Weight.	Brand of Powder.	Nature and Length.	Weight.	Diameter.				
			lbs.			lbs.	in.	feet.			
1186	Box target, 6-in. plates on 18 in. of teak, and a skin of two ½-in. plates, ribs, &c. No. 27.	7-in. M. L. rifled gun of 130 cwt.	22	Rifle L. G.	Chilled shot, E., 13·62 in.	117·0	6·9	1328·5	1432·0	66·06	Struck fair between two ribs; penetrated armour-plate, and stuck in backing. Shot broke up and remained in the hole; indent 15 in.
1190	,,	,,	,,	,,	,,	116·5	,,	1331·4	,,	,,	Struck near 1186 between two ribs. Shot penetrated to inner skin, which it bulged slightly; shot remained broken up in the hole.
1191	,,	,,	,,	,,	Chilled shot, Belgian head, 14·79 in.	,,	,,	,,	,,	,,	Struck fair between two ribs; *penetrated* the target completely. Shot broke to pieces; skin of target was here only half-inch.
1192	,,	,,	,,	,,	Chilled shot. Ogival head, 13·86 in.	115·0	,,	1340·0	,,	,,	Struck partly on a rib and armour-plate bolt; *penetrated* the target completely. Shot broke to pieces.
1196	,,	,,	,,	,,	,,	116·25	,,	1332·8	,,	,,	Struck fair between two ribs; *penetrated* the target completely. Shot broke to pieces.
1189	,,	,,	,,	,,	Steel, Ogival head, 13·25 in.	115·5	6·92	1360·0	1481·4	68·14	Struck fair between two ribs; *penetrated* the target completely. Shot picked up inside whole, having rebounded off armour-plate at rear of box.
1198	,,	,,	,,	,,	Hemispherical-headed steel shot, 12·0 in.	115·0	6·92	1380·0	1518·6	69·85	Struck on rib. Shot penetrated plate, and lodged in the backing; total indent about 18 in.; rib at back cracked, and forced back slightly.
1201	,,	,,	,,	,,	,,	,,	,,	1371·0	1498·9	68 95	Struck fair between two ribs. Shot penetrated the plate, and lodged in the backing; total indent about 14·5 in.
1175	4·5-in. unbacked plates, inclined at an angle of 38° with the ground.	,,	,,	,,	Chilled, Belgian head, 14·79 in.	115·0	6·9	1339	1432·0	66·06	Struck fair; made a jagged hole 1 ft. long, and a small hole through the plate. Shot broke into pieces.
1202	,,	,,	,,	,,	Chilled shell, Ogival head, 15·6 in.	115·5	,,	1349	1457·4	67·23	Struck fair between two ribs; *penetrated* the target, bursting and setting fire to the backing.
1205	,,	,,	,,	,,	,,	115·0	,,	1346	1444·7	66 65	Struck full on a rib; burst in the backing; rib broken, and inner skin bulged and cracked.
1176	,,	,,	,,	,,	Chilled, Ogival head, 13·86 in.	115·19	,,	1339	1432·0	66·06	Struck fair; *penetrated* the plate. Shot broke to pieces.
1178	,,	,,	,,	,,	,,	113·0	,,	1372	1475·9	68·04	Struck fair; *penetrated* the plate. Shot broke into pieces.
1180	,,	,,	16	,,	,,	113·5	6·9	1253	1235·6	57·00	Struck top of plate, scooping out a hole 4½ in. deep. Shot broke to pieces.
1183	,,	,,	,,	,,	Steel, Ogival head, 13·25 in.	114·0	6·92	1240	1215·4	55·91	Struck on edge of former hit; scooped out a piece of plate, and glanced off. Shot broke.
1184	,,	,,	,,	,,	Chilled, Ogival head, 13·86 in.	114·0	6·9	1277	1289·0	59·47	Struck fair; scooped a hole 11 × 6¼ in., and about 2 in. deep. Shot broke to pieces.

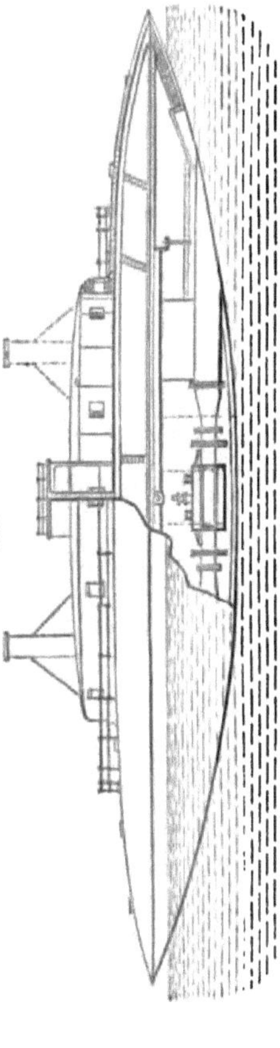

324.

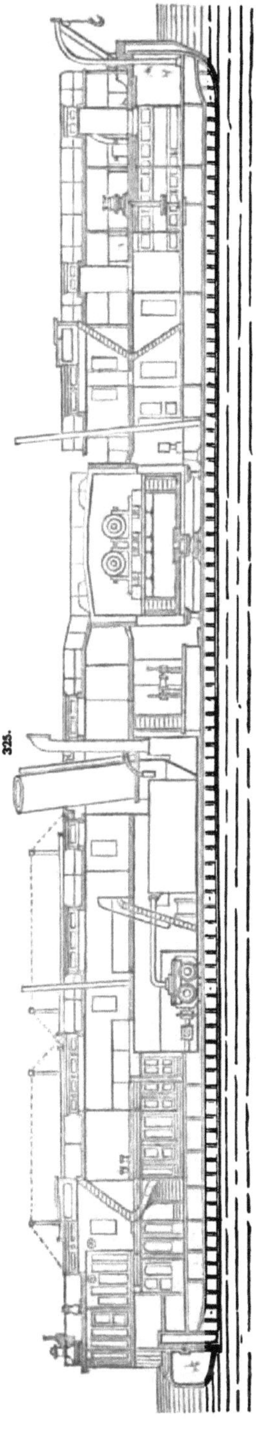

325.

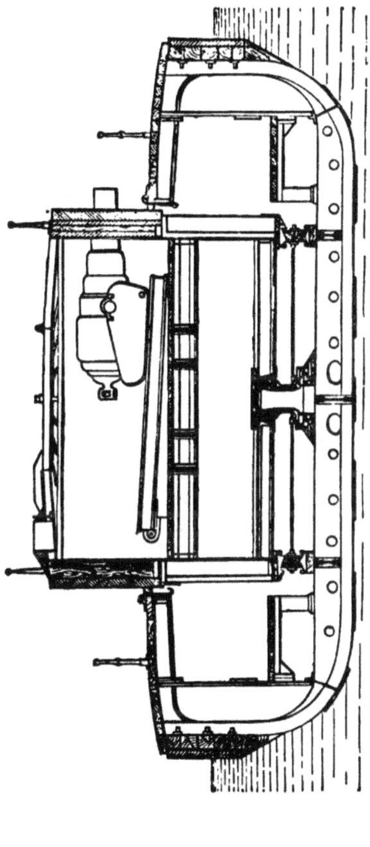

326.

The experiments against iron-plated targets seem to demonstrate the superiority of elongated over spherical projectiles, when the shot or shell are made of a hardened material.

The principal objections to the spherical, as compared with the elongated form, may be enumerated as follows:—The form is ill adapted for penetration, either in the case of steel or chilled iron projectiles, which require a pointed cylindrical form to develop their full power. The diameter being larger in proportion to the weight, the projectile experiences a greater resistance, both from the air in its flight and from the plate on its impact. The range and accuracy is considerably inferior. The capacity of the projectile, as a shell, is much less. Elongated projectiles have been found to be less liable to alter their shape on impact; and the cylindrical form is much better adapted for steel or chilled shells, which, as spherical, would be almost worthless.

But it is to be remembered that neither has the target been fully worked out as a scientific application of strength, nor have any satisfactory experiments been made as to the effects of a different mode of attack—that is, that which would rack in the fullest degree, by simultaneous blows—against the present system of penetration by separate shot. And this is the more important, since no great naval battle has been decided except at close quarters, where range and accuracy are thrown aside, and the victory belongs to those who hit hardest at short ranges with whole broadsides delivered at once. Here the superior velocity obtainable in all cases with spherical shot, up to 800 yds., would not fail to give more work, particularly if one-third of the shot's weight of powder be used as a battering-charge, which may well be done in modern guns. See ARTILLERY.

It should be taken into account, that if armour-plate be driven in or penetrated at short ranges by even spherical shot, shell effects follow,—the pieces of armour-plate, nuts, splinters, &c., doing all that could be expected from any shell; and that at long ranges all shot will strike at an angle with the horizontal, and any shot may strike the deck or masts even more probably than the armour-plate. The most complete armour and the cheapest that a ship can have is the power of sinking, so as to expose but little of her hull out of water.

Figs. 324, 325, 326, represent the construction of the 'Buffalo' and the 'Tiger,' ships recently built for the Dutch Government; the former being a *ram* and *monitor* combined, and the latter a *monitor*. The dimensions of the 'Buffalo' are:—breadth 40 ft., length 205 ft., depth 24 ft., tonnage 1472, load-draught 15 ft. 6 in. The dimensions of the 'Tiger' are:—breadth 44 ft., length 187 ft., and depth 11 ft. 6 in. Both vessels carry turrets of the same construction and dimensions, 6 ft. above the upper deck, and 22 ft. in diameter; pierced for two guns, and fixed upon a circular platform upon the main deck. This platform is moved by means of machinery similar to a railway turn-table, and can make one revolution in 45 seconds. The wall of the turret consists of 8-in. plates of malleable iron, 12 in. of teak, and 1-in. plates of wrought iron, making a total thickness of 21 in. Each of the 8-in. plates is 14½ ft. by 3 ft., and forms one-fifth of the circle. There are ten of these plates in each turret; they are secured to teak-backing by nut and screw bolts with "elastic cup-washers." In the 'Buffalo,' Fig. 324, a low wall rises out of the deck: it is of the same composition and thickness as the turret, the bottom of which it surrounds, and which, as well as the turning machinery, it protects. The outer skin of the vessel, for about 100 ft. amidships, is composed of 6-in. armour-plates, tapering off to 4½ in. forward and 3 in. aft, laid on 10 in. of teak amidships, tapering off at each end to 8 in.

Works relating to Armour:—'Report of the Secretary of the Navy on Armoured Vessels,' Washington, 1864. Norton and Valentine, 'Report on the Munitions of War at the Paris Exhibition,' royal 8vo, 1868. Dislère Note, 'Sur la Marine des États-Unis,' 8vo, Paris, 1867. Captain Noble's 'Report on the Penetration, &c., of Armour Plates,' fol., 1866.

See also:—Humber's 'Record of Modern Engineering,' fol., 1863. 'Revue Maritime et Coloniale.' 'Journal of the Royal United Service Institute.'

ARRASTRE, *in Gold-mining*.

ARRASTRE. FR., *Moulin à mulet mexicain*; GER., *Mexicanische Quartz Mühle*; ITAL., *Aia da quarzo*; SPAN., *Arrastre*.

The arrastre consists of a circular pavement of stone, about 12 ft. in diameter, on which the quartz is ground by means of two or more large stones, or mullers, dragged continually over its surface, either by horses or mules, but more frequently by the latter. The periphery of the circular pavement is surrounded by a rough kerbing of wood or flat stones, forming a kind of tub about 2 ft. in depth, and in its centre is a stout wooden post, firmly bedded in the ground, and standing nearly level with the exterior kerbing. Working on an iron pivot in this central post is a strong upright wooden shaft, secured at its upper extremity to a horizontal beam by another journal, which is often merely a prolongation of the shaft itself. This upright shaft is crossed at right angles by two strong pieces of wood, forming four arms, of which one is made sufficiently long to admit of attaching two mules for working the machine. The grinding is performed by four large blocks of hard stone, usually porphyry or granite, attached to the arms either by chains or thongs of raw hide, in such a way that their edges, in the direction of their motion, are raised about an inch from the stone pavement, whilst the other side trails upon it. These stones each weigh from 300 to 400 lbs., and in some arrastres two only are employed, in which case a single mule is sufficient to work the machine.

Fig. 327 is a sectional view of a Mexican arrastre, in which A is the upright shaft; B, arms to which mullers, C, are attached; and D, the central block of wood in which the lower bearing works.

Some of the arrastres used by Mexican gold miners, and for the purpose of testing the value of quartz veins, are very rudely put together, the bottom being made of unhewn flat stones laid down in clay; but in a well-constructed arrastre, intended to be permanently employed, the stones are carefully dressed and closely jointed, and, after being placed in their respective positions, are grouted in with hydraulic cement.

The charge for an ordinary arrastre is 450 lbs. of quartz, previously broken into pieces of about the size of pigeons' eggs. The machine is now set in motion, a little water being from time to time added, and at the expiration of from four to five hours the quartz has become reduced to a finely-divided state, and more water is added, until the contents of the arrastre assume the consistency of tolerably thick cream. Quicksilver is then sprinkled over its surface to the amount of 1½ oz. for every ounce of gold supposed to be contained in the finely-divided rock which is generally known, with a considerable degree of accuracy, from the results obtained from previous charges. The grinding is after this continued for another two hours, during which time the mercury is divided into minute globules, and becomes disseminated throughout the mass, which should be of such a consistency as not to allow it to sink to the bottom, but be so held in suspension as to meet, and amalgamate with, all the particles of gold. At the expiration of this time the amalgamation is considered complete, and the process of settling the amalgam from the ground silicious matter is commenced. Water is now let into the paste so as to render it very thin, and perfectly mobile, the mules being driven very slowly, in order to allow the particles of gold and amalgam to yield to the influence of their densities, and to sink to the bottom. After having in this way slowly agitated the mixture for about half-an-hour, the thin mud is allowed to run off, leaving behind it, in the bottom of the arrastre, the gold combined with mercury in the form of amalgam. Another charge of broken quartz is now put in, and the operation is repeated, time after time, until it is thought desirable to stop for the purpose of cleaning up. In the roughly-constructed arrastre, having a bottom of uncut stones laid in clay, the run is seldom less than ten days, and is sometimes extended to three weeks or a month. In this case the amalgam settles in the crevices between the paving-stones, which have to be dug up, and all the sand and mud between them carefully washed. If, however, the machine be well constructed, and provided with a closely-paved bottom, the cleaning up is more frequently repeated, since the quicksilver and amalgam do not find their way so readily between the stones, but remain on the surface, from which they are easily collected in an iron vessel, for subsequent treatment by straining and retorting.

327.

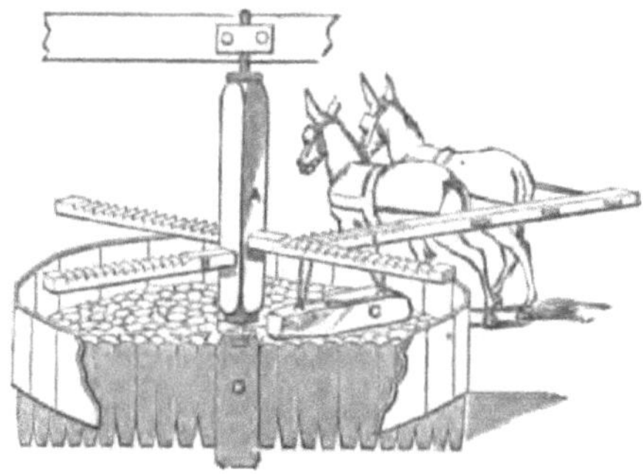

The arrastre does its work slowly, and consumes a large amount of power in proportion to the quantity of rock crushed, but is an excellent amalgamator, and is often valuable for the purpose of testing newly-discovered veins, and ascertaining their approximate yield. It is also the arrangement most commonly adopted by a miner, who, having found a rich pocket in his vein, is desirous of converting a portion of it into money, and of ascertaining whether it be likely to continne productive, before incurring the expense of erecting more costly and complicated apparatus. A modification of the arrastre is not unfrequently employed for the treatment of pyrites separated from tailings by washing, and is generally considered to be well adapted for that purpose.

ARRIS. Fr., *Arête;* Ger., *Kante;* Ital., *Spigolo, Risega;* Span., *Vivo, arista.*

The angle formed by the meeting of two surfaces which are not in the same plane. In builders' work any angle which retains its original sharpness. With workmen the term "arris" has only one dimension, that of length, in which it differs from the word "edge," which usually has two dimensions, namely, length and thickness. The junction of two sides of a square stone is called its *arris*, but the side of least dimensions of a board or a stone slab is called its *edge*.

ARRIS FILLET. Fr., *Nervure angulaire;* Ger., *Scharfkantiger Vorsprung;* Ital., *Corrente triangolare;* Span., *Filete de arista.*

An arris fillet is a slight piece of timber, triangular in section, used to raise the slates on a roof where they are cut by a chimney-shaft, skylight, or wall, when it cuts the slates obliquely; the object being to throw off the wet which would otherwise find its way under the flashing. See Fillet.

ARRIS GUTTER. Fr., *Gouttière angulaire;* Ger., *Scharfkantige Dachrinne;* Ital., *Gorna a* V; Span., *Gotera de forma* V.

An arris gutter is a gutter formed to a V-shape, usually of wood, and fixed to the eaves of a building.

ARRIS RAIL. Fr., *Ornière angulaire;* Ger., *Kantenschiene;* Ital., *Regolo triangolare;* Span., *Rail de forma* V.

328.

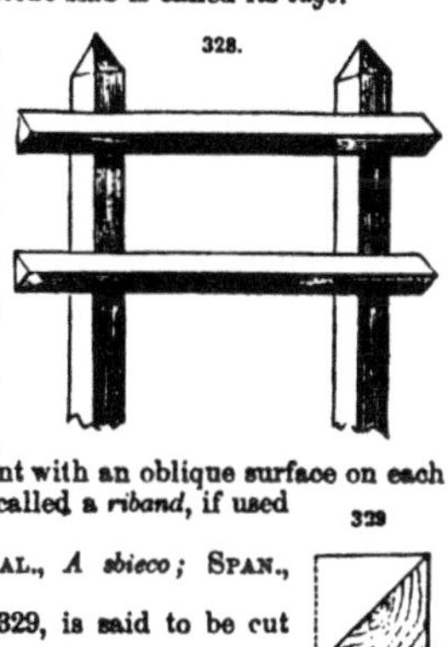

An *arris rail*, in carpentry, is a rail cut to a triangular section, and when fixed to posts, as Fig. 328, shows the arris in front with an oblique surface on each side. A flat rail, or one with a rectangular section, would be called a *riband*, if used in a palisade.

ARRIS-WISE. Fr., *En Diagonale;* Ger., *Eckweise;* Ital., *A sbieco;* Span., *Diagonalmente.*

329

A balk or piece of squared timber sawn diagonally, Fig. 329, is said to be cut "arris-wise."

In bricklayers' work, tiles laid diagonally.

ARSENIC. Fr., *Arsenic;* Ger., *Arsenik;* Ital., *Arsenico;* Span., *Arsénico.*

Arsenic is a metal of a steel-grey colour, and brilliant lustre, though usually dull from tarnish. This metal is usually obtained from arsenious acid, and the latter by calcination from native arseniurets, such as those of iron, copper, and other metals. Hence it is, in most cases, a secondary product. Yet much of the arsenious acid of commerce is manufactured from iron pyrites, which, when the arsenic is extracted, serves no other purpose. Arsenic enters as an important agent into many branches of art, and is a useful metal in forming fusible alloys.

The metal arsenic is not poisonous, but one of its oxides (arsenious acid), formed by its combustion in air, is extremely so; and in operating either with the metal or the acid some caution is required on the part of the operator. When arsenious acid is operated on, if we moisten it, the inhalation of the dust is prevented. And when an alloy of arsenic is melted and operated on, the vapours of the metal are made harmless when the operator fills his mouth with grains of charcoal, renewing them from time to time. This charcoal will absorb any arsenic which may accidentally enter the organs of respiration. Arsenical pyrites is the common ore of this metal. It is here combined with iron, silver, gold, bismuth, and antimony. In all cases of its application in practice, we may consider the arsenious acid as the only ore: and as this is obtained as a secondary product in the calcination of cobalt ores, we shall include the description of its manufacture in the article on that substance.

The arsenious acid of commerce is white, glassy when fresh, but generally opaque when brought into market for sale. For metallurgical purposes no arsenious acid in powder ought to be used, for it is frequently adulterated with gypsum or other matter. The commercial article is always more or less perfectly glassy or milky, or transparent in the interior of the flat pieces, while on the exterior it appears opaque; it is generally vitreous throughout its whole mass. It is slightly soluble in water. About ten parts may be dissolved in boiling water: this quantity, however, depends on the amount of acid present. Water never dissolves the whole of it, even when less than the above quantity is exposed to its action; it will dissolve more when a large quantity of acid is afforded. Arsenious acid consists of 75·8 metal, and 24·19 oxygen. It sublimes in open vessels at 380°; it is decomposed by hydrogen, carbon, sulphur, phosphorus, and some metals, such as lead, iron, silver, &c.

The metal arsenic is easily obtained pure when arsenious acid is mixed with fatty oil, or a compound of carbon and hydrogen, or finely-pulverized soft charcoal, and heated gently in a glass tube. It evaporates at 356°, and is therefore easily smelted, and the metal condenses in the cold parts of the heated tube. In large quantities it may be obtained by mixing arsenious acid with coarse charcoal-powder, or what is better still, *culm*—small fragments of bituminous coal—and exposing it in a large crucible to a red heat. This crucible is covered by a second one, as shown in Fig. 330, and well luted; the lower pot is exposed to a red heat, while the upper one is kept cool. The metal thus formed and evaporated will condense in the upper pot, from which it is easily separated, when cold. The same operation may be performed on arsenical pyrites, without carbon; and the metal is obtained in a similar manner. Iron, nickel, and other permanent metals remain in the lower pot, combined with some arsenic.

330.

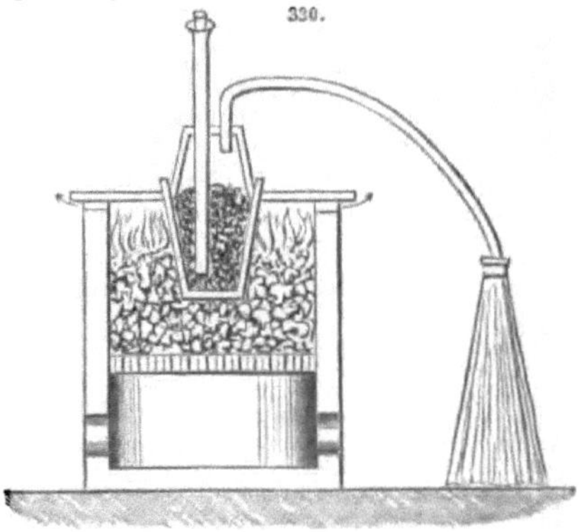

The metal is of a high lustre, and greyish-white; its specific gravity is 5·70. Its weight and lustre increase with its purity. It evaporates without melting; and its vapours, which smell strongly of garlic, are sometimes confounded with those of phosphorus. Arsenious acid does not smell; it is the metal only which emits this odour. It is not ductile, nor malleable, and may be converted into fine powder in a mortar. It is highly combustible, and deflagrates when either mixed or heated gently with saltpetre.

If this metal, in its pure state, is of little interest to the metallurgist, its alloys are of much value. All metals, without an exception, are made more fusible by the addition of arsenic; in some instances its influence is remarkably distinct. The alkaline metals combine with it with great facility, even when it is simply heated with the oxides of those metals—such as potassa or soda. It requires extreme caution to operate on these alloys; that is, on those of the alkaline metals and arsenic, because they decompose rapidly in damp air, and evolve arseniuretted hydrogen—a virulent poison—the effect of which resists the most refined skill of the physician. In combination with lead—in shot—arsenic is harmless; and also in all compounds of the proper metals, when its quantity is not too large. Aluminum, and all the metals of this class, combine very readily with arsenic. In fact, all metals combine easily with arsenic, but they are quite as easily decomposed. The decomposition of arsenical alloys is effected by merely continued heat, and, with the exception of silver, in a short time. The higher the degree of heat is, so much shorter is the time in which the act is accomplished. When it is desirable to retain arsenic in the composition, it is necessary to melt the metals at as low a heat as possible. The combination of arsenic with other metals is as easily performed as the decomposition. Metallic arsenic and lead cannot be combined directly; but when melted lead is covered by arsenious acid, some lead is oxidized, and in its place arsenic is absorbed. In the same manner, other metals, which melt at or near the heat at which arsenious acid volatilizes, may be combined with arsenic. Iron, chromium, copper, and others, cannot be alloyed by these means, but they may be effectually combined with arsenic in a manner described in our articles on these metals; and there is no doubt that all alloys of this kind are most safely and correctly compounded by that manner—namely, cementing the metals directly, or their oxides, with arsenious acid and carbon, at a heat at which neither the refractory metals nor the alloy is melted, and then melt the alloy thus formed at the lowest heat at which it will dissolve in a crucible, with the exclusion of oxygen; that is, under a cover of fusible glass.

Alloys of arsenic cannot be converted into vessels in which food for men or animals is prepared, but it finds extensive applications in other cases; and when its properties are more thoroughly understood, it will be still more generally used. In virtue of its property of causing the fluidity of metals, when present in small quantities, it promotes the union of those metals which, without its assistance, do not unite. Zinc and lead do not unite very readily; but with the assistance of a little arsenic, both form a firm combination. Iron has no affinity for lead, but when arsenic is present it forms an alloy with it. Thus we may form combinations which, without the assistance of arsenic, cannot so easily be accomplished. Iron and alumina may be formed by melting grey-iron and pure alumina together; in this case all the impurities of the cast iron are in the compound. When pure iron filings or turnings are cemented in alumina, arsenious acid, and carbon, and then melted in a crucible so as to expel the arsenic, an alloy of iron and aluminum of great purity is formed, which, however, contains traces of arsenic.

Arsenic, like *antimony*, has a remarkable tendency to cause metals to crystallize; but it does not make quite as brittle alloys as the latter. In producing a high degree of fluidity, it admits the melting of metals at a low heat, and consequently the formation of small crystals and fine grain, and enables the metals to contract into a small compass, which causes them to be close and to assume a high polish. With the closeness of grain, the hardness and brittleness increases.

Arsenic causes all metals to be whiter than they naturally are.

ARTESIAN WELL. Fr., *Puits artésien*; Ger., *Artesischer Brunnen*; Ital., *Pozzo artesiano*; Span., *Pozo artesiano*.

An artesian well is a shaft sunk or bored through impermeable strata, until a water-bearing stratum is tapped, when the water is forced upwards by means of the hydrostatic pressure due to the superior level at which the rain-water was received.

When comparing the operations and tools of artesian well-borers, George Rowdon Burnell, in a paper, given in the Minutes of the Institution of Civil Engineers, "On the Machinery Employed in Sinking Artesian Wells on the Continent," takes three systems, namely,—the Chinese, or Fauvelle's system; the French well-borers', or, rather, the usual well-borers' system; and Kind's system. Of these, the system of Fauvelle was at first much patronized by Arago and by Dr. Buckland, but it is now very little practised on the Continent, and not at all in Great Britain. The principles upon which it was founded were, first, that the motion given to the tool in rotation was simply derived from the resistance that a rope would oppose to an effort of torsion; and, therefore, that the limits of application of the system were only such as would provide that the tool should be safely acted upon; and, secondly, that the injection of a current of water, descending through a central tube, should wash out the detritus created by the cutting tool at the bottom. The difficulties attending the removal of the detritus were enormous; and, though the system of Fauvelle answered tolerably well when applied to shallow borings, it was found to be attended with such disadvantages when applied on a large scale, that it has been generally abandoned. The quantity of water required to keep the boring-tool clear is a great objection to the introduction of this system, especially as in the majority of cases artesian wells are sunk in such places as are deprived of the advantage of a large supply.

In the ordinary system of well-boring, the motion of the tools used for the comminution, or for the removal of the rocks, is effected by the use of solid iron rods, connected with the upper parts of the machinery, so that the weight of the rods, and the weight of the tools themselves, increase in proportion to the depth of the excavation. It follows from this fact, that where the excavation is very deep, there is considerable difficulty in transmitting the blow of the tool, in consequence of the vibration produced in the long rod, or in consequence of the torsion; and, for the same reason, there is a danger of the blows not being equally delivered at the bottom. It has been attempted to obviate this difficulty by the use of hollow rods, presenting greater sectional area than was absolutely necessary for the particular case, in order to increase their lateral resistance to the blows tending to produce vibration. The trepan is made so as to fall from a height of 2 feet, but the disengagement of the machinery is effected by the reaction of the column of water that the trepan works in. The majority of well-borers have, in many instances, used the hollow rods filled with cork, or with similar substances, and they have also tried the Euyenhausen joint; but they do not appear to have made so much use of Kind's system for removing the products of the excavation, and they more frequently resort to augers and chisels.

The first well that was executed of great depth, and which gave rise to the adoption of tools which directed public attention to the art of well-boring, was that for the city of Paris by Mulot, at the Abattoir of Grenelle. This was commenced in the year 1832; and, after more than eight years' incessant labour, water rose, on the 26th of February, 1842, from the total depth of 1798 feet. Subsequent to this, many wells have been sunk on the Continent, with the hope of attaining the brine springs so often met with in the Rhine provinces, or the springs destined for the supply of towns, and which are even deeper than the well of Grenelle, reaching in some cases to the extraordinary depth of 2800 feet; but all of them, like the Grenelle well, of small diameter. In their construction, however, the German engineers introduced some important modifications of the tools employed; and, amongst other inventions, Euyenhausen imparted a sliding movement to the striking part of the tool used for comminuting the rock, so as to fall always through a certain distance; and thus, while he produced a uniform action upon the rock at the bottom, he avoided the jar of the tools. Kind also began to apply his system to the working of the large excavations for the purpose of winning coal. Whilst the art was in this state, and when he had already executed some very important works in Germany, Belgium, the North of France, Creuzot, Seraing, &c., the Municipal Council of Paris determined to intrust him with the execution of a new well they were about to sink at Passy. The well of Passy was intended to be executed in the Paris basin, which it was to traverse with a diameter, hitherto unattempted, of 1 mètre, 3 2809 ft.; that of the Grenelle well being only 20 centimètres, 8 in. It was calculated that it would reach the water-bearing stratum at nearly the same depth as the latter, and would yield 8000 mètres or 10,000 cubic mètres in 24 hours, or about 1,786,240 gallons to 2,232,800 gallons a-day.

The operations were undertaken by Kind under a contract with the Municipality of Paris, by which he bound himself to complete the works within the space of twelve months from the date of their commencement, and to deliver the above quantity of water for the sum of 300,000 francs, 12,000*l*. On the 31st of May, 1857—after the workmen had been engaged nearly the time stipulated for the completion of the work, and when the boring had been advanced to the depth of 1732 ft. from the surface—the excavation suddenly collapsed in the upper strata, at about 100 ft. from the ground, and filled up the bore. Kind would have been ruined had the engineers of the town held him to the strict letter of his contract; but it was decided to behave in a liberal manner, and to release him from it, the town retaining his services for the completion of the well, as also the right to use his patent machinery. The difficulties encountered in carrying the excavation through the clays of the upper strata were found to be so serious that, under the new arrangement, it required six years and nine months of continuous efforts to reach the water-bearing stratum, of which time the far larger portion was employed in traversing the clay beds. The upper part of this well was finally lined with solid masonry, to the depth of 150 ft. from the surface; and beyond that depth tubing of wood and iron was introduced. This tubing was continued to the depth of 1804 ft. from the surface, and had at the bottom a length of copper pipe pierced with holes to allow the water to enter. At this depth the compound tubing could not be made to descend any lower; but the engineers employed by the city of Paris were convinced that they could obtain the water by means of a preliminary boring; and therefore they proceeded to sink in the interior of the above tube of 3 2809 ft. diameter, an inner tube 2 ft. 4 in. diameter, formed of wrought-iron plates 2 in. thick, so as to enable them to traverse the clays encountered at this zone. At last, the water-bearing strata were met with on the 24th of September, 1861, at the depth of 1913 ft. 10 in. from the ground-line; the yield of the well being, at the first stroke of the tool that pierced the crust, 15,000 cubic mètres in 24 hours, or 3,349,200 gallons a-day; it quickly rose to 25,000 cubic mètres, or 5,582,000 gallons a-day; and as long as the column of water rose without any sensible diminution, it continued to deliver a uniform quantity of 17,000 mètres, or 3,795,000 gallons a-day. The total cost of this well was more than 40,000*l*., instead of 12,000*l*., at which Kind had originally estimated it.

In sinking the well of Passy, the weight of the trepan for comminuting the rock was about 1 ton 16 cwt., 1800 kilog.; the height through which it fell was about 60 centimètres; and its diameter was 3 ft. $3\frac{3}{8}$ in., 1 mètre. The rods were of oak, about 8 in. on the side, and the dimensions of the cutting tool were limited to 3 ft. $3\frac{3}{8}$ in. because it worked the whole time in water; but generally the class of borings Kind undertook were of such a description as justified resorting to tools of great dimensions. When sinking the shafts for winning coal, his operations required to be carried on with the full diameters of 10 ft. or 14 ft.; and he then drove a boring of 8 ft. 4 in. diameter in the first instance, and subsequently enlarged this excavation. There can be no objection to executing artesian borings of this diameter, other than the probable exhaustion of the supply; particularly as it is now known that the yield of water by these methods is proportionate to the diameter of the column; though, strange as it may appear, the first opposition to Kind's plan of sinking the well of Passy was founded upon the assumption that he would not meet with a larger supply of water from the sub-cretaceous formations than had been met with at Grenelle, where the diameter of the boring was at the bottom not more than 8 in. It is now, however, proved that there is a direct gain in adopting the larger borings, not only as regards the quantity of water to be derived from them, but also in their execution, arising from the fact that the tools can be made more secure against the effects of torsion or of concussion against the sides of the excavation, which is the cause of the most serious accidents met with in well-sinking.

331.

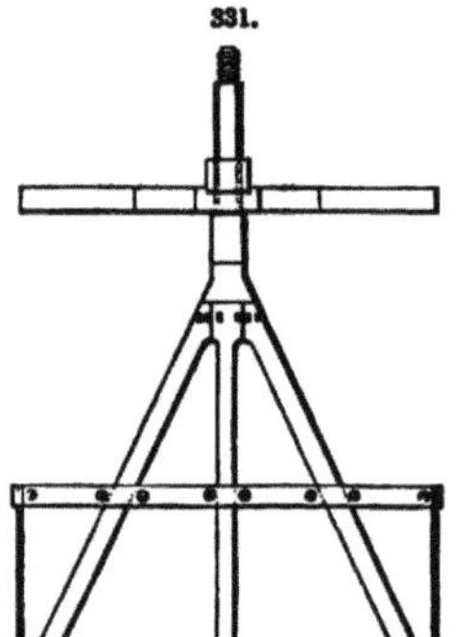

332.

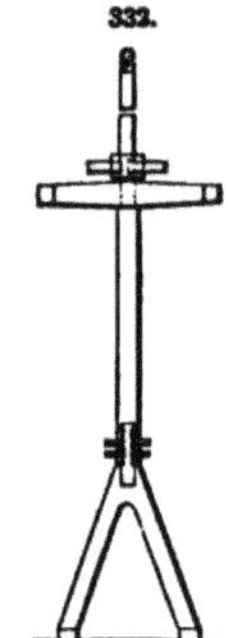

333.

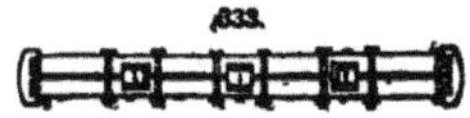

The trepan of M. Kind contains some peculiar details, which are shown in Figs. 331, 332. The trepan is composed of two principal pieces, the frame and the arms, both of wrought iron, with the exception of the teeth of the cutting part, which are of cast steel. The frame has at the bottom a series of holes, slightly conical, into which the teeth are inserted, and tightly wedged up, Fig. 333. These teeth are placed with their cutting edges on the longitudinal axis of the frame that receives them; and at the extremity of the frame there are formed two heads, forged out of the same piece with the body of the tool, which also carries two teeth, placed in the same direction as the others, but double their width, in order to render this part of the tool more powerful. By increasing the dimensions of these end-teeth, the diameter of the boring can be augmented, so as to compensate for the diminution of the clear space caused by the tubing, necessarily introduced for security in traversing strata disposed to fall in, or for the purpose of allowing the water from below to escape at an intermediate level.

Above the lower part of the frame of the trepan is a second piece, composed of two parts bolted together, and made to support the lower portion of the frame. This part of the machinery also carries two teeth at its extremities, which serve to guide the tool in its descent, and to work off the asperities left by the lower portion of the trepan. Above this, again, are the guides of the machinery, properly speaking, consisting of two pieces of wrought iron, arranged in the form of a cross, with the ends turned up, so as to preserve the machinery perfectly vertical in its movements, by pressing against the sides of the boring already executed. These pieces are independent of the blades of the trepan, and may be moved closer to it or farther away from it, as may be desired. The stem and the arms are terminated by a single piece of wrought iron, which is joined to the frame with a kind of saddle-joint, and is kept in its place by means of keys and wedges. The whole of the trepan is finally jointed to the great rods that communicate the motion from the surface, by means of a screw-coupling, formed below the part of the tool which bears the joint; this arrangement permits the free fall of the cutting part, and unites the top of the arms and frame, and the rod, Fig. 334. It has been proposed to substitute for this screw-coupling a keyed joint, in order to avoid the inconvenience frequently found to attend the rusting of the screw, which often interposes great difficulties in cases where it becomes necessary to withdraw the trepan. Kind introduced some modifications in the trepans employed in carrying out the large borings for the coal mines in Belgium. These modifications were rendered necessary by the large dimensions he was obliged to give the borings, and by the preliminary use of a smaller trepan.

The sliding joint is the part of Euyenhausen's invention most unhesitatingly adopted by Kind, and it is one of the peculiarities of his system as contrasted with the processes formerly in use. So long as his operations were confined to the small dimensions usually adopted for artesian borings, he contented himself with making a description of joint with a free fall; a simple movement of disengagement regulating the height fixed by the machinery itself, like the fall of the monkey in a pile-driving machine; but it was found that this system did not answer when applied to large excavations, and it also presented certain dangers. Kind then, for the larger class of borings, availed himself of sliding guides, so contrived as to be equally thrown out of gear when the machinery had come to the end of the stroke, and maintained in their respective positions by being made in two pieces, of which the inner one worked upon slides, moving freely in the piece that communicated the motion to the striking part of the machinery. The two parts of the tool were connected with pins, and with a sliding joint, which, in the Passy well, was thrown out of gear by the reaction of the column of water above the tool unloosing the click that upheld the lower part of the trepan, Figs. 335, 336, 337. The changes thus made in the usual way of

334.

335.

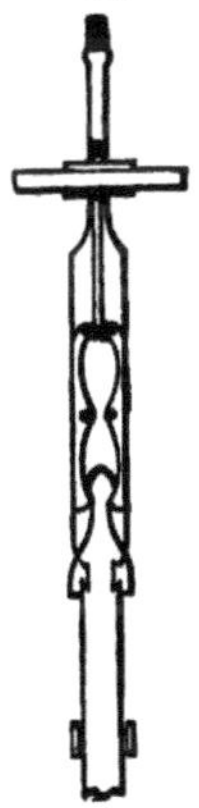

336.

337.

releasing the tool, and in guiding it in its fall were, however, matters of detail; they involved no new principle in the manner of well-boring; and the modern authorities upon the subject consider that there was something deficient in Kind's system of making the column of water act upon a disc by which the click was set in motion. This system, in fact, required the presence of a column of water, not always to be commanded, especially when the borings had to be executed in the carboniferous series.

The rods used for the suspension of the trepan, and for the transmission of the blows to it, were of oak; and this alone would constitute one of the most characteristic differences between the system of tools introduced by Kind and those made by the majority of well-borers, but which, like the disengagement of the tool intended to comminute the rock, depended for its success upon the boring being filled with water. The resistance that the wood offers, by its elasticity, to the effects of any sudden jar, is also to be taken into account in the comparison of the latter with iron, for the iron is liable to change its form under the influence of this cause. The resistance to an effort of torsion need not, however, be much dwelt on, for the turn given to the trepan is always made when the tool is lifted up from its bed. For the purpose of making the rods, Kind recommended that straight-grown trees, of the requisite diameter, should be selected, rather than that they should be made of cut timber, as there is less danger of the wood warping, and the character of the wood is more homogeneous. He generally used these trees in lengths of about 50 feet, and he connected them at the ends with wrought-iron joints, fitting one into the other, Fig. 338. The ironwork of the joints is made with a shoulder underneath the screw-coupling, to allow the rods to be suspended by the ordinary *crow's-foot* during the operation of raising or lowering them. In the works executed at Passy there was a kind of frame erected over the centre of the boring, of sufficient height to allow of the rods being withdrawn in two lengths at a time, thus producing a considerable economy of time and labour.

All the processes yet introduced for removing the products of the excavation must be considered to be, more or less, defective, because all are established on the supposition that the comminuting tool must be withdrawn, in order that the spoon, or other tool intended to remove the products of the working of the comminutor, may be inserted. This remark applies to Kind's operations at Passy and elsewhere, as he removed the rock detached from the bottom of the excavation by a spoon, Figs. 339, 340, which was a modification of the tool he invariably employs for this purpose. It consisted of a cylinder of wrought iron, suspended from the rods by a frame, and fastened to it, a little below the centre of gravity, so that the operation of upsetting it, when loaded, could be easily performed. This cylinder was lowered to the level of the last workings of the trepan, and the materials already detached by that instrument were forced into the tool, by the gradual movement of the latter in a vertical direction. Some other implements, employed by Kind for the purpose of removing the products of the excavation in the shafts for the coal-mines of the North of France, were ingenious, and well adapted to the large dimensions of the shafts; but they were all, in some degree, exposed to the danger of becoming fixed, if used in the small borings of artesian wells, by the minute particles of rocks falling down between their sides and the excavation from above. Their use was therefore abandoned, and the well of Passy was cleared out with the spoon, the bottom of which was made to open upwards, with a hinged flap, which admitted the finer materials detached by the trepan. There were also several tools for the purpose of withdrawing the broken parts of the machinery from the excavation, or whatever substances might fall in from above; and all were marked by a great degree of simplicity, but they did not differ enough from those generally used by well-borers for the same purposes to merit further remarks. In fact, the accidents intended to be guarded against or remedied are so precisely alike in all cases, that there can be little variety in the manufacture of these instruments. But G. R. Burnell believed that Kind deprived himself of a valuable appliance in not using the ball-clack, *la soupape à boulet*, that other well-borers employ, Fig. 341. The tools used by Mulot, Laurent, Dru, and other French well-borers, are admirably adapted for the extraction of the materials from excavations of small diameter, but would be of no avail if applied to the well of Passy. They seem to have been designed for wells which rarely exceed the diameter of that executed at Grenelle.

338.

339.

341.

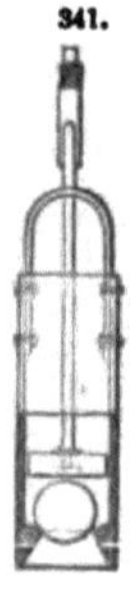

342.

343.

340.

At Passy, great strength was given to the head of the striking-tool, and to the part of the machinery applied to turn the trepan, because the great weight of the latter superinduced the danger of its breaking off under the influence of the shock, and because the solidity of this part of the machinery necessarily regulated the whole working of the tool. The head of the "assemblage" was connected with the balance-beam of the steam-engine by a Vaucanson chain, with a screw-coupling, admitting of being lengthened as the trepan descended, Figs. 342, 343. The balance-beam, in order to increase its elastic force in the upward stroke, is in Kind's works made of wood, in two pieces; the upper one being of fir and the lower one of beech. The whole of the machinery is put in motion by steam, which is admitted to the upper part of the cylinder, and presses it down, and thus raises the tool at the other end of the beam to that part in connection with the cylinder. The counterpoise to the weight of the tools is also placed upon the cylinder-end of the beam. The cylinder receives the steam through ports that are opened and closed by hand, like those of a Nasmyth's hammer which was invented by Robert Wilson, of Patricroft; so that the number of the strokes of the piston may be increased or diminished, and the length of the strokes may be increased, as occasion may require.

The balance-beam is continued beyond the point where the piston is connected with it, and it goes to meet the blocks placed to check the force of the blow given by the descent of the tool. The guides of the piston-head are attached to the part of the machinery that acts in this manner; but at Passy, Kind made the balance-beam work upon two free plummer-blocks, or blocks having no permanent cover, that they might be more easily moved whenever it was necessary to displace the beam, for the purpose of taking up or letting down the rods, or for changing the tools; for the balance-beam was always immediately over the centre of the tools, and it therefore had to be displaced every time that the latter were required to be changed. This was effected by allowing the beam to slide horizontally, so as to leave the mouth of the pit open. The counter-check, above mentioned, likewise prevented the piston from striking the cylinder-cover with too great a force when it was brought back by the weight of the tools to its original position. The operation of raising and lowering the rods, or of changing the tools, was performed at Passy by a separate steam-engine, and the spoon was discharged into a special truck, moving upon a railway expressly laid for this purpose in the great tower erected over the excavation. All these arrangements were in fact made with the extreme attention to the details of the various parts of the work which characterizes the proceedings of foreign engineers, and conduces so much to their success.

The beating, or comminution of the rock, was usually effected at Passy at the rate of from fifteen strokes to twenty strokes per minute. The rate of descent, of course, differed in a marked manner, according to the nature of the rock operated upon; but, generally speaking, the trepan was worked for the space of about eight hours at a time, after which it was withdrawn, and the spoon let down, in order to remove the detritus. The average number of men employed in the gang, besides the foreman, or the superintendent of the well, was about fourteen: they consisted of a smith and hammerman, whose duty it was to keep the tools in order; and two shifts of men intrusted with the excavation, namely, an engine-driver and stoker, a chief workman, or sub-foreman, and three assistants. The total time employed in sinking the shafts executed upon this system in the North of France, where it has been applied without meeting with the accidents encountered in the Passy well, was found to be susceptible of being divided in the following manner: from 25 per cent. to 56 per cent. was employed in manœuvring the trepan; from 11 per cent. to 14½ per cent. in raising and lowering the tools; from 19 per cent. to 21 per cent. in removing the materials detached from the rocks, and cleaning out the bottom of the excavation; and from 8 per cent. to 10½ per cent. was lost, owing to the stoppage of the engines, or to the accidents from broken tools, or to other causes always attending these operations. In the well of Passy there was, of course, a considerable difference in the proportions of the time employed in the various details of the work; and the long period occupied in obviating the effects of the slips which took place in the clays, both in the basement beds of the Paris basin and in the sub-cretaceous strata, would render any comparison derived from that well of little value; but it would appear that, until the great accident occurred, the various operations went on precisely as Kind had calculated upon.

The essential difference between the systems of Kind and those of the ordinary well-borers appears to consist in the use of a trepan of considerable diameter, falling through a certain height; and in the rods of oak, working in water, which were thus easily counterbalanced. All the other tools employed are in point of fact quite irrespective of the merits of the various processes. The advantages of these inventions would be very great, provided the well were sunk in such circumstances as would allow of its being worked under water, and the results of the Passy boring demonstrated that these inventions had a real existence, for the accidents attending that work were such as might have been expected in the course of boring artesian wells. It may be questioned whether the engineers of the town were justified in passing the contract with Kind to finish the work within the time, and for the sum at which he undertook it, but they certainly treated him with kindness and consideration, in allowing him to conduct the work at the expense of the city of Paris, for so long a period after the expiration of his contract. It seems, however, that the French well-borers could not at the time have attempted to continue the well upon any other system than that introduced by Kind; that is to say, upon the supposition that it should be completed of the dimensions originally undertaken. The patent has now expired, and the processes he introduced have consequently been open to the imitation of all the world, and no doubt they will be adopted by the French well-borers. Experience has shown, in fact, that this detail was badly executed at the well of Passy. The masonry lining was introduced after his contract had expired, and when he had ceased to have the control of the works; the wrought-iron tubing at the lower part of the excavation being a subsequent idea. It has followed from this defective system of tubing—the wood necessarily yielding in the vertical joints—that the water in its upward passage escaped through the joints, and went to supply the basement beds of the Paris basin, which are as much resorted to as the London sand-beds for an artesian supply; and, in fact, the level of the water has been raised in the neighbouring wells by the quantity let in from below, and the yield of the well itself has been proportionally diminished, until it has fallen to 450,000 gallons per day. That the increased yield of the neighbouring wells is to be accounted for by the escape of the water from the artesian boring is additionally proved by the temperature of the water in them; it is found to be nearly 82° Fah., or nearly that observed in the water of Passy. This was an unfortunate complication of the bargain made between Kind and the Muni-

BORING AT KENTISH TOWN, LONDON.

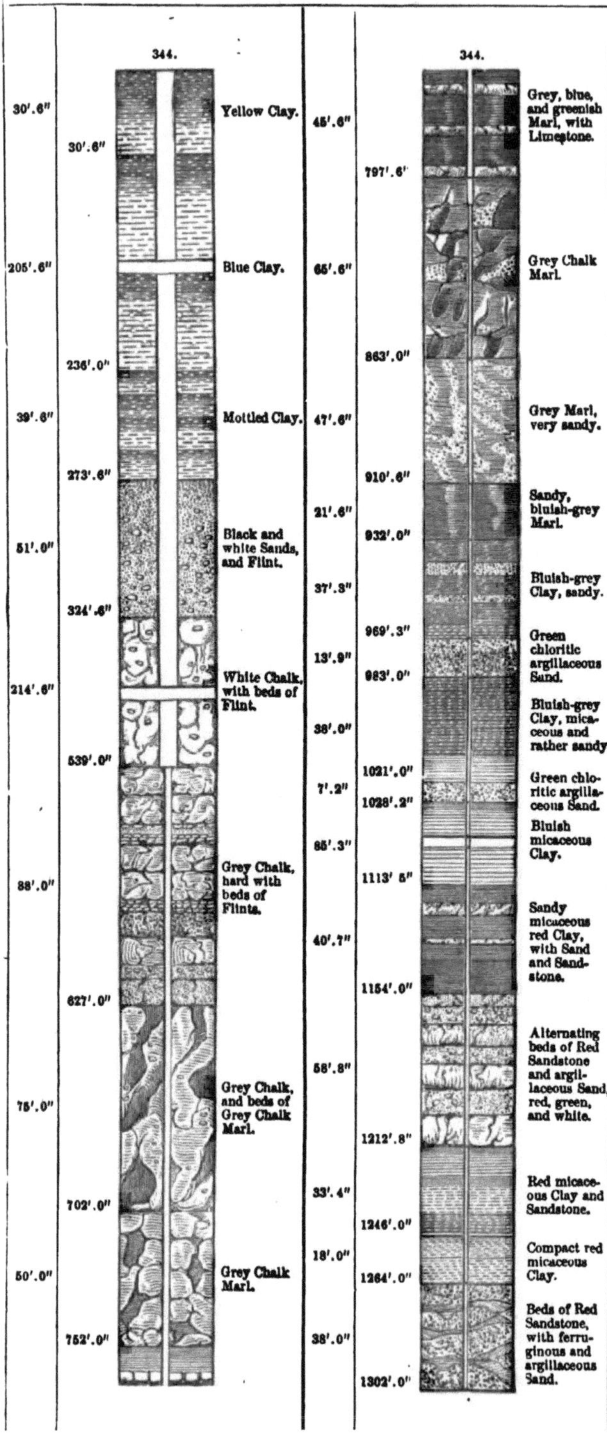

BORING AT PASSY, PARIS.

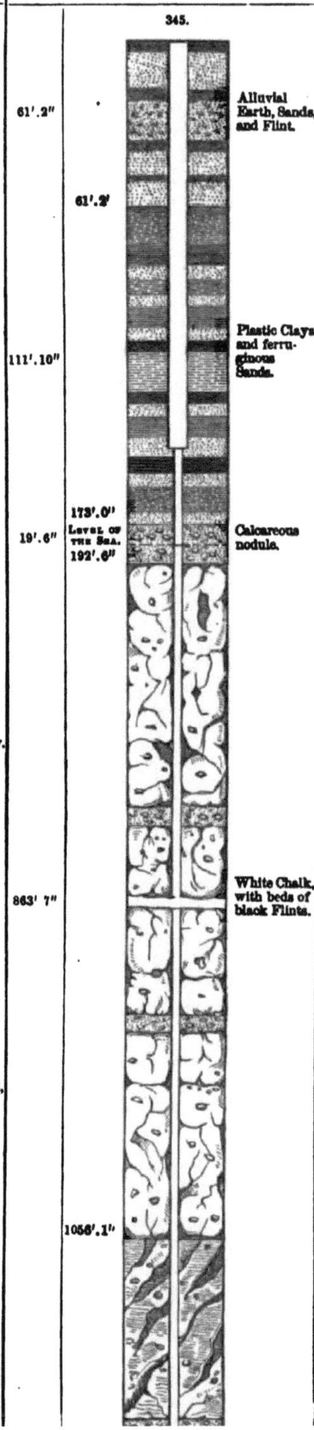

Boring at Passy, Paris—*continued.*

345.

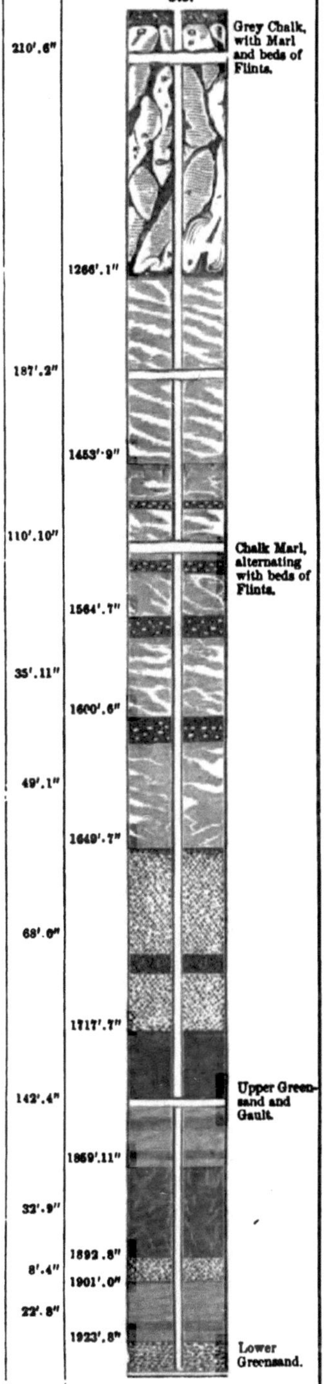

Boring at Grenelle, Paris.

346.

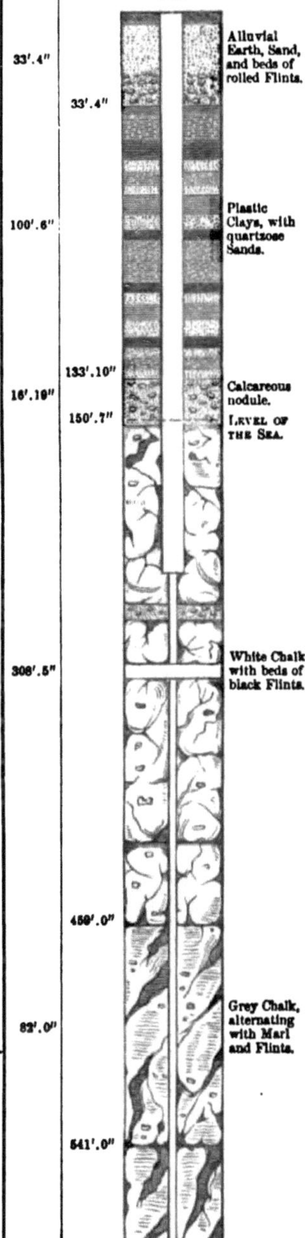

316.

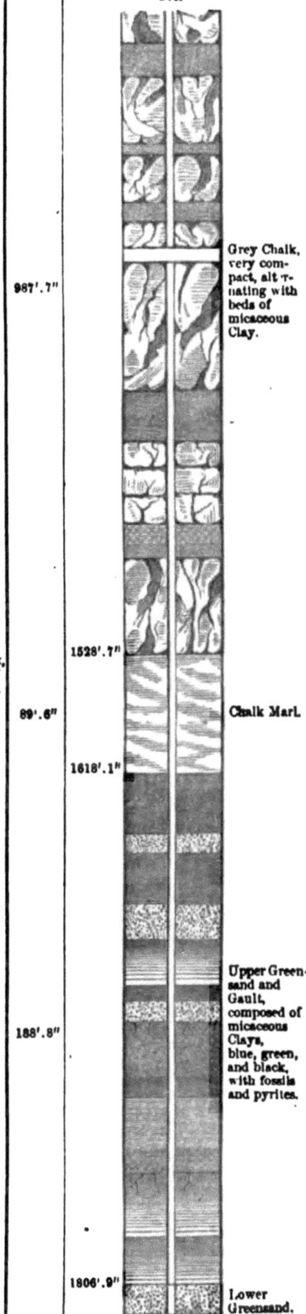

cipal Council; but it in no respect affects the choice of the boring machinery, which seems to have complied with all the conditions it was designed to meet. The descent of the tubes and their nature ought to have been the subject of special study by the engineers of the town, who should have known the nature of the strata to be traversed better than Kind could be supposed to do, and should have insisted upon the tubing being executed of cast or wrought iron, so as effectually to resist the passage of the water. At any rate, this precaution ought to have been taken in the portions of the well carried through the basement beds of the Paris basin, or through the lower members of the chalk and the upper greensand. It may also be observed that a remarkable change has been noticed in connection with the subterranean disturbances which have lately taken place in France. The water in this well is at these times rendered thick, cloudy, and totally unfit for human consumption. The discolouration of the water, however, is not dependent upon this cause, but is owing to the strata through which the boring passes being washed out, which is in itself a serious objection to the use of the water of this well.

The system applied by Dru is worthy of attention, not so much on account of the novelty of the invention, or of any new principle involved in it, as on account of the contrivances it contains for the application of the tool, "*à chute libre*," or the free-falling tool, to artesian wells of large diameters. It has been already explained, that under Kind's arrangements the trepan was thrown out of gear by the reaction of the water which was allowed to find its way into the column of the excavation; but that it is not always possible to command the supply of the quantity necessary for that purpose; and even when possible, the clutch Kind adopted was so shaped as to be subject to much and rapid wear. Dru, with a view to obviate both these inconveniences, made his trepan in the manner shown in Fig. 335, in which it will be seen that the tool was gradually raised until it came in contact with the fixed part of the upper machinery, when it was thrown out of gear. The bearings of the clutch were parallel to the horizontal line, and were found in practice to be more evenly worn, so that this instrument could be worked sometimes from eight days to fourteen days without intermission; whereas, on Kind's system, the trepan was frequently withdrawn after two days' or three days' service. Another great recommendation of Dru's system, if applied in cases where water is scarce, is that there is no necessity for a column of water with the trepan; but in all other essential respects the details of his machinery are the same as those employed by Kind, who must be considered to have advanced the science of well-boring by the introduction of timber-rods, the manner of balancing them, and by the use of the trepan, all of which were first applied at the artesian boring of Passy, and enabled it to be executed in a comparatively short space of time. The tools introduced by Dru and other modern well-borers are doubtlessly better fitted for the artesian borings of small diameter, and for such as are free from water from the upper strata; but the advantage ceases when these conditions are reversed.

The nature and depths of the different strata bored through in sinking an artesian well at Kentish Town, London are shown in Fig. 344; Fig. 345 shows the nature and extent of the strata that were met in boring the artesian well at Passy, Paris; and Fig. 346, in a similar manner, exhibits the range and nature of the strata perforated in boring the artesian well at Grenelle, to which we have so often referred.

We take an account of some recent operations in artesian well-boring by M. Dru, of Paris, from a paper read by him at the Conservatoire des Arts et Métiers, Paris, 6th June, 1867, and published in the 'Proceedings of the Institute of Mechanical Engineers.'

The artesian wells at present sunk in the tertiary formation of the Paris basin, Fig. 347, range in size generally from about 8 in. down to 2 in. diameter, with a depth of about 230 to 350 ft.

347.

Géological Section from Niort to Verdun, through the Paris basin.

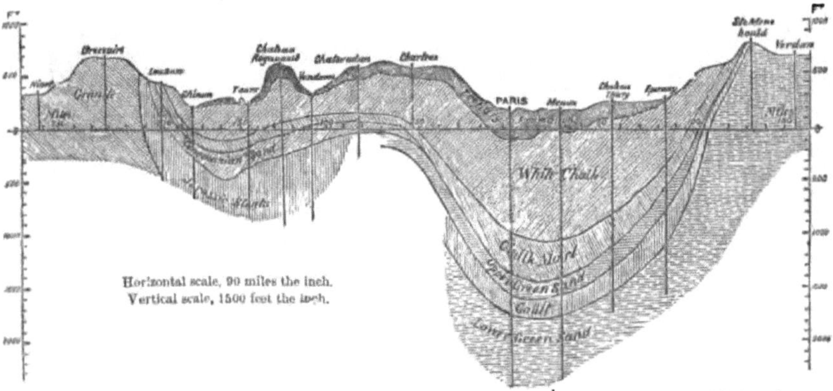

The bore-hole is usually lined with copper, in order to make the wells water-tight and bring the water to the surface without loss. These works often present considerable difficulties in their execution, from the frequent changes in the nature of the ground passed through, and from the impediments that are so often encountered in driving the tubes through the beds of sand and clay.

Borings of a much greater depth and larger size are now in process of execution in Paris, for the purpose of bringing to the surface a large supply of the artesian waters from the lower greensand underlying the chalk, in the secondary formation; the existence of a supply of water in that stratum has already been proved by the well sunk at Grenelle of 3¾ in. diameter, and the subsequent one of 27½ in. diameter sunk at Passy. Each of these two artesian wells required six or seven years' work for its completion; their situation in Paris is shown in the plan, Fig. 348.

348.

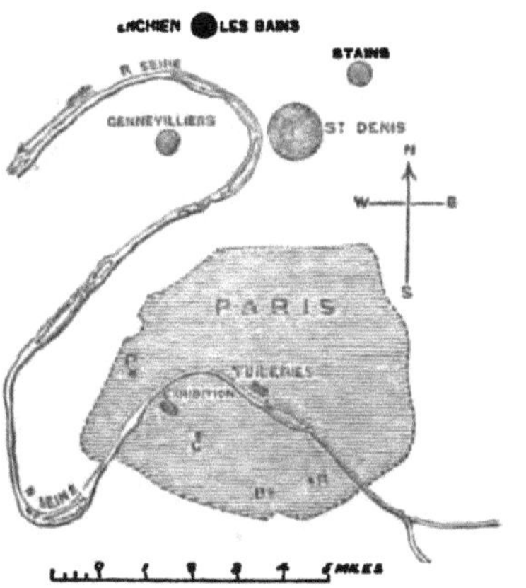

349.

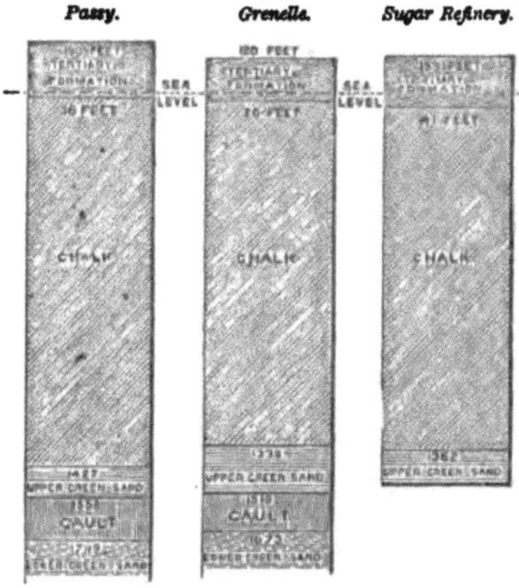

Reference.—P, Passy Well. G, Grenelle. B, Butte-aux-Cailles. R, Sugar Refinery.

A large artesian well was, in 1867, being constructed by Dru at Butte-aux-Cailles, Fig. 348, for the supply of the city of Paris, which is intended to be carried down through the greensand to a depth of 2600 or 2900 ft. to reach the Portland limestone. The boring in 1867 was 490 ft. deep, and its diameter 47 in.

During the previous 2½ years, M. Dru was engaged in sinking a similar well of 19 in. diameter for supplying the Sugar Refinery of M. Say, in Paris, Fig. 349; 1570 ft. deep of this well had been bored in 1867, see Fig. 349.

For the smaller wells, hand-boring tools are in use; but these are limited to borings of inconsiderable depth and small diameter. For borings of the diameter of these large wells, it is necessary to make use of special tools, worked entirely by steam-power; and in some cases of sinking mine-shafts, tools of as large a diameter as 14¾ ft. have been used. The boring is effected by a rotary motion in the case of the small diameters; but in borings of a large diameter and considerable depth, percussive action alone is employed, which is effected by raising the tool and letting it fall with successive strokes.

The apparatus employed by M. Dru in boring the large wells that have been mentioned, is shown in Figs. 350, 351. The boring-rod A is suspended from the outer end of the working beam B, which is made of timber hooped with iron, working upon a middle bearing, and is connected at the inner end to the vertical steam cylinder C, of 10 in. diameter and 39 in. stroke. The stroke of the boring-rod is reduced to 22 in., by the inner end of the beam being made longer than the outer end, serving as a partial counterbalance for the weight of the boring-rod. The steam cylinder is shown enlarged in Fig. 352, and is single-acting, being used only to lift the boring-rod at each stroke, and the rod is lowered again by releasing the steam from the top side of the piston; the stroke is limited by timber stops both below and above the end of the working beam B.

The boring-tool is the part of most importance in the apparatus, and the one that has involved most difficulty in maturing its construction. The points to be aimed at in this are,—simplicity of construction and repairs· the greatest force of blow possible for each unit of striking-surface; and freedom from liability to get turned aside and choked.

The tool used in small borings is a single chisel, as shown in Figs. 353, 354; but for the large borings it is found best to divide the tool-face into separate chisels, each of convenient size and weight for forging. All the chisels, however, are kept in a straight line, whereby the extent of striking-surface is reduced; and the tool is rendered less liable to be turned aside by meeting a hard portion of flint on a single point of the striking-edge, which would diminish the effect of the blow.

The tool is shown in Figs. 355, 356, 357, 358, 359, 360, 361; and is composed of a wrought-iron body D, connected by a screwed end, E, to the boring-rod, and carrying the chisels, F F, fixed in separate sockets and secured by nuts above; two or four chisels are used, or sometimes even a greater number, according to the size of the hole to be bored. This construction allows of any broken chisel being easily replaced; and also, by changing the breadth of the two outer chisels, the diameter of the hole bored can be regulated exactly as may be desired. When four chisels are used, the two centre ones are made a little longer than the others, as shown in Fig. 359, to form a leading hole as a guide to the boring-rod. A cross-bar G, of the same width as the tool, guides it in the hole in the direction at right angles to the tool; and in the case of the larger and longer tools a second cross-bar is also added higher up, at right angles to the first and parallel to the striking-edge of the tool.

350.

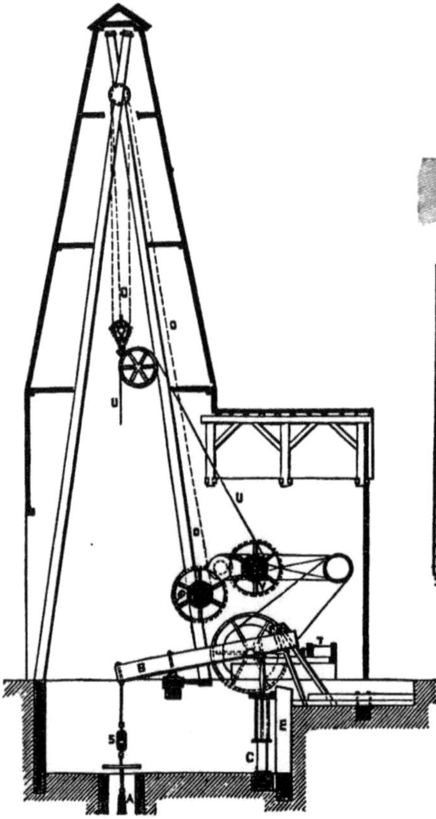

351.

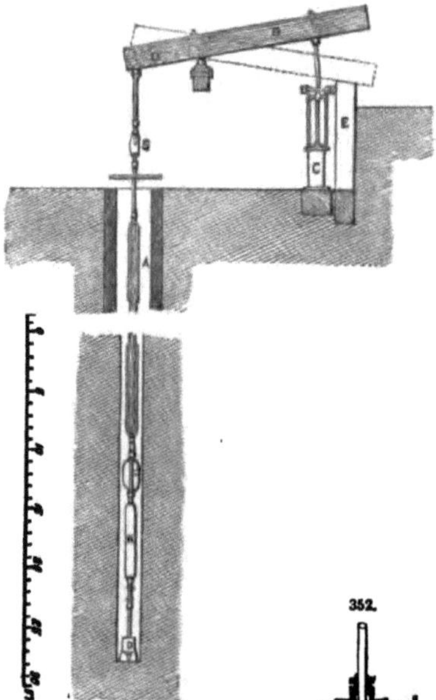

352.

353.

354.

355.

356.

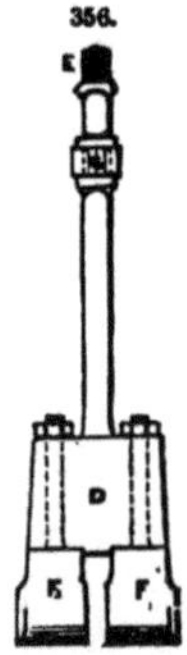

357.

If the whole length of the boring-rod were allowed to fall suddenly to the bottom of a large bore-hole at each stroke, frequent breakages would occur; it is therefore found requisite to arrange for the tool to be detached from the boring-rod at a fixed point in each stroke, and this has led to the general adoption of *free-falling tools*. There have been several contrivances for effecting this object; and M. Dru's plan of self-acting free-falling tool liberated by reaction, is

358.

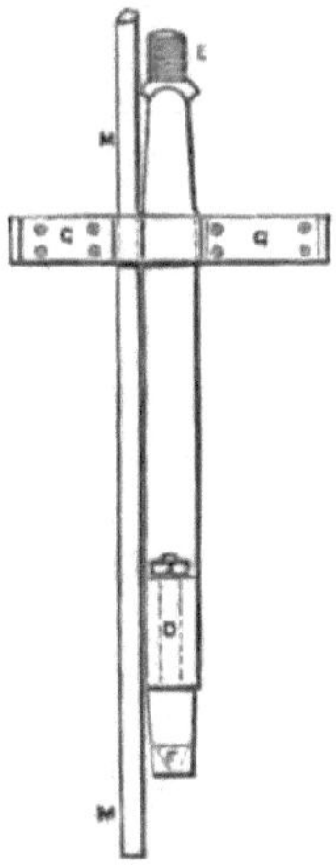

359.

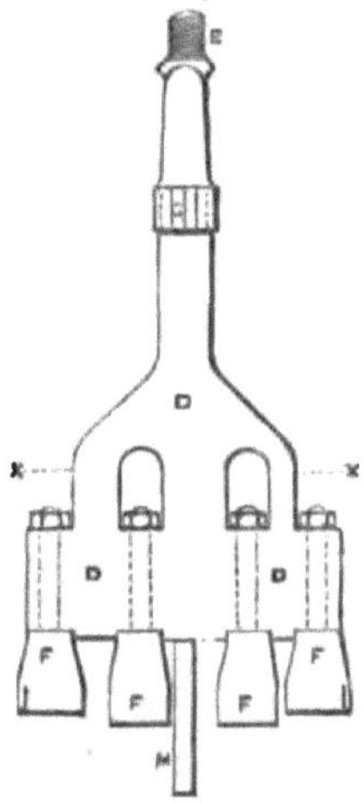

360.

361.

shown in side and front view in Figs. 362, 363. The hook H, attached to the head of the boring-tool D, slides vertically in the box K, which is screwed to the lower extremity of the boring-rod; and the hook engages with the catch, J, centred in the sides of the box K, whereby the tool is lifted as the boring-rod rises. The tail of the catch, J, bears against an inclined plane, L, at the top of the box K; and the two holes carrying the centre-pin, I, of the catch, are made oval in the vertical direction, so as to allow a slight vertical movement of the catch. When the boring-rod reaches the top of the stroke, it is stopped suddenly by the tail end of the beam B, Fig. 351, striking upon the wood buffer-block E; and the shock thus occasioned causes a slight jump of the catch, J, in the box K; the tail of the catch is thereby thrown outwards by the incline L, as shown in Fig. 364, liberating the hook H, and the tool then falls freely to the bottom of the bore-hole, as shown in Fig. 365. When the boring-rod descends again after the tool, the catch, J, again engages with the hook H, enabling the tool to be raised for the next blow, as in Fig. 363.

Another construction of self-acting free-falling tool, liberated by a separate disengaging-rod, is shown in side and front view in Figs. 366, 367. This tool consists of four principal pieces, the hook H, the catch J, the paul I, and the disengaging-rod M. The hook H, carrying the boring-tool D, slides between the two vertical sides of the box K, which is screwed to the bottom of the boring-rod; and the catch, J, works in the same space upon a centre-pin fixed in the box, so that the tool is carried by the rod, when hooked on the catch, as shown in Fig. 367. At the same time the paul I, at the back of the catch, J, secures it from getting unhooked from the tool; but this paul is centred in a separate sliding hoop, N, forming the top of the disengaging-rod M, which slides freely up and down within a fixed distance upon the box K; and in its lowest position the hoop, N, rests upon the upper of the two guides P P, Fig. 366, through which the disengaging-rod, M, slides outside the box K. In lowering the boring-rod, the disengaging-rod, M, reaches the bottom of the bore-hole first, as shown in Figs. 366, 367, and being then stopped it prevents the paul, I, from descending any lower; and the inclined back of the catch, J, sliding down past the paul, the latter forces the catch out of the hook H, as shown in Fig. 368, thus allowing the tool, D, to fall freely and strike its blow. The height of fall of the tool is always the same, being determined only by the length of the disengaging-rod M.

The blow having been struck, and the boring-rod continuing to be lowered to the bottom of the hole, the catch, J, falls back into its original position, and engages again with the hook H, as shown in Fig. 369, ready for lifting the tool in the next stroke. As the boring-rod rises, the tail of the catch, J, trips up the paul, I, in passing, as shown in Fig. 370, allowing the catch to pass freely; and the paul before it begins to be lifted returns to the original position, shown in Fig. 367, where it locks the catch J, and prevents any risk of its becoming unhooked either in raising or lowering the tool in the well.

The boring-tool shown in Figs. 355, 356, which is employed in boring the well of 19 in. diameter at the Sugar Refinery, weighs ¾ ton, and is liberated by reaction, by the arrangement shown in Figs. 362 to 365; and the same mode of liberation was applied in the first instance to the larger tool, shown in Figs. 358 to 361, employed in sinking the well of 47 in. diameter at Butte-aux-Cailles. The great weight of the latter tool, however, amounting to as much as 3½ tons, necessitated so violent a shock for the purpose of liberating the tool by reaction, that the boring-rods and the rest of the apparatus would have been damaged by a continuance of that mode of working; and M. Dru was therefore led to design the arrangement of disengaging-rod for releasing the tool, as shown in Figs. 366, 367. This mode of liberation is consequently the one now in use at Butte-aux-Cailles, where it has proved completely successful in practical working. In this case the cross-guide, G, fixed upon the tool is made with an eye for the disengaging-rod, M, to work through freely, as shown in Figs. 358, 360. For borings of small diameter, however, the disengaging-rod cannot supersede the reaction system of liberation, as the latter alone is able to work in borings as small as 3¼ in. diameter; and a bore-hole no larger than this diameter has been successfully completed by M. Dru with the reaction-tool to a depth of 750 ft.

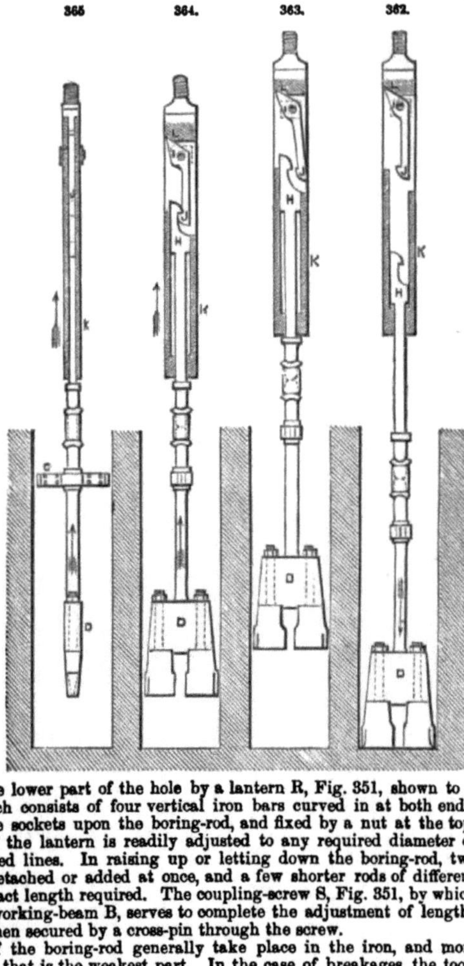

The boring-rods employed are of two kinds, wrought iron and wood. The wood rods, shown in Figs. 351, 371, are used for borings of large diameter, as they possess the advantage of having a larger section for stiffness without increasing the weight; and also when immersed in water the greater portion of their weight is floated. The wood for the rods requires to be carefully selected, and care has to be taken to choose the timber from the thick part of the tree, and not the toppings. In France, Lorraine, or Vosges, deals are preferred.

The boring-rods, whether of wood or iron, are screwed together either by solid sockets, as in Fig. 372, or with separate collars, as in Fig. 373. The separate collars are preferred for the purpose, on account of being easy to forge; and also because, as only one-half of the collar works in coupling and uncoupling the rods, while the other half is fixed, the screw-thread becomes worn only at one end, and by changing the collar, end for end, a new thread is obtained when one is worn out, the worn end being then jammed fast as the fixed end of the collar.

The boring-rod is guided in the lower part of the hole by a lantern R, Fig. 351, shown to a larger scale in Figs. 371, 374, which consists of four vertical iron bars curved in at both ends, where they are secured by movable sockets upon the boring-rod, and fixed by a nut at the top. By changing the bars, the size of the lantern is readily adjusted to any required diameter of bore-hole, as indicated by the dotted lines. In raising up or letting down the boring-rod, two lengths of about 30 ft. each are detached or added at once, and a few shorter rods of different lengths are used to make up the exact length required. The coupling-screw S, Fig. 351, by which the boring-rod is connected to the working-beam B, serves to complete the adjustment of length; this is turned by a cross-bar, and then secured by a cross-pin through the screw.

In ordinary work, breakages of the boring-rod generally take place in the iron, and more particularly at the part screwed, as that is the weakest part. In the case of breakages, the tools usually employed for picking up the broken ends are a conical screwed socket, shown in Fig. 375, and a claw, shown in Fig. 376; the socket being made with an ordinary V-thread for cases where the breakage occurs in the iron; but having a sharper thread, like a wood screw, when used where the breakage is in one of the wood rods. In order to ascertain the shape of the fractured end left in the bore-hole, and its position relatively to the centre line of the hole, a similar conical socket is first lowered, having its under-surface filled up level with wax, so as to take an impression of the broken end, and show what size of screwed socket should be employed for getting it up. Tools with nippers are sometimes used in large borings, as it is not advisable to subject the rods to a twist.

When the boring-tool has detached a sufficient quantity of material, the boring rod and tool are drawn up by means of the rope O, Fig. 350, winding upon the drum Q, which is driven by straps and gearing upon the steam-engine T. A scoop is then lowered into the bore-hole by the wire-rope U, from the other drum V, and is afterwards drawn up again from the excavated material. A friction-break is applied to the drum Q, for regulating the rate of lowering the boring-rod down the well. The scoop, shown in Figs. 377, 378, consists of a riveted iron cylinder, with a handle at the top, which can either be screwed to the boring-rod or attached to the wire-rope; and the bottom is closed by a large valve, opening inwards. Two different forms of valve are used, either a pair of flap-valves, as shown in Fig. 378, or a single-cone valve, Fig. 379; and the bottom ring of the cylinder, forming the seating of the valve, is forged solid, and steeled on the lower edge.

366. 367 368. 369. 370.

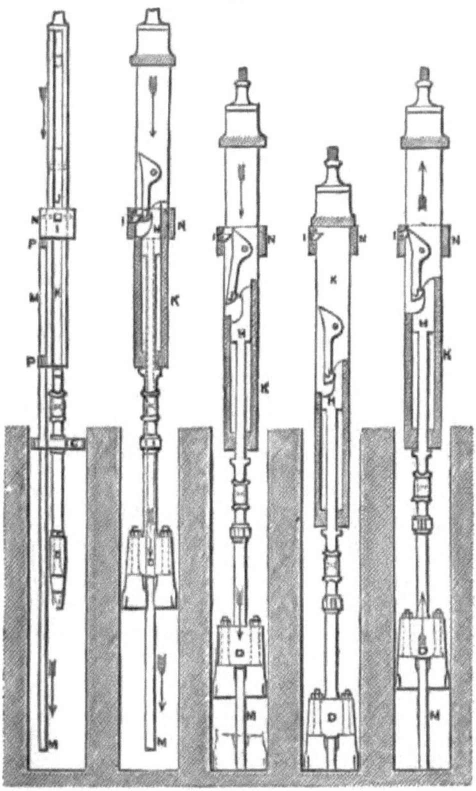

371.

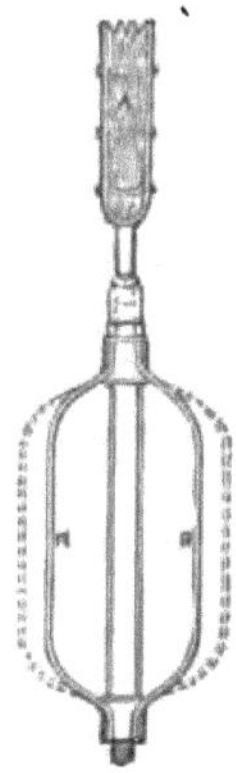

373.

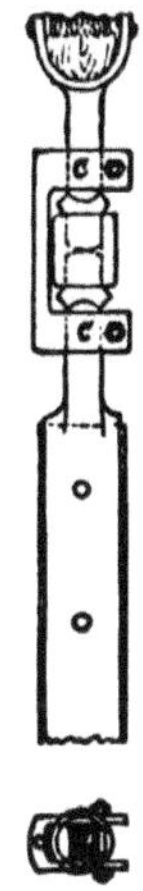

372.

373.

On lowering this cylinder to the bottom of the bore-hole, the valve opens, and the loose material enters the cylinder, where it is retained by the closing of the valve, whilst the scoop is drawn up again to the surface. In boring through chalk, as in the case of the deep wells in the Paris basin, Fig. 347, the hole is first made of about half the final diameter for 60 to 90 ft. depth, and it is then enlarged to the full diameter by using a larger tool. This is done for convenience of working; for if the whole area were acted upon at once, it would involve crushing all the flints in the chalk; but, by putting a scoop in the advanced hole, the flints that are detached during the working of the second larger tool are received in the scoop and removed by it, without getting broken by the tool.

374.

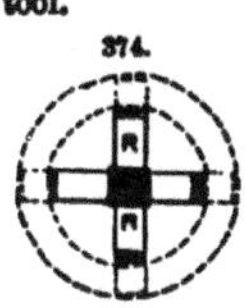

375.

376.

377.

378.

379.

The resistance experienced in boring through different strata is various; and some rocks passed through are so hard, that with 12,000 blows a-day of a boring-tool weighing nearly 10 cwt., with 19 in.

height of fall, the bore-hole was advanced only 3 to 4 in. a-day. As the opposite case, strata of running sand have been met with so wet, that a slight movement of the rod at the bottom of the hole was sufficient to make the sand rise 30 to 40 ft. in the bore-hole. In these cases M. Dru has adopted the Chinese method of effecting a speedy clearance, by means of a scoop closed by a large ball-clack at the bottom, as shown in Fig. 377, and suspended by a rope, to which a vertical movement is given; each time the scoop falls upon the sand a portion of this is forced up into the scoop, and retained there by the ball-valve.

An artesian well is always some time in settling down to its permanent working state, generally one or two months; and when the water first reaches the surface it undergoes considerable fluctuations, being charged from time to time with the substances at the bottom of the bore-hole. In the Grenelle well there were fluctuations at starting of 300 ft. in the height of the water. The velocity of the flow of water from the artesian wells varies considerably, and the following are some examples of the delivery at the surface by those already completed in the Paris basin :—

	Depth of Bore-hole.	Diameter of Bore-hole.	Gallons delivered per minute.	Velocity of discharge. Feet per second.
	feet.	inches.		
St. Denis, Hôtel Dieu ..	262	2·28	28	2·68
Gennevilliers	312	2·28	31	2·89
Stains	266	2·95	176	9·87
Elbeuf	492	2·95	66	3·71
Paris, Grenelle	1795	3·74	484	16·63
„ Passy	1923	27·56	1980	1·64

The localities of these wells are shown in the plan, Fig. 348; and Fig. 347 is a geological section passing through the Paris basin. Sections of the strata bored through in the wells at Passy and Grenelle are shown in Fig. 349; and a section in the boring now in progress at the Sugar Refinery is shown in the same Figure. In the case of the artesian well at Grenelle, the water is carried up to a height of 128 ft. above the ground by a stand-pipe of 3·74 in. bore, from the top of which the water overflows at the rate of 100 gallons a-minute, with a velocity of 3·94 ft. a-second.

Borings of large diameter, for mines or other shafts, are also sunk by means of the same description of boring-tools, only considerably increased in size, extending up to as much as 14 ft. diameter. The well is then lined with cast-iron or wrought-iron tubing, for the purpose of making it water-tight; and a special contrivance, invented by Kind, has been adopted for making a water-tight joint between the tubing and the bottom of the well, or with another portion of tubing previously lowered down. This is done by a stuffing-box, shown in Fig. 380, which contains a packing of moss at A A. The upper portion of the tubing is drawn down to the lower portion by the tightening-screws B B, so as to compress the moss-packing when the weight is not sufficient for the purpose. A space, C, is left between the tubing and the side of the well, to admit of the passage of the stuffing-box flange, and also for running in concrete for the completion of the operation. The moss-packing rests upon the bottom flange D; but this flange is sometimes omitted. The joint is thus simply made by pressing out the moss-packing against the sides of the well; and this material, being easily compressible and not liable to decay under water, is found to make a very satisfactory and durable joint.

380.

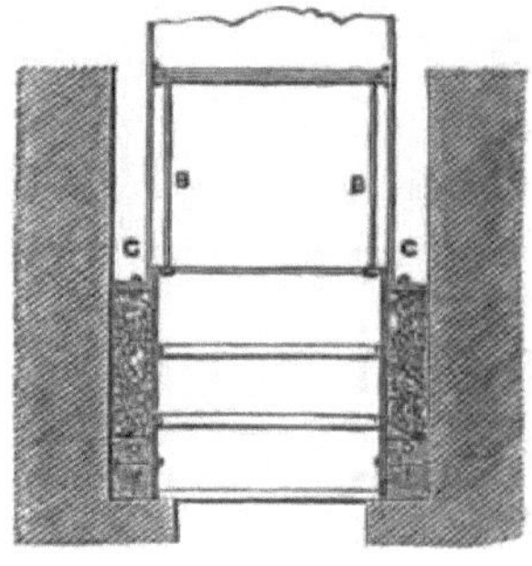

M. Dru states that the reaction-tool has been successfully employed for borings up to as large as about 4 ft. diameter, as in the case of the well at Butte-aux-Cailles of 47 in. diameter; but beyond that size he considers the shock requisite to liberate the larger and heavier tool would probably be so excessive, as to be injurious to the boring-rods and the rest of the attachments; and he therefore designed the arrangement of the disengaging-rod for liberating the tool in borings of large diameter, whereby all shock upon the boring-rods was avoided and the tool was liberated with complete certainty.

In practice it is necessary to turn the boring-tool partly round between each stroke, so as to prevent it from falling every time in the same position at the bottom of the well; and this was effected in the well at Butte-aux-Cailles by manual-power at the top of the well, by means of a long hand-lever fixed to the boring-rod by a clip bolted on, which was turned round by a couple of men through part of a revolution during the time that the tool was being lifted. The turning was ordinarily done in the right-hand direction only, so as to avoid the risk of unscrewing any of the screwed couplings of the boring-rods; and care was taken to give the boring-rod half a turn when the tool was at the bottom, so as to tighten the screw-couplings, which otherwise might shake loose. In the event of a fracture, however, leaving a considerable length of boring-rod in the hole, it was sometimes necessary to have the means of unscrewing the couplings of the portion left in the hole, so as to raise it in parts, instead of all at once. In that case a locking-clip was added at each screwed joint above, and secured by bolts, as shown at C in Fig. 373, at the time of putting the rods together for lowering them down the well to recover the broken portion; and by this means the ends of the rods were prevented from becoming unscrewed in the coupling-sockets, when the rods were turned round backwards for unscrewing the joints in the broken length at the bottom of the bore-hole.

M. Dru states that in his own experience, owing to the difficulties attending the operation, the occurrence of delays from accidents was the rule, while the regular working of the machinery was the exception. He also states that, although the chisel-shaped tools previously described were the form principally employed, he considers it would be a mistake to attempt to use any one form of tool exclusively for all descriptions of ground. For passing through granite, or any other primary rock, a percussive action is indispensable, and the force of the blow is required to be concentrated upon a small extent of cutting-edge, in order to produce any effect by the blow; but in softer ground a greater number of cutting-chisels are used in the falling-tool.

As long ago as 1842 ropes have been used in France for boring purposes, but they have not been found to answer in boring through clay, because the tool becomes choked and sticks fast, and the rope breaks, leaving the tool imbedded at the bottom of the bore-hole · it is then necessary to have recourse to rods for raising the boring-tool, and M. Dru therefore preferred to use rods in the first instance. In boring through sand, however, ropes have been successfully employed, and by this means borings have been carried down with great rapidity, as much as 60 ft. depth having been accomplished in a fortnight through a bed of sand in boring a well in the upper stratum of the Paris basin. The section of the Paris basin shows that, owing to the variable strata to be passed through, no one form of boring-tool, such as has been referred to, could be used for all parts of the bore-hole, but different tools were required, according to the particular stratification at the bottom of the hole.

When running sands are met with, the plan adopted is to use the Chinese ball-scoop, Fig. 377, described for clearing the bottom of the bore-hole; and where there is too much sand for it to be got rid of in this way, a tube has to be sent down from the surface to shut off the sand. This, of course, necessitates diminishing the diameter of the hole in passing through the sand; but on reaching the solid rock below the running sand, an expanding tool is used for continuing the bore-hole below the tubing with the same diameter as above it, so as to allow the tubing to go down with the hole.

In the case of meeting with a surface of very hard rock at a considerable inclination to the bore-hole, M. Dru employs a tool, the cutters of which are fixed in a circle all round the edge of the tool, instead of in a single diameter line; the length of the tool is also considerably increased in such cases, as compared with the tools used for ordinary work, so that it is guided for a length of as much as 20 ft. He uses this tool in all cases where from any cause the hole is found to be going crooked, and has even succeeded by this means in straightening a hole that had previously been bored crooked.

The cutting action of this tool is all round its edge; and therefore in meeting with an inclined hard surface, as there is nothing to cut on the lower side, the force of the blow is brought to bear on the upper side alone, until an entrance is effected into the hard rock in a true straight line with the upper part of the hole.

381.

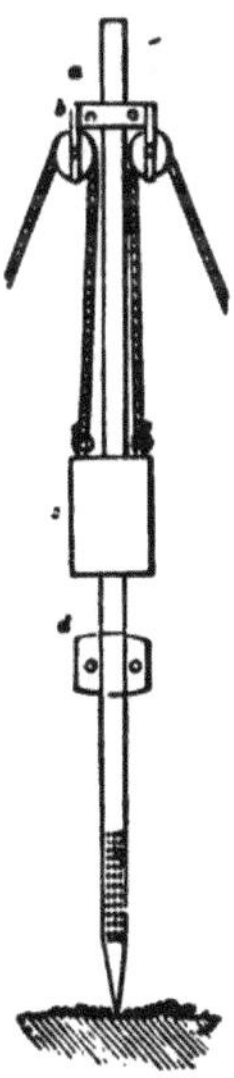

382.

Norton's patent Tube Well.—This well consists of a hollow wrought-iron tube about 1¾ in diameter, composed of any number of lengths from 3 to 11 ft., according to the depth required. The water is admitted into the tube through a series of holes, which extend up the lowest length to a height of 2½ ft. from the bottom.

The position for a well having been selected, a vertical hole is made in the ground with a crow-bar to a convenient depth; the well-tube *a*, having the clamp *d*, monkey *c*, and pulleys *b*, Fig. 381, previously fixed on it, is inserted into this hole.

The clamp is then screwed firmly on to the tube from 18 in. to 2 ft. from the ground, as the soil is either difficult or easy; each bolt being tightened equally, so as not to indent the tube.

The pulleys are next clamped on to the tube at a height of about 6 or 7 ft. from the ground, the ropes from the monkey having been previously rove through them.

The monkey is raised by two men pulling the ropes at the same angle. They should stand exactly opposite each other, and work together steadily, so as to keep the tube perfectly vertical, and prevent it from swaying about while being driven. If the tube shows an inclination to slope towards one side, a rope should be fastened to its top and kept taut on the opposite side, so as gradually to bring the tube back to the vertical. When the men have raised the monkey to within a few inches of the pulleys, they lift their hands suddenly, thus slackening the ropes and allowing the monkey to descend with its full weight on to the clamp. The monkey is steadied by a third man, who also assists to force it down at each descent. This man, likewise, from time to time, with a pair of gas-tongs, turns the tube round in the ground, which assists the process of driving, particularly when the point comes in contact with stones.

Particular attention must be paid to the clamp, to see that it does not move on the tube; the bolts must be tightened up at the first appearance of any slipping.

When the clamp has been driven down to the ground, the monkey is raised off it, the screws of the clamp are slackened, and the clamp is again screwed to the tube, about 18 in. or 2 ft. from the ground. After this, the monkey is lowered on to it, and the pulleys are then raised until they are again 6 or 7 ft. from the ground.

The driving is continued until but 5 or 6 in. of the well-tube remain above the ground, when the clamp, monkey, and pulleys are removed, and an additional length of tube screwed on to that in the ground. This is done by first screwing a collar on to the tube in the ground, and then screwing the next length of tube into the collar, till it buts against the lower tube; a little white-lead must be placed on the threads of the collar before the ends of the tubes are screwed into it.

The driving can thus be continued until the well has obtained the desired depth. Soon after another length has been added, the upper length should be turned round a little with the gas-tongs, to tighten the joints, which have a tendency to become loose from the jarring of the monkey. Care must be taken, after getting into a water-bearing stratum, not to drive through it, owing to anxiety to get a large supply. From time to time, and always before screwing on an additional length of tube, the well should be sounded, by means of a small lead attached to a line, to ascertain the depth of water, if any, and character of the earth which has penetrated through the holes perforated in the lower part of the well-tube. As soon as it appears that the well has been driven deep enough, the pump is screwed on to the top and the water drawn up. It usually happens that the water is at first thick, and comes in but small quantities; but after pumping for some little time, as the chamber round the bottom of the well becomes enlarged, the quantity increases and the water becomes clearer.

When sinking in gravel or clay, the bottom of the well-tube is liable to become filled up by the material penetrating through the holes; and before a supply of water can be obtained, this accumulation must be removed by means of the cleaning-pipes.

The cleaning-pipes are of small diameter, ¼-in. externally, and the several lengths are connected together in the same way as the well-tubes, by collars screwing on over the adjoining end of two pipes.

To clear the well, one cleaning-pipe after another is lowered into the well, until the lower end touches the accumulation; the pipes must be held carefully, for if one were to drop into the well it would be impossible to get it out without drawing the well. A pump is then attached to the upper cleaning-pipe by means of a reducing-socket; the lower end of the cleaning-pipe is then raised and held about an inch above the accumulation by means of the gas-tongs: water is next poured down the well outside the cleaning-pipe, and, being pumped up through the cleaning-pipe, brings up with it the upper portion of the accumulation; the cleaning-pipe is gradually lowered, and the pumping continued until the whole of the stuff inside the well-tube is removed. The pump is then removed from the cleaning-pipe, and the cleaning-pipes are withdrawn piece by piece; and finally the pump is screwed on to the upper end of the tube-well, Fig. 382, which is then in working order.

The tube being very small, is in itself capable of containing only a very small supply of water, which would be exhausted by a few strokes of the pump, the condition, therefore, upon which alone these tube-wells can be effective, is that there shall be a free flow of water from the outside through the apertures into the lower end of the tube. When the stratum in which the water is found is very porous, as in the case of gravel and some sorts of chalk, the water flows freely; and a yield has been obtained in such situations as great and rapid as the pump has been able to lift, that is 600 gallons an hour. In some other soils, such as sandy loam, the yield in itself may not be sufficiently rapid to supply the pump; in such cases, the effect of constant pumping is to draw up with the water from the bottom a good deal of clay and sand, and so gradually to form a reservoir, as it were, around the foot of the tube, in which water accumulates when the pump is not in action, as is the case in a common well. In dense clays, however, of a close and very tenacious character, the American tube-well is not applicable, as the small perforations become sealed, and water will not enter the tube. When the stratum reached by driving is a quicksand, the quantity of sand drawn up from the water will be so great, that a considerable amount will have to be pumped before the water will come up clear; and even in some positions, when the quicksand is of great extent, the effect of the pumping may be to injure the foundations of adjoining buildings on the surface of the ground.

The tube-well cannot itself be driven through rock, although it might be used for drawing water from a subjacent stratum through a hole bored in the rock to receive it.

Subject to these conditions, these tube-wells afford a ready and economical means for drawing water to the surface from a depth not exceeding 27 or 28 ft.

Works and Papers on Artesian Wells:—Garnier, F., 'Traite sur les Puits Artésiens,' 4to, Paris, 1826. Héricourt de Thury. 'Considérations Géologiques et Physiques sur la cause du Jaillissement des Eaux des Puits Fores, 8vo, Paris, 1829. Degoussee et Saurent, 'Guide du Sondeur,' 3 vols., Paris, 1861. G. R. Burnell, 'On Artesian Wells,' Transactions Inst. C. E., 8vo, 1864. B. Latham, 'Papers on Water Supply,' Part I., 8vo, 1865. Van Ertborn, 'Memoire sur les Puits Artésiens,' 8vo, Anvers, 1866. Dru, 'On the Machinery for Boring Artesian Wells,' Proceedings Inst M. E., 1867.

ARTIFICIAL STONE. Fr., *Pierre artificielle*; Ger., *Künstlicher Stein*; Ital., *Pietra artifiziale*; Span., *Piedra artificial*.

See Stone.

ARTILLERY. Fr., *Artillerie*; Ger., *Artillerie*; Ital., *Artiglieria*; Span., *Artilleria*.

The term artillery is applied to all descriptions of ordnance, whether light or heavy, and everything required for their service.

A chief point in the science of gunnery, and one which limits its future, is the construction of guns of sufficient strength to curb and govern the utmost force of any explosive compound which may be used in them.

No modern theory of constructing guns can be called new, since guns are in existence that have been either recovered from wrecks, or preserved in other ways, showing every variety of coils, hoops, casting, wire-binding, and so on, as far as the appliances then in use could furnish the quondam inventors with means of carrying their inventions into effect. That in which novelty has been attained, is the improvement of processes by which large castings or forgings, accurate turning and boring, can be secured, or by which chemical knowledge can be brought to bear on the manipulation of metals: but no such progress can make a built-up gun, or machine of any sort, stronger than a perfectly homogeneous one, in which the varying strains are closely calculated, and properly met by the scientific disposition of the necessary strength. Thus, while we admire the ingenuity of the methods by which guns have been built up, we cannot think that such processes will now, more than in former times, continue to be preferred to well cast guns of good material.

Guns are burst by two forces—the rending action of the powder gas, and the unequal heating of a nearly homogeneous metal; but it is not, as supposed by some authors, by interior cracks caused by contraction, since metals do not contract but expand by heat; and guns are often burst without such rapidity of firing as would induce sufficient heat to be felt on the outside, scarcely even on the inside of the gun. If cracks on the interior, and all observers are agreed on this point, are the first indications of the bursting of a gun—if this be accompanied by a sensible dilatation of the bore, it is clear that, as the outside of the gun has not increased, though the inside has cracked, some of the metal has been compressed into a smaller space than it previously occupied, and this compression being greatest on the inside, as shown by the greater width of crack, ought to be resisted by a metal at that point harder, that is, having greater strength to resist impact, than at any other portion of the gun. In cannon cast solid in sand and afterwards bored out, the reverse is the case. Increased thickness of metal, as trunnions or other projections, will determine in many cases the line of fracture. A muzzle swell is beneficial, as tending to prevent vibration, which is begun at the point where the projectile leaves the gun, and the true concussive explosion takes place. The cascable, also, is not arbitrary, nor alone necessary for supporting a breeching; but, like the trunnions, reinforces the gun, though, unlike them, it does so

383.

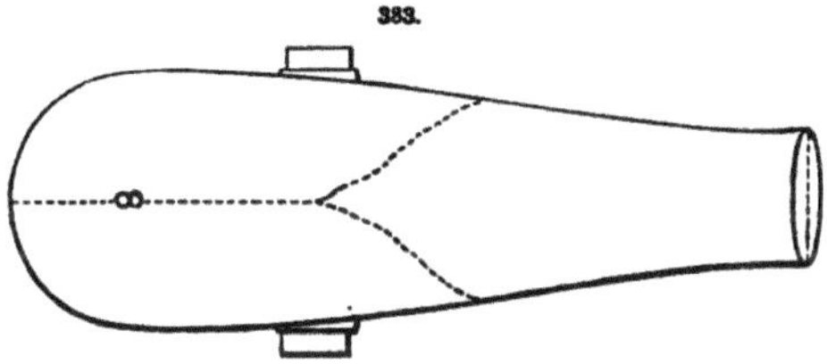

American Gun.

384.

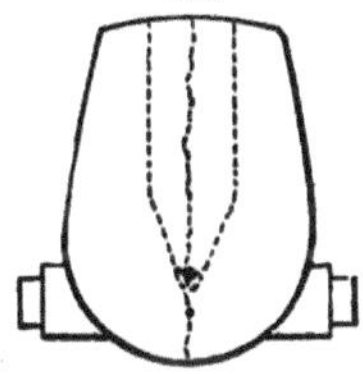

385.

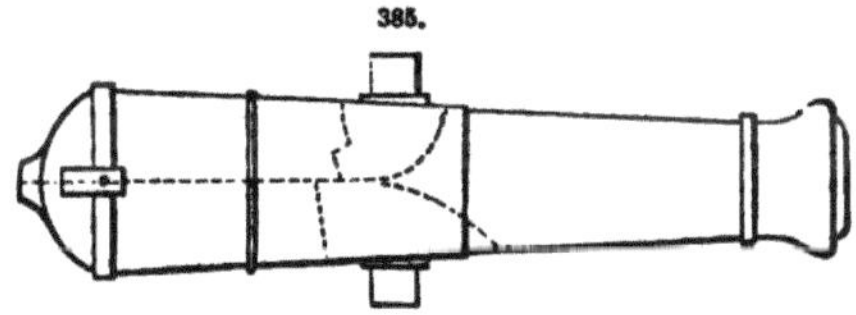

386.

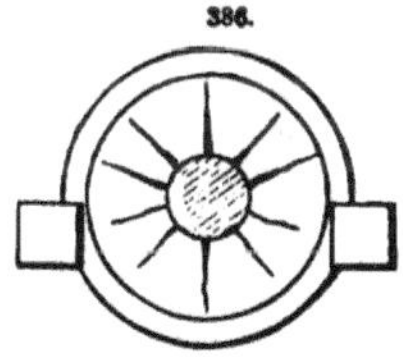

usefully, in consequence of being in the prolongation of the axis of the piece. We give two practical illustrations of the bursting of guns, Figs. 383, 385; and one of that of mortars, Fig. 384.

Fig. 386 shows the comparative intensity of strains in a gun when the trunnions are under the plane, or when there is no cascable.

387

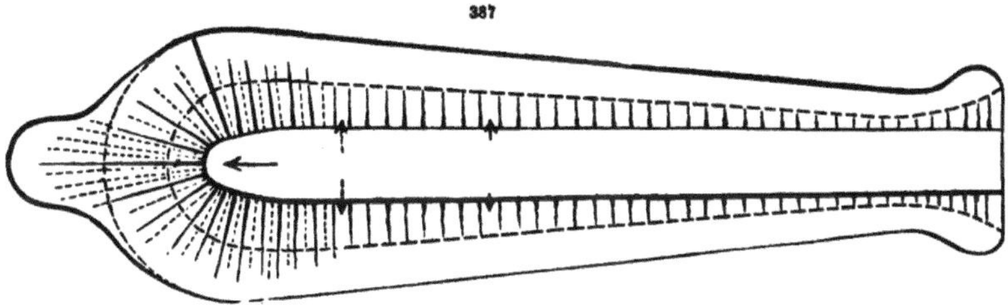

Now let us examine the parts in which the lines of least resistance are to be found. In the gun, a radial strain, Fig. 387, is exerted on all parts of the bore equally, except to the rear, where it is no longer radial but direct; no longer resisted by wedges, as hereafter shown, but by a cylinder of iron which terminates in the cascable. At the trunnions, too, which are not often placed in modern guns in the horizontal plane of the axis, but below it, there will be another line of least resistance where the trunnions are not, the necessary consequence of a line of greatest resistance where they are.

Still more is this seen in the mortar, Fig. 384, which has an enormously greater lateral resistance at the trunnions than at the muzzle. This would feel the want of a cascable also, were it not that the powder-chamber is conical, and smaller than the bore; but as it is, it splits open as if wedges had been driven into the muzzle—the guns as if wedges had been driven into the breech.

J. A. Longridge, in a paper "On the Construction of Artillery," read before the Institution of Civil Engineers, came to the conclusion that the required object could be attained by constructing a gun in such an initial state of equilibrium, that when the varying strain, caused by internal pressure, should come upon it, the initial strain should be equivalent to the induced strain, and the sum of the two strains constant throughout. With the view of proving the truth of these conclusions, J. A. Longridge caused a series of experiments to be made, the first of which was to ascertain the ultimate force of gunpowder. A number of cast-iron cylinders 1 in. diameter and 0·1 in. thick were prepared. They were bored and turned accurately, and purposely made of very hard, brittle iron, so as of themselves to give the minimum of strength. These cylinders were then wrapped round with iron wire, the tensile strength of which had been previously ascertained. The number of coils varied with the degree of strength required, and each coil was laid on with the initial strain, which the experimenter's calculations led him to believe would, at the moment of bursting, cause all the coils to give way together. The cylinders were filled with Government cannon-powder, and the ends secured, leaving no vent but a touch-hole the size of a small pin, through which the powder was exploded. Several of these burst; but it was found that a cylinder with ten coils of wire upon it could not be burst. The diametral section of the cast iron was 0·2 in.; and taking its strength at 8 tons the square inch, the result is

		Tons.
$\frac{2}{10}$ths of an inch × 8 tons per square inch	=	1·6
There were ten coils of wire, each wire by experiment broke with 60 lbs., and was $\frac{1}{28}$th inch in diameter; therefore the tensile strength of the two sides = 2 × 28 × 60 × 10 ÷ 2240	=	15·0
Total strength		16 6

The internal diameter of the cylinder was 1 inch, consequently the ultimate strength of the powder did not exceed 17 tons a square inch. This must not, however, be held as the maximum effect that may be produced. The explosion of powder is more of an impact than a pressure; and although, strictly speaking, impact is only a pressure of short duration, the length of this duration is generally admitted to have great influence upon the effect produced. It has been stated that a bar to resist, with safety, the sudden application of a given pull, requires to have twice the strength that is necessary to resist the gradual application and steady action of the same pull. From this follows that, the more rapid the explosion, the greater the strain upon the gun. Not that the ultimate pressure is greater; but that, being produced in less time, its effect is greater.

The cause of weakness in a cast-iron gun is, in the first place, that the actual strength of the interior of a large gun, or mortar, is far below that of the average of ordinary castings, and always must be so whilst guns are cast solid. So long as this is the case, the outside must cool and solidify first; whilst the interior, cooling more slowly, must be drawn and rendered less dense, and consequently less resisting. This cannot be obviated by any care in selecting material.

The worst part of this iron, in the chase of the gun, is afterwards bored out; but still the metal around the internal circumference is weakened below the average; and at the bottom of the powder-chamber it is in the worst possible condition. This is fully accounted for by the law of cooling. Whenever a variation in thickness occurs, a difference in the rate of cooling must also take place; this alone gives rise to a state of varied stress amongst the particles of the metal, diminishing the effect of the metal as a resisting substance. In ordinary castings this is well understood; but the same law operates in guns, though in a smaller degree; take, for instance, the accompanying sketch of a gun, Fig. 388, distorted in its proportions, for the sake of illustration, and suppose it to have cooled down after casting. Although, in the present state of our knowledge of the subject, it would be impossible to determine the absolute position of the isothermal lines at any period of cooling, yet it is certain they must approximate to the dotted lines shown in Fig. 388; and following these lines according to some definite law, would be the lines of equal stress of the particles of the gun, when cold. When, therefore, the gun is bored out, it is evident that the inner circumference of the bore must be in a state of varying strain, and that strain is one of tension. Consequently, the internal part of the gun is, throughout, in an initial state of more or less tension, and, as regards its power to resist a tensile strain, it is inferior to the normal, or average strength of the material. But beyond this, whenever a change of dimensions occurs, the cooling will give rise to varying strains, which may account for fracture taking place at those particular parts. To obviate this, wrought iron and steel have been tried, and in point of workmanship great results have been arrived at; but the same objections apply as to guns of cast iron; the inner part of the gun must be in a state of initial tension. The expense

388.

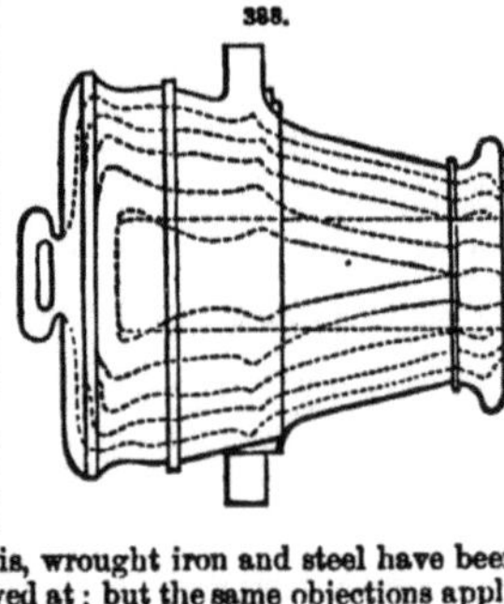

and difficulty of manufacture are also very great. As a specimen of steel manufacture, the gun forged by Krüpp, and afterwards bored and mounted in a cast-iron jacket at Woolwich, may be instanced. It was bored out to 8 in., and was from 4 to 4·5 in. thick. Taking the tensile force of hammered cast steel at 40 tons to the square inch, the resistance would be from 320 to 360 tons, which, if the strain had been uniform throughout, would have been equal to between 40 and 45 tons to the square inch on the diameter; yet the gun burst at the first discharge, with 25 lbs. of powder and a 260-lb. shot. Mallet mentions a wrought-iron 8-in. gun, forged at the Gospel Oak Iron-works, and proved at Woolwich in 1855, which burst into several pieces at the first discharge. The thickness at the breech end of this gun, which was stated to be of nearly the same dimensions as the established cast-iron guns of the same calibre, was about 9 in.; and, taking the tensile force at 20 tons to the square inch, the material, provided it had been uniformly strained, ought to have resisted a diametral strain of 360 tons, or about 45 tons to the square inch. This gun, which appeared in every respect sound to the eye and of perfect material, burst with a proof-charge of 28 lbs. of powder and two spherical 8-in. shot. The conclusion has been arrived at, that the manufacture of large forged wrought-iron guns is an operation of great difficulty, expense, and uncertainty; and however the difficulty and expense may be decreased, the uncertainty must still remain; at the best, it is but substituting for cast iron a material of higher tensile strength; the radical defect of a homogeneous mass still remaining, namely, the unequal distribution of the strain from the inner to the outer circumference. The same remarks apply with still greater force to guns of hammered cast steel, of large dimensions. The principle, which appears to be the basis of a sound and reliable construction, is that of manufacturing the gun of successive layers, laid on with an original increasing strain, from the centre to the circumference.

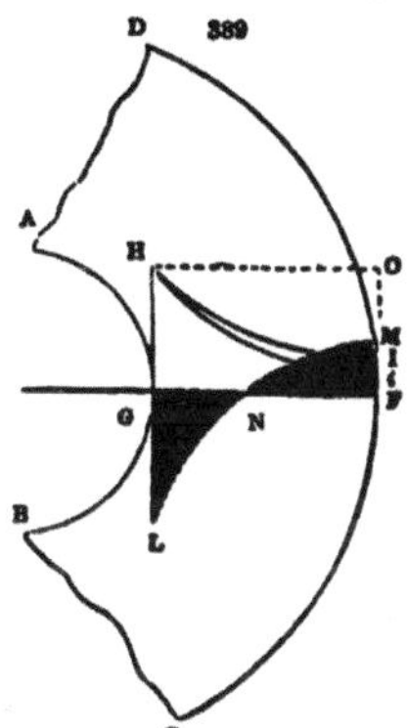

There is an objection to the use of hoops, from the want of continuity. The special requirement is, that each layer of the gun shall be in a definite initial state of tension or compression previous to explosion.

If, in Fig. 389, A B C D represents a portion of a section of an 8-in. gun, of which A G B is the inner and D F C the outer circumference, the state of tension of any particle between G and F may be denoted by ordinates drawn at the points in question, those above G F representing tension, and those below, compression. If, now, the gun is of any homogeneous material, such as cast iron, the state of tension at the time of explosion, and when the gun is about to burst, will be denoted by a curve H I, or H i. Then, supposing the tensile force of the material to be 12 tons to the square inch, and the thickness of the gun 6·5 in., when the strain at G is G H, or 12 tons, at F it is F I = 3 tons, or F i = 1·75 ton, according as the one or the other formula is adopted. The areas of these curves give, of course, the total strengths of the gun at the bursting point, and are found to be 36·72 tons and 30·871 tons respectively, instead of 78 tons, which it would have been if uniformly strained at 12 tons to the square inch. Now the object sought to be attained in the method of construction under consideration is, that each particle, such as K, shall, when explosion takes place, be equally strained with G. In order that this may be so, the initial state of the tension must be such as represented by the curve L N M, those between G and N being in compression, while those particles between N and M are in tension.

If, now, it is attempted to accomplish this by means of hoops, it will be found impossible, inasmuch as each hoop is a homogeneous cylinder, and follows the same law throughout its thickness, as is represented by the curve H I. Figs. 390, 391, and 392, represent the successive states of stress of rings, put on so as that when explosion takes place they shall be all equally strained at their inner circumferences. Fig. 390 shows two rings; Fig. 391 shows three rings; Fig. 392 shows four rings. The numbers denote the strains in tons per square inch. From this it will be seen that when the four rings are put on, instead of the curve L N M of Fig. 389, there are a series of abrupt changes, the two inner rings being in compression, and the two outer in tension. When the explosion takes place, the state of maximum strain is represented by Fig. 393.

The area between the dotted and full lines shows the work done by the explosion, and, taking the total thickness of the gun, it amounts to 10·1 tons to the inch of thickness; whereas, had the construction been of very thin rings or of small wire, it would have been represented by the area between the dotted line L N M O H, Fig. 389, and would have been = 12 tons to the inch of thickness, showing a superiority of about 20 per cent. in favour of the wire over the hoops. This is upon the supposition that the workmanship of the hoops is perfect, which is not the case in practice. To afford some idea of the accuracy required, the radii of the several rings, shown in Fig. 393, are given in the following Table:—

No. of Ring.	Inner Radius.	Outer Radius.	Thickness.	Difference.
1	4·0000	5·3222	1·3222	$R_1 - \rho_2 =$ ·0031
2	5·3191	7·2928	1·9737	$R_2 - \rho_3 =$ ·0035
3	7·2893	9·4633	2·1740	$R_3 - \rho_4 =$ ·0035
4	9·4598	11·8247	2·3649	

Thus it appears, that in order to give the requisite amount of initial stress, the external radius of the first ring must be $\frac{31}{10000}$ of an inch, or about $\frac{1}{300}$ of an inch larger than the internal radius of the second; the external radii of the second and third $\frac{34}{10000}$ of an inch greater than the internal radii of the rings next to them. Therefore, whilst the whole effect depends upon so small a quantity as about $\frac{1}{300}$ of an inch, it is evident that a very small error in workmanship will materially affect the result, and tend to deviations from the proper initial strains.

390.

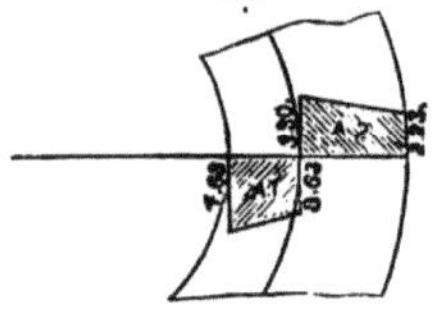

391.

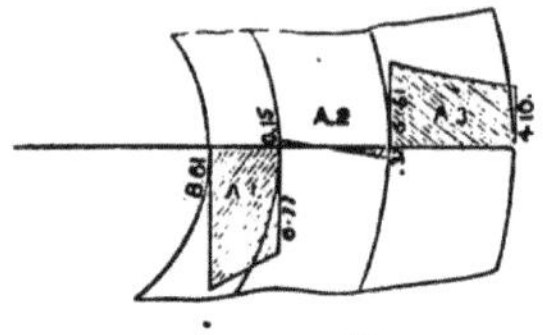

392.

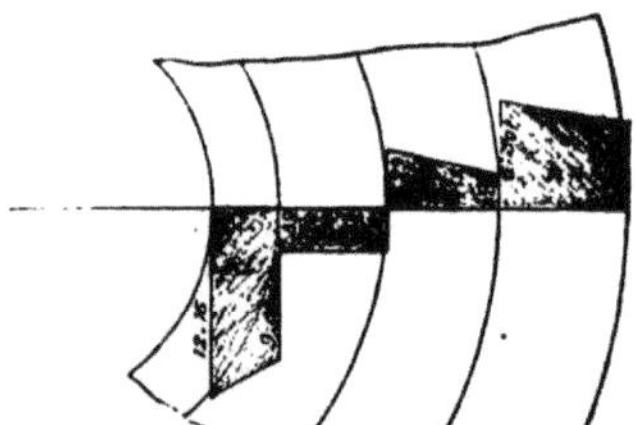

393.

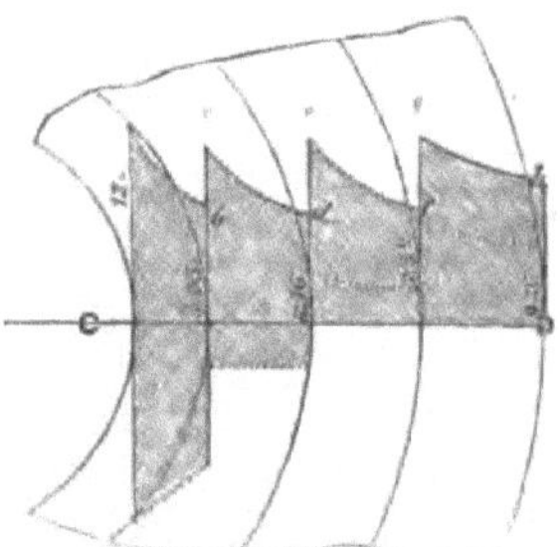

Fig 394 represents the states of stress of the rings before explosion and at the instant of maximum strain, when the rings are accurately put on.

Fig. 395 represents the states of stress of the same gun when the outer ring has been made $\frac{1}{100}$ of an inch too small.

394.

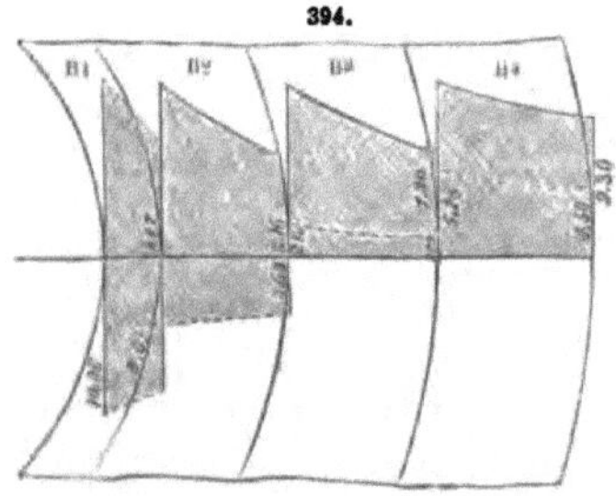

395.

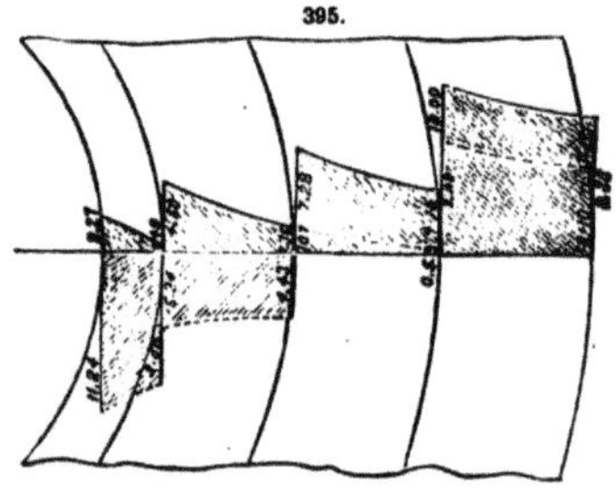

The result is, that before explosion the maximum compression of the inner ring is increased from 10·086 tons to 11·244 tons, and the maximum tension of the outer ring from 5·778 tons to 7·823 tons to the square inch; whilst at the time of maximum strain during explosion the tension of the same ring is only 2·268 tons, although the outer ring is strained to 12 tons, its assumed ultimate strength. The absolute strength of the gun is thus reduced from an average of 10·5 tons to 6 tons per inch thickness, or about 40 per cent., by an error of only $\frac{1}{100}$ of an inch in a ring about 17 inches in diameter. Rings, therefore, present practical difficulties which are entirely avoided by the use of wire, as it may be coiled on with the exact strain indicated by theory. The method adopted by J. A. Longridge was to coil a quantity of wire on a drum, fixed with its axis parallel to that of a lathe on which the gun was placed. On the axis of this drum there was another drum, to which was applied a break, similar in principle to Prony's dynamometric break, so adjusted as to give the exact tension required for each successive coil of the wire. The whole apparatus was extremely simple, and the wire laid on with great regularity. It is evident the apparatus might be so arranged that the process would proceed with the same ease and regularity as winding a thread on to a bobbin, and at the same time with the greatest accuracy as regards the initial tension. No such facility attends the use of hoops. They must be accurately bored; and after each layer is put on, the gun must be placed in the lathe, and the hoops turned on the outside. Great accuracy is indispensable; and not only is the amount of labour much greater, but it must be of a far higher and consequently of a more expensive class. Then, as to the accuracy of tension with hoops, its atttainment is almost impracticable, while the process of shrinking on is not to be depended upon. Not only is there a difficulty in insuring the exact temperature required, but scarcely any two pieces of iron will shrink identically; and when the

$\frac{1}{100}$ part of an inch of contraction would give rise to a great variation in tension, the necessity of perfect accuracy is apparent. It has been proposed to force the hoops in longitudinal sections by hydrostatic pressure into a gun slightly conical. Captain Blakely, in a lecture given by him at the United Service Institution, in 1859, gave an account of his experiments. His first gun was an 18-pounder, Fig 396, consisting of one series of wrought-iron rings shrunk on a cast-iron cylinder,

396.

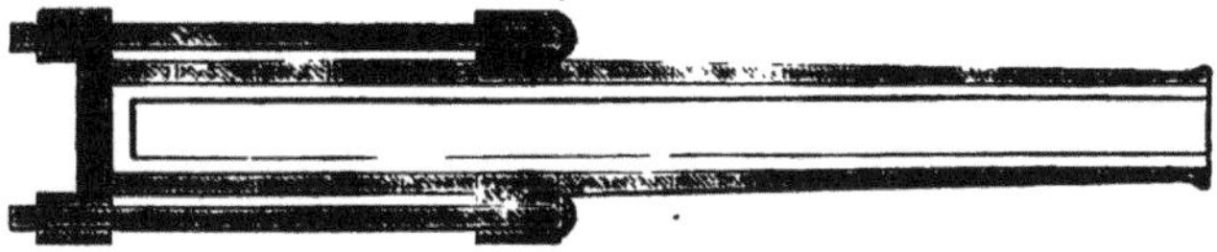

5·5 in. in diameter inside, and $1\frac{3}{8}$ in. thick. The wrought-iron rings were from 2 in. thick downwards; the total thickness of the breech was $3\frac{3}{4}$ in., that of the ordinary 18-pounder service-gun being $5\frac{3}{8}$ in. This gun was fired frequently, and stood well. It was then bored out as a 24-pounder; but not being truly bored, the cast iron was reduced on one side to only $\frac{1}{2}$ in. thick. In this state it sustained without injury several hours' firing, with charges varying from one shot and 4 lbs. of powder, to one shot, two wads, and 8 lbs. of powder. At the third round with this latter charge it burst. This gun had a thickness of only 2 5 in. round the charges, as compared with a service 24-pounder 6 in. in thickness.

397.

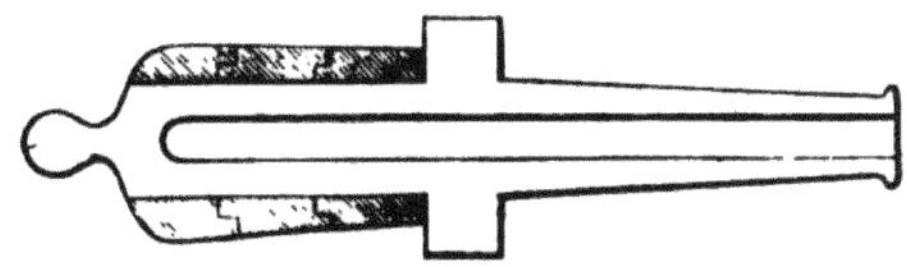

Capt. Blakely next got a 9-pounder, Fig. 397, turned down from the trunnions to the breech, and on this part he put wrought-iron rings of such a size as to replace the metal removed. This gun was fired, round for round, with a cast-iron service-gun of the same size and weight. The following Table gives the result:—

No. of Shot. Blakely.	Charge of Powder.	No. of Shot.	No. of Rounds fired. Blakely's.	Service.	No. of Shot Fired from Service-Gun.
	lbs.				
4	8	2	2	2	4
86	3	1	86	86	86
26	4	1	26	26	26
5	5	1	5	5	5
10	5	2	5	5	10
636	6	2	318	110	Burst 220
3	6	3	1	..	..
4	6	4	1	..	..
5	6	5	1	..	..
6	6	6	1	..	..
7	6	7	1	..	..
8	6	8	1	..	..
9	6	9	1	..	..
1580	6	10	158	..	..
2389	..	..	607	234	351

Thus it appears that Captain Blakely's gun stood 607 rounds, and the Government service-gun only 234 rounds; the number of shot thrown being 2389 and 351 respectively, or nearly as 7 to 1.

Proceeding with Longridge's experiments, the first point to be settled is the amount of initial strain to be put on each coil. The formula adopted by Longridge was $t = \mathrm{T} \cdot \frac{x^2 - r^2}{x^2}$.

According to Professor Hart's investigation, the formula is $t = \mathrm{T} \cdot \mathrm{R}\, \frac{x^2 - r y}{x (r + y)}$, which, in the case of small wire is nearly $t = \frac{\mathrm{T\,R}}{x} \cdot \frac{x - r}{x + r}$.

These, however, are general formulæ which require modification, according to the varying circumstances of each case, before they can be applied to practice.

Experiments on cylinders prepared according to the first formula were conducted as follows. A number of brass cylinders were prepared of exactly the same dimensions, namely:—

Internal diameter	1 in.
External "	$1\frac{3}{16}$ in.
Thickness of brass	$\frac{3}{32}$ in.

These cylinders were accurately turned and bored, and had a flange $\frac{1}{4}$ in. in depth and $\frac{1}{8}$ in. in thickness, at each end. Each end was widened out, so as to afford seating to two gun-metal balls, which were accurately ground to fit them. The total content of each cylinder, with the balls in their places, was 300 grains of the best sporting-powder. When the powder was put into the cylinder, and the balls were placed at each end, the whole was bound together by a very strong wrought-iron strap, similar to the strap of a connecting-rod, with a jib and cotter. The cotter was driven tightly home, and the powder was then fired through a small touch-hole left in the side-seating, Fig. 398.

398.

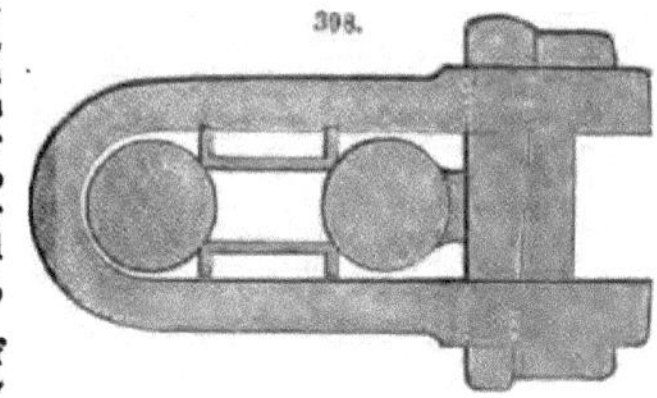

The first experiments were to ascertain the effect of the powder on the cylinders, without any wire. They were commenced with charges of powder, beginning at 50 grains, and increasing until the cylinder burst.

The results were as follows :—

No. of Experiment.	No. of Cylinder.	Condition.	Charge of Powder.	Effect.
			grains.	
1	1	Without wire	50	Slightly bulged.
2	"	"	60	Bulged a little more.
3	"	"	70	" " external diam. $1\frac{5}{16}$.
4	"	"	80	" " " $1\frac{7}{16}$.
5	"	"	90	Burst.
6	2	Two coils of wire, $\frac{1}{64}$ in.	90	No effect.
7	"	" " one end loose	100	Bulged at loose end.
8	3	Without wire	70	Bulged to $1\frac{11}{32}$.
9	4	Six coils of $\frac{1}{64}$ wire	100	No effect.
10	"	" "	110	"
11	"	" "	120	" one end of wire came loose.
12	"	Same cylinder, with one coil of $\frac{1}{32}$ wire	100	Burst, the end of the wire being badly fastened; wire uninjured.
13	5	Two coils of $\frac{1}{32}$ wire	100	No effect.
14	"	" "	120	"
15	"	" "	130	"
16	6	Four coils of $\frac{1}{32}$ wire	120	"
17	"	" "	130	"
18	"	" "	140	"
19	"	" "	150	"
20	"	" "	160	"
21	"	" "	170	"
22	"	" "	180	"
23	"	" "	200	"

The strength of the wire used in these experiments was ascertained, by trial, to be as resisting a dead tension :—

$\frac{1}{64}$ .. 23 lbs. = 120,000 lbs. the sq. in.
$\frac{1}{32}$ .. 70 lbs. = 92,000 lbs. the sq. in.

If now the expansive force of the powder is taken to be inversely as the volume, its ultimate strength may be approximately arrived at from the last experiment. The powder then could not burst the cylinder. Now the strength of the cylinder, supposing all the material to be equally strained, could not exceed the following to the lineal inch of cylinder :—

Wire	17,920 lbs.
Brass	3,136 "
	21,056 lbs., or 9·4 tons.

As the internal diameter was exactly 1 in., it shows that the ultimate force of the material in Experiment 23 did not exceed 9·4 tons the sq. in. Assuming the law as above, the ultimate pressure, supposing the cylinder to have been full, could not exceed $9·4 \times \frac{300}{200}$, or 13 tons the sq. in.

The enormous strain to which these cylinders were subjected is evinced by the effects upon the gun-metal balls, which were more or less cut away by the gases where they touched the cylinders.

A subsequent experiment was conducted in the following manner :—

A brass cylinder, Fig. 399, was constructed of nearly the same internal dimensions as a 3-pound mountain-gun, about 3 in. diameter and 36 in. long. The thickness of the brass was $\frac{1}{4}$ in.; at the breech end it was covered with six coils of steel wire, square in section, and of No. 16 wire-gauge = $\frac{1}{16}$ of an inch. These coils extended about 15 in. along the cylinder, and were reduced towards the muzzle to two coils. Consequently, the thickness of the cylinder was,

At the breech $\frac{1}{4}$-in. brass + $\frac{3}{8}$-in. iron = $\frac{5}{8}$ in.
" muzzle $\frac{1}{4}$-in. brass + $\frac{1}{8}$-in. iron = $\frac{3}{8}$ in.

The thickness of the 3-pounder gun, with which it may be compared, being,

At the breech .. .. 2·37 in.
muzzle .. .. 0·75 in.

This cylinder was not mounted as a gun; it had no trunnions; it was cleated with wood; and the object of a deep steel ring, which was screwed on the muzzle, was simply to cover the ends of the cleating. This cleating had nothing to do with the principle involved, and was only used to screen the construction from general observation.

399.

This cylinder was proved with repeated charges, varying from ¼ lb. of powder and one round shot, to 1½ lb. of powder and two shots. The cylinder was simply laid on the ground, with a slight elevation, its breech abutting against a massive stone wall, so as to prevent recoil. It stood the proof without injury. Another trial was made with this gun, before the Committee of Ordnance, in the following manner:—

The gun was clamped on a block of oak with iron clamps, and allowed to recoil on a wooden platform. Two rounds were fired; first with a charge of 1 lb. of powder, one shot, fixed to wood bottom, and one wood over the shot; the recoil was 7 ft.; the gun was found to have slightly shifted its position on the block; a trifling expansion of the wire had also taken place at the breech. At the second round the gun was fired with 2 lbs. of powder, one shot, and one wad, and burst; the separation took place about 2 in. in front of the base-ring; the breech was completely separated from the rest of the gun, and was blown 90 yds. directly to the rear. The wire was unravelled to the length of 3 or 4 ft., and the brass cylinder burst in a peculiar manner, turning its ends upwards and outwards. It also opened slightly at the centre of the gun, but the wire did not give way at that point. The ordinary proof-charge for a gun of this diameter would be 1¼-lb. shot and one wad.

In order to try more particularly the effect of the wire in giving strength to the cylinder, this gun was, after bursting, sawn in two at the centre, and one end of each portion was plugged with a brass plug, which was secured in its place by iron bands and several coils of wire: these guns were then secured to slides of wood, as in the former instance; they were placed opposite the proof-butt, and that made from the breech end was loaded with ¾ lb. powder, and shot. It burst; the breech being blown out, and the wire uncoiling to a considerable extent.

The muzzle portion was then loaded with a similar charge. It did not burst; but was much shaken by the discharge, and portions of the iron bands gave way. It was then loaded with 1 lb. of powder and one shot, which, on discharge, burst in two places, the breech being completely separated from the gun, and the slide on which it had been fired was rent into several pieces.

The bursting of this cylinder was not due to its construction, but to the manner in which it was mounted, shown in Figs. 400, 401.

400.

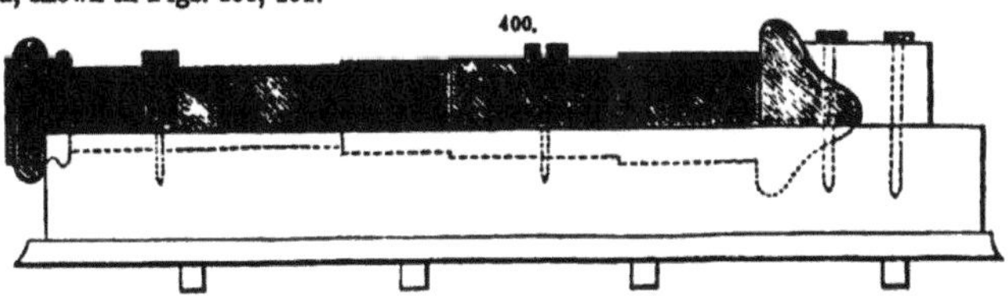

401

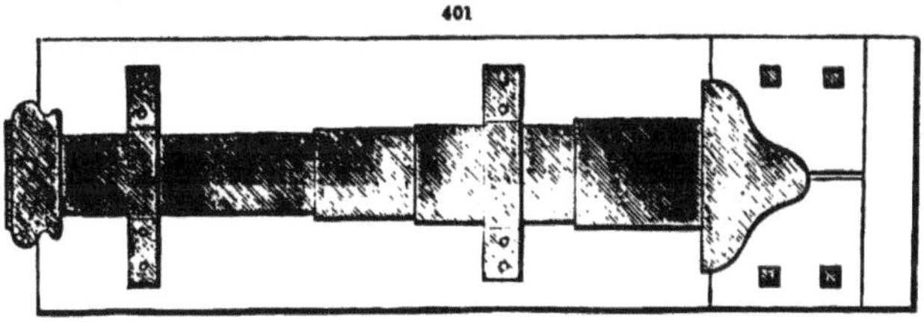

Experiments were afterwards made with a piece of the broken cylinder about 2 ft. in length, stripped of all the wire, with the exception of two coils. It was then a brass tube 2 ft. long and

$\frac{1}{4}$ in. thick, with two coils of square steel wire, each $\frac{1}{16}$ in. thick, making together $\frac{1}{4}$ in. brass and $\frac{1}{8}$ in. wire. In the middle of this was placed $1\frac{1}{2}$ lb. of Government cannon-powder, and the ends were filled with closely-fitting wood plugs, fixed tightly with iron wedges. A trench 3 ft. deep was then dug in stiff clay, and the cylinder laid at the bottom. At each end a railway sleeper was driven firmly into the clay; the trench was then filled in, and the clay well pounded with a heavy beater. The powder was then fired by means of a patent fuse. The whole of this arrangement is shown in Fig. 402. The wooden plugs and sleepers were thrown out with great violence, and a large mass of clay blown out at each end, but the cylinder remained uninjured.

402.

403.

It was, after this, charged with 2 lbs. of powder, the ends filled with closely-fitting iron plugs, and the whole bound together, Fig. 403, by an iron strap of a sectional area of 5 sq. in. The powder was then ignited, and the iron strap torn asunder, but the cylinder remained intact, except at the ends, where, from the wire being imperfectly fastened, it uncoiled, and the cylinder was torn open. Taking the tensile force of the iron strap at 18 tons the square inch, the force of the powder must have been above 13 tons the square inch; yet this was resisted by $\frac{1}{4}$ in. of brass and $\frac{1}{8}$ in. of steel wire.

The diametral strain must have been 39 tons; and, taking the brass at 10 tons the square inch, it leaves 34 tons for the steel wire, which, divided over the two sides, or $\frac{1}{4}$ in., would give for the ultimate resisting strength of the wire not less than 136 tons the square inch of section. The wire used was of the finest quality.

404

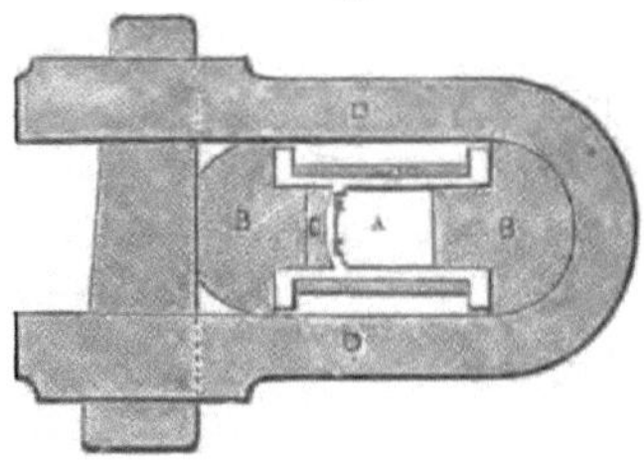

Further experiments were now instituted; firstly, to try the effect of wire in enabling hard cast iron to resist a bursting strain; secondly, with a view of ascertaining whether it was possible to transmit the force of the powder through a thin breech of cast iron to a yielding substance placed between that breech and the carriage of the gun.

Two sets of cylinders were prepared; the first set arranged as shown in Fig. 404, where A is the powder-chamber; B B, cast-iron plugs; C, the space between the bottom of the powder and the plug B, filled up with a soft material; D, a wrought-iron strap, with jib and cotter for keying up the plugs B B. The object was to ascertain whether the diaphragm at E E would be shattered by the force of the explosion. Six cylinders were thus prepared, and charged with from 50 to 250 grains of Government cannon-powder, the total contents of the cylinder being 310 grains.

The following were the results:—

No. of Cylinder.	Wire.	Charge.	Results.	Material behind the Diaphragm.
		grains.		
0	Two coils ..	50	No effect	Lead.
"	" " ..	50	"	"
"	" " ..	100	"	"
"	" " ..	120	"	"
"	" " ..	150	Burst	"
1	Four coils ..	150	No effect	"
"	" " ..	180	Top flange burst ..	"
3	Six coils ..	180	No effect	"
"	" " ..	200	"	"
"	" " ..	220	"	"
"	" " ..	240	Flange burst	"
6	Eight coils ..	240	"	"
8	" " ..	200	No effect	Gutta-percha.
"	" " ..	220	Burst	Gutta-percha softened by heat.
9	Ten coils ..	240	No effect	Lead.
"	" " ..	250	Flange burst.	

Iron wire, No. 21 gauge, $\frac{1}{32}$ in. diameter, was used; its breaking strain was 60 lbs.

In no case was the bottom of the cylinder injured, except in the second experiment with cylinder No. 8, when the gutta-percha was softened by the heat of the first explosion. The lead transmitted the force perfectly in every case; showing that there is no practical difficulty in transmitting the force through even so thin a diaphragm as $\frac{1}{10}$ of an inch, even when of so brittle a material as cast iron.

405.

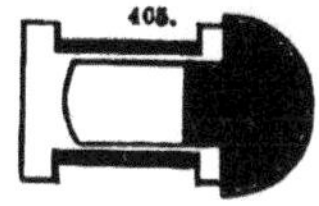

The second set of cylinders, Fig. 405, each contained 305 grains when full to the plug. The plugs fitted accurately, and the powder was fired through a small touch-hole, the size of a pin, with the following results:—

No. of Cylinder.	Wire.	Charge.	Results.	Remarks.
		grains.		
0	None	40	No effect.	
"	"	50	"	
"	"	60	"	
"	"	70	"	
"	"	80	Burst.	
2	Four coils ..	130	No effect.	
"	" " ..	150	Flange burst.	
7	Eight coils ..	200	No effect .. .	A wrought-iron flange $\frac{1}{4}$ in. deep, contracted on flange.
"	" " ..	220	"	
"	" " ..	240	"	
"	" " ..	250	"	
"	" " ..	260	"	
"	" " ..	270	"	
"	" " ..	280	"	
"	" " ..	290	"	Hoop on flange shifted.
5	" " ..	200	"	
"	" " ..	220	"	
"	" " ..	230	"	
"	" " ..	240	.. ..	Flange cracked.
4	Four coils ..	200	No effect.	
"	" " ..	250	.. ..	Flange cracked.
10	Ten coils ..	310	No effect.	

In these experiments the same wire as in the last was used. Its breaking strain was 60lbs.; consequently, the actual strength of the material in the cylinder to the lineal inch was:—

			Tons.	Tons.
No. 0.	Cast iron 0·10 × 2 × tons	=	1·76	1·76
	Nil.			
" 2.	Cast iron, as above	=	1·76	
	Wire $4 \times 28 \times 2 \times \frac{60}{2240}$	=	6 00	
				7·76
" 7.	Cast iron, as above	=	1·76	
	Wire $8 \times 28 \times 2 \times \frac{60}{2240}$	=	12 10	
				13·76
" 5.	Same as No. 7	=	..	13·76
" 4.	" " 2	=	..	7 76
" 10.	Cast iron	=	1·76	
	Wire $10 \times 28 \times 2 \times \frac{60}{2240}$	=	15·00	
				16·76

The enormous force of the expansive gases in these experiments was shown by their action on the plugs, which, although accurately fitted and of hard iron, were chiselled and grooved out in an extraordinary manner; the rents, too, were rapidly enlarged. The results obtained as regards strength were so conclusive, that Longridge proceeded to construct a small gun, represented in Fig. 406. This gun was 2·96 in. in bore, and 36 in. long in the clear; it had on it twelve coils of No. 16 W. G. iron wire at the breech, decreasing to four coils at the muzzle. The thickness of cast iron was $\frac{11}{16}$ in. at the breech and $\frac{1}{2}$ in. at the muzzle. The gun was cast hollow, and a recess left in the thick part of the breech, in which an india-rubber washer, $\frac{3}{4}$ in. thick, was placed. The trunnions formed no part of the gun, but consisted of a strap passing round the breech, with two side-rods extending about one-third the length of the gun, and terminating in the trunnions themselves. Thus the whole force of the recoil was transmitted through the heavy mass at the breech, then through the caoutchouc, and along the side-rods to the trunnions. The whole was mounted on a wood carriage on four roller-wheels, about 8 in. in diameter. The weight of the gun and wrought-iron trunnion-strap was 3 cwt., and the carriage 2 cwt. 0 qr. 15 lbs., making a total of 5 cwt. 0 qr. 15 lbs. The shot were cast as nearly the size of the bore as possible, so as to move freely, but with very little windage. The spherical shot weighed 3¾ lbs., and the conical shot

406.

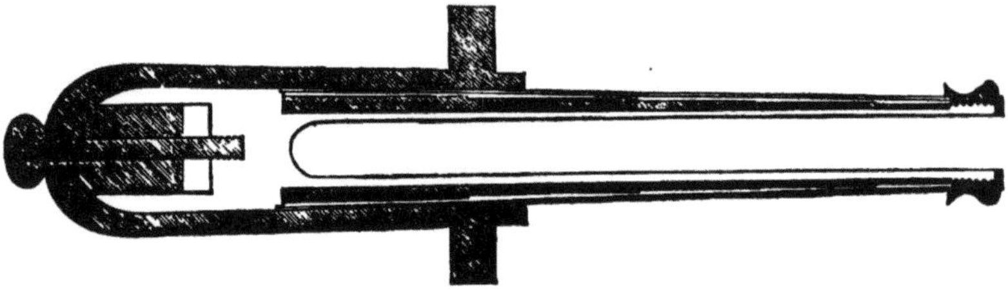

from 6 to 7½ lbs. The following Table exhibits the results of the trial of this gun with 70° elevation, Government cannon-powder being used.

No.	Description of Shot.	Weight.	—	Charge of Powder.	Range to First Graze.
		lbs.		oz.	yds.
9	Round	8¾	7°	11	1400
4	Elongated ..	6½	7	11	1200
5	" ..	6	7	8	1220
6	" ..	7½	7	11	1542
8	" ..	7	7	11	Lost beyond 1500
7	" ..	7	7	16	" 1800
10	" ..	6½	7	16	" 1500
11	" ..	6½	7	16	Lost beyond 1800

The variations in range were due, partly, to not having very exact means of adjusting the elevations, and partly to differences in the form of the shot. The trials just described were, moreover, only intended as preliminary, it being intended to carry out a more complete series at another time. Unfortunately, this intention was frustrated by an accident which destroyed the gun. Longridge had an idea that it might be possible to obtain more accuracy of flight by using shot somewhat on the principle of an arrow, Fig. 407, with a long, light shaft, and heavy head. The head was of cast iron, and weighed about 8¾ lbs.; the shaft was of fir, fitted tightly into the iron head. When fired by mistake with a heavy charge of powder, the wood was driven forward with great force, entering and splitting the iron head. This was wedged so tightly in the chase, that it never left the gun, but tore it asunder about 12 in. from the muzzle. The muzzle with the shot in it were thrown forward about 15 yds., and the wire was uncoiled, but not broken. This accident was due to the action of the shot, and had nothing to do with the principle upon which the gun was constructed. Enough, however, had been done to show that, with a gun weighing only 3 cwt., a shot of 7½ lbs. could be thrown from 1500 to 1800 yds., a result, it is believed, not attainable by any 6-pounder in the service.

407.

We now proceed to describe the guns recently adopted by the French navy. These guns are of cast iron, strengthened by steel hoops up to the trunnions, or even a little further on the chase; they are all rifled and breech-loading. The shot used are of two kinds:—1. Oblong, or elongated hollow shot, containing gunpowder and an arrangement for firing it at the moment of impact; 2. Cast steel cylindrical, or cylindrical ogival-headed shot, to be used against iron-clad vessels; the former at short, the latter at long distances. Both kinds of shot are provided with two rows of projections, fitting in the rifled grooves of the gun, and made of zinc, copper, or bronze. The powder-cartridges are made of parchment-paper, while a wad of cork or dry sea-weed is placed between the powder and the projectile. The calibre of these guns is 0·16 m.; 0 19 m.; 0·24 m.; 0·27 m. The following are the chief dimensions of each of these guns:—

Rifled Gun of 0·16 m. Calibre.

Total length	3·385 m.
Diameter at the breech	0·634 m.
Diameter of bore	0·1647 m.
Weight of gun	5000 kilos.

The bore is made with three parabolic grooves, the inclination of which varies from 0° at the beginning to 6° at the mouth of the gun. With a charge of 5 kilos. of powder, and elongated projectile weighing 31·5 kilos., and a wad 0·16 m. in length, the range of this gun was as follows:—

950	mètres,	at an	angle of	2°
3500	"	"	"	10°
7250	"	"	"	35°

At this last distance the lateral deviation is 16 mètres, and the longitudinal deviation, on an average, 44 mètres. With a charge of 7·5 kilos., and a steel cylindrical, or cylindrical ogival-headed shot weighing 45 kilos., the range of the ogival-headed projectile at 4° was about 1700 mètres; the correctness and length of range were about the same as when the elongated shot and a charge of 5 kilos. were used. The last-mentioned steel shot must not be fired at iron-clad vessels at a greater distance than 600 mètres; but at a distance of 300 mètres this steel shot perforates iron plates of 15 centimètres thickness.

Rifled Gun of 0·19 m. Calibre.

Total length	3·800 m.
Diameter at the breech	0·772 m.
Diameter of bore	0·194 m.
Weight of gun	8000 kilos.

The bore is made with five parabolic grooves, the inclination of which varies from 0° at the beginning to 6° at the mouth of the gun. With a charge of 8 kilos., a cast-iron projectile weighing 52 kilos., and a wadding of sea-weed 190 millimètres long, between the powder and shot, the range of this gun was:—

900 mètres, at an angle of	2°
3300 " " "	10°
7000 " " "	35°

At this last distance the lateral deviation is about 14 mètres, and the average longitudinal deviation about 42 mètres. With a charge of 12·5 kilos., and a cylindrical, or cylindrical ogival-headed projectile weighing 75 kilos., the range of this gun is, under the same angle of inclination, nearly the same for a distance of from 800 to 1000 mètres. The cylindrical projectile is, however, only intended to be fired to a distance of 300 mètres. The solid steel projectiles are formidable weapons against iron-clad vessels covered with armour-plates of 15 centimètres, at distances varying from 300 mètres for the cylindrical and 800 mètres for the cylindrical ogival-headed shot.

Rifled Gun of 0·24 m. Calibre.

Total length	4·560 m.
Diameter of the breech	0·980 m.
Diameter of the bore	0·240 m.
Weight of the gun	14,000 kilos.

The bore is made with five parabolic grooves, the inclination of which varies from 0° to 6°. With a charge of 16 kilos. this gun throws an elongated shot, weighing on an average 100 kilos., as follows:—

1000 mètres, at an angle of	2°
3600 " " "	10°
7800 " " "	35°

With a charge of 20 kilos. this gun projects a steel cylindrical ogival-headed shot weighing 144 kilos. At an angle of 3° the range is 1120 mètres with the ogival-headed and 1020 mètres with the cylindrical shot. This gun has the greatest effect within 1000 mètres, at which distance a few shots fired from it would destroy the heaviest and strongest walls in existence.

A cylindrical projectile, weighing 144 kilos., fired against a shield constructed of 80 centimètres of wood and armour-plate of 15 centimètres, not only perforated that shield, but also carried with it 140 to 150 kilos. of the iron plate and about a cubic mètre of wood.

Rifled Gun of 0·27 m. Calibre.

This gun is of the same construction as the three last mentioned,—*viz.* a cast-iron breech-loading gun strengthened by steel hoops. Its dimensions are:—

Total length	4·660 m.
Diameter at the breech	1·133 m.
Diameter of the bore	0·275 m.
Weight of gun	22,000 kilos.

With a charge of 24 kilos. it throws an elongated projectile weighing 144 kilos.; with a charge of 30 kilos. it throws a solid steel cylindrical, or cylindrical ogival-headed shot weighing 216 kilos.

Two cannons of 42 centimètres calibre were lately cast at Ruelle. The material of which they were composed was cast iron strengthened by steel hoops. These guns weigh each 37,000 kilos.; the diameter of the bore is 0·424 m. The extreme external diameter is 1·360 m.; diameter at the breech, 1·300 m.; diameter at the end of the hooped part, 1·050 m. This gun will throw:—

1. A solid spherical shot 0·42 m. diameter, and weighing 300 kilos., with a powder-charge of 50 kilos.

2. A hollow spherical projectile, weighing 210 kilos., containing 9 kilos. of gunpowder, with a charge of 33 kilos.

The Fraser gun is made according to Armstrong's system, with bar-iron wound round a solid mandril. This description of gun is made of three different calibres,—9 in., 0·228 m.; 8 in., 0·203 m.; 7 in., 0·177 m. The wrought bar-iron used is submitted to a strain of 21 to 23 tons the square inch; if it does not stand this test, its strength is insufficient; if it is stronger, it becomes too unmanageable. The bars are welded together, after testing, in lengths varying from 50 to 200 ft. = 15 m. to 60 m. These long bars are then placed in a reverberatory furnace, through which they are dragged, as they become sufficiently hot, to be rolled on a mandril; they are then submitted to the blows of a steam-hammer.

Fig. 408 represents the 24-centimètre gun on its carriage.
Fig. 409 the 24-centimètre gun in course of construction and partly hooped.
Fig. 410, hoop with trunnions.
Fig. 411, breech screw and plug for gun of 24 centimètres calibre.

408.

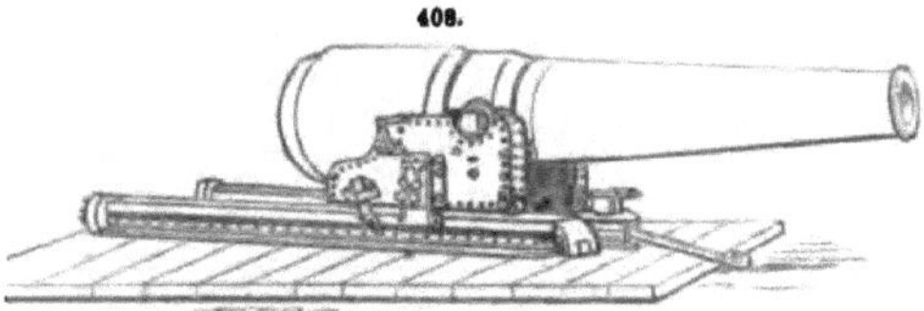

410

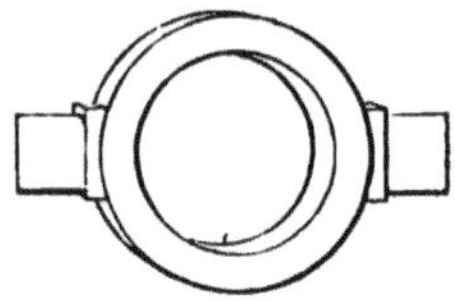

409.

411.

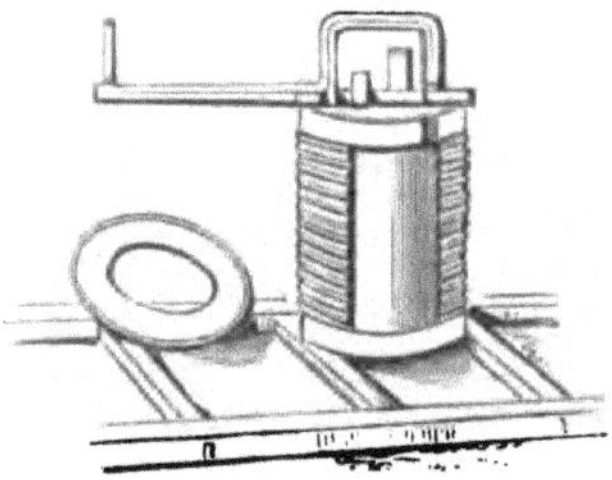

American Cast-iron Guns.—Although the United States' Government has made little progress in the adaptation of wrought iron and steel to cannon-making, it has certainly attained to a remarkable degree of perfection in the figure, material, and fabrication of its cast-iron guns. While constructors in Europe have carefully preserved the traditional shapes and ornamentation of early times, shapes that once had a significance, but are now only sources of weakness, the aim in America has been to ascertain the exact amount and locality of *strain*, and to proportion the parts with this reference, to the entire abandonment of whatever is merely fanciful and traditional.

The consequent saving of weight with a given strength at the point of maximum strain, is well illustrated by placing a section of the British 8-in. gun, 68-pounder, over that of the United States' army 8-in. columbiad, Fig. 412.

412

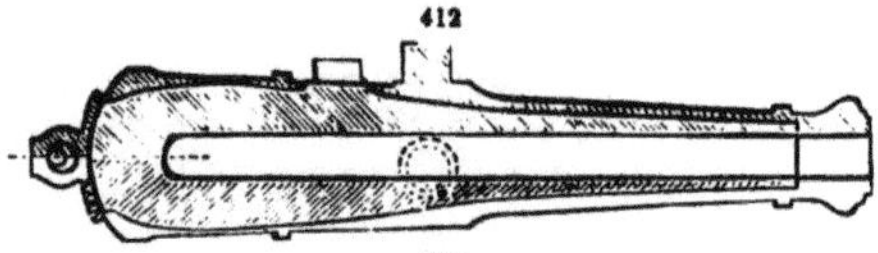

413.

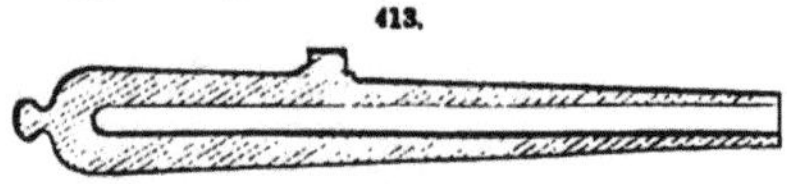

414.

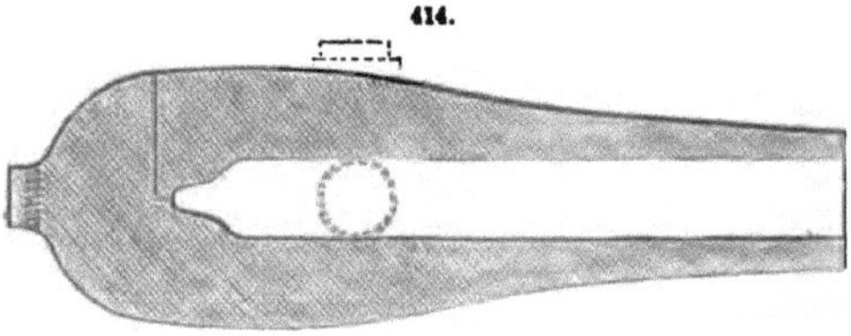

Rodman's process of casting guns hollow and cooling them from within, for the purpose of modifying the initial strains, when added to the advantages of good proportion and strong material, produces nearly or quite the best result attainable with simple cast iron. But the tension of this material at its elastic limit is so low, that it will not alone endure the pressure necessary to give the highest velocities to the heavy projectiles demanded by iron-clad warfare.

Considering, however, the failure of such a large proportion of the heavy wrought-iron guns, both built-up and solid, and the present scarcity and enormous cost of steel masses of the proper quality, it is by no means certain that the cast-iron barrel lined with steel, or, as so largely and successfully used in America, France, and Spain, strengthened by hoops, is not the best temporary resort.

Hollow casting, the most obvious means of improvement, is not deemed important for heavy ordnance alone. The 4 2-in. rifled United States' siege-gun, Fig. 413, is cast hollow and cooled from within. Indeed, the advantages of the process can be better realized in the 8 or 10 in barrel cast for hooping, than in the 15-in. columbiad.

All United States' army guns down to 4 2-in. bore are hollow-cast. The 20-in., 15-in., and 13-in. navy guns have been cast hollow. Fig. 414 shows the 15-in., and Fig. 415 the 13-in. gun.

The following abstract of official reports will explain the conduct and results of the hollow-casting process.

On the 4th of August, 1849, two 8-in. columbiads were cast at the Fort Pitt Works, from the same iron. No. 1 was cast solid, in the usual manner; No. 2 was cast on a hollow core, through which a stream of water passed while the metal was cooling. The iron for both castings was melted at the same time in two air-furnaces, each containing 14,000 lbs. After melting, the liquid iron remained in the furnaces, exposed to a high heat, for one hour; it was then discharged into a common reservoir, whence it issued in a single stream, which, after proceeding a few feet, separated into two branches, one leading to each mould.

415.

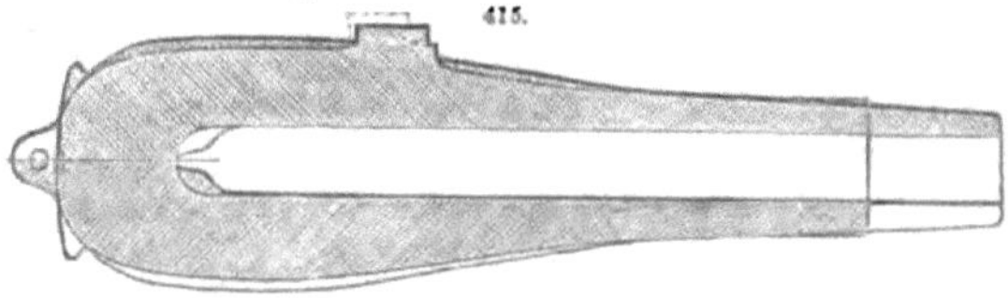

The solid casting was cooled as usual in an open pit. The hollow casting was cooled in the interior by passing a stream of water through the core for a period of 40 hours, when the core was withdrawn; after which the water passed through the interior cavity formed by the core, for 20 hours. The average quantity of water passed through during the whole period was 1·66 cubic ft. a minute, or 100 ft. an hour; making in all 6000 cubic ft., weighing 187 tons. The temperature of the water was increased 20° during the 1st hour; 13° during the 20th hour; 8° during the 40th hour; and 3° during the 60th and last hour. The weight of the water passed through was 30 times the weight of the casting; and the heat imparted by the casting to the water, and carried off by the latter, was equal to 10° on the whole quantity of water used. The mould for this casting was placed in a covered pit, which had previously been heated to about 400°; and this heat was kept up as long as the stream of water was supplied. Both columbiads were completed and inspected September 6th, and were found to be accurate and uniform in their dimensions and weights.

The charges used in testing the guns were as follows:—

Proof Charges.—1st fire, 12 lbs. powder, 1 ball, and 1 wad; 2nd fire, 15 lbs. powder, 1 shell, and 1 sabot.

Service Charges.—10 lbs. powder, 1 ball, and 1 sabot; mean weight of balls used, 63½ lbs.; mean weight of shells used, 49 lbs.; mean proof range of powder used, 298 yds.

The guns were fired alternately, up to the 85th round, at which columbiad No. 1, cast solid, burst. Then the proof proceeded with No. 2, which burst at the 251st round, having endured nearly three times as much service as the other.

In 1851, two more 8-in. columbiads were cast at the same foundry, and under similar circumstances; the one was cast solid, and the other hollow. The iron for both remained in fusion 2½ hours, exposed to a high heat.

The core for the hollow gun was formed upon a water-tight cast-iron tube closed at the lower end. The water descended to the bottom of this tube by a central tube open at the lower end, and ascended through the annular space between the tubes. The water passed through the core at the rate of 2½ cubic ft. a-minute, or 150 ft. an hour. At 25 hours after casting, the core was withdrawn, and the water thereafter circulated through the interior cavity formed by the core, at the same rate for 40 hours; making 65 hours in all. The whole quantity of water passed through the casting was nearly 10,000 cubic ft., weighing about 300 tons, or about 50 times the weight of the casting. The heat imparted by the casting to the water, and carried off by the latter, was equal to 6° on the whole quantity of water used.

A fire was kindled in the bottom of the pit directly after casting, and was continued 60 hours. The pit was covered, and the iron case containing the gun-mould was kept at as high a temperature as it would safely bear, being nearly to a red heat, all the time.

In the same year, two other 10-in. columbiads were cast, of the same iron, the one solid, and the other hollow. Both moulds were placed in the same pit, and all the space in the pit outside of the moulds was filled with moulding-sand, and rammed. This was done because the iron cases of the moulds were not large enough to admit the usual thickness of clay in the walls of the mould. It was apprehended that the heat of the great mass of iron within would penetrate through the thin mould, and heat the iron cases so much as to cause them to yield and let the iron run out of the mould. The external cooling of the 10-in. hollow gun, by the contact of the flask with green sand, was therefore much more rapid than that of the 8-in. hollow gun.

Water was passed through the core at the rate of about 4 cub. ft. a-minute, or 240 ft. an hour, for 94 hours; amounting in all to 22,560 ft., weighing about 700 tons, or 70 times the weight of the casting. The mean elevation of the temperature of all the water passed through the core in 94 hours, was about 3½°. At the end of this period an attempt was made to withdraw the core from the casting, which proved unsuccessful. The contraction of the iron around it held it so firmly, that the upper part of it broke off, leaving the remainder imbedded in the casting. The stream of water was then diminished to about 2 ft. a-minute, which continued to circulate through the core for 48 hours. The supply of water allotted to and circulated through both the 8-in. and 10-in. guns was equal, in weight, to the weight of each casting, in about 1 hour and 20 minutes.

When proving these guns, 80 rounds a-day were easily made with 7 men in 5 hours, from the 8-in. gun, and with 9 men, 60 rounds were made in the same time from the 10-in. gun. 15 rounds were sometimes made from the 8-in. gun in 30 minutes. The two guns making the pair to be compared were fired alternately, one discharge from each, in regular succession, until one of them burst, when the firing of the survivor was continued by itself. The powder of the cartridges of each pair was of the same proof range, and taken from the same cask.

Proof Charges.—8-in., 1st round, 12 lbs. powder, one ball and sabot, and 1 wad; 8-in., 2nd round, 15 lbs. powder, 1 shell with sabot: 10-in., 1st round, 20 lbs. powder, 1 ball and sabot. and 1 wad; 10-in., 2nd round, 24 lbs. powder, 1 shell with sabot.

Service Charges.—8-in., 10 lbs. powder, 1 ball with sabot; 10-in., 18 lbs. powder, 1 ball with sabot. Weight of 8-in. balls, 63½ lbs.; of shells, 48½ lbs.: weight of 10-in. balls, 124 lbs.; of shells, 91 lbs.

The number of rounds fired from each gun, including proof charges, were as follows:—8-in. gun, No. 3, cast solid, 73 rounds; 8-in. gun, No. 4, cast hollow, 1500 rounds: 10-in. gun, No. 5, cast solid, 20 rounds; 10-in. gun, No. 6, cast hollow, 249 rounds.

Each of them, excepting the 8-in. gun, No. 4, cast hollow, burst at the last round; and that remains unbroken, and apparently capable of much further service.

On comparing the enlargements of the bores, made by an equal number of rounds of the guns cast solid with those cast hollow, it will be seen that, in both pairs of guns, the enlargement is least in those cast hollow.

The less endurance of the 10-in. hollow gun than that of the 8-in. hollow one, is accounted for by the fact that the 10-in. gun had no fire on the exterior of the flask while cooling, it having been rammed up in the pit, where it was supposed, at the time of casting, the heat of the gun would have been retained by the sand until the interior should have been cooled by the circulation of water through the core-barrel. This supposition was found to be erroneous on digging out the sand, as its temperature was found to be much lower than had been expected.

One of the 15-in. American naval guns was fired 900 times at elevations from 0 to 5°. The charge commenced at 35 lbs. It was then increased to 50 lbs. With 60 lbs., 220 rounds were fired. The gun at length burst with 70 lbs. The shot in all cases was 440 lbs. After the first 300 rounds, the chamber, Fig. 414, was bored out to a nearly parabolic form, and the chase was turned down 3 in., so as to fit the port designed for the 13-in. gun.

Columbiads.—The columbiads, Figs. 416, 417, are a species of sea-coast cannon, which combine certain qualities of the gun, howitzer, and mortar; in other words, they are long, chambered pieces, capable of projecting solid shot and shells, with heavy charges of powder, at high angles of elevation, and are therefore equally suited to the defence of narrow channels and distant roadsteads.

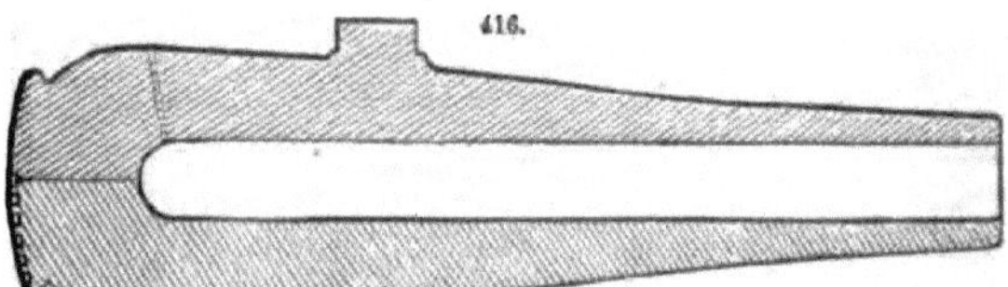

416.

The columbiad was invented by Bumford, and used in the war of 1812 for firing solid shot. In 1844 the model was changed, by lengthening the bore and increasing the weight of metal, to enable it to endure the increased charge of powder, or ⅛th of the weight of the solid shot. Six years after this, it was discovered that the pieces thus altered did not always possess the requisite strength. In 1858 they were degraded to the rank of shell guns, to be fired with diminished charges of powder, and their places supplied with pieces of improved model.

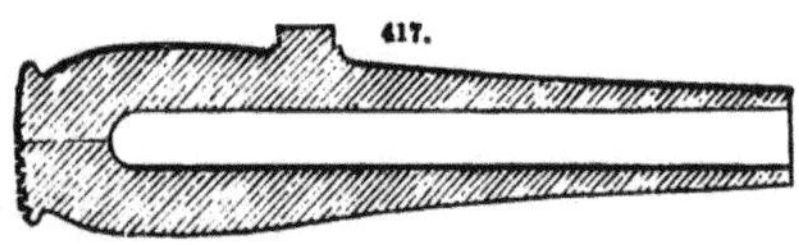

417.

The changes made in forming the new model, consisted in giving greater thickness of metal in the prolongation of the axis of the bore, which was done by diminishing the length of the bore itself; in substituting a hemispherical bottom to the bore and removing the cylindrical chamber; in removing the swell of the muzzle and base ring; and in rounding off the corner of the breech. The present model, as illustrated, was proposed by Rodman, in 1860.

In addition to the heavy ordnance before mentioned, the Navy Department has introduced a superior gun of 10-in. calibre, called a 125-pounder. The exterior dimensions are nearly the same as those of the 11-in. gun, except that the maximum diameter of the reinforce is continued farther forward, 3 calibres. The first of these guns was cast solid, and endured 47 lbs. of powder and 125-lb. balls for some hundred rounds. The new 10-in. un is cast hollow; charge, 40 lbs.; shot, 125 lbs.

The chambers of the Navy 13 and 15 in. guns, as shown in Figs. 414, 415, have recently been changed to a shape nearly parabolic.

The Navy Department has four 12-in. rifles, cast hollow, of about the exterior dimensions of the 15-in. gun. It is believed that they will have satisfactory endurance with 50-lb. charges and 600-lb. bolts.

Twenty-inch guns for the army and navy have recently been cast at Pittsburg. The following are the particulars of the metal and the fabrication of the first 20-in. army gun:—

The iron was high No. 2, warm blast, 200° hematite, from Blair county, Pennsylvania. The smelted pigs were remelted and cast into pigs, which were again melted in three air-furnaces. The weight of iron was 172,000 lbs.; the time of melting, 7½ hours; the time of casting, 23 minutes. Water, run through the core at the rate of 30 gallons a-minute, during the first hour was heated from 36° to 92°; during the second hour, at the rate of 60 gallons a-minute, water emerged at 61°. From the 15th to the 20th hour after casting, the water was heated 21·5°. After the 26th hour the core-barrel was removed, and air was forced into the bore at the rate of 2000 cubic feet a-minute. The metal was considered too high to be cooled by the direct contact of water. At the 50th hour after casting, the air emerging from the gun was 130 seconds in rising, 60° to 212°. The gun was cast on the 11th of February, 1864. On the 17th, the difference in the temperature of the entering and emerging air was 100°; on the 20th it was 33°. Air circulated through the bore till the 24th.

418.

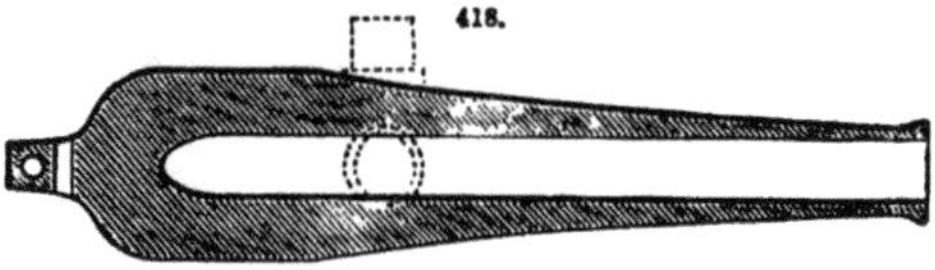

The mould, 5 to 6 in. in thickness, was made in a two-part iron flask, 1½ in. thick. On the 23rd the upper part of the mould was removed; on the 24th the lower part was removed; on the 25th the gun was removed from the pit.

The density of the metal taken from the casting was 7·3028. The tenacity was 28,737 lbs. per square in.

Fig. 418 shows a section of the 11-in. Dahlgren gun used in the United States' navy.

PARTICULARS AND CHARGES OF U. S. HOLLOW-CAST IRON ARMY ORDNANCE.

The Heavy Guns have no Preponderance.

Name of Gun.	Length.	Length of bore.	Max. diam.	Weight.	Service charge.	Bursting charge Shell.	Weight of Shot.	Weight of Shell.	Remarks.
Smooth-bores.	in.	in.	in.	lbs.	lbs.	lbs.	lbs.	lbs.	
20-in. columbiad ..	243·5	210	64	115,200	100	..	1000	..	Weight of shell not determined.
15-in. „	190	165	48	49,100	50 / ·6 grain	17	440 / 425	330	Cored shot.
13-in. „	177·6	155·94	41·6	32,731	30 No. 5	7	300 / 280	224	
10-in. „ of 1860	136·66	105·5	32·	15,059	15 for shell / 18 for shot	3	127½	100	
8-in. „ „	123·5	110	25·6	8,465	10	1½	68	48	
Rifles.									
4½-in. siege-gun of 1860	133	120	16	3,450	3½	..	30	30	Twist uniform. 1 turn in 15 ft. Preponderance 300 lbs.
3-in. field-gun of 1861	72·65	65	9·7	830	1	..	10	10	Twist uniform. 1 turn in 10 ft. Preponderance 40 lbs.

PARTICULARS AND CHARGES OF U. S. CAST-IRON NAVY ORDNANCE IN SERVICE.

Name of Gun.	Length of bore.	Max. diam.	Weight.	Service charge.	Max. charge.	Weight of Shot.	Weight of Shell.	Remarks.
Smooth-bores.	in.	in.	lbs.	lbs.	lbs.	lbs.	lbs.	
20-in. gun ..	163	64	100,000	Probably 100	..	1000	..	Shell not determined. Cored shot, and gun cast hollow.
15-in. „ ..	130	48	42,000	35	60	400	330	Cast hollow. Cored shot.
13-in. „ ..	130	44·7	36,000	40	..	280	224	Cast hollow.
11-in. „ ..	132	32	16,000	15	20	170	130	Lately cast hollow.
10-in. „ ..	119½	29·1	12,000	12½	16	125	100	Cast solid.
9-in. „ ..	107	72·2	9,200	10	13	93	70	Cast solid.
125-pdr. (10-in.)	117¾	33·25	16,500	40	..	125	100	Cast hollow.
Rifles.								
Parrott 10-in. ..	144	40	26,500	25	..	230 to 250	250	The Parrott guns are hooped with wrought iron, and are lately cast hollow.
„ 8-in. ..	136	32	16,300	16	..	132 to 175	152 to 175	
„ 100-pdr. (6·4-in.) ..	130	25·9	9,700	10	..	70 to 100	100	

Bronze Guns.—An alloy of about 90 parts of copper and 10 parts of tin, commonly known as "gun-metal" in Europe, is popularly called "brass" in America, when used for cannon, and named "bronze" by recent American writers. A strong cast iron is also known in America as "gun-metal."

The "work done" in stretching to the elastic limit and to the point of fracture, is less for ordinary bronze than for wrought iron of maximum ductility, and for low steel. This defect, added to the costliness of bronze, to the various embarrassments experienced in the casting of large masses, to its softness, and consequently rapid wear and compression, and to its injury by heat, has not warranted its employment for large calibres and high charges. The increase of cost, with increase of weight, would probably be greater for bronze than for cast iron, and much greater than for steel or wrought iron, because bronze must be cast under great pressure, to be sound and tenacious. So that, were it the proper metal in other particulars, an unnecessarily large and actually immense non-paying capital would be tied up in a national bronze armament. The high value of the old material would not offset this cost to the extent that it does in railway matters, for obvious reasons.

The mean ultimate cohesion of gun-metal, according to European authorities and the experiments of the United States' Government, is about 33,000 lbs. per sq. in. Mallet states it from 32,334 lbs. to 43,536 lbs. Major Wade states it from 17,698 lbs. to 56,786 lbs.

Benton says, that "the density and tenacity of bronze, when cast into the form of cannon, are found to depend upon the pressure and mode of cooling. This is exhibited by the means of observations made on five guns cast at the Chicopee Foundry, namely:—

Density.			Tenacity per Square Inch.	
Breech square.	Gun-head.	Finished gun.	Breech square.	Gun-head.
8·765	8·444	8 740	46,509	27,415

The guns were cast in a vertical position, with the breech square at the bottom. In consequence of the difference in the fusibility of tin and copper, the perfection of the alloy depends much on the nature of the furnace and the treatment of the melted metal. By these means alone, the tenacity of bronze has been carried, at the Washington Navy Yard Foundry, as high as 60,000 lbs.

The fabrication of bronze ordnance appears to be far better understood in Spain, and more especially in Turkey, than in America or England. Some bronze guns of 20 tons weight have been cast in Spain, but they cannot be rapidly fired.

According to American and British authorities, the want of uniformity, even in different parts of the same gun, is a striking defect. For instance, for light pieces, especially for field-cannon, bronze is much used, but there are many objections even to this alloy. As the tin is much more fusible than the copper, and must be introduced when the latter is in fusion, it is difficult to seize the precise moment when the alloy can be properly formed: part of the tin is frequently burned and converted into scoria.

Major Wade, after calculating the results of experiments on a lot of bronze guns, cast at Chicopee, says, "The most remarkable feature of the above table is the irregular and heterogeneous character of the results which it exhibits in samples taken from different parts of the same guns. By an examination of the results obtained from the heads of all the guns cast, it will appear that the density varies from 8 308 to 8 756, a difference equal to 28 lbs. in the cubic foot; and that the tenacity varies from 23,529 to 35,484, a difference in the ratio of 2 to 3. These differences occur in samples taken from the same part of different castings, the gun-head; the part which, in iron cannon, gives a correct measure of the quality of the metal in all parts of the gun. The materials used in all these castings were of the same quality: they were melted, cast, and cooled in the same manner, and were designed to be similarly treated in all respects. The causes why such irregular and unequal results were produced, when the materials used and the treatment of them were apparently equal, are yet to be ascertained."

The authorities generally agree that the tin in bronze guns is gradually melted by the heat of successive explosions. If this is the case with field-guns, the heavy charges and projectiles, and the quick firing demanded in iron-clad warfare, would soon destroy this material. Colonel Wilford stated, at a meeting of the United Service Institution, that iron mortars were introduced because holes were burned in the chambers of bronze mortars by the immense heat of the powder-gas. Heat also causes the drooping of the parts of a bronze gun that overhang the trunnions.

As to decomposition, Captain Benton says, "Bronze is but slightly corroded by the action of the gases evolved from gunpowder, or by atmospheric causes " but Captain Simpson remarks, that the gases produced by the combustion of gunpowder also produce an injurious effect upon this kind of piece, by acting chemically on the bronze.

All these defects of bronze for the bore of a gun, irrespective of strength, namely, the melting of the tin, the change of figure, the conversion. abrasion, and compression, obviously aggravate each other; and, when taken in connection with rifling and excessive pressures, are conclusive evidence as to the unfitness of the material to meet the conditions of greatest effect under consideration.

The average ultimate tenacity of bronze is so low—in fact, little above that of the best average cast gun-iron—that the loss of strength, due to want of regulated initial tension and compression, becomes a very serious defect when calibres are large and pressure high. To remedy it by hooping bronze with steel or iron, would not avoid the defective surface of the bore just considered.

The Dutch, however, have lined cast-iron guns with bronze, and Blakely has constructed some experimental guns in the same way for another reason: bronze can safely elongate more than cast iron, without permanent change of figure; and when it is put in a position where it must be more elongated by internal pressure, the strength of the whole structure is thus brought into service—the principle of varying elasticity, already considered, is approximately realized.

Bronze hoops upon steel or iron barrels would avoid the defect of a soft bore, but they would increase the defect just considered, due to the unequal stretching of the layers of a tube by internal pressure. A principal advantage of bronze hoops is, that with the little heat they would get from the powder, they would expand to the same extent, approximately, as the more highly-heated iron barrel, thus reducing the danger of bursting by rapid firing.

Other Alloys.—Phosphorus is known to improve the strength of copper, and to make it cast soundly. Abel, chemist to the British War Department, stated before the Institution of Civil Engineers, that he had made some experiments upon the combinations of phosphorus with copper, and "had found that by the introduction of a small proportion, say from 2 to 4 per cent., of phosphorus into copper, a metal was produced remarkable for its density and tenacity, and superior in every respect to ordinary gun-metal. He believed the average strain borne by gun-metal might be represented by 31,000 lbs. upon the square inch; whilst the material obtained by adding phosphorus to copper bore a strain of from 48,000 to 50,000 lbs. But the increased tenacity was not the only beneficial result obtained by this treatment of copper. The material was more uniform throughout, which was scarcely ever the case with gun-metal. The experiments alluded to were merely preliminary, and had been, to a certain extent, checked by the improvements since introduced in the construction of field-guns, which had led to a discontinuance of the employment of gun-metal."

Aluminium has been found to add great strength to copper. The compound formed of these two metals is called Aluminium Bronze. John Anderson, superintendent of the Royal Gun-factory, Woolwich, found the tensile strength of an alloy of 90 per cent. of copper and 10 per cent. of aluminium was 73,181 lbs. the square in., or twice that of gun-metal, and its resistance to crushing 132,146 lbs., that of gun-metal being 120,000 lbs. The aluminium bronze did not begin to change its form until the pressure exceeded 20,384 lbs. In transverse strength or rigidity it was also found superior to gun-metal, in the ratio of 44 to 1. Its tenacity and elasticity depend on a particular number of meltings: at the first melting aluminium bronze is very brittle, a state to which it again returns after fusion.

The first melting appears to produce an internal mechanical mixture, rather than a chemical combination of the metals; as, in the proportion of 10 of aluminium and 90 of copper, an alloy of a very brittle character is produced by the first melting; but by repeated meltings a more uniform combination seems to take place, and a metal is produced free from brittleness, and having about the same hardness as iron. The alloy, containing rather less than 10 per cent. of aluminium, is said to possess the most uniform composition and the best degree of hardness; but it is not always an easy thing to produce this desirable uniformity of texture, as patches of extreme hardness sometimes occur, which resist the tools, and are altogether intractable to the action of the rollers.

Aluminium bronze, composed of 9 parts by weight of copper and 1 of aluminium, was found by J. Anderson to have a tensile strength of about 43 tons, 96,320 lbs.; but two other specimens, which were not quite sound, had only a mean tensile strength of 22½ tons, 50,400 lbs. So that the metal is liable to great variations in strength.

The cost of this alloy would of course prevent its extensive introduction as a cannon-metal.

Sterro-metal, a recent invention of Baron de Rosthorn, of Vienna, is described by a correspondent of the 'Times.' The mechanical properties of the alloy were carefully examined at the Polytechnic Institution, Vienna, with the following results:—

Tensile Strength of Sterro-Metal.

	Tensile Strength, in Tons.	Reduced to Pounds.
Sterro-Metal.		
After simple fusion	27	60,480
„ forging red-hot	34	76,160
„ drawn cold	38	85,120
Gun-Metal—Bronze.		
After simple fusion	18	40,320

The same copper, from Boston, U.S., was used in making both the sterro-metal and the gun-metal, and for the latter the best English tin was employed. Both alloys were cast under precisely similar conditions, and run into the same mould. Similar tests were made at the Arsenal, Vienna, and the results are as follows:—

Tensile Strength of Sterro-Metal.

	Tensile Strength, in Tons.	Reduced to Pounds.
Sterro-Metal.		
After simple fusion	28	62,720
„ forging red-hot	32	71,680
Drawn cold and reduced from 100 to 77 of transverse sectional area	37	82,880

The specimens of metal operated on in the preceding experiments were analyzed at the Austrian mint. The results are as under:—

	Polytechnic Metal.	Arsenal Metal.
Copper	55·04	57·63
Spelter	42·36	40·22
Iron	1·77	1·86
Tin	0·83	0·15
	100·00	99·86

Experience has shown that the proportion of spelter may vary from 38 to 42 per cent., without materially affecting the quality of the alloy. The specific gravity of the forged metal is 8·37, and that of the same metal, drawn cold into wire, 8·40. But sterro-metal possesses another quality, which, in reference to its application for guns, is regarded as more important than its high tenacity, namely, great elasticity. It is not permanently elongated until stretched beyond $\frac{1}{400}$ of its length. Sterro-metal, it should be stated, is from 30 to 40 per cent. cheaper than gun-metal. Field-guns, from 4 to 12 pounders, have been made of single pieces of metal, worked by the action of a hydraulic press, whereby expense in forging is avoided; but reliable experiments have demonstrated that the metal thus treated has precisely the same properties and the same tensile strength as bars of it drawn out under the steam-hammer.

The following is the official report of experiments made by John Anderson, upon this metal, variously compounded and treated:—

Composition of this alloy, as made in the Arsenal at Vienna, is,—copper, 60; zinc, 41·88; iron, 1·94; tin ·156. And, as made at the Polytechnic, Vienna, its composition is,—copper, 60; zinc, 46·18; iron, 1·93; tin, ·905.

Alloys of similar composition to that of the Austrian metal have been prepared in the Royal Gun-factories, from which a better result has been obtained than from mixtures of the Austrian metals, also prepared in the Royal Gun-factories. The subjoined Table shows the results of the experiments with these different specimens.

This alloy is said to be the invention of Baron de Rosthorn, of Vienna. It derives its name from a Greek word signifying "firm." It consists of copper and spelter, with small portions of iron and tin; and to these latter its peculiar properties are attributed.

It has a brass-yellow colour, is close in grain, is free from porosity, and has considerable hardness, whereby it is well adapted to bearing-metal, or other purposes, where resistance to friction is needed.

The inventor proposes that, in heavy ordnance, the interior should consist of a tube of sterro-metal, and, over this, wrought or cast iron should be shrunk, from the breech to beyond the trunnions.

COMPOSITION AND STRENGTH OF STERRO-METAL, WOOLWICH.

Composition	Treatment.	Strain at Permanent Elongation of ·002 the inch.	Breaking Weight.	Ultimate Elongation the inch.
		tons.	tons.	inch.
Austrian mixture	As received ..	6·75	26·75	·1
R. G. factories' mixture of copper, 60; zinc, 39; iron, 3; tin, 1·5	Cast in sand ..	11·	21·5	·05
R. G. factories' mixture of copper, 60; zinc, 44; iron, 4; tin, 2	"	13·75	19·25	·015
" " "	Cast in iron ..	17·25	24·25	·016
" " "	Cast in iron and annealed	15·25	23·25	·02
" " "	Forged red-hot ..	17·	28·	·045

In a discussion before the Institution of Civil Engineers, Charles Fox said that "he believed it would eventually be found that the best gun could be constructed with some extremely dense and homogeneous alloy, cast and used without being drawn under the hammer. If a gun was made of an alloy possessing very great density, the detonating force of the powder would be resisted by a greater quantity of the metal employed than it could be by making use of one with greater elasticity. He thought, therefore, the best guns would be made of iron, mixed with some other metals, such as wolfram and titanium, so as to insure the greatest strength and density. Mushet had obtained great density, by mixing with iron a small percentage of wolfram, and great strength by the use of titanium. Therefore, he was inclined to believe, that guns cast of the densest alloys would have greater effect, in proportion to their thickness, than could be obtained by any complicated and expensive mode of construction."

It is obviously impossible, in the absence of further experiments, to predict either great success or failure for the alloys considered, as compared with steel. The field for discovery and improvement is certainly broad and promising; but no more so than in the case of steel. Although the alloying of copper, especially for cannon, has been practised for more than five hundred years, and should, therefore, be in advance of steel-making, which, for the purposes of artillery, is the work of the last decade, both metals—in fact, all metals—are undeveloped, because their chemical relations, and especially their elongation, within and beyond the elastic limit, and the corresponding pressures, have not been properly investigated.

While certain alloys, of both iron and copper, have one important feature in common—homogeneity, due to fusibility, at practicable temperatures—the alloys of iron have this grand advantage; iron is everywhere cheap and abundant; and the other necessary ingredients and fluxes—carbon, manganese, zinc, and silicium—are equally abundant, and, in some localities, already mixed, which would appear to be, on the whole, advantageous, although the mixtures are not found in proper proportions.

The fitness of metals for cannon depends chiefly on the amount of their elongation within the elastic limit, and the amount of pressure required to produce this elongation; that is to say, upon their elasticity.

It also depends, if the least possible weight is to be combined with the greatest possible preventive against explosive bursting, upon the amount of elongation and the corresponding pressure, beyond the elastic limit; that is to say, upon the ductility of the metal.

Hardness, to resist compression and wear, is the other most important quality.

Cast iron has the least ultimate tenacity, elasticity, and ductility; but it is harder than bronze and wrought iron, and more uniform and trustworthy than wrought iron, because it is homogeneous.

The unequal cooling of solid castings leaves them under initial rupturing strains; but hollow casting, and cooling from within, remedies this defect, and other minor defects.

Wrought iron has the advantage of a considerable amount of elasticity, a high degree of ductility, and a greater ultimate tenacity, than cast iron; but, as large masses must be welded up from small pieces, this tenacity cannot be depended upon: this defect, however, is more in the process of fabrication than in the material, and may be modified by improved processes. Another serious defect of wrought iron is its softness, and consequent yielding, under pressure and friction.

Low cast steel has the greatest ultimate tenacity and hardness: and, what is more important, with an equal degree of ductility, it has the highest elasticity.

It has the great advantage over wrought iron, of homogeneity, in masses of any size.

It is, unlike the other metals, capable of great variation in density, by the simple processes of hardening and annealing, and, therefore, of being adapted to the different degrees of elongation that it is subjected to, in either solid or built-up guns.

Bronze has greater ultimate tenacity than cast iron, but it has little more elasticity, and less homogeneity; it has a high degree of ductility, but it is the softest of cannon-metals, and is injuriously affected by the heat of high charges.

The other alloys of copper are very costly, and their endurance, under high charges, is not determined.

In view of the duty demanded of modern guns, simple cast iron is too weak, although it can be used to advantage for jackets over steel tubes—a position where mass, small extensibility, and the cheap application of the trunnions and other projections, are the chief requirements. And, although cast-iron barrels, hooped with the best high wrought iron, and with low steel, cannot fulfil all the theoretical conditions of strength, and do not endure the highest charges, they have thus far proved trustworthy and efficient.

Wrought iron, in large masses, cannot be trusted, and is, in all cases, too soft.

Bronze is soft, and destructible by heat.

Low steel is, therefore, possibly in connection with cast iron, as stated above, by reason of the associated qualities which may be called strength and toughness, the only material from which we can hope to maintain resistance to the high pressures demanded in modern warfare.

See Armour. Gunpowder. Materials of Construction, *strength of.*

Works relating to this subject:—Struensee, 'Aufangsgründe v. Artillerie,' 1788. Meineke, 'Anleitung zum Guss des Bronzirten Geschützes,' 1817. 'Report of Experiments on the Strength and other Properties of Metals for Cannon,' 4to, Philadelphia, 1856. R. Mallet, 'On the Construction of Artillery,' 4to, 1856. 'Report of Experiments upon British and Foreign Ores for the Purposes of Cast-Iron Ordnance,' fol., 1859. J. A. Longridge, 'On the Construction of Artillery,' Transactions Inst. C. E., 1860. Captain Rodman's 'Reports of Experiments on the Properties of Metals for Cannon,' 4to, Boston, 1861. 'Reports from the Select Committee on Ordnance,' 1862–63. D. Treadwell, 'On the Construction of Hooped Cannon,' royal 8vo, New York, 1864. Holley's 'Treatise on Ordnance and Armour,' 8vo, New York, 1865. Aloncle, 'Études sur l'Artillerie,' 8vo, Paris, 1866. 'Reports of the Whitworth and Armstrong Committee,' 2 vols., fol., 1866. H. L. Abbott, 'Siege Artillery in the Campaigns against Richmond,' 8vo, New York, 1868.

See also:—'Journal of the United Service Institution.' 'Aide-Mémoire to the Military Sciences.'

ASHLERING. Fr., *Maçonnerie de moëllon*; Ger., *Schalwerk, Bruchstein Mauerwerk*; Ital., *Ritti*; Span., *Obra de silleria y sillarejo.*

419.

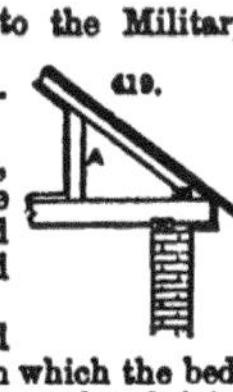

Ashlering, in carpentry, are the short, upright pieces of timber or quartering, as A in sketch, fixed in *garrets* to the *floor* and *rafters* to cut off the acute angle formed by the rafters and floor. They are usually 2½ in. thick by 4 in. wide, and from 2½ to 3 ft. long, spaced about 12 in. apart. They are lathed over and plastered as in ordinary *partitions.*

Ashlar or Ashler Work.—In masonry, where each stone is squared and dressed to given dimensions. It is usually applied to the squared stone-facing of walls in which the beds are dressed horizontal and the joints vertical and disposed at uniform distances, so as to break joint with the stone in the course above and below. The face may be worked in any way. It may be left rough from the quarry, when it is called "rock-ashlar;" or it may be dressed in a variety of ways, in which case it is called "dressed ashlar."

In the neighbourhood of London, the term ashlar is applied to a thin facing of squared stones laid in courses, with close-fitting joints, and set in fine mortar or putty. See MASONRY.

ASH-PAN. FR., *Cendrier;* GER., *Aschkasten;* ITAL., *Ceneracciolo;* SPAN., *Cenicero*

See BOILERS.

ASH-PIT. FR., *Fosse à cendre;* GER., *Aschloch;* ITAL., *Cenerario;* SPAN., *Cenicero.*

See BOILERS.

ASPHALTE. FR., *Asphalte;* GER., *Asphalt;* ITAL., *Asfalto;* SPAN., *Asfalto.*

Asphalte is a bituminous limestone found in the Jura Mountains and other localities, which is used in the formation of pavements and in the manufacture of bituminous cement.

Bitumen is found in nature in various conditions, and is met with in many parts of the world. It is supposed to be the substance mentioned in Genesis, chap. xi., ver. 3, as having been used, instead of mortar, in building the Tower of Babel; and there are numerous proofs of its having been used in ancient buildings in Egypt and Assyria. It is found, more or less pure, in large quantities washed on the shores from the surface of the Dead Sea or Lake Asphaltites, and is supposed to be derived from bituminous springs in the neighbourhood of that lake. Immense quantities of bitumen exist in the island of Trinidad, at a place called the Tar Lake, where the ground, for an unknown depth, contains so large a proportion of bitumen, that in hot weather it becomes too soft to walk upon. In some localities there are beds of shale so highly impregnated, that upon wells or pits being dug they become filled with bitumen. In other localities there are bituminous sands. In Auvergne, in France, are many beds of this description; and near Clermont bitumen exudes from the ground into a kind of wells, which have received the name of Fountain of Pitch.

It is, however, from beds of bituminous sandstone that, next to the bituminous limestone, the best description of bitumen is obtained. It is from these beds, which have been technically termed molasses, that most of the bitumen, or mineral tar, is obtained for mixing with the bituminous limestone in the manufacture of asphaltic mastic.

Bitumen is composed of carbon, hydrogen, and oxygen, in the proportion of about 85 carbon, 12 hydrogen, and 3 oxygen. The colour is a deep black, with a very slight tinge of redness. It has a peculiar aromatic odour, somewhat resembling, but still very different from that of tar and pitch. The odour is very strong when at a boiling temperature, but at ordinary temperatures it is scarcely perceptible. At a temperature under 50° Fahrenheit it is solid and brittle; from 50° to about 70° it is soft and plastic; from 70° to 90° it has a pasty consistence; from 90° to 110° or 120° it is glutinous; and above 120° it is liquid. The specific gravity is about 1·03.

The geological origin of bitumen is somewhat uncertain. The most probable hypothesis appears to be that it was produced from beds of coal while subject to heat and pressure at great depths below the surface of the earth, and that it was afterwards forced upwards through the superincumbent strata during some convulsions of nature. In its progress to, or on its arrival at the surface, it impregnated the limestone and sandstone rocks, and became mixed with the other strata in which it is now found. Here it may be necessary to observe, that the vague conjectures upon which geology is founded, and such matter as rest rather upon a speculative than a substantial, philosophical basis, are neither examined nor discussed in this work, and they receive but little of our attention.

For the purpose of obtaining bitumen, or mineral tar, the sandstone is broken into pieces about the size of the stones used for macadamizing roads, and placed in caldrons and boiled in water. In about an hour the bitumen becomes liquid and rises to the surface, and the stone falls to the bottom in grains of sand. The bitumen is then skimmed off. If, however, the sand be in very fine grains, a considerable quantity of it becomes mixed in the boiling with the bitumen, and rises with it to the surface of the water. A second operation, therefore, becomes necessary in order to render the bitumen sufficiently free from sand. For this purpose it is placed in another caldron, and heated to such a degree as to render it quite liquid. The water remaining in the skimmings soon evaporates, and the sand falling to the bottom, the pure bitumen is drawn off. In this second operation a considerable quantity of bitumen is lost in consequence of the impracticability of separating it from the sand at the bottom of the caldron. Of late years bitumen is sometimes extracted by chemical solution, and the liquid in which it has been dissolved drawn off by evaporation. A small admixture of pure sand is of very little detriment for most purposes to which bitumen is applied; but it is essential that it should be free from earthy or vegetable matter. In extracting bitumen from such soils as that in which it is found in the island of Trinidad, it is necessary to resort to complex chemical processes, and even then the result is inferior to the products of the bituminous sandstones.

Bitumen has been used from remote antiquity, and probably asphalte also may have been known to the ancients; but it does not appear to have been applied to its present uses until the beginning of the eighteenth century. The first mine of asphalte was that of the Val-de-Travers, near Neuchâtel, in Switzerland. It was discovered by Dr. d'Eyrinis, who published in the year 1721 a small volume, in which the nature and uses of asphalte are very fully explained, and its adoption for various purposes enthusiastically advocated. It was not, however, until 1838 that the first pavements of asphalte were laid in the streets of Paris.

The valuable properties of asphalte now became fully appreciated; and not only so, but they were greatly exaggerated, and the material applied to purposes for which it was not adapted. About this time asphalte produced an industrial fever not unlike the celebrated South-Sea Bubble or the more recent railway mania. Societies were formed in Paris, whose shares increased in price within a few months to ten times their original cost; and then in a very short time a reaction occurred, and one-tenth of the original cost could not be obtained for the same shares. The evils of excessive speculation disappeared in course of time, and at present the production and application of asphalte is an extensive and well-regulated branch of industry. It is much more used on the continent of Europe than in England. It is the material generally adopted for the pavements of Paris, and indeed for very many of the carriage-roads. During the period of excitement several companies were formed in England for the supply and application of asphalte, and amongst others, one promoted by Claridge, and one known as the Metropolitan Mineral Rock and Asphalte Company, which still maintain the highest character for excellence of materials and workmanship.

The best description of asphalte consists of a limestone composed of nearly pure carbonate of lime, impregnated with from six to ten parts by weight of bitumen. The principal source of supply at the present time is from the Jura Mountains, in the neighbourhood of Seyssel, on the west bank of the river Rhone.

The colour is a dark chocolate, approaching to black. In cold weather the stone is hard, and easily broken; but in warm weather it softens under the blow of the hammer. The specific gravity is about 2·25.

The bituminous limestone, after being quarried, is broken to a size that would pass through a ring about 2½ inches in diameter. It is then reduced to powder, either by grinding, or by crushing, or by exposure to heated steam. The first method is generally preferred; but when the stone is hard and deficient in bitumen, it is more convenient to crush it between revolving cylinders; on the other hand, with stone which is soft and contains an excess of bitumen, the hot vapour process may be more advantageous, especially in warm weather. The powder requires to be sifted, and all the large particles again passed through the mill, or otherwise reduced to the requisite degree of fineness. The next process is to place the powder in a caldron, in which a small quantity of pure bitumen has been previously melted, and to bring it into a state of fusion and stir it by machinery until the whole is thoroughly mixed, when it is drawn off and run into moulds to form blocks of about 1 cwt. These blocks, when cold, are ready for exportation as asphaltic mastic. The finest mastic contains 87 per cent. of the carbonate of lime and 13 per cent. of bitumen, but several qualities of mastic are prepared by adding to the other ingredients various proportions of very coarse-grained sand, termed "grit." The mastic requires to be remelted on the spot, where it is to be used in the laying of pavements or other works; and for this purpose portable caldrons are provided, and a small quantity of the mineral tar is added, as a flux, with every melting.

Mastic of pure asphalte is seldom used without some admixture of grit, a certain proportion of which adds to its power of enduring wear and tear, and of resisting the tendency to soften at high temperatures. Asphalte containing about 15 per cent. of fine grit is best adapted for covering roofs, and that containing about 25 per cent. of coarse grit for the laying of footpaths. Asphalte pavements should not become appreciably soft when subjected to a temperature of 160° Fahrenheit.

The principal uses of asphalte are for covering roofs; for protecting underground vaults, magazines, railway-arches, &c., from the percolation of water; for pavements; and for damp-courses, to prevent moisture from rising in walls by capillary attraction. A layer of asphalte ½-inch in thickness may be considered impervious to moisture, and, therefore, an effectual damp-course. As a covering to roofs, it is exposed to variations of temperature, which are destructive to cement and most other compositions, and frequently cause cracks even in lead; but good asphalte, ¾-inch in thickness, firmly supported on a thick bed of concrete, will form a substantial water-tight covering through all alternations of climate. As a pavement it has the advantage of being free from joints, of presenting a very smooth surface, and of being capable of enduring a great amount of wear. A good paving of asphalte, ¾-inch thick, will last longer under heavy traffic than Yorkshire flagging 3 in. thick. As a covering for underground arches it may not be considered very much superior to Portland cement, but it has the advantage of possessing a certain degree of elasticity, which renders it less liable to crack upon slight settlements occurring in the brickwork or masonry underneath it. One great recommendation for asphalte is, that it is easily repaired; and that the materials from old work may be melted and, with a small addition of mineral tar, re-used in new work.

The dark colour of asphalte pavements is somewhat to its disadvantage; and the closeness of its texture, and its property of readily condensing on its surface the moisture of a damp atmosphere, renders it unsuitable for the floors of inhabited apartments. Its property of becoming soft at a high temperature renders it unfit for floors of furnace-rooms, forges, &c., or for any close proximity to fires. On the other hand, it is to be observed that, though liable to melt, it never takes fire, and that it is, therefore, well adapted for fire-proof construction.

Asphalte has been used as a cement in the brickwork of tanks, &c.; but for such purposes it appears to possess very little, if any, advantage over Portland cement, which is a much cheaper material. It has very little adhesion for brick or stone; and, consequently, however superior the asphalte itself may be in regard to strength and impermeability, it does not, in combination, form a solid homogeneous mass of masonry. The absence of adhesion also prevents it from being applicable as a covering or lining to vertical surfaces.

Good asphalte is not much affected by the ordinary variations of temperature; but still it becomes somewhat soft under a summer sun, and is liable to slide down if laid upon a steep incline. Asphalte roofs and pavements should, therefore, be laid nearly horizontal, and when steep inclines are unavoidable, they should not be more than about two or three feet in width. Vertical surfaces above 6 inches in height should be avoided when practicable; but, with care, greater heights may be executed with safety, especially in short lengths.

The preparation of asphaltic mastic and mineral tar, as above described, is evidently somewhat expensive; and the carriage of the material adds greatly to the cost, especially in this country, which has no source of supply nearer than the Jura Mountains. The bitumen is known to be very similar to some cheaper substances, and, among others, to common coal-tar; and carbonate of lime of similar quality to that of the bituminous limestone is easily obtainable in almost every locality. There is, therefore, every inducement to endeavour to produce an artificial compound, which may supply the place of natural asphalte. The effort to invent such a compound merits approbation, and has been successfully made by the Metropolitan Asphalte Company. Up to the present time, every artificial asphalte has proved to be a failure in cases where strength was required, or where exposed to great alternations of temperature. Preparations of coal-tar have been advantageously employed in protecting walls, arches, &c., from damp, when the artificial asphalte could be itself protected from the weather and the air excluded; and tar pavements are extremely cheap, and well adapted for many situations.

ASSAYING. Fr., *Essai des Métaux*; Ger., *Probirkunst*; Ital., *Saggio dei Metalli*; Span.. *Ensayo de metales.*

The term "assaying" is frequently used in the general sense of chemical analysis; but, strictly, it is only applicable to that mode of separating metals from their ores, or *gold* and *silver* from the baser metals, in which no wet reagents, generally speaking, are employed, and the action of heat is called into play. We shall, in this article, give concise methods for enabling any one to detect in commercially valuable minerals and ores those constituents of which they are composed.

The forms of *blowpipe* generally used for assaying are shown in Figs. 420, 421. They consist of a tube made of *brass* or of *german-silver*, bent near the end, and terminated with a finely-pointed nozzle. The best form of blowpipe is represented in Fig. 422. The tube and nozzle are made of

420.

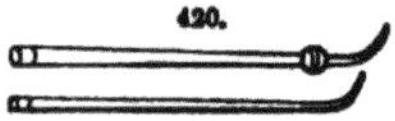

421.

422.

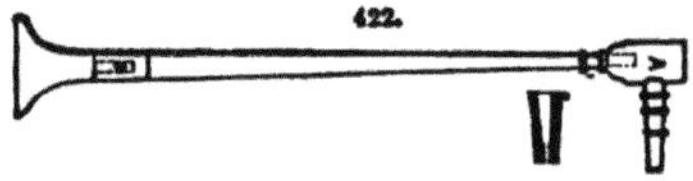

the same material as the common blowpipe, the point of platinum, and the mouthpiece of horn, wood, or ivory. The air-chamber A serves to partially regulate the blast, and receives the tube and nozzle, which are ground to fit accurately, each of these pieces being movable. The point, which enters the flame, as before observed, is made of platinum.

In using the blowpipe, the lips are pressed against the mouthpiece, and the stem firmly held; the cheeks are inflated with air, which is expelled from the mouth through the pipe, by contracting the muscles of the cheeks, care being taken to inhale only through the nostrils; by this means a continuous flame is kept up.

When a flame is propelled by a current of air blown into or upon it, the flame produced may be divided into two parts, possessing respectively the properties of reduction and oxidation. The reducing-flame is produced by a weak current of air acting upon the flame of a lamp or candle; the carbon contained in the flame is thus brought in contact with the substance to be examined, which it reduces. The oxidizing-flame is formed by blowing strongly into the interior of the candle-flame. Combustion is thus thoroughly effected; and if a small piece of an oxidizable body be held at the point of the flame, the former speedily acquires an intense heat, and combines freely with the oxygen of the surrounding air. The substance to be analysed should, when exposed to the flame of the blowpipe, be supported upon some infusible, and, in many cases, incombustible material.

When it is required to reduce an oxidized substance, to fuse a body without oxidizing it, or to oxidize a body on which the reducing action of carbon alone is unimportant, that body is placed in a small hollow in a piece of charcoal. The best kind of charcoal for this purpose is made from closely-grained pine-wood, free from knots, and should be cut by a small, fine saw into convenient pieces.

For holding in the flame substances which would be affected by charcoal, platinum-wire, 0·012 in. diameter, is formed into a small hook. The hook is heated and dipped into borax or microcosmic salt, which adheres to it, forming a small globule in which the substance to be tested is placed. Platinum-foil is used for the same purposes as the wire. Platinum spoons, shaped as in Fig. 423, are used for fusing the mineral with reagents, as carbonates of potash and soda,

423.

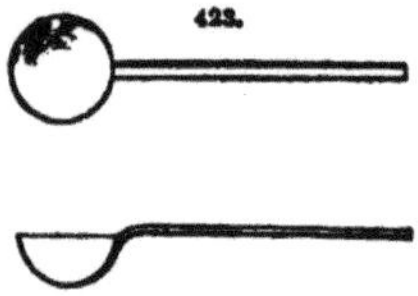

424.

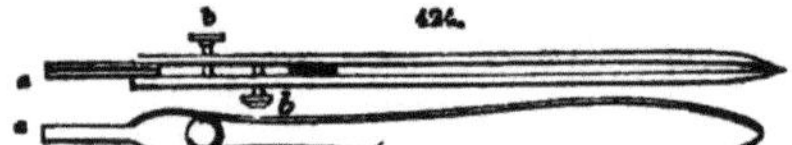

bisulphate of potash, and saltpetre. When the substance can be determined by the colour it gives to the blowpipe-flame, it is held in the latter by brass or steel forceps with platinum tips, Figs. 424, 425. To take up the mineral, the knobs *b b*, Fig. 424, are pressed, the platinum points *a*, *a*

425.

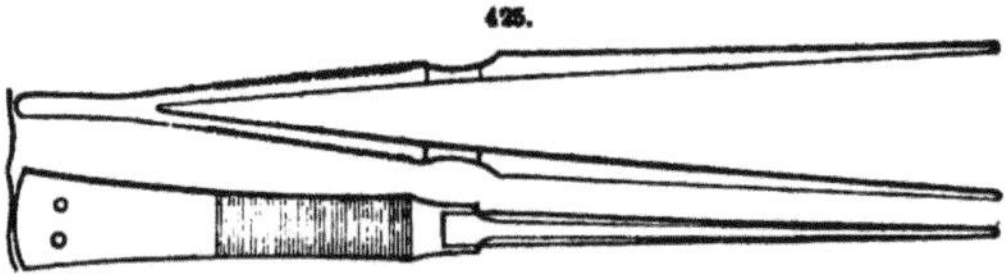

then open, and close, when required, by their own elasticity. For manipulating in acids, forceps with glass points, Fig. 426, are used.

Glass tubes of various diameters, in lengths of 5 or 6 in., open at both ends, are used for roasting substances containing *sulphur*, selenium, *arsenic*, *antimony*, and tellurium. These, when heated with an access of air, evolve characteristic fumes. They are generally heated by a spirit-lamp. Small test-tubes are also required, in order to detect the presence of water, mercury, or other bodies which are volatilized by heat without access of air.

The reagents most commonly used in assaying are carbonate of soda, borax, and microcosmic salt. The carbonate of soda must be anhydrous and perfectly pure. It is chiefly used to reduce metallic oxides and sulphides, to decompose silicates, and to determine the fusibility of different substances.

426.

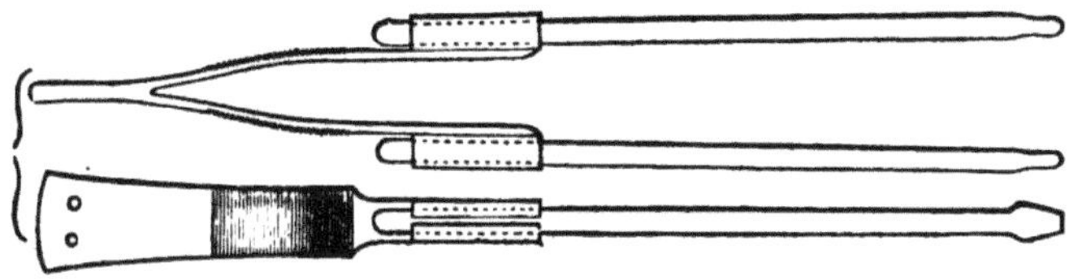

Pure borax is heated below its melting-point to expel its water of crystallization, and is then pulverized. In using borax, a small quantity is formed into a bead on the end of a platinum-wire, to this bead is then added a minute quantity of the powdered substance to be examined. The whole is then held in the blowpipe-flame, and the following results observed:—whether the borax dissolve the substance or not; the colour of the bead formed in the oxidizing and reducing flames respectively; and whether the colour of the bead alter when cooling. Only sufficient of the substance should be added to give a colour to the bead; if this be too intense to be clearly distinguished, more borax may be added. When microcosmic salt, which is a combination of phosphate of soda and ammonia, is used, it should be fused upon platinum-foil, to expel the water and excess of ammonia contained in it. It is then used upon platinum-wire in the same way as borax.

The following reagents are required in certain cases:—nitrate of potash, also called saltpetre, for oxidizing certain substances by fusing with them either on platinum-foil or in the platinum spoon; bisulphate of potash, for eliminating certain volatile matters, as lithia, boracic acid, hydrofluoric acid, bromine, iodine, also for decomposing salts of titanic, tantalic, or tungstic acids; nitrate of cobalt, chemically pure and in solution, for detecting the presence of alumina, magnesia, oxide of zinc, oxide of tin, and titanic acid, which, when moistened with this reagent and strongly heated, assume certain characteristic colours; silica, for various purposes; fluoride of calcium, known as fluor-spar, which, mixed with bisulphate of potash, is used for ascertaining the presence of lithia and boracic acid; oxide of nickel or oxalate of nickel, which latter is a salt of oxide of nickel with oxalic acid, for the detection of potash in large quantity in salts which also contain soda and lithia; protoxide, black oxide, of copper, for detecting chlorine, bromine, and iodine; tin-foil, for reducing various metallic oxides dissolved in borax or microcosmic salt (the hot fluid is touched on charcoal with a piece of tin-foil, and then strongly heated for some seconds under the reducing-flame); fine silver, for discovering sulphur and sulphuric acid. The reagents should be kept in glass-stoppered bottles.

In addition to the apparatus already described, the following articles are desirable, though, with some exceptions, not indispensable; a steel hammer, a small anvil, a steel crushing-mortar, an agate mortar, two or three files, a pair of scissors, a magnet, a pocket-lens, some porcelain capsules, a spirit-lamp, and a good pocket-knife; blue litmus-paper, turmeric-paper, small quantities of strong sulphuric, nitric, and hydrochloric acids, and a few glass rods.

Methods of Analysis.—*In the Test-tube.*—The tube being thoroughly dry and clean, a small portion of the pulverized substance under examination is placed in it, and heated over a spirit-lamp until the glass softens; it must then be noticed whether any vapour or sublimate is collected in the upper portion of the tube. This vapour may be water, mercury, sulphur, selenium, tellurium, or arsenic. If the product be liquid, its alkaline or acid reaction should be tested by litmus-paper. Organic substances may be detected by their odour. Quicksilver can be discovered in the sublimate by means of a lens. The sublimate of selenium is reddish brown, of tellurium grey, and of arsenic black; that of the latter being sometimes metallic. If these substances do not appear as sublimates, it must not be concluded that they are not present, as they may exist in combinations not readily destroyed by heat alone. Oxygen and ammonia are sometimes evolved; the former may be recognized by introducing an incandescent splinter of wood, which will immediately burst into a flame; and the latter by its alkaline reaction upon moistened red litmus-paper. Often, however, ammonia exists in such a state of combination, that heat alone will not disengage it. When any substance is supposed to contain such a combination, it must be mixed with caustic soda, or caustic lime, and heated in a clean test-tube, when free ammonia will escape.

In the Open Tube.—The substance in a state of powder is placed in the tube half an inch from one end, and heated by degrees, the tube being slightly inclined in order to produce a current of air. The constituents of the substance thus combine with the oxygen of the air, and are volatilized. Sulphur forms sulphurous acid, which is detected by its pungent smell. Selenium forms a steel-grey deposit, and also a vapour, characterized by its smell. Arsenic volatilizes as arsenious acid, antimony as antimonious acid, and tellurium as tellurous acid, all forming dense white fumes. The deposit from arsenic is crystalline, from the others amorphous; that from tellurous acid forms small beads. The tubes used should be made of difficultly fusible German or Bohemian glass.

On Charcoal.—The action of most substances when heated on charcoal is similar to their action in the test-tube. The experimenter should make himself familiar with the incrustations formed by different substances when heated on charcoal. We will only describe the action of those substances which are of practical importance.

Reactions of some of the more common Metallic Oxides, with Borax and with Microcosmic Salt, on Platinum-Wire, in both Flames.

Metallic Oxides.	With Borax.		With Microcosmic Salt.	
	Oxidizing-Flame.	Reducing-Flame.	Oxidizing-Flame.	Reducing-Flame.
Oxide of Manganese, Mn_2O_3.	Affords an intense amethyst colour, which on cooling becomes violet. A large quantity of the oxide produces an apparently black bead, which, however, if pressed flat, is seen to be transparent.	The coloured bead becomes colourless. With a large amount of the oxide, this reaction is best obtained upon charcoal, and is facilitated by the addition of tin-foil.	With a considerable quantity of oxide, an amethyst colour is obtained, but never so dark as in borax. A small quantity of oxide only having been taken, a colourless bead is obtained, in which, however, the amethyst colour may be brought out by the addition of a little nitre. While the bead is kept fused, it froths and gives off bubbles of gas.	The coloured bead immediately loses its colour, either on platinum-wire or on charcoal. After the reduction the fluid bead remains still.
Oxide of Iron, Fe_2O_3.	With a small proportion of oxide, the bead is of a yellow colour while warm, and colourless when cold; with a larger proportion, red while warm, and yellow when cold; and with a still larger amount, dark red while warm, and dark yellow when cold.	Treated alone on platinum-wire, the bead becomes of a bottle-green colour, and if touched with tin, it becomes a pale sea-green. On charcoal with tin, it assumes at first a bottle-green colour, which by continued blowing changes to a sea-green.	With a certain amount of oxide, the bead is of a yellowish-red colour, which on cooling changes to yellow, then green, and finally becomes colourless. With a large addition of oxide, the colour is, when warm, dark red, and passes, while cooling, into brownish-red, dark green, and finally brownish-red. During cooling, the colours change more rapidly than is the case with borax.	With a small proportion of oxide there is no reaction. With a larger amount, the bead is red, while warm, and becomes, on cooling, successively yellow, green, and russet. With the addition of tin the bead becomes, during cooling, first green, and then colourless.
Oxide of Cobalt, CoO.	Colours the bead of an intense blue, both whilst hot and when cold. When much oxide is present, the colour is so deep as to appear black.	As in the oxidizing-flame.	As with borax, but less intensely coloured. During cooling the colour becomes somewhat paler.	As in the oxidizing-flame.
Oxide of Nickel, NiO.	Colours intensely. A small amount of oxide affords a bead which, while warm, is violet: and becomes of a pale reddish-brown on cooling. A larger addition produces a dark violet colour in the warm, and reddish-brown in the cold bead.	The oxide is reduced, and the metallic particles give the bead a turbid grey appearance. If the blast be continued, the metallic particles fall together without fusing, and the bead becomes colourless. This reaction is readily obtained with tin upon charcoal, and the reduced nickel fuses to a bead with the tin.	Dissolves into a reddish bead, which becomes yellow on cooling. With a large addition of the oxide, the bead is brownish while hot, and orange when cold.	On platinum-wire the nickeliferous bead undergoes no change. Treated with tin upon charcoal, it becomes at first opaque and grey; and, after long-continued blowing, the reduced nickel forms a bead, which remains colourless.

Oxide of Zinc, $Zn\,O$.	Dissolves easily into a clear, colourless bead, which, when much oxide is present, may be rendered opaque and flocculent by an intermittent flame, and become so spontaneously with a still larger addition of the oxide. When a considerable quantity is dissolved, a bead is obtained, which is pale yellow while hot, and colourless when cold.	On platinum-wire the saturated bead becomes at first opaque and grey, but by a sustained blast is again rendered clear. On charcoal the oxide is gradually reduced; the metal is volatilized, and encrusts the charcoal with oxide.	As with borax.	As with borax.
Oxide of Lead, $Pb\,O$.	Dissolves readily to a clear yellow bead, which loses its colour upon cooling, and, when containing much oxide, can be rendered dull under an intermittent flame. With a still larger addition of oxide, it becomes opaque yellow on cooling.	The plumbiferous bead spreads out on charcoal, becomes turbid, bubbles up, until the whole of the oxide is reduced, when it again becomes clear. It is, however, difficult to bring the lead together into a bead.	As with borax; but a larger addition of oxide is required to produce a yellow colour in the warm bead.	On charcoal the plumbiferous bead becomes grey and dull. With an excess of oxide, a part is volatilized, and forms an incrustation on the charcoal beyond the bead. The addition of tin does not render the bead opaque, but somewhat more dull and grey than in its absence.
Oxide of Tin, $Sn\,O_2$.	In small quantity dissolves slowly into a clear, colourless bead, which, when cold, remains clear, and cannot be rendered opaque with an intermittent flame. If a saturated bead, which has been allowed to cool, be reheated to incipient red heat, it loses its rounded form, and exhibits imperfect crystallization.	A bead containing but little oxide undergoes no change. If much of the latter be present, a part may be reduced upon charcoal.	In small quantity dissolves very slowly to a colourless bead, which remains clear on cooling.	The bead undergoes no change, either on charcoal or platinum-wire.
Oxide of Bismuth, $Bi\,O_3$.	Dissolves readily to a clear bead, which, with a small amount of the oxide, is yellow while warm, and becomes colourless on cooling. With a large addition, the bead is, in the hot state, of a deep orange colour, which changes to yellow, and finally becomes opaline, in process of cooling.	The bead becomes at first grey and turbid, then begins to effervesce, which action continues during the reduction of the oxide, and it finally becomes perfectly clear. If tin be added, the bead becomes at first grey, from reduced bismuth, but, when the metal is collected into a bead, the bead is again clear and colourless.	Dissolves in small quantity to a clear, colourless bead. A large addition of oxide affords a bead which, while warm, is yellow, and becomes colourless on cooling. When in sufficient proportion, the bead may be rendered opaque, and, under an intermittent flame, a still larger addition of oxide renders the bead spontaneously opaque on cooling.	On charcoal, and especially with the addition of tin, the bead remains colourless and clear while warm, but becomes, on cooling, of a dark grey colour, and opaque.

REACTIONS OF SOME OF THE MORE COMMON METALLIC OXIDES, WITH BORAX AND WITH MICROCOSMIC SALT, ON PLATINUM-WIRE, IN BOTH FLAMES—*continued.*

Metallic Oxides.	With Borax.		With Microcosmic Salt.	
	Oxidizing-Flame.	Reducing-Flame.	Oxidizing-Flame.	Reducing-Flame.
Oxide of Copper, Cu O.	Produces an intense colouration. If in small quantity, the bead is green, while warm, and becomes blue on cooling. If in large proportion, the green colour is so intense as to appear black. When cool, this becomes paler, and changes to a greenish-blue.	If not too saturated, the cupriferous bead soon becomes nearly colourless, but immediately on solidifying assumes a red colour, and becomes opaque. By long-continued blowing on charcoal the copper in the bead is reduced, and separates as a small metallic bead, leaving the remainder colourless. With the addition of tin, the bead becomes of an opaque, dull red, on cooling.	With an equal proportion of oxide as applied to borax this salt is not so strongly coloured. A small amount imparts a green colour in the warm, and a blue in the cold. With a very large addition of oxide, the bead is opaque in the hot state, and, after cooling, of a greenish-blue.	A tolerably saturated bead assumes a dark green colour under a good flame, and, on cooling, becomes of an opaque brick-red, the moment it solidifies. A bead containing but a small proportion of the oxide becomes equally red and opaque on cooling, if treated with tin upon charcoal.
Oxide of Mercury, Hg O.	No reaction.	No reaction.	No reaction.	No reaction.
Oxide of Silver, Ag O.	The oxide is partly dissolved, and partly reduced. In small quantity it colours the bead yellow while warm, the colour disappearing on cooling. In large quantity the bead is yellow while warm, but during cooling becomes paler to a certain point, and then again deeper. If re-heated slightly, the bead becomes opalescent.	On charcoal the argentiferous bead becomes at first grey, from reduced metal, but afterwards, when the silver is collected into a bead, it becomes clear and colourless.	Both the oxide and metal afford a yellowish bead, which, when containing much oxide, becomes opaline, exhibiting a yellow colour by daylight, and a red one by artificial light.	As with borax.
Oxide of Antimony, $Sb\ O_3$.	Even when in large proportion dissolves to a clear bead, which is yellow while hot, but almost entirely loses its colour on cooling. On charcoal the antimonious acid may be almost expelled, so that tin produces no further change.	A bead that has only been treated for a short time in the oxidizing-flame, when submitted to the reducing-flame becomes grey and turbid, from the reduced antimony. This soon volatilizes, and the bead again becomes clear. The addition of tin renders the bead ash-grey, or black, according to the amount of oxide it contains.	Dissolves, with ebullition, to a bead of a pale yellow colour while warm.	On charcoal the saturated bead becomes at first dull, but as soon as the reduced antimony is volatilized it again becomes clear. With tin the bead is at first rendered grey by the reduced antimony, but by continued blowing is restored to clearness. Even when the bead contains but little oxide, tin produces this reaction.

Arsenic, when heated upon charcoal by the blowpipe-flame, covers the former with a coating, white in the centre and grey at the edges, of arsenious acid. This coating is immediately volatilized, when brought in contact with the flame, and gives off the odour of garlic which characterizes arsenic. The vapour evolved is highly poisonous, and should not be inhaled. Metallic arsenic dissolves readily in nitric and hydrochloric acids; in the first case, if heat be applied and an excess of acid used, arsenic acid will be formed; and in the second, arseniuretted hydrogen, a very poisonous gas, is evolved, leaving chloride of arsenic.

Antimony melts easily, coating the charcoal under both the oxidizing and reducing flames, with an incrustation—white where thick, and bluish where thin—of antimonious acid. Antimonious acid is less volatile than arsenious acid, and tinges the reducing-flame blue; but is simply melted by the oxidizing-flame. Antimonious acid, when moistened with a solution of nitrate of cobalt, and gradually brought to a high temperature in the oxidizing-flame, after cooling, presents a dusky green appearance. The best solvent for antimony is aqua regia, nitro-muriatic acid, which converts it into chloride of antimony.

Bismuth melts readily, and coats the charcoal, under both flames, with its oxide. The colour of this coating resembles that of an orange, and becomes paler on cooling. The edges of the oxide, which have been more exposed to the action of the charcoal, become converted into carbonate of bismuth, which is white. By applying either flame the oxide is driven from place to place, being first reduced by the charcoal to metallic bismuth, which is volatilized and re-oxidized. The colour of the flame undergoes no alteration. Bismuth dissolves in nitric acid, from its solution in which it may be precipitated as a white sediment by dilution with pure water.

Copper.—This metal, when unalloyed, melts easily before the blowpipe. When placed in the oxidizing-flame it becomes coated with black oxide of copper, while the flame is strongly tinged with green. Metallic copper is easily obtained from its oxide in the reducing-flame, without incrusting the charcoal. Many compounds of copper may be reduced to the metallic state by mixing them with carbonate of soda, and then heating in the reducing-flame. Copper dissolves readily in nitric acid, giving off nitrous fumes, and forming a deep azure-blue solution on the addition of ammonia. A polished surface of iron or steel, immersed in a solution of copper, soon becomes coated with this metal.

Gold melts easily before the blowpipe, is not acted on by fluxes, and is soluble in aqua regia.

Lead fuses readily, covering the charcoal with oxide of a dark yellow colour, which becomes paler on cooling. Beyond the oxide, carbonate of lead is formed, of a bluish-white colour. The oxide, when heated in the oxidizing-flame, acts in the same manner as the oxide of bismuth; but in the reducing-flame it volatilizes, tinging the flame blue. Lead readily dissolves in nitric acid; and its oxide, litharge, is soluble even in vinegar.

Platinum.—Infusible, not affected by borax or microcosmic salt, except in a state of fine dust, when reactions for iron or copper, which occur in small quantities, as impurities, take place. It is soluble only in boiling aqua regia.

Silver, when fused alone upon charcoal, covers it with a thin coating of dark brown oxide. If lead be present, it first forms a yellow oxide; then, as the silver becomes purer, the silver forms a dark red deposit beyond. Antimony, when present, forms a white crust of antimonious acid, which, on further exposure to the heat, becomes red on the exterior. If antimony and lead are simultaneously present in the silver, a crimson incrustation forms upon the charcoal after the former metals have been volatilized. Rich silver ores sometimes produce the same result, when fused upon charcoal. Silver dissolves readily in nitric acid, and may be re-deposited by a plate of copper.

Tin melts readily, and oxidizes in the inner flame. The melted metal, when exposed to the reducing-flame, becomes covered, as well as the charcoal, with oxide, which is pale yellow while hot, and becomes white when cool. This oxide cannot be reduced by either flame. The best solvent for tin is hydrochloric acid; nitric acid oxidizes this metal, but has no effect upon its oxide.

Zinc melts with facility, and burns with a bright greenish-white flame in the oxidizing-flame. The product of this combustion is oxide of zinc, evolved in dense white fumes, which coat the charcoal. This coating, while hot, is yellow, and turns white on cooling; it shines brilliantly when heated with the oxidizing-flame, but cannot be volatilized. The reducing-flame volatilizes it but slowly. Zinc is readily soluble in dilute sulphuric acid, hydrogen being evolved, and sulphate of zinc formed.

In Platinum Forceps.—If the substance to be examined does not attack platinum, a small fragment held in the forceps is exposed to the oxidizing-flame; if its action upon platinum is feared, it should be placed upon charcoal or refractory porcelain. In this method of examination the substance is recognized by the colour it imparts to the flame. The following are a few substances of frequent occurrence classified according to the colours which they give to the blowpipe-flame:—

Blue Flame.	Green Flame.	Red Flame.	Violet Flame.
Arsenic .. light.	Copper .. emerald.	Lime .. purplish.	Potash .. clear.
Antimony .. greenish.	Baryta .. pale.	Lithia .. crimson.	
Bromide of copper } mixed with green.	Boracic acid dark.	Strontia .. dark crimson.	
Chloride of copper } intense.	Ammonia .. very dark.		
Lead .. .. pale, clear.	Iron .. .. dark.		
Selenium .. azure.	Iodide of copper } intense emerald.		
	Phosphoric acid } pale.		
	Zinc .. .. very pale.		

With Borax.—As this method serves to distinguish metallic oxides, all substances containing unoxidized metals must be previously roasted, in order to convert them into oxides. The same treatment is necessary when microcosmic salt is substituted for borax.

With Carbonate of Soda.—The substance to be analyzed is powdered and made into a paste with carbonate of soda and water. A small portion of the paste is then gradually heated upon the charcoal until the temperature is as high as possible. Three reactions may then take place: either the substance will fuse with effervescence, it will be reduced, or the alkali will be absorbed into the charcoal, leaving the substance on the surface unchanged. Silica, titanic and tungstic acids fuse with effervescence.

The oxides of gold, silver, tungsten, antimony, arsenic, copper, mercury, bismuth, tin, lead, zinc, iron, nickel, and cobalt, when mixed with carbonate of soda and heated upon the charcoal in the reducing-flame, are reduced. Lead, zinc, antimony, and bismuth, volatilize partially, forming incrustations on the charcoal. Mercury and arsenic are volatilized as soon as reduced, leaving no marks upon the charcoal.

That part of the charcoal upon which the assay has rested must be pulverized in a mortar, when any metal which may be contained in it will be found in the form of a shining metallic powder, if brittle, or flakes, if malleable.

Sulphur may be detected by heating the substance with double its weight of carbonate of soda, upon charcoal, in the reducing-flame. The assay and that portion of the charcoal which has absorbed any alkali are powdered and placed upon a moistened surface of polished silver, which, if sulphur be present, receives a black stain.

Manganese is detected by fusing the substance with carbonate of soda and nitre, upon platinum, in the oxidizing-flame. The bead thus formed is of a turquoise colour when cool.

In order to detect quantities of phosphorus too minute to give any reaction in the blowpipe-flame, part of the substance is pulverized with five times its bulk of a mixture of 3 parts carbonate of soda, 1 nitrate of potash, and 1 silica, and the whole fused in a platinum spoon or crucible. The resulting mass is mixed with water, filtered, and a few drops of carbonate of ammonia added; the silica is precipitated by boiling, and removed by filtration. A small quantity of acetic acid is then added to the filtrate, which is boiled to expel the carbonic acid, and to which pure nitrate of silver is added. If phosphorus or phosphoric acid be present, a yellow precipitate appears; if, on the contrary, the solution contain arsenic, or any compound of that metal, the precipitate is of a reddish-brown colour.

Assay of Fuel.—To estimate approximately the amount of carbon in any particular fuel, a portion of the fuel should be dried, weighed, and heated in a platinum crucible until further increase of temperature causes no reduction in weight. The difference between the weight of the remaining ashes and that of the substance previous to heating gives the desired result.

Assay of Gold and Silver.—In assaying gold, the metal is wrapped, with three times its weight of fine silver and twelve times its weight of pure lead, in a piece of thin paper, and melted in a bone-ash *cupel*, Fig. 427. This cupel is either heated in a muffle, shown at *m*, Fig. 428, or by the oxidizing-flame of a blowpipe. The lead and copper become oxidized, the fused oxide of lead dissolves that of the copper, and both are absorbed by the cupel, leaving the gold and silver combined in the form of a button. This button should be rolled into a thin plate and boiled with nitric acid, spec. grav. 1·18, which extracts the greater part of the silver. The remainder is then washed with pure water, and boiled in nitric acid, spec. grav. 1·28, to extract the last traces of silver; after which it is washed, heated to redness in a crucible, and weighed.

427.

428.

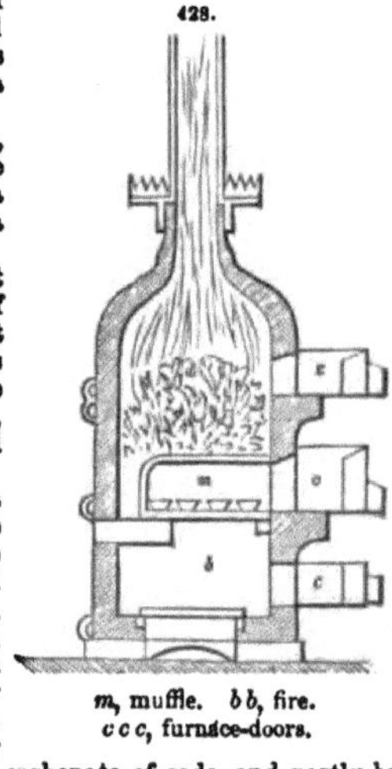

m, muffle. *b b*, fire.
c c c, furnace-doors.

The assay of silver is generally conducted by the wet process, and is based on the fact that chloride of silver is an insoluble salt, and that a solution of common salt can be made of such a strength as to precipitate a certain weight of pure silver from a solution of that metal in nitric acid.

Lead may be extracted from galena, its sulphide, and its most common ore, by mixing 300 grains of galena with 450 grains of dried carbonate of soda and 20 grains of charcoal, and placing it in a crucible with two large iron nails, heads downwards. This crucible is covered, and heated moderately for half-an-hour. The remainder of the nails is carefully removed from the liquid mass, which is then allowed to cool, the crucible broken, and the lead extracted and weighed.

To ascertain if it contains silver, the button is placed in a small bone-ash cupel, heated in a muffle, until the whole of the lead is oxidized and absorbed by the bone-ash the cupel is made of, leaving the minute globule of silver. Small globules of lead may be conveniently cupelled on charcoal before the blowpipe, by pressing some bone-ash into a cavity scooped in the charcoal, placing the lead upon its surface, and exposing it to a good oxidizing-flame as long as it decreases in size. If any copper be present, the bone-ash will show a green stain after cooling. Pure lead gives a yellow stain. In the above process the sulphur of the lead ore, galena, is removed partly by the sodium of the carbonate of soda, and partly by the iron of the nails, the excess of carbonate of soda serving to flux any silica with which the galena may be mixed. (See Articles on the various Metals.)

Works relating to Assaying:—Berthier, 'Traité des Essais par la voie sèche,' 2 vols., 8vo, Paris, 1848. 'Records of Mining and Metallurgy,' by J. Arthur Phillips and John Darlington, crown 8vo, 1857. Scheerer and Blandford, 'An Introduction to the Use of the Mouth Blow-pipe,' 12mo, 1864. J. Silversmith's 'Handbook for Miners, Metallurgists, and Assayers,' 12mo, New York, 1866. G. Kurstel, 'Nevada and California Processes of Gold and Silver Extraction,' 8vo, San Francisco, 1866. B. Kerl, 'Metallurgische Probirkunst,' royal 8vo, Leipzig, 1866. J. Arthur Phillips, 'Mining and Metallurgy of Gold and Silver,' royal 8vo, 1867. Mitchell's 'Manual of Practical Assaying,' 3rd edition, 8vo, 1868.

ATOM. Fr., *Atome;* Ger., *Atom;* Ital., *Atomo;* Span., *Atomo.*

Bodies are not composed of one continuous substance, but—as is evidenced by their porosity and their faculty of increasing or diminishing their volume under certain influences, and even of changing their actual state—they are formed of an aggregation of small particles, called *molecules*, placed at specific distances from each other, and maintained in equilibrium by the powers of attraction and repulsion which they reciprocally exercise.

These molecules, however, are not the final limit of subdivision of which matter is susceptible. By bringing other forces into play, it is possible, in most cases, to divide them into yet smaller masses.

It is to these last that the name of *atom* has been given.

We have said "*in most cases*," because there are some exceptional ones when the molecule is *not* divisible.

The bodies to which they appertain are then said to have an atom and a molecule that are homologous.

By knowing the atomic weights of all the simple bodies, and the molecular weights of either the elements or the compounds which they form, we arrive at far more correct notions regarding the constitution of bodies, than by trusting solely to the rough fact of equivalents. See Molecule.

How the first of these are attainable will be shown in the next article.

ATOMIC WEIGHTS. Fr., *Poids atomiques;* Ger., *Atomgewicht;* Ital., *Pesi atomichi;* Span., *Pesos atomicos.*

Higgins and Dalton were the first to think of explaining chemical combinations by the hypothetical juxtaposition of atoms. Dalton argued that, those atoms being insecable, the various quantities of a body A, which unite with an invariable quantity of another body B, must bear to each other ratios that are rational and commensurable. From this atomical hypothesis he then deduced, *a priori*, the law of multiple proportions, which, after receiving the sanction of experience, has become one of its most solid foundations.

The atomical theory affords a satisfactory explanation of equivalents, that is to say, of the fact that bodies enter into combinations in quantities bearing the same ratio, though they vary between themselves.

For example, let us suppose the weight of one atom of potassium to be 39 times greater than that of one atom of hydrogen, and that one atom of the one or the other of those bodies to one atom of chlorine is required in order to form a definite combination. As the weight of the atom of chlorine remains the same in both cases, it is evident that, to saturate it, it will take 39 times the weight of potassium that would be required of hydrogen. Moreover, as these proportions cannot alter, although, instead of two atoms, an indefinite number enter into combination, the result is, in general, that to saturate any given quantity of chlorine, it will take 39 times more potassium than hydrogen. This is what is meant when we say that the relative equivalent of potassium to that of hydrogen, taken as unity, is 39. In the atomical theory, the equivalents of bodies thus become the weight of their atom compared to the weight of an atom of hydrogen taken as unity, and are known under the name of *Atomic Weights.*

The notion of atomic weight carries with it, however, something more precise than that of equivalent. It is another ratio, but one more fully determined.

For instance, let us suppose an atom of oxygen to play the same part as an atom of hydrogen, that the two bodies, in short, may be substituted the one for the other, and atom for atom; as experiment proves that 8 parts, in weight, of oxygen go to 1 of hydrogen, we are bound to conclude that the atom of oxygen weighs 8 times heavier than that of hydrogen;—that the atomic weight of oxygen is 8.

We will now assume that it takes 2 atoms of hydrogen to replace one of oxygen. As one of hydrogen is replaced by 8 of oxygen, 2 of hydrogen will require 16 of oxygen; this will lead us to the admission that the atom of oxygen weighs 16 times heavier than that of hydrogen;—that the atomic weight of oxygen is 16.

It thence follows that—according as the atom of oxygen is substituted for 1 or for 2 atoms of hydrogen—the atomic weight of the first of those two bodies is 8 or 16; whereas the equivalent—which only represents a simple relation of weight, irrespective of atoms—remains always equal to 8.

It is necessary to bear in mind that the numbers whereby the atoms of the different bodies are expressed have reference solely to their relative weights—not to their bulks, which are supposed to be equal in all cases.

Another important distinction to which we must call attention is this, that *compound bodies can have no atomic weight:* they have only a molecular weight.

Simple bodies have both a molecular weight and an atomic weight.

These two weights may be used indiscriminately in special cases when the molecule contains only one atom.

Two methods are commonly used to determine atomic weights: the one is founded on the fact that an atom is the smallest portion of a body which can exercise a reaction, the other is based upon the different specific heats. Both these methods are indispensable, as they cannot be always used indiscriminately.

First Method.—In order to determine the atomic weight of a simple body, it is necessary, in the first place, to know the molecular weights of that body in a free or uncombined state, and of all —or, at least, the greater number of the compounds which it forms; it is, moreover, requisite to ascertain the relative quantities which enter into the composition of these latter. We then choose, as the weight of the atom, the largest number that will exactly divide the weights of that body contained either in its free molecule or in that of its various compounds. For, in fact, a single molecule must contain a whole number of atoms, since these are indivisible; therefore, the weight of any number of atoms is necessarily always capable of being divided by that of a single atom.

One example will suffice to make this clearly understood. In comparing the weights of equal volumes of free hydrogen, hydrochloric acid, hydrobromic acid, hydriodic acid, hydrocyanic acid, hydrosulphuric acid, hydroselenic acid, hydrotelluric acid, ammonia, and so on, we find that the *molecular weights* of these different bodies, as compared with that of the *molecular* weight of hydrogen, taken as unity (and not with that of its atom, which we still suppose to be unknown), are as follows:—

Names of Bodies.	Weights of Molecules compared with the Weight of a Molecule of Hydrogen = 1.	Quantitative Composition of the Molecule.	
		Amount of Hydrogen in a Molecula.	Quantities of:
Pure hydrogen	1	1	0 other bodies.
Hydrochloric acid	18·25	$\frac{1}{2}$	17·75 chlorine.
Hydrobromic acid	40·50	$\frac{1}{2}$	40 bromine.
Hydriodic acid	64·00	$\frac{1}{2}$	63·5 iodine.
Hydrocyanic acid	13·5	$\frac{1}{2}$	13 carbon and nitrogen combined.
Water	9	1	8 oxygen.
Hydrosulphuric acid ..	17	1	16 sulphur.
Hydroselenic acid	40·75	1	39·75 selenium.
Hydrotelluric acid	65·5	1	64·5 tellurium.
Formic acid	23	1	22 carbon and oxygen combined.
Ammonia	8·5	$1\frac{1}{2}$	7 nitrogen.
Phosphoretted hydrogen ..	17	$1\frac{1}{2}$	15·5 phosphorus.
Arseniuretted hydrogen ..	39	$1\frac{1}{2}$	37·5 arsenic.
Acetic acid..	30	2	28 carbon and oxygen combined.
Ethylene	14	2	12 carbon.
Propionic acid	37	3	34 carbon and oxygen combined.
Alcohol	23	3	20 " "
Ether	37	5	32 " "

From this Table it will be seen that the greatest common divisor of the numbers, $\frac{1}{2}$, 1, $1\frac{1}{2}$, 2, 3, 5, which express the weights of hydrogen contained in the molecules of the different bodies, is $\frac{1}{2}$; therefore $\frac{1}{2}$ represents the atomic weight of hydrogen. All the weights in the foregoing Table refer to the molecule of hydrogen. If, however, we take as unity the weight of this atom instead of that of the molecule, the numbers would be doubled, as follows:—

Names of Bodies.	Molecular Weights compared with the Weight of One Atom of Hydrogen.	Quantitative Composition of the Molecule.	
		Amount of Hydrogen in a Molecule.	Quantities of:
Pure hydrogen	2	2	0 other bodies.
Hydrochloric acid	36·5	1	35·5 chlorine.
Hydrobromic acid	81	1	80 bromine.
Hydriodic acid	128	1	127 iodine.
Hydrocyanic acid	27	1	26 cyanogen.
Water	18	2	16 oxygen.
Hydrosulphuric acid ..	34	2	32 sulphur.
Hydroselenic acid	81·5	2	79·5 selenium.
Hydrotelluric acid	131	2	129 tellurium.
Formic acid	46	2	44 carbon and oxygen.
Ammonia	17	3	14 nitrogen.
Phosphoretted hydrogen ..	34	3	31 phosphorus.
Arseniuretted hydrogen ..	78	3	75 arsenic.
Acetic acid	60	4	56 carbon and oxygen.
Ethylene	28	4	24 carbon.
Propionic acid	74	6	68 carbon and oxygen combined.
Alcohol	46	6	40 carbon and oxygen.
Ether	74	10	64 " "

And 1, being the greatest common divisor, would be the true atomic weight of hydrogen.

In the same manner, we can determine the atomic weights of other simple bodies; for instance, nitrogen. For this purpose we must first examine the molecular weights and compositions of the different volatile compounds of nitrogen, as protoxide and binoxide of nitrogen, hyponitrous acid, hydrous and anhydrous nitric acid and ammonia; we can then form the following Table:—

Names of Bodies.	Weight of Molecule compared with an Atom of Hydrogen = 1.	Amount of Nitrogen.	Amount of:
Protoxide of nitrogen ..	44	28	16 oxygen.
Binoxide of nitrogen ..	30	14	16 "
Hyponitrous acid	46	44	32 "
Hydrated nitric acid ..	63	14	49 " and hydrogen combined.
Anhydrous nitric acid ..	108	28	80 "
Ammonia	17	14	3 hydrogen.
Nitrogen	28	28	0 other bodies.

14, being the greatest common divisor of the numbers 14 and 28, becomes the atomic weight of nitrogen, and will remain so unless a new combination of that body be discovered, the molecule of which shall contain a quantity of that metalloid equal to a submultiple of 14.

Second Method.—This method is due to Dulong and Petit. The atomic weights of several bodies being already known, those savants found that the same amount of heat is always requisite in order to raise by 1 degree the weights of various simple bodies proportional to their atomic weights. Thus, to increase by 1 degree 25 grammes of sodium, 32 grammes of sulphur, 118 grammes of tin, 31 grammes of phosphorus, &c., one same quantity of heat is required, which, for the present, we will represent by the letter P.

P raises 23 grammes of sodium 1 degree. It is evident, then, that to raise 1 gramme—that is, 23 times less of that element—also 1 degree, 23 times less heat will be required, or $\frac{P}{23}$. Therefore, $\frac{P}{23}$ represents the calorific capacity of sodium.

In a like manner it will be found that the calorific capacity of sulphur is $\frac{P}{32}$, that of tin $\frac{P}{118}$, and that of phosphorus $\frac{P}{31}$.

It will be seen that the specific heats decrease when the atomic weights increase, and that in the same ratio; so that the atomic weights being 1, 2, 4, 8, 16, &c., the specific heats will be $\frac{1}{2}$, $\frac{1}{4}$, $\frac{1}{8}$, $\frac{1}{16}$, &c.

We are taught by arithmetic that if the two factors of a multiplication be so modified that the one becomes 2, 3, 4, 5 times less while the other becomes 2, 3, 4, 5 times greater, the product is invariable. We must, therefore, always obtain sensibly the same number when we multiply the specific heats of various bodies by their atomic weights.

Thus, the product of the atomic weight of sodium by its specific heat is $\frac{P \times 23}{23} = P$. The product of the atomic weight of sulphur by its specific heat is $\frac{P \times 32}{32} = P$. The constant number P has been numerically determined, and is sensibly equal to 6·666.

If it be required to find the atomic weight of a simple body, its specific heat must be ascertained. Let C represent the heat, and x its unknown atomic weight, we have: $C \times x = 6{\cdot}666$; whence we derive $x = \frac{6{\cdot}666}{C}$. The atomic weight is found, therefore, by dividing the number 6·666 by the specific heat derived from experiment.

Dulong and Petit have enunciated this law by saying that the specific heats are inversely proportional to the atomic weights.

To enable the use of this method, it is necessary that the bodies whose specific heat we wish to ascertain, exist under similar conditions. Thus the specific heat of gases cannot be used to determine their atomic weights. But, in this case, the desired result is arrived at in a different manner.

M. Vœstyn discovered that, in compound bodies, each atom retains its specific heat. If the molecule of a compound body contains 2, 3, 4 simple atoms, the product of its specific heat by its molecular weight will be 2, 3, 4 times the constant number 6·666.

So that, supposing that the atomic weight of a gas be required, it must be made to enter into such a combination as will assume the solid state, and of which the specific heat must be ascertained. By multiplying the number representing that calorific capacity by the molecular weight of the compound, and dividing the product by 6·666, the quotient gives the number of atoms of which the molecule is composed. The analysis of the compound being made, and the atomic weight of one of its elements known, the atomic weight of the other can be readily deducted.

Let us assume, as an example, that it is sought by this process to find the atomic weight of oxygen: we combine it with hydrogen, and then determine the specific heat of the water thus formed, or rather, we know that it is equal to 1, since the specific heat of water has been taken as the unity: on the other hand, the molecular weight of the water must be ascertained, and it will be found equal to 18. Now, as the number 18 contains 3 times 6, we arrive at the conclusion that the water contains 3 atoms; finally, it is shown by analysis that 18 of water contain 16 of oxygen, and 2 of hydrogen. We are, moreover, aware that the weight of an atom of hydrogen is 1; we must thence conclude that the atom of oxygen weighs 16. For, if it weighed less than 16, there would be more than one atom in a molecule of water, and, as the latter already contains 2 atoms of hydrogen, it would contain altogether more than 3 atoms, which would be in contradiction with the conclusions deduced from the calorific capacity of water.

It has been seen that, in lieu of dividing 18 by 6·666, we divided it only by 6. That is because the number P is not absolutely constant, but varies between 6 and 7: so that in selecting 6·666 we only chose the mean. This, however, does not in any way weaken either the law or the results derived from it. The specific heats are only approximate, because it is not possible to tell what quantity of caloric a body absorbs in order to expand as well as to become heated, a quantity which increases the specific heat found, and falsifies the result; but this slight divergence between theory and experiment offers no inconvenience: it tends, it is true, to render the atomic weights likewise only approximate; fortunately that approximation is sufficiently great to render the analysis of the compounds, into which enter the bodies of which the atomic weights are required, all-sufficient in completely establishing the latter.

If the atomic weight of silver were required, we should divide the number 6·666 by 0·05701, the specific heat of the metal, and thus we should obtain $\frac{6 \cdot 666}{0 \cdot 05701} = 117$; if, on the other hand, we combine silver and chlorine, and analyze the chloride of silver, we shall find that that compound contains 35·5 of chlorine to 108 of silver.

As 35·5 represents the atomic weight of chlorine, we may consider this latter to be combined with 1, 2, 3, 4 . . . atoms of silver. And, in these several hypotheses, the atomic weight of silver would be 108, 54, 27, . . &c.

Again, we may suppose the chloride of silver to contain only 1 atom of silver, to 2, 3, 4, 5, . . . atoms of chlorine, so that the quantity of silver in combination with 35·5 of chlorine would represent but $\frac{1}{2}$, $\frac{1}{3}$, $\frac{1}{4}$, $\frac{1}{5}$, . . . the weight of its atom. In these several hypotheses the atomic weight of silver would be 216, 324, 432, 540, . . . &c.

Other hypotheses may be added to the above; but, be they what they may, they will always give for the atomic weight of silver values that differ considerably from the number 117, found by means of the specific heat. There is but one supposition which yields a value in approximate harmony with that number, it is the one whence we deduced 108 as the atomic weight: 108 must, therefore, be regarded as the true atomic weight of silver.

The object of chemical notation is to represent the various bodies known by means of brief formulæ which shall indicate at once their molecular weight, and their composition both as to quality and quantity; thus enabling the sense of the different reactions that take place to be better understood.

In the construction of these formulæ a symbol has been adopted that represents the atom —not the molecule—of each simple body; such are the symbols used in the Table. These symbols are generally obtained by taking the initial letter of the name of the body, thus: Θ for oxygen, S for sulphur, and so on. When the names of several bodies commence with the same letter, the first letter is then taken to designate that body only that has been longest known; whereas, for the symbol of the others, the *two* first letters of their name are used. For instance, sulphur, selenium, silicium, strontium, beginning each with an S, S signifies sulphur, while Se represents selenium, Si silicium, and St strontium.

There are, however, some exceptions to this rule: occasionally, instead of using the two first letters, the first and one of the letters in the body of the word are taken. Thus, arsenic is expressed by As, stannum (tin) by Sn, stibium (antimony) by Sb, and hydrargyrum (mercury) by Hg.

Finally, in the same way that some of the symbols are taken from the Latin, like the three last, others are borrowed from the German. Thus, the symbol of tungsten is W, from the German Wolfram.

All simple bodies being indicated by a symbol that expresses not only their nature, but also their atomic weight, nothing is easier than to represent a compound molecule. The only thing needed is to write down, side by side, the several constituent atoms, adding above each one an exponent indicative of its number. When that number is equal to 1, the exponent is dispensed with. Thus, $\text{S}\,\Theta^3$ implies a compound molecule formed of 1 atom of sulphur and 3 atoms of oxygen.

In symbolic writing it has been agreed to place the most electro-positive of the several components always first. But this rule is only strictly followed for those compounds that contain but two elements.

It is clear that the formulæ of which we are now speaking represent the qualitative composition of bodies. It is also clear that they represent their molecular weights. Since a molecule can have no other weight than the sum of the weights of the atoms contained in it, in order to know how much it weighs, it will suffice to multiply the atomic weight of each element by its exponent, and add together the products. Thus, the formula of glycerine being $\text{C}^3\,\text{H}^8\,\Theta^3$, its molecular weight will be equal to

	the weight of 3 atoms of carbon	3 × 12 = 36
+	the weight of 8 atoms of hydrogen	8 × 1 = 8
+	the weight of 3 atoms of oxygen	3 × 16 = 48
	Total	92

Finally, these formulæ represent likewise the centesimal composition of bodies. Knowing the quantity of the various elements contained in a certain weight of the compound—that of its molecule, we arrive, by means of a simple proportion, at the knowledge of its centesimal composition.

For example, supposing that we wish to find the centesimal composition of acetic acid; we deduce from its formula, $C^2 H^4 \Theta^2$, in the first place, that the molecule of this acid weighs 60, and that it contains

2 atoms = 24 of carbon,
4 atoms = 4 of hydrogen,
and 2 atoms = 32 of oxygen.

We next lay down the three proportions :—

$$\text{1st. } 60 : 24 :: 100 : x, \text{ where } x = \frac{24 \times 100}{60} = \frac{24 \times 10}{6} = 40.$$

$$\text{2nd. } 60 : 4 :: 100 : x, \text{ where } x = \frac{4 \times 100}{60} = \frac{4 \times 10}{6} = 6 \cdot 666.$$

$$\text{3rd. } 60 : 32 :: 100 : x, \text{ where } x = \frac{32 \times 100}{60} = \frac{32 \times 10}{6} = 53 \cdot 333.$$

We now know in what manner, by the aid of a formula, it is possible to learn the quantitative and qualitative composition, as well as the molecular weight, of the compound which it represents. It remains to be seen how, with a given body, the formula is to be established; it is the other side of the problem.

To establish the formula of a compound body, we first of all ascertain by analysis its centesimal composition; then we determine its molecular weight. Our next step is, by a series of proportions, to find out the composition of a certain weight of that substance known to represent its molecular weight. After which we divide the quantities of its several elements by their atomic weights; the quotient shows how many atoms there are of each. Finally, we only have to write down, side by side, the symbols expressing the different atoms, beginning with the most electro-positive, and to surmount those symbols by an exponent indicating the number of the atoms.

Let us apply this rule to an example, and suppose that it be required to establish the formula for propionic acid. We analyze the acid, and we find that it contains 48·648 centesimals of carbon, 43·243 of oxygen, and 8·108 of hydrogen.

We next look for its molecular weight, and find it equal to 74. Having done that, we lay down the three proportions :—

1st. 100 : 48·648 :: 74 : x, where x = 35·999, or nearly 36.
2nd. 100 : 43·243 :: 74 : x, where x = 31·999, or nearly 32
3rd. 100 : 8·108 :: 74 : x, where x = 5·999, or nearly 6.

So that one molecule of propionic acid weighs 74, and contains 36 of carbon, 32 of oxygen, and 6 of hydrogen.

The weight of one atom of carbon is 12; if, then, we divide the weight of that body contained in one molecule of propionic acid, that is 36, by 12, we shall have the number of its atoms; and, as $\frac{36}{12} = 3$, we conclude that it contains 3 atoms of carbon.

In like manner, the weight of one atom of oxygen being 16, we divide the weight of oxygen contained in the molecule by that number; that is, $\frac{32}{16} = 2$: therefore, propionic acid contains 2 atoms of oxygen.

Finally, one atom of hydrogen weighs 1, and as there are 6 of hydrogen, and $\frac{6}{1} = 6$, we conclude that propionic acid contains 6 atoms of that element.

Hence the formula for propionic acid is $C^3 H^6 \Theta^2$.

It is sometimes necessary to indicate that a certain number of molecules of a same body take part in a reaction. It is then customary to place at the left of the formula a coefficient, to express that number. Thus, to signify 3 molecules of propionic acid, we write $3\ C^3 H^6 \Theta^2$.

Lastly, in order to render an exact account of the reactions, it is the practice to represent them by means of equations. In these equations the first side contains the formulæ of the different bodies entering into reaction, preceded by a coefficient indicating how many molecules react; and the second side, which is separated from the first by the sign =, contains the formulæ of the products formed by the reaction. As nothing is lost during chemical action, it is clear that the second side of the equation must contain strictly all the atoms that existed in the first, only differently grouped.

To give an example of a chemical equation, we will represent the reaction which gives rise to chloride of potassium, K Cl, by means of hydrochloric acid, H Cl, and of potassium K H Θ.

$$\underset{\text{Potassium.}}{KH\Theta} + \underset{\text{Hydrochloric acid.}}{HCl} = \underset{\text{Chloride of potassium.}}{KCl} + \underset{\text{Water.}}{H^2\Theta}.$$

The atom of potassium, the two atoms of hydrogen, the atom of oxygen, and the atom of chlorine that compose the first side, are all found in the second side, but grouped in a different way.

The following Table contains a complete list of the elementary bodies known at present (1869). Every description of matter which has been examined is made up of these 65 elements, either combined together to form compounds, or in an uncombined or free state.

Names of Elements.	New Symbols.	Old Symbols.	Hydrogen = 1. New Atomic Weights.	Hydrogen = 1. Atomic Weights of Berzelius.	Hydrogen = 1. Atomic Weights of Gerhardt.	Atomic Volumes.	Weight of a Cubic Foot in Ounces Avoirdupois. Water = 1000 oz.	Specific Gravity. Water = 1. When water = 1000, the decimal point is to be placed to the Right, three places of Figures.
Aluminium	Al	Al	27·5	27·39	13·75	5·3	2670	2·5—2·67, Wöhler; 2·67, Deville.
Antimony	Sb	Sb	122	129·24	122	17·9	6720	6·72, Marchand and Scheerer; Kopp.
Arsenic	As	As	75	75·22	75·22	13·3	5670	5·63, Karsten; 5·67, Herapath.
Barium	Ba	Ba	137	137·06	68·5	..	..	
Bismuth	Bi	Bi	210	213·20	210	21·2	9800	9·80, Marchand and Scheerer; 9·78, Kopp.
Boron	B	B	11	21·82	10·9	..	..	
Bromine	Br	Br	80	80·10	80	25·8	3190	*Liquid:* 3·19, Pierre; 2·99, Löwig.
Cadmium	Cd	Cd	112	111·66	56	6·5	8690	8·69, Stromeyer; 8·45, Kopp.
Cæsium	Cs	Cs	133	..	..	..	..	
Calcium	Ca	Ca	40	40·32	20	12·6	1580	1·58, Bunsen.
Carbon	C	C	12	12·04	22	3·4	3520	*Diamond:* 3·52, Brisson.
Cerium	Ce	Ce	92	..	..	..	..	
Chlorine	Cl	Cl	35·5	35·52	35·5	26·7	1330	*Liquid:* 1·33, Faraday.
Chromium	Cr	Cr	52·5	52·70	26·26	3·8	7010	7·01, Bunsen and Frankland.
Cobalt	Co	Co	59	59·07	29·5	3·5	8510	8·49, Brunner; 8·51, Berzelius.
Copper	Cu	Cu	63·5	63·39	31·75	3·6	8950	8·95, Marchand and Scheerer; 8·93, Kopp.
Didymium	Di	Di	95	..	..	..	..	
Erbium	Erb	E	..	..	..	..	..	
Fluorine	Fl	F	19	18·70	19	..	..	
Glucinum	Gl	Gl	14	..	..	2·2	2100	2·1, Debray.
Gold	Au	Au	197	196·98	..	10·2	19340	19·34. G Rose; 19·26, Brisson.
Hydrogen	H	H	1	1	1	..	..	
Indium	In	In	35·9(?)	..	..	..	..	
Iodine	I	I	127	127·08	127	25·7	4950	4·95, Gay-Lussac
Iridium	Ir	Ir	198	197·44	98·56	4·5	21800	21·80, Hare.
Iron	Fe	Fe	56	56·17	28	3·6	7840	7·84, Broling; 7·79, Karsten.
Lanthanum	La	La	92·8	..	..	..	..	
Lead	Pb	Pb	207	207·47	103·5	9·2	11390	11·39, Karsten; 11·33, Kopp.
Lithium	Li	Li	7	13·08	7	11·9	590	0·59, Bunsen.
Magnesium	Mg	Mg	24	25·34	12	6·9	1740	1·74, Bunsen; 1 70, Kopp.
Manganese	Mn	Mn	55	55·23	27·5	3·5	8030	8·03, Bachmann; 8·01, John.
Mercury	Hg	Hg	200	200·52	100	7·4	13600	*Liquid:* 13·60, Regnault, Kopp.
Molybdenum	Mo	Mo	96	95·53	48	5·3	8630	8·62—8·64, Buchholz.
Nickel	Ni	Ni	59	59·19	29·5	3·4	8820	8·60. Brunner; 8·82, Tupputi.
Niobium	Nb	Nb	94 (?)	..	..	..	..	
Nitrogen	N	N	14	14·02	14	..	..	
Norium	No	No	..	..	..	..	..	
Osmium	Os	Os	199	199·13	..	..	..	
Oxygen	O	O	16	16	16	..	..	
Palladium	Pd	Pd	106·5	106·64	..	4·6	11800	11·80, Wollaston.
Phosphorus	P	P	31	31·41	31	16·8	1840	*Yellow:* 1·84, Schrötter; 1·83, Kopp.
Platinum	Pt	Pt	197·5	197·44	98·5	4·6	21500	21·5, Wollaston, Berzelius.
Potassium	K	K	39	78·47	39	45·6	860	0·86, Gay-Lussac and Thenard.
Rhodium	Rh	Rh	104·4	104·48	..	4·7	11200	11·0, Wollaston; 11·2, Cloud.
Rubidium	Rb	Rb	85·4	..	..	..	..	
Ruthenium	Ru	Ru	104·4	..	..	..	..	
Selenium	Se	Se	79·5	79·37	79·5	18·4	4280	*Amorphous:* 4·28, Schaffgotsch.
Silicon	Si	Si	28	44·51	..	11·2	2490	2·49, Wöhler.
Silver	Ag	Ag	108	216·29	108	10·2	10570	10·4, Karsten; 10·57, G. Rose.
Sodium	Na	Na	23	46·43	23	23·7	970	0·97, Gay-Lussac and Thenard.
Strontium	Sr	Sr	87·5	87·48	43·75	17·2	2540	2·54, Bunsen.
Sulphur	S	S	32	32·17	32	15·2	2070	*Trimetric:* 2·07, Marchand and Scheerer, Kopp.
Tantalum	Ta	Ta	137·6	..	..	..	..	
Tellurium	Te	Te	129	128·48	129	20·6	6240	6·24, Berzelius; 6·18, Löwe.
Terbium	Tr	Tr	..	..	..	..	..	
Thallium	Tl	Tl	204	..	..	..	..	
Thorium	Th	Th	..	..	..	..	..	
Tin	Sn	Sn	118	117·83	59	16·2	7300	7·29, Karsten; 7·30, Kopp.
Titanium	Ti	Ti	50	48·3	25	..	..	
Tungsten	W	W	184	190·44	92	5·3	17900	17·2, Allen and Aiken; 17·5—18·3, Wöhler.
Uranium	U	U	120	118·88	60	3·3	18400	18·4, Péligot.
Vanadium	V	V	137	137·32	..	..	..	
Yttrium	Yt	Y	68	..	..	..	..	
Zinc	Zn	Zn	65·2	65·16	32·6	4·6	7130	7·13, Kopp; 7·1—7·2, Bolley.
Zirconium	Zr	Zr	89·6	67·26	..	..	..	

The succeeding tabulated form shows at one view where the ordinary substances, not agreeing with the required alphabetical order, are placed, and from what words their respective symbols are taken.

Aluminium	Al	Hydrargyrus (mercury)	Hg	Rhodium	Rh
Argentum (silver) ..	Ag	Hydrogen	H	Rubidium	Rb
Arsenic	As	Indium	In	Ruthenium	Ru
Aurum (gold)	Au	Iodine	I	Selenium	Se
Barium	Ba	Iridium	Ir	Silicon	Si
Bismuth	Bi	Kalium (potassium) ..	K	Stannum (tin)	Sn
Boron	B	Lanthanum	La	Stibium (antimony) ..	Sb
Bromine	Br	Lithium	Li	Strontium	Sr
Cadmium	Cd	Magnesium	Mg	Sulphur	S
Cæsium	Cs	Manganese	Mn	Tantalum	Ta
Calcium	Ca	Molybdenum	Mo	Tellurium	Te
Carbon	C	Natrium (sodium) ..	Na	Terbium	Tr
Cerium	Ce	Nickel	Ni	Thallium	Tl
Chlorine	Cl	Niobium	Nb	Thorium	Th
Chromium	Cr	Nitrogen	N	Titanium	Ti
Cobalt	Co	Norium	No	Uranium	U
Cuprum (copper)	Cu	Osmium	Os	Vanadium	V
Didymium	Di	Oxygen	O	Wolfram (tungsten) ..	W
Erbium	Erb	Palladium	Pd	Yttrium	Yt
Ferrum (iron)	Fe	Phosphorus	P	Zinc	Zn
Fluorine	Fl	Platinum	Pt	Zirconium	Zr
Glucinum	Gl	Plumbum (lead)	Pb		

Prefixes and Affixes.—Hypo-. This term is used to indicate that the substance to which it is applied contains less oxygen than the other substance from which the name is derived; thus *Hyponitric acid* contains less oxygen than nitric acid. *-ous* is used to imply less oxygen than the termination, *-ic*; thus nitr*ous* oxide contains half as much oxygen for its nitrogen as nitr*ic* oxide. Many of the elements are capable of uniting with other elements in several different proportions to form chemical compounds. Sulphur, for example, is specially apt to form more than one compound with a single element. When sulphur unites with a metal, the compound formed is called a sulph*ide*, just as a compound of oxygen and a metal is called an oxide, or one of chlorine and a metal a chloride,—the termination, *-ide*, which always indicates combination, being added to the first syllable of the word sulphur, or oxygen, or chlorine, and the new word ending in *ide* being then connected with the name of the metal, as in the case of sulphide of copper. But when, as in the case of calcium, there are several distinct sulphides, it is customary to distinguish one from the other by means of various Latin and Greek prefixes. Thus the compound which contains one atom of sulphur and one atom of calcium is the proto-sulphide, or simply the sulphide of calcium, the prefix *proto-* being derived from the Greek word for first; the compound which contains two atoms of sulphur to one of calcium is the bisulphide of calcium, from the Latin for twice; and in like manner we have a tersulphide, containing three atoms of sulphur to one of calcium, and a quinquisulphide containing five atoms of sulphur. The compound containing the highest proportion of sulphur is often called the *per*sulphide. A good custom is to designate the compounds which contain more sulphur than the protosulphide by prefixes of Latin origin, and to distinguish those which may contain less sulphur than the protosulphide by means of Greek prefixes; thus, if there were a compound of two atoms of calcium and one of sulphur, it would properly be called a *di*-sulphide of calcium, the prefix being from the Greek *δις*. The same prefixes are used in an analogous manner in connection with the words oxide, chloride, bromide, iodide, and the similar words ending in *ide*.

Many modern writers on chemistry employ a notation which we append. Some of the symbols and contractions of this notation are embodied in the notation we have just explained.

· One equivalent of oxygen;—written above a symbol representing an element, and repeated to indicate two, three, or more equivalents; thus, $\dot{Fe}$ denotes a compound of one equivalent of oxygen with one of iron; $\dddot{S}$ a compound of three equivalents of oxygen with one of sulphur.

' One equivalent of sulphur;—used in the same manner as the preceding; thus, $\overset{,,}{Fe}$ denotes a compound of two equivalents of sulphur and one of iron.

A dash drawn across a symbol having either of the foregoing signs above it, denotes that two equivalents of the substance represented by the symbol are joined with the number of equivalents of oxygen or sulphur indicated by the dots or commas; thus, $\overset{\cdots}{\text{F̶e}}$ represents a compound of two equivalents of iron and three of oxygen, forming sesqui-oxide of iron.

+ indicates, in organic chemistry, a base or alkaloid, when placed above the initial letter of the name of the substance; as, $\overset{+}{M}$, morphine; $\overset{+}{Q}$, quinine.

— indicates, in organic chemistry, an acid, when placed above the initial letter of the name of the acid; as, $\bar{C}$, citric acid; $\bar{T}$, tartaric acid.

Each symbol, when used singly, always indicates a single atom or equivalent of the substance represented by it; thus, O stands for one atom or equivalent of oxygen, C for a single equivalent of carbon, and the others in like manner. A compound body made up of single equivalents of its constituents is represented by the two symbols of the respective constituents written side by side;

as, HO, a compound of one equivalent of hydrogen with one of oxygen, forming water. To express more than one atom or equivalent of a substance, a number is used, either prefixed to the symbol, or, more commonly, written after it, below the line; as, $2\,O$, or O_2, two equivalents of oxygen.

A secondary compound, as a salt, is indicated by writing the symbols of the constituent compounds one after another, with the sign + between them, the symbol of the base being always placed first; thus, $CaO + CO_2$ represents carbonate of lime. A comma is frequently used instead of the sign +, commonly to express a more intimate union than would be expressed by that sign. The period is also sometimes used to indicate a union more intimate than that denoted by the sign +, but less so than that implied by a comma. A number written before the symbol of a compound designates a corresponding number of equivalents of that compound; as, $3\,SO_3$, three equivalents of sulphuric acid. When the formula of the quantity contains several terms, those to which the figure applies are included in parentheses or brackets, to which the figure is prefixed; as $3\,(CaO + SO_3)$, three equivalents of sulphate of lime. See Equivalents. Heat. Isomorphism. Molecular Volume.

ATTEMPERATOR, in Brewing. Fr., *Attempérateur;* Ger., *Eine Vorrichtung um eine gleichmässige Temperatur zu behalten;* Ital., *Regolatore della temperatura;* Span., *Regulador de temperatura.*

The attemperator invented by A. B. Walker is shown at A, Figs. 429, 430, and consists of a rectangular frame, in which a series of flattened tubes *b*, *b*, are arranged side by side, with spaces

429.

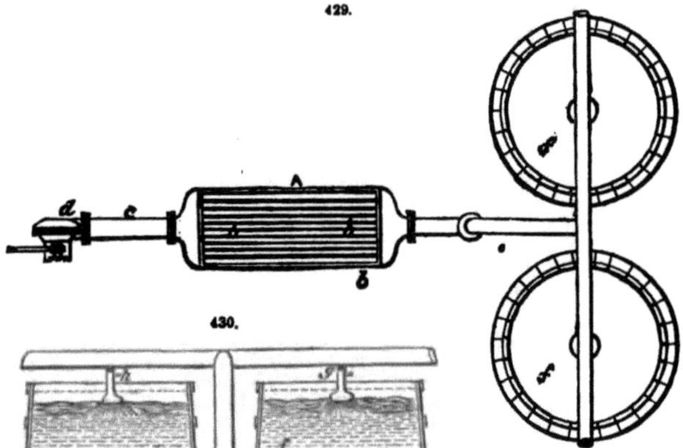

430.

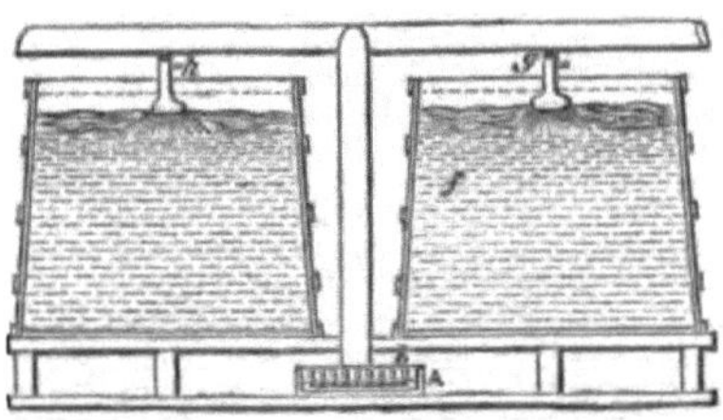

of about 5 in. between them. The spaces between the tubes are filled with spring-water, ice, or any freezing mixture of salts or acids. At one end the tubes are connected by the pipe *c*, to the fan *d;* at the opposite end they are connected to the pipe *e* which leads up to the fermenting-tuns *f*, branching off directly over each tun, and having a vertical tube *g*, pierced at the bottom with small holes for the exit of air. The supply of air to each tun is regulated by the slide *h*, so as to suit the progress of the fermentation. As soon as the fermentation commences, and the yeast begins to form on the surface of the wort, the fan is put in motion, and a current of air is thus driven through the tubes of the attemperator, and caused to pass over the surface of the wort until the fermentation is completed.

This atmospheric attemperator has been applied to the manufacture of malt, so that malt may be made in summer as well as in winter. Hitherto it has been impracticable to make malt in

431.

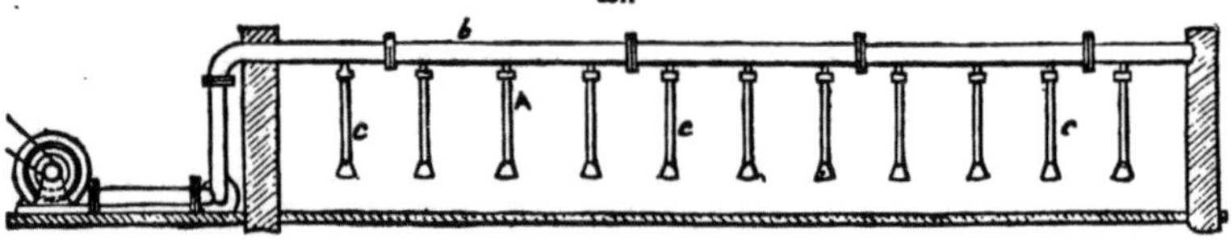

the summer equal to that made in winter, or, in fact, to make it at all in the summer, owing to the unsuitability of temperature. Fig. 431 shows the application to malting-rooms: A is the malting-room; *b*, pipes leading from the attemperator directly through the malting-room, a few feet above the floor; *c c*, flexible tubes attached to the pipes, *b*, with rose perforated caps at the ends. Whilst the malt is on the floor in process of germination, the temperature of the room is regulated to that degree of temperature best suited for malting, by currents of attemperated air from the attemperator distributed over the surface of the malt on the floor.

Fig. 432 represents the attemperator invented by Joshua Crockford, as fixed to a wooden mash-tun with sliding side-doors and fixed roof, such as is largely used in Burton breweries. It consists

432.

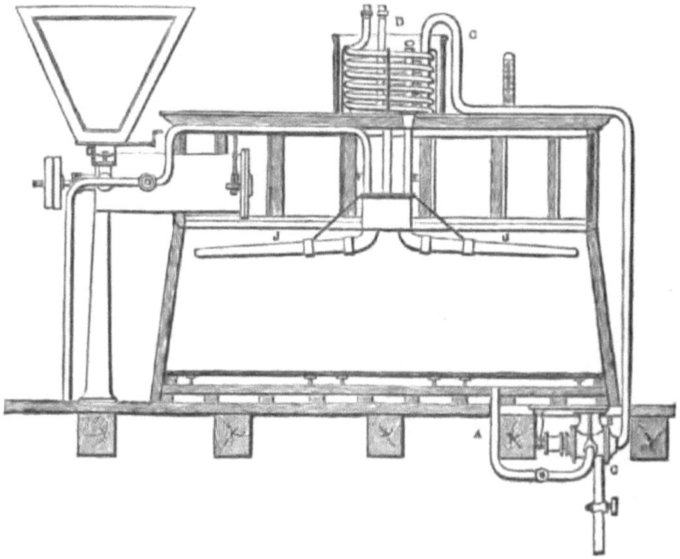

of a circular vessel, D, fixed on the top of the mash-tun, this vessel containing a coiled steam-pipe, and being fitted with suitable pipes for conducting the wort into and out of it. When it is desired to raise the temperature of the mash, the wort is drawn from the tun by the pipe, A; and the centrifugal pump, B, being set in action, it is raised into the vessel, D, through the pipe, C. There it is heated by the action of the steam in the coiled pipes; and then it is led down through the pipe, E, to the central vessel of the sparger, J, which distributes it over the goods. It will be noticed that the pipe, C, conducts the wort to the bottom of the vessel, D, whilst the pipe, E, draws off the wort from near the surface, where it is of course hottest. The pipe, F, is for admitting the ordinary supply of hot liquor to the *sparger*.

So long as the pump, B, is in action, a constant current is maintained through the goods, the wort being drawn off at the bottom, heated, and again sparged on the top continually. By the use of this attemperator, the temperature of the mash can be kept at any desired point for any length of time; and in the case of small brewings, where the loss of heat, from radiation and other causes, is proportionately very great, it is particularly valuable.

ATTIC. Fr., *Grenier, Mansarde*; Ger., *Der Uebersatz*; Ital., *Attico*; Span., *Baxo tejado*.

See Building.

ATWOOD'S MACHINE. Fr., *Machine d'Atwood*; Ger., *Atwood Maschine für Anzeige von freian Fall der Körper*; Ital., *Macchina d'Atwood*; Span., *Maquina Attwood*.

See Acceleration.

AUGER. Fr., *Tarière*; Ger., *Zimmermannsbohrer*; Ital., *Trivella*; Span., *Haladra*.

An *auger* is an instrument for boring holes, chiefly in wood. It consists of a long shank or axis, having a cutting-edge at one end, and usually a handle placed crosswise at the other, by which it is turned with both hands. A *pod-auger* has a straight channel or groove, like the half of a bean-pod. A *screw-auger* has a twisted blade, by the spiral groove of which the chips are discharged. Peculiar augers are employed, termed *ground-augers*, for boring *artesian wells*, or perforating soils or rocks, instruments of this sort consist of a handle for working, a rod which may be lengthened as the perforation extends, and a bit, mouth, or cutting-piece, resembling the bit of a common auger, for soils or soft rocks; and a chisel for harder rocks. A *bit* is a name common to all exchangeable boring tools; for wood, bits are generally applied by means of a crank-formed handle, known as the carpenter's *brace*. The smaller tools used for metal, and applied by the *bow-drill*, *ratchet-brace*, lathe, or drilling-machine, are termed drills, or *drill-bits*. The particular names used to designate *bits* are, in most cases, derived from their forms and the purposes for which they are employed. For wood, bits termed *shell-bits* are numerous; the simplest form is shown in Fig. 433, which is shaped like a *gouge*, with the piercing end sharpened to a semicircular edge, for shearing the fibres around the circumference of the hole. When large, the *shell-bit* is termed a *gouge-bit*. Sometimes the piercing end is drawn to a radial point, and it is then termed a *spoon-bit*, of which the cooper's *dowel-bit* and the *table* or *furniture bit* are examples. When the

433.

434.

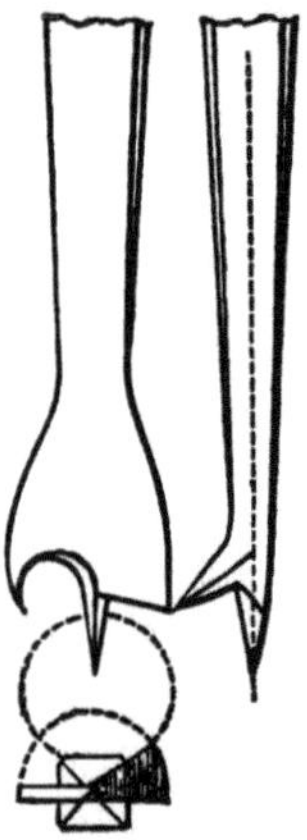

435.

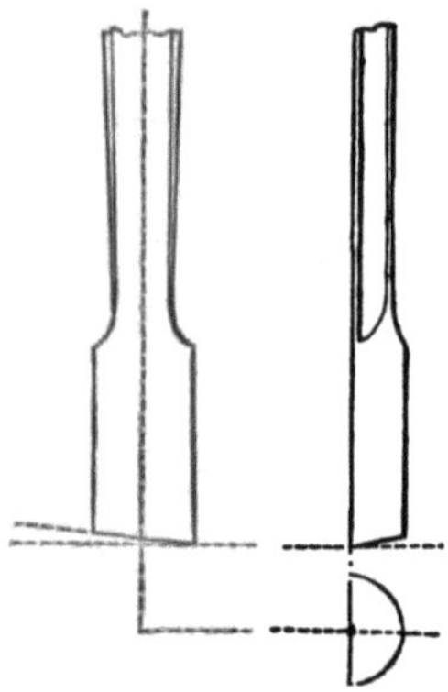

436.

end is bent into a semicircular form horizontally, it is then termed a *duck-nose bit*. The *centre-bit*, shown in Fig. 434, is another typical form, in which the end is flat, provided with a centre point or pin, filed triangularly, which serves as a guide; the *centre-bit* has also a shearing-edge, or *nicker*, serving to cut the fibres round the margin of the hole, and a broad chisel-edge, or cutter, to pare away and remove the wood within the circle defined by the nicker. The *plug centre-bit*, used chiefly for making countersinks for cylinder-headed screws; the *button-tool*, which retains only the centre-pin and nicker, and is used for cutting out discs of leather, and of similar substances; the *flute-drill*, the *cup-key drill*, the wine-cooper's bit, are all modifications of the *centre-bit*. The *half-round bit*, shown in Fig. 435, is employed in enlarging holes in metal, and is usually fixed vertically, or worked by a lathe. The cutting end of this bit is ground with an inclined plane, from 3° to 6° from the perpendicular, according to the hardness of the metal to be bored. The *rose-bit* is shown in Fig. 436; it is cylindrical, and terminates in a truncated cone, the oblique surface of which is cut into teeth, like the *rose-countersink*, of which it is a modification. The *rose-bit* is often used for enlarging holes of considerable depth in metals.

437.

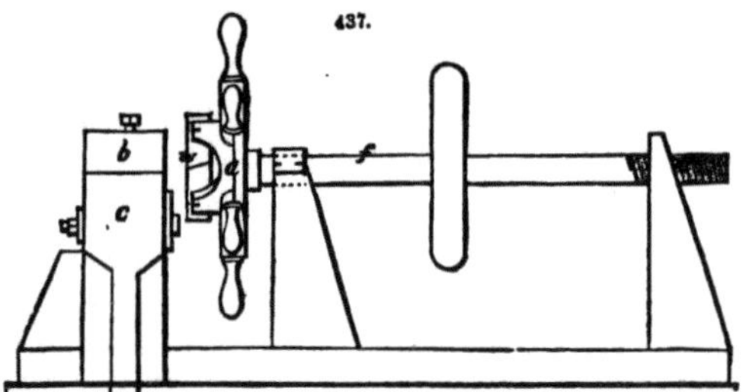

438.

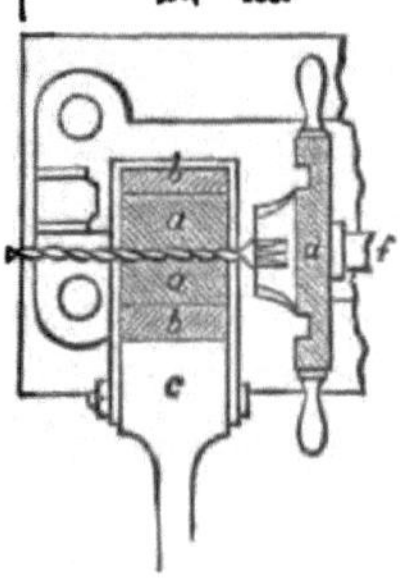

439.

440.

441.

Figs. 437, 438, 439, 440, 441, refer to Ransom Cook's machine for turning the lips of augers. In using this machine the cam-lever *c* is first raised to nearly a vertical position, thus allowing the clamps *b* to open. An auger or bit, with the lips shaped as shown in Fig. 441, being red-hot, is placed in one of the crimping-dies *a*, with the lips projecting beyond the die, and towards the wrench, Figs. 437, 438. The upper end of the cam-lever is then brought quickly down, thus forcing the other crimping-die against the auger, and firmly holding it between the two dies. A quick turn is given to the screw-shaft *f*, which brings the wrench in the hub of the wrench-wheel *d*

into an embrace with the end of the auger, the centre of which enters the hole in the wrench, while the lips pass into the slots on its side. One of the handles of the wrench-wheels is then seized by the operator, and turned towards himself, when the wrench, keeping the auger straight by means of its hold of the centre, turns or bends the lips into the desired position, both being turned at the same time and to the same angle, while the shoulders of both are left in the same line.

442.

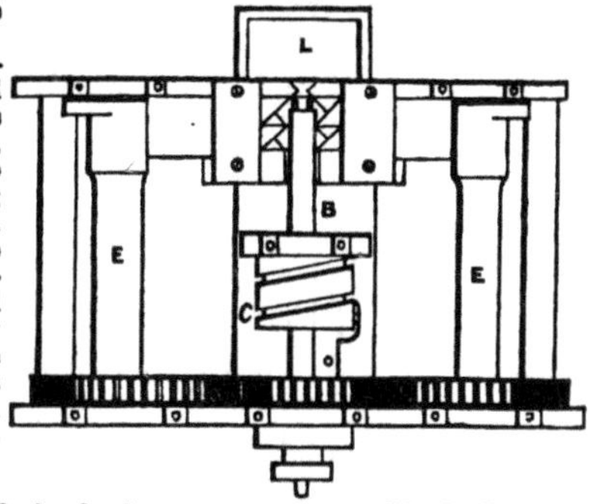

Grier and Boyd's machine for making auger-bits is shown in Fig. 442. The revolving and longitudinally-moving shaft L B, has a recess in its end for holding and twisting the blank, in connection with a series of dies arranged to clasp and hold the auger as fast as it is twisted; thus completing the process in one operation. The rotating-shaft B has a hole in it made longitudinally, of proper size and form, to receive the blank and hold it while being twisted and drawn out. The tongs L, having suitably constructed jaws, act in combination with the shaft B. C is a screw so arranged as to impart to the shaft B an intermittent longitudinal movement during the operation of twisting the blank, so that both twisting the blank and setting the lips are done in one operation.

Fig. 443 represents a machine for swaging the heads of screw-augers, invented by R. Jennings. The jaws of the gripping-dies are arranged vertically—one stationary, and the other movable; their faces grasp the twisted auger, and present the upper end to a hammer, which *swages* the points and lips at one blow. The heading-die is operated by the rotating-shaft B, through the medium of the loose driving-wheel E, provided with the pins *g*, the sliding-wheel F, placed on a sliding-rod H, and the fixed inclined lip *a*. The forming-die consists of the portions L L D, constructed and arranged as shown in Fig. 443. The heading-die D, mould L, toggle M, and operating mechanism, are so arranged that the driving-shaft B may, at the will of the operator, be connected with, or disconnected from, the continually rotating driving-wheel E.

443.

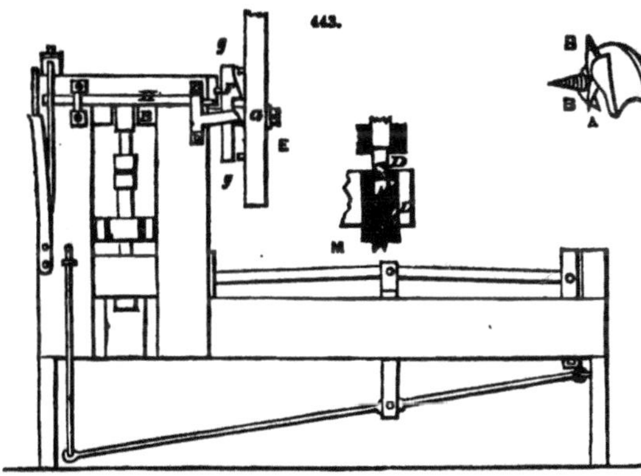

445.

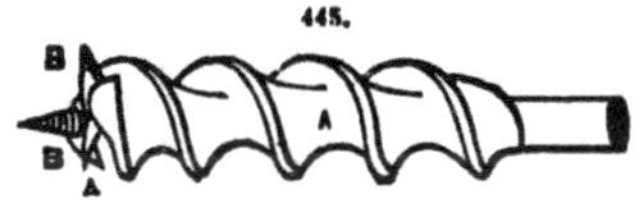

444.

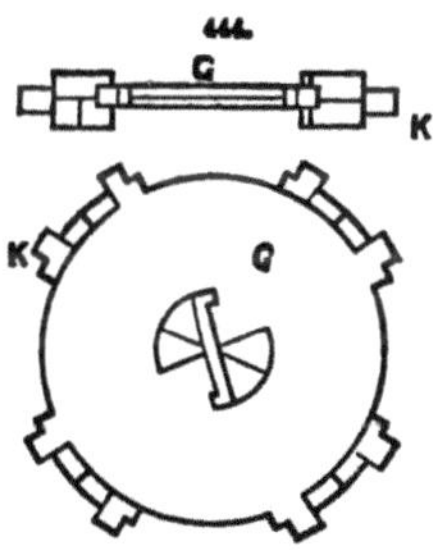

Fig. 444 illustrates the principles upon which a simple and ingenious machine, for forming the twist of auger-blanks, is constructed. This device, patented by Mary Tower, consists of a series of pairs of circular metallic plates, G, K, superimposed on each other, each plate having a peculiarly shaped mortice through the centre, and provided also with projecting and overlapping studs upon its periphery. When these plates are arranged in certain positions, the central mortices correspond with each other, and admit the flat bar of metal, previously heated, of which the auger is to be made; the upper plate is then turned round, with a continuous movement, until each plate has, by means of contact of the projecting studs with each other, been forced to assume the position required for imparting to the metal the necessary twist.

Jennings' double-twisted pod-auger is shown in Fig. 445; this instrument has two *floor-lips* B, B, which project beyond the positions in which they have heretofore been placed; and the spur *a*, instead of being situated at the outer front corner of the cutting-edge of the floor-lip, where the latter, from its necessary thinness, is weakest, is projected from its hinder part, or heel, where it is strongest. A new relation between the cutting-edge of the floor-lip and the spur is thus effected, so that the one does not interfere with the operation of the other. Hence the cutting-edges of a double-twist auger-bit are so formed, that the cutting-edges of the scorers and chisel do not intersect the worm or helix of the shaft A at the same point.

The ship-auger, shown in Figs. 446, 447, has the cutting portion B of the auger attached to the screw portion by means of a *dove-tail notch*, formed by the shoulder *b*, inclined end *d*, dowel *f*, and screw *e*. This method of construction was invented by J. W. Hoagland.

The expanding auger-bit of L. H. Gibbs is shown in Figs. 448, 449, 450, 451. In this instrument the plate B is received into the slot D in the auger A; and by inserting the lower pin *g* into one of the series of holes in the plate B, the cutting-lip *j* on the plate B can be set further out or in, so as to bore larger or smaller holes. This invention consists of the adjustable plate B, with the rib *d*, lip *j*, index holes *c, c, c, c*, in B, combined with the auger A, with slot D, tapering-pins *g, g*, and *set-screw h*, as shown in the figures.

446.

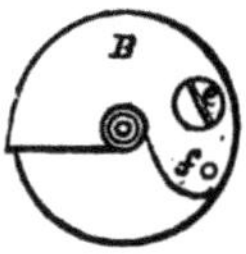

447.

448.

449.

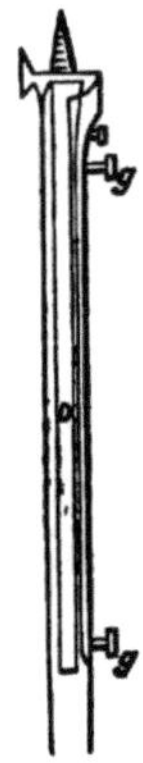

450.

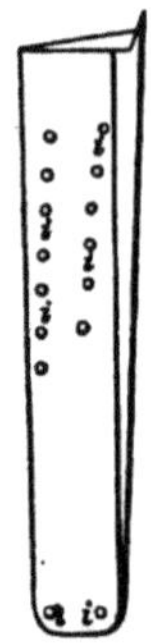

451.

452.

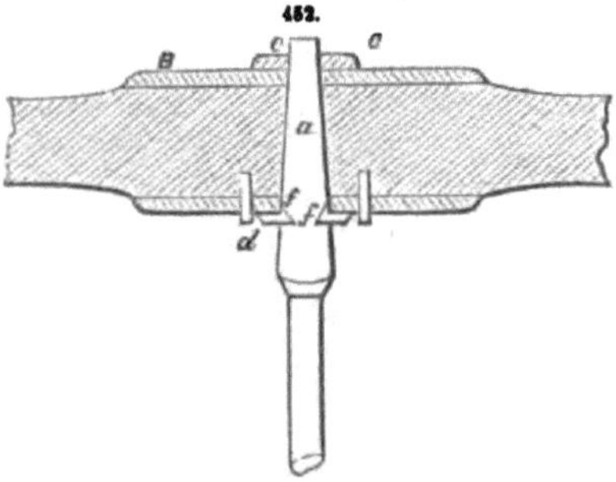

453.

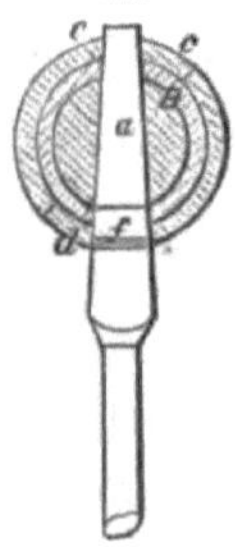

Figs. 452, 453, illustrate C. W. Cotton's method of securing augers to handles by having a metallic tube B placed around the centre of the handle, a transverse rectangular taper-hole *a*, made through the handle and tube, and a metallic band C, placed around the tube B, and turning loosely on it; the band C having slots *c d*, made through it, a part of the slot *d* being taper; then, the shank of the auger being placed in the hole *a*, and through the slots *c d* in the band, the shank is secured to the handle by turning the band, which causes the edges of the taper portion of the slot *d* to pass into the notches *f, f*, of the shank.

Figs. 454, 455. 456, 457, represent a *gimlet* or auger handle, invented by G. H. Talbot. This inventor does not claim the method of giving a revolving action, in either direction, to the boring-

454.

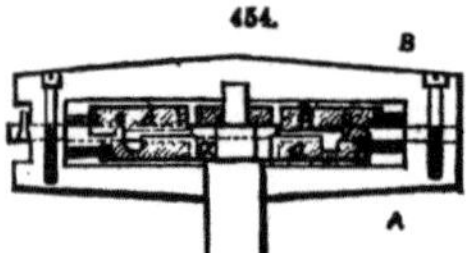

456.

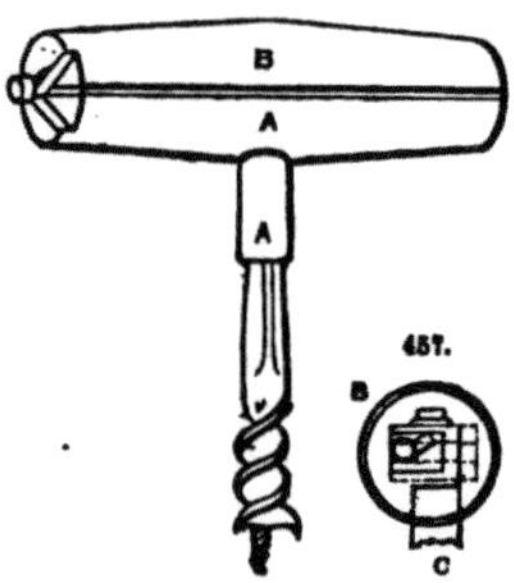

458.

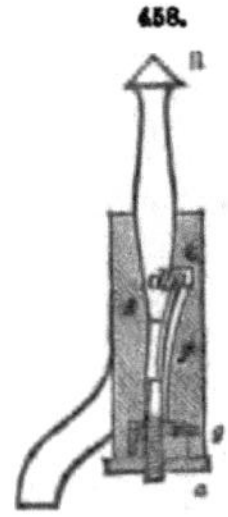

459.

455.

457.

tool by reversible pawls and ratchets, as such arrangement is common to *drill-stocks*. His invention consists in forming gimlet or auger handles so internally arranged, that when the pressure of the hand is applied on both sides of the axial line of the bit, and under the usual clutch of the hand over the centre line, the accompanying devices present no obstruction, but are protected from injury or derangement, as shown in Figs. 456, 457.

A mode of securing brace-bits in their sockets, proposed by E. W. Nichols, is shown in Figs. 458, 459. The bit R is represented as being fastened to the stock S. To loosen the bit, the nut *a* is turned, thus screwing down the shank of the spring, until the projection B of the spring springs back into the recess *c*, in the stock, when the bit may be drawn out. The pin *g* enters a groove in the outside of the nut *a*, which serves to confine *a* to the end of the stock.

Fig. 460 shows a ground-auger, invented by David Ring. It consists of two semicircular discs, D and E, of steel, each furnished at the ends of its straight-edge with cutters R and *j*, one pointing upwards, the other downwards. The cutter can be expanded by means of slots *f* in the discs, through which the confining-screws are passed into the cutter-heads at the bottom of the shaft: the latter is furnished with a gimlet-point, to facilitate its entrance into the earth.

Fig. 461 represents a hollow auger, invented by Wychoff. The cutters A A are each only half the width of the kerf, and are provided with two or more projecting cutting-points C and D, so arranged as to cut in parallel but concentric planes.

Another hollow auger, invented by J. McClure, is shown in Fig. 463. In this the cutters can be adjusted to different sizes of tenons by means of inclined cutter-bearing pieces, sliding upon inclined supporting-pieces *f f*. It is used in combination with a receding centre *b*.

460.

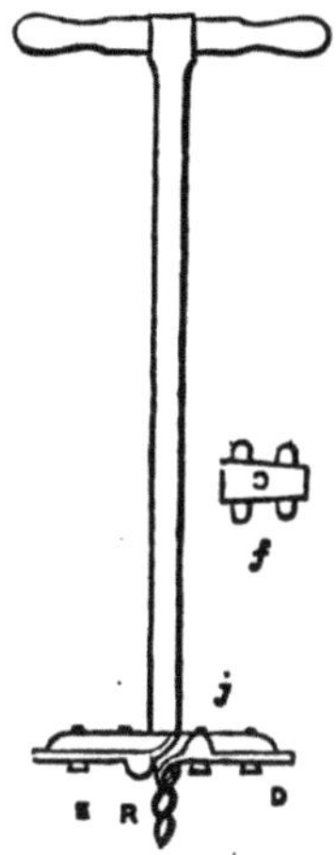

462.

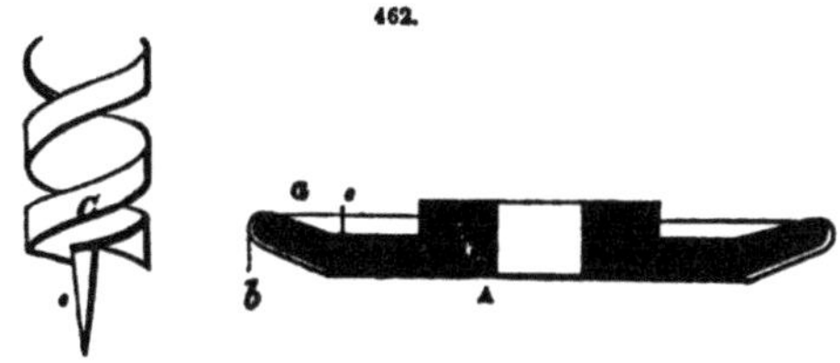

461.

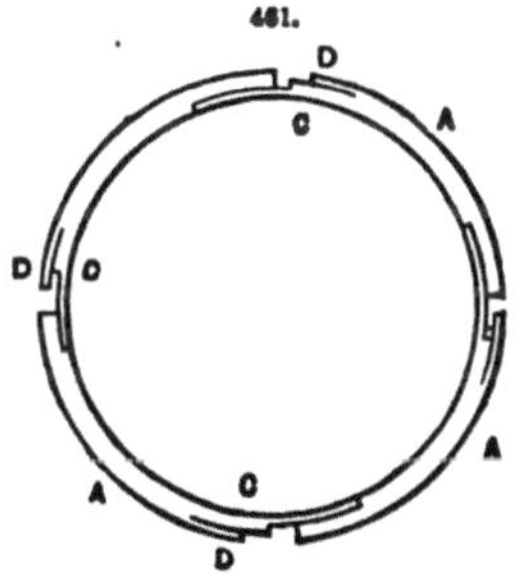

463.

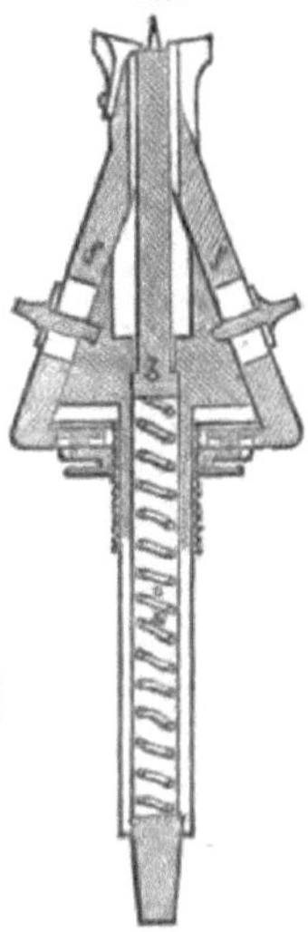

464.

465.

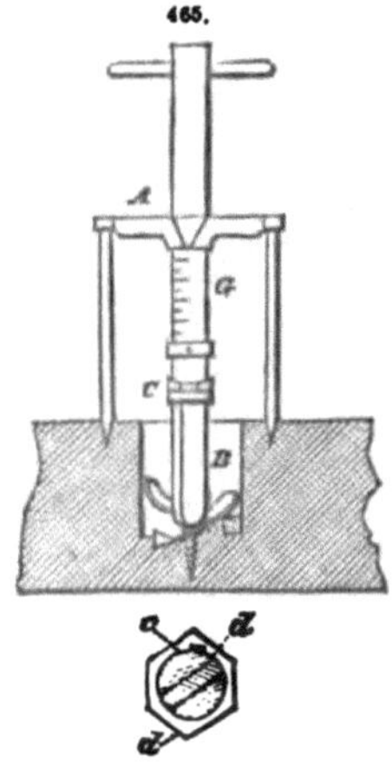

Fig. 462 shows a tool for finishing augers, invented by R. Jennings. It consists of a wheel or burr, having a bevelled surface *a* on one side, and a semicircular edge, corrugated to form a series of cutters *b*, which have a radial or nearly radial position, and extend from the inner edge of *a* to the outer edge of the same. These cutters are placed entirely around the semicircular periphery of the wheel, and act in combination with the concave surface *c* at the opposite side of the wheel.

Fig. 464 represents an auger invented by H. T. Love. This differs from common augers, in having the cutting floor-lips semicircloid in form on their cutting-edges, these edges forming an obtuse angle with the axis of the auger.

The auger invented by E. C. Gillett consists of a cam-sleeve C, Fig. 465, in combination with a slotted shaft G and the flat, notched shank *d* of an auger B. It is so arranged, that the shank of the auger will slide into the long slot in the shaft, through the cam-sleeve, which is then turned till it catches into the notches in the flat shank of the auger, and holds it fast in the shaft.

Hardening Steel-cutting Tools.—Any substance which combines chemically with iron will impart hardness to it; but the presence of carbon is required in order to produce that peculiarity, softness after tempering, which we observe in so high a degree in steel. Other substances, which have sufficient cohesion, produce a similar, but weaker, effect on iron than carbon; such are sulphur, or arsenic, and perhaps others; but no other substance is more available and more perfectly suited for this purpose than carbon. When the difference between hardened and tempered steel is caused by carbon or any other substance, in chemical union in the hardened, and in mechanical admixture in tempered steel, it is evident that this condition must be dependent upon temperature or other agencies. German steel manufactured at a high heat requires a white heat for hardening, and its carbon is so firmly united to the iron that this kind of steel may be welded to it with great facility. Refined shear steel, which has been much heated and hammered, will bear, next to German steel, the highest heat in hardening and welding. Ordinary cast steel will bear less than shear steel, and is welded with difficulty; the finest cast steel will bear the least degree of it, and cannot be welded to iron by the common process. This shows that a large quantity of foreign substances cause the steel to be more fusible, and to bear less heat in hardening. The degree of heat by which the chemical union of carbon and iron is accomplished is, therefore, not permanent; it varies with the fusibility of the metal. This we observe as well in cast iron as in all other metallic alloys. The change in the constitution of steel may be produced at very low temperatures, indeed at almost any temperature. The finest edge and a high degree of hardness are produced by the mere hammering of steel; and, if the above theory of the constitution of hardened and tempered steel is correct, the mechanical mixture of carbon and iron in tempered steel is converted into a chemical compound by mere compression. We recognize this fact in drawing wire, sheet iron, or in hammering iron, or any other metal or alloy. In these operations a large quantity of heat is liberated by compression, which may in some measure account for the metamorphosis; but we find that when the compression is very rapid, and brought about by great force, the phenomenon attending its application is similar to that of a strong heat, —the metal becomes extremely brittle. The best edge, and consequently the highest degree of cohesion and compactness in steel, is produced by striking a small bar of tempered steel with a small steel hammer, on a cold polished steel anvil, so as to avoid any considerable or perceptible increase of heat. If this operation is performed on a square bar of steel, by means of a hammer of the smallest size, of one or two ounces weight, and the corner of the steel thus compressed is ground down so as to remove the surface which has been touched by the hammer, we obtain an edge which cannot be surpassed for fineness by any other means of hardening. This shows that hardening may be performed by a variety of means, and that heat is not absolutely necessary. The strong cohesion of the metal is the cause of this phenomenon; and by whatever means we produce the close contact of the particles of metal and carbon, we secure hardness. These reflections serve to explain the manipulations which are employed in the manufacture of steel.

The common means by which steel is hardened are well known; these form no part of our investigations; but we may remark here that it is not so much the degree of heat to which steel is exposed before *chilling* it, as the difference of temperature between the cooling medium and the heat of the metal, together with the heat-conducting capacity of the *refrigerator*. Experiments have shown that little is gained by substituting other fluids than pure water for hardening steel. This is an entirely practical operation: the temperature of water may be in all instances the same, and fresh common spring water, or river water, is as good as any other fluid. But as the liability of steel to lose some of its component parts increases with the heat to which it is exposed, and as, near its smelting point, it assumes the nature of cast iron, it is found necessary, in order to preserve its original character, to perform the hardening operation at the lowest possible heat; for these reasons, water as cold as possible is used; and as, by plunging the hot metal into it, an atmosphere of steam is formed around it, which is a bad conductor of heat, either the metal or the water ought to be moved, to expose the hot metal to renewed action of the cold particles. The application of acids or salts for hardening is injurious to steel, however good conductors of heat such solutions are. Some of the fluid will always penetrate the metal and cause its decomposition. The use of oil or fat for this purpose is, if not equally prejudicial, at least of little benefit. That fluids penetrate metals, and particularly iron, is shown in wire factories: when iron wire is cleaned, after annealing, in diluted sulphuric acid, which is, by neutralization and washing, as far removed as practicable, it retains always some of the acid, which causes the wire to be brittle when fresh. An exposure of the wire to the atmosphere for some time removes the acid, and therefore the cause of brittleness. Wire thus cleaned by acids is often permanently injured in its strength, which does not happen with wire which is cleaned by the old method, with sand and water.

Annealing Steel for Cutting-tools.—It has been recommended, and it is also practicable in some instances, to modify the heat of the metal and the cooling medium for hardening, so that the contact of the two produces the required degree of hardness. A uniform degree of heat cannot be applied to all kinds of steel; and since the mode and time of heating is also important, and the fluid refrigerator cannot be uniform in composition and temperature, it is easily understood that this method of hardening cannot be universal. The common mode, and perhaps the best one for hardening, is to expose the steel to such a degree of heat, and so to cool it in water, that it assumes the highest degree of hardness, and then temper by exposure to a moderate heat.

A variety of means have been proposed for tempering hardened steel, such as melted fusible metals, lead and alloys of lead, heated fat or oil. When we reflect on the nature of steel, we soon find that the various kinds require different degrees of heat, by which they assume a definite texture or hardness, and that neither a certain degree of heat nor a certain colour of its tempered surface will indicate the actual condition of the steel. The operations on steel are of so delicate a nature that they cannot be brought under general rules—they are entirely dependent on the skill of the practical man; no language can impart that information which is applicable in all cases. The following statements are for these reasons to be considered as relatively true, and not as generally applicable.

Steel which has been hardened to the extreme should be exposed to a heat of 400° for chirurgical instruments, such as lancets—it assumes then a faint yellow colour on its polished surface; to 425° for razors, which tempers it yellow; to 432° for penknives, which it also tempers yellow; to 468° for scissors and cold chisels—the colour is brown-yellow; to 490° for edge-tools and common cutlery—colour purple; to 508° for table-knives—colour also purple; to 530° for small springs and weapons, to 537° for large springs, saws, *augers*—the colour is blue; to 580° for large saws, the colour of which it tempers to dark blue; beyond 600° steel becomes black, is annealed and soft. Steel which requires a high degree of heat for hardening, demands also more heat for tempering than that which hardens at a lower degree of heat; and steel exposed to a tempering heat of a certain degree for a length of time, increases in softness, even when the intensity of heat is not increased.

The colour of steel in the fresh fracture is white, like deadened silver. When tempered, it becomes more grey; and when annealed, it is grey, but the intensity depends on the kind of steel or amount of carbon. In all cases, the hardened, as well as the annealed steel, should not exhibit to the eye, even when aided by a lens, any crystallization. Hardened steel shows a little higher lustre than annealed steel. In making these distinctions, we should always regard hammering as equal to hardening by heat and refrigeration.

Steel is not very ductile. When cold, it will not bear much alteration of form, and when heated, it is, in some kinds, equal to iron. Cast steel will not bear much bending. German steel, shear steel, and all kinds of hammered steel, are more ductile when hot than those kinds which have been less subjected to compression.

The degrees of heat at which steel melts vary considerably: German steel requires the highest heat for melting—this may be about 3600°; the best kinds of cast steel will melt at considerably less, or about 2800°.

Oxygen has little effect on hardened steel; still, white cast iron, with much carbon, is superior to steel in resisting oxidation. The pure white colour of hardened steel, and its susceptibility of being oxidized when heated, cause the beautiful colours of tempered steel. When heated under a cover of oil, these colours do not appear; or, when such coloured and heated steel is brought into an atmosphere of hydrogen, the colour also disappears. This shows that colour is caused by oxidation, and that, in tempering steel under a coating of oil, extreme caution is required to hit the proper point of heat. These colours proceed from a coating of oxide, which, in its extreme thinness, causes the yellow, and when thick, blue; and finally, it becomes opaque and black. Acids facilitate the formation of these colours; alkalies delay or prevent them altogether. Pure iron requires a higher heat, to show the same colour, than steel: the best steel, and also white cast iron, show the series of colours at comparatively the lowest degree of heat. These tests may be arranged in the following order:—

Very faint yellow, for lancets	temp. from 400° to 430°	Fah.
Pale straw yellow, for razors and scalpels	„ 430° to 450°	„
Full yellow, for penknives and chisels	„ 470° to 480°	„
„ for cast iron	470°	„
Brown, for scissors and chisels	temp. about 490°	„
„ for wrought iron	495°	„
Red, for carpenters' tools in general	temp. 510° to 520°	„
Purple, for fine watch-springs and table-knives	temp. about 530°	„
Bright blue, for swords and lock-springs	„ 550°	„
Full blue, for daggers, fine saws, and needles	temp. 560°	„
Dark blue, for common saws	temp. about 600°	„

See HAND-TOOLS.

AWL. FR., *Alène*; GER., *Ahle, Pfrieme*; ITAL., *Lesina*; SPAN., *Leina*.

An awl is a pointed instrument for piercing small holes, as in leather or wood. The blade is differently shaped and differently pointed for different uses, as in the brad-awl, saddler's-awl, shoemaker's-awl, and so on.

466.

Fig. 466 shows the method of fastening an awl into the haft, invented by R. Egan. A metal socket-piece B forms the end of the haft A, to which it is attached by means of a stout tongue screwed into the haft. This socket has an internal screw, which is formed to receive the threaded shank of the awl C.

AXE. FR., *Hache*; GER., *Axt*; ITAL., *Accetta, Scure*; SPAN., *Hacha*.

An axe is an instrument, usually of iron, with a steel edge or blade, for hewing and chopping wood. See HAND-TOOLS.

AXIS. FR., *Axe*; GER., *Achse*; ITAL., *Asse*.

The straight line, real or imaginary, passing through a body, on which it revolves, or may be supposed to revolve, is termed an axis. See MECHANICAL POWERS.

AXLE-GREASE. FR., *Graisse d'essieu*; GER., *Lagerschmiere*; ITAL., *Untume*; SPAN., *Grasa, ó gordura pora eje*.

See ANTIFRICTION METAL. FRICTION.

AXLES. FR., *Essieu*; GER., *Welle*; ITAL., *Sala*; SPAN., *Eje*.

An axle or axle-tree is a transverse bar or shaft, connecting the *naves* of the opposite wheels of a car or carriage. Railway axles are termed *leading* and *tracing*, from their position in the front or in the rear of a car or truck, respectively. The journal-box of an axle is called an *axle-box*.

Thomas Thorneycroft, in a paper read before the Institute of Mechanical Engineers, considers an axle as having certain relations to a girder in principle. Girders generally have their two ends resting on two points of support, and the load is either located at fixed distances from the props, or dispersed over the whole surface: just so with the axle; it has its points of support and its loaded parts; but it is not clearly evident which are the loaded parts and which the props. It is found that the inclined surface of the wheel-tire ranges from 1 in 12 to 1 in 20, and, consequently, the direct tendency of the wheels under a load is to descend that incline, so that every vertical blow which the wheels may receive is compounded of two forces, namely, the one to crush the wheels in the direction of their vertical plane, and the other to move the lower parts of the wheels together: it will be seen that these two forces have a direct tendency to bend the axle somewhere between the wheels; should that yielding, or bending, extend no farther than one-half the elastic limit, if long continued, fracture will ultimately take place; but should the elastic limit be exceeded, the axle takes a permanent bend, the wheels are then diverted from their vertical plane, and, as a matter of course, leave the rails. To demonstrate this, an axle reduced in the middle to $3\frac{3}{4}$ in. diameter was placed upon two props 4 ft. 9 in. apart, and loaded in the middle: the utmost of its deflection, without a permanent set, was ·232 in., the load carried 7 tons. An axle reduced to 4 in. in the middle was then placed upon the props 4 ft. 9 in. apart: its utmost deflection, without a permanent set, was ·281 in., the load carried 9 tons. Another axle, but parallel, $4\frac{5}{16}$ in. diameter, was next placed upon the props 4 ft. 9 in. apart: its utmost deflection, without a permanent set, was ·343 in., the load carried 14 tons. Hence, by reducing an axle of $4\frac{5}{16}$ in. diameter in the middle to $3\frac{3}{4}$ in., its limit of elasticity is reduced from ·343 in. to ·232 in., and the load required to produce that elasticity is from 14 to 7 tons. Fig. 467 shows the position of the wheels to the rails when the bending of the axle has exceeded its elastic limit.

467.

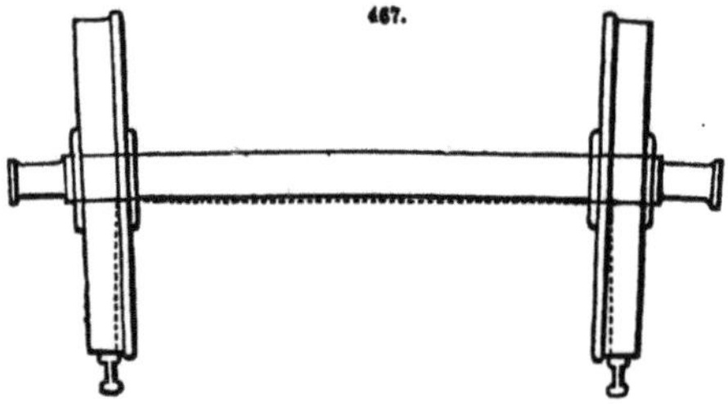

470.

471.

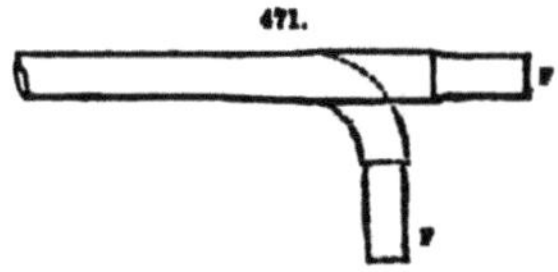

468.

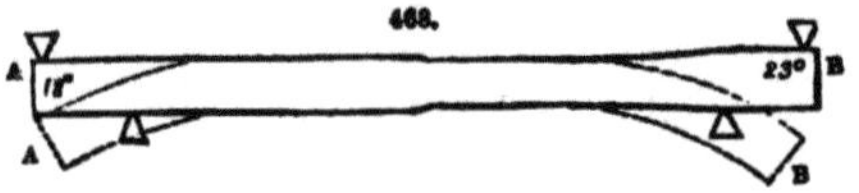

469.

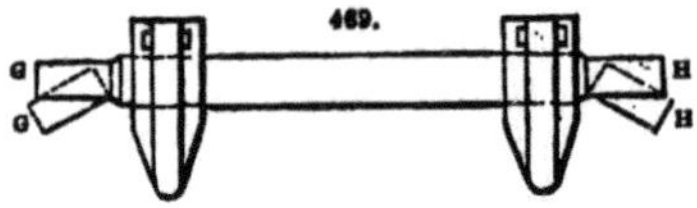

472.

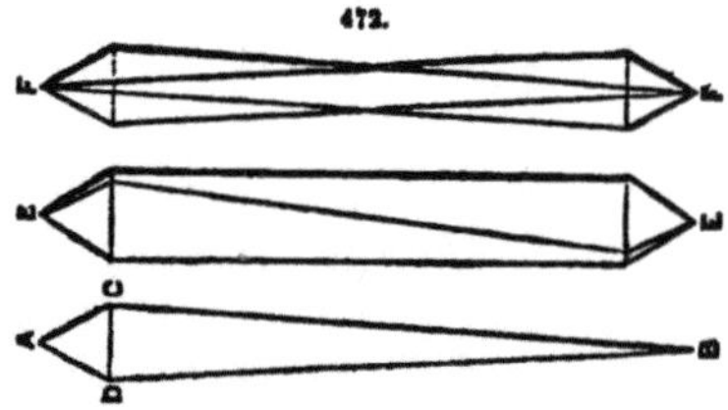

To ascertain what influence the reduction of an axle in the middle would have on its strength to resist sudden impact, compared to an unreduced one, an axle was made as represented by Fig. 468, which shows the end A parallel to the centre $4\frac{1}{2}$ in. diameter, and the end B drawn down from the back of the wheel towards the centre, where it is 4 in. diameter. The end A was subjected to impact—the relative position of prop and ram was the back of the wheel and the neck of the journal; this end received forty-six blows of the ram, and bent to an angle of 18°. The end B was then subjected to impact—the prop and ram in the same relative position, when it bent back to an angle of 23° with only sixteen blows of the ram, as shown by the dotted lines in Fig. 468. To ascertain what influence a shoulder behind the wheel would have on the strength of the axle at that part compared to one without a shoulder, an axle was cut in two, Figs. 470, 471, the end E turned from the neck of the journal, leaving a shoulder $\frac{1}{8}$th in. deep as a stop to the wheel; and the end F turned from the neck of the journal to the same diameter, but no shoulder left. The end E was subjected to hydraulic pressure, the load being in a direct line with the shoulder, when it broke in two with a load of 60 tons. The end F was subjected in the same way to hydraulic pressure, when it bent into the form shown by the dotted lines with 84 tons. To ascertain what influence the position of the wheel in relation to the neck of the journal would have on the strength of the latter under impact, a piece of an axle, Fig. 469, was prepared, with a journal taken down at each end, the end G keyed into a cast-iron frame, and the face of the frame in a line with the neck of the journal; the journal was then subjected to the impact of a ram falling 10 ft., when it broke off at the seventh blow. The end H was keyed into the cast-iron frame in the same way, but with the neck of the journal projecting $1\frac{1}{4}$ in. from the face of the frame; the journal was then subjected to the impact of the same ram falling 10 ft., when it broke at the twenty-fourth blow.

From these experiments, and from the acknowledged deteriorating influence of vibration or bending on iron, especially when continued any great length of time, it is Thorneycroft's opinion that neither shafts nor railway-axles ought to be reduced in the middle, but rather, if there is to be a departure from the parallel form, they should be made thickest in the middle, and thus effectually prevent any vibration or bending whatever.

In the discussion which followed upon Thorneycroft's paper, E. A. Cowper said that Thorneycroft's conclusions were arrived at by experiment, unaided by theory. If A B, Fig. 472, is taken as the axis of a railway-axle, A and B as the centres of the journals, and C as the centre of one wheel, we should have the case of a girder weighted at A and B, and supported at C; the proportionate *strength* ought, therefore, to be as a triangle A C D, and a triangle B C D. Now if we put two of these triangles together, as at E E, we shall at once arrive at the result that the strength of the axle should be uniform between the wheels, and consequently parallel, and from the wheels to the journals the strength should be as a triangle; and if we had merely to do with a strain due to the weight on the journals of the axle, these proportionate strengths would be strictly correct. *But* we have to provide against the lateral strains, from the flanges of the wheels suddenly striking the switches and crossings in passing through them; and this is so much greater than the mere weight on the axle, that it must be considered chiefly in determining the form of an axle; if, therefore, the Fig. A C B D were reversed on itself, the outline thus given would represent the proportionate *strength* of an axle, as at F F, which is fully in accordance with the usual tapered form of axles. *The actual diameters* can be easily arrived at, by taking the cube root of the width.

J. Ramsbottom, in the same discussion, considered that whether in the case of axles or machinery of any kind, there must be an error in the proportions if any one could say beforehand at what point a fracture was likely to occur. He could only conceive one instance in railway practice where axles should be parallel, and that was in the case where the forces were applied in a line directly parallel to the axle; as at A A, Fig. 473. If a pair of wheels are running, for

473.

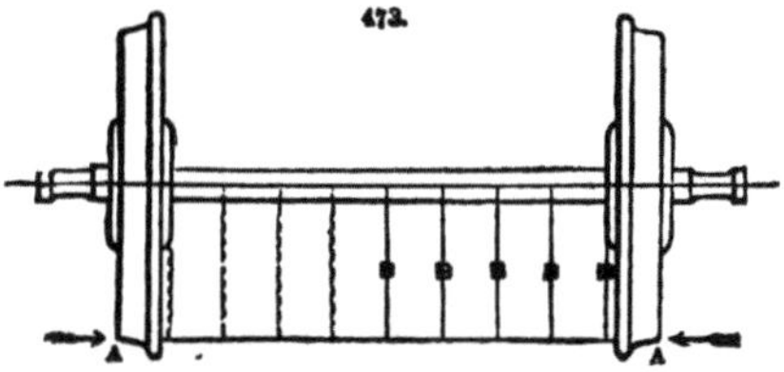

474.

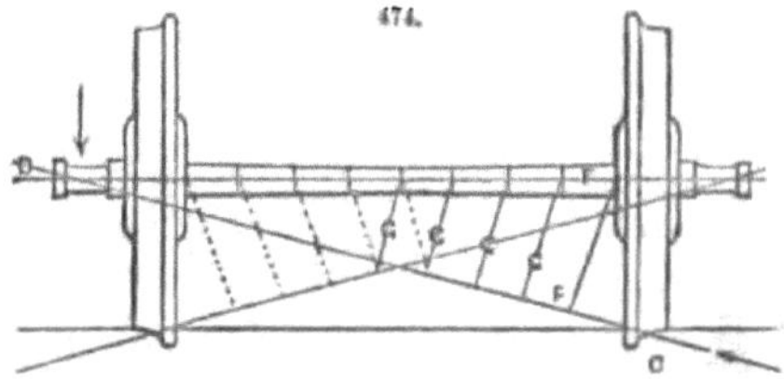

instance, between rails converging to a point, the axle should be parallel, since the effect of the leverage is the same at all points B B between the wheels. But there is another and more important force resulting from a lateral blow upon one wheel only, coupled with the load of the vehicle on the axle, the direction of which will be tolerably well indicated by a line C D, Fig. 474, drawn from the circumference of the wheel on one side to the centre of the journal on the other side; and if the line F F is drawn perpendicular to this, from the axle close to the wheel, it will represent the greatest effective leverage, tending to break or strain the axle at that point; and the strain upon any other point of the axle may be found by drawing lines G G parallel to this; in fact, the cube roots of these lines will give the diameter of the axle at the points where they fall.

To obtain the advantage of the shoulder at the inner side of the boss without weakening the axle, A. Slate has suggested the introduction of a small shoulder inside the boss of the wheel near the outer side, see A, Fig. 475. The wheel has to be bored out a tight fit, and forced on to the axle by a hydraulic press, but the first, 1½ in. from the inner side, must be slightly coned, as shown at B, so as to remove the grip on the axle from the extreme edge of the boss, and prevent the tendency of the axle to break at that point.

475.

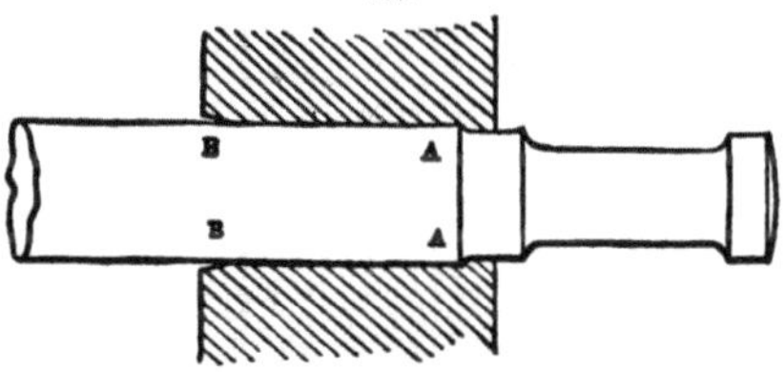

Fig. 476 is a longitudinal section, and Fig. 477 a transverse section, of the axle-boxes employed on the North-Kent line. The top of the box is circular, for a peculiar arrangement of springs; the box is cast open-fronted, with a movable front A to attach by screws, a grease-tight joint being maintained by an elastic substance between. From this arrangement the interior of the box can be inspected, and a new brass applied, without lifting the carriage. The back of the box round the axle is cast with a round-edged projecting lip B B. A plate of metal C C, with centre hole fitting the axle, is secured by bolts to the back of the box, with a piece of leather D D, the orifice of which is enlarged into a partial pipe-form round the axle, to give increased bearing-surface. This leather presses equally on the axle and on the lip of the box, and thus a tight joint is maintained, which preserves the grease without overflowing above the level of the bottom of the axle. The bolt-holes in the metal plate are vertical and oblong, so that when the upper bearing-brass wears, and causes a corresponding wear in the plate and leather, and consequent leak below, the two latter may be drawn upwards to fit the lower part of the axle. At the top of the box there is a screw-tap E, for feeding, with holes through it to admit the ingress and egress of air. This tap serves to feed an upper chamber F, with holes to the axle as usual; this tap also feeds the lower chamber G, which, in addition, catches the grease that falls through from the upper chamber by the working of the axle. A piece of hard wood H is applied between the end of the axle and the front of the box, to prevent wear of the shoulder-collar of the bearing-brass. Two collars of light wood I, I, float on the oil or grease in contact with the lower side of the axle, and thus carry up the lubricating material, if it happens to be below the level of the axle. Figs. 478, 479 are a longitudinal section and back view, showing an upper and lower feed of an axle and axle-box. To retain the oil or grease, a conical metal spring C C is inserted in a corresponding circular groove at the back of the box; by its elastic expansion it presses against a strip of leather lining the groove, and thus forms a tight joint. The small end of the conical spring clips a leather pipe-collar D D, fitted on the axle, which collar may either revolve with the axle in the small end of the spring, or it may be fixed to the spring, and the axle revolve within

476.

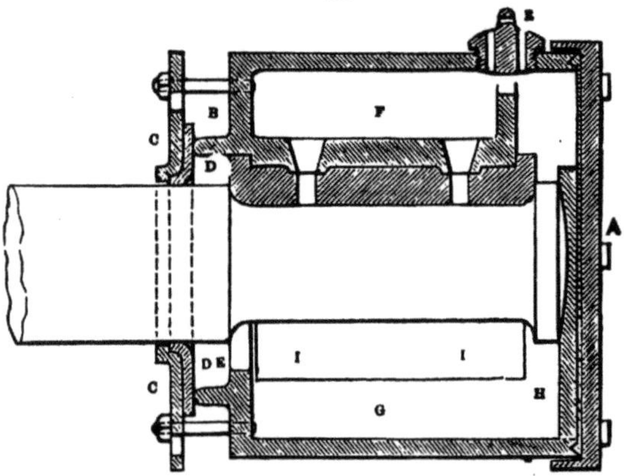

477

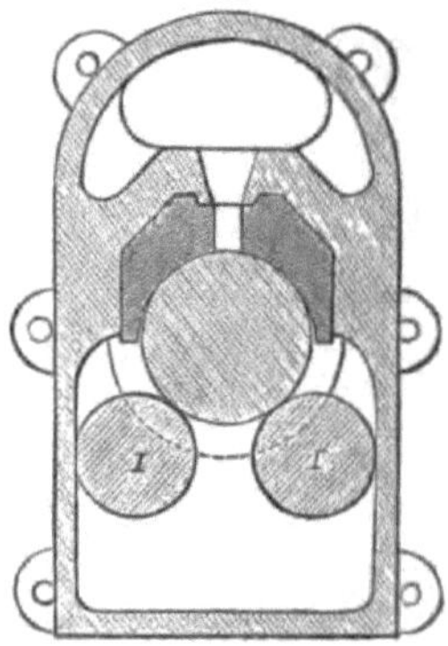

the leather collar; as the spring expands against the groove in the box, it has no tendency to press the axle or leather too tightly, so as to cause much friction. The conical spring is prevented from

478.

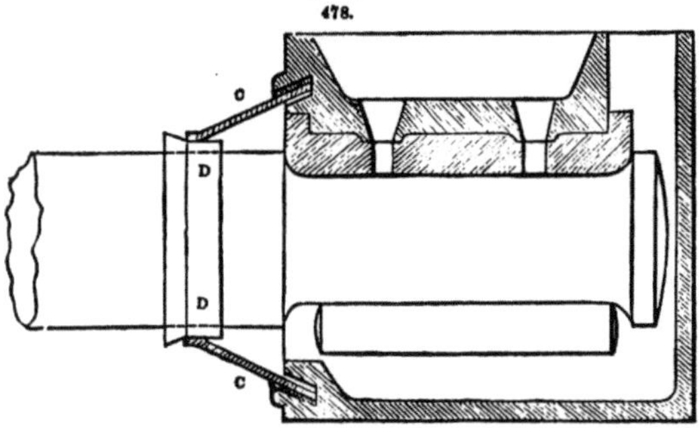

479.

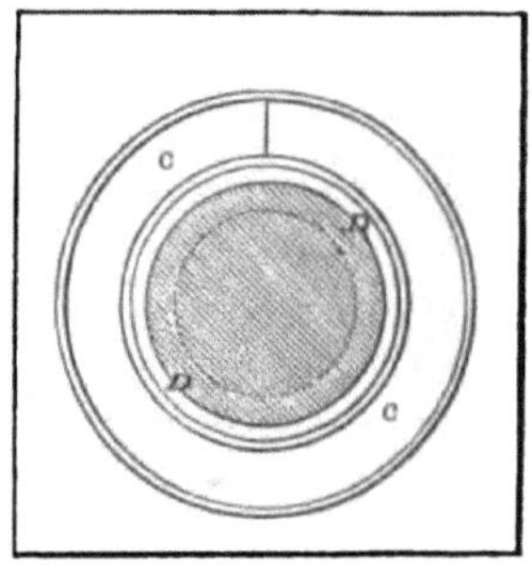

turning, by a stud; the edges of the spring overlap each other, to keep out dirt, and the hollow space between the spring and the axle may be filled with sponge or cotton-waste. Figs. 480, 481,

480.

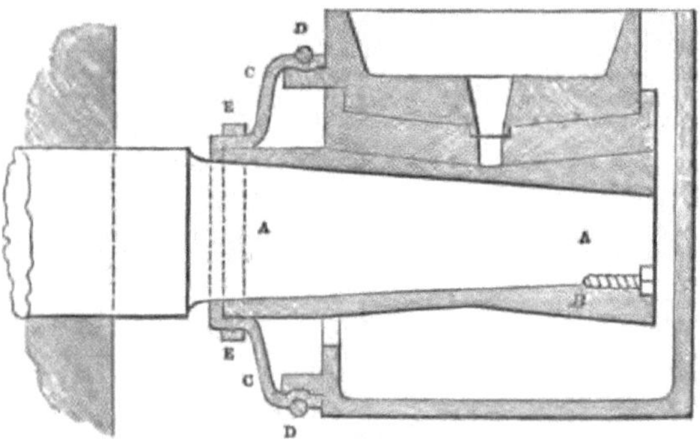

481.

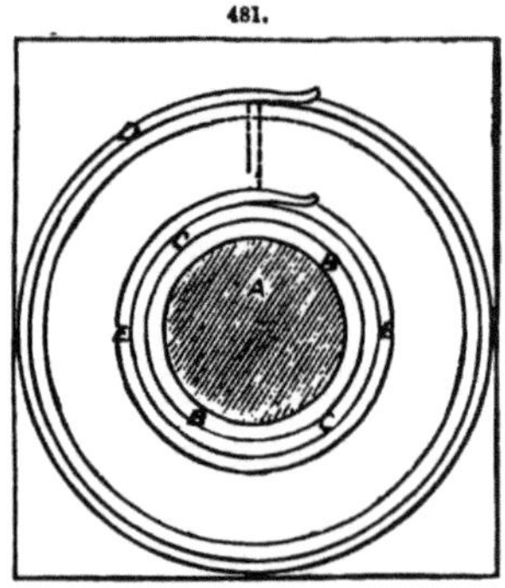

show a longitudinal section and back view of an axle-box on a similar arrangement, in which a conical pipe of blocked leather C C, is secured to the box-lip by an elastic ring D D, similar to a key-ring, and clipped to the axle by a second ring E E. Both the spring cone and leather cone will, by their free action, accommodate any irregular movement of the box, and prevent loose wear between the metal plate and leather. In all cases where any material comes in contact with the revolving axle, it is essential that the surface be properly smoothed, that the pressure be as light as may be convenient, and the lubrication certain. Fig. 480, in addition to the axle-box arrangement, shows a mode of applying movable journals to axle-arms, either new or old. Thus the journals A A, may be forged down to a taper, with the object of extending the distance of the bearing from the wheel, or of increasing the diameter of the axle-bearing. The movable bearing B B, may be of wrought iron or of cast iron, well got up and *case-hardened;* and manufacturers might be enabled to supply cheaply a superior class of axle-boxes and axle-bearings. This arrangement could be well adapted to the hollow axle shown in Figs. 482, 483, and it also enables

482.

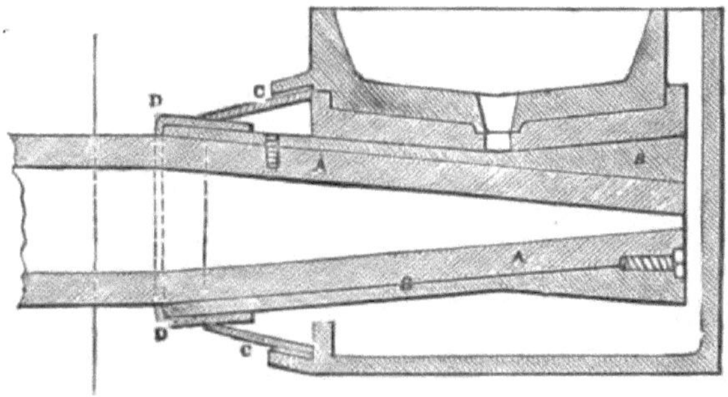

483.

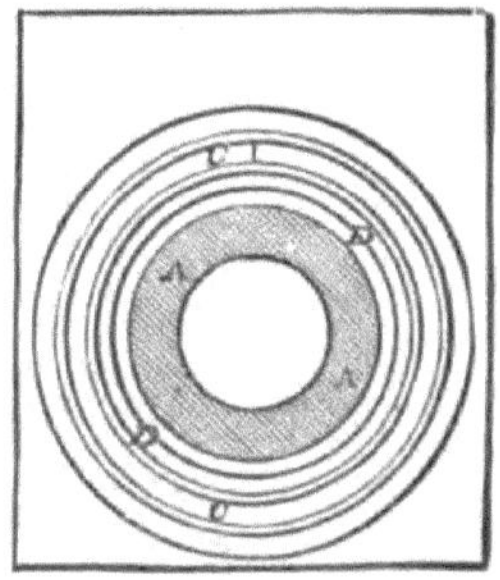

constructors, at comparatively small cost, to replace axles, when rendered unsafe by long running and much vibrating.

It should be remarked that it is desirable in all cases to get the axle-bearings as long as is convenient, even when not required for bearing-surface, for the following reasons:—The points of the springs which support the frames are at a considerable elevation above the axle-bearing, and act with mischievous leverage to tilt the axle-boxes laterally to the carriage, when the wheel-flanges strike the rail. It is evident that unless there be some proportion in the length of bearing to the height of the spring, there will be a great strain upon the guards of the axle-boxes to keep them steady.

The axle-box described above is the first application of the principle of retaining the grease or oil, by closing in the back of the box.

Many similar contrivances have since been brought out by various other persons for the purpose of retaining the grease in the axle-boxes, but it appears that the original application of the principle was due to W. B. Adams, in May, 1847. (See 'Proceedings of the Institution of Mechanical Engineers,' April 27, 1858.) There does not appear to be very material variation in any of the subsequent plans.

As to the practical mechanism for keeping out dirt and preserving the bath of grease, it must vary with circumstances. Many axle-boxes are so close to the wheel-bosses, that the leather pipe-collar is the only practical arrangement; and having come into general use, it is difficult to change it; but W. B. Adams prefers the elastic metal collars, C C, shown in Figs. 478 and 482, pressing upon leather pipe-collars D D.

The object sought is to form a tight joint between the box and the axle, which are both exposed to rough jolts and a tilting movement of the box on the axle; therefore the medium to from the joint should be flexible, and not liable to be put out of order.

The mode of lubrication from above the bearing has one objection, in the liability to accident by dirt getting on the arm, and from the holes wasting a most important part of the bearing-surface; but W. B. Adams thinks it preferable to retain this arrangement, keeping the holes small, as it renders a security in case of any accident happening to the lower reservoir.

Two forms of journal are shown in the diagrams; one the double cone, Figs. 480 and 482; the other the ordinary cylindrical journal with collars, Figs. 476 and 478. There is an advantage in the double cone with the small diameter in the centre of the bearing, that it has a tendency to cause the lubricating fluid to press outwards from the centre while in rapid motion. The cylindrical bearing between collars has also this disadvantage, that the box is not kept in its position by gravity, but by a very small collar-surface, which, being vertical, does not retain the lubricating fluid so easily as the horizontal surface; and, moreover, by its larger diameter has a tendency to throw it off by centrifugal action. Where the boxes fit tightly to the guards, the collar-bearings are frequently subject to rapid wear; and lateral thrust is more destructive than the downward pressure of the load. The small rounded corners next the collars, intended to prevent the "nicking" or breaking of the axle, are of little service to give the box a centripetal tendency. The cylindrical bearing has the advantage, that the bearing-surface is not lessened by end play; and with the axle working in a bath of lubricating material, the collars will at all times be safe. In either case, of the cone or the cylinder, it is clear that the lubricating bath below will be the safest precaution against heating. As regards the strength of the axle, the coned journal has the advantage by its gradual tapering form, supposing an equal amount of metal in both cases. The fitting of the bearing-brass to the journal is a matter of greater nicety with the cone than with the cylinder. With cylindrical journals the usual practice is, to make the bearing-brass of considerably larger radius than the journal, so that it bears on a very small surface, which wears to a polish, and gradually extends to the half diameter. In point of fact, railway bearings are made to grind themselves to a true fit or work, instead of being accurately ground and fitted beforehand, as is the case with nicer machinery.

In the axle-box, Fig. 484, the lid vibrates on a hinge, and is secured above by a disc on the axis of the handle, engaging a recess on the upper edge of the box. The floor of the oil-chamber is inclined upwards and outwards, and has a loose perforated frame to hold the packing. In this arrangement of J. Montgomery, shown in the figure, the lid A of the journal-box C, is hinged below and secured by a handle B and jamming-cam *b*; C′ is the loose or hinged rack frame, placed inside the axle-box C.

484.

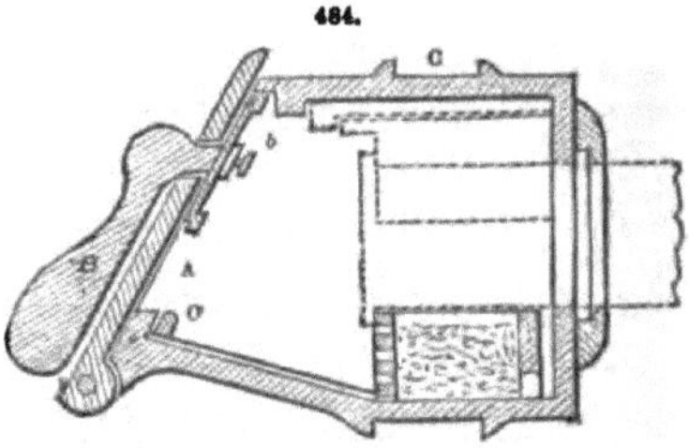

W. G. Craig, of Manchester, in a paper published in the 'Proceedings of the Institution of Mechanical Engineers,' observes that in axle-boxes the principal defect is in the seats for the gun-metal bearings. These are arranged in various ways, but all involving expensive fitting of the bearings into their places to secure them from getting loose, notwithstanding which they frequently turn on one side, as represented at A, in Fig. 485, and are also liable to drop out of their places entirely and remain in the bottom of the axle-box, causing the journal to come in contact with the cast-iron box, which is soon destroyed, as in the example shown in Fig. 486,

485.

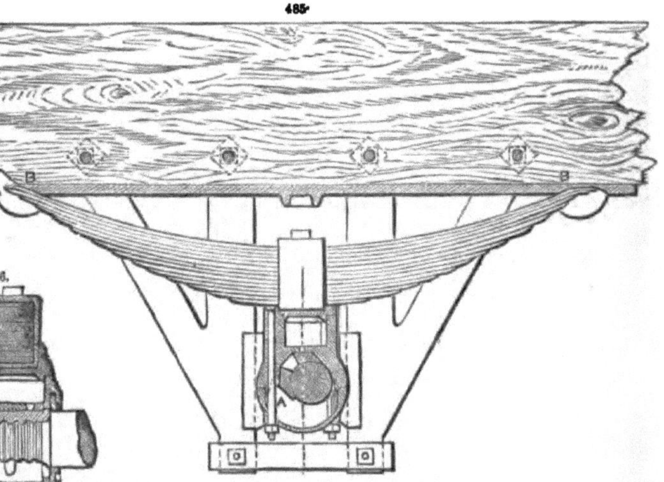

486.

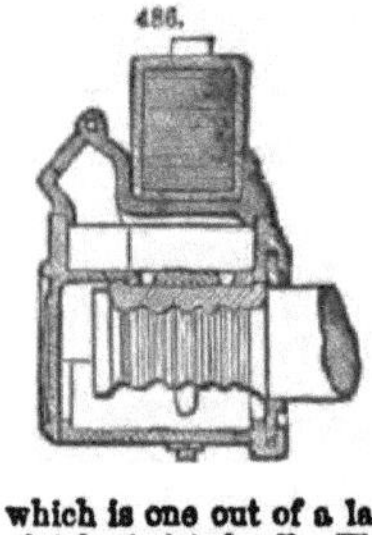

which is one out of a large number of similar cases. In many instances the journals are completely twisted off. These occurrences necessarily cause great delays to trains, and require the carriage or wagon to be sent into the workshops for repair. In order to prevent the brasses from coming out of their places, they are cast with deep sides, and not unfrequently riveted to the axle-boxes; but the consequence of this is, that when the brass warms, its sides grip the journal, and soon cause the bearing to heat.

Another defect in axle-boxes is in the mode of lubricating, which tends to create the evil of hot journals before the remedy can act. This arises from the thickness of the metal that forms the seat of the bearing and bottom of the grease-chamber, which in many cases amounts, together with the bearing, to two inches thickness above the journal, and must be heated through by friction before the grease can be brought down to lubricate the bearing. In numerous instances, from the metal becoming overheated, the grease is rendered fluid, and runs out and is wasted.

Another disadvantage is that, from the carelessness of workmen, the holes for the passage of the grease through the axle-box and bearing are often not made to correspond, and the flow of grease to the journal is consequently obstructed. It is also difficult to clean out such holes when required, to do which properly involves a considerable loss of time, and the evil arising from not frequently picking them out is a total stoppage of the lubricating material, causing cut journals.

Another difficulty is that of getting the grease to the back of the grease-chamber to lubricate the inner collar. This is now done by the greaser pressing back the grease with his knife, when time admits, and is generally only imperfectly accomplished whilst the carriages are temporarily stopped at a station.

Further, the number of parts belonging to the axle-boxes now in general use, bolts, nuts, leather packing, and so on, all involve expensive fitting, and are subject to frequent displacement.

In the axle-box grease-covers, again, there is a constant rattling noise with all iron covers that are without springs, and they often require repair from being made slight, and the hinges being strained; and with the sliding arrangement, the covers get lost in transit, and grit and other foreign matter then mix with the grease, and produce hot axles, either by stopping up the lubricating holes, or by coming in contact with the journal. An objection also to iron covers is, that during the summer months they quickly conduct the sun's heat to the grease and cause it to run to waste.

In the spring fittings, W. G. Craig states that all arrangements are defective where bolts, nuts, or screwed spring-clips are used for fixing the spring to the axle-box, as these are not found sufficient to retain the spring in position in ordinary concussions, entailing the frequent loss of bolts and nuts, and at times of the whole spring. It is also objectionable for the ends of the spring to bear against fixed shoes fastened to the sole-bars, since the spring, on elongating, causes the ends to rub hard upon the shoe, which wears a recess in it, and also wears the top plate of the spring; the result is that a shoulder is formed upon the shoe, and in many instances the spring cuts the body of the shoe entirely away and bears upon the wood, as represented at B B, Fig. 485, which is an exact copy of one of the many so found on the Manchester, Sheffield, and Lincolnshire Railway, in all such cases the spring is prevented from acting beyond the limit of the shoulder formed on the shoe. Amongst the methods adopted to remedy this defect, the spring ends have been softened and bent downwards to fit the shoe, and the top plate of the spring has been thickened to allow for wear; the former plan has been abandoned as inefficient, and the latter, which is now generally in use, is subject to the objection before named.

The failures of the axle-boxes, bearings, grease-covers, and spring fittings described above, involve a serious item in the working expenses; and from his experience of the causes producing those failures, W. G. Craig was led to adopt the construction of axle-box and spring fittings, which we shall now describe.

The improved axle-box is cast in one piece to avoid all expensive fitting of leather packing, bolts, nuts, and so on; the mode of lubricating through the axle-box seat by means of holes is dispensed with, and the lubrication is effected direct through the gun-metal bearing. All expensive fitting of the bearing is avoided by casting an oblong hole in the axle-box seat, to receive a projection C, Figs. 487 and 488, cast upon the top of the bearing; these can then be put

487.

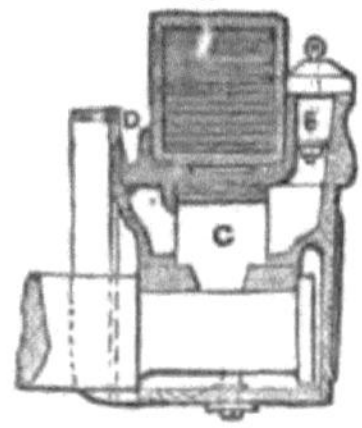

488.

489.

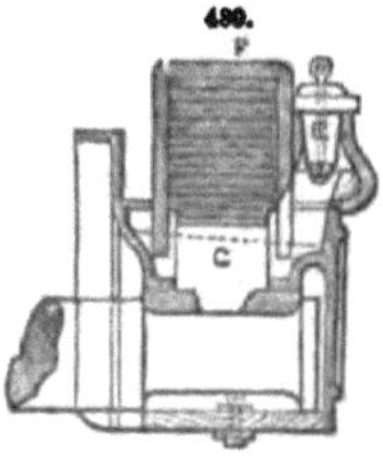

together as they come from the foundry, a slight looseness being of little consequence, since the projecting piece completely prevents the possibility of the bearing turning on one side, as at A, Fig. 485. This axle-box is also much lighter than the ordinary construction, and all bolts and nuts are dispensed with, excepting the two bolts that fasten the under-keep.

The projection cast on the top of the bearing is made hollow, and, passing through the oblong hole in the seat of the axle-box into the grease-chamber above, supports either the cap of the axle-box, as in Fig. 487, or the under-side of the spring itself, as in Fig. 488. By this means is ensured not only the impossibility of the brass turning, but also the perfect lubrication of the journal through the large aperture in the projection, which brings the grease in direct communication with it, so that a very small increase of temperature is sufficient to cause its perfect lubrication. In this case the cast iron of the axle-box is never required to get hot in order to soften the grease, that being one of the chief causes of hot journals. A further advantage obtained by the projecting piece on the top of the bearing is, that the brass may be made of a shallow form, so that it will never require its sides to be reduced to prevent them from clipping the journal.

In the grease-covers, W. G. Craig has adopted two plans to suit different forms of springs. In the first, shown in Fig. 487, the cover D clasps the top of the axle-box, and prevents the admission of grit, dust, &c., which cannot be too carefully excluded from the top of the box. It is cast with an oblong recess upon it, extending upwards, which receives the grease above the level of the grease-chamber, and is fitted with a wooden plug or plunger E, secured by a chain to the wagon, in place of the ordinary iron cover or slide; the grease being thus introduced above the level of the grease-chamber ensures the perfect lubrication of the whole bearing, being forced by the pressure of the plunger into the recesses at the back of the brass.

The second modification of improved grease-cover, shown in Figs. 487 and 489, is combined with an improved spring-fixing; a wrought-iron plate is employed, about $\frac{1}{16}$ths in. thick, and of sufficient size to project over the axle-box top about ½ in. all round; an opening is made in the plate to admit the grease, and two other openings to receive the ends of the spring-clip F, to which the plate is secured by a *cotter*. On the top of the plate is fixed a piece of wood, about 1 in. thick, on which is screwed an oblong casting, similar to the oblong recess in Fig. 487, having a wooden plug E to fit the opening: the advantage of this arrangement is, that this portion of the grease-cover can be replaced when broken, without having occasion to lift the wagon.

The improved spring-fixing consists of a simple and cheap contrivance in combination with the axle-box; a piece of iron F, ½ in. thick, and the same width as the spring, is simply bent over the spring to form a clip, Fig. 489, allowing the ends to project to form cotter-holes, and to fit into the cover of the grease-chamber. Two cotters fasten the clip tightly to the spring, and are cut so as to correspond with the length of the grease-chamber, by which means they are prevented from getting loose. The cotters form the two sides of an oblong recess on the under-side of the cover, and receive between them the projecting hollow piece C on the top of the brass, as in Fig. 488. A piece of wood or india-rubber is inserted between the under-side of the spring and the top of the projection C, to form a cushion for the spring.

A modification of the improved spring-fixing, for application to existing stock, is made by fixing two pieces of iron, about 2 in. wide, and the length of the width of the spring, and about ⅝ths in. thick, by means of the bolt through the centre of the spring; on each side of these pieces is fixed a welded hoop of iron, 1 in. by ½ in., as a support for the centre of the spring, in addition to the ordinary small rivet passing through the spring, and a corresponding recess is made on the cast-iron cover of the grease-chamber.

The fittings at the ends of the springs are improved by making the shoes G G, Figs. 490 and 491, to slide on a metal bed H, about ¾ in. thick, and of sufficient length to allow of the play of the spring with the sliding-shoes attached.

490.

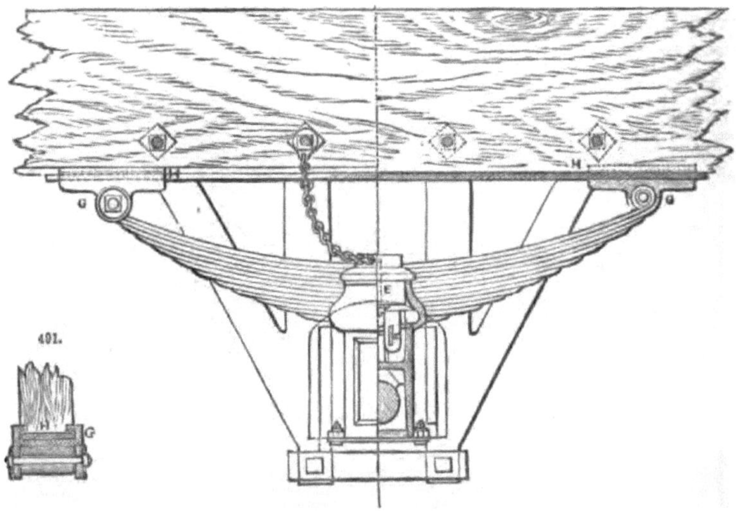

491.

The plate H is fixed on the under-side of the sole-bar, materially strengthening the wagon, and the sliding-shoes have grooved sides to secure them to the plate. The spring is connected to the shoe by a bolt passing through the side cheeks of the shoe and the spring-eye, the object being to retain the spring attached to the wagon, when the journals require examination, thus saving considerable expense and time in lifting.

The sliding-shoe maintains the perfect action of the springs, and their working is, in consequence, so easy, as to be almost equal to that of a first-class carriage fitted with scroll-irons, if the same length of spring is used in each.

AXLE-BOX. Fr., *Boîte à graisse;* Ger., *Schmierbüchse;* Ital., *Serbatoio dell'untume;* Span., *Caja dell eje.*

An axle-box is the journal-box of an axle, especially a railway axle. See Axles.

AXLE-TREE. Fr., *Tourillon;* Ger., *Wagenachse*, Ital., *Sala;* Span., *Eje.*

An axle-tree is a fixed transverse bar of wood or iron supporting a carriage, on the rounded ends of which the opposite wheels revolve. See Axles.

BABBIT'S METAL. Fr., *Métal de Babbit;* Ger., *Babbit's Metall;* Ital., *Lega Babbiti.*

See Alloys.

BACK. Fr., *Arrière;* Ger., *Rücken;* Ital., *Il dietro;* Span., *Atras.*

Back.—Of a hand-rail or rafter, the upper part. Of a stone, the part opposite to the face. When the stone is used as a facing to a brick wall, it is squared at the back; but if the wall is built of *rubble* work the back of the stone is left rough, as it comes from the quarry.

Back, in Brewing.—See Sparger.

Back-flap.—A leaf hinged to a window-shutter, in order that the shutter may fold back to enable it to fit into the side-boxings. Back-flaps, in the best work, are framed in panels, but in inferior work they are formed of one piece; in either case they are made of less thickness than the shutter, usually an inch when the latter is from 1¼ to 1½ in.

Back-day is a term applied to a day, the wages of which are usually kept in arrear. If the workmen are paid on Saturday, their wages are made up to the previous evening. When a workman quits his employment, he is said to take his "back-day," that is, to require his account to be closed.

492.

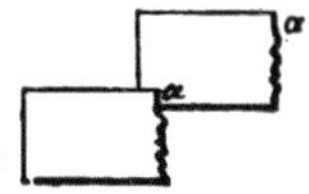

Back-hearth.—The part under the grate, of a stone or an iron hearth to a fireplace.

Back-joint.—That part of the back of a stone step, dressed to fit into the rebate of the upper step, as *a a*, Fig. 492. When there is no rebate, the part of the under-side of the step, dressed to fit over the edge of the lower step, is called the back-joint. It is usually from 1 to 2 in. wide.

The term is also applied to the part of the back-edge of a landing or step against which a wood or paved floor abuts, and which is dressed to make a close joint.

Back-lining of a Boxing.—The piece of framing which forms the back part of the *boxing*. It should be tongued to the architrave grounds at one edge, and to the inside lining of the sash-frame at the other.

493.

Back-lining of a Sash-frame.—The part of a cased *sash-frame* next the wall, opposite the pulley-stile.

Back of a Window.—The part between the sill of the sash-frame and the floor. When this part is recessed into the wall, the sides are called elbows. See Fig. 493.

BACKER. Fr., *Ardoise à dossier*; Ger., *Unterschiefer*; Ital., *Lavagna stretta.*

A narrow slate laid on the back of a broad, square-headed slate, at the spot where a course of slates begins to diminish in width, is called a backer.

BACK-LASH. Fr., *Jeu*; Ger., *Todter Gang, Spielraum*; Ital., *Retrocolpo.*

See Reaction.

BACK-LINKS.

See Parallel Motion.

BACKING. Fr., *Remplage*; Ger., *Füllung*; Ital., *Ripieno*; Span., *Ripio.*

Backing is the hearting or internal mass of a wall faced with material of a superior quality. The filling over the haunches of a vault or arch is also termed the backing. Filling with earth or other material behind a wall to strengthen it against heavy shocks, as behind a *sea-wall* or between the walls of a *traverse*, is termed backing up.

Backings are slips of wood, fixed against rough walls, to which are attached wood linings or other finishings.

BACKWATER. Fr., *Eau dormante*; Ger., *Stauwasser*; Span., *Remanso.*

Water held back in a stream or reservoir by some obstruction, or water thrown back by the turning of a water-wheel, is called backwater.

BADIGEON. Fr., *Badigeon*; Ger., *Kitt aus Steinstaub und Gyps*; Ital., *Stucco*; Span., *Betun.*

Badigeon is the name given to any mixture for stopping holes in stone or wood work, for the purpose of hiding defects. The badigeon used for stonework is composed of plaster and freestone, well mixed and ground together. The badigeon employed for woodwork is sawdust and strong glue, or putty and chalk.

BAG.

A bag is a measure reckoned equal to a Winchester bushel striked, or 2150·42 cub. in. Twenty-five bushels or bags make a hundred, or ton, of lime. A bag of plaster of Paris is reckoned at 14 lbs., and eight bags equal a bushel nearly.

BALANCE. Fr., *Balance*; Ger., *Wage*; Ital., *Bilancia*; Span., *Balanza.*

An apparatus for weighing bodies is termed a *balance*; in its original and simplest form, Fig. 494, it consists of a beam or lever A B, suspended exactly in the middle, O, with a scale or basin of precisely equal weight, hung to each extremity A, B.

In the balance, Fig. 494, where any two equal weights whatever, placed one at each extremity, are to be in equilibrium, it is requisite, first, that the balance preserve its equilibrium when it carries no weights; secondly, that the two arms be of equal length. For, let M, N, be the moments of the two parts of the balance when it has no weights appended; a, b, the lengths of the two arms; P, P, the weights which are in equilibrium upon it. Then, we must have $M + P a = N + P b$; whence, $P (b - a) = M - N$. And, since the equation must hold, whatever be the value of P, we have, necessarily, $M = N$, $a = b$. If either of the two specified conditions is wanting, the balance is said to be *false*; but it may still be made to serve justly to determine the weight.

494.

For the first defect is easily corrected, by giving an equipoise to the naked balance, through the addition, in one of the scales, of such a weight as will produce an equilibrium. The second defect is obviated by placing the body to be weighed, first in one scale and then in the other; its just weight will be a mean proportional between the two different marked weights, that have served to balance it. For if the weight P, placed successively in the two scales, is balanced by the two weights Q, Q′, we shall have $\left\{\begin{matrix} P \times a = Q \times b \\ P \times b = Q' \times a \end{matrix}\right\}$, whence, by multiplying these equations together, we have $P^2 = Q \times Q'$, and hence the weight of P becomes known in terms of the registered weights Q, Q′.

There is also another mode of making a false balance serve exactly to ascertain the equality of two weights. This method, invented by Borda, consists in placing the two weights successively in the same scale of the balance, and counterpoising them by the same weight placed in the other scale. If the two weights have, in the two successive weighings, been in equilibrium with the same counterpoise, they are necessarily equal to one another. Let the first weight be P, and let it be in equilibrium with the marked weight Q. We cannot conclude at once that $P = Q$, because we are supposed not to be sure of the justness of the balance. Let the second weight be P′, and let this, also, be in equilibrium with the same marked weight Q; and here, too, for the same reason, we cannot say that $P' = Q$. But we may, with certainty, affirm that $P = P'$; because these two weights, being in equilibrium with the same marked weight Q, under the same circumstances, cannot but be equal. If P be a known standard weight, say of 100 lbs., and P′ an irregular piece of metal, the weight of which was sought, then we know that $P' = 100$ lbs., without knowing the weight of Q. Since it is physically impossible to secure, with the utmost degree of precision, the perfect equality of the lengths of the arms, A O, O B, Fig. 494, and of the moments M, N, it will always be well to employ this method of two weighings, when it is wished scrupulously to ascertain the true weight of a quantity according to a given stand of weights.

The first method gives also the relations of the arms of the beam to each other. The body to be weighed is placed on one of the scale-boards, and counterbalanced by marked weights; let P be the weight thus obtained. The body is then placed on the other scale-board; let Q be the weight necessary to produce equilibrium. If Q be equal to P, then it is exactly the weight of the body. If this be not the case, let x be the weight of the body to be weighed and a and b the lengths of the arms. In the first instance $x a = P b$; [1] and in the second $x b = Q a$, [2].

If these equations be multiplied member by member, and the common factors a and b cancelled, the result will be $x^2 = P Q$, whence $x = \sqrt{P Q}$, [3]; that is to say, that the real weight of a body is the geometrical mean between the weights found in the above manner. We obtain in addition, by dividing equation [1] by equation [2], the result $\frac{a}{b} = \frac{P b}{Q a}$, or $\frac{a^2}{b^2} = \frac{P}{Q}$, or lastly $\frac{a}{b} = \frac{\sqrt{P}}{\sqrt{Q}}$, [4]; therefore the arms of the beam stand in the same relation to each other as the square roots of the weights which were placed at the extremities of the opposite arms. From the third and fourth equations can be deduced $\frac{a}{b} = \frac{x}{Q}$, or $\frac{a}{b} = \frac{P}{x}$. If, for instance, $P = 1^k{\cdot}350$ and $Q = 1^k{\cdot}362$, then $x = 1^k{\cdot}35598$, or nearly $x = 1^k{\cdot}356$, and $\frac{a}{b} = {\cdot}995594$, or $\frac{a}{b} = {\cdot}996$ nearly.

In order that a balance *be just*, it is not sufficient that the two equal weights are in equilibrium; but the inequality of weight must also have the power to overcome the friction of the fulcrum. For this purpose the beam must encounter no resistance in its revolution round the fixed point,

495.

496

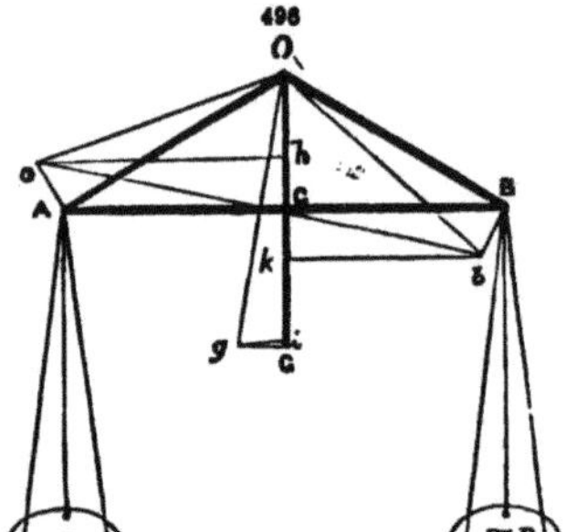

or there must be no friction between the beam and its supports; for any friction would necessitate an increased weight of the body to produce equilibrium. In exact balances, the beam is supported, upon polished surfaces of steel or agate, by steel blades, Fig. 495. By this means the friction is reduced to a minimum.

The perfection of the balance, and the facility of managing it, require other conditions. In the first place, for every, the smallest, inequality of the two weights, the needle of the balance should slowly incline itself, stopping in a position oblique to the horizon, without going completely down. Further, if the balance is in a state of equilibrium, and if the needle is accidentally made to incline itself, it ought not to stop in that inclined position, nor completely to go down, but slowly to reassume its vertical position.

The fulfilment of these conditions depends, above all, upon the relative position of the three points O, C, G, Fig. 496, of which O is the centre of motion, G the centre of gravity of the beam of the balance, C the intersection of the vertical line O G with the horizontal line A B, which joins the points of suspension of the two scales.

The most advantageous situation of the three points O, C, G, is readily determined from the succeeding derived equations. The equilibrium of a balance having been disturbed, by the addition of a very small weight p to one of the equal weights P, P, that are weighed in it, to determine the inclination, which will thence ensue, in the needle of the balance.

Suppose that, by the addition of the small weight p to the weight P, hanging from the point B, the beam passes into the position a O b; and let the angle A O a = B O b = G O g = θ, and the angle A O C = B O C = α; and from the points a, b, g, let there be drawn $a\,h$, $b\,k$, $g\,i$, perpendicular to O G, and let M be the weight of the beam.

Then, for the balance to be in equilibrium in the position a O b, we must have

$$P \times a h + M \times g i = (P + p) \times b k; \qquad [Y].$$

$$\text{Now, } a\,h = \text{A O sin. } (a + \theta) = \text{A O sin. } a \text{ cos. } \theta + \text{A O sin. } \theta \text{ cos. } a$$
$$= \text{A C cos. } \theta + \text{O C sin. } \theta;$$
$$b\,k = \text{A O sin. } (a - \theta) = \text{A O sin. } a \text{ cos. } \theta - \text{A O sin. } \theta \text{ cos. } a$$
$$= \text{A C cos. } \theta - \text{O C sin. } \theta;$$
$$g\,i = \text{O G sin. } \theta.$$

Substituting these values in [Y], the equation of equilibrium, we shall have,

$$\tan.\ \theta = \frac{p \times \text{A C}}{(2\,\text{P} + p)\,\text{O C} + \text{M} \times \text{O G}};$$

or putting M′ for the weight of the loaded balance, so that $M' = M + 2\,P + p$, we shall have more briefly

$$\tan.\ \theta = \frac{p \times \text{A C}}{\text{M}' \times \text{O C} + \text{M} \times \text{C G}}.$$

The mobility of the balance, therefore, will be the greater, the longer are its arms, the nearer together are the three points O, C, G, and the less it is loaded. If O C = o, then, $\tan.\ \theta = \dfrac{p \times \text{A C}}{\text{M} \times \text{C G}}$; and consequently the mobility of the balance is the same, whatever be the load which it carries. If, at the same time, O C = zero, and G C = zero, so that the three points O, C, G, coincide, then $\tan.\ \theta = \infty$ or infinity; and then for every, the smallest, derangement of the equilibrium, the needle of the balance will go down, and describe a quadrant of a circle, so as itself to become horizontal. And if O C and C G were negative, tan. θ would become negative, and the needle, still going down, would describe more than the quadrant of a circle. It is evident that these constructions of the balance are faulty.

If the needle or pointer of a balance, which is in perfect equilibrium, be made to incline at the angle θ, to find the force with which the beam tends to resume the position of equilibrium. Let S be the *moment of* INERTIA; of the loaded balance, with respect to the axis of motion. The accelerating force, with which the beam will tend to turn round in the direction a A b B, to place itself again in the position A O B, will be

$$= \frac{\text{P} \times a\,h - \text{P} \times b\,k + \text{M} \times g\,i}{\text{S}}.$$ See INERTIA, moment of.

And substituting the values of $a\,h$, $b\,k$, $g\,i$, before found, the force sought is found to be

$$= \frac{\sin.\ \theta}{\text{S}}\,(2\,\text{P} \times \text{O C} + \text{M} \times \text{O G})$$
$$= \frac{\sin.\ \theta}{\text{S}}\,(\text{M}' \times \text{O C} + \text{M} \times \text{C G}).$$

It must be observed, in computing the moment of inertia S, that the masses of the weights P, P, which hang freely from the points A, B, must be understood to be concentered in these points. If O C = O G = zero, the restoring force vanishes; in that case, the balance is said to be indolent; indeed it stops altogether, the needle resting indifferently in any position. But if O C and C G are negative, the restoring force is negative; and then the balance is said to be unstable, because, for every, the smallest, inclination of the pointer, instead of resuming the position of equilibrium, it deviates more and more from it, and goes completely down.

The weight of the beam may be considered as applied to the centre of gravity G, Figs. 497, 498. If this point be removed from the perpendicular to G′, by a slight movement of the beam, it

497.

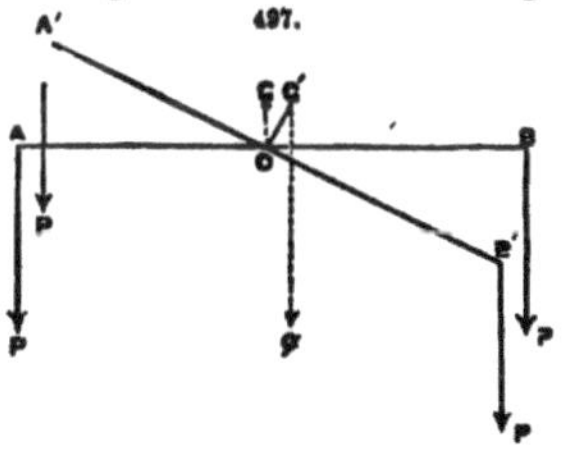

498.

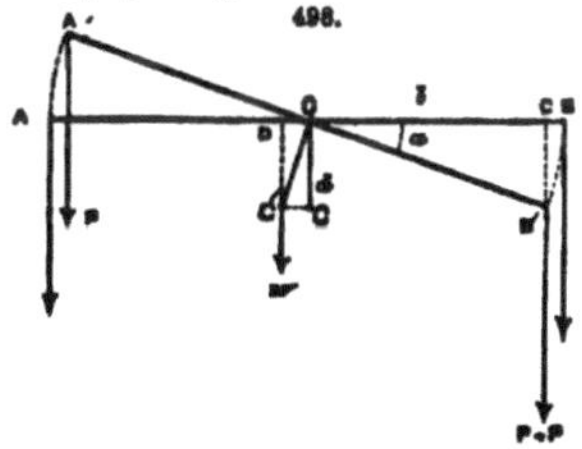

will be raised. The force of gravity causing it to descend, the beam will incline to return to its first position, which it will retain after a series of oscillations. This third position is the one which should be adopted.

The centre of gravity of the beam must not, however, be situated too far below the axis of suspension; for we should then have a so-called *inert* balance. This is easily accounted for. Let A B be the points of suspension of the scale-boards, and let them be placed in a horizontal line passing through the point of suspension O, or, more exactly, through the axis of suspension.

Suppose a weight P be applied to A, and a weight P + p to B. Let q be the weight of the beam, which may be considered as applied to G. The excess of weight p applied to B will firstly lower the arm O B of the beam; but if this excess be not too great, the beam will soon attain a certain position A′ B′, where it will finally remain in equilibrium after several oscillations; this position can be determined.

If G′ be the new position of the point G, draw the horizontal G′ I, and the perpendiculars A′ C and B′ D. Considering the moments of the forces with regard to the point O, and taking into account the direction of the rotation which each force tends to produce, the result will be

$$(\text{P} + p) \times \text{O D} = \text{P} \times \text{O C} + q \times \text{G}'\text{I}. \qquad [5]$$

But A′ O B′ being a straight line, the same as A O B, and the arms A′ O and B′ O of the beam being equal, it follows that O C is equal to O D; the terms P × O D and P × O C disappear from the above equation, leaving $p \times OD = q \times G'I$. If δ be the distance G O or G′O; b the length O B or O A of the arm of the beam, and α the angle B O B′ or its equal G O G′, then

$$OD = b \cos. \alpha \text{ and } G'I = \delta \sin. \alpha.$$

The condition of equilibrium then becomes $pb \cos. \alpha = q\,\delta \sin. \alpha$, from which is deduced $\tan. \alpha = \frac{pb}{q\delta}$; [6]. It is, therefore, evident that if the distance G O or δ be too great, the value of tan. α, and consequently of α, for the same excess of weight p, might be very small; the beam would then remain in equilibrium, in a position but little different from its primitive position, with which it might be confounded. The balance would, therefore, not be sensible enough.

The angle α may be regarded as the measure of sensibility of the balance; it will increase as the weight q of the beam diminishes, as the arm b lengthens, and as the distance δ from the centre of gravity of the beam to the axis of suspension becomes shorter. In precise balances, the distance from the centre of gravity of the beam to the axis of suspension can be changed by means of a nut E moving up and down a screw, the axis of which is perpendicular to the direction of the beam, Fig. 495. The angle described by the beam is indicated by a needle, fixed perpendicularly; the end of this needle moves round a graduated arc. It is now evident that, by placing the points of support A and B in a horizontal line, passing through the axis of suspension, the sensibility of the balance does not become affected by the weight to be estimated. It is by virtue of this fact that the terms P × O D and P × O C disappear from the equations of the moments. The case would be altered if the point O were above or below the horizontal A B: the lengths O D and O C would then no longer be equal; the weight P would remain in the equation, consequently the angle α, and with it the sensibility of the balance, would depend upon the weight of the scale-boards. To diminish as much as possible the friction, the scale-boards are suspended by hooks resting on blades at the extremity of the beam, as seen in Fig. 494.

The lever-balance, Fig. 499, is also a lever of the first order. The materials to be weighed are placed in a scale-board, suspended at the extremity B of one of the arms of the beam; and equilibrium is established by means of a slider A, which moves along the other arm. If the beam be suspended in such a manner that the friction may be omitted, as shown in the figure, and if the centre of gravity of the beam lie in a perpendicular from the point of suspension O, equilibrium will ensue when P is the weight of the materials to be weighed and p that of the slider,

$$P \times OB = p \times OA, \text{ whence } P = p \times \frac{OA}{OB}.$$

The quantities P and O A being constant, it is evident that the weight Q is proportional to the distance O M between the slider and the point of suspension. For convenience, the arm O B is divided into a certain number of equal parts, at each of which is marked the weight which produces equilibrium when the slider is at that spot. This kind of balance has sometimes two rings for suspension, and consequently two corresponding systems of blades; one or the other ring is used, according to the nature of the substances to be weighed; if these be light, the ring farther from the point B, to which the scale-board is attached, is used. For heavy matters, the ring nearer B is employed. The arm O A, in this case is divided in two different manners.

499.

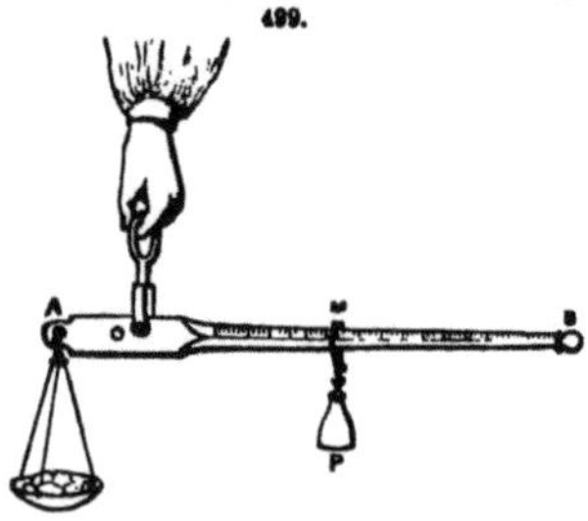

500.

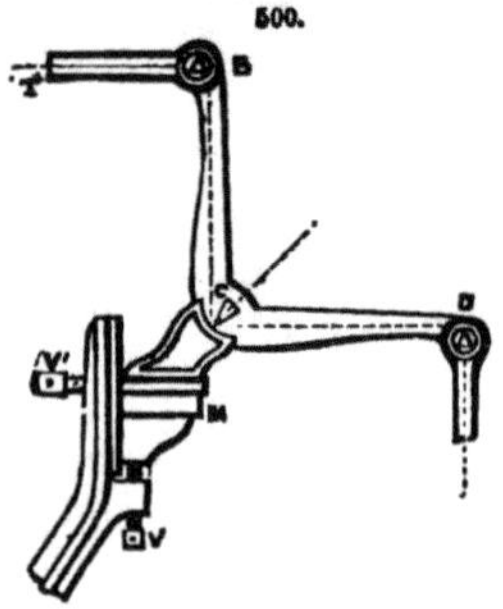

501.

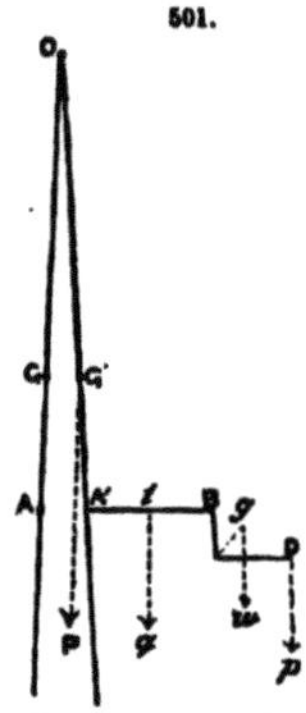

Balance, *Moment*.—An apparatus invented by Didion to measure directly the moment of the weight of a *ballistic pendulum* with regard to the axis of suspension. This apparatus was adopted in 1857 for testing powder in powder manufactories in France. It consists of a bent lever B C D, Fig. 500, in which the arms C B and C D are equal, and form an angle slightly greater than a right angle. At the extremity B of one of the arms is joined, by means of blades, an inclined rod T, which can be fixed or jointed to the ballistic pendulum; at the extremity D of the other arm is jointed, also by means of blades, a second rod, supporting a scale-board for the reception of weights. The bent lever rests at C upon a support M, which can be fixed to the structure bearing the pendulum. This support can be moved either in a direction parallel to the arm C B, or in a direction perpendicular to it, by moving the screws V or V′. Let O A, Fig. 501, be the perpendicular from the point of suspension of the ballistic pendulum, containing the centre of gravity G. Suppose the straight line O G A brought to the position O G′A′, forming an acute angle α; a rod A′ B, jointed at A′ to the pendulum and at B to one of the arms of the balance, is placed perpendicularly to O A′, and makes consequently an angle α with the horizon. The bent lever is so constructed that when the arm C B is placed parallel to O A′, the other arm C D is horizontal, or the angle B C D is equal to $90° + a$. In the scale-board suspended to D a weight is placed sufficient to maintain equilibrium in the position just indicated; this weight will give the weight P of the pendulum, applied at G′, in relation to the point of suspension O.

The theory of this apparatus is easily understood. Take, firstly, the equilibrium of the pendulum and the moments of the forces acting upon it, with regard to the axis through O. The suspension being effected by means of blades, the reaction of the supports may be considered as zero; there then remain but the moment of the weight P and the moment of the reaction R exercised at A′ by the rod, in the direction A′B; if we make O A = L and O G = l, we get

$$P\,l \sin. a = R\,L. \qquad [1]$$

Take, secondly, the equilibrium of the rod A′B; let R′ be the reaction exercised on the rod, at B, by the bent lever, and let q be the weight of the rod, applied to its centre of gravity i; if we move the forces R, R′ and q in the direction A′B, we shall obtain

$$R + q \sin. a = R'. \qquad [2]$$

Then the rod receives from the pendulum a reaction R equal and opposite to that which it exerts on the pendulum.

Take, lastly, the equilibrium of the bent lever, and take the moments with regard to the axis through C; the moments of the reaction of the supports may here also be considered as zero, as well as the friction of the blades at B and D. The moment of the reaction R′, exercised at B, by the rod A′B, must be rendered equal to the sum of the moments of the weight p of the scale-board suspended at D, and of the weight π of the levers, applied to the centre of gravity g. If λ be the length of the lever-arm C B, λ' that of the arm C D, very little different from λ, and δ the distance g C, the point g lying in the bisecting line of the angle B C D, then

$$R'\lambda = p\lambda' + \pi\,\delta \cos. (45° + \tfrac{1}{2} a). \qquad [3]$$

By eliminating R and R′ from equations [1], [2], and [3], and seeking the value of the moment Pl, we find :—

$$Pl = \frac{L}{\sin. a}\left[p\,\frac{\lambda'}{\lambda} + \pi\,\frac{\delta}{\lambda} \cos. (45° + \tfrac{1}{2} a)\right] - q\,L. \qquad [4]$$

In the balance for moment, constructed according to Didion's design, $a = 5° \; 44' \; 21''$, whence $\sin. a = 0 \cdot 1$ and $\cos. (45° + \tfrac{1}{2} a) = 0 \cdot 6708$. Also $\lambda = \lambda' = 0^m \cdot 2$, therefore

$$Pl = 10\,L\,(p + 3 \cdot 354\,\pi\,\delta) - q\,L. \qquad [5]$$

This apparatus gives the moment Pl within at least one ten-thousandth of its value. But it is of the greatest importance that the arms of the lever be exactly equal, and that their length be known, for an error of one-tenth of a millimètre in one of their lengths would give rise to an error of one five-hundredth in the calculation of the moment sought. In order to test the equality of the lever-arms, the bent lever should be placed in two positions, so that each arm occupies alternately a horizontal position; if the arms were unequal, two different weights p and p' would be necessary to produce equilibrium in each case. If this be the case, Didion substitutes for p, $\sqrt{p\,p'}$; but this rule, which is exact when applied to the common balance, is here only approximative. See 'Traité d'Artillerie,' par Didion.

A *Weighing Machine* is a balance employed chiefly in shops, warehouses, and railways for weighing heavy weights. The beam A B C, Fig. 502, moves round a horizontal axis passing through O; at the extremity C is suspended the scale-board in which the weight p, destined to produce equilibrium with the weight P, is placed. The weight to be determined is placed on a horizontal plane D E, to which is attached on one side a vertical partition E F, to protect the balance from injury; this is fastened to a horizontal bar T, the end of which is shown in the figure, by oblique bars G G. T is suspended from the short arm of the beam by a rod B; D E also rests at I upon a lever K L, movable at I and suspended by the rod L A from another point A of the same arm of the beam. The lower part of the machine is enclosed in a case, which is omitted in the figure. The lower stage M N supports the blade K, upon which the lever K L turns; and the vertical partition N Q supports the axis of suspension of the beam. When there is no weight either upon E D or upon $b\,c$, the point m, fixed upon the arm O C, must be exactly opposite the point n, fixed to the frame-work of the machine. As this is, in reality, seldom the case, a small weight is placed in the cup a, suspended to C, till m is opposite n; this weight is called the tare. A weight to be weighed must be placed on D E; weights are then placed on the scale-board $b\,c$ till the points m and n are opposite. The weight p on the scale-board is exactly a tenth of the weight P on the stage D E.

502.

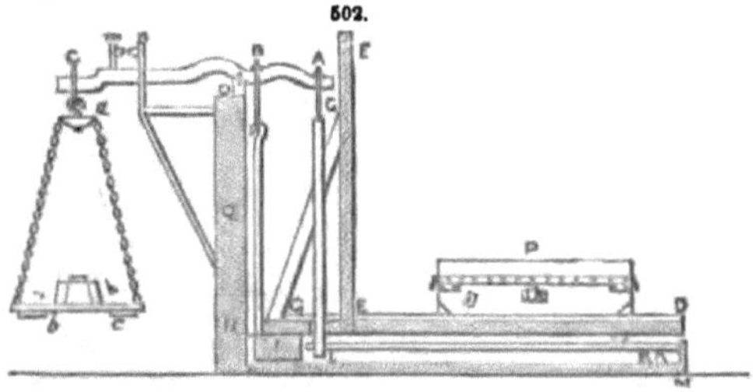

503.

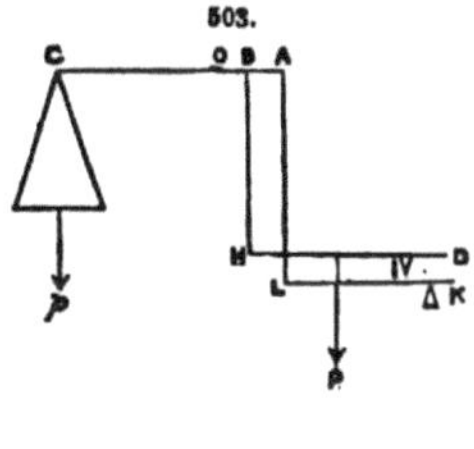

In Fig. 503 this balance is shown in its most simple form. The dimensions of the different parts are such as to produce the proportion O B : O A = K I : K L = 1 : n. [1]

The relation of O B to O C is generally $\frac{1}{10}$. When the apparatus is set in motion, D H remains horizontal. If the points B and H are lowered a short distance h, by virtue of the above proportion the point A lowers itself $n\,h$. It will be the same with the point L; consequently, in the same proportion, the point I will be lowered the distance h, like the point H; and I H will remain horizontal. The actual work of P will be $+\,\mathrm{P}h$, as the stage is lowered h. But the point C will be raised $10\,h$, as $\mathrm{OC} = 10\,\mathrm{OB}$; the actual work of p will therefore be $-10\,p\,h$. The actual work of the reactions exercised by the points of support I K O is zero, if the friction be omitted. The other forces that influence the system are mutual, equal in pairs, and having opposite signs, and their virtual actions cancel one another in the summing up. There then remains $\mathrm{P}h - 10\,p\,h = 0$, whence $p = \frac{\mathrm{P}}{10}$. In reality, the four points A, B, C, O, Fig. 502, are not in the same horizontal plane; but if they are projected on a horizontal plane passing through O, as in Fig. 503, A B C, the same reasoning holds good.

Conditions of Equilibrium in Roberval's Balance.—A n, $a\,k$, Fig. 504, are two equal and parallel bars turning at their middle points about pivots C, c, which are in the same vertical line, and connect with the stand H C. The extremities of these bars are connected by means of pins A, a; n, k, with two other bars A a, $n\,k$, so as to allow of free motion about these points. Then, since A C, $a\,c$ are equal and parallel, A a is equal and parallel to C c; and therefore A a, $n\,k$ will always be vertical in every position. In this balance, two horizontal rods are fixed to the vertical bars A a, $n\,k$, on which the weights P and Q are placed, and if these balance each other in one position, the equilibrium will remain undisturbed in every other position. On the same principle the balances now in general use are constructed, and we will prove that if the weights P, Q are equal they will balance each other in whatever part of the scale-pans, E, F, they are placed. Suppose the weight P to be placed in the scale E, immediately over the rod A a; and an equal weight Q in the other scale F acting at any distance C m from the centre. At n introduce two equal and opposite forces R, r, acting vertically in opposite directions, each equal to Q. Also introduce two pairs of equal and opposite forces S, s; T, t; each of these forces having the same proportion to Q as $m\,k$ to $n\,m$. These six forces will manifestly balance each other, and not affect the equilibrium. Now the forces Q and s being proportional, $m\,h$, $m\,n$ are equivalent to a force represented by the diagonal $m\,k$. In like manner, r and T are equivalent to the same force $k\,m$ acting in the opposite direction. Hence the forces Q, s; r, T, mutually destroy each other; and, therefore, the force Q is equivalent to the equal force R acting in the direction n R, and the two forces S and t acting in the directions C n, $k\,c$. But the forces S, t only produce pressure on the pivots C, c, and have no effect in turning the bars round. And since P and R are equal and at equal distances from C, they will balance each other.

504.

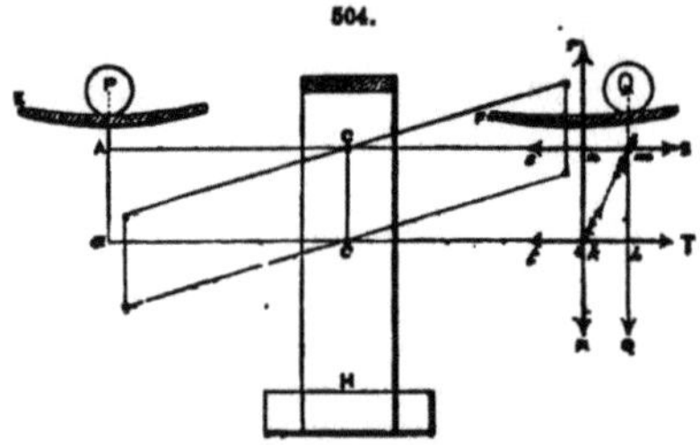

505.

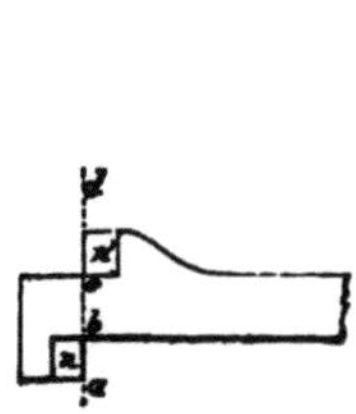

506.

507.

Fig. 505 represents the extremity of this beam seen from above; $n\,n$ are two blades, inclined in contrary directions, their edges $a\,b$ and $c\,d$ are placed in a straight line in the lower plane of the beam. This extremity enters a stirrup, forming the lower end of the side A A′, and which is represented in Fig. 506 in perspective, and Fig. 507 in plan. Its lower cross-bar is a blade, the edge of which, $b\,c$, is turned upwards; the two prongs m and m' are so arranged that this edge is on the outer side at m and on the inner at m'. When the end of the beam is in connection with the stirrup, the edge $a\,b$ of the blade n rests on the outer face of the prong m, the edge $c\,d$ of the blade n' rests on the inner face of m', and the lower face of the beam rests on the edge $b\,c$ of the blade forming the cross-bar of the stirrup. The edges of the three knives thus form a prolongation of each other, and each bears on a corresponding plane. It follows, that in all portions of the parallelogram the contact takes place along this same edge, which thus forms the axis of articulation, without producing an appreciable amount of friction. The opposite end of the beam is jointed in the same manner to the stirrup, which forms the lower end of the side B B′. At O′ the jointing is of the same nature, with this difference, that the stirrup is fixed. This system of jointing gives to the parallelogram the desired sensibility. The lower beam and the greater part of the lateral sides are generally concealed by the body of the balance, as shown in Fig. 508. The conditions of stability and sensibility are otherwise the same as in the ordinary balance; the centre of gravity of each beam, and particularly of the upper beam, must be a short distance below the axis of suspension of that beam.

508.

Fig. 509 represents an *assay balance* of great delicacy and precision, with the glass case enclosing it. In front, and slightly projecting from the stand, is seen the end of a lever, by means of which the beam can be raised in order to relieve the knife-edged fulcrum, upon which it rests, from

509.

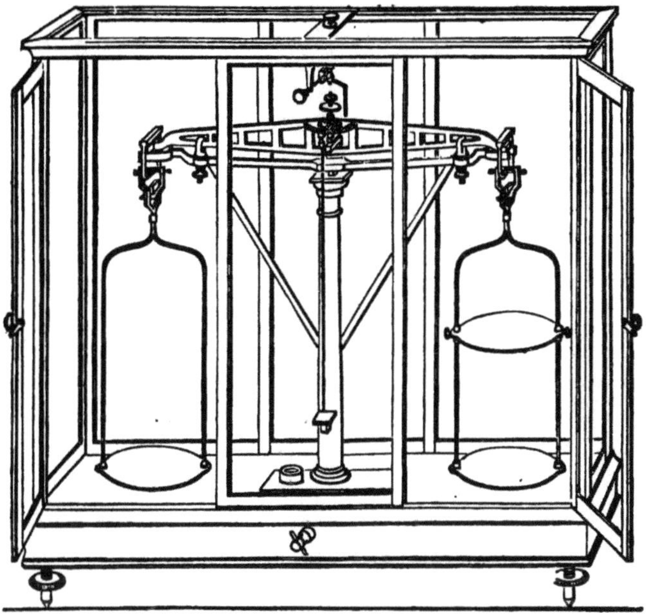

fatigue and wear when the instrument is not in use. At the upper part will be observed a small mechanical contrivance resorted to by some makers, though not in general use. In front of the beam is a horizontal graduated semicircle, over which a little metallic needle can be made to travel with the aid of a pin, worked by a button on the exterior upper surface of the case. By shifting this needle to the right or to the left, the equilibrium may be obtained with much accuracy, while the readings on the graduated semicircle show the increments or decrements of weight. See Hydraulic Machines. Lever. Weighing Machines.

BALANCE-VALVE. Fr., *Soupape à bascule*; Ger., *Entlasteter Schieber*; Ital., *Valvola bilanciata*; Span., *Válvula autamatora.*

See Valves.

510.

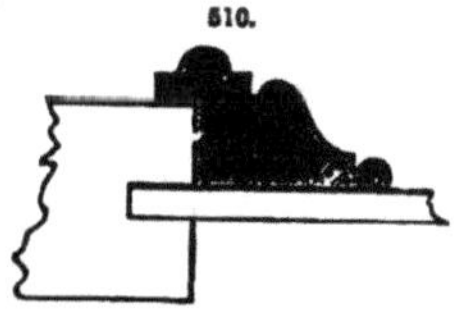

BALECTION. Fr., *Moulure*; Ger., *Fenster oder Thürverkleidung*; Ital., *Modanatura.*

Balection, bolection.—A term applied to mouldings which project beyond the face of the work which they decorate, Fig. 510. They are principally used for external doors.

BALK, BAULK. Fr., *Poutre, Guéret*; Ger., *Balken*; Ital., *Trave*; Span., *Viga.*

In engineering, the ridge or bank between two excavations; also a term applied to timber in the log, roughly squared, and before it is sawn up for use; and sometimes used to designate the horizontal timbers in an open roof.

BALL-AND-SOCKET JOINT. Fr., *Genou de charnière*; Ger., *Kugelgelenk*; Ital., *Snodatura sferica, Nocella*; Span., *Juego de nuez.*

See Joints.

BALL-COCK. Fr., *Robinet modérateur*; Ger., *Zulass und Absperrhahn*; Ital., *Chiave a galleggiante*; Span., *Válvula flotante.*

See Water-Supply.

BALL-LEVER. Fr., *Levier à bascule*; Ger., *Regulirungs brebel*; Ital., *Leva a contropeso.*

See Water-Supply.

BALL-VALVE. Fr., *Ventouse*; Ger., *Ablass*; Ital., *Valvola pallata.*

See Water-Supply.

BALLAST. Fr., *Lest*; Ger., *Ballast*; Ital., *Stiva, Ghiaia*; Span., *Lastre.*

Ballast is a term applied to gravel, stone, or other material used for loading ships, to give them steadiness in the water when sailing without a cargo, or when the cargo is not of sufficient weight.

In *civil engineering*, a layer of broken stone or coarse gravel, from 6 to 18 in. in thickness, laid above the formation-level of roads and railways, where it answers the purpose of a foundation for the road-metal in the former, and for the sleepers and permanent way in the latter. Originally the material used for this purpose was of the same description as that used for ballasting ships, hence the term.

Thames ballast is the gravel dredged from the bed of the river Thames, and is used for ships' ballast, making concrete, formation of roads, &c.

Where stone and gravel are scarce, clay burned to the hardness of at least ordinary bricks is used instead of the ordinary gravel or stone ballast, the process being as follows:—

Clay as used for the manufacture of bricks is dug and slightly tempered to make a homogeneous mass; a log of wood is fixed upright in the ground, from which a flue, technically termed a "live hole," of 8 or 9 in. in diameter, and about 7 ft. long, according to the size of the heap to be burned, is formed of bricks on edge, or old drain-pipes, on the upper side of which, at the end next the wood, is left an opening. Pieces of wood are next piled around the upright log, in a conical form, about 4 or 5 ft. in diameter at the base, and about 4 ft. high. Over this cone of wood is placed about half-a-ton of coal, and on the coal a coating of clay 6 to 8 in. thick. The whole is then set on fire through the horizontal flue.

As the wood burns down, coals are thrown in until all becomes in a state approaching incandescence. It is then ready to form the core for the reception of another coating of clay; and as the burning proceeds, coal and breeze are added to feed the fire, then another coating of clay, and so on, until the size of the heap becomes such as to render it inconvenient to add more. The burning should proceed gradually at first, as in brick-making, to expel the moisture, after which the heat may be urged. When the mass is thoroughly burned through, it should be left to cool, and afterwards broken into pieces of the size required. From $\frac{3}{4}$ of a cwt. to 1 cwt. of coal is required to burn a cubic yard of clay; or, according to some authorities, about 11 cub. yds. of breeze, and 4 tons of coal, including small coal or slack, is required to burn 100 cub. yds. of clay.

A cubic yard of clay, measured in the solid, before digging, will, when burned and broken up, occupy a space of $1\frac{1}{4}$ to $1\frac{1}{2}$ cub. yd., and weighs about a ton.

About 22 cub. ft. of Thames ballast, 21 of broken granite, and from 22 to 27 cub. ft. of other stones, when broken, weigh a ton.

BALLAST-WAGON. Fr., *Waggon à lest;* Ger., *Ballast Wagen, Kieswagen;* Ital., *Carro da Ghiaia;* Span., *Wagon para lastre.*

A ballast-wagon is a wagon employed in removing earth in excavations and similar works. They are usually made to hold 2 to 3 cub. yds. Ballast-wagons should be made with springs, as without springs they increase the wear and tear of the rails considerably, and add to the expense of maintaining the way. Figs. 511, 512, show forms of ballast-wagons generally in use.

511.

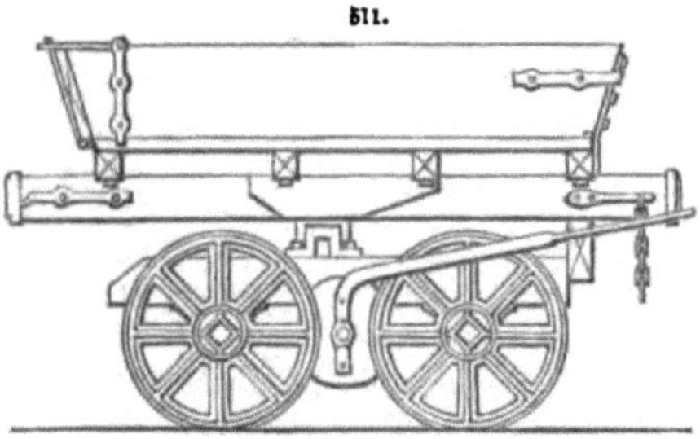

512.

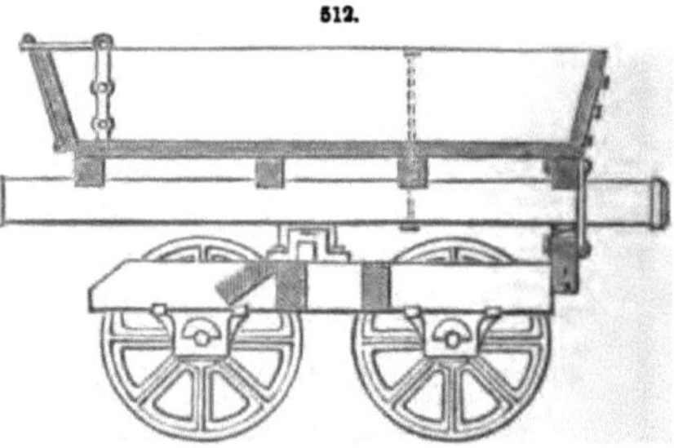

BALUSTER. Fr., *Balustre;* Ger., *Docke einer Treppenlehne;* Ital., *Balaustro;* Span., *Balaustre.*

A small pillar, a series of which, fixed under the coping of the parapet of a bridge, or otherwise, is called a balustrade. It also applies to the pillars of wood or iron which support the hand-rail at the side of a stairs.

BALUSTRADE. Fr., *Balustrade;* Ger., *Balustrade;* Ital., *Colonnata di balaustri;* Span., *Balaustrada.*

See Baluster.

BANDING-PLANE. Fr., *Rabot à rainure;* Ger., *Nuthhobel;* Ital., *Pialla da intorsiatore;* Span., *Cepillo de Bocel.*

A banding-plane is a plane used for cutting out grooves and inlaying strings and bands in straight and circular work.

BANKER. Fr., *Etabli;* Ger., *Werkbank;* Ital., *Ceppo;* Span., *Banco de cantero.*

The bench or table upon which bricklayers and stonemasons prepare and shape their materials.

BANK-NOTE PRINTING MACHINE. Fr., *Machine à imprimer les billets de Banque;* Ger., *Presse zum Drucken der Cassenscheine;* Ital., *Macchina per stampare i biglietti di Banco ed altri.*

The Bank-Note Printing Machine, invented by Thomas Grubb, of Dublin, and described by him in a paper read before the Institution of Mechanical Engineers, is shown in Figs. 513 and 514; Fig. 514 is a side elevation, and Fig. 513 a vertical section taken at right angles to Fig. 514.

The machine consists of a horizontal polygonal cylinder A, Figs. 513 and 514, of twenty sides, on ten of which are held the ten engraved plates from which the notes are to be printed. These engraved plates are held in position in dovetails formed by ten plain plates screwed upon the ten intermediate sides of the polygon. The plates and polygon are maintained at the proper temperature for working, by steam admitted inside the hollow polygon through the back-bearing B, Fig. 513.

513.

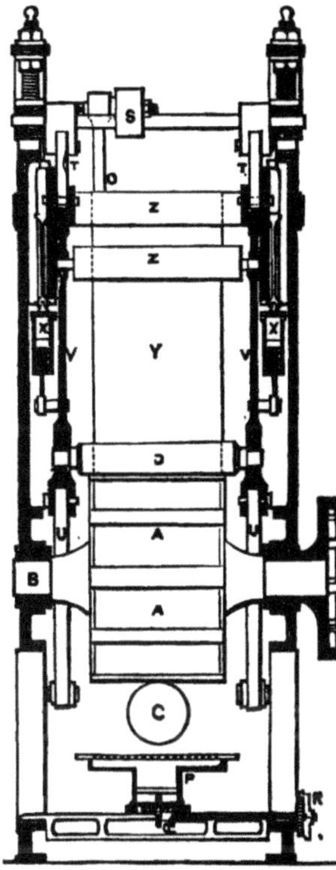

514.

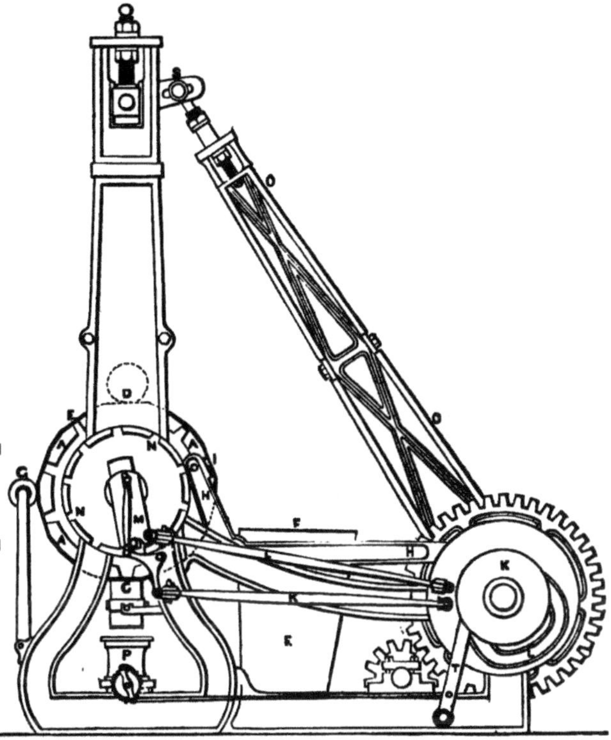

At the ordinary rate of working, each impression occupies ten seconds; and the polygon is held stationary during eight seconds of the time, and during the remaining two seconds it is turned round through $\frac{1}{10}$th of a revolution, in the direction shown by the arrow in Fig. 514, so as to bring the next engraved plate to the top for being printed from. During the eight seconds that the polygon is stationary, the undermost plate at C is inked by the machine, the uppermost plate at D is printed from, and the plate at E has the wiping of its surface completed. The machine requires two attendants, one standing in the box F, who lays the paper, removes it after being printed, and observes that the work is proceeding satisfactorily; while the other gives the final wiping by hand to the surface of each plate as it comes into the position E, Fig. 514. The greater portion of the superfluous ink is wiped off by the machine, thereby lessening the labour of the final wiping by hand: this is done by means of the wiping-roller G, on which is wound a length of cotton cloth. This wiping-roller is stationary while the polygon is at rest; but on the polygon beginning to move, the roller presses against the surface of the previously inked plate, and, by turning in the opposite direction, wipes off the superfluous ink. When the surface of this roller becomes surcharged with ink, a length of the cotton cloth upon it equal to one round is torn off. By the revolution of the polygon, therefore, each plate on arriving at the bottom at C is inked, and as it proceeds upwards is first partially wiped by the machine at G, and afterwards finally wiped by hand at E, and on arriving at the top is printed from by the printing-roller D.

Several special contrivances are required for producing the respective actions of the machine, the principal of which are the apparatus for locking and turning the polygon, and for the motions of the printing-roller, and the inking apparatus.

At each printing the polygon is subjected to a heavy strain by the vertical pressure of the printing-roller D, Fig. 514, tending to turn it first in one direction and then in the opposite direction alternately, as the printing-roller passes over from one side of the centre to the other; it therefore requires to be very firmly locked during the time that it has to be held stationary for printing. This is effected by the locking-lever H, which, being lowered by its cam, inserts the steel tooth I into one of the ten spaces round the circumference of the polygon. The turning of the polygon through $\frac{1}{10}$th of a revolution between each printing is accomplished by the levers and cams K K and L L, one of these levers, K, alternately withdraws and protrudes the tooth J, while the other, L, moves the carrier-box M through the required arc. The action of the levers H and K is so arranged, that as the tooth of the locking-lever is withdrawn, that of the turning-lever enters the toothed circular plate N, and *vice versâ*. The polygon is thus sure to be duly acted upon, and always under restraint, thereby ensuring against risk of accident, which might occur if ordinary means were employed for turning the polygon.

For working the printing-roller D, Figs. 513 and 514, the ordinary method of causing it to roll up an inclined plane into contact with the plate was inadmissible in the present machine, and the following arrangement has been adopted for the purpose. The roller is first caused to descend vertically at the proper time and with a sufficient pressure upon the plate, on which the paper to be printed is previously laid; next, the roller is carried horizontally over the surface of the plate, to produce the impression; then it is raised from the plate and withdrawn to its first position. The motion of the centre of the roller is, therefore, as shown by the dotted line P Q R in Fig. 515; and this motion is produced in the following manner. The vertical descending movement from P to Q is effected by the long trussed rod O and its double cam, acting on the bell-crank S and toggle-lever joint T. The horizontal traverse from Q to R is produced by the printing cams and their levers U acting upon the sectors V, Figs. 513 and 515. These sectors roll against inverted straight-edges W, and the bearings of the printing-roller D being concentric with the arcs of the sectors V, the roller is traversed horizontally over the engraved plate. The curved motion from R to P, in withdrawing the printing-roller and returning it to its original position, is produced by the joint action of the two sets of cams. The blocks W, against which the sectors V roll, slide vertically in grooves in the side frames of the machine; the under-side of these blocks is faced with a flat surface of hardened steel, against which rolls the upper segmental surface of the sectors, which is also made of hardened steel. These steel bearing-surfaces are retained in contact by the spiral springs X, Fig. 513, which carry the weight of the sectors, the printing-roller, and the levers U. The sectors are prevented from slipping in rolling against the blocks W, by means of bridle-levers, which admit of motion in other directions, but prevent slipping. An endless sheet of flannel Y passes round the printing-roller D and over the guide-rollers Z above, travelling in the direction shown by the arrows in Fig. 515; the upper rollers Z are carried by the sectors V and slide-blocks W.

515.

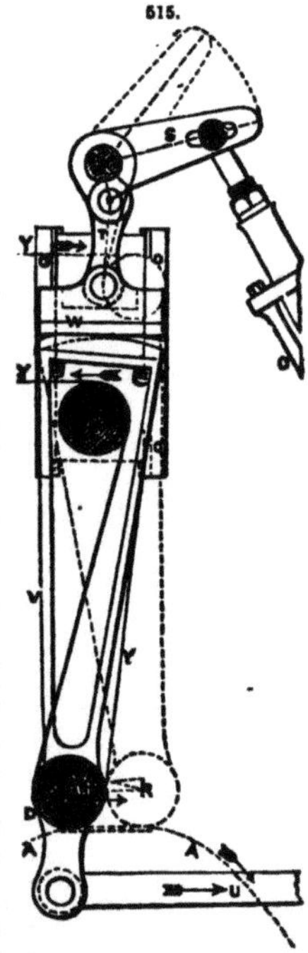

The inking of the engraved plates is performed by the inking-roller C, Figs. 513 and 514. Upon the capability of the machine to ink the engraved plates effectively depends its usefulness; and herein lay the main difficulty, and the reason of the failure of previous attempts at improvement. The difficulty was much enhanced by the circumstance that, in using the ink applicable to engraved plates, a portion of it becomes thickened and unfit for immediate use. In the ordinary hand-inking, this thickened ink collects in a ring on that part of the dabber which barely comes into contact with the plate during the inking; and this requires to be removed occasionally from the dabber by a knife or scraper, otherwise, by its mixing with the fresher ink, the quality of the work would be injured. In experiments made preliminary to constructing the present machine, a cylindrical roller was used for inking the plates by hand, and was supplied with ink from a perforated surface, through which the ink was caused to exude; and it was found that the thickened ink, technically termed "gatherings," adhered at first to the perforated surface between the perforations, and afterwards, when the film had arrived at a certain thickness, it was transferred bodily to the surface of the inking-roller. These experiments led to the adoption of the following inking apparatus:—

The short cylinder P, Figs. 513 and 514, containing the supply of ink, is fitted with a piston, the rod of which is a screw that projects below the cylinder, as shown in Fig. 513. The upper end of the cylinder expands into a horizontal rectangular tray, rather larger in size than the plate to be inked; and this tray is covered with a flat plate of steel, perforated with a number of small holes. The piston is slowly raised by the screwed piston-rod and the bevil-wheels Q, and ratchet-wheel R. In order to charge the cylinder with ink, the piston is lowered, and the perforated top plate removed for filling in the ink; and on replacing the top the piston is raised until the ink exudes through the perforations in the top plate, the lowering and raising of the piston being effected by hand by a winch upon the spindle of the ratchet-wheel R. The supply of ink during the working of the machine is kept up by the lever T from the cam-shaft, Fig. 514, acting on the ratchet-wheel R, a small but sufficient quantity of ink being in this way forced up through the perforated surface for every plate inked by the inking-roller C.

The inking-roller C, shown enlarged in Figs. 516, 517, and 518, is formed of a number of discs of woollen cloth, screwed tightly together upon a spindle and finished in the lathe. It is worked to and fro continually without intermission by the rack and sector B. This rack slides on a cylindrical rod D, which allows the frame E, carrying the roller, to be raised and lowered sufficiently to cause the inking-roller either to apply ink to the engraved plate on the polygon A above, as shown in Figs. 516 and 517, or to descend and roll on the inking-table P below, for obtaining a fresh supply of ink, as shown in Fig. 519. The roller is held up for inking the polygon by the weighted lever G, having an adjustable weight; and on the pressure of this weight being removed at the proper moment by a cam, the roller C drops and rolls upon the ink-table P with its own weight, and that of the frame E, as shown in Fig. 516. It thus takes up a fresh supply of ink, while the polygon is turned through one-tenth of a revolution so as to bring the next engraved plate round into position, ready to be inked when the inking-roller is raised again by the lever G.

516.

517.

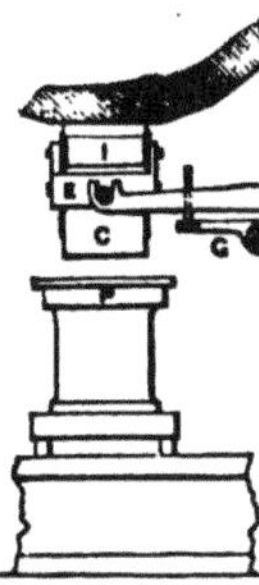

518.

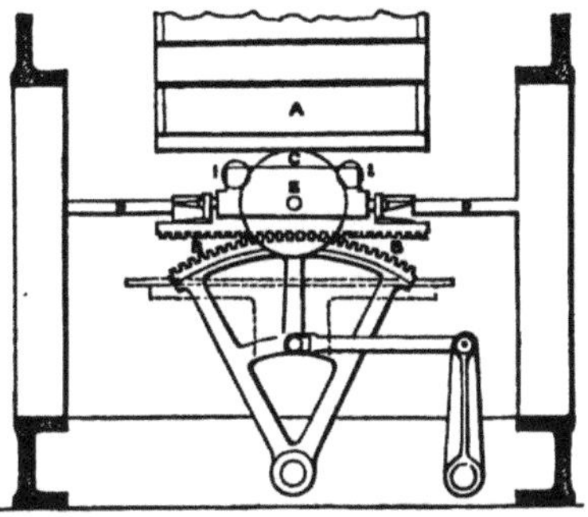

As the ink is transferred from the perforated inking-table P to the inking-roller C in a series of dots through the holes in the inking-table, the engraved plates would be inked in a similar and therefore imperfect manner; but this is prevented by the application of the two small distributing-

519.

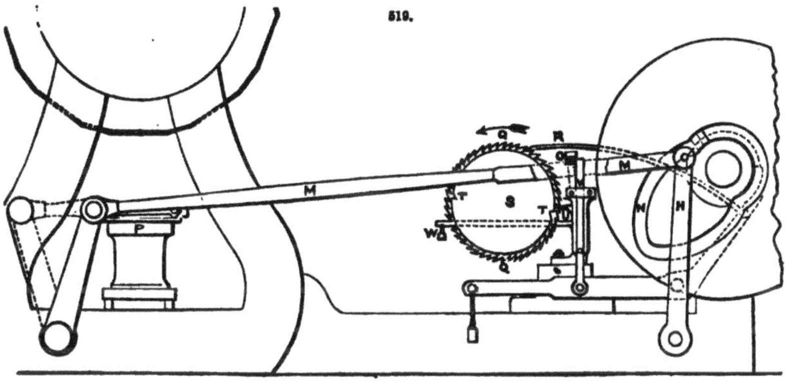

rollers I I, which are pressed in contact with the inking-roller by springs. One of these distributing-rollers simply revolves in contact with the inking-roller; but the other has on one of its bearings a screw, which gives it an end-traversing motion in addition to the rotary motion; and thus the ink coming to the roller in dots is equalized over its entire surface, and the inking of the engraved plates is rendered uniform.

The occasional removal of the thickened ink from the surface of the inking-table P is provided for by the arrangement shown in Fig. 519. The ink-scraper J is shown in the position when out of action, and H is the lever and M the connecting-rod for communicating the required motion to it from the cam N. On the scraper beginning to move across the inking-table P from the position shown at J in the drawing, a small cam at each end causes it to rise and so to pass clear over the table without scraping; while in the return motion the scraper descends into contact with the table, and pushes before it the thickened ink, which falls into a trough placed to receive it.

If the scraping were to be performed once for every impression printed, then it would only be required to allow the cam-roller at the end of the lever H to remain continually in the groove of the cam N. But as it is required that the scraping should be performed only occasionally, or once for every thirty-six impressions or revolutions of the cam-shaft, the following arrangement is adopted. At O is a latch, which so long as it is not raised holds the connecting-rod M pushed outwards laterally, so that the cam-roller is out of the cam-groove N. The ratchet-wheel Q has seventy-two teeth, and is carried forwards one tooth for every impression by the pawl R, which is actuated by the cam shown by the dotted line. Attached to the ratchet-wheel is a disc S, having only two notches T T in its circumference; and U is a rocking-frame, rocking on a pin at its lower end, and having on the upper part a projecting finger, which is caused by the weight W to press continually against the edge of the disc S. V is a bolt, worked up and down at each revolution of the cam-shaft by the same lever and cam that work the pawl R of the ratchet-wheel. The upper end of this bolt slides in the rocking-frame U; and so long as the finger is on the edge of the disc

S, the bolt V rises and falls just clear of the latch O, as shown in the drawing, so that it produces no action.

But once for every thirty-six impressions of the machine, the finger of the rocking-frame U enters one of the notches T in the disc, and the rocking-frame falling forwards, tilts the bolt V into the position indicated by the dotted line, when the next rising of the bolt lifts the latch O, as shown by the dotted line. The cam-roller is now thrown into the cam-groove N by the lateral pressure of the spring lever H, and the scraping of the inking-table P is performed once by the revolution of the cam N. At the end of the revolution the cam-roller is thrown out of the cam-groove N by the inclined stop X at the end of the groove, and is instantly held out by the dropping of the latch O; and before the bolt V again rises, the ratchet-wheel Q has been turned forwards one tooth, so that the finger of the rocking-frame U is again out of the notch T, and therefore the bolt V rises clear of the latch O, and ceases to lift it for the next thirty-six impressions.

BAR. Fr., *Barre*; Ger., *Stab*; Ital., *Scanno, Banco di rend*; Span., *Barra*.

A bar is a piece of wood, metal, or other solid matter, long in proportion to its diameter or width. A bank of sand, gravel, or other matter forming a shoal at the mouth of a river or harbour, obstructing entrance or rendering it difficult, is called a bar.

The term is also applied to an ingot, lump, or wedge of metal from the mines, run in a mould and unwrought.

BARBETTE. Fr., *Barbette*; Ger., *Geschutzbank, Pritsche*; Ital., *Barba, Parapetto a barba, Artiglieria in barba*; Span., *Barbetta*.

See Ordnance.

BARGE-BOARD. Fr., *Flache dosse*; Ger., *Schwarte, Schillstück*; Ital., *Fasciame d'aggetto del tetto*; Span., *Guarda Malletas*.

A barge-board, or verge-board, is a board, more or less ornamental, attached to the gable ends of a roof, chiefly in old English houses, to hide the ends of the horizontal timbers, and to protect from the weather the under-side of the *barge-course*, which is usually boarded or plastered.

BARGE-COUPLES. Fr., *Chevron extérieur*; Ger., *Aussere Sparren*; Ital., *Puntoni che aggettono*.

Barge-couples are the rafters placed under the *barge-course*, which serve as grounds for the *barge-boards*, and carry the plastering or boarding of the soffits.

BARGE-COURSE. Fr., *Battellement*; Ger., *Traufziegelreihe*; Ital., *Filare d'aggetto del tetto*; Span., *Boquillas*.

The barge-course is that part of the tiling or slating of a roof which projects over the gable of a building. The term is also applied, when there is no projection, to a coping of stone or brick laid above the tiling or slating along the sloping sides of the gable. A fillet of cement is usually run along the inside, to cover the junction between it and the slating.

BARLEY-DRESSING MACHINE. Fr., *Machine à préparer l'orge*; Ger., *Maschine zum Zurichten der Gerste*; Ital., *Macchina a preparare l'orzo*.

Figs. 520, 521, represent a machine, designed by J. T. Poyser, for screening and dressing barley and removing the broken grains. The barley to be dressed is placed in the hopper X,

520. 521.

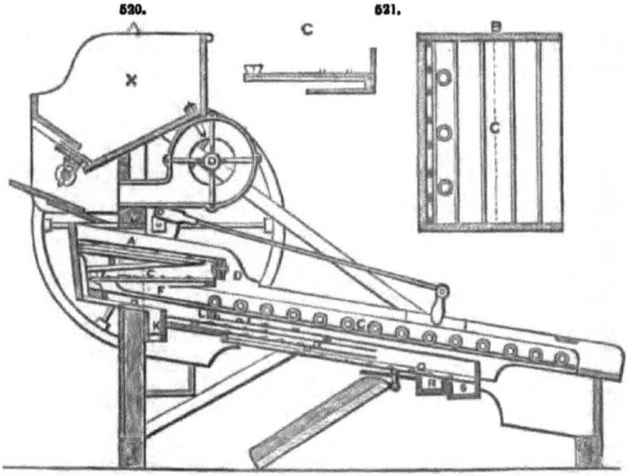

Fig. 520; this hopper has an opening at the bottom provided with a slide, by which the amount of grain allowed to pass through can be regulated. In falling from the hopper, the grain is subjected to the blast of a fan, which removes the dust, smut, &c. The grain is received on an inclined board which delivers it on to the screen A; this screen is mounted in a frame, which has a rapid reciprocating motion imparted to it in an inclined direction, by being coupled by rods to a crank on a transverse shaft.

The grain is delivered on to the screen A, and any beans, peas, stones, &c., which may be mixed with it, pass over this screen and down on to a second screen C, over which they also pass in a passage leading to the shoot K, where they are delivered. The shoot K is provided with an inclined perforated plate, which effects the separation of any stray corn which may have passed over the screens, A and C, with the refuse, the latter passing over the inclined perforated plate in the shoot, and being delivered on one side of the machine, whilst the grain passes through the plate and is delivered on the opposite side. The refuse which passes over the plate C is prevented from clogging the passages leading from the lower end of that plate to the spout K, by causing it to separate and pass on either side of circular blocks of wood placed opposite these passages and between the plates C and B. These blocks are shown in the plan and section of the perforated plate C, Figs. 520 and 521, and they not only serve the purpose we have mentioned, but, by retarding the progress of the refuse, cause the corn to be more effectually separated from it.

The screen-plates, A and C, have shallow transverse bars on their upper sides, and they are perforated with openings of sufficient size to permit of the passage through them of the grains of barley. The grain which passes through the screens is received by the plain zinc trays, B and E, and by them delivered on to the perforated plate F. This plate is provided with transverse bars on its upper side, and over it the grain passes down to the screen G, which is without transverse bars, but which is subjected to the action of a series of rollers. A portion of the broken and damaged corn and refuse matters of a smaller size than the grains of barley pass through the plate F and fall upon the inclined board H, down which they are led to the perforated plate L, which is provided with transverse bars. The damaged corn and small refuse matters fall through this plate on to the plate N, which leads them to the perforated and ribbed plate O, through which they fall on to the last perforated screen P, any refuse passing through this screen being delivered beneath the machine

Any good barley which may have been delivered through the screen F, and over the plate H, to the screen L, passes over that screen to the plain plate M, which carries it forward and delivers it on to the screen Q. This screen also receives any good grain which may have been separated from the light and broken grain during its passage over the inclined plate N, and the perforated plates O and P. The best barley passes over the perforated plate Q, into the spout S, whilst that which passes through the perforated plate is received by a plain plate, which conducts it to the spout R, by which it is delivered at the side of the machine as a second sample.

We must now return to the grain which, passing over the perforated plate F, is delivered to that marked G. Whilst passing over this latter screen, it is subjected to the action of a series of small wooden rollers covered with india-rubber, the axes of these rollers being carried in slotted holes in a pair of side-rails in such a manner that the rollers are free to rise and allow the grain to pass beneath them. The side-rails just mentioned are supported by brackets, which pass through long slots in the sides of the movable frame carrying the screens, and are bolted to the main frames of the machine, so that the rollers have no reciprocating motion, the screen-plate G passing to and fro under them.

BARN MACHINERY. Fr., *Machines employées dans les fermes;* Ger., *Landwirth schaftliche Apparate;* Ital., *Macchine impiegate nelle cascine.*

Barn machinery may be divided into the following classes,—threshing and dressing apparatus for grain; mills and flour-dressing apparatus; corn-crushing, straw-chopping, and root-cutting implements.

Fig. 522 shows a longitudinal section of a Portable Threshing Machine. The sheaves of corn having been unbound, are passed to the feeder, who stands in the dickey A, and shakes the straw

522.

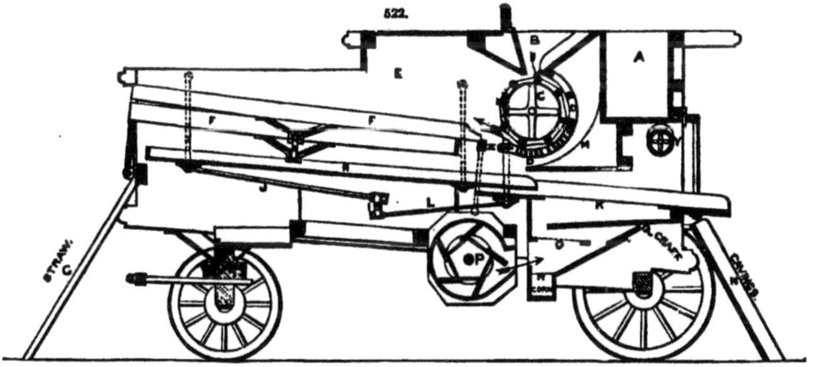

loose into the mouth of the machine B; here it meets the drum C revolving at a high velocity, which carries it over the concave or breastwork D, beating out the grain from the ears. The grain falls through the breastwork D, while the straw is thrown off into the chamber E, whence it is taken forward by the shakers F F, and thrown over the straw-board G. The shakers F by their motion also separate the straw, and allow any grain that may have been thrown off from the drum with the straw to fall on to the shock-board or shogging-board H, whence, by means of the motion given to the shogging-board from the crank-shaft I, through the connecting-rod J, it is passed down to the riddle-board K, to which an opposite motion is communicated by the other connecting-rod L from an opposite crank on the shaft I, the two boards, H and K, advancing to each other or receding simultaneously, both being suspended by links to the framing, and swinging loosely between the sides of the machine.

The grain that passes through the breastwork D is guided by the bent plate M, so as to fall upon the top end of the riddle-board K. In working the machine, many heads and short straws fall from the breastwork and shakers, and a coarse riddle or perforated plate is put on the top of the riddle-board K to take them out from the corn; these "cavings," as they are called, pass over the end of the board K, and fall down the caving-board N. The corn and all smaller substances, chaff, and so on, pass through the top riddle, and down an inclined board on to a second riddle O, of a finer mesh or perforation; here they meet a blast of air from the fan or blower P, which blows away the chaff through the opening Q, while the grain which is heavier falls into the spout R.

The grain is thence carried away by the elevator S, Figs. 523, 525. This is composed of a number of tins or cans T fixed on a belt U, which works over two pulleys V V, one at the top and

523. 524. 525.

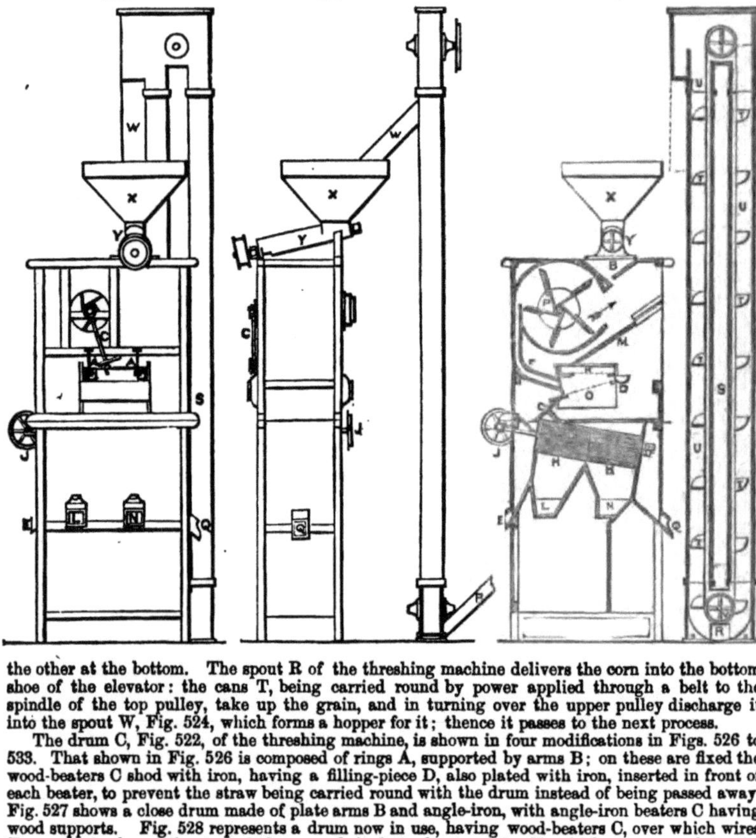

the other at the bottom. The spout R of the threshing machine delivers the corn into the bottom shoe of the elevator: the cans T, being carried round by power applied through a belt to the spindle of the top pulley, take up the grain, and in turning over the upper pulley discharge it into the spout W, Fig. 524, which forms a hopper for it; thence it passes to the next process.

The drum C, Fig. 522, of the threshing machine, is shown in four modifications in Figs. 526 to 533. That shown in Fig. 526 is composed of rings A, supported by arms B; on these are fixed the wood-beaters C shod with iron, having a filling-piece D, also plated with iron, inserted in front of each beater, to prevent the straw being carried round with the drum instead of being passed away. Fig. 527 shows a close drum made of plate arms B and angle-iron, with angle-iron beaters C having wood supports. Fig. 528 represents a drum now in use, having wood-beaters C, over which wires E are strained; on the top of these are bolted the beater-plates F, which are grooved in the face, as shown in Fig. 529, the object of the grooves being to allow the grain to pass into them, and prevent it from being broken, as was the case with plain straight beaters.

The breastwork D, Fig. 522, is shown in detail in Fig. 528. It consists of a frame made in two portions, the upper part from G to H being of wood, and the lower part from H to I of iron; these are carried by iron rods passing across the machine at the three points G, H, and I, which afford the means of adjusting the position of the breastwork by means of screw adjusting-pieces at the sides of the machine, so as to bring it nearer to or farther from the drum, as may be required. The adjustments at G and H being horizontal sliding adjustments, and that at I vertical, a slot J is required at the outer extremity of the frame, in which the rod I works. The distance between the breastwork and drum is gradually contracted from the top towards the bottom, to ensure the whole of the straw being exposed to the action of the drum; this distance is sometimes about 1½ in. at the top, and reduced to only ¼ in. at the bottom, but the adjustment is dependent upon the state of the grain. The upper half of the breastwork is furnished with transverse strips or ribs of wrought iron K, about $1\frac{1}{2}$ in. wide by $\frac{1}{4}$ or $\frac{3}{8}$ in. thick, let in to the sides of the framework; these are perforated with holes, through which a set of wires are passed, forming a concave grating to prevent the straw from getting through. The lower half of the breastwork is a frame of iron bars D, also

526.

527.

528.

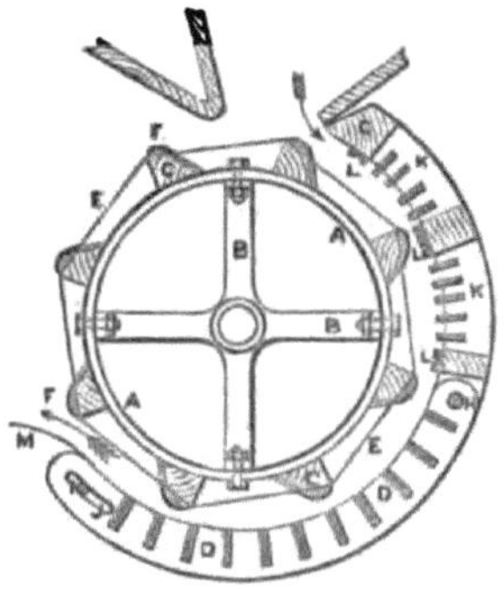

529.

530.

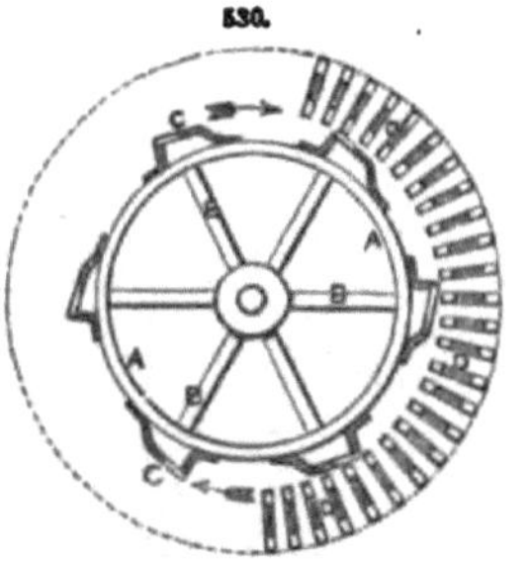

531.

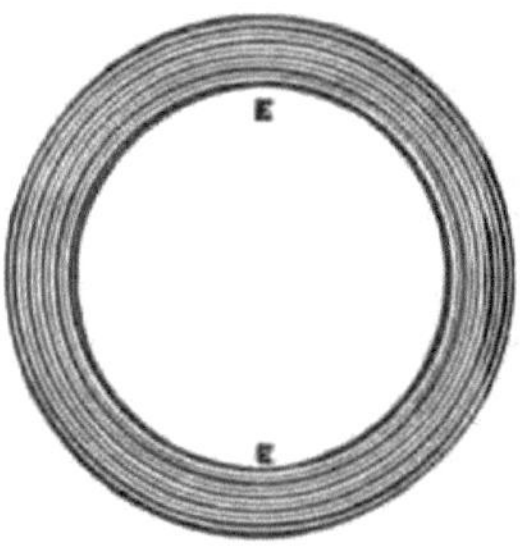

532.

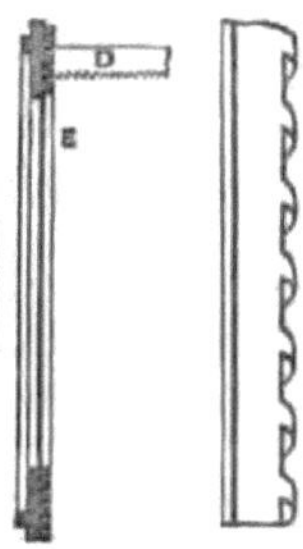

533.

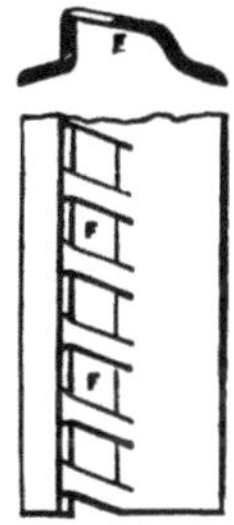

covered with wire. Strips of bar-iron L rolled with one or two longitudinal grooves are fastened on the wooden frame of the breastwork, projecting slightly to prevent the straw from passing too quickly without being thoroughly threshed. The straw is guided from the breastwork to the shakers by the plate M.

Another plan of breastwork, invented by Barrett, Exall, and Andrewes, is shown in Fig. 530, formed by a series of flat bars D, held in their places by the sides of the machine in which slots are cast; the bars are all moved together, to or from the drum A, by a worm-plate E, Figs. 531, 532, in which their ends rest, and are thus adjusted to the required distance from the drum. The plan of drum adopted in connection with this breastwork is also worth notice, as being new, simple, and efficacious. Three or more sets of arms B are keyed on the spindle, each set being completed by a hoop A on the outside, on which are fixed the wrought-iron beater-plates C, of the form shown in Fig. 530. The beater-plates are made of flat iron, rolled or pressed into the shape, with the spaces F cut in, as shown in Fig. 533, so that any grain they may collect may pass into the drum until it can find its way again to the breastwork D. This beater seems to be more perfect in its mechanical construction than the others, where wood and iron are used together.

Figs. 534 to 539 show three modifications of the straw-shaker F, Fig. 522. Figs. 534, 535, represent side-elevations and end-elevations of the two positions of a "rail-shaker," which is made of a

534.

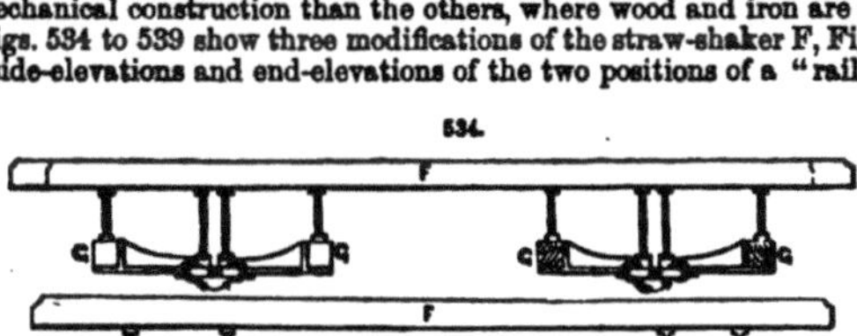

535.

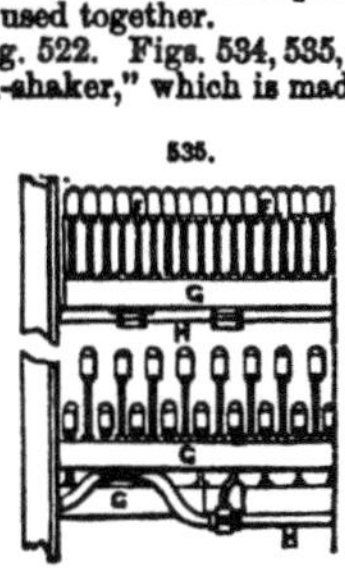

number of rails or bars F F attached alternately to two frames G G, actuated at both ends by the two crank-shafts H; the shafts H are also coupled together by means of cranks on their outer ends and coupling-rods. Figs. 536, 537, show a "box-shaker," consisting of a set of boxes F F, covered with perforated sheet iron, or with sloping strips of thin hoop-iron, as shown dotted; the boxes

536.

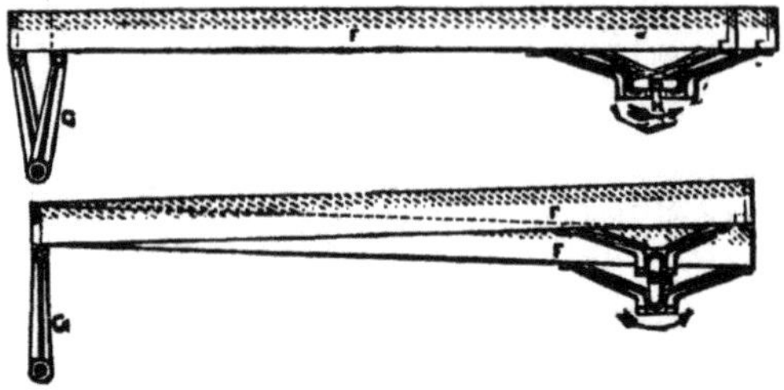

37.

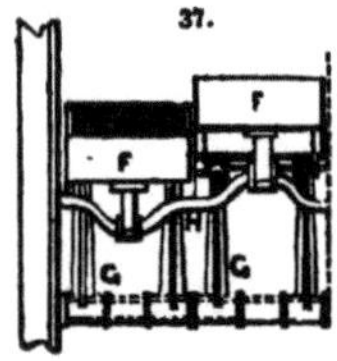

are carried at one end by radius-links G, and the other ends are attached to double cranks on the shaft H, which give the alternating movement, the boxes thus rising and falling to receive or deliver the straw the one to the other, and the rotary motion of the cranks also carrying it forwards, as in the former case.

An improvement upon these two methods is shown in Figs. 538, 539. The great fault of the rail-shaker consists in its passing so many straws through; while the objection to the box-shaker

538.

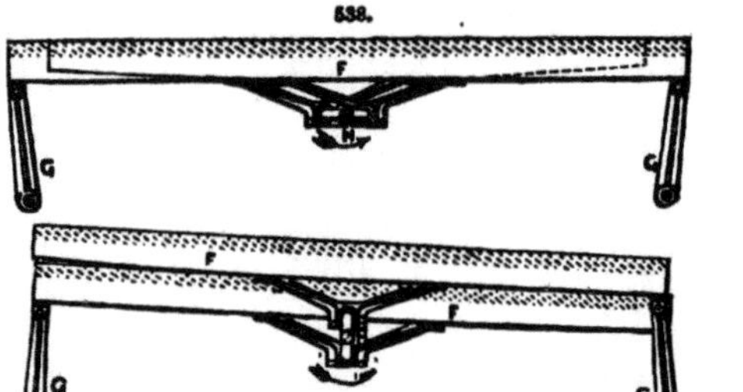

539.

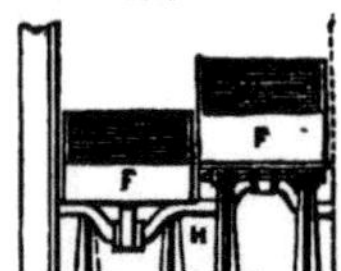

is, that any grain that has been thrown by the drum to the outer end of the chamber E, Fig. 522, is carried along with the straw over the end of the shaker, in consequence of the vertical motion of the crank being reduced at that point, and becoming only a horizontal or longitudinal motion at the end. To obviate these defects, the new shaker, Figs. 538, 539, was invented by James Good. The principle on which this is constructed differs from that of the ordinary box-shaker, Figs. 536, 537, in this respect, that while the latter has the crank-bearings at the same end for all the boxes and the radius-links G at the other end, in Good's shaker the boxes are supported by the links alternately at opposite ends, the crank-shaft H being thus between the links G; or, in other words, the crank-shaft H may be said to be moved to the middle of the length of the boxes, and half of the boxes to be then turned round end for end. Thus, while the ends that are attached to the links close to the drum rise only so much as is due to the vibration of the rocking-links, the ends of the other boxes have a considerable lift imparted to them from the crank; in the centre of the length of the boxes, just over the crank-shaft H, the lift and throw are the same as in the most effective part of the old shaker; and again at the outer end the straw is tossed by the loose ends of the boxes, while only a passing or horizontal motion is given by the others. This is known as the "cross-shaker," owing to the boxes moving crosswise, or alternately up and down.

There is another shaker, invented by T. and H. Brinsmead, the features of which are different from those of every other. Immediately over an inclined plane of wood, and sufficiently above it to allow them to revolve without touching it, are placed transversely, and therefore horizontally, a series of triangular prisms of wood, armed at their edges with curved iron teeth, so arranged that, as the prisms revolve, the teeth of each pass between those of the two adjoining prisms. The prisms being made all to revolve simultaneously in one direction, either by cranks on the ends of the spindles coupled together by one rod, or by a train of wheels, the straw which falls on those at the lower end of the plane is tossed and carried upwards by the action of the teeth as they rise, which also, as they pass down again between the teeth of the prism next above, deliver the straw on to these last, and sweep down to the bottom of the plane the corn and any short straw that has fallen through. At the bottom of the plane is a curved wire-netting, through which the corn readily passes, but which stops the passage of any straw; and the revolving teeth of the lowest prism rake up the straw again as it accumulates, and toss it upwards and onwards as at first.

In the shaker shown in Fig. 522 there are five boxes, two working on links at the end of the machine, and the three alternate boxes vibrating on links at the end nearest the drum; the inner ends of the latter are prolonged, to allow for the throw of the crank, and are attached by means of an angle-iron to a cross-bar Z, extending across the machine, and carried by two links outside the framing, one on each side, as shown dotted, thus avoiding any links inside the machine.

The slogging-board H and the riddle-board K, Fig. 522, were originally in one length; and the vibration then caused by this single piece moving backwards and forwards impeded the introduction of portable machines. They were, however, afterwards parted into two lengths, and in the present machine the motion of the two boards in opposite directions neutralizes the disturbing effect of the reciprocating weight of both.

The elevator S, Figs. 523, 525, has been already described in a form generally in use. Another kind of elevator also in use is formed of sheet iron bent round a spindle in a spiral form, working in a trough curved to the radius of the outside of the worm, and touching it, thus winding the corn, &c., that falls into the trough, by means of the worm, up the spout; but as it must necessarily be kept at only a slight inclination, this elevator is not so general in its application as that previously described, nor nearly so cheap in construction.

From the spout W, Figs. 523, 524, the grain is delivered into the hopper X of the corn-dressing machine, and thence into the barley-horner Y, which is shown more fully in Figs. 540 to 543. It is here subjected to the action of a number of knives fixed on a spindle, which loosen the

540.

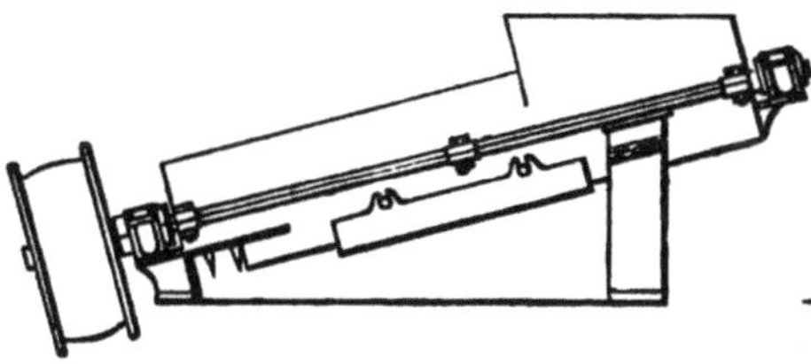

541.

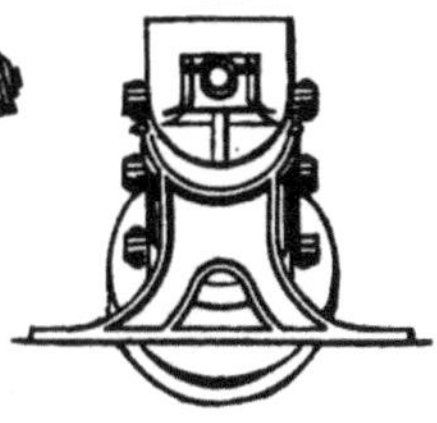

542.

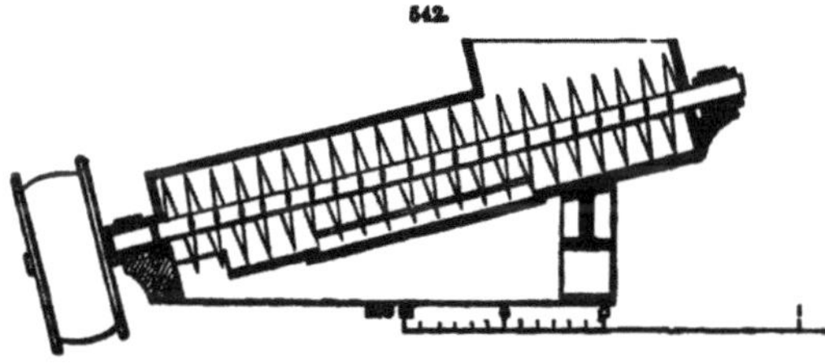

543.

husks or whites of the wheat and cut the ears or horns off the barley. Independently of the inclined position of the barley-horner, the grain is kept in motion by the "set" of the knives, which are in a spiral form. Motion is given to the spindle by a pulley fixed on the end.

The grain and loose ears pass from the barley-horner into the mouth B of the dressing machine, and are met in their descent by a current of air from the fan or blower P, which clears the grain of all chaff and ears that may have been left in it; the grain falling upon the inclined board M is conducted to the riddle K, which is carried upon the links A A, Fig. 523, and has motion communicated to it by the crank at the end of the spindle C. Any stones or ears that may have got in are here taken out, and pass over the riddle K into the spout D, whence they are conducted into the delivery-spout E. The grain, on falling through the riddle, is caught by a fine wire-sieve O, through which all small seeds pass, and are carried also to the spout E. The grain passing over this sieve, is swept by a current of air taken from the back of the blower P by the passage F, by which it is effectually cleaned of any lighter seeds that may be too large to pass through the sieve, and also of any chaff that may have been loosened from the grain by the riddling. The board G carries the grain to the mouth of the revolving sieve or screen H, which receives a rotary motion from the wheels J. The small imperfect corn, or "light" corn, falls through the first meshes into the spout L; the mesh, being then altered and widened, allows the broken corn and a larger size, or "tail" corn, to fall through into the spout N; while nothing but the best corn can find its way to the spout Q. A simple apparatus is here fixed, consisting of a *weighing machine* with rods and bell-cranks, so arranged as to shut off the delivery and ring a bell when the scale falls. A bushel of corn weighs 60 lbs.; four bushels make a sack; and the weight of the sack itself being 7 lbs., 247 lbs. is therefore the weight to be put in the scale; the empty sack is held open to the spout Q by means of rods fixed on this scale of the machine. When the four bushels of corn are delivered, the scale falls, loosing the catch of the slide, which immediately shuts, and allows a bell that has hitherto rested upon the top of it to swing clear, and ring, thus calling the attendant to put a fresh sack on and reopen the slide.

The whole process, from the time when the corn in the straw is fed to the threshing machine, Fig. 522, to the time when the grain, dressed and sorted for market, is sacked in half-quarter quantities from the spout Q, Fig. 524, is thus entirely self-acting; and in this time the following separations are made:—straw, cavings, chaff, seeds, light corn, tail-corn, best corn; besides dust, which must inevitably be mixed up with the straw at first, and which is blown away in the process.

The portable threshing machine is now so arranged as to combine the dressing and separating process with the threshing and winnowing in one machine.

The riddle-boards K and O, Fig. 522, are divided longitudinally and vertically into two parts, as is also the blower or fan P, thus forming two distinct sets of blowers and riddles. The corn passing through the first set, arrives at the spout R, as before described, from which it is taken up by elevators, and passed into the hopper of a barley-horner Y, placed under the dickey A of the portable machine. It is thus passed to the other side of the machine, where it falls down in front of the second half of the divided blower, corresponding to the blower P in Fig. 524, upon the new or second set of riddles, corresponding to the riddles K and O in the same figure, down to a second spout corresponding to the spout R on the other side of the machine. It is here again taken up by another set of elevators, and discharged into the hopper B of the separator, Figs. 544, 545, passing through the revolving screen H, and being delivered as before, the light corn by the spout L, the tail at N, the best at Q, and the seeds and dirt at E E.

In this separator a blower P is fixed, either above the machine and screen-case H, to blow all chaff and seeds out before passing into the screen, as in Fig. 544, or at the end, as in Fig. 545. In

544.

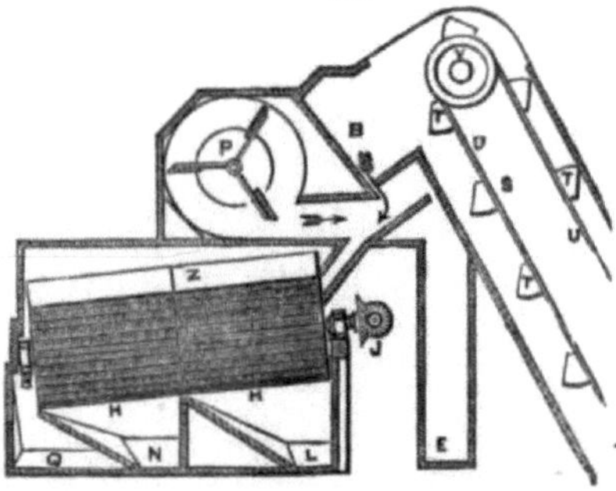

545.

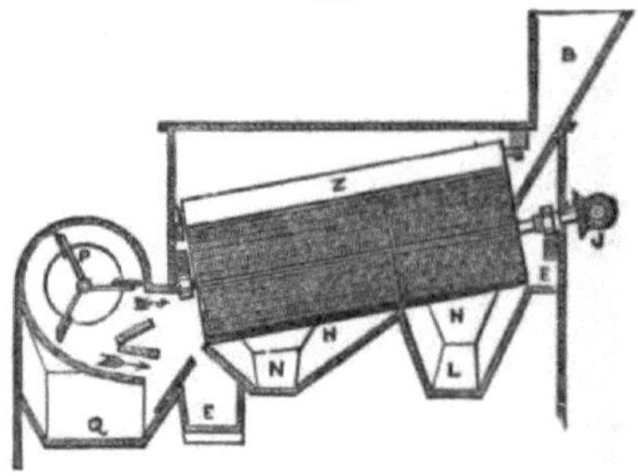

the latter plan, shown in Fig. 545, the corn is subjected to the action of the blast while on the screen and in falling from it. The arrangement, however, shown in Fig. 544, introduced by Clayton, Shuttleworth, and Co., is found to be an improvement, and has been adopted in place of the other arrangement. The plan first tried was to blow up the screen; then under; then up and under; but in the last plan the chaff and light corn are blown out before reaching the revolving-screen H, consequently power is saved; and experience shows that a better sample is obtained by this arrangement. In both the plans, shown in Figs. 544 and 545, brushes Z Z are fixed above the revolving screens H to keep the wires clear. The separating apparatus is fixed on the side of the portable threshing machine, and the grain is delivered in the same separations as enumerated previously.

546.

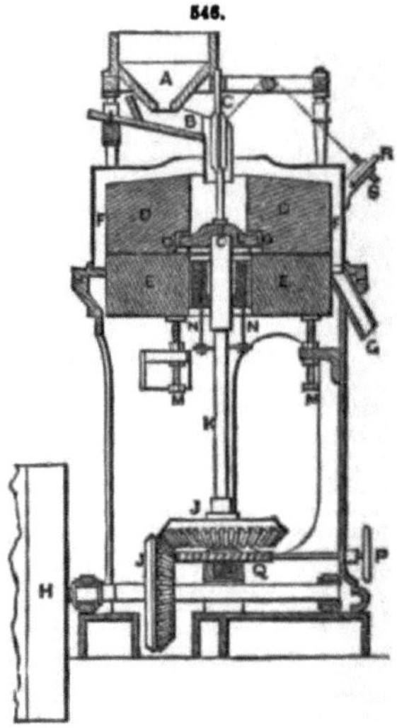

The action of threshing is still supposed by many to be a continuous series of blows; by others to consist of rubbing between the beaters and the surface of the breast-work; and by others, again, to be the combination of the two actions.

In the present machines in this country the straw is fed across the drum, so as to allow the drum to "boult" it, or carry it through without twisting or breaking the straw, which is with many farmers a serious consideration. In the old and in the present American machines, the straw is broken up by means of pegs on the drum and breastwork; but as the Royal Agricultural Society take notice of the state of the straw, whether it is broken or not, it has become an object to preserve it.

With regard to the riddles, the top riddle-plate K, Fig. 522, is coarse in the openings, to allow the corn to pass through freely; the second riddle O is finer; and in the combined machines with the split-blower and riddles, the third and fourth riddles, corresponding to K and O in Fig. 525, are each finer than the one before it. The riddles are sometimes made of wire netting, and sometimes of wood perforated at an inclination; but more frequently of punched sheet iron.

The next process through which the grain has to pass is grinding, breaking, or kibbling, as it is called, according to the degree of fineness of the meal required. Several methods have been proposed, but the old plan of one stone revolving above another which is fixed is still found as good and economical as any. Fig. 546 shows a vertical section of a portable corn-mill constructed on this plan. The corn being fed into the hopper A, is shaken down the spout B by the damsel C, working against the spout, into the eye of the upper or running stone D, whence it gets into the furrows of the two stones D and E, passing out from them as meal into the casing F, from which it is carried off by the spout G. Motion is given to the runner D from the pulleys H through the shaft I by means of the mitre-wheels J to the vertical spindle K; at the top of the spindle are two studs, on which rides the casting L, which in its turn carries the running-stone D by the two steps O. The damsel C being fastened in the casting L, receives motion from it. The bed-stone E is adjusted by the set screws M M; and the screwed wedges N keep the wood-packing against the spindle K. The coarseness or fineness of the meal is regulated by the hand-wheel P working a worm-gearing in the worm-wheel Q, which is keyed on the top of a brass bush resting in an outer bush or seating, with a square thread cut in the two bushes; thus, by turning the wheel P, the worm-wheel Q causes the inner bush to turn in its seat on the thread, thereby raising or lowering the spindle K, and with it the upper stone D. The hand-wheel R acts as a nut upon the screw S, raising or lowering the spout or shoe B, and thus diminishing or increasing the feed to the stones. The object of these mills on farms is to break wheat, barley, oats, &c., into meal for food for man and beast; it is only worked occasionally, and therefore the arrangements for cooling the flour and meal are not required.

The Flour-dressing Machine consists of a case containing an inclined cylinder of wire gauze of various degrees of fineness; on a spindle passing along the centre of this cylinder are fixed by means of arms keyed on it a set of brushes, which revolve at about 300 to 500 revolutions per minute. The meal or broken corn is passed into one end of the cylinder, and the fine flour falls at once through the wire into the first compartment; by means of the brushes and alterations in the gauge of the wire five or six separations are made.

The smaller implements forming part of the barn works are of modern introduction, and have been brought forward by science to assist the practical agriculturist. Those in most general use are—

Linseed and Corn crushers; Chaff-cutters, or Straw-choppers; Turnip-cutters and Root-cutters; Gorse crushers or cutters; Oilcake crushers or breakers.

The Linseed and Corn crushers have been introduced to effect a saving in the quantity of corn necessary for animals, as the crushing or bruising ensures the whole of the nutriment contained in the grain being rendered available, instead of the animal swallowing the food without properly masticating it. The process is simple; the grain is merely passed between plain or grooved rollers, crushing, not grinding, being the object; the bulk is thereby increased at least one-third, and its nutritive power in the same ratio. This idea is really of very old date, though only of recent adoption, having been recommended by Hartlib in 1650.

The Chaff-cutters are made with two or more knives shaped concave or convex towards the edge, and fixed on a shaft carrying a fly-wheel. A feed-motion is attached to bring the straw or hay up to the knives, the straw being placed in a box, and the knives working across the end of the box and close against it. The length of the cut is variable, and may be altered from about ¼ inch to 3 inches by adjusting the amount of the feed.

Turnip-cutters are discs, arms, or plates of metal with knives or cutters to pare or slice turnips or other roots, which lie against the knives by their own weight; the roots are cut in slices for cattle, and in strips for sheep, cross-cutters being then introduced.

Gorse-crushers are made with toothed rollers to bruise the gorse for feeding beasts, which eat it with avidity; it is crushed by the machine to a harmless pulp, and cut into short lengths.

Oilcake-breakers are made with toothed rollers, by which the cake is taken hold of and broken, the cut being adjusted by set screws, so as to regulate the degree of fineness required, according as the cake is being broken for cattle or sheep; the dust from the cake passes through a grating.

The results of experience with the several machines have led to the adoption of the following speeds of working as the most eligible for the purpose:—

The speed of the drum of the threshing machine is found to be best at about 5000 ft. of the circumference a-minute.

The straw-shaker should pass the straw at the rate of 75 to 80 ft. a-minute.

The shogging-board and riddle-board should be worked at about 200 revolutions of the crank a-minute.

The blowers should run at about 2000 ft. of the circumference a-minute.

The barley-horner spindle should make 400 to 500 revolutions a-minute.

The elevators should work at 100 or 150 ft. a-minute, but the rate is dependent upon the quantity to be taken up, and it may sometimes be found necessary to quicken the speed.

The best speed of a 3-ft. stone for a mill similar to the one described is found to be about 140 to 150 revolutions a-minute, or about 1400 ft. a-minute of the circumference, instead of 1550 to 1600 ft. a-minute, as given by the ordinary rule, the lower speed giving the greatest quantity of work done for the least amount of power expended.

The smaller machines are not so delicate in their operations, and are more dependent upon the kind of stuff they are fed with and the state it is in, and therefore do not allow of any fixed rule.

The growth of the threshing machine having been traced from the simple threshing drum and breastwork to the complete machines now in use, an interesting experiment may be mentioned, which was tried at the meeting of the Yorkshire Agricultural Society, at Ripon, in 1854. to ascertain the power consumed by the several parts of the machine.

A combined fixed machine with a dressing apparatus as described, required 6·15 horse-power to drive it when at work, and 1·77 horse-power when empty, leaving 4·38 horse-power available for doing the work. This machine threshed 200 sheaves of wheat in 13 80 minutes, and the power expended was accordingly 6·15 horse-power for 13·80 minutes, equivalent to 84·87 horse-power for one minute: or, multiplying by 33000 and dividing by 200, the power expended was—

14004 units of work to thresh 1 sheaf of wheat

(one unit of work being one pound weight raised one foot high). The 1·77 horse-power required to drive the machine, when empty, was divided as under:—

	Horse-power.
Dressing Machine	·37
Elevator	·11
Shaker and Riddle	·28
Blower	·20
Drum and Shafting	·81
Total	1·77

A similar machine required 4·23 horse-power to drive it when at work, and 2·70 horse-power when empty, leaving 1·53 horse-power available for doing the work. This machine threshed 200 sheaves of wheat in 22·68 minutes, and the power expended was accordingly 4·23 horse-power for 22·68 minutes, equivalent to 95·94 horse-power for one minute; or, multiplying by 33000, and dividing by 200, the power expended was—

15830 units of work to thresh 1 sheaf of wheat.

The 2·70 horse-power required to drive the machine when empty was divided as under:—

	Horse-power.
Dressing Machine	·34
Elevator	·28
Shockboard and Pulley	·39
Blower and Drum	1·46
Main Shaft and Shaker	·23
Total	2·70

The power expended in threshing 1 sheaf of wheat has been gradually increased from about 6000 units in the earlier machines by the additions in successive years of further apparatus to render the process more complete, several operations being combined in the one machine.

Taking a similar basis of calculation, the power required to work the portable corn-mills and smaller barn implements, as reduced from the average results of the trials at the show of the Royal Agricultural Society in 1855, is as follows:—

Portable Corn-mills,	about	9000	units	to grind	1 lb. of corn.
Corn-crushers,	„	3600	„	to crush	1 lb. of linseed or oats.
Chaff-cutters,	„	2200	„	to cut	1 lb. of chaff.
Turnip-cutters,	„	150	„	to cut	1 lb. of turnips.
Oilcake-breakers,	„	180	„	to break	1 lb. of cake for cattle.
		350	„	to break	1 lb. of cake for sheep.

A trial of threshing machines took place in Kent, in April, 1856, when one machine threshed, without finishing, about 21½ quarters of wheat with 350 lbs. of coal in 3½ hours; while another machine, having extra machinery attached to it for finishing, threshed and finished for market in the same time about 25½ quarters with 563 lbs. of coal, and that under disadvantage, owing to very high wind, and the windy side of the stack having fallen by lot to it. A stack of barley was threshed and finished by the second machine in 7¾ hours, including stoppages amounting to 1¼ hour, making the actual time 6½ hours; in this time the machine was found to have yielded 73 quarters of barley, or at the rate of 11·23 quarters per hour; the engine employed was of 7 horse-power.—(Taken from a paper by W. Waller, given in 'Proc. Inst. of Mec. Eng.,' 1856.)

BARKER'S MILL. Fr., *Roue hydraulique de Barker*; Ger., *Segner's Wasserrad*; Ital., *Mulinello idraulico a reazione*; Span., *Molino de Barker*.

The simplest form of reaction water-wheel is that of Barker's Mill, Fig. 547. Water-wheels are divided into groups according to the form of the part which receives immediately the action of the water.

The following synoptical Table exhibits at one view the different kinds of water-wheels and machines of rotation:—

- Water-wheels and machines of rotation.
  - Vertical ..
    - with floats ..
      - plane, in
        - a water-course
          - rectilinear.
          - circular.
        - an indefinite fluid.
      - curved. Wheels of Poncelet.
    - with buckets receiving the water
      - at summit. Overshot wheels.
      - below summit. Breast or undershot.
  - Horizontal ..
    - with floats ..
      - struck by an isolated vein.
      - placed in a cylinder. Tub-wheels.
      - outside cylinder. Turbines of Fourneyron.
    - with conduits. Turbines of Burdin, of Boyden, of Francis.
    - reaction. Barker's Mill. Wheels of Segner, of Euler.

The water-mill, shown in Fig. 547, invented by Barker, performed the operation of grinding corn without either wheel or trundle; A is a pipe or channel that brings water from a reservoir to the upright tube B.

The water runs down the tube, and thence into the horizontal trunk C, which has equal arms; and, lastly, runs out through holes at *d* and *e*, opening on contrary sides near the ends of those arms. These orifices, *d*, *e*, have sliders fitted to them, so that their magnitude may be increased or diminished at pleasure.

The upright spindle D is fixed in the bottom of the trunk, and screwed to it below by the nut

g, and is fixed into the trunk by two cross-bars at f; so that, if the tube B and trunk C be turned round, the spindle D will be turned also.

The top of the spindle goes square into the rynd of the upper mill-stone H, as in common mills; and as the trunk, tube, and spindle turn round, the mill-stone is turned round thereby. The lower or quiescent mill-stone is represented by I, and K is the floor on which it rests, in which is the hole L to let the meal run through, and fall down into a trough which may be about M. The hoop or case that goes round the mill-stone rests on the floor K, and supports the hopper in the common way. The lower end of the spindle turns in a hole in the bridge-tree G F, which supports the mill-stone, tube, spindle, and trunk. This tree is movable on a pin at h, and its other end is supported by an iron rod N fixed into it, the top of the rod going through the first bracket O, and having a screw-nut o upon it, above the bracket. By turning this nut forward or backward the mill-stone is raised or lowered at pleasure.

547.

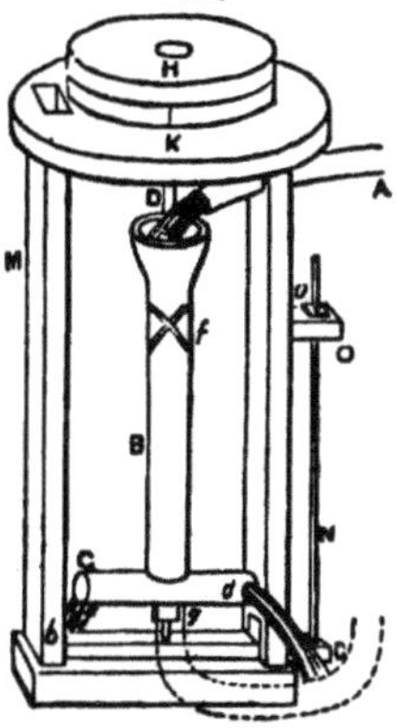

Whilst the tube B is kept full of water from the pipe A, and the water continues to run out from the ends of the trunk, the upper mill-stone H, together with the trunk, tube, and spindle, turn round. But if the holes in the trunk were stopped, no motion would ensue, even though the tube and trunk were full of water. For, if there were no hole in the trunk, the pressure of the water would be equal against all parts of its sides within. But when the water has free egress through the holes, its pressure there is entirely removed: and the pressure against the parts of the sides which are opposite to the holes turns the machine.

James Rumsey improved this machine, by conveying the water from the reservoir, not by a pipe, as A D B, in great part of which the spindle turns, but by a pipe which descends from A, without the frame L N, till it reaches as low or lower than G, and then to be conveyed by a curvilinear neck and collar from G to g, where it enters the arms, as is shown by the dotted lines at the lower part of the figure.

Most of the authors who have attempted to lay down the theory of this mill have fallen into error. The most ingenious theory we have yet seen is by William Waring (given in the 'American Transactions,' vol. iii.), which, with some such corrections as appeared necessary to adapt his rules to practical purposes, is nearly as follows:

The first inquiry relates to the *magnitude of the pipe* which conveys the water from the reservoir to the centre of the horizontal tube $e\ d$, at g. To this end, let $A =$ the area of the orifice by which the water is admitted at g; $h =$ the perpendicular height of the surface of the water in the reservoir above g; $d =$ the vertical depth of any horizontal section of the pipe below the same surface; $S =$ the surface or area of the horizontal section of the pipe, at the depth d. Then, since the areas in the several parts of the pipe should be inversely as the velocities, and the velocities are in the subduplicate ratio of the depths below the head, those areas must be inversely in the subduplicate ratio of the depths; consequently, $\frac{S}{A} = \frac{\sqrt{h}}{\sqrt{d}}$, and $S = A\sqrt{\frac{h}{d}}$. So that the pipe must have its bore increased from the level of g upwards in the ratio of 1 to $\sqrt{\frac{h}{d}}$; and if a section in any part be less than would be assigned by this ratio, the water will be obstructed in its passage.

Of the Initial Force with which the Machine commences its Motion.—If we conceive the water pressing in the tube from g towards e, previous to the opening of the aperture, there will manifestly be no motion occasioned, because the forces on the opposite sides of the tube balance each other, and the force against the end C is resisted by the fixed axle Dg; or, if we consider both arms, it is balanced by the equal force acting upon the equal end at d in an opposite direction. But if one of the apertures, as d (its area being $= a$), is opened, the pressure upon that portion of the tube is taken away, and the equal and opposite pressure upon an equal portion of the contrary side of the tube, having now nothing to keep it in equilibrio, tends to move the arm Cg about the axis Dg. In like manner, when the aperture e (also $= a$) is opened, the pressure, which was previously counterbalanced by the opposite pressure on the orifice e, now exerts its tendency to produce a rotatory motion about the axis Dg; so that, combining together the effects of both these unbalanced pressures, and considering that the pressure of water upon any point is proportional to the depth of that point below the upper surface of the fluid, we shall have $2\,a\,h\,w$ for the force which causes the rotatory motion to commence; the values of a and h being taken in feet, and w representing $62\frac{1}{2}$ lbs. avoirdupois, the weight of a cubic foot of water. But as the velocity of rotation increases, the pressure depending upon the relative velocities of the water and the sides of the tube diminishes, and consequently the power is diminished, notwithstanding what is gained by that which we now proceed to consider.

The Centrifugal Force.—This may be found in a similar manner to that which is adopted when considering the theory of the *centrifugal pump*. Thus, if, besides the preceding notation, we take l for the length of each arm $g\,d$, $g\,e$, t for the time of rotation, g for $32\frac{1}{6}$ ft., the measure of the force of gravity, and π for 3·141593, since a is the section of the flowing water at right angles to its motion, we shall have, by proceeding as in the article just referred to, $\frac{2\,\pi^2\,l^2}{g\,t^2} =$ the length of a column of water, whose pressure is equal to the centrifugal force, or $\frac{4\pi^2 a w l^3}{g t^2} = 76\cdot70625 \frac{a l^3}{t^2}$ the weight of a column of water in lbs., which is equivalent to the centrifugal force of the fluid in both arms. And this is equivalent to the augmentation of power at the apertures, because fluids press equally in all directions.

The inertia of the fluid greatly counteracts the effects of the centrifugal force. The inertia of the rotatory tube with the contained fluid would not continue to resist the moving power after the velocity became uniform, were the same fluid retained in it as was in it when the motion was first imparted; but as this passes off, and there is a continual succession of new matter acquiring a motion in the direction of the rotatory, there must be a constant reaction against the sides of the tube equal to the communicating force. Now this reaction is very different from that of a fluid confined in the tube when it begins to move, because a particle at the extremity of the tube is not to receive its whole circular motion there, but gradually acquires it by a uniform acceleration during its passage along the tube; so that we must here inquire what force will give to the quantity of water $a l w$, in the time $\frac{l}{v}$ of its passing through its respective horizontal arm, the velocity $\frac{2\pi r}{t}$, in the direction of the aperture. Then, according to the rules given for forces in dynamics, we shall have $\frac{12\cdot273\, a l v}{t} \times \frac{8\cdot0208}{5} = 19\cdot6878 \frac{a l v}{t}$ for the resistance in lbs. opposed to each arm, such resistance being estimated as if accumulated at the distance $\frac{1}{2} l$ from the centre of motion.

Acquired Velocity of the Water.—According to the theory of hydraulics, the velocity of water issuing through an aperture at the depth h below the upper surface of a reservoir, is expressed by $8\cdot0208 \sqrt{h}$, which when reduced, in conformity with the experiments, becomes $5\sqrt{h}$ very nearly; and this is the velocity of the water passing out of the tubes at the commencement of the rotation. Then, as

$$\sqrt{2ahw} : 5\sqrt{h} :: \sqrt{\left(2ahw + 76\cdot70625 \frac{a l^2}{t^2}\right)} : 5\sqrt{\left(h + 38\cdot35312 \frac{l^2}{w t^2}\right)}$$
$$= 5\sqrt{\left(h + \cdot61365 \frac{l^2}{t^2}\right)} = v,$$

the required velocity of the water.

Ratio of the Central Force to the Inertia.—This will be ascertained by substituting for v in the expression $19\cdot6878 \frac{a l v}{t}$, its value just found; so that we have $98\cdot439 \frac{a l^2}{t^2} \times \sqrt{\left(\cdot61365 + \frac{h t^2}{l^2}\right)}$ for the inertia, while the centrifugal force is measured by $76\cdot70625 \frac{a l^2}{t^2}$. Now we find that $76\cdot70625 \frac{a l^2}{t^2} : 98\cdot439 \frac{a l^2}{t^2} \times \sqrt{\left(\cdot61365 \frac{h t^2}{l^2}\right)} :: 1 : 1\cdot2833 \sqrt{\left(\cdot61365 \frac{h t^2}{l^2}\right)}$, or as $1 : \sqrt{\left(1 + \frac{1\cdot646\, h t^2}{l^2}\right)}$ very nearly; which is the ratio of the power gained by the centrifugal force to the obstruction arising from inertia. Whence it appears that the latter is greater than the former, except when $t = 0$, $h = 0$, or $l = \infty$, cases never occurring in practice; and that the longer the arms, the less the fall of water, and the greater the velocity of rotation, the nearer these forces approach to the ratio of equality.

Adjustment of the Parts and Motion.—Here it must be particularly observed that the centrifugal force should not exceed the gravity of the water revolving in the arms $g d$, $g e$; for in that case the water would be drawn into the tube faster than it could be naturally supplied at its entrance, by the velocity proper to that depth, and of consequence a vacuum must be occasioned; nor should the velocity of the apertures be greater than half that of the water through them; for the apertures being still adapted in point of magnitude to the velocity, the effluent quantity or number of acting particles is as the time, the momentum is in the simple ratio of the relative velocity, and therefore the greatest effect will be produced when the velocity of the apertures is equal to half that due to the head of water. These two conditions expressed algebraically will furnish the equations,

$$76\cdot70625 \frac{a l^2}{t^2} = 2 a l w \;\ldots\; \frac{2\pi l}{t} = \frac{5}{2}\sqrt{h + l};$$

from which equations we deduce the following,

$$\text{namely,} \begin{cases} h = 9\cdot29345\, l = 15\cdot1446\, t^2, \\ l = 1\cdot6296\, t^2 = \cdot1076\, h, \\ t = \sqrt{\cdot61365\, l} = \sqrt{\cdot06603\, h}. \end{cases}$$

Whence it appears that h, l, and t^2, are nearly in the constant ratio of 15, $9\frac{1}{4}$, and 1.

Still it should be observed, that while l and t are preserved in a constant ratio, the values of $76\cdot70625 \frac{a l^2}{t^2}$, and of $12\cdot273 \frac{a l v}{t}$, that is, of the central force and of the inertia, must remain the same; so that the brachia may be made of any length at pleasure (not less than $\cdot1076\, h$) if the time of revolution be taken in a corresponding proportion, or so that the velocity of the apertures undergo no variation, which will be ensured by making $t = \sqrt{\cdot61365\, l}$; for a double or triple radius, revolving in a double or triple time, or with half or a third the angular velocity, has the same absolute velocity at the extremity; and with the same power there applied, will produce the same effect. Hence,

The moving force and velocity of the machine, when the effect is a maximum, may be found. For, if we put $\cdot 61365\, l$ for t^2, and $9\cdot 29345\, l$ for h, in the expression $\sqrt{\left(1 + \frac{1\cdot 646\, h\, t^2}{l^2}\right)}$, it becomes $\sqrt{1+8} = 2$; in which case the resistance of inertia is just double the central force, or the gravity of the water in the tube $= 125\, a\, l$, which, taken from the impelling force, leaves $62\cdot 5\, (a\, h + l) - 125\, a\, l = 62\cdot 5\, a\, (h - l) = 55\cdot 775\, a\, h$ lbs. avoirdupois = the real moving force, at the distance of the centres of the apertures from the centre of motion, l being taken $= \cdot 1076\, h$. And by a like substitution, the velocity $\frac{5}{2}\sqrt{h + l}$ becomes $\frac{5}{2}\sqrt{1\cdot 1076\, h} = 2\cdot 63205\sqrt{h}$, feet per second.

Area of the Apertures.—If A = the area of a section of the race perpendicular to the direction of its motion, V = its velocity per second, both in feet, a and h as before; then it will be $A\,V = 10\, a\sqrt{\left(h + \cdot 61365\, \frac{l^2}{t^2}\right)}$ cubic feet = the quantity of water emitted per second by both apertures: hence $a = \frac{A\,V}{14\cdot 2722\sqrt{h}} = \frac{\cdot 070066\, A\,V\sqrt{h}}{h}$, the area proper for one of the apertures.

From the preceding investigation we may deduce the following rules.

1. Make each arm of the horizontal tube, from the centre of motion to the centre of the aperture, of any convenient length, not less than ¼th of the perpendicular height of the water's surface above these centres.

2. Multiply the length of the arm in feet by ·61365, and take the square root of the product for the proper time of a revolution in seconds, and adapt the other parts of the machinery to this velocity; or,

3. If the time of a revolution be given, multiply the square of this time by 1·6296 for the proportional length of the arm in feet.

4. Multiply together the breadth, depth, and velocity per second of the race, and divide the last product by 14·27 times the square root of the height, for the area of either aperture; or, multiply the continual product of the breadth, depth, and velocity of the race, by the square root of the height, and by the decimal ·07; the last product, divided by the height, will give the area of the aperture.

5. Multiply the area of either aperture by the height of the head of water, and the product by 55·775 (or 56 lbs.), for the moving force, estimated at the centres of the apertures in pounds avoirdupois.

With respect to different forms and developments of reaction water-wheels, we give, with some slight, but necessary, alterations, the following general observations from a treatise on Hydraulics, by J. F. D'Aubuisson de Voisins.

We designate by the appellation *reaction water-wheels* machines in which the water contained in them, and which issues from them with a certain effort, reacts upon the parts of the machine opposite the orifices of issue with an equal effort; in consequence of which it constrains these parts to recoil, and so occasions the motion of rotation. The following example will enable us to appreciate this mode of action; but before giving this example, it is necessary to revert to a principle.

The equality between action and reaction, which is regarded nearly as an axiom in mechanics, has been directly demonstrated by Daniel Bernoulli, in the case of a jet issuing from a vase ('Hydrodynamica,' pp. 279 and 303). He found, by calculation and experiment, that the effort exerted upon the vase by the reaction of the jet was equal to the weight of a prism which had for its base the orifice, and for its height twice the height due to the velocity of issue; and we know that such is the measure of the effort of which the jet is capable.

Let there be a vase or great vertical tube, of which A is the base, Fig. 548, which is movable around its axis C, at the foot of which is fixed a horizontal tube B D, open at B, and closed through its remaining extent. If this apparatus be filled with water, the fluid will exert an equal pressure on all parts of the tube; that which takes place at any point will be destroyed by the pressure upon the point diametrically opposite, and there will be an equilibrium. But if we make an orifice at a, for example, there will no longer be a pressure upon this point; that exerted upon the opposite side will be no longer counterbalanced, and it will drive the tube in the direction from a to e; the jet issuing at a, acting by its *reaction*, will cause the machine to turn around its axis C, and in a direction opposite to its own, in the same manner as the elastic fluid arising from igniting the powder contained in the charge of a squib or rocket, issuing downwards, drives it rapidly upwards.

548.

If, at the lower part of the great vertical tube A, we have radiating from it many tubes similar to B D, and similarly pierced, we shall have the machine of reaction designed, towards the middle of the last century, by Segner, professor of mathematics at Göttingen, which the Germans consequently name *Segner's Wheel*.

Euler, having made this an object of his studies ('Académie de Berlin,' 1750), proposed, 1st, to give a curved form to the horizontal tubes, so as to obtain a pressure resulting from the centrifugal force; 2nd, to cause the water to issue through the extremities of the tubes, which extremities he curved so as to make them perpendicular to the radius of the wheel drawn to them.

M. Manouri d'Ectot, profiting by the indication of these improvements, planned a machine such as we see in Fig 549. Its tubes, swelling in the middle, and curved like an S, were united and held by iron bars. The motive water is conveyed to them by means of a great vertical tube, which is bent horizontally at B, and, passing under the wings or revolving arms, rises vertically, and terminates at the common centre O.

549.

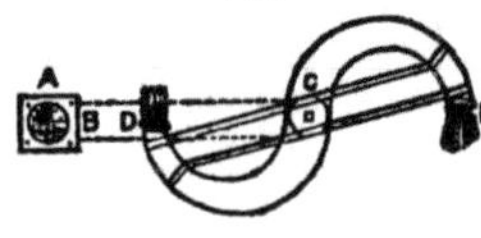

These wheels have been successfully established in the mills of Brittany, of Normandy, and of the environs of Paris; "from authentic experiments, they produced an effect superior to that of the best executed 'pot-wheels,'" says Carnot, in the name of the commission of the Institute appointed to the examination of this machine ('Journal des Mines,' 1813, tom. XXXIII.). However, in common practice, we cannot, without difficulty, keep tight the junction of the stationary part, the tube conducting the water, with the movable part, the wings or arms of the wheel. Otherwise, this wheel seems better fitted than any other to transmit the action of a current of water directed from below upwards, such as issues from certain artesian wells.

Euler, whose ideas upon these reaction machines were derived from Segner's, designed one which seemed to him better fitted to reap the full advantage of this mode of the action of water. It had the form of a great bell, or rather, it was a truncated cone, hollow in the middle, consisting of two concentric surfaces, made of sheet-iron plates, with a space between them, open at the top and closed at the bottom; small bent pipes were fitted vertically all around and at the bottom, their extremities being horizontal and in the direction of the motion, or rather, in a direction opposite to it. The motive water entering at the top of the machine, filled the space between the two conical envelopes, and issued through the small tubes. Though unwieldy, this machine has been used advantageously in France.

550.

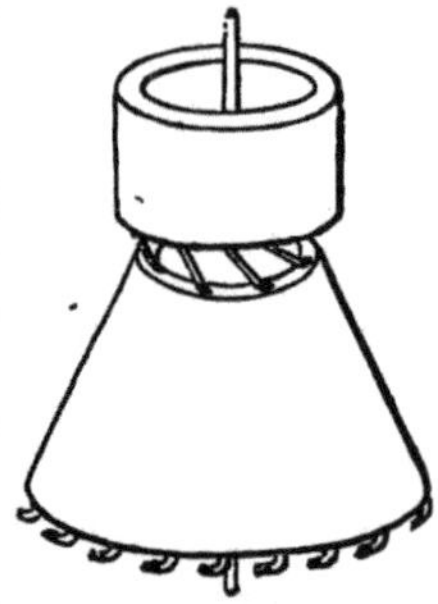

Three years after, Euler gave a more complete theory of reaction wheels; and on this occasion he projected a second, which is described in the 'Mémoires de l'Académie de Berlin,' 1754. It consisted of two parts, placed one above the other as shown in Fig. 550. The upper was immovable, and formed a cylindrical and annular reservoir, with small tubes fixed to the bottom, rectilinear, but inclined at an angle determined by calculation, and delivering the water upon the lower part. The latter, movable around its axis, presented at the top an annular trough, from the bottom of which projected twenty tubes, diverging in their descent, the ends of which, bent horizontally, delivered the water in the air. All of these pipes were covered, as far as the bending, by a smooth sheet-iron surface, designed to lessen the resistance of the air.

Such a machine, with tubes uniformly curved, not being obstructed at their extremity, and not being entirely full of water, has a close resemblance to the duct-wheels of M. Burdin; and the theory of Borda would be equally applicable to it.

The learned engineer whom we have just named, and to whom the works of Euler were unknown, also made a *reaction turbine*, which bears a great resemblance to that of the illustrious geometer. We give a short description of one which he established at the mill of Ardes, in the department of Puy-de-Dôme.

The fall is 6·56 ft. Under a wooden basin, where the water is maintained at a constant height of 3·28 ft., is placed the machine of rotation represented in Fig. 551. Three injecting orifices, fitted to the bottom of the basin, deliver the water horizontally in the crown, or small annular basin, which forms its upper part. It then enters into three pyramidal enclosures, with vertical axes, whose extremities are bent horizontally, having an orifice of issue. The height of the machine is 3·28 ft.; and generally it is one-half the fall.

551.

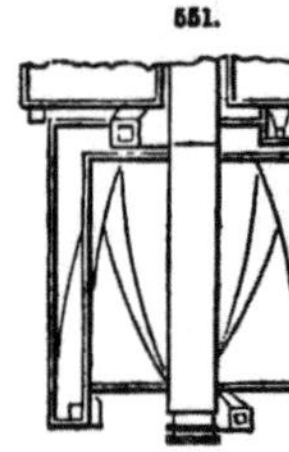

It is contrived so that the turbine, under the injecting orifices, may have a velocity of 14·53 ft., that due to the height of 3·28 ft. The water arriving upon the machine with a velocity equal to that of the points which receive it, there is no shock. Moreover, the head upon the orifices of the conduits being 3·28 ft., the water will issue from them also with the relative velocity of 14·53 ft.: and as that of the orifices in an opposite direction is the same in value, the absolute velocity of the fluid will be zero. The two conditions necessary for the *maximum* of effect are thus fulfilled, and the dynamic effect of the turbine will be P H. The total fall of the water being put = H. This fall, when it is taken to measure the entire force of the current, is the difference of level between the fluid surfaces of the upper and lower reaches. But for hydraulic wheels it is reckoned from the upper level, or that of the reservoir, to the lowest point of the wheel, as this point may be lowered to the level of the lower reach, when this level is constant. P = the weight of water furnished in one second by the motive current.

But in practice, many circumstances always occur to change the conditions of this greatest effect. Still, M. Burdin has never seen the useful effect of his reaction turbines below 0·65 P H, and sometimes it has been as high as 0·75 P H ('Annales des Mines,' tom. III., 1828).

Nearly a century has elapsed since the theory of reaction machines was the object of Euler's researches: his memoirs upon this subject, which, however, I am not in a situation to properly appreciate, bear, according to competent judges, the impress of his analytical genius. But since their publication, and partly in consequence of the works of this great man, the theory of machines in motion, especially in all pertaining to their dynamic effect, has reached a much greater degree of generality and simplicity.

For a summary application to reaction wheels of this theory, I will suppose, with M. Navier, that the water enters them without shock, and runs through them without a sudden change of velocity; I shall only then, have to consider its absolute velocity immediately after its exit from the machine. We have remonstrated (see TURBINE WATER-WHEELS) that when water issues through orifices made in the circumference of a wheel in motion around its vertical axis, its velocity, relatively to that of the machine, is, upon the last element of the orifices, $\sqrt{2gh+v^2}$, h being the height of the reservoir above these orifices, and v their velocity of rotation. We suppose their extremity to be horizontal, and perpendicular to the radius of the circumference described; then, their velocity v is found directly opposed to that which the fluid possesses upon this extremity, and its absolute velocity, immediately after quitting it, is then $\sqrt{2gh+v^2}-v$. But the dynamic effect is equal to the force of the motor, minus the half of the *vis viva* which the water possesses after issuing from the machine. and we shall thus have

$$E = Ph - \frac{P}{2g}\left(\sqrt{2gh+v^2}-v\right).$$

This equation shows that the effect is greater, as the complex factor of the second term in the second member is smaller, and that it will be at its *maximum* and equal to Ph when this factor is zero; now, we cannot have $\sqrt{2gh+v^2}-v=0$, except v is infinite. Whence we conclude that in reaction machines the effect can never be, even in theory, equal to the force of the motor, and that it is greater in proportion as the velocity of rotation is the more considerable.

Finally, in the year 1838, M. Combes, mining engineer, took up the theory of reaction machines, and extended it to all the circumstances of motion; after having studied carefully that of Euler, he established a more general one, which he presented to the Academy of Sciences; but as yet it has not been published. From the short notice upon this subject, inserted in the reports of the sessions of the Academy of Sciences (session of 6th August), the formulæ of M. Combes indicate in reaction machines what those of M. Poncelet have shown for turbines, that the velocity of the wheel may experience great variations, either increasing or decreasing, from that giving the *maximum* of effect, without a marked diminution in this effect. "It is necessary," observes the author, "that the gates of the reaction wheel should be fixed upon the wheel itself; and in order that the useful effect may remain always the same, notwithstanding the variations in the volume of water, it is requisite that the gates should act at once upon the whole of the orifices of entry and issue of the movable pipes, which should have between them a constant ratio, determined by the equation of motion."

See FLOAT WATER-WHEELS. OVERSHOT WATER-WHEELS. TURBINE WATER-WHEELS. UNDERSHOT WATER-WHEELS.

BAROMETER. FR., *Baromètre*; GER., *Barometer*; ITAL., *Barometro*; SPAN., *Barómetro*.

A barometer is an instrument for determining the weight or pressure of the atmosphere, and hence the actual and probable changes of the weather, or the height of any ascent.

The form commonly used was invented by Torricelli at Florence in 1643. It consists of a glass tube, 33 or 34 in. in length, closed at top, filled with mercury, except the vacuum at the top, and inverted in an open cup of mercury. A graduated scale is attached to the tube to note the variations of the column of mercury.

552.

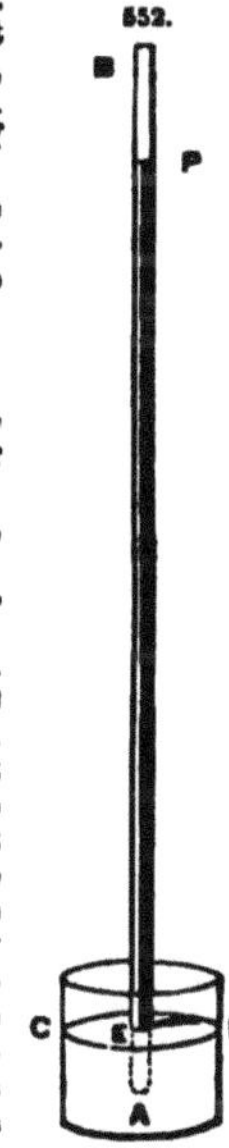

The *Aneroid barometer* is a form of the instrument in which the atmosphere acts upon the elastic top of a thin metallic box, which has previously been partially exhausted of air, and furnished with levers and an index to note the changes produced by atmospheric pressure.

The *Siphon barometer* is another form of the common barometer.

The *Sympiesometer* is another form of barometer.

A *Marine barometer* is a barometer with tube contracted at the bottom to prevent rapid oscillations of the mercury; it is suspended in gimbals from an arm or support on shipboard.

A *Mountain barometer* is a portable mercurial barometer, with tripod support, and a cong, for measuring heights.

A *Wheel barometer* is a barometer with recurved tube and a float, from which a cord by passing over a pulley moves an index.

Experiments show that if a vessel be exhausted of air it will be lighter than when it was full, hence air has weight. And we show (see HYDROSTATIC BALANCE) how the weight and *specific gravity* of air may be accurately determined. The weight of a column of the atmosphere is shown by the barometer; for, let A B, Fig. 552, be a glass tube, 34 in. long, open at A and closed at B. Fill the tube with mercury, and placing the finger firmly on the end A, so as to prevent the mercury from escaping out of the tube, invert it, and plunge the end A into a vessel C D, of mercury. If the finger be now removed, it will be found that the mercury will stand at about 29 or 30 in. in the tube above the level of the mercury in the basin. That the mercury is sustained in the tube by the pressure of the air upon the surface of the mercury in the basin may be proved by placing the barometer under the receiver of an air-pump. As the air is exhausted, the mercury sinks in the tube; and when the exhaustion is so complete that very little pressure is exerted on the surface of the mercury in the basin, the mercury in the tube and in the basin are nearly on the same level. On the air being again admitted into the receiver, the mercury rises in the tube to its former height. Since the pressure of a fluid on any portion of the surface is the weight of the superincumbent column of the fluid, the pressure of the mercury upwards against the surface C D, is the weight of a column of mercury, whose base is C D and altitude P E; and this pressure is balanced by the pressure of the air downwards on the surface C D. From numerous experiments it has been found that the density of air is proportional to the force that compresses it. For let A B C D, Fig. 553, be a bent cylindrical tube of glass, having one end A open and the other D closed. Suppose the communication between the two branches to be cut off by a small quantity of mercury poured in at A until it just fills the base B C. Then the air in C D will be of the same density as the air in A B. If more mercury be poured in at A, it will force the mercury to rise in D C; and if this pouring be continued until the mercury stands at S, as high above the point T, to which it has risen in D C, as the altitude of the mercury in the barometer, then that column of mercury, from what we have before shown, is equivalent to the weight of the column of air incumbent upon it. Hence the pressure against the air in D T, arising from the pressure of the atmosphere and the mercury in S B, is twice as great as it was against the air in C D; and it has been observed that D T = ½ C D; therefore, the air being compressed into half the space, the density is doubled. In like manner, if another column of mercury be added, so that the altitude of the mercury in A B, above the mercury in C D, shall be twice the altitude of the mercury in the barometer, the pressure of the air in D T, arising from the weight of the atmosphere and the mercury in S B, will now be three times as great as it was against the air in C D. Also D T has been found to be = ⅓ C D, and, therefore, the density in D T is = 3 times the density of the atmosphere. In this way the density has been found in all cases, within a moderate extent, to be proportional to the compressing force. And since the force that compresses the air is balanced by the elasticity of the air, it is evident that the elastic force of the air is equal to the compressing force, and may be measured by it. Also it follows, that the elasticity of the air is proportional to its density. This is the law of Boyle.

553.

Gay-Lussac found by experiment, that all gases, under the same pressure, expand uniformly for equal increments of temperature; this is true, at least, from the freezing to the boiling point of the thermometer. Suppose a column of the atmosphere to rest on a base whose area is 1; and suppose this column to be divided into an infinite number of strata, of equal thickness, parallel to the horizon. Then, since each stratum of air is compressed by the weight of those above it, the lower strata will be more compressed, and, therefore, will be denser than those above them.

The nature and properties of both atmospheric air and mercury have been carefully ascertained; so far, the determination of the heights of mountains by the barometer presents no difficulty; but, to solve the ultimate equation, has, up to the present time, defied the skill of mathematicians. The formulas presented by writers on mechanics to effect this object, only gave approximate results. This ultimate equation, which may be presented under the form [1], can be solved with the greatest ease by dual arithmetic, a new art.

$$z = \frac{k\,(1 + \alpha\,\tau)}{g}\left(1 + \frac{z}{r}\right)\log._\epsilon\left[\frac{H}{h}\left(1 + \frac{z}{r}\right)^2\right]. \qquad [1]$$

In this equation all the quantities, except the required height z, are known; the logs. are hyperbolic, and this equation may be put under the form

$$\sqrt{\frac{H}{h}}\left(1+\frac{z}{r}\right) = \sqrt{\frac{H}{h}} + \frac{k\,(1+\alpha\,\tau)}{r\,g}\sqrt{\frac{H}{h}}\left(1+\frac{z}{r}\right)\log._\epsilon\left[\frac{H}{h}\left(1+\frac{z}{r}\right)\right];$$

in this last equation put $\sqrt{\frac{H}{h}} = A$; and suppose A = hyperbolic log. of a, that is, suppose $\log._\epsilon a = A$; put $x = \sqrt{\frac{H}{h}}\left(1 + \frac{z}{r}\right)$, and $B = \frac{k\,(1+\alpha\,\tau)}{r\,g}$.

These substitutions being made, [1] becomes

$$x = A + B\,x\,\log._\epsilon x^2. \qquad [2]$$

The base of the hyperbolic system of logarithms $\epsilon = 2\cdot718281828$ and the dual logarithm of $\epsilon = 100000000, = 10^8$, the dual logarithm of ϵ is written $\downarrow, \epsilon = 10^8$. Equation [2] may be put under the form $\epsilon^x = a\,x^{2Bx}$; taking the xth root of both sides of the last equation, we have $\epsilon = a^{\frac{1}{x}}\,x^{2B}$.

Put $n^{2B} = \epsilon$, then $\frac{n^{2B}}{x^{2B}} = a^{\frac{1}{x}}$; therefore, $\left(\frac{n}{x}\right)^x = a^{\frac{1}{2B}}$; or $\left(\frac{x}{n}\right)^x = \frac{1}{a^{\frac{1}{2B}}}$; consequently $\left(\frac{x}{n}\right)^{\frac{x}{n}} = \left(\frac{1}{a^{\frac{1}{2B}}}\right)^{\frac{1}{n}}$. If c be put for the known quantity $\left(\frac{1}{a^{\frac{1}{2B}}}\right)^{\frac{1}{n}}$ and y for $\frac{x}{n}$, the last equation becomes $y^y = c$, hence in a dual form we have $y\downarrow, (y) = \downarrow, (c)$; [3]. A general solution of all equations of the form [3], is given in Part II. of the 'Art and Science of Dual Arithmetic,' by Oliver Byrne.

In measuring heights by the barometer, it is necessary to know, to the greatest nicety, the ratio of the density of mercury to that of air. The accurate and indefatigable M. Regnault found, at Paris, that a litre of air at 0° centigrade, under a pressure of 760 millimètres, weighed 1·293187 grammes; and at the level of the sea, in latitude 45°, it weighed 1·292697 grammes. He also found that a litre of mercury, at the temperature of 0° cent., weighed 13595·93 grammes.

A litre of water, at its maximum density, weighs 1000 grammes; therefore the ratio of the density of mercury to that of air, in latitude 45°, will be $= \frac{13595 \cdot 93}{1 \cdot 292697} = 10517 \cdot 49$.

At Paris, a litre of atmospheric air weighs 1·293187 grammes; but this number is only correct for the locality in which the experiments were made—that is, the latitude of 48° 50′ 14″, and a height of 60 mètres above the level of the sea. Taking 1·292697 grammes for the weight of a litre of air under the parallel of 45° latitude, and at the same distance from the centre of the earth as that at which the experiments were tried, to be 1·292697 grammes, then, putting w for the weight of the litre of air, in any other latitude, any other distance from the centre of the earth,

$$w = \frac{1 \cdot 292697\,(1 \cdot 00001885)\,(1 - \cdot 002837)\cos.\lambda}{1 + \frac{2q}{n}}. \qquad [4]$$

$n = 6366198$ mètres, the mean radius of the earth, q the height of the place of observation above the mean radius, and λ the latitude of the place. In applying formula [4] to a particular example, the author found that at Philadelphia, U.S., lat. 39° 56′ 51″·5 N., the weight of a litre of air was 1·2914392 grammes; the ratio of the density of mercury to that of air at the level of the sea was 10527·735; and $n = 6367653$ mètres at Philadelphia.

Regnault also found, by experiments, that 1·36706 represents the volume of air at 100° centigrade thermom., the volume at 0° being supposed = 1. Before the time of Regnault, these and many other constants were greatly in error. Experiments show that air, under the same pressure, expands uniformly for equal increments of temperature; that the expansion due to the same increase of temperature is not the same for all gases, as many scientific men have supposed. However, in air the expansion for a unit of bulk is ·36706, according to Regnault, from the freezing to the boiling points; and therefore the expansion for each degree of Fahrenheit is $\frac{1}{180}$ of ·36706.

Let a = the expansion of air for each degree of the thermometer, k = the ratio of the elasticity of air to its density at the temperature of melting ice; then the bulk at the temperature x will be increased, and therefore the density diminished in the ratio of $1 + a\,x$ to 1; consequently $k\,(1 + a\,x)$ = the ratio of the elasticity to the density at the temperature x.

For the air, let
- p = the elasticity, g = the force of gravity, } at the surface of the earth;
- P = the elasticity, D = the density at temp. x, G = the force of gravity, } at the altitude s above the surface;

$$\text{then } \frac{P}{D} = k\,(1 + a\,x), \text{ and } \therefore P = D\,k\,(1 + a\,x).$$

Then, for what is conventionally termed the differential of P, in the notation of the differential calculus, put $d\,P$;

$$\text{therefore, } P - d\,P = \text{the pressure at the altitude } (s + d),$$
$$P = \text{the pressure at the altitude } s.$$

$\therefore - d\,P$ = the difference of pressures $= D\,G\,d\,s = \frac{D\,g\,r^2\,d\,s}{(r+s)^2}$; r being put for the radius of the earth where the observation is made.

$$\therefore -\frac{d\,P}{P} = \frac{g\,r^2}{k\,(1 + a\,x)}\left(\frac{1}{(r+s)^2}\right), \text{ and hence, by integrating, we find}$$

$$\log._e P = \frac{g\,r^2}{k\,(1 + a\,x)}\left(\frac{1}{(r+s)}\right) + \text{constant},$$

$$\therefore \log._e p = \frac{g\,r^2}{k\,(1 + a\,x)}\left(\frac{1}{r}\right) + \text{constant};$$

$$\text{consequently, } \log._e p - \log._e P = \log._e \frac{p}{P} = \frac{g\,r^2}{k\,(1 + a\,x)}\left(\frac{1}{r} - \frac{1}{r+s}\right);$$

$$\therefore \log._e \frac{p}{P} = \frac{g\,s}{k\,(1 + a\,x)}\left(\frac{r}{r+s}\right). \qquad [5]$$

From experiment it also appears that mercury contracts uniformly as its temperature decreases.

For mercury, let
- H = height of barometer, M = density of mercury, T = temperature of mercury, } at the surface of the earth;

and H_1, M_1, T_1, the same quantities respectively at the altitude s; and $\frac{1}{\beta}$ = the condensation of mercury for one degree of the thermometer; therefore, $M_1 = M\left(1 + \frac{T - T_1}{\beta}\right)$; but, $p = g\,H\,M$; and $P = G\,H_1\,M_1 = \frac{g\,r^2}{(r+s)^2}\,(H_1)\,M\left(1 + \frac{T - T_1}{\beta}\right)$;

$$\therefore \frac{p}{P} = \left(\frac{r+s}{r}\right)^2 \frac{H}{\left(1 + \frac{T - T_1}{\beta}\right)H_1}.$$

Putting h for $\left(1 + \frac{T - T_1}{\beta}\right)H_1$, and equating the hyperbolic logarithms of the last value of

$\frac{p}{P}$ with the value before found [5], we have $\frac{g z}{k(1+a x)}\left(\frac{r}{r+z}\right) = \log._e \left\{\frac{H}{h}\left(1+\frac{z}{r}\right)^2\right\}$;

$$\therefore z = \frac{k(1+a x)}{g}\left(1+\frac{z}{r}\right)\log._e\left[\frac{H}{h}\left(1+\frac{z}{r}\right)^2\right].$$

The temperature x has been supposed to remain the same throughout the whole column z, whereas it always decreases as we ascend from the surface of the earth; but, being ignorant of the law of this change, a mean value (τ) between the values at the two stations is taken and considered constant: the mean τ, being substituted for x in the last equation, [1] is established.

We append the barometrical Tables of M. Mathrew, from which the heights of mountains may be found by mere inspection. This method will be found useful when great accuracy is not required, or when an observer wishes to avoid the labour of calculating.

These Tables, based upon Laplace's formula, are sufficiently extended to enable heights—or rather, differences of level—to be calculated to nearly 9000 mètres.

Let us suppose the following observations to have been made:—

At the lower station { H, height of the barometer; T, temperature of the barometer; t, temperature of the air.

At the upper station { h, height of the barometer; T′, temperature of the barometer; t', temperature of the air.

The height h of the barometer, at the temperature T′, observed at the upper station, becomes h' when brought to the temperature T of the barometer at the lower station. Now, for every degree centigrade, the expansion of mercury is 0·00018002; that of brass, according to the barometric scale, is 0·00001878; and the difference of these two expansions is $0{\cdot}00016124 = \frac{1}{6200}$; therefore we have $h' = h\left(1+\frac{T-T'}{6200}\right)$.

Let s be the height of the lower station above the level of the sea, and L the latitude of the place. The difference of level Z between the two stations is

$$Z = 18336^{m} \log.\frac{H}{h'} \times \left\{\begin{array}{l}\left\{1+\frac{2(t+t')}{1000}\right\}\\ (1+0{\cdot}00265\cos.2L)\\ \left(1+\frac{Z+15926}{6366198}+\frac{s}{3183099}\right)\end{array}\right\}$$

This is the formula to which the equation of *Celestial Mechanics* is brought by introducing the term $\frac{s}{3183099}$, which is relative to the height s of the lower station above the sea.

But we have just found $h' = h\left(1+\frac{T-T'}{6200}\right)$; therefore, by calling M = 0·4342945 the modulus of the logarithms, we shall have $\log. h' = \log. h + M\frac{T-T'}{6200}$,

$$\text{then}\qquad 18336^{m}\log. h' = 18336^{m}\log. h + 1^{m}{\cdot}2843\,(T-T'),$$

$$\text{and lastly,}\quad 18336^{m}\log.\frac{H}{h'} = 18336^{m}\log.\frac{H}{h} - 1^{m}{\cdot}2843\,(T-T'),$$

and we shall be enabled, after the substitution, to put the foregoing equation in the following form:—

$$Z = \left(18336^{m}\log.\frac{H}{h} - 1^{m}{\cdot}2843\,(T-T')\right) \times \left\{\begin{array}{l}\left\{1+\frac{2(t+t')}{1000}\right\}\\ \left(1+0{\cdot}00265\cos.2L+\frac{Z+15926}{6366198}\right)\\ \left(1+\frac{s}{3183099}\right)\end{array}\right\}$$

It is from this complete formula, with all the data of the observations H, h, T, T′, t and t', that the following barometrical Tables have been constructed.

After having calculated the first approximate value of Z,

$$a = 18336^{m}\log.\frac{H}{h} - 1^{m}{\cdot}2483\,(T-T'),$$

$$\text{and the second,}\ A = a\,\frac{2(t+t')}{1000},\ \text{we shall have}$$

$$Z = A\left\{\begin{array}{l}\left(1+0{\cdot}00265\cos.2L+\frac{A+15926}{6366198}\right)\\ \left(1+\frac{s}{3183099}\right)\end{array}\right\}$$

Table I. gives the metrical values of $18336^{m}\log. H$ and of $18336^{m}\log. h$ for barometrical heights from 265 to 801 millimètres; only, all those values are diminished by the constant $44428^{m}{\cdot}128$, which neither alters the value of the term $18336^{m}\log.\frac{H}{h}$ nor of the difference $18336^{m}\log. H - 18336^{m}\log. h$.

Table II. gives the correction $-1^m \cdot 2843\,(T - T')$ dependent upon the difference $T - T'$ in the temperatures of the barometer at the two stations. It is generally subtractive. It would be additive if $T - T'$ were negative, that is, if the temperature T' of the barometer, at the upper station, were greater than the temperature T at the lower station.

If the barometrical scale were divided upon glass or upon a wooden mounting, the correction, which then would become $-1^m \cdot 43\,(T - T')$, would be directly obtainable by calculation.

Table III. gives, for an approximate height A, and the latitude L, the correction, always additive,

$$A\left\{0 \cdot 00265 \cos. 2\,L + \frac{A + 15926}{6366198}\right\}.$$

The first term $A\,0 \cdot 00265 \cos. 2\,L$ arises from the variation of gravity between the latitude of 45 degrees and the latitude L of the place of observation. It is positive between the equator and 45 degrees, and negative between 45 degrees and the pole.

The second term $\frac{A + 15926}{6366198}$ A is due to the diminution of gravity in the vertical between the two stations; it is always positive and larger than the first. The sum of these two terms has, therefore, the advantage of being always positive.

The small correction $A\,\frac{s}{8183099}$ is owing to the height s of the lower station above the sea. That height is known, but, with an approximation amply sufficient, we may take

$$s = 18336^m \log. \frac{760}{H} \cdot$$

The correction then becomes $\quad A\,0 \cdot 00576 \log. \frac{760}{H}$

It is always additive, and is given in Table IV. It is obtained at the lower station, together with A and the height H of the barometer.

Method of performing the Calculation.—Take from Table I. the two numbers corresponding to the barometrical heights H and h, obtained by observation. From this difference subtract the correction $1^m \cdot 2843\,(T - T')$, which will be found in Table II., together with the thermometrical difference $T - T'$ of the barometers. The approximate height a will thus be got.

The correction $a\,\frac{2\,(t + t')}{1000}$, for the temperature of the air, has next to be calculated by multiplying the thousandth part of a by twice the sum of the temperatures t and t'. It bears the same sign as $t + t'$. A second approximate height A is then obtained.

Having A and the latitude L of the place, find, in Table III., the correction, always additive, $A\left\{0 \cdot 00265 \cos. 2\,L + \frac{A + 15926}{6366198}\right\}$, which arises from the variation of gravity in latitude, and its diminution in the vertical between the two stations.

When the height of the lower station is rather great, or when the height H of the barometer at that station is below 750 millimètres, Table IV. will give the additive correction

$$A\,0 \cdot 00576 \log. \frac{760}{H} \cdot$$

This Table has a double entry, but the correction, which never varies very much, may be easily gathered at sight when it is wished to take it into account.

Example.—Measurement of the height of Mont Blanc, by MM. Bravais and Martins, on the 29th August, 1844. Mean latitude, 46 degrees.

At the lower station:

Height of the barometer at the Observatory of Geneva	$H = 729^{mm} \cdot 65$
Barometrical thermometer	$T = 18° \cdot 6$
Free thermometer	$t = 19° \cdot 3$

At the upper station, 1 mètre below the summit:

Height of the barometer	$h = 424^{mm} \cdot 05$
Barometrical thermometer	$T' = -4° \cdot 2$
Free thermometer	$t' = -7° \cdot 6$
Table I. gives for $H = 729^{mm} \cdot 65$	$8069^m \cdot 9$
Table I. gives for $h = 424^{mm} \cdot 05$	$-3748^m \cdot 1$
Difference	$4321^m \cdot 8$
Table II. gives for $T - T' = 22° \cdot 8$	$-29^m \cdot 3$
First approximate height a	$4292^m \cdot 5$
Correction $\frac{a}{1000}\,2\,(t + t') = 4 \cdot 292 \times 23 \cdot 4$	$+100^m \cdot 4$
Second approximate height A	$4392^m \cdot 9$
Table III. gives for $A = 4392 \cdot 9$ and $L = 46°$	$+13^m \cdot 4$
Table IV. gives for $H = 729^{mm}$ and 4400^m	$+0^m \cdot 4$
Difference of level between the two stations	$4406^m \cdot 9$

By adding 408 mètres to this difference of level—the height of the Observatory of Geneva above the sea, and 1 mètre more for the distance of the upper station below the summit we find the height of Mont Blanc to be $4815^m \cdot 9$ above the level of the sea.

TABLE I.

Value of 1833^{m} log. H, and the value of 18336^{m} log. h, in mètres, diminished by the constant 44428^{m}·128.
H, h, millimètres, height of mercury in the barometer tube, at the lower and upper stations, respectively.

H or h.	Mètres.	Difference.	H or h.	Mètres.	Difference.	H or h.	Mètres.	Difference.	H or h.	Mètres.	Difference.
265	4·5	30·0	330	1751·3	24·1	395	3183·1	20·1	460	4396·2	17·3
266	34·5	29·9	331	1775·4	24·0	396	3203·2	20·1	461	4413·5	17·3
267	64·4	29·7	332	1799·4	24·0	397	3223·3	20·0	462	4430·8	17·2
268	94·1	29·7	333	1823·4	23·9	398	3243·3	20·0	463	4448·0	17·1
269	123·8	29·6	334	1847·3	23·8	399	3263·3	19·9	464	4465·1	17·2
270	153·4	29·4	335	1871·1	23·7	400	3283·2	19·9	465	4482·3	17·1
271	182·8	29·3	336	1694·8	23·7	401	3303·1	19·8	466	4499·4	17·1
272	212·1	29·2	337	1918·5	23·6	402	3322·9	19·8	467	4516·5	17·0
273	241·3	29·2	338	1942·1	23·5	403	3342·7	19·8	468	4533·5	17·0
274	270·5	29·0	339	1965·6	23·5	404	3362·5	19·7	469	4550·5	17·0
275	299·5	28·9	340	1989·1	23·4	405	3382·2	19·6	470	4567·5	16·9
276	328·4	28·8	341	2012·5	23·3	406	3401·8	19·6	471	4584·4	16·9
277	357·2	28·7	342	2035·8	23·2	407	3421·4	19·5	472	4601·3	16·8
278	385·9	28·6	343	2059·0	23·2	408	3440·9	19·5	473	4618·1	16·8
279	414·5	28·5	344	2082·2	23·1	409	3460·4	19·5	474	4634·9	16·8
280	443·0	28·3	345	2105·3	23·1	410	3479·9	19·4	475	4651·7	16·8
281	471·3	28·3	346	2128·4	23·0	411	3499·3	19·3	476	4668·5	16·7
282	499·6	28·2	347	2151·4	22·9	412	3518·6	19·3	477	4685·2	16·7
283	527·8	28·1	348	2174·3	22·8	413	3537·9	19·3	478	4701·9	16·6
284	555·9	28·0	349	2197·1	22·8	414	3557·2	19·2	479	4718·5	16·6
285	583·9	27·9	350	2219·9	22·7	415	3576·4	19·2	480	4735·1	16·6
286	611·8	27·8	351	2242·6	22·7	416	3595·6	19·1	481	4751·7	16·5
287	639·6	27·7	352	2265·3	22·6	417	3614·7	19·1	482	4768·2	16·5
288	667·3	27·6	353	2287·9	22·5	418	3633·8	19·0	483	4784·7	16·5
289	694·9	27·5	354	2310·4	22·5	419	3652·8	19·0	484	4801·2	16·4
290	722·4	27·4	355	2332·9	22·4	420	3671·8	18·9	485	4817·6	16·4
291	749·8	27·3	356	2355·3	22·3	421	3690·7	18·9	486	4834·0	16·4
292	777·1	27·2	357	2377·6	22·3	422	3709·6	18·8	487	4850·4	16·3
293	804·3	27·2	358	2399·9	22·2	423	3728·4	18·8	488	4866·7	16·3
294	831·5	27·0	359	2422·1	22·1	424	3747·2	18·8	489	4883·0	16·3
295	858·5	27·0	360	2444·2	22·1	425	3766·0	18·7	490	4899·3	16·2
296	885·5	26·8	361	2466·3	22·0	426	3784·7	18·7	491	4915·5	16·2
297	912·3	26·8	362	2488·3	22·0	427	3803·4	18·6	492	4931·7	16·2
298	939·1	26·7	363	2510·3	21·9	428	3822·0	18·6	493	4947·9	16·1
299	965·8	26·6	364	2532·2	21·9	429	3840·6	18·5	494	4964·0	16·1
300	992·4	26·5	365	2554·1	21·8	430	3859·1	18·5	495	4980·1	16·1
301	1018·9	26·4	366	2575·9	21·7	431	3877·6	18·5	496	4996·2	16·0
302	1045·3	26·3	367	2597·6	21·7	432	3896·1	18·4	497	5012·2	16·0
303	1071·6	26·2	368	2619·3	21·6	433	3914·5	18·4	498	5028·2	16·0
304	1097·8	26·2	369	2640·9	21·5	434	3932·9	18·3	499	5044·2	16·0
305	1124·0	26·1	370	2662·4	21·5	435	3951·2	18·3	500	5060·2	15·9
306	1150·1	26·0	371	2683·9	21·5	436	3969·5	18·2	501	5076·1	15·9
307	1176·1	25·9	372	2705·4	21·3	437	3987·7	18·2	502	5092·0	15·8
308	1202·0	25·8	373	2726·7	21·3	438	4005·9	18·2	503	5107·8	15·8
309	1227·8	25·7	374	2748·0	21·3	439	4024·1	18·1	504	5123·6	15·8
310	1253·5	25·6	375	2769·3	21·2	440	4042·2	18·1	505	5139·4	15·8
311	1279·1	25·6	376	2790·5	21·2	441	4060·3	18·0	506	5155·2	15·7
312	1304·7	25·5	377	2811·7	21·1	442	4078·3	18·0	507	5170·9	15·7
313	1330·2	25·4	378	2832·8	21·0	443	4096·3	18·0	508	5186·6	15·7
314	1355·6	25·3	379	2853·8	21·0	444	4114·3	17·9	509	5202·3	15·6
315	1380·9	25·2	380	2874·8	20·9	445	4132·2	17·9	510	5217·9	15·6
316	1406·1	25·2	381	2895·7	20·9	446	4150·1	17·8	511	5233·5	15·6
317	1431·3	25·1	382	2916·6	20·8	447	4167·9	17·8	512	5249·1	15·5
318	1456·4	25·0	383	2937·4	20·8	448	4185·7	17·8	513	5264·6	15·5
319	1481·4	24·9	384	2958·2	20·7	449	4203·5	17·7	514	5280·1	15·5
320	1506·3	24·8	385	2978·9	20·7	450	4221·2	17·7	515	5295·6	15·4
321	1531·1	24·8	386	2999·6	20·6	451	4238·9	17·6	516	5311·0	15·4
322	1555·9	24·7	387	3020·2	20·5	452	4256·5	17·6	517	5326·4	15·4
323	1580·6	24·6	388	3040·7	20·5	453	4274·1	17·6	518	5341·8	15·4
324	1605·2	24·6	389	3061·2	20·4	454	4291·7	17·5	519	5357·2	15·3
325	1629·8	24·4	390	3081·6	20·4	455	4309·2	17·5	520	5372·5	15·3
326	1654·2	24·4	391	3102·0	20·4	456	4326·7	17·4	521	5387·8	15·3
327	1678·6	24·3	392	3122·4	20·3	457	4344·1	17·4	522	5403·1	15·2
328	1702·9	24·3	393	3142·7	20·2	458	4361·5	17·4	523	5418·3	15·2
329	1727·2	24·1	394	3162·9	20·2	459	4378·9	17·3	524	5433·5	15·2
330	1751·3		395	3183·1		460	4396·2		525	5448·7	

TABLE I.—*continued.*

H or h.	Mètres.	Difference.	H or h.	Mètres.	Difference.	H or h.	Mètres.	Difference.	H or h.	Mètres.	Difference.
525	5448·7	15·2	594	6432·0	13·4	663	7307·1	12·0	732	8095·5	10·9
526	5463·9	15·1	595	6445·4	13·4	664	7319·1	12·0	733	8106·4	10·9
527	5479·0	15·1	596	6458·8	13·4	665	7331·1	12·0	734	8117·3	10·8
528	5494·1	15·1	597	6472·2	13·3	666	7343·1	12·0	735	8128·1	10·8
529	5509·2	15·0	598	6485·5	13·3	667	7355·1	11·9	736	8138·9	10·8
530	5524·2	15·0	599	6498·8	13·2	668	7367·0	11·9	737	8149·7	10·8
531	5539·2	15·0	600	6512·0	13·3	669	7378·9	11·9	738	8160·5	10·8
532	5554·2	14·9	601	6525·3	13·3	670	7390·8	11·8	739	8171·3	10·8
533	5569·1	15·0	602	6538·6	13·2	671	7402·6	11·9	740	8182·1	10·8
534	5584·1	14·9	603	6551·8	13·2	672	7414·5	11·9	741	8192·9	10·7
535	5599·0	14·8	604	6565·0	13·2	673	7426·4	11·8	742	8203·6	10·7
536	5613·8	14·9	605	6578·2	13·1	674	7438·2	11·8	743	8214·3	10·7
537	5628·7	14·8	606	6591·3	13·1	675	7450·0	11·8	744	8225·0	10·7
538	5643·5	14·8	607	6604·4	13·1	676	7461·8	11·8	745	8235·7	10·7
539	5658·3	14·7	608	6617·5	13·1	677	7473·6	11·7	746	8246·4	10·7
540	5673·0	14·8	609	6630·6	13·1	678	7485·3	11·7	747	8257·1	10·6
541	5687·8	14·7	610	6643·7	13·0	679	7497·0	11·7	748	8267·7	10·7
542	5702·5	14·7	611	6656·7	13·0	680	7508·7	11·7	749	8278·4	10·6
543	5717·2	14·6	612	6669·7	13·0	681	7520·4	11·7	750	8289·0	10·6
544	5731·8	14·6	613	6682·7	13·0	682	7532·1	11·7	751	8299·6	10·6
545	5746·4	14·6	614	6695·7	13·0	683	7543·8	11·7	752	8310·2	10·6
546	5761·0	14·6	615	6708·7	12·9	684	7555·5	11·6	753	8320·8	10·6
547	5775·6	14·6	616	6721·6	12·9	685	7567·1	11·6	754	8331·4	10·5
548	5790·2	14·5	617	6734·5	12·9	686	7578·7	11·6	755	8341·9	10·5
549	5804·7	14·5	618	6747·4	12·9	687	7590·3	11·6	756	8352·4	10·6
550	5819·2	14·4	619	6760·3	12·9	688	7601·9	11·6	757	8363·0	10·5
551	5833·6	14·5	620	6773·2	12·8	689	7613·5	11·5	758	8373·5	10·5
552	5848·1	14·4	621	6786·0	12·8	690	7625·0	11·5	759	8384·0	10·5
553	5862·5	14·4	622	6798·8	12·8	691	7636·5	11·5	760	8394·5	10·4
554	5876·9	14·3	623	6811·6	12·8	692	7648·0	11·5	761	8404·9	10·5
555	5891·2	14·4	624	6824·4	12·7	693	7659·5	11·5	762	8415·4	10·4
556	5905·6	14·3	625	6837·1	12·7	694	7671·0	11·5	763	8425·8	10·5
557	5919·9	14·3	626	6849·8	12·7	695	7682·5	11·5	764	8436·3	10·4
558	5934·2	14·2	627	6862·5	12·7	696	7694·0	11·4	765	8446·7	10·4
559	5948·4	14·2	628	6875·2	12·7	697	7705·4	11·4	766	8457·1	10·4
560	5962·6	14·2	629	6887·9	12·7	698	7716·8	11·4	767	8467·5	10·4
561	5976·8	14·2	630	6900·6	12·6	699	7728·2	11·4	768	8477·9	10·3
562	5991·0	14·1	631	6913·2	12·6	700	7739·6	11·4	769	8488·2	10·4
563	6005·1	14·2	632	6925·8	12·6	701	7751·0	11·3	770	8498·6	10·3
564	6019·3	14·1	633	6938·4	12·6	702	7762·3	11·3	771	8508·9	10·3
565	6033·4	14·1	634	6951·0	12·5	703	7773·6	11·3	772	8519·2	10·3
566	6047·5	14·1	635	6963·5	12·6	704	7784·9	11·3	773	8529·5	10·3
567	6061·6	14·0	636	6976·1	12·5	705	7796·2	11·3	774	8639·8	10·3
568	6075·6	14·0	637	6988·6	12·5	706	7807·5	11·3	775	8550·1	10·3
569	6089·6	14·0	638	7001·1	12·4	707	7818·8	11·3	776	8560·4	10·2
570	6103·6	14·0	639	7013·5	12·5	708	7830·1	11·2	777	8570·6	10·3
571	6117·6	13·9	640	7026·0	12·4	709	7841·3	11·2	778	8580·9	10·2
572	6131·5	13·9	641	7038·4	12·4	710	7852·5	11·2	779	8591·1	10·2
573	6145·4	13·9	642	7050·8	12·4	711	7863·7	11·2	780	8601·3	10·2
574	6159·3	13·8	643	7063·2	12·4	712	7874·9	11·2	781	8611·5	10·2
575	6173·1	13·9	644	7075·6	12·4	713	7886·1	11·2	782	8621·7	10·2
576	6187·0	13·8	645	7088·0	12·3	714	7897·3	11·1	783	8631·9	10·1
577	6200·8	13·8	646	7100·3	12·3	715	7908·4	11·2	784	8642·0	10·2
578	6214·6	13·8	647	7112·6	12·3	716	7919·6	11·1	785	8652·2	10·1
579	6228·4	13·7	648	7124·9	12·3	717	7930·7	11·1	786	8662·3	10·2
580	6242·1	13·7	649	7137·2	12·3	718	7941·8	11·1	787	8672·5	10·1
581	6255·8	13·7	650	7149·5	12·2	719	7952·9	11·0	788	8682·6	10·1
582	6269·5	13·7	651	7161·7	12·2	720	7963·9	11·1	789	8692·7	10·1
583	6283·2	13·6	652	7173·9	12·2	721	7975·0	11·0	790	8702·8	10·0
584	6296·8	13·6	653	7186·1	12·2	722	7986·0	11·0	791	8712·8	10·1
585	6310·4	13·6	654	7198·3	12·2	723	7997·0	11·0	792	8722·9	10·0
586	6324·0	13·6	655	7210·5	12·1	724	8008·0	11·0	793	8732·9	10·1
587	6337·6	13·6	656	7222·6	12·1	725	8019·0	11·0	794	8743·0	10·0
588	6351·2	13·5	657	7234·7	12·1	726	8030·0	11·0	795	8753·0	10·0
589	6364·7	13·5	658	7246·8	12·1	727	8041·0	10·9	796	8763·0	10·0
590	6378·2	13·5	659	7258·9	12·1	728	8051·9	10·9	797	8773·0	10·0
591	6391·7	13·5	660	7271·0	12·1	729	8062·8	10·9	798	8783·0	10·0
592	6405·2	13·4	661	7283·1	12·1	730	8073·7	10·9	799	8793·0	9·9
593	6418·6	13·4	662	7295·1	12·0	731	8084·6	10·9	800	8802·9	9·9
594	6432·0		663	7307·1		732	8095·5		801	8812·8	

TABLE II.—Correction; $-1^{m}\cdot2843\,(T-T')$.

T − T'.	Correction.	T − T'.	Correction.	T − T'.	Correction.	T − T'.	Correction.	T − T'.	Correction.	T − T'.	Correction.
°	m.	°	m.	°	m.	°	m.	°	m.	°	m.
0·0	0·0	4·2	5·4	8·2	10·5	12·2	15·7	16·2	20·8	20·2	25·9
0·2	0·3	4·4	5·7	8·4	10·8	12·4	15·9	16·4	21·1	20·4	26·2
0·4	0·5	4·6	5·9	8·6	11·0	12·6	16·2	16·6	21·3	20·6	26·5
0·6	0·8	4·8	6·2	8·8	11·3	12·8	16·4	16·8	21·6	20·8	26·7
0·8	1·0	5·0	6·4	9·0	11·6	13·0	16·7	17·0	21·8	21·0	27·0
1·0	1·3	5·2	6·7	9·2	11·8	13·2	17·0	17·2	22·1	21·2	27·2
1·2	1·5	5·4	6·9	9·4	12·1	13·4	17·2	17·4	22·3	21·4	27·5
1·4	1·8	5·6	7·2	9·6	12·3	13·6	17·5	17·6	22·6	21·6	27·7
1·6	2·1	5·8	7·4	9·8	12·6	13·8	17·7	17·8	22·9	21·8	28·0
1·8	2·3	6·0	7·7	10·0	12·8	14·0	18·0	18·0	23·1	22·0	28·3
2·0	2·6	6·2	8·0	10·2	13·1	14·2	18·2	18·2	23·4	22·2	28·5
2·2	2·8	6·4	8·2	10·4	13·4	14·4	18·5	18·4	23·6	22·4	28·8
2·4	3·1	6·6	8·5	10·6	13·6	14·6	18·8	18·6	23·9	22·6	29·0
2·6	3·3	6·8	8·7	10·8	13·9	14·8	19·0	18·8	24·1	22·8	29·3
2·8	3·6	7·0	9·0	11·0	14·1	15·0	19·3	19·0	24·4	23·0	29·5
3·0	3·9	7·2	9·2	11·2	14·4	15·2	19·5	19·2	24·7	23·2	29·8
3·2	4·1	7·4	9·5	11·4	14·6	15·4	19·8	19·4	24·9	23·4	30·1
3·4	4·4	7·6	9·8	11·6	14·9	15·6	20·0	19·6	25·2	23·6	30·3
3·6	4·6	7·8	10·0	11·8	15·2	15·8	20·3	19·8	25·4	23·8	30·6
3·8	4·9	8·0	10·3	12·0	15·4	16·0	20·5	20·0	25·7	24·0	30·8
4·0	5·1										

The correction is subtracted when T − T' is positive, and added when T − T' is negative.

TABLE III.

Correction; $A\left\{\cdot00265\ \cos.\ 2\,L + \dfrac{A + 15926}{6366198}\right\}$; always to be added.

Height approximating to A.	LATITUDE L.										
	0°	3°	6°	9°	12°	15°	18°	21°	24°	27°	30°
m.	m.	m.	m.	m.	m.	m.	m.	m.	m.	m.	m.
100	0·5	0·5	0·5	0·5	0·5	0·5	0·5	0·4	0·4	0·4	0·4
200	1·0	1·0	1·0	1·0	1·0	1·0	0·9	0·9	0·9	0·8	0·8
300	1·6	1·6	1·6	1·5	1·5	1·5	1·4	1·4	1·3	1·2	1·2
400	2·1	2·1	2·1	2·0	2·0	1·9	1·9	1·8	1·7	1·7	1·6
500	2·6	2·6	2·6	2·5	2·5	2·4	2·4	2·3	2·2	2·1	2·0
600	3·2	3·1	3·1	3·1	3·0	2·9	2·8	2·7	2·6	2·5	2·4
700	3·7	3·7	3·6	3·6	3·5	3·4	3·3	3·2	3·1	2·9	2·7
800	4·2	4·2	4·2	4·1	4·0	3·9	3·8	3·7	3·5	3·3	3·2
900	4·8	4·8	4·7	4·6	4·6	4·5	4·3	4·1	4·0	3·8	3·6
1000	5·3	5·3	5·3	5·2	5·1	5·0	4·8	4·6	4·4	4·2	4·0
1100	5·9	5·8	5·8	5·7	5·6	5·5	5·3	5·1	4·9	4·7	4·4
1200	6·4	6·4	6·3	6·2	6·1	6·0	5·8	5·6	5·4	5·1	4·8
1300	7·0	6·9	6·9	6·8	6·7	6·5	6·3	6·1	5·8	5·5	5·2
1400	7·5	7·5	7·4	7·3	7·2	7·0	6·8	6·6	6·3	6·0	5·7
1500	8·1	8·1	8·0	7·9	7·7	7·5	7·3	7·1	6·8	6·4	6·1
1600	8·6	8·6	8·5	8·4	8·3	8·1	7·8	7·6	7·2	6·9	6·5
1700	9·2	9·2	9·1	9·0	8·8	8·6	8·4	8·1	7·7	7·4	7·0
1800	9·8	9·8	9·7	9·5	9·3	9·1	8·9	8·6	8·2	7·8	7·4
1900	10·4	10·3	10·2	10·1	9·9	9·7	9·4	9·1	8·7	8·3	7·8
2000	10·9	10·9	10·8	10·7	10·5	10·2	9·9	9·6	9·2	8·7	8·3
2100	11·5	11·5	11·4	11·2	11·0	10·8	10·4	10·1	9·7	9·2	8·8
2200	12·1	12·1	12·0	11·8	11·6	11·3	11·0	10·6	10·2	9·7	9·2
2300	12·7	12·6	12·5	12·4	12·1	11·8	11·5	11·1	10·7	10·2	9·6
2400	13·3	13·2	13·1	13·0	12·7	12·4	12·1	11·6	11·2	10·6	10·1
2500	13·9	13·8	13·7	13·5	13·3	13·0	12·6	12·2	11·7	11·1	10·5
2600	14·5	14·4	14·3	14·1	13·9	13·5	13·1	12·7	12·2	11·6	11·0
2700	15·1	15·0	14·9	14·7	14·4	14·1	13·7	13·2	12·7	12·2	11·5
2800	15·7	15·6	15·5	15·3	15·0	14·7	14·2	13·8	13·2	12·6	12·0
2900	16·3	16·2	16·1	15·9	15·6	15·2	14·8	14·3	13·7	13·0	12·3
3000	16·9	16·8	16·7	16·5	16·2	15·8	15·3	14·8	14·2	13·6	12·9
3500	20·0	19·9	19·8	19·5	19·2	18·7	18·2	17·6	16·9	16·1	15·3
4000	23·1	23·1	22·9	22·6	22·2	21·7	21·1	20·4	19·6	18·7	17·8
5000	29·7	29·6	29·4	29·0	28·5	27·9	27·2	26·3	25·3	24·2	23·1
6000	36·6	36·5	36·2	35·8	35·2	34·4	33·5	32·5	31·3	30·0	28·6
7000	43·8	43·7	43·4	42·9	42·2	41·3	40·2	39·0	37·6	36·1	34·5

TABLE III.—*continued.*

Height approximating to A.	LATITUDE L.										
	33°	36°	39°	42°	45°	48°	51°	54°	57°	60°	63°
m.	m.	m.	m.	m.	m.	m.	m.	m.	m.	m.	m.
100	0·4	0·3	0·3	0·3	0·2	0·2	0·2	0·2	0·1	0·1	0·1
200	0·7	0·7	0·6	0·6	0·5	0·5	0·4	0·3	0·3	0·2	0·2
300	1·1	1·0	0·9	0·9	0·8	0·7	0·6	0·5	0·4	0·4	0·3
400	1·5	1·4	1·3	1·1	1·0	0·9	0·8	0·7	0·6	0·5	0·4
500	1·8	1·7	1·6	1·4	1·3	1·2	1·0	0·9	0·8	0·6	0·5
600	2·2	2·1	1·9	1·7	1·6	1·4	1·2	1·1	0·9	0·8	0·6
700	2·6	2·4	2·2	2·0	1·8	1·6	1·4	1·3	1·1	0·9	0·7
800	3·0	2·8	2·5	2·3	2·1	1·9	1·7	1·4	1·2	1·0	0·9
900	3·4	3·1	2·9	2·7	2·4	2·1	1·9	1·6	1·4	1·2	1·0
1000	3·7	3·5	3·2	2·9	2·7	2·4	2·1	1·8	1·6	1·3	1·1
1100	4·1	3·8	3·5	3·2	2·9	2·6	2·3	2·0	1·8	1·5	1·2
1200	4·5	4·2	3·9	3·6	3·2	2·9	2·6	2·2	1·9	1·6	1·4
1300	4·9	4·6	4·2	3·9	3·5	3·2	2·8	2·5	2·1	1·8	1·5
1400	5·3	5·0	4·6	4·2	3·8	3·4	3·0	2·7	2·3	1·9	1·6
1500	5·7	5·3	4·9	4·5	4·1	3·7	3·3	2·9	2·5	2·1	1·8
1600	6·1	5·7	5·3	4·9	4·4	4·0	3·5	3·1	2·7	2·3	1·9
1700	6·5	6·1	5·6	5·2	4·7	4·2	3·8	3·3	2·9	2·5	2·1
1800	7·0	6·5	6·0	5·5	5·0	4·5	4·0	3·5	3·1	2·6	2·2
1900	7·4	6·9	6·4	5·8	5·3	4·8	4·3	3·8	3·3	2·8	2·4
2000	7·8	7·3	6·7	6·2	5·6	5·1	4·5	4·0	3·5	3·0	2·5
2100	8·2	7·7	7·1	6·5	5·9	5·4	4·8	4·2	3·7	3·2	2·7
2200	8·6	8·1	7·5	6·9	6·3	5·7	5·0	4·5	3·9	3·3	2·8
2300	9·1	8·5	7·8	7·2	6·6	5·9	5·3	4·7	4·1	3·5	3·0
2400	9·5	8·9	8·2	7·6	6·9	6·3	5·7	5·1	4·3	3·7	3·2
2500	9·9	9·2	8·6	7·9	7·2	6·5	5·9	5·2	4·5	3·9	3·3
2600	10·4	9·7	9·0	8·3	7·6	6·8	6·1	5·4	4·8	4·1	3·5
2700	10·8	10·1	9·4	8·6	7·9	7·1	6·4	5·7	5·0	4·3	3·7
2800	11·3	10·5	9·8	9·0	8·2	7·5	6·7	5·9	5·2	4·5	3·9
2900	11·7	11·0	10·2	9·4	8·6	7·8	7·0	6·2	5·5	4·7	4·1
3000	12·2	11·4	10·6	9·8	8·9	8·1	7·3	6·5	5·7	4·9	4·2
3500	14·4	13·5	12·6	11·6	10·7	9·7	8·8	7·8	6·9	6·0	5·2
4000	16·8	15·8	14·7	13·6	12·5	11·4	10·3	9·2	8·2	7·2	6·3
5000	21·8	20·5	19·2	17·8	16·4	15·0	13·7	12·3	11·0	9·8	8·7
6000	27·1	25·6	24·0	22·3	20·7	19·0	17·4	15·8	14·2	12·7	11·3
7000	32·8	30·9	29·1	27·1	25·2	23·3	21·4	19·5	17·7	15·9	14·3

TABLE IV.—Diminution of weight in the vertical due to the heights of the lower station.

Correction; $A\left(\cdot 00576 \log. \frac{760}{H}\right)$; always to be added.

Height approximating to A.	HEIGHT OF BAROMETER AT LOWER STATION.									
	460	490	520	550	580	610	640	670	700	730
m.	m.	m.	m.	m.	m.	m.	m.	m.	m.	m.
100	0·1	0·1	0·1	0·1	0·1	0·1	0·0	0·0	0·0	0·0
200	0·3	0·2	0·2	0·2	0·1	0·1	0·1	0·1	0·0	0·0
300	0·4	0·3	0·3	0·2	0·2	0·2	0·1	0·1	0·1	0·0
400	0·5	0·4	0·4	0·3	0·3	0·2	0·2	0·1	0·1	0·0
500	0·6	0·5	0·5	0·4	0·3	0·3	0·2	0·2	0·1	0·1
600	0·8	0·7	0·6	0·5	0·4	0·3	0·3	0·2	0·1	0·1
700	0·9	0·8	0·7	0·6	0·5	0·4	0·3	0·2	0·1	0·1
800	1·0	0·9	0·8	0·6	0·5	0·4	0·3	0·3	0·2	0·1
900	1·1	1·0	0·9	0·7	0·6	0·5	0·4	0·3	0·2	0·1
1000	1·3	1·1	0·9	0·8	0·7	0·6	0·4	0·3	0·2	0·1
1200	1·5	1·3	1·1	1·0	0·8	0·7	0·5	0·4	0·2	0·1
1400	1·8	1·5	1·3	1·1	0·9	0·8	0·6	0·4	0·3	0·1
1600	2·0	1·8	1·5	1·3	1·1	0·9	0·7	0·5	0·3	0·2
1800	2·3	2·0	1·7	1·5	1·2	1·0	0·8	0·6	0·4	0·2
2000	2·5	2·2	1·9	1·6	1·4	1·1	0·9	0·6	0·4	0·2
2200	2·8	2·4	2·1	1·8	1·5	1·2	0·9	0·7	0·5	0·2
2400	3·0	2·6	2·3	1·9	1·6	1·3	1·0	0·8	0·5	0·2
2600	3·3	2·9	2·5	2·1	1·8	1·4	1·1	0·8	0·5	0·3
2800	3·5	3·1	2·7	2·3	1·9	1·5	1·2	0·9	0·6	0·3
3000	3·8	3·3	2·8	2·4	2·0	1·6	1·3	0·9	0·6	0·3
4000	5·0	4·4	3·8	3·2	2·7	2·2	1·7	1·3	0·8	0·4
5000	..	5·5	4·7	4·0	3·4	2·8	2·1	1·6	1·0	0·5
6000	..	..	..	4·9	4·1	3·3	2·6	1·9	1·2	0·6
7000	..	..	..	..	..	..	3·0	2·2	1·4	0·7
8000	..	..	..	..	..	..	..	..	1·6	0·8

BARRACKS. FR., *Casernes*; GER., *Caserne*; ITAL., *Caserne*; SPAN., *Casernas*.

Barrack.—The word barrack is probably derived from the Saxon *Parruc*, an enclosure, or from the Spanish *Barraccas*, small huts for fishermen, and is the general term employed in this country for a building or collection of buildings of a permanent nature used for the residence of troops; while buildings of a less permanent kind, such as at Aldershot or the Curragh, we generally call camps. On the Continent, however, the term equivalent in sound to our barracks appears limited to huts or field-huts, while the more permanent buildings are called casernes, or caserme, derived, no doubt, from the Spanish or Italian. *Casa*, a house. The Germans generally have adopted the French term.

In the mediæval ages, and before standing armies formed a part of the institution of the country. soldiers were generally lodged in the different feudal or royal fortresses, which occupied the principal strong positions, both for offence and defence, throughout the kingdom; but after the Revolution, and more particularly in Ireland, where it was necessary to keep up a very numerous force, we begin to find large sums expended in the construction of buildings for the purposes of our modern barracks. In the reign of the first George, the Royal Barracks, Dublin, were thus erected, affording accommodation at that period for perhaps 5000 men, on an area of about 14 acres, although, owing to the enlarged cubical space now allotted to each soldier, they will accommodate no more than 1800. Few, therefore, if any of our barracks present any archæological interest extending further back than the middle or latter end of the last century, if we except the generally incongruous alterations effected in some of our old fortresses to accommodate our soldiers, such as those in Edinburgh and Stirling Castles, the Tower, Dover Castle, Yarmouth, &c., &c. The officers' quarters at Dover Castle are a creditable adaptation of the architecture of the existing buildings. A very good idea of what military lodging in the middle ages was, may be obtained by inspecting the lower stories of Roslin Castle, the keep at Newcastle, and other *souterraines* of castles of that period.

With the exception of a few on a large scale, built during the last century, or in the early part of the present one—such as those of Dublin, Cork, and Fermoy, in Ireland: the cavalry barracks of Piershill, near Edinburgh; those of York and Canterbury—the greater part of our barracks were composed of mere makeshifts, many of them in the heart of large towns. cribbed and confined in space, overcrowded. and quite devoid of any sanitary arrangements, drainage, or even a proper supply of water.

It was not until the General Report of the Commission appointed for improving the Sanitary Condition of Barracks and Hospitals appeared in 1861, that any idea could be formed of the real state of the existing dwellings of the British army in the United Kingdom, extending in number to 243 distinct barracks, and 167 hospitals. Many of them are described as being built in densely-peopled neighbourhoods, and closely surrounded by dwellings of the civil population, often of the very lowest classes. Some of the Metropolitan barracks, such as St. George's, behind the National Gallery, and Portman Barracks, are especially notorious in this respect, and also many of those in Portsmouth; Knightsbridge Barrack is also specially condemned. In numerous cases, existing barracks consist of blocks of private buildings, or ordinary dwelling-houses, often in low neighbourhoods, in which, owing to political or other causes, it was at some period deemed desirable to station troops, such as the Ship Street and Linen Hall Barracks in Dublin, the barracks in Cashel, Galway, and many other towns in Ireland, and in the manufacturing districts in England and Scotland. In such cases, no sanitary improvements short of entire demolition and reconstruction can ever afford accommodation adequate to what modern hygiene would require. Even in more recent times, when money has not been spared, and it would be naturally supposed the faults of older buildings would be avoided, our barrack architects or engineers have not been more successful—witness, the Guards' Barracks in St. James's Park, the new barracks in the Tower and in Edinburgh Castle, the Cavalry Barracks, Hounslow, and many others we could mention. As the unit of 600 cub. ft. space a man is now generally adopted as the minimum for barrack accommodation, it will be sufficient to state, that when the late Commission inspected the barracks of the United Kingdom, they found under the existing regulations provision made for 75,800 men, giving each a cubic space varying from 290 ft. to 550 ft. a man; while at the rate of 600 ft. a man, only 53,800 could be accommodated, showing a deficiency of barrack room for 22,000 men. In the Chatham district alone, 100 men were accommodated in the space sufficient only for 57.

The Barrack Commissioners, however, do not involve all existing barracks under a sentence of condemnation. Some of the block arrangements of the Irish barracks, especially those of Parsonstown, Templemore, Naas, Dundalk, Island Bridge, and Beggars' Bush Barracks, near Dublin, are commended, as well as the more modern barracks at Bury and Ashton, in Lancashire, and York Cavalry Barracks.

The recommendation of the Commission generally as to drainage, warming, water-supply, ablution, and other sanitary arrangements, have been gradually carried out wherever possible during the last seven or eight years, although, from economical motives connected with the War Department administration, much remains yet to be done.

The most recently built and improved barrack in London is that of Chelsea, near the Royal Hospital, constructed for a battalion of the Guards, from the designs of G. Morgan, who obtained the appointment of architect by public competition in 1858–9. It possesses all the required necessaries and accessories for modern military dwellings, including gymnasium, married soldiers' barracks, chapel, provost, stores, hospital, and so on, and cost on the whole a sum of 296,000*l.*, or about 245*l.* a head, including all ranks; but which sum included the purchase of a very expensive site.

The subjoined sketch, Fig. 554, shows two units of barrack-rooms, complete, with the passage between them, or one floor of a barrack-house, in the new Chelsea Barracks. Each house, as it consisted of two or three floors, would contain four or six units, to accommodate twenty men each.

Chelsea Barracks consist simply of a number of units such as those shown in Fig. 554, extending in a straight line of about 1000 ft. in length, with a detached building for the staff-officers at one end, and for the staff-sergeants at the other. In the rear are placed the miscellaneous buildings.

554.

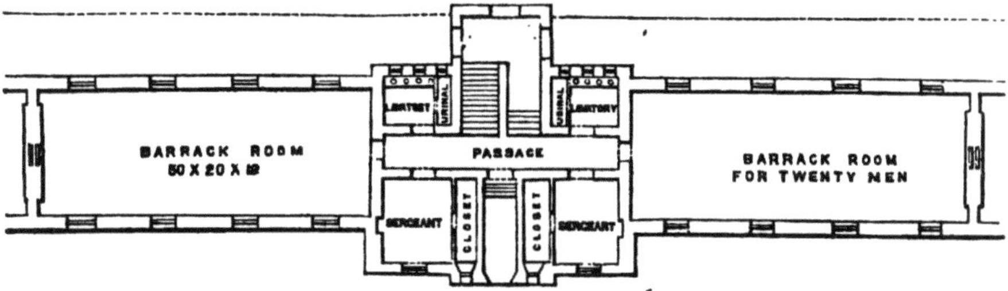

The synopsis, given page 247, shows the different buildings deemed necessary for the accommodation of a battalion of infantry 1200 strong.

Married Soldiers' Quarters.—During the last few years, quarters for married soldiers, Fig. 555, affording one room of about 220 superficial area to each married soldier, or about 6 per cent. on the regimental strength, have been built at the principal barracks in the United Kingdom, with all necessary out-offices, washing and ablution rooms, at a cost of about 180*l*. each man.

555.

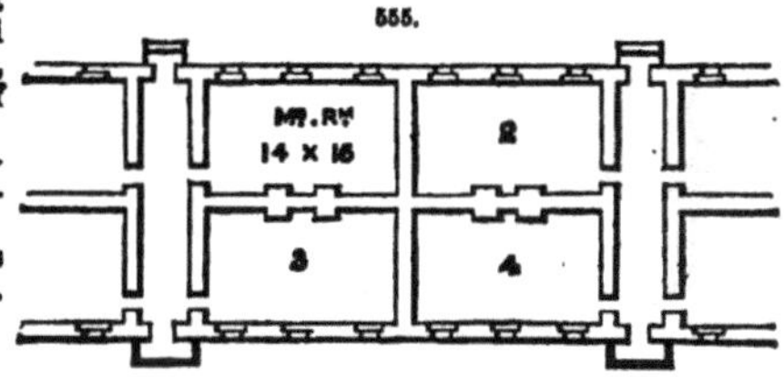

Ground-plan of Married Soldiers' Quarters.

Constructive Elements of Healthy Barrack-rooms.—The elements of healthy barrack-room construction are:—

1. Accommodation for 20 to 30 beds in each room, at 600 cub. ft. per head.
2. Height of room, 11 to 12 ft.
3. Breadth of room, 19 to 20 ft.
4. Windows equal to one-half the number of beds, arranged on opposite sides of the room.
5. No more than two rows of beds in each room; and 5 ft. in breadth, at least, allowed for each bed.
6. No barrack-room to contain a sergeant's bunk.

Each barrack-room should have a sergeant's room opening from the landing or passage; and connected with the barrack-room, and opposite the entrance, there should be a well-lighted and ventilated lavatory, with fixed ablution-basins, and proper water-service laid on. One basin will be sufficient for every ten men. In the same room should be placed a night-urinal.

The barrack-room unit would then consist of (1) barrack-room, (2) sergeant's room, (3) ablution-room, (4) night-urinal; and a barrack would consist of a number of these units arranged as the available space may allow.

Lighting.—Home barracks are now generally lighted with gas, except in very remote situations. No gas, at the public expense, is allowed for officers' quarters, except for the mess establishment and passages. Gas-lights, when properly arranged, afford great facilities for the improvement of the ventilation of soldiers' rooms.

Articles of Regulation-pattern, &c.—To ensure uniformity as well as economy, articles of regulation-pattern, such as iron shelving, skirting, latrines, and ablution apparatus, ash bins, sinks, cooking apparatus of every kind, wash-house fitments, urinals, and so on, are now used generally both in construction and repair of all barracks in the United Kingdom, as well as those in the Colonies when circumstances will admit. Rules are also laid down for the thickness of floors description of doors, gates, &c., to be used generally. The external painting of barracks is performed every four, and the internal every seven years.

Although the majority of our barracks have been constructed without any regard to their defence in case of attack by a mob or insurrectionary body, yet, within the last few years, it is laid down as a general rule that they should be sufficiently fortified, by loop-holed flanking defences or otherwise, to resist a *coup de main* at least. Many of the Irish barracks are so constructed, and those at Ashton and Bury, in Lancashire, have flanking defences at the angles. On the other hand, we find many barracks so commanded by surrounding buildings, as to be hopelessly untenable in case of a resolute attack.

In addition to the buildings themselves, the War Department provides the necessary articles of furniture, bedding, and cooking utensils for all soldiers' and non-commissioned officers' quarters. For officers' quarters, large fixtures only, such as presses, kitchen tables and dressers, racks and pins, curtain-cornices, and so on, are provided at the public expense. All wilful damage committed by the occupants must be made good by them without cost to the public.

Maintenance.—Both the designing and repairs of all barracks are now carried on under the supervision of the Director of Works, who has under him the officers of the corps of Royal Engineers and the Civil Branch of Royal Engineers' Department. Formerly there was a Board of Commissioners for Barracks, but their office was abolished by the late Duke of Wellington in 1818, when Master-General of the Ordnance, and their duties transferred to the corps of Royal Engineers. An officer named barrack-master is placed in local supervision and charge of all our principal barracks.

The total sum voted for the construction and repairs of barrack buildings in the United Kingdom, for the year ending 31st March, 1869, was ..	£319,229
For those in the Colonies	142,905
Total	£462,134

Faulty Construction.—The principal faults in the construction of existing barracks are those of site, defective drainage, and ventilation; the crowding of the blocks of buildings too close on each other, or piling floors too high; back-to-back barrack-rooms with windows only on one side, as in the Wellington Barracks and Edinburgh Castle, are especially condemned; also long, narrow, dark corridors, as in Hounslow Barracks; defective water-supply and the use of wells generally; and placing the soldiers' rooms over the stables, as is the case in too many modern cavalry barracks.

Approved Construction.—The approved construction of a barrack may be comprised in a few words. It should be as simple, yet as durable as possible; the walls built hollow, to preserve the rooms from damp, and the spaces under the floors properly ventilated; the floors and stair-cases to be fire-proof, the former constructed of wrought-iron joists, bedded in concrete, and the latter on as easy inclines as possible. Provision should be made for collecting the rain-water falling on roofs, which is always valuable for washing and cooking purposes. The site should have sufficient elevation to afford easy drainage, and every care should be taken to make all parades and exercising-grounds as solid and dry as possible. Fresh air, warmed by proper stoves, should be introduced into each room, so as to keep the temperature as near as possible steady to 60° Fahr.; while up-cast shafts, to remove all foul air, should be formed from each room at opposite corners. The latrines, urinals, ablution-rooms, and baths, should be plentifully supplied with water laid on with a proper head of pressure, either by direct service or from tanks or cisterns placed at a proper height. The introduction of wrought and cast iron in joists, sashes, &c., recommended wherever possible, as the wear and tear of material in barracks is enormous. The floors should be of wood, but those of ablution and bath rooms of asphalte; and of cook-houses, &c., flagging or tiles of a durable nature.

Regimental Hospital Establishment.

[*Part of the tabulated arrangement, page* 247.]

7 per cent. to be provided on barrack accommodation.
1200 ft. of cubic space a bed, and 80 to 100 sq. ft. a bed.

Ground Floor.

	Feet.		Feet.
Wards:—No. 2 large 1020 ×	26 × 14	No. 2 nurses' rooms	11 × 9.4
" small ..	20 × 13	Surgery	18 × 11
Waiting-room	18 × 11	Scullery	11 × 9
Day-room	18 × 15	Water-closets, sinks, &c.	

First Floor.

Orderlies' room	26.9 × 18	Clean-linen store	20 × 13
Lavatory	8 × 5	Bedding-store	15 × 10
Water-closet and urinal.		Pack-store	18 × 11

Kitchen Building.

Quarters for hospital-sergeant and steward, one room each	20 × 13
Kitchen	15 × 14
Scullery and beer-cellar; Cook's room; Room for medical comforts; Larder	Size according to available space.

Yard.

	Feet.		Feet.
Foul-bedding store	10 × 8	Dead-house	12 × 12
Coal and wood stores ..	10 × 8	No. 4 latrines.	
Wash-house and laundry, each	14 × 16	" urinals.	

Cavalry Barracks.—The general arrangements, and cubical space, for the officers and men, mess establishment, &c., of a barrack for a regiment of cavalry, do not differ much from the accommodation provided for an equal number of infantry. The open areas for parades, exercising-grounds, and the like, however, require to be larger.

In modern cavalry barracks several important sanitary improvements have been recently made. In many of the older existing barracks, including Knightsbridge, Regent's Park, Hounslow, Hampton Court, Brighton, York Old Barracks, Hulme, Preston, Sheffield, Canterbury, Piershill near Edinburgh, Dublin Royal Barracks, Island Bridge, and several others, the men's rooms are situated over the stables, an arrangement which possesses some conveniences, but which is strongly condemned in the Report of the Barrack Commission of 1861. We may mention among cavalry barracks in which this objectionable arrangement does not exist, those of Dundalk, see Fig. 556, which is one of the best of all our cavalry barracks; Newbridge, near the Curragh Camp; Cahir; the New York Barracks; Maidstone, and other places.

SYNOPSIS OF A BARRACK TO CONTAIN A BATTALION OF INFANTRY 1200 STRONG, OR 12 COMPANIES.

1 Lieut.-colonel.	1 Paymaster.	7 Staff-sergeants, Class A.	50 Corporals.	1 Mess-man.
2 Majors.	1 Adjutant.	19 ,, ,, ,, B.	76 Married Soldiers.	1 Mess-waiter.
12 Captains.	1 Quartermaster.	23 Married Sergeants.	874 Privates.	1 Cook.
14 Lieutenants	1 Surgeon.	12 Single ,,	1 Canteen Sergeant	2 Com. Officers' Servants.
10 Ensigns.	1 Assistant-surgeon.	24 Drummers.	1 Librarian.	Stabling for 8 horses.

ACCOMMODATION.

(1) Commanding Officers Quarters.

	Feet.
2 Sitting-rooms ..	20 × 18
2 Bed-rooms	20 × 18
2 Servants' rooms..	16 × 15
Kitchen, scullery, larder, water-closets, and other out-offices.	

(2) Field Officers' Quarters.

	Feet.
1 Sitting-room	18 × 16
1 Bed-room	18 × 16
1 Servant's room	16 × 15
Water-closet, small yard, and out-house.	

(3) Officers' Quarters.

	Feet.
1 Room each	18 × 16
Servant's room, for 2 officers	18 × 16
Water-closets, 1 for 4 officers.	
Sink.	
Yard.	
Out-house.	
Latrine, 1 for 10 servants.	

(11) Regimental Stores.

	Feet.
Bread-store	18 × 16
Meat-store	18 × 16
Clothing-store	35 × 18
Expense magazine.	
Shifting-room.	
Yard.	

(4) Mess Establishment.

	Feet.
Mess-room ..	45 × 25 × 16
Ante-room	30 × 25
No. 3 water-closets.	
Urinals and lavatory.	
Mess-waiter's room	18 × 14
,, ,,	18 × 12
Pantry, wash-up room.	
Plate-closet.	
Wine-cellar and smaller offices at rear.	

(5) Mess-man's Quarters.

	Feet.
2 Living-rooms	18 × 16
Store	10 × 12
Beer-cellar	18 × 16
Mess-kitchen..	25 × 20
Scullery	18 × 12
Larder	18 × 8
Coal-store.	
Water-closet.	
Latrine for servants.	
Urinal.	

(6) Non-com. Officers' Quarters.

	Feet.
1st class Staff-sergeant, 2 rooms	16 × 14
2nd class ditto, 1 room	16 × 14
Water-closet, 1 on each floor.	
Sinks ,,	
Ablution-room for men	10 × 10
,, ,, women	10 × 10
Latrines and urinals.	

(7) Sergeants' Mess Establishment.

	Feet.
Mess-room	40 × 20 × 13
Store	12 × 10
Kitchen	20 × 20
Scullery	15 × 12
Larder	15 × 7.6
Beer-cellar.	
Coal-cellar.	
Cook's living-room ..	20 × 15
Yard.	
Latrines, No. 2.	
Urinals, No. 2.	
Shed	10 × 8

(8) Married Soldiers' Quarters.

	Feet.
1 Room each	16 × 14
Water-closet and sink, 1 to each floor.	
Men's ablution-room.	
Women's ,,	
4 Men's latrines.	
4 Women's latrines.	
6 Children's ,,	
Urinals.	
Yard.	
Wash-house	46 × 20
Laundry	24 × 20
Bath-room	20 × 6
Drying-ground according to available space.	

(9) Soldiers' Barracks.

	Feet.
Non-com. officers' mess, 1 room	14 × 14
Soldiers' rooms, No. 38. for 34 men each	70 × 20 × 12
Ablution-rooms, No. 38	14 × 10
Night-urinals, No. 38.	
Cook-houses, No. 2	40 × 20 × 14
Latrines, No. 48 compartments.	
Urinals, No. 48.	
Bath-house, Non com. officers'	17 × 5.6
Bath-house, soldiers' ..	27.6 × 17
Attendant's room ..	17 × 9
Shed for tools	20 × 10

(10) Orderly, Guard Rooms, &c.

	Feet.
Orderly-room	18 × 12
Clerk's room	24 × 18
Paymaster's office ..	18 × 14
Guard-house	21 × 18
Guard-room cells, No. 5	10 × 8
Prisoners' room ..	24 × 18
Night-urinal.	
Ablution-room.	
No. 2 latrines.	
No. 2 urinals.	
Armourer's shop ..	13 × 13
Armourer's store ..	13 × 13
Shoemaker's shop ..	20 × 18
Tailor's shop	24 × 18
Master tailor	18 × 16

Means of Recreation and Exercise.

	Feet.
Game-room ..	65 × 30 × 16
Reading-room	50 × 30
Library	14 × 11.7
Coffee-room	16 × 12.6
Lavatories	16 × 8
Librarian's room ..	14 × 11.7
2 Bed-rooms	14 × 11.7
Latrine, 3 compartments.	
No. 2 urinals.	
No. 2 fives' courts ..	55 × 30
Cricket-ground.	
Skittle-ground.	
Gymnasium	80 × 40
Instructor's and dressing room	15 × 12
Drill-ground, as space permits.	
Drill-shed.	
Practice-room for band	27 × 18

Divine Worship.

	Feet.
Chapel school ..	80 × 58
Infant school.. ..	25 × 15
*Teacher's room ..	14 × 11
Bed-room	14 × 11
Kitchen and offices.	
Playground.	
Schoolmasters' quarters as Staff-sergeants.	
Latrines for boys and girls.	

* Female.

Canteen Establishment.

	Feet.
Canteen shop	78 × 16
,, bar	30 × 8
,, store	16 × 10
,, cellar	16 × 7
,, ,, No. 2 ..	18 × 16
Tap-room	50 × 26
No 2 living-rooms for canteen-man	16 × 14
Kitchen and scullery.	
Yard.	
Latrines and urinals.	

Cell Establishment.

	Feet.
(Proportion 2 per cent. on accommodation)	
No. 20 cells ..	12 × 7.6 × 10
1 Sink and urinal on each floor.	
1 Water-closet ditto.	
Latrine in yard.	
Shed for shot-drill ..	40 × 30
Exercise shed	50 × 10
Kitchen	16 × 14
Yard.	
Provost-sergeant's quarters as for Class A.	
Staff.	
Coal-store.	
Water-closet.	

Barrack-master's Establishment.

	Feet
Barrack office	18 × 14
Barrack-sergeants' quarters as Staff-sergeants.	
Foul-bedding store ..	30 × 20
Unserviceable stores	33 × 20
Meat-store	30 × 20
Coal-yard	60 × 60
Coal-shed	60 × 30
Engine-house.	
Ladder-shed.	
Meter-house.	
Latrine and urinal.	
Hospital establishment. See *Hospital*.	

The shelves and other fittings of cavalry barracks are nearly identical with those of infantry, except that some modification and provision must be made for racks for lances, swords, pistols, and carbines, with which the cavalry soldier is armed.

556.

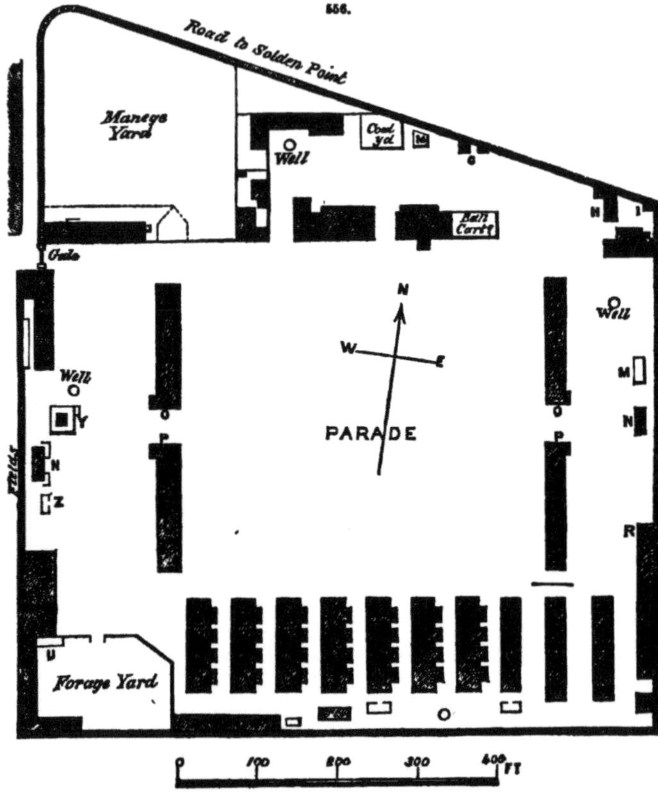

Block-plan of Cavalry Barracks, Dundalk.

References.—*A*, Officers' quarters. *B*, Barrack-master's house. *C*, Bedding store. *E*, Barrack store. *F*, Straw store. *G*, Officers' privies. *H*, Hospital privy. *I*, Dead-house. *K*, Hospital. *L*, Soldiers' quarters. *M*, Ash-pits. *N*, Soldiers' privies. *O*, Cook-house. *P*, Wash-house and sheds. *R*, Armourer's shop. *S*, Infantry stables. *T*, Troop stables. *W*, Soldiers' wash-house. *X*, Forge and shop. *Y*, Magazine. *Z*, Dung-pits. *V'*, Engine-house. *W'*, Canteen. *X'*, Cells. *Y'*, Barrack-sergeants' quarters and cells. *Z'*, Guard-room and regimental store.

Cavalry stables, when they do not form a part of the range as above mentioned, are generally placed behind the main buildings, and are either built double, that is, with two rows of stalls and a passage between them, or single, with one row of stalls—in either case, the stall being considered as the unit. The cubical space allotted for each horse is about 1200 ft. The double stables have a width of 30 ft., the single 17 or 18 ft., with an average height of 10 ft. The size of the single stall averages 9 ft. 6 in. from the wall to outside of heel-post, and the width 5 ft. 7 in. The width of a stall for an officer's horse being 6 ft. Officers' stables have saddle-rooms, and separate hay and straw stores, with doors sufficiently large to admit a carriage. The stable accommodation provided for officers is that of the number of chargers they are entitled to draw forage for. Hospital or infirmary stables are also provided, in the proportion of 6 boxes and 14 stalls to every regiment, at the present establishment of 253 horses.

The drainage and ventilation of our cavalry stables are now carefully attended to. Ceilings to stables were formerly considered as indispensable, yet many are in favour of open roofs well ventilated at the ridge, and the under-side of rafters lathed and plastered to prevent the fall of dust. The stalls are generally paved with granite pitchers, 6 in. × 3 in., laid diagonally in Portland cement to a slope from front to rear of 1 in 80, and falling from the centre of stall to each side 1 in 40, with a dressed channel of stone or terra-cotta emptying into a trapped underground drain outside, and quite clear of stables. Local materials, however, in many cases, may be used.

The fitments of all cavalry stables are of cast and wrought iron, of a uniform established pattern, Figs. 557, 558, 559. The horses of the privates are separated by swing bales of hollowed wrought iron, 2⅜ in. diam. The hay-racks and mangers are horizontal, supported by cast-iron or stone corbels built into walls. The use of the old-fashioned over-head hay-rack and wooden manger is discontinued in all military stables.

557.

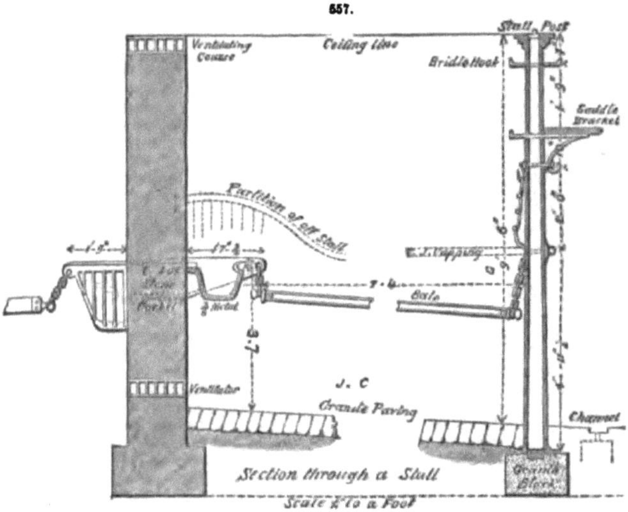

The walls of officers' stables are boarded to a height of 7 ft., and the stalls separated by 1½ in. tongued oak partitions, let into cast-iron capping and sole-pieces. The external doors are double, hung in two heights, and a swing window is provided for every stall in both single and double stables. A perforated course of air-bricks is built under the eaves, and an air-brick, 9 × 9, in walls 6 ft. above floor between every two stalls. Wrought-iron corn-bins are provided in each hay and straw store; and in the troop stables the saddlery is kept on wrought-iron brackets, screwed into the heel-posts of cast iron. Officers' saddle-rooms are provided with a small stove; and gas and water are generally laid on to all stables where practicable.

559.

558.

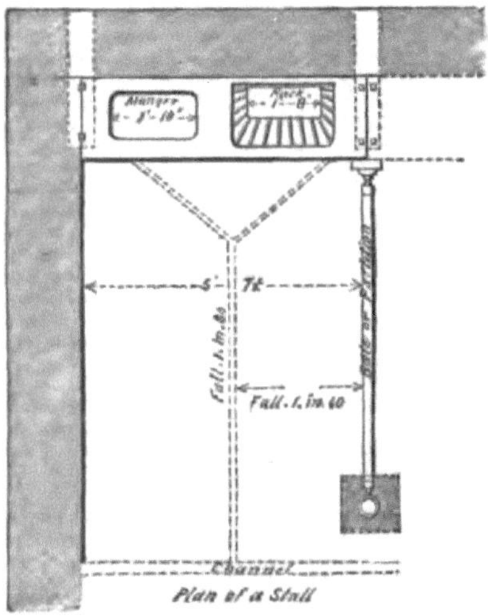

Stables should be placed from 30 to 40 ft. in rear of the men's quarters, and large corridors, covered with glass roofs, should be formed between them, so as to afford shelter to the men attending the stables in wet weather. At other times they can be utilized as drill-sheds. The litter-stalls, manure-pits, &c., placed in rear of stables.

The following buildings are generally included in a project for a cavalry barrack:—a riding-school; stores for forage, according to local circumstances; forges; shoeing-sheds; medicine-room; litter-stalls, &c. (and where artillery or military train are quartered, provision must be made for gun and carriage sheds, &c., &c.); workshops for saddlers, harness-makers, &c.

In 1858, when prizes were offered for the best designs for barracks for infantry and cavalry, Mr. H. Wyatt obtained that for the latter, and afterwards prepared plans for a cavalry barrack at Nottingham, which, however, owing to some local difficulties as to site, &c., has not as yet been carried out. The synopsis used by him and all the other competitors was from the Blue-Book containing the Report of the Committee on Barrack Accommodation in the Army, with the Minutes of Evidence, dated 1855.

Casemated Barracks.—In nearly all our recently erected fortifications, convenient bomb-proof barracks or casemates for the garrisons are provided, with all the necessary store and other accommodation, and forming a vast improvement on the old casemates at Chatham, Cork Harbour,

Plymouth, and other places, which were so justly reprobated by the Barrack Commission. The new casemated barracks thus constructed will allow the War Department in some degree to make up for the loss of accommodation caused by the enlarged cubical space now allowed to each man, without any particular outlay for this specific object. Properly arranged, a casemated barrack is quite as dry and healthy as one of ordinary construction.

Temporary Barracks.—For some time prior to the Crimean war, the formation of large camps, composed not of tents, but of buildings of a more permanent nature, and capable of accommodating not one or two regiments, but a large *corps d'armée* of from 10,000 to 15,000 men of all arms, and necessary war material, had found much favour in the eyes of the military authorities. Commencing with the encampment on Chobham Heath in 1853, the great military camp at Aldershot was next projected in 1854, the site being on a vast expanse of waste land or heath on the borders of Hampshire and Surrey, within a convenient distance of the metropolis (about 35 miles) and our great naval arsenal of Portsmouth, and easily reached by the South-Western Railway, which passes within a short distance of it. Aldershot can now accommodate 20,000 men; and probably up to this date not less than 1½ million sterling has been spent on it.

Another grand training-camp for the troops stationed in Ireland is that of the Curragh, situated on the vast plain of that name in the county of Kildare; it was formed in 1855 for 10,000 men, but is now capable of accommodating many more. There are also large temporary barracks at Colchester; Parkhurst, in the Isle of Wight; Chichester; and Shorncliffe, near Dover. The huts of these camps are principally of framed fir, clap-boarded, and covered with asphalted felt, with a central nucleus of brick fire-places. The framing is raised off the ground by brick sleeper-walls in every case.

The original temporary buildings are being now, especially at Aldershot, gradually replaced by others of a more permanent nature.

The sanitary arrangements, water-supply, roads, &c., of our large camps are generally satisfactory, and the health of the troops better than in barracks situated in large towns.

The principle of construction in our temporary barracks is simply the arrangements of the different units of accommodation for soldiers' and officers' huts round a series of squares; in the Curragh Camp these squares have an area of 380 ft. × 360 ft. for drilling purposes. The soldiers' huts are of a uniform size, 40 ft. × 20 ft., and accommodate 25 men each. The officers' huts are divided into 8 small rooms about 9 ft. square, and the sergeants' something similar; so that space is economized as much as possible. The officers of higher rank are of course better accommodated; but married officers have a just cause of complaint in the very limited space allotted to them, and the wooden buildings in winter afford anything but comfortable lodging to their occupants. The cost of a soldier's hut may be estimated at 85*l*. each.

In Parkhurst Barracks the huts are covered with a rebated tile, which extremely resembles brickwork, and forms a very warm and durable covering.

The arrangement of the old Roman camps, as described by Polybius and Hyginus, might be studied with improvement by modern military engineers.

Indian Barracks.—Very extensive improvements have of late years taken place in our Indian barracks, involving indeed nearly a complete reconstruction. A cubical space of 1500 ft. per man is now allowed, and all the buildings are raised some feet from the ground, and surrounded on all sides with a verandah, 10 ft. 6 in. wide. A barrack-room for 24 men—and they seldom hold less—is 100′ × 24′ × 15′. They never exceed two stories in height. Ventilation and drainage are carefully looked after. They are wanting in none of the conveniences and luxuries (if they may be termed so) of our modern home-barracks. A sum of not less than 10 millions sterling has been lately provided by the Indian Government for the purposes of barrack improvement and reconstruction, showing how vastly important this subject is considered in our Indian empire. See 'Suggestions for Improving Barracks at Indian Stations,' issued by the Secretary of State for India, 1864.

Foreign Barracks.—On the Continent, barracks are much more numerous than in England. They are generally on a much larger scale and with greater architectural pretensions, although very often deficient both in material comforts and sanitary arrangements: their position, too, is more influenced by political considerations than English barracks are. The men's quarters in Continental cavalry barracks are, like many of our own, often placed over the stables. Generally speaking, the soldiers' quarters are airy enough, though cold and comfortless in winter; and our expensive and well fitted up officers' quarters and mess establishment are altogether wanting.

The great foreign camps at Chalôns, in Rhenish Prussia, Silesia, &c., are on a much more vast scale that any of ours, and have their huts generally formed of *sod* or wattle, thatched with reeds or straw; but these are only occupied during the exercise months in summer and autumn. A French hut to hold 20 men is 6·50 mètres long and 4·35 wide, and 3 30 high to ridge. See Laisnes' 'Aide Mémoire,' p. 598.

In France, sailors as well as soldiers and marines are provided with barracks at all the large naval stations—a system which might be very advantageously adopted by this country, instead of allowing our seamen to disperse over the face of the land when their ships are paid off.

BARRAGE. Fr., *Barrage*; Ger., *Damm Schlagbaum*; Ital., *Chiusa*; Span., *Estacado.*

Barrage is a French term, and signifies, in an engineering sense, the barring of a river or other watercourse by artificial means, in order to facilitate navigation or irrigation in parts where the incline is too rapid, and the quantity of water—from that or other causes—would be insufficient for those purposes were it left to spread freely and in waste over its normal bed.

In mountainous districts and hot countries, but more especially in tropical climes, the rivers are all subject periodically either to a great excess of water or to an almost total want thereof. To-day they are raging torrents, flooding and devastating the neighbouring country; to-morrow, mere streamlets, often fordable, and frequently reduced to the 400th part of their ordinary average volume.

Such irregularities exercise a most detrimental influence over the interests of the populations where they occur. The agricultural prosperity of India, for instance, suffers greatly from these causes; for there, unless aided by artificial irrigation, all cultivation must necessarily cease during the dry season.

In Demerara, Surinam, some parts of Georgia, and a few other places, the evil is in a measure guarded against by the facility which those countries possess of cutting canals and obtaining water from the interior; but, as a rule, it may be said that there is a total absence in all tropical lands of that due provision for regulating the supply of water which is of such vital importance to the welfare and prosperity of every country.

It is not our province to dwell upon the physical causes that determine this state of things; but, having referred in a cursory manner to the very serious damage to commerce and agriculture arising therefrom, we propose to lay before our readers, as concisely as possible, some of the most successful remedies which the engineer's art has from time to time suggested in counteraction of the evil. There is, however, a peculiarity appertaining to the great tropical rivers running through countries having dry seasons, and crossing deltas or alluvial plains, which it may be interesting to mention. It is the fact that the beds of those rivers are sometimes as high as remote parts of the neighbouring country, while their borders are much higher; so that the overflow diverges at nearly right angles to the direct flow. The cause of the bed and sides of these rivers rising above the natural level of the surrounding plains is due to the earthy matter, held in suspension by the natural flow, being deposited in proportion as velocity is diminished.

There are two kinds of barrage, the *Barrage* and the *Barrage-mobile*: early examples of the former are found in the permanent dams placed across streams and watercourses, so as to increase or maintain their depth, for the purpose either of rendering them navigable or obtaining a fall with the view of propelling machinery; the surplus water, in such cases, being conducted through sluices, or over by-washes, dams, tumbling-bays, or overfalls, prepared to carry it off; and, in the event of floods arising in the river, additional sluices being opened, in order to prevent it overflowing and injuring the side-dams and adjacent property. Works of this nature may be seen on the Thames, on the Arno at Florence, at the reservoir of Gros-Bois, on the Canal de Bourgogne, on the Vesoult, at Courbeton, at Conflans-sur-Seine, and on numerous other canalized European rivers; and although they will hardly bear comparison, in point of magnitude, with those executed in tropical regions, nevertheless many of them are of sufficient importance to deserve special mention in the course of this article.

BARRAGE-FIXE is the term applied to permanent dams, built of masonry.

BARRAGE-MOBILE, or *movable barrage*, is that which can be raised, lowered, or removed at will, and is formed partly of masonry, partly of timber.

The most simple form of *barrage-mobile* is that represented in Fig. 560, where the current of water passes between two lateral walls, whose intervening space is partially closed by a certain number of small beams A, A′, A″, A‴, superposed, and forming an overfall by which the liquid flows into the trough below. The advantage of this plan is, that it enables the ridge of the weir to be heightened or lowered at pleasure, and with very great facility.

560.

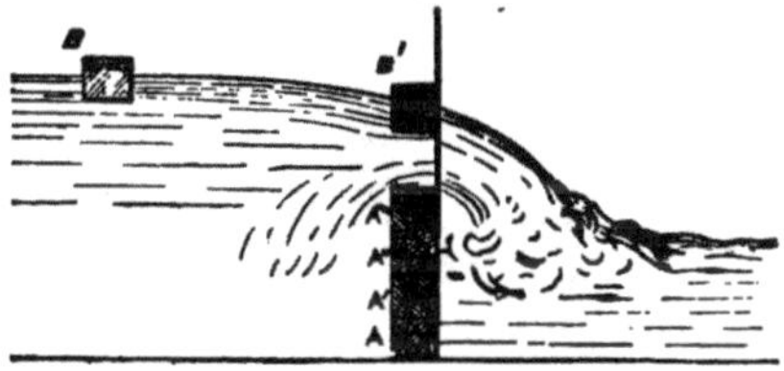

A rather remarkable phenomenon is connected with this sort of barrage. When it is required to raise it, another beam B, floating in the upper trough, is borne by the stream till it reaches B′, where its extremities rest against the rabbets that secure those already fixed; but no sooner has it attained this point than it is seen suddenly to sink, falling straight on to the beam A‴, and thus taking up its allotted position as if by instinct.

This fact is easily explained. The space that separates the beams A‴ and B′ forming a kind of adjutage, the pressure upon the under surface of B′ becomes less than the atmospheric pressure exerted upon its upper surface (see HYDRAULICS). The beam acts, therefore, in obedience to the difference of those two forces augmented by its own weight.

It must be observed that the first beams A, A′, which have to be put beneath the level of the lower trough, do not sink thus naturally into position: in order that the phenomenon may take place, they must already be in sufficient number to rise above the surface of the water in the nether basin, so as to produce a fall.

A difficulty arises in ascertaining the exact coefficient of the expenditure of fluid, by a dam or overfall, which it is important to point out before proceeding further.

An *Overfall* is an orifice, open at the top, and the lower part of which presents a flat, horizontal surface called a *Sill*. The lateral edges being generally vertical, the opening may be considered as a rectangle, of which the upper side has been removed; this assimilation would still be admissible even in a case where the length of the sill, in relation to the thickness or depth of the sheet of water passing over it, was very great.

Let L be the length of sill;

y the vertical distance between the sill and the surface of the liquid at some point of the reservoir where it would be comparatively stagnant;

η the thickness of the sheet of water passing the overfall;

Q the expenditure in a second of time.

The water being supposed to flow freely into the open air, we find that for one molecule, starting from the reservoir without any sensible initial velocity, and actually traversing the vertical plane of the sill, the load varies from y to $y - \eta$, which shows at once that η must be smaller than y; for, by virtue of Bernoulli's theorem (see HYDRAULICS), no flow can take place where the load is negative. The velocity corresponding to the mean load will therefore be

$$\sqrt{2g\left(y-\frac{1}{2}\eta\right)},$$

and, as the section of the orifice is represented by $L\eta$, the theoretical expenditure will be expressed by $L\eta\sqrt{2g\left(y-\frac{1}{2}\eta\right)}$. If, then, we designate by m the coefficient of expenditure as applied to the flow in the present instance, we shall have $Q = mL\eta\sqrt{2g\left(y-\frac{1}{2}\eta\right)}$.

In the foregoing expression η and m are unknown auxiliaries that no theory has been able yet to determine. The only thing we know is that η must be smaller than y, and experience shows that the ratio $\frac{\eta}{y}$ is inconstant, but that it rarely descends below 0·72 for overfalls having a narrow sill: we will suppose, therefore, its mean value to be equal to 0·86. As regards m, since it varies very little with the load or the dimensions of a narrow-edged orifice, we may reasonably assume it to be equal to the mean 0·62. We thus obtain

$$Q = 0\cdot 62 \times 0\cdot 68\, L y \sqrt{2g \times 0\cdot 57 y}$$
$$= 0\cdot 403\, L y \sqrt{2gy}.$$

In reality, if we put $Q = rLy\sqrt{2gy}$, r being a ratio that can only be determined experimentally, it is evident that it has not a constant value. MM. Poncelet and Lesbros, in experimentalizing upon a narrow-edged overfall, $0^{m}\cdot 20$ in length, and sufficiently distant from the bottom and lateral sides of the reservoir, found that r varied from 0·385 to 0·424; its greatest value corresponding to the smallest loads. The mean of these two numbers is 0·405, which differs but little from the result 0·403 obtained above. So that for narrow-edged overfalls, placed at a sufficient distance from the bottom and sides of the reservoir, and flowing freely into the open air, we find $Q = 0\cdot 405\, Ly\sqrt{2gy}$ nearly: but this formula may give a result either a little too great or a little too small, according as the ratio $\frac{y}{L}$ is great or small.

It is very difficult, even by approximate valuations, to keep an account of all the circumstances and local conditions that may tend to influence the value of r. We may observe, however, that if a canal be barred across its entire width by a narrow-edged overfall whose sill is tolerably distant from the bottom, and if, at the same time, $\frac{y}{L}$ be small, it would be advisable to make r a little larger, and to put $Q = 0\cdot 45\, Ly\sqrt{2gy}$, or, which is the same thing, $Q = 2\,Ly^{\frac{3}{2}}$.

We will now cite a particular and rather remarkable case, inasmuch as it shows how theory may give an increased limit to the coefficient r and the corresponding value of η. It is that where the sill B of the overfall, Fig. 561, widening at its junction with the reservoir, is prolonged by means of an open channel, slightly inclined, wherein the liquid acquires a sensibly uniform motion. Then the common velocity of all the streams passing A B is $\sqrt{2g(y-\eta)}$; and, as there is no further contraction beyond the above section, the expenditure Q is given by the formula $Q = L\eta\sqrt{2g(y-\eta)}$.

561.

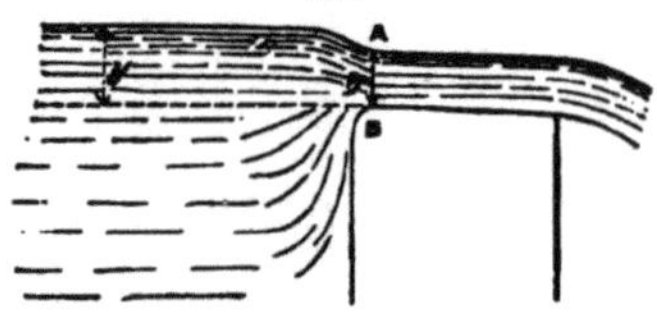

When L and y are invariable, Q becomes a function of η only, and its maximum is easily found. For, in fact, $\frac{Q^2}{2gL^2} = \eta^2(y-\eta) = y\eta^2 - \eta^3$: the maximum of the second member, and consequently that of Q, is found by making the derivation taken in relation to η equal to zero, which gives $\eta(2y - 3\eta) = 0$, whence $\eta = \frac{2}{3}y$, since $\eta - 0$ would lead to an expenditure that would be nul. Then, in the expression of Q, by making $\eta = \frac{2}{3}y$, we get

$$Q = \frac{2}{3\sqrt{3}} Ly\sqrt{2gy} = 0\cdot 385\, Ly\sqrt{2gy}.$$

The surface depression $y - \eta$ corresponding to the maximum expenditure is, therefore, one-third of the height y, and the value of the corresponding ratio r is 0·385. As the theoretical hypotheses are never completely realized in practice, if the sill of the overfall be followed by a channel, r will very rarely attain the superior limit 0·385. According to MM. Castel and Lesbros, the mean expression for an overfall like the present would be $Q = 0\cdot 35\, Ly\sqrt{2gy}$; but here again there may be a very great variation in the ratio r between one overfall and another.

When, in lieu of there being a comparatively stagnant reservoir above the overfall, there is a current with an appreciable velocity U_0, the expressions $\sqrt{2g\left(y-\frac{1}{2}\eta\right)}$ and $\sqrt{2g(y-\eta)}$, considered above, no longer represent the velocity of the streams of liquid passing over the sill. If we call that velocity U, Bernoulli's theorem then gives $\frac{U^2 - U_0^2}{2g} = y - \frac{1}{2}\eta$, or $\frac{U^2 - U_0^2}{2g} = y - \eta$, according as the question has reference to an orifice discharging freely into the air, or to one followed by an open channel.

Although, in a general sense, it is possible to tell, with a sufficient degree of exactitude, the velocity with which a liquid flows from a narrow-edged orifice—whether that orifice be plane, or widening towards the interior of the reservoir, or followed by an open channel slightly inclined—there yet remains one quantity of far greater practical importance, which, unfortunately, it is not so easy to ascertain, and that is the expenditure. This last depends not only upon the velocity with which the molecules pass the plane of the orifice, but also upon the angles at which the several liquid streamlets intersect that plane, angles that vary from one point of the orifice to another according to laws at present unknown. The only positive assertion that can be made is, that the real expenditure of water is *less* than the product of the area of the orifice by the velocity of the streams traversing it, which quantity has been improperly termed the *theoretical expenditure*. We have indicated several numbers which, in certain special cases, will afford an approximate solution of the question. In reality, however, in each of those particular cases the coefficient of expenditure is inconstant, and varies according to secondary circumstances, whose influence is very imperfectly known, such as the dimensions of the orifice, and its position in relation to the bottom and sides of the reservoir. Consequently, the only advice we can give, when the approximation obtainable by the mean coefficients of expenditure which we have indicated is not deemed sufficiently satisfactory, is to select from known collections of experiments those that bear the closest relationship to the case under investigation, and to borrow therefrom the coefficient that appears to be the best applicable.

The Tables at the end of this article, and which are taken from the more complete and extended ones published by MM. Lesbros and Poncelet, give the coefficients of expenditure which we consider the most useful in practice.

The abrupt sectional changes that take place in watercourses give rise to various problems that are of great interest to the engineer. Unfortunately, the actual state of science renders it impossible at present to solve them in so satisfactory a manner as would be desirable. In the following example, theory supplies us with a few data—incomplete and inaccurate, no doubt,—but capable, nevertheless, of being utilized in practice.

Under ordinary circumstances, the sill of the barrage is higher than the level of the water in the lower trough; but, where the current is variable, it sometimes happens that the latter rises above the weir, as shown in Fig. 562; the barrage is then said to be *noyé*, or *submerged*.

Let us suppose a barrage, or overfall, to be thrown across a watercourse, of which the level—and consequently the expenditure—are variable: in order to simplify as much as possible the calculations that have to be made, we will imagine the channel in the vicinity of the barrage to be rectangular and the bottom horizontal. The expenditure having a definite and known value, it is required to find: 1, the greatest height to which the lower level of the water can rise without in any way affecting the upper level: 2, in the case of that limit being exceeded, what would be the nature of the fall that would ensue from the upper to the lower trough.

Let L be the width of the current; h_0 its depth above and a few mètres from the weir; U_0 its mean velocity at that point; h and U analogous quantities for a section taken a little below the fall; c the height between the crest of the weir and the bottom; v the velocity of the sheet of liquid passing over the crest: η the thickness of that same sheet of liquid.

There being no lateral contraction, if the barrage acts as a narrow-edged overfall, the expenditure Q will be given by the formula $Q = 0\cdot45\, L \left(h_0 - c + \frac{U_0^2}{2g}\right) \sqrt{2g\left(h_0 - c + \frac{U_0^2}{2g}\right)}$.

If the crest, instead of being narrow-edged, happened to be of some considerable length, with a slight incline, then the coefficient 0·45 ought to undergo a certain reduction, and descend to 0 385, or even to a lower number, such as 0·36 or 0·37. The formula holds good so long as the overfall empties itself freely into the open air, to accomplish which it is necessary only that the level of the water in the lower trough be beneath the crest of the weir. When it exceeds that point, but only by a quantity less than η, it appears evident that the formula must not be modified: the very utmost that can happen is that the streamlets, losing their parabolic form, and becoming parallel as they cross the overfall as in the case when this latter is of any great thickness or width, the numerical coefficient would have to be reduced, as just stated: which would tend slightly to raise the level above the weir, the expenditure remaining the same. In examining the question more closely, it is seen that the level below the weir may be raised even higher yet without any perceptible alteration taking place in the expenditure.

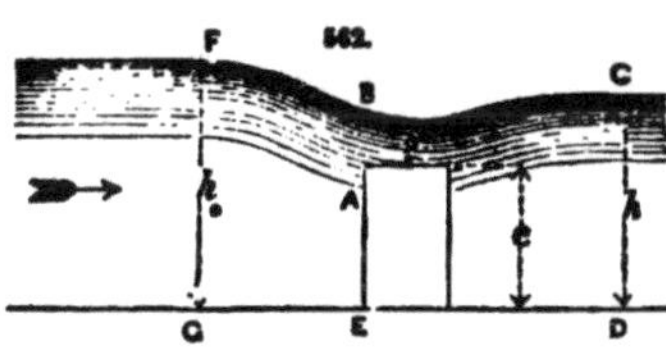

Effectually, if we apply the general theorem of the quantities of motion projected, to the liquid contained between the vertical plane A B, Fig 562, which passes through the sill of the barrage, and the section C D where the velocity is U, it will at once be found that the algebraical increment of the quantity of motion of the system projected along the horizontal line during a very short space of time, θ, is $\frac{\Pi Q \theta}{g}(U - v)$, admitting that all the molecules of the same section possess the same velocity. As to the impulses, we need only keep account of those produced by the pressure on the surfaces B A E and C D, disregarding the atmospheric pressure which acts uniformly upon the entire system. These pressures must accord very nearly with the hydrostatic law, in the first place because the streams are ostensibly parallel as they traverse A B and C D, and because the motion of the liquid in contact with A E is comparatively slow; hence the value of the projected impulse is $\frac{1}{2}\,\Pi\,\theta\,L\left\{(c+\eta)^2-h^2\right\}$. We have, therefore,

$$\frac{\Pi\,Q\,\theta}{g}(U-v)=\frac{1}{2}\,\Pi\,\theta\,L\left\{(c-\eta)^2-h^2\right\};$$

or, by simplifying, $\frac{2\,Q}{L\,g}(U-v)=(c+\eta)^2-h^2$. We moreover have $\frac{Q}{L}=v\,\eta=U\,h$; and, by eliminating Q and U, the former expression becomes $\frac{2\,v^2}{g}\,\eta\left(\frac{\eta}{h}-1\right)=(c+\eta)^2-h^2$, equation of the third degree in h, whence that quantity might be deduced if v and η were known. In order to find those unknown auxiliaries, it must be admitted, in conformity with what has been seen in the theory of the overfall, that the surface depression $h_0-\eta-c$, above the fall, is connected with the total load upon the sill by the equation $3\left(h_0-\eta-c+\frac{U_0^2}{2g}\right)=h_0-c+\frac{U_0^2}{2\,g}$, whence we derive $3\,\eta=2\left(h_0-c+\frac{U_0^2}{2\,g}\right)$, which, in combination with

$$Q=0{\cdot}37\,L\left(h_0-c+\frac{U_0^2}{2g}\right)\sqrt{2\,g\left(h_0-c+\frac{U_0^2}{2\,g}\right)},$$

in order to find η, would give $Q=0{\cdot}37\,L\times\frac{3}{2}\,\eta\sqrt{2\,g\times\frac{3}{2}\,\eta}=0{\cdot}68\,L\,\eta\sqrt{2\,g\,\eta}$.

Having found η, we calculate $v=\frac{Q}{L\,\eta}$, and we thus obtain the necessary elements to arrive at the numerical value of h. If, on the other hand, the depth of water below the fall were to exceed the aforesaid limit, what would be the depth above the weir?

In order to answer this question—supposing the expenditure Q to remain constantly the same—let us indicate the alteration that takes place in η, v, h_0, U_0, h, by η'. v', h'_0, U'_0, h': we then get the following equations:—

$$\frac{2\,\eta'\,v'^2}{g}\left(\frac{\eta'}{h'}-1\right)=(c+\eta')^2-h'^2,\quad \frac{v'^2}{2\,g}-\frac{U'^2_0}{2\,g}=h'_0-c-\eta',\quad \frac{Q}{L}=\eta',\ v'=h'_0U'_0,$$

whereof the first and two last are known; the second is an immediate application of Bernoulli's theorem to the passage of a molecule from the section F G, where the velocity is U'_0, to the section A B. By means of these four equations the unknown quantities η', v', h'_0, U'_0, may be determined, when Q and h' are given. The thickness η' of the sheet of water A B cannot be calculated by the same expression as η, because the barrage acts no longer as an overfall.

The influence of a barrage, when submerged, becomes less and less perceptible; that is to say, the fall h'_0-h' grows smaller and smaller as the level of the water below the weir rises. This will be understood *a priori*, without any mathematical demonstration; for, if the barrage be covered by a sheet of water much higher than itself, it will then occupy but a small fractional portion of the transverse section, and thus, in a measure, may be compared to a slight undulation of the bottom.

Bateau-vanne, or *floating-gate*, is the name given to a sort of self-acting, movable barrage, invented by M. Sartoris, and which, in point of simplicity, comes next under our notice. It is used chiefly for purposes of irrigation, and for regulating the supply of water in mill-ponds, and is as follows:—

A caisson, or *boat* as the French term expresses it, A A, whose transverse section is rectangular, rests against two stone piers, the space separating these being partially closed by a platform B B', Fig. 563, raised above the bottom C C of the river. The water passing between A and B has a certain velocity U, while the streams that pass beneath the boat have a less velocity U'. If we call h the height of A below the level N, we have $\frac{U^2}{2\,g}=h$. On the other hand, if, within a given section D D', all the streams could be reckoned as parallel and having an equal velocity, and a and b were taken to designate the heights A B and D D'—the opening being supposed to be rectangular—the incompressibility of water would give $U\,a=U'\,b$, which relationship is necessary in order that the mass of water comprised between A B D D' may remain always the same. Finally, by applying Bernoulli's theorem to a molecule passing from the point D, with a pressure p' and a velocity U', to the point A where those quantities become p and U, we have the equation

563.

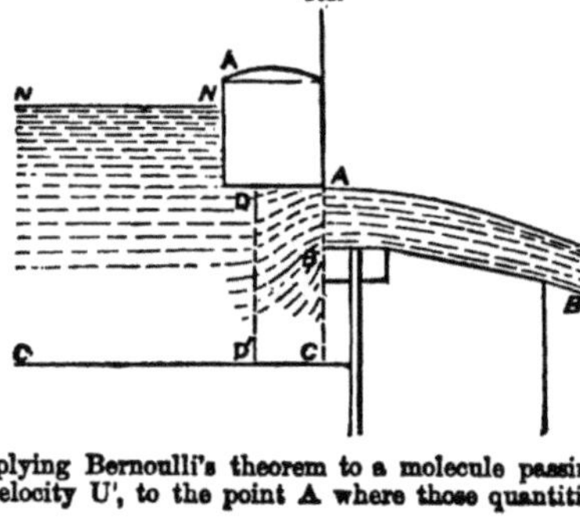

$$\frac{p'-p}{\Pi}=\frac{U^2-U'^2}{2\,g}.$$

From these three relations it is not difficult to deduce

$$\frac{p'-p}{\Pi}=h\left(1-\frac{a^2}{b^2}\right).$$

This calculation of the pressure p' upon the bottom of the float may be a little uncertain, especially as the velocities of the molecules that traverse the section D D′ are not all equal, and the converging of the streams towards the orifice A B prevents them also being horizontal; but it is quite sufficient to show that p' must be greater than p, and that the excess increases with h. The weight of the boat and its friction against the piers may sometimes be inadequate to establish equilibrium with the vertical force produced by the excess $p'-p$; in that case this is what is done: on the side facing the upper trough are several cocks, by means of which water enough is let into the caisson to balance it in the position it is intended it shall occupy. If it be required to sink it lower, more water is let in; if, on the contrary, it has to be raised, another set of cocks, facing the lower trough, are opened, the water flows from them, and the boat rises as it becomes lighter.

The two following styles of flood-gate are by M. Chaubart. They have been tested on the canal that borders the Garonne, with very satisfactory results, and for which the inventor received great praise. It would seem, therefore, that they may be serviceably employed for purposes of navigation and irrigation, as well as in regulating the level of mill-ponds. We have thought proper to introduce them here because they bear a rather close relationship, in principle, to a system of barrage-mobile now getting into very general use in France, and of which we shall presently have to speak at some length.

First Plan.—A canal, whose section is rectangular, is closed by an inclined gate A B, Fig. 564, occupying its entire width. To this gate is permanently fixed a quadrant C D, made of cast iron, which rolls on the horizontal plane E F. When the gate is in its initial position, the level N of the water touches its summit, and it must be so arranged that the resultant of the forces of gravity and the pressure of the water, acting upon the apparatus, shall pass through G, the actual point of contact of C D and E F. Equilibrium is then established. But if, from any unforeseen cause, it so happens that more water comes into the trough above, and that the level N rises, the centre of pressure rises with it, and the resultant advances in front of G. The gate then swings *en bascule*, till it assumes the position A′ B′, and the point of contact of the curve is removed from G to G′, while the surplus water runs off both at A′ and B′.

564.

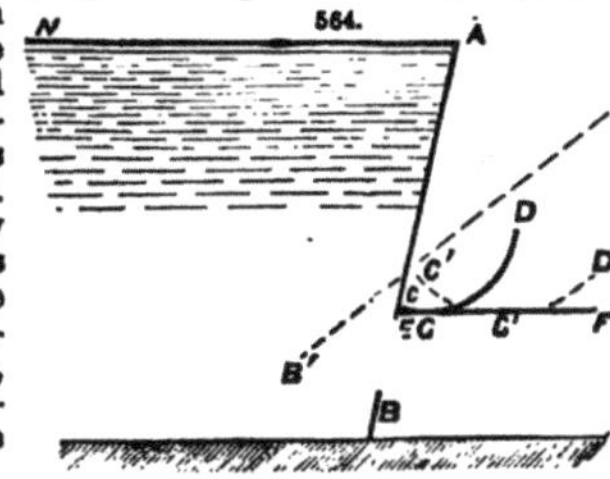

It may so happen, when in this new position A′ B′—if the curve C D has been properly determined—that the resultant will pass through G after the level has returned to N. In that case it is clear that the gate will remain in the position A′ B′, and that it will only quit it when a further rise causes its angle of inclination to be increased, or when an additional fall in the level forces it to right itself. In a word, whatsoever position it may assume, it will only remain in equilibrium so long as the level of the water in the tank reaches its normal height:—a greater height will widen the outlet, a less height will contract or close it. The apparatus may consequently be used to ensure a constant level of water in a reservoir where the supply is variable.

Second Plan.—By the aid of the gate just described, M. Chaubart has solved, as we have seen, the problem of obtaining a constant level in a reservoir where the *supply* is variable. Now, on the contrary, it is wished to get a constant and equal supply of water through a rectangular opening in a tank where the *level* varies. We will explain in what manner M. Chaubart has effected this object.

When in its natural position, the gate A B, Fig. 565, leaves a free outlet for the water between its lower edge and the bottom of the basin. The level N of the water, being now supposed to have attained its maximum height, nearly touches the top A of the gate, while a certain given quantity Q of water is discharged, in the unity of time, by the above rectangular opening whose dimensions are known. The gate is kept in equilibrium by the line A B resting against a fixed curve C C′ C″, and the resultant of the actions exercised by the pressure of the water and by gravity passes through the point of actual contact C. When the level falls to N′, the total pressure necessarily diminishes in proportion, and the gate swings *en bascule* till it finds another position of equilibrium A′ B′, which line rests upon a different point, C′, of the fixed curve; while the point B, advancing to B′, the section of the outlet, on the other hand, is proportionately increased. It may easily be conceived, then, that if the curve C C′ C″ has been properly traced, it is quite possible for the enlargement of the orifice to compensate for the diminution of pressure. If, now, we call l the width of the outlet, y the height of B′ above the bottom F F′, Y the height of the level N′ above the same horizontal line, m the coefficient of expenditure applicable in the present instance, the invariability of the expenditure will be expressed as follows:

565.

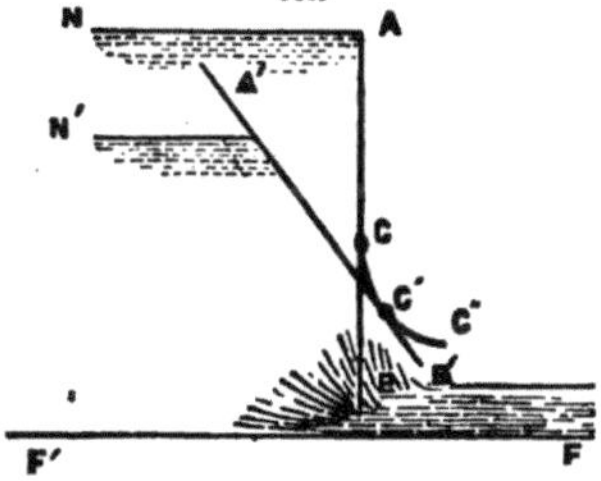

$$Q=mly\sqrt{2g\left(Y-\frac{1}{2}y\right)}.$$

from which equation may be deduced the values of Y in relation to y, and inversely.

Unfortunately, it is very difficult to ascertain the exact shape of the required curve; therefore, as the limits and nature of this work will not allow us to enter into a long theoretical investigation, we will content ourselves with laying before our readers an approximate graphical method.

It must be admitted that during the displacement of the gate the straight line A B turns upon a certain curve, which has to be determined in such a manner as to fulfil the essential conditions of the problem. In the first place, the initial point C of the curve, Fig. 565, will be known by taking the intersection of A B with the resultant R of the weight of the gate and the pressure of the water. Let us next imagine the line A B to undergo a slight change of position in the direction of A′ B′; during this movement it will have turned round an instantaneous and variable centre of rotation between C and C′, and, consequently, the approximate supposition will be admissible, that its entire rotation took place round the point D, where the tangents D C and D C′ to the curve C C′ meet. But, as that point is not known, it must be taken, by guess, a little below C′, and the line A B is made to move at a slight angle round the centre D; in this fresh position, the extremity B, now arrived at B′, being at a certain height y above the bottom F F′, Y has to be calculated by aid of the last equation; we then draw the new resultant R′ of the weight and pressure, and find its intersection C′ with the last position of A B. We shall ascertain that the point D, around which A D has been made to turn, was properly chosen if it happens to be equidistant from C and C′; if that be not the case, we shift the point D primitively adopted, and a second attempt will give, in a sufficiently approximate manner, both the line A′ B′ and the second point C′ of the curve sought. From that one we pass to another, and so on, till the required curve is ultimately traced by a series of points or tangents.

Components of a Movable Barrage.—A movable barrage, established across a navigable river, comprises two essential parts, namely, the navigable way and the overfall, see Fig. 566.

566.

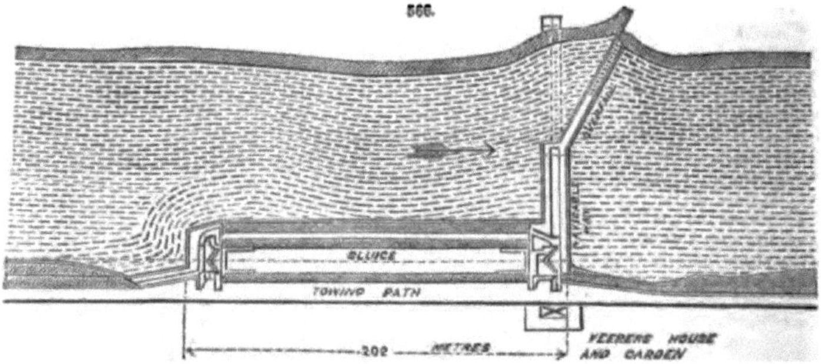

The former is used for purposes of navigation when there is a sufficient natural draught of water in the river for ships to pass; the movable lifts which serve to close the way are then laid flat on their platform.

The overfall serves to maintain the level of the river at a determinate height when the barrage is in use; it likewise serves as an outlet for the water while the lifts of the navigable way are being raised.

In addition to these two essential parts, there is generally, also, a lock through which the navigation takes place when the barrage is closed, see Fig. 566; when there is no lock adjoining, then the navigation can only be performed by removing the barrage and releasing the water at certain fixed periods.

The *sill* or *platform* of a navigable way should be placed at a depth not less than that of the bottom of the river above the weir.

On the Upper Seine these sills are $0^{m} \cdot 60$ below low-water mark.

The sill of the overfall should be so raised that—having due regard to economy and facility of construction—its section, added to that of the navigable way, shall offer an outlet proportional to the quantity of water that flows down the river at its different periods. Moreover, it must be at such a height that it may give free passage to the waters of the river while the lifts are being raised without producing too heavy a fall from the upper to the lower basin; these conditions are most important.

There necessarily exists, therefore, a relation between the section of the navigable way and that of the overfall; and another, moreover, between the width of the latter and the height of its sill.

The sills of the overfalls on the Upper Seine have been placed at $0^{m} \cdot 50$ above low-water mark.

The establishment of a navigable way is a costly work; its breadth should, consequently, not be greater than is absolutely necessary for the requirements of navigation.

On the Yonne the breadth of the navigable ways is 35 mètres; on the Upper Seine, between Montereau and Paris, it varies from 40 to 55 mètres, measured perpendicularly to the course of the river.

The width of the overfalls ranges between 60 and 70 mètres.

When the breadth of the river will not admit of the overfall being placed perpendicularly to its course, and in a line with the navigable way, it may be put obliquely, as shown in Fig. 566; in that case the angle of inclination must not be less than 60 degrees.

The platform of the navigable way should be of sufficient width to receive all the various components of the lifts, dams, &c., when the foundations are laid upon concrete.

On the Upper Seine, that width is 9ᵐ·50, divided into three zones, namely: two measuring 1ᵐ·75 for the piers, and the third 6 mètres for the pavement destined to receive those portions called the movable parts of the barrage. The platform, moreover, must be of such a thickness as will enable it to resist the different pressures and forces to which it has to be subjected.

On the Upper Seine that thickness is 2 mètres, irrespectively of the pavement.

When the difference of level between the upper and lower basin is as much as 2ᵐ·40, as is the case at the several barrages of the Upper Seine, each movable lift of the navigable way is capable of exercising a vertical strain upon the platform equal to about 2200 kilogrammes. If the sill were, at the same time, subjected to an under-pressure, which may very easily occur, it would be necessary to fix at the foot of each lift a block of stone 2 mètres cube, in order to secure it. As stones of similar dimensions, however, are not easily obtainable, and as they would entail, moreover, many difficulties of construction, as well as of repair, recourse has been had to disc anchors, which firmly bind the sill to the bed of concrete beneath the foundation.

567.

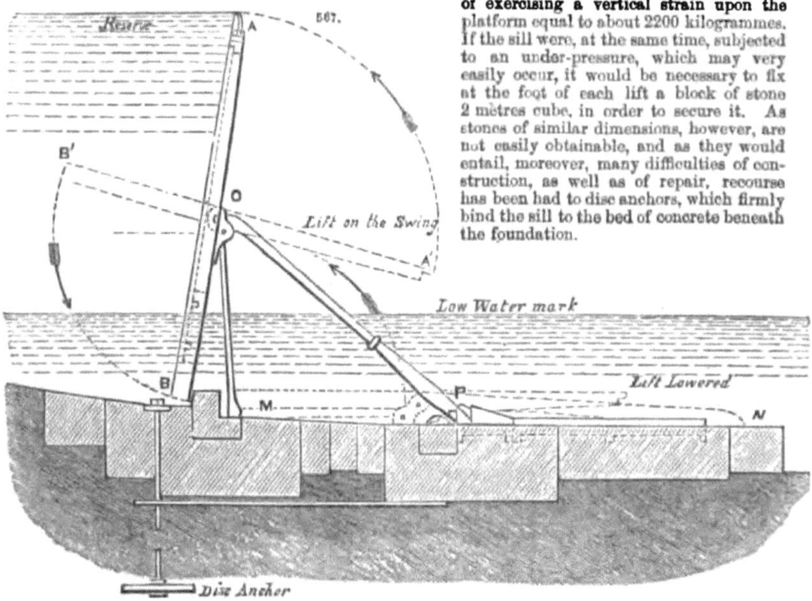

A glance at the above cut, Fig. 567, will make this arrangement clearly intelligible. There is a separate anchor for every lift, and they are fixed in position before the concrete is poured in, their rods or shanks being kept vertical during the operation.

In the navigable ways of the Seine, the pressure in the direction of the buttress of each lift is equal to about 4500 kilogrammes, producing a horizontal component equal to about 3500 kilogrammes. This component tends to make the piles of the foundations give way, and cause the mass of concrete to pivot towards the lower basin. When, however, the piles are well driven in and bound together, no such disturbance need be feared; but, as the layers of concrete below the fall may possibly be undermined by the water, it is requisite that the pavement of the platform be able to resist, by its power of adhesion alone, a sliding force equal to 3500 kilogrammes for every lift. If, then, we take 1100 kilogrammes only as representing the adhesive force of each square mètre of masonry, we find that the adhesion added to the friction gives a power of resistance greatly exceeding 3500 kilogrammes, so that little danger need be apprehended.

The lift of a navigable way is composed of three principal parts, namely:

1st. Of a framework of timber susceptible of moving upon a horizontal axis placed perpendicularly to the direction of the current. When this framework is raised, it is supported by its axis, while its base rests against a sill attached to the platform of the barrage.

2nd. Of a *chevalet* or stay, made of iron, and bearing the horizontal axis mentioned above. The lower part of the chevalet is terminated by two spindles working in sockets that are attached to the sill against which the foot of the lift rests, Fig. 568; so that this chevalet is able to turn upon its base, carrying with it, as it moves, the framework of the lift.

3rd. Of a buttress of iron, the head of which forms an articulation with that of the chevalet, its foot resting against a cast-iron shoe, firmly cemented in the platform

These three pieces are all that constitute the lift, and the whole arrangement presents very much the appearance of a painter's easel with a picture upon it, as will be seen by the foregoing illustration. Fig. 567, already referred to, shows the lift raised; A B being the framework of timber, O the centre on which it oscillates, A' B' its position when giving way to the water, and M N its position when lying flat on the platform to allow the passage of vessels. But, in order to make the arrangement better understood, we give an enlarged cut representing the chief features of the system, Fig. 569, wherein A B shows the framework or wooden swing-gate, *a b* its position when raised, and A' B' its position when lowered, C the chevalet, and D the buttress or prop.

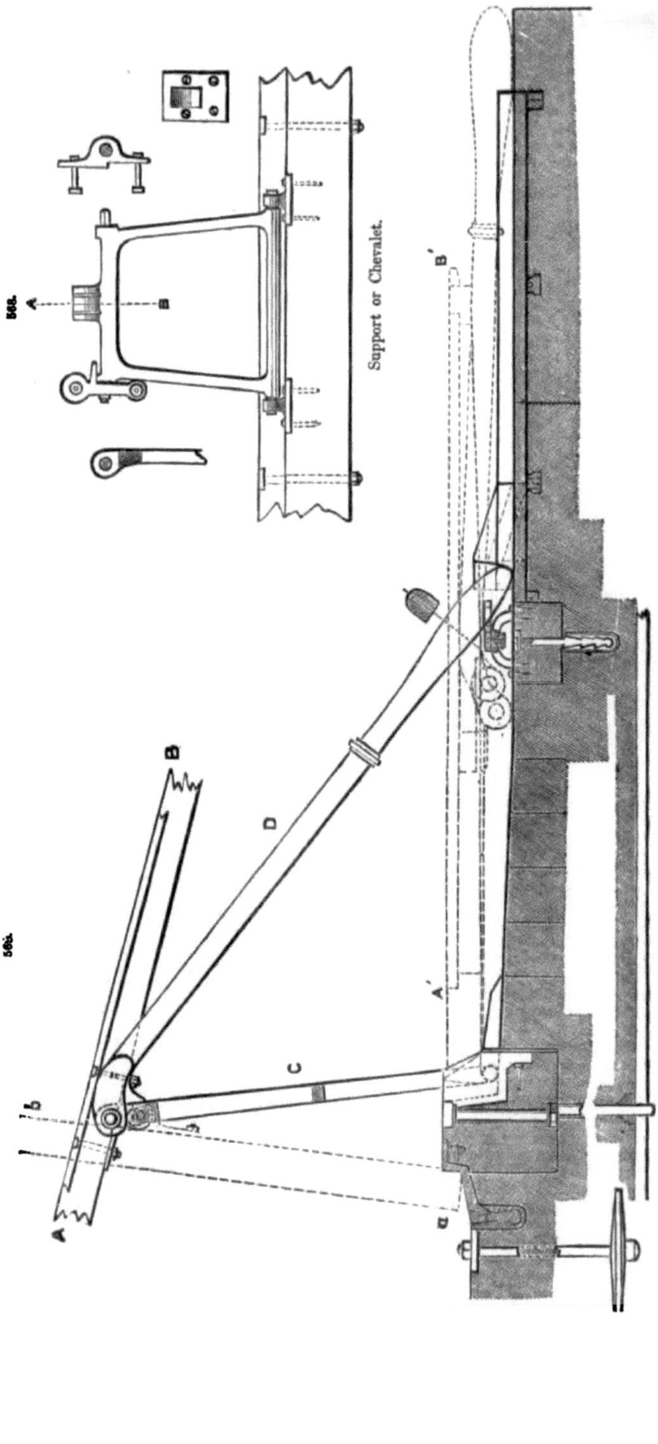

563.

Support or Chevalet.

564.

This is the system that has been so successfully and ingeniously put into practice by M. Chanoine at the celebrated barrage of Conflans-sur-Seine—of which a general view is here shown, Fig. 570—and at various other places.

Besides the three principal components of a lift, above described, there remains yet one addition of some import:

The *Counterpoise*.—Upon referring to the several illustrations, it will be observed that that portion of the lift which is above its axis of suspension, and which is called the *volée*, or *fly*, is wedge-shaped, getting thinner towards the top; whereas that which is below the axis of suspension is uniform in its thickness, which is equal to the thickest part of the fly. This is done in order to nearly balance the gate, giving the lower portion, however, a slight preponderance over the upper, which has a longer radius. The moment of the weight of the timbers forming the lower part, called the *culasse*, or *breech*, is about 110 out of water; and that of the timbers of the fly is nearly equal to it: but when the breech is entirely immersed, the moment of its weight is destroyed by the very fact of immersion; so that, in manœuvring, the weight of the fly becomes an obstacle to the lowering of the breech: to remedy this it was necessary to append a counterpoise to the latter, Figs. 571 and 572, composed of a mass of cast iron, movable, and held and guided by three parallel iron bars, Fig. 572, along which it may slide, and weighing about 66 kilogrammes. The moment of this counterpoise, together with the rest of the ironwork of the breech, is about 96.

In order to give some idea how the system operates, let us suppose a lift to be raised and in position, and then observe by what means it is lowered.

If the end of the buttress, which is rounded, be drawn on one side from the shoe, it is evident that, losing its point of support, it will slide upon the platform in the direction of the pressure exerted against the lift; that the *chevalet* will necessarily follow the buttress, turning upon its base; and that the gate itself, in rotation, will follow the *chevalet*: so that the two former will be stretched upon the platform in prolongation of one another, while the latter rests on the top of both, covering them. See Fig. 569.

The buttresses are made to slip from their respective shoes by means of an iron bar, placed horizontally upon the platform, and furnished with catches, so disposed at distances that they draw aside the buttresses one by one, in succession, and in the order in which it is intended to lower the lifts. This bar must be easy of management, and arranged in such a manner

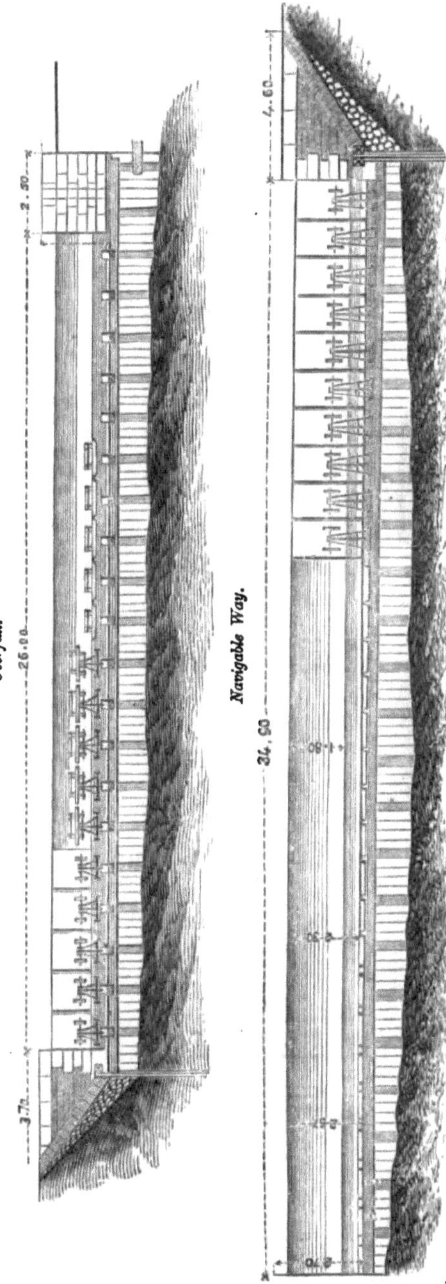

570.

that its action may not be impeded by gravel, sand, or any foreign matter carried down by the current. It is terminated at one end by a rack worked by a vertical wheel, by the aid of which its motion is imparted and thence transmitted from buttress to buttress. Upon being released from the shoe, the buttresses slip into guiding rails, or grooves, in which they slide till they reach the bottom.

571.

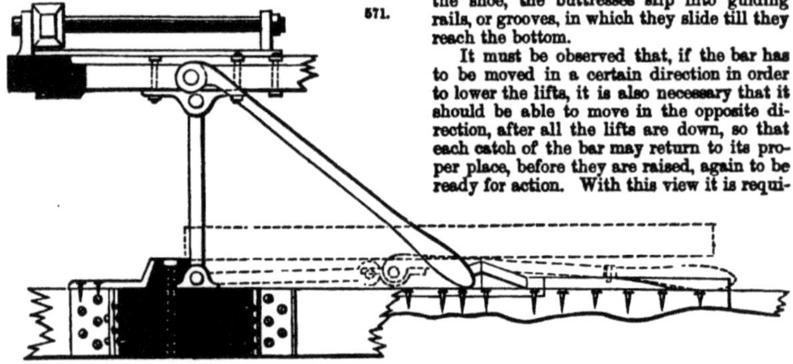

It must be observed that, if the bar has to be moved in a certain direction in order to lower the lifts, it is also necessary that it should be able to move in the opposite direction, after all the lifts are down, so that each catch of the bar may return to its proper place, before they are raised, again to be ready for action. With this view it is requisite so to arrange that there may be a chamber reserved beneath the articulation of the chevalet and buttress wherein the bar may freely work.

572.

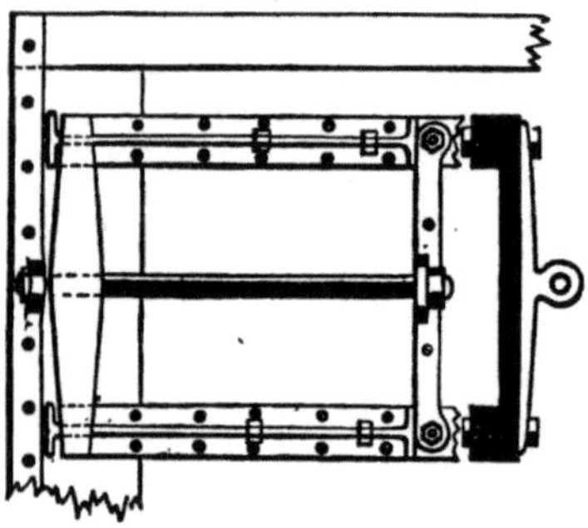

We have already implied that the lift proper is divided by its axis of rotation into two distinct parts: the lower part it has been agreed to call the *breech*, and the upper the *fly*. It is necessary to bear in mind this distinction. We will now describe the

Method of Raising the Lifts.—If the base of the *breech* be fixed to the sill of the barrage, and we attempt to raise the lift by seizing the top of the *fly* with a hook, a resistance is at once experienced, that increases rapidly with the height of the fall in the navigable way, and becomes almost insurmountable when the fall attains a height of $0^{m} \cdot 30$.

With the movable lifts actually in use, the operation is performed in a totally different manner. Instead of proceeding as above, it is the lower part

573.

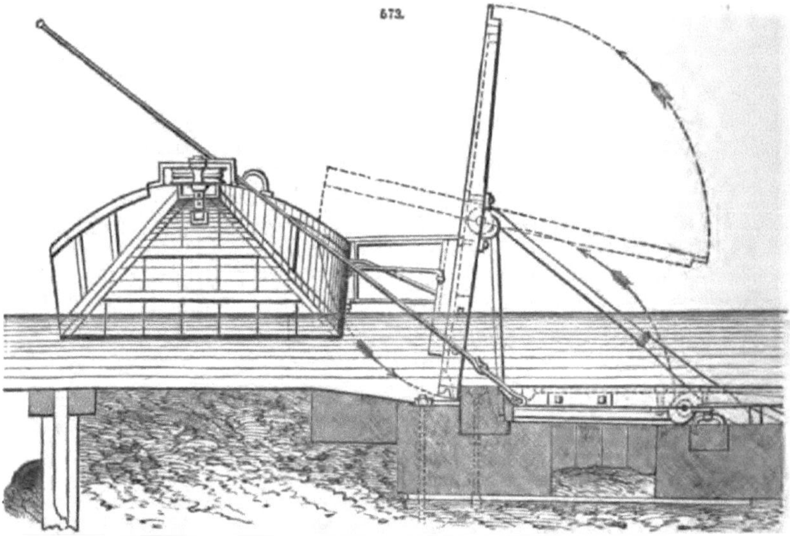

of the *breech* that is first raised; whereby the fall of water which, in the former instance, was an impediment, becomes, to a certain extent, an auxiliary, because it raises the woodwork of the lift as soon as the water has made its way beneath it. To this effect, the base of the *breech* is provided with a stout iron handle. The keeper, entering a boat fitted for the purpose, seizes this handle with a hook; then, pulling, by degrees the *breech* of the lift rises from the platform, dragging with it its *chevalet*, and the latter its *buttress*, Fig. 573. When these three have arrived at the end of their course, the extremity of the buttress comes and rests against the shoe, and the gate remains suspended on its axis of rotation, while the *breech* is upheld by the boatman's hook. As soon as the hook is detached, if the *breech* be a little heavier than the *fly*, or if it be slightly pushed, the gate immediately turns upon its axis, and the *breech* rests against the sill of the weir. This is effectually what takes place; but in order to ensure precision and regularity in the working of the different parts, many other accessories are needed, the details of which our limits will not allow us to particularize.

There are twelve *barrages* between Paris and Montereau, of which the normal heights of the falls are as follows:—

		Mètres.
Barrage of	Port-à-l'Anglais	H = 2·40
"	Albion	H = 1·85
"	Evry	H = 1·54
"	Coudray	H = 1·82
"	La Citanguette	H = 1·43
"	Vives-Eaux	H = 1·46
"	Melun	H = 1·44
"	La Cave	H = 1·97
"	Samois	H = 2·00
"	Champagne	H = 1·59
"	La Madeleine	H = 1·64
"	Varennes	H = 1·62

The extreme values are $H = 2^{m}\cdot40$ and $H = 1^{m}\cdot43$.

When the reserve is on a level with the top of a barrage, the real fall is equal to the normal fall, just given, less the surface incline of the water between that barrage and the one immediately below it. That incline may vary from 0 to $0^{m}\cdot15$. Moreover, it is necessary to deduct the thickness of the sheet of water surmounting the crest of the lower weir.

We regret that space will not allow us to enter into further details upon so interesting and important a subject. As we shall have occasion, however, to return to it again when speaking of flood-gates and canals, we must refer to those heads for more minute particulars touching the construction and working of the movable lift and the self-acting bascule gate. The lifts of a navigable way ought never to be self-acting, though the inconveniences likely to arise from their being so constructed would not be of any very serious nature. As regards the overfall, the case is different: there, on the contrary, it is of great importance that the lifts, swung *en bascule*, should be self-acting.

At the barrage of Conflans the overfall is composed of twenty lifts, each $1^{m}\cdot35$ high, $1^{m}\cdot20$ wide, and separated by spaces measuring $0^{m}\cdot10$. They are all self-acting, each gate swinging *en bascule*, and being regulated to resist a certain pressure by means of a counterpoise; so that, when by reason of a sudden increase of water the pressure becomes too great, they immediately yield, and, presenting a wider opening for the flood, prevent inundations, very much after the manner of the gates invented by M. Chaubart. It is, however, but an act of justice to state that the idea of a barrage with movable lifts was first due to M. Thénard, about the year 1840.

That gentleman, for a great many years chief engineer of the canal operations on the river Isle, had been unceasingly occupied in search of, and experimenting upon, the means of arriving at some efficient and practical mode of regulating, controlling, and utilizing the supply of water in rivers. He so far succeeded in this object, that he was enabled to sustain the waters of the river Isle at 7 ft. 4 in. above the level of the bed, procure a convenient draught of water to get boats up during dry weather, maintain them at this level sufficiently long so that the free flowing of the river was incapable of drawing them away, and, having arrived at this point, to restore the waters to their natural course in order not to expose the valleys to an overflow that would be prejudicial.

The first report, addressed to the Administration of Bridges and Highways, on the trials made by M. Thénard, is dated in 1831: it announced the good opinion formed of them by the inspector of the division. In 1839, for the purpose of verifying it, another commission, composed of Inspectors, general and divisional, of Bridges and Highways, was appointed by the Government. M. Thénard, having perfected with skill and success a happy idea of a provisional flood-gate, suggested to him by the divisional inspector, M. Mesnager, was able to render his system of *Barrage* more complete. On the 4th of July, 1841, the commission concluded their experiments and reported thereon.

Up to this time M. Thénard had only had occasion to apply his system to fixed existing barrages, raising the level of the water about 2 ft. 6 in. only. Confiding in the certainty of his system, however, he obtained authority to make a further trial, in which the retained body of water above the lower level was raised to a height of nearly 9 ft.

Among the authors who may be consulted are MM. Bresse, 'Cours de Mécanique Appliquée;' Lesbros, 'Expériences Hydrauliques sur les Lois de l'Ecoulement de l'Eau;' Chanoine, 'Notice sur les Barrages Mobiles;' Dubuat, 'Principes d'Hydraulique;' Chanoine and Lagrené, 'Mémoire sur les Barrages à Hausses Mobiles;' Mari; Graeff, 'Rapport sur la Forme et la Mode de Construction du Barrage d'Enfer sur le Furens,' Mémoires des Ponts et Chaussées, No. 134, 4e série; Gibbs, 'Cotton Cultivation, and the Barrage of Great Rivers,' crown 8vo, 1862; Breton, 'Mémoire sur les Barrages,' 4to, 1867.

TABLE I.—COEFFICIENTS OF EXPENDITURE.

Narrow-edged rectangular orifices, 0m·20 wide, and varying in height, discharging freely into the open air.

	Loads upon the ridges of the Orifices.	Coefficients of Expenditure for Orifices whose Heights are					
		m. 0·20	m. 0·10	m. 0·05	m. 0·03	m. 0·02	m. 0·01
1. Orifices completely isolated from the bottom and sides of the reservoir.	m. 0·02	0·572	0·596	0·616	0·639	0·660	0·695
	0·03	0·578	0·600	0·620	0·641	0·659	0·689
	0·04	0·582	0·603	0·623	0·640	0·659	0·684
	0·06	0·587	0·607	0·626	0·639	0·657	0·677
	0·10	0·592	0·611	0·630	0·637	0·655	0·667
	0·20	0·598	0·615	0·631	0·634	0·649	0·655
	0·30	0·600	0·616	0·630	0·632	0·645	0·650
	0·40	0·602	0·617	0·629	0·631	0·642	0·646
	0·60	0·604	0·617	0·627	0·630	0·638	0·641
	1·00	0·605	0·615	0·625	0·627	0·632	0·629
	1·50	0·602	0·611	0·619	0·621	0·620	0·617
	2·00	0·601	0·607	0·613	0·613	0·613	0·613
	3·00	0·601	0·603	0·606	0·607	0·608	0·609
2. Orifices whose lower surface is not contracted; *i.e.* is on a level with the bottom of the reservoir.	0·02	0·599	0·624	0·664	0·691	0·703	0·756
	0·03	0·603	0·629	0·665	0·687	0·702	0·747
	0·04	0·605	0·633	0·666	0·686	0·701	0·741
	0·06	0·610	0·637	0·667	0·686	0·699	0·732
	0·10	0·615	0·643	0·669	0·684	0·698	0·722
	0·20	0·621	0·648	0·670	0·681	0·696	0·712
	0·30	0·622	0·648	0·670	0·681	0·695	0·709
	0·40	0·623	0·648	0·669	0·681	0·695	0·706
	0·60	0·624	0·648	0·668	0·679	0·693	0·703
	1·00	0·624	0·647	0·666	0·676	0·692	0·701
	1·50	0·624	0·644	0·665	0·675	0·687	0·697
	2·00	0·619	0·641	0·664	0·675	0·683	0·693
	3·00	0·614	0·639	0·662	0·675	0·680	0·689
3. Orifices whose vertical sides are not contracted; *i.e.* are in the same plane as the sides of the reservoir.	0·02	..	..	0·655	..	..	0·715
	0·03	..	..	0·653	..	..	0·706
	0·04	0·649	..	0·651	..	..	0·699
	0·06	0·647	..	0·648	..	..	0·691
	0·10	0·645	..	0·645	..	..	0·683
	0·20	0·641	..	0·642	..	..	0·675
	0·30	0·639	..	0·642	..	..	0·671
	0·40	0·639	..	0·641	..	..	0·668
	0·60	0·638	..	0·639	..	..	0·665
	1·00	0·638	..	0·634	..	..	0·658
	1·50	0·637	..	0·627	..	..	0·651
	2·00	0·636	..	0·621	..	..	0·647
	3·00	0·634	..	0·614	..	..	0·644
4. Orifices without any contraction, either of the sides or of the bottom.	0·02	..	..	..	..	..	..
	0·03	..	..	..	..	..	..
	0·04	..	..	..	..	..	..
	0·06	..	..	0·699	..	..	..
	0·10	..	..	0·696	..	..	..
	0·20	0·708	..	0·693	..	..	..
	0·30	0·687	..	0·691	..	..	..
	0·40	0·682	..	0·690	..	..	..
	0·60	0·679	..	0·688	..	..	..
	1·00	0·676	..	0·685	..	..	..
	1·50	0·672	..	0·681	..	..	..
	2·00	0·668	..	0·680	..	..	..
	3·00	0·665	..	0·678	..	..	..

TABLE II.

Narrow-edged rectangular orifices, $0^{m}\cdot20$ in width, and varying in height, and continued outwards by a rectangular, horizontal, and open channel, of the same width as the orifice.

	Loads upon the ridge of the orifices.	COEFFICIENTS OF EXPENDITURE FOR ORIFICES WHOSE HEIGHTS ARE					
		m. 0·20	m. 0·10	m. 0·05	m. 0·03	m. 0·02	m. 0·01
1. Orifices completely isolated from the bottom and sides of the reservoir.	m. 0·02	0·480	0·484	0·488	0·501	..	0·599
	0·03	0·493	0·507	0·525	0·551	..	0·626
	0·04	0·503	0·527	0·555	0·598	..	0·645
	0·06	0·518	0·557	0·594	0·632	..	0·667
	0·10	0·542	0·586	0·624	0·633	..	0·671
	0·20	0·574	0·606	0·631	0·632	..	0·664
	0·30	0·591	0·612	0·629	0·631	..	0·658
	0·40	0·597	0·615	0·626	0·630	..	0·652
	0·60	0·600	0·615	0·625	0·628	..	0·644
	1·00	0·601	0·613	0·624	0·625	..	0·631
	1·50	0·601	0·612	0·619	0·620	..	0·618
	2·00	0·601	0·607	0·613	0·613	..	0·613
	3·00	0·601	0·603	0·606	0·607	..	0·609
2. Orifices whose lower surface is not contracted.	0·02	0·480	..	0·487	..	..	0·616
	0·03	0·493	..	0·526	..	..	0·642
	0·04	0·502	..	0·552	..	..	0·660
	0·06	0·517	..	0·583	..	..	0·676
	0·10	0·538	..	0·605	..	..	0·682
	0·20	0·566	..	0·617	..	..	0·679
	0·30	0·580	..	0·622	..	..	0·676
	0·40	0·587	..	0·625	..	..	0·673
	0·60	0·595	..	0·627	..	..	0·670
	1·00	0·600	..	0·628	..	..	0·665
	1·50	0·602	..	0·627	..	..	0·657
	2·00	0·602	..	0·623	..	..	0·654
	3·00	0·601	..	0·618	..	..	0·652
3. Orifices whose lower surface is isolated from the bottom, and whose vertical sides project $0^{m}\cdot02$ from the sides of the reservoir.	0·02	0·496	..	0·557	..	..	0·675
	0·03	0·510	..	0·577	..	..	0·683
	0·04	0·522	..	0·592	..	..	0·688
	0·06	0·539	..	0·611	..	..	0·693
	0·10	0·563	..	0·628	..	..	0·694
	0·20	0·591	..	0·637	..	..	0·684
	0·30	0·607	..	0·636	..	..	0·677
	0·40	0·615	..	0·635	..	..	0·673
	0·60	0·625	..	0·635	..	..	0·669
	1·00	0·628	..	0·635	..	..	0·663
	1·50	0·627	..	0·634	..	..	0·656
	2·00	0·626	..	0·634	..	..	0·651
	3·00	0·624	..	0·632	..	..	0·648
4. Orifices whose lower surface is not contracted, and whose vertical sides project $0^{m}\cdot02$ from the sides of the reservoir.	0·02	..	..	0·512	..	..	0·625
	0·03	..	..	0·543	..	..	0·651
	0·04	0·518	..	0·566	..	..	0·667
	0·06	0·536	..	0·595	..	..	0·686
	0·10	0·560	..	0·621	..	..	0·697
	0·20	0·589	..	0·637	..	..	0·698
	0·30	0·603	..	0·643	..	..	0·696
	0·40	0·613	..	0·646	..	..	0·694
	0·60	0·623	..	0·648	..	..	0·690
	1·00	0·630	..	0·649	..	..	0·685
	1·50	0·633	..	0·647	..	..	0·679
	2·00	0·632	..	0·644	..	..	0·674
	3·00	0·630	..	0·639	..	..	0·670

TABLE III.

Narrow-edged rectangular overfalls, $0^m \cdot 20$ wide, { discharging freely into the open air .. .. (A)
outwardly extended by means of an open horizontal channel of equal section with the orifice .. .. (B)

The arrangements A and B may present the variations defined at 1, 2, and 3, Table I., and 3 and 4, Table II., when the rectangular orifices are closed at the top.

Loads upon the Sill of the Overfall.	Coefficients of Expenditure for overfalls presenting the arrangement A, with its variation.				Coefficients of Expenditure for overfalls presenting the arrangement B, with its variation.			
	Tab. I. 1	Tab. I. 2	Tab. I. 3	Tab. II. 4	Tab. I. 1	Tab. I. 2	Tab. II. 3	Tab. II. 4
m.								
0·01	0·424	0·384	0·492	0·292	..	..	0·395	..
0·02	0·417	0·402	0·473	0·318	0·196	0·208	0·383	0·175
0·03	0·412	0·410	0·459	0·337	0·234	0·232	0·373	0·205
0·04	0·407	0·411	0·449	0·352	0·263	0·251	0·365	0·234
0·05	0·404	0·411	0·442	0·362	0·278	0·268	0·360	0·260
0·06	0·401	0·410	0·437	0·370	0·286	0·281	0·355	0·276
0·07	0·398	0·409	0·435	0·375	0·292	0·288	0·352	0·285
0·08	0·397	0·409	0·434	0·379	0·297	0·294	0·349	0·291
0·09	0·396	0·409	0·434	0·380	0·301	0·298	0·347	0·295
0·10	0·395	0·408	0·434	0·382	0·304	0·302	0·345	0·299
0·12	0·394	0·408	0·434	0·383	0·309	0·308	0·343	0·306
0·14	0·393	0·408	0·434	0·383	0·313	0·312	0·341	0·311
0·16	0·393	0·407	0·433	0·384	0·316	0·316	0·340	0·315
0·18	0·392	0·406	0·432	0·383	0·317	0·319	0·339	0·319
0·20	0·390	0·405	0·432	0·383	0·319	0·323	0·338	0·322
0·22	0·386	0·405	0·430	0·382	0·320	0·325	0·337	0·325
0·25	0·379	0·404	0·428	0·381	0·321	0·329	0·336	0·329
0·30	0·371	0·403	0·424	0·378	0·324	0·332	0·334	0·332

BARREL. FR., *Tonneau;* GER., *Fass;* ITAL., *Barile, Bariglione;* SPAN., *Barril.*

A barrel is a round vessel or cask, of more length than breadth, and bulging in the middle, made of staves and headings, and bound with hoops.

The term is also applied to a tube, or to any hollow cylinder, as the barrel of a *gun*, the barrel of a *pump*, and so on.

BARREL-CURB. FR., *Margelle;* GER., *Senkrahmen;* ITAL., *Appoggio impiegato nella costruzione dei pozzi;* SPAN., *Mardelle.*

A barrel-curb, or well-curb, is an open cylinder, about 3 ft. 6 in. or 4 ft. in length, formed of strips of wood nailed round horizontal ribs of elm, and used as a mould in well-sinking to keep the well cylindrical during the process of sinking. When the required depth has been attained, this cylinder is usually left in the bottom of the well, under the steming, brickwork being built up under the horizontal ribs.

BARREL-DRAIN. FR., *Tranchée en berceau;* GER., *Tonnenförmiger Absugscanal;* ITAL., *Fogna cilindrica;* SPAN., *Alcantarilla Cilíndrica.*

A barrel-drain is a brick or stone drain of cylindrical form.

BARRIER. FR., *Barrière;* GER., *Barriere;* ITAL., *Palizzata;* SPAN., *Barrera.*

In *fortification* a barrier is a kind of fence made in a passage or retrenchment to stop an enemy. It is usually a palisade or stockade.

BARROW. FR., *Brouette;* GER., *Schubkarren;* ITAL., *Carriuola;* SPAN., *Carretilla de mano.*

A barrow is a light, small carriage, borne or moved by hand.

The body of the excavator's barrow, Figs. 574, 575, is spread wide open, and the sides are much inclined; the centre of gravity of the load is therefore situated much lower, with respect to the

574.

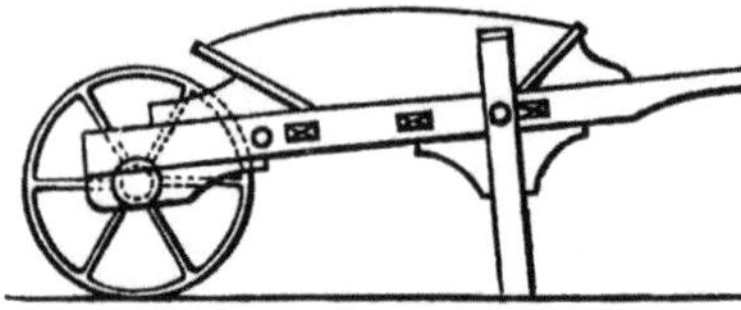

575.

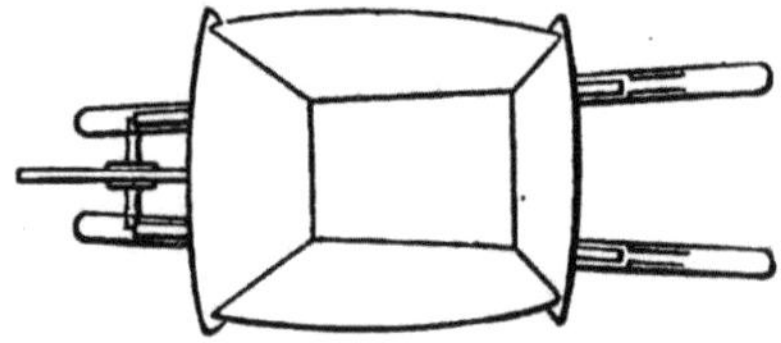

handles, than in the ordinary barrow, which renders it steadier and easier to wheel. The contents are discharged by inclining the barrow at an angle of 45°, and supporting it constantly on the wheel. The nave of the wheel is prolonged on each side, and serves as an axle, the periphery of which is about 1 in. in thickness and rounded on the edge.

BARS, Guard. Fr., *Garde-fou*; Ger., *Fenster Vergitterung*; Ital., *Sbarra*.

Usually wrought-iron bars in front of windows, about ⅝ in. in diameter, if round, or ¾ in. if square, spaced about 5 in. apart, and fixed perpendicularly through horizontal rails of flat bar-iron, 1¾ in. wide by ⅜ in. thick, built into the jambs of the window: the ends of the bars at the bottom should be let into the stone sill, and run with lead.

BASCULE BRIDGES. Fr., *Pont-levis*; Ger., *Zugbrücke*; Ital., *Ponte levatoio*; Span., *Puente-báscula*.

See Drawbridges.

BASEMENT. Fr., *Soubassement*; Ger., *Fundament, Untere Theil*; Ital., *Basamento*; Span., *Bajamente, Planta baja*.

See Building.

BASE-LINES. Fr., *Lignes de base*; Ger., *Grundlinie*; Ital., *Base di triangolazione*; Span., *Base trigonométrica*.

See Geodesy.

BASE-PLATE. Fr., *Plaque de fondation*; Ger., *Grundplatte*; Ital., *Piastra di fondazione*; Span., *Plancha de cimiento*.

The foundation-plate of heavy machinery, as of the steam-engine, is termed the base-plate; the *bed-plate*.

BASIN. Fr., *Bassin*; Ger., *Schiffsdocke*; Ital., *Bacino*; Span., *Pantano, dique*.

Any hollow place containing water, as a dock for ships, is called a basin. See Docks.

BASKET-HANDLED ARCH. Fr., *Arc en anse de panier*; Ger., *Stichbogen*; Ital., *Arco scemo*; Span., *Arco elíptico*.

Any arch less than a semicircle on the same chord is called a basket-handled arch, hence all semi-elliptic arches are included in the term.

BASTARD ASHLAR. Fr., *Moëllon gisant*; Ger., *Pläner*; Ital., *Pietra rozza*.

Bastard ashlar are stones intended for ashlar-work, which are merely rough scabbled to the required size at the quarry; or the face-stones of a rubble wall, which are selected, squared, and dressed, to resemble ashlar.

BASTARD STUCCO. Fr., *Stuc mêlé de mastic*; Ger., *Kalkmörtel mit feinem Sande versetzt*; Ital., *Stucco rozzo*.

The finishing coat in plastering when prepared for paint is termed bastard stucco. It is composed of similar stuff to that used for trowelled stucco, with the addition of a small portion of hair, but is accompanied with less labour, not being floated; it is generally employed in three-coat work.

BASTARD-TOOTHED FILE. Fr., *Grosse lime*; Ger., *Bestossfeile*; Ital., *Lima bastarda*; Span., *Lima*.

See Hand-tools.

BASTION. Fr., *Bastion*; Ger., *Bastion, Bollwerk*; Ital., *Bastione*; Span., *Baluarte*.

See Fortification.

BAT. Fr., *Morceau de brique*; Ger., *Schieferstein*; Ital., *Pezzo di mattone*; Span., *Medio ladrillo*.

The half of a brick is usually termed a bat. Other portions are named according to the size, as a quarter bat, three-quarter bat, and so on.

BATEA. Fr., *Cuvette pour laver l'or*; Ger., *Waschschüssel für Gold*; Ital., *Bacinella per lavare l'oro*; Span., *Batea*.

A batea is a conical-shaped dish, Fig. 576, employed for washing gold and pulverized samples of gold quartz.

576.

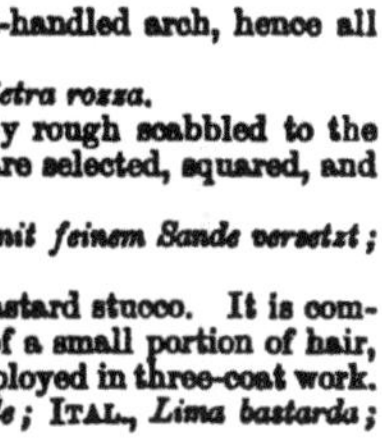

From the general irregularity of the produce of quartz in gold mines, it is impossible to ascertain the average yield of vein-stuff without crushing and experimenting on large quantities; but the most usual method of judging approximately of the value of rock, is to pulverize a small quantity and wash the resulting powder in a batea or horn spoon. In selecting the rock for this purpose, it is evidently of the greatest importance that it should represent a fair average of the vein or streak from which it is taken, and consequently several hundred-weights should be broken from the whole area of the exposed surface, taking care that every part be represented by samples of nearly equal weights. The whole must now be broken by a hammer on an iron plate, into pieces of about the size of walnuts. The resulting heap is then carefully mixed, by turning over with a shovel, and subsequently cut through the middle, so as to leave a trench through its centre, extending to the floor on which it has been placed. The two sides are afterwards carefully scraped down, and removed as a representative sample on which the yield of the vein is to be estimated. For the purpose of a rough approximation, this may be at once pulverized in a mortar or otherwise, and its contents judged of in accordance with the results obtained by washing. Where, however, greater accuracy is aimed at, and the original heap contained a large quantity of broken rock, at least a hundred-weight should be scraped from the sides of the cutting, and this, after being further reduced to the size of peas, must again be cut through, and a sample of about 4 lbs. obtained, by the means employed in the first instance, as the final sample. This is pulverized in a mortar, and the whole passed through a sieve of wire gauze, of forty holes to the lineal inch, after which it is ready for treatment, either by washing or assay.

The most accurate results are obtained by carefully washing a 4-lb. sample in the batea, Fig. 576, which is about 20 in. in diameter, and 2¼ in. in depth.

BATH. Fr., *Bain*; Ger., *Bad*; Ital., *Bagno*; Span., *Baño*.

See Washhouses and Baths.

BATH-METAL. Fr., *Tombac*; Ger., *Tomback*; Ital., *Lega di Bath*; Span., *Tumbaga*.

Bath-metal is an alloy consisting of 4½ oz. zinc and 1 lb. copper. See Alloys.

BATTEN. Fr., *Volige;* Ger., *Dünnes Brett;* Ital., *Tavola stretta, Listello;* Span., *Astilla.*

Batten is a term applied to sawn timber under 3 in. in thickness, when the width is 7 in., to distinguish it from other widths, such as *deals* and *planks.*

The term *batten* is also applied to boards in long lengths of less than 7 in. wide, though seldom more than 2 or 3 in. in width—the thickness varying from $\frac{1}{2}$ in. to $1\frac{1}{2}$ in., according to the purpose for which they are intended.

BATTENING. Fr., *Construction en volige;* Ger., *Schalwerk;* Ital., *Costruzione di listelli.*

Narrow boards or battens fixed to walls, intended to be papered over canvas or to receive the laths for plastering—also battens nailed on the rafters of a roof to receive the slating—are called battening.

Wall-battens should be spaced about 12 in. apart, and are generally $2\frac{1}{2}$ or 3 in. wide, and $\frac{3}{4}$ to $1\frac{1}{4}$ in. thick.

Slate-battens must be spaced to correspond with the gauge of the slate. They should be from $2\frac{1}{2}$ to 3 in. wide, and from $\frac{3}{4}$ to $1\frac{1}{4}$ in. in thickness, according to the strength required.

BATTER. Fr., *Talus;* Ger., *Verjungung;* Ital., *Inclinazione di muro Scarpa;* Span., *Inclinacion, talud.*

In building, batter is a term employed to signify leaning back; it is usually expressed by the ratio of the departure from the perpendicular to the height, as 1 in 10, Fig. 577, which means that for every 10 ft. in height the wall batters 1 ft.

577.

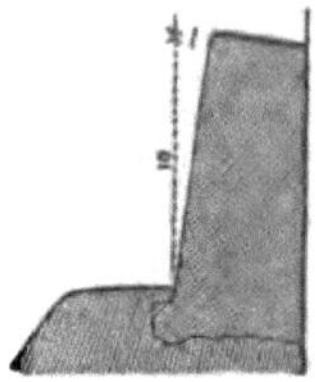

Retaining walls are sometimes battered on the face to the extent of 1 in 5; latterly, however, 1 in 10 has become more general, as when the batter is great the joints of the sloping wall hold the wet, which soon finds its way into the work: it is on this account that most engineers prefer walls with a vertical face, or at most with a very slight batter.

BATTERY. Fr., *Batterie galvanique;* Ger., *Galvanische Batterie;* Ital., *Batteria Galvanica.*

When a series of voltaic elements, cells, couples, or pairs, are arranged in such a manner that the zinc of one element is in connection with the copper of another element; the zinc of this with the copper of a third, and so on; such an arrangement is termed a Galvanic or Voltaic Battery.

The earliest galvanic battery was constructed by Volta in 1800. It consisted of an insulated plate, Fig. 578, upon which was placed a series of discs of copper and of zinc soldered together. Above the copper of the first disc was placed a disc of cloth *d,* saturated with acidulated water; upon the disc of cloth was then laid another metallic disc. These discs were thus alternately laid one upon another, until a pile, Fig. 579, had been built up, care being taken to lay them the same way. To the ends of this pile were attached wires *p* and *n,* which, when connected in any way, set in motion a current of electricity.

578.

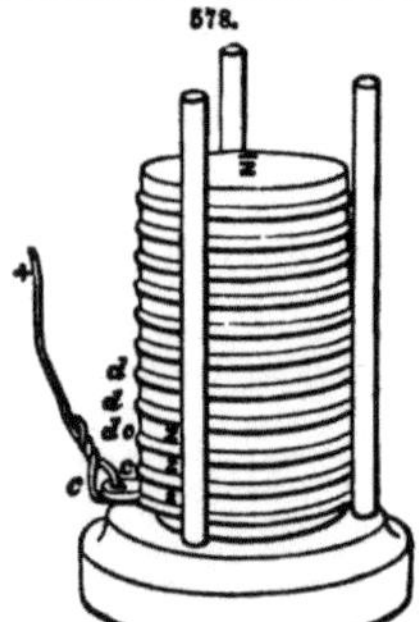

580.

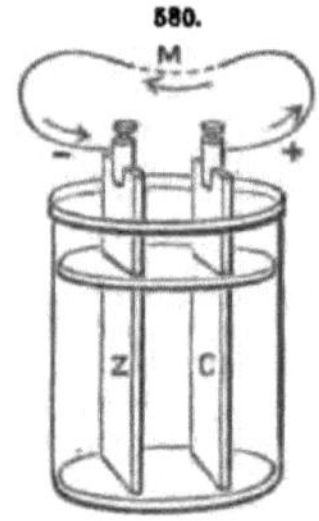

579.

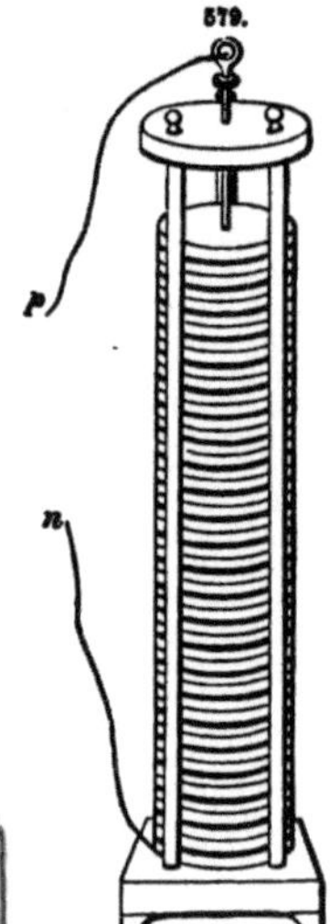

581.

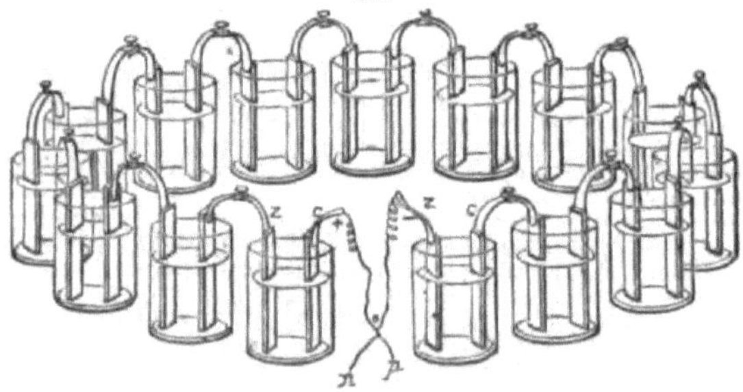

The piles constructed in this way were, however, but weak, and useless for experiments which lasted any length of time; for as the number of elements was augmented, the weight of the upper discs pressed the liquid from the lower discs, which became dry, and so lost their conductibility.

This led Volta to invent the improved modification shown in Fig. 581, which he called the *couronne de tasses*, or crown of cups. Instead of the damp cloth of the pile, a number of jars, arranged in a circle, and filled with acidulated water, are employed. The jars communicate successively one with the other by means of metallic arches formed of a plate of zinc soldered to a plate of copper; the copper of each arch being plunged into the jar which precedes it, and the zinc into the jar which follows it. The two jars which form the extremities of the series receive respectively a plate of zinc Z, and a plate of copper C, to each of which is fastened a conducting-wire. The first corresponds to the negative pole and the second to the positive pole.

To comprehend the principle of this battery, we will suppose that two plates, cut, one Z from a sheet of zinc, and the other C from a sheet of copper, Fig. 580, are placed, without contact with each other, in a jar containing slightly acidulated water. To the upper edges of the plates let two pieces of wire be fastened. In this state the apparatus will manifest no development of the electric fluid; but if the ends of the wires be brought into contact at M, an electric current will be set in motion, passing through the wires from the point where the wire is fastened to the copper C, to the point where the other wire is soldered to the zinc Z. The current will continue to flow so long as the ends of the wires are in contact, but the moment the ends are separated the current ceases.

It will be seen that the electric fluid is evolved by the combination of three bodies—the zinc, the copper, and the acidulated solution in which they were immersed. The production of the current depends on the chemical action of the solution on the zinc. That metal being very susceptible of oxidation, decomposes the water which is in contact with it. One constituent of the water combining with the zinc produces a compound called the oxide of zinc, and this oxide entering again into combination with the acid which the water holds in solution, forms a soluble salt. If the acid, for example, be sulphuric acid, this salt will be the sulphate of the oxide of zinc; and as fast as it is produced it will be dissolved in the water in which the plates were immersed. The copper not being as susceptible of chemical action as the zinc, remains comparatively unaffected by the solution; but the hydrogen evolved in the decomposition of the water collects upon its surface, after which it rises and escapes in bubbles at the surface of the solution.

It is to this chemical action upon the zinc that the production of the electric current is due. If a similar action had taken place in the same degree on the copper, a similar and equally intense electric current would be produced in the opposite direction; in that case the two currents would neutralize each other, and no electric effect would ensue. From this it will be seen that the efficacy of the combination must be ascribed to the fact that one of the two metals immersed in the solution is more oxidizable than the other, and that the energy of the effect and the intensity of the current will be so much the greater as the susceptibility of oxidation of one metal exceeds that of the other.

It appears, therefore, that the principle may be generalized, and that electricity will be developed and a current produced by any two metals similarly placed, which are oxidizable in different degrees. And, indeed, if two pieces of the same metal are differently acted upon, either by heat or chemically, a current of electricity will be produced on their being connected together.

Zinc being one of the most oxidizable metals, and being also sufficiently cheap and plentiful, is generally used for voltaic combinations. Silver, gold, and platinum are severally less susceptible of oxidation, and of chemical action generally, than copper, and would therefore answer voltaic purposes better, but are excluded by their greater cost, and by the fact that copper is found sufficient for all practical purposes. It is not, however, absolutely necessary that the inoxidizable plate C of the combination should be a metal. It is only necessary that it be a good conductor of electricity.

In certain voltaic combinations, charcoal properly solidified has therefore been substituted for copper, the solution being such as would produce a strong chemical action on copper. Each combination of two metals, or of one metal and charcoal, is called either a cell, a couple, an element, or a pair.

A series of jars, Fig. 582, when arranged in a similar manner to Volta's *couronne de tasses*, that is, the zinc of one jar in connection with the copper of the next jar, the order being zinc, acid, copper, zinc, acid, copper, and so on, is termed a battery, and by its means the effects produced by a single element are capable of being greatly increased. If, however, only one element is employed, it is in itself a battery.

582.

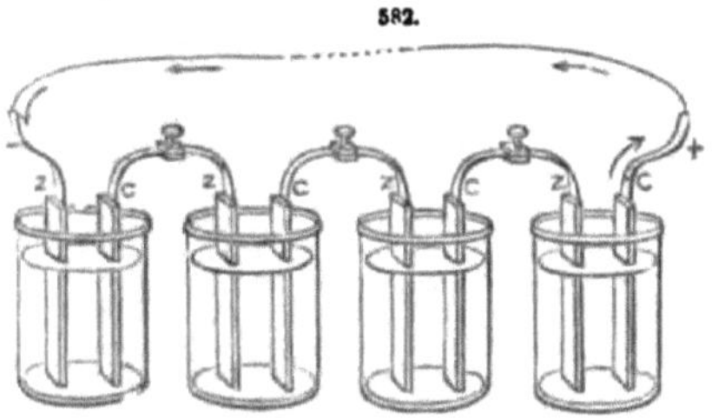

583.

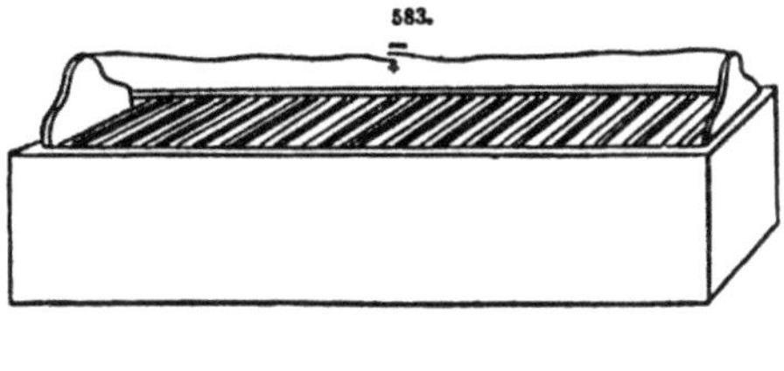

The part of a battery from which the current is supposed to proceed is called the positive pole, and the part towards which the current flows, the negative pole. These poles, shown at + and −, Fig. 582, are often termed electrodes, so that + would be the positive and − the negative electrode.

The arrangement of the *couronne de tasses*, and of batteries similarly constructed, was so cumbersome that they were soon superseded by the Trough battery, which is shown in Fig. 583. This battery was invented by Cruickshank, and consists of a rectangular water-tight trough divided into cells by plates formed of zinc and copper soldered together. The cells are filled with acidulated water or a solution of salt and water; and two plates of copper, furnished with conducting-wires, are immersed in the last cell at each end. It was with a battery of this kind, composed of 3000 couples, that at the commencement of the present century Davy succeeded in decomposing potash and soda, and thus discovered potassium and sodium. The trough battery is rather inconvenient on account of its weight, and also through the wood of the case warping under the action of the acids.

The arrangement introduced by Wollaston is more convenient. An element of Wollaston's battery, Fig. 584, is composed of a plate of zinc Z, round which is bent a plate of copper C, actual contact being prevented by placing small pieces of wood or cork between the plates. When the element is to be put in operation, it is immersed in a vase containing acidulated water; the negative pole then establishes itself at the wire connected with the zinc, and the positive pole at the wire fastened to the copper. To unite several elements into a battery, it is necessary to connect the copper of each element with the zinc of the following. The elements thus connected can then be mounted on a stand, Fig. 585, and, when the battery is to be operated, plunged separately into vases containing the exciting liquid. The necessity of a vase for each of the elements of Wollaston's battery renders it cumbersome.

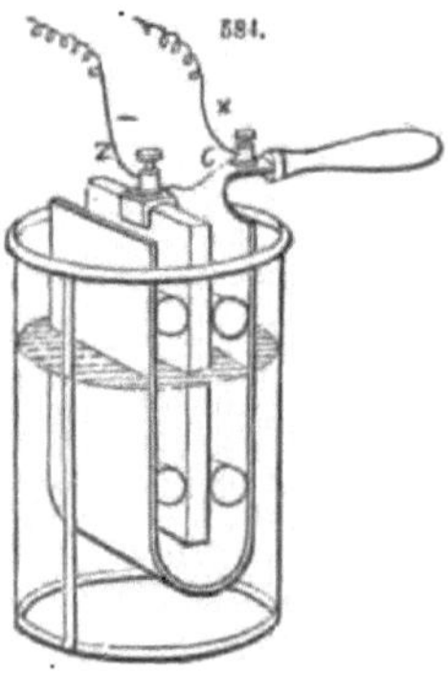

584.

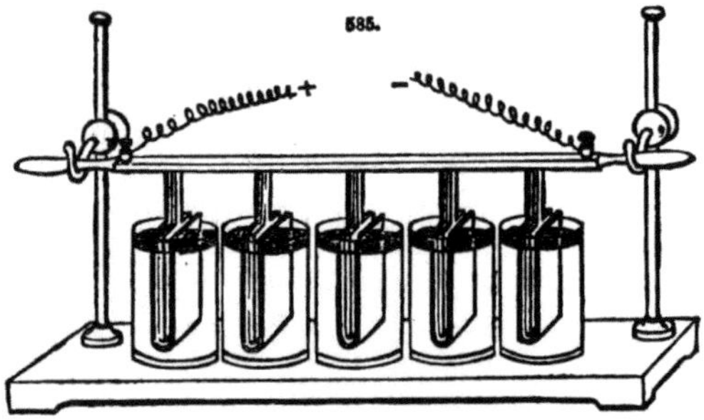

585.

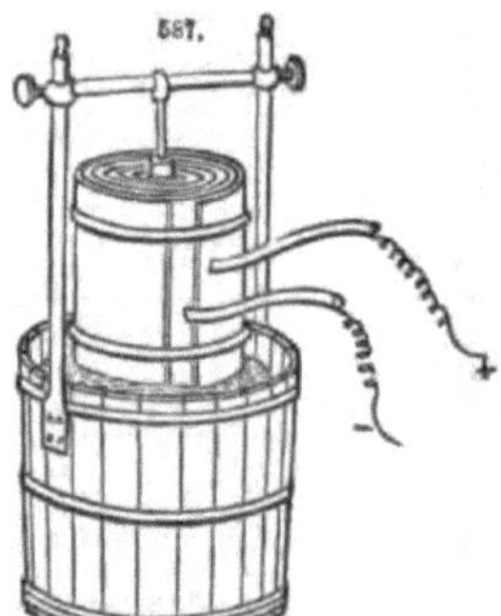

587.

586.

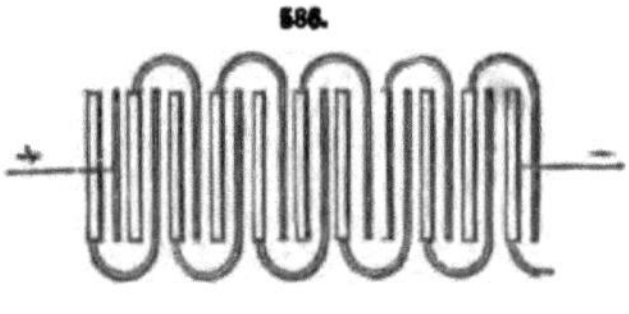

Muncke's battery removes this inconvenience, as by its means we can have a considerable number of elements in a small space. The plates of zinc and of copper are soldered together vertically, bent into the form of the letter U, and then fitted alternately one into the other, Fig. 586, in such a manner that the alternation of the metals is complete. Together the elements form a single system, which is fixed in a wooden frame, and, when required for use, immersed in a stone trough filled with acidulated water.

Batteries composed of a number of elements, as Muncke's battery, are especially applicable when the current encounters in its polar circuit any great resistance. When this resistance is weak, it is preferable to have the advantage of increasing the area of the surface of the elements rather than their number. This condition is realized by Dr. Hare's cylindrical battery, Fig. 587, which consists of a large sheet of zinc and one of copper, soldered together at one end. The sheets are rolled, without touching each other, round a cylinder of wood, and each is attached to metallic conducting-wires; the negative wire is that in connection with the zinc, and the positive wire that fastened to the copper.

When the apparatus is to be used, it is plunged into a tub containing acidulated water. When several of these elements are united in a battery, an apparatus is obtained, of which the caloric power is so great as to have obtained for it the name of a calorimotor.

The batteries we have described present practically several serious inconveniences. The water being decomposed by the zinc, liberates hydrogen, which, charged with acid particles, is released into the air, rendering in a very short time the surrounding atmosphere irrespirable. Besides this, the liberated hydrogen adheres as a film on the surface of the copper, presenting a great resistance to the current, and sensibly diminishing its intensity. Lastly, this film has a variable thickness, from which results a perpetual variation in the intensity of the current itself. These inconveniences, however, disappear in a greater or less degree in batteries operated by two liquids.

Daniell's Battery.—An element of Daniell's battery, Fig. 588, consists of an outer jar containing acidulated water in which is immersed a cylinder of zinc Z, in the interior of which is placed a porous earthenware pot, filled with a solution of sulphate of copper which surrounds a cylinder of copper C. The conducting-wire attached to the zinc, Fig. 588, corresponds to the negative pole of the element.

It is easily seen that in this arrangement the film of hydrogen which acts so disadvantageously in ordinary batteries, no longer exists; and the metal, following the same direction as positive electricity, deposits itself upon the copper cylinder. The disengagement of the hydrogen is replaced by a decomposition of the copper, which does not change the physical condition of the system. The action of this battery is very regular, it lasts a long time without renewal, and the surrounding atmosphere is not sensibly affected by its fumes.

588.

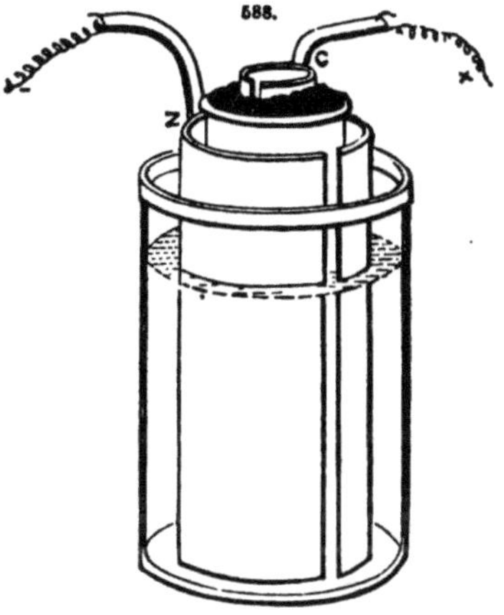

589.

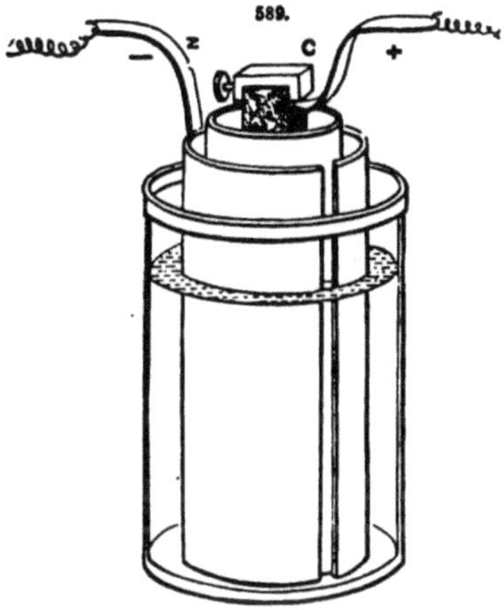

Amongst the causes which vary the intensity of the current in this battery is the change in the nature of the liquids, the acidulated water becoming charged more and more with sulphate of zinc; and the decomposition of the sulphate of copper. Experience has shown that the first circumstance does not affect the intensity of the current in any sensible degree. To remedy the second, in proportion as the sulphate of copper is decomposed, fresh portions of that salt are added; and for this purpose there is a sort of perforated flange on the upper part of the copper cylinder, which is kept filled with crystals of sulphate of copper.

Bunsen's Battery consists of a vase, Fig. 589, containing dilute sulphuric acid, in which is placed a plate of zinc Z as in Daniell's battery; but in the porous pot is poured nitric acid at 40°, and, instead of the copper, a prism of charcoal, made from the residuum taken from the retorts of gas-works, is used.

To form a battery, Fig. 590, the carbon of one element is united with the zinc of the following element, by means of clamps, the positive pole evidently corresponding to the last carbon, and the negative pole to the last zinc.

590.

In Bunsen's battery, the hydrogen liberated by the decomposition of the water decomposes in its turn the nitric acid in which the carbon is immersed; the result is that hyponitric acid is formed, which gradually dissolves, and never shows itself in the form of bubbles.

Bunsen's battery is only a modification of a battery previously invented by Grove, in which the place of the carbon was occupied by a plate of platinum; but the high price of this metal caused the use of Grove's battery to be very limited. Several inventors had thought of replacing the platinum plate by charcoal; and a few batteries of this description were in use at the time Bunsen's arrangement was introduced.

It should be noticed that a great inconvenience attends all batteries in which nitric acid is employed, owing to the diffusion of nitrous vapour which vitiates the surrounding air.

When a battery is in action, the work produced at the poles corresponds to the oxidation or consumption of the zinc, in a similar manner to the caloric engine, where the work performed is in proportion to the amount of coal consumed. If ordinary zinc is used, in batteries furnished with acidulated water, the zinc is always found attacked, whether the battery is in operation or not; but this does not occur if chemically pure zinc is used. However, there is no occasion to use pure zinc if amalgamated zinc, which is easily made, is employed.

The amalgamation of the zinc is best effected by dipping it in dilute sulphuric acid, and then rubbing it over with mercury, or by immersing it in a solution of a salt of mercury, and afterwards rubbing it briskly, when the amalgamation will be complete.

There have been numerous batteries contrived, which we do not notice, as their description would be of no interest. We only mention the Bichromate of Potash battery, which is frequently made use of in the current for the purpose of exciting various electrical machines. The element consists of a spherical-shaped bottle, with a broad neck, containing a solution of $\frac{1}{10}$ bichromate of potash, added to an equal quantity of sulphuric acid. Into the liquid is plunged a double plate of zinc, Fig. 591, in the interior of which is arranged a plate of charcoal, which answers to the positive pole. When it is desired to suspend the action of the battery, the plate of zinc is raised, so as to prevent its contact with the liquid.

The work produced by chemical action is not the only means of obtaining a galvano-electric current; the same result may be arrived at by the action of heat. This development of electricity by heat was discovered by Seebeck, of Berlin, in 1821, and is called *thermo-electricity*.

591.

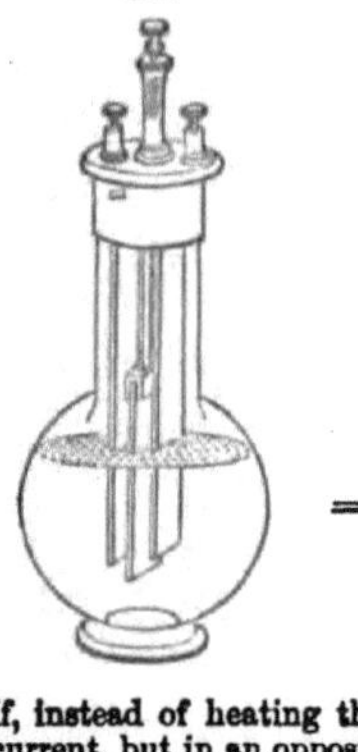

592.

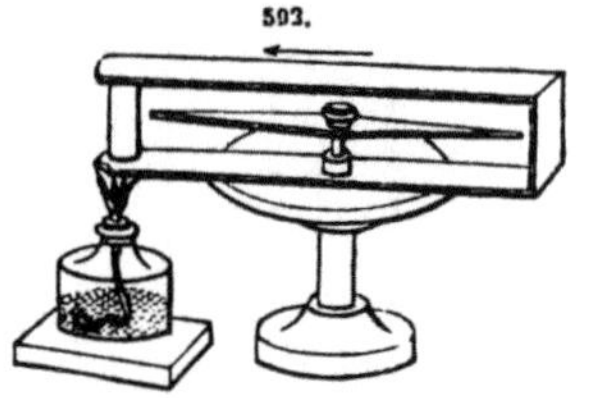

593.

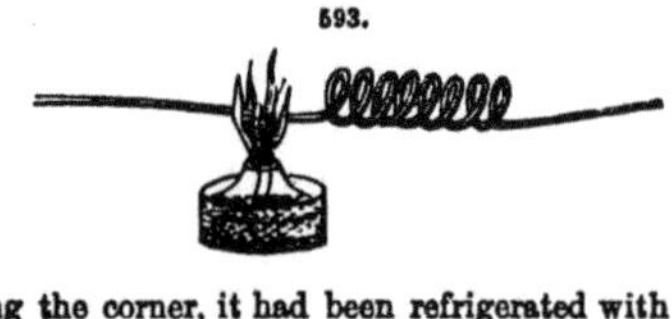

To make a thermo-electric couple, a plate of copper, Fig. 592, is bent round and soldered at its two ends to a cylinder of bismuth; we have thus a sort of rectangular figure, three sides of which are made by the plate of copper, and the fourth by the bismuth cylinder. The apparatus must be arranged in such a manner that the long sides of the rectangle are nearly in the magnetic meridian; and in the interior of the apparatus is placed a magnetic needle. If one of the compound corners of the circuit, where the two metals are joined, is heated with a spirit-lamp, Fig. 592, the needle is deflected; this denotes the production of an electric current which is directed in the copper plate from the warm corner to the cold corner. If, instead of heating the corner, it had been refrigerated with ice, it would also have produced a current, but in an opposite direction.

Although this experiment is especially sensible with bismuth and antimony, it will succeed with any of the metals. It is also not absolutely necessary to heat the point in the circuit where the two metals join, since if an elevation of temperature takes place at any point where a perfect similarity of structure does not exist, a current immediately manifests itself. This important fact is demonstrated by various experiments. Take, for example, a piece of platinum wire which has been twisted into a knot, and heat it near the twist; it will produce a current of electricity, the force of which may be observed by placing the extremities of the wires in connection with a galvanometer. The current, it will be seen, changes from one side to the other of the knot as either side is heated.

The same result will occur if a portion of the wire is coiled into a spiral, Fig. 593, and treated in the same manner.

In metals which are, similar to bismuth, not homogeneous, it is very common to observe thermo-electric currents.

It should be noticed that to a difference in structure, and to no other circumstance, as, for example, a change of dimensions, is this phenomenon attributed. When the molecular construction is the same on both sides of the heated point, there will be no current manifested. The two following experiments made by M. Magnus are decisive on this head. He reduced a copper cylinder in the middle until it was only the thickness of a fine wire; on heating the metal at the place where the sudden change in its diameter occurred, he did not observe any current, although there must have been a difference in the diffusion of the electricity from one side to the other of this point.

For his second experiment, Magnus took two tubes, A B and C D, Fig. 594, and filled them with mercury: the extremities A and D being connected with the wires of a galvanometer. The mercury contained in C was heated, and the extremity of the tube B plunged into it. In this case also no current was observed. Yet if a difference of structure is necessary to produce a current, this difference may be very insignificant. So that if the extremities of the two platinum wires were bent into hooks, and one of them after being heated laid upon the other, no effect would be produced, because the action of heat upon the platinum could not make it undergo any modifica-

594.

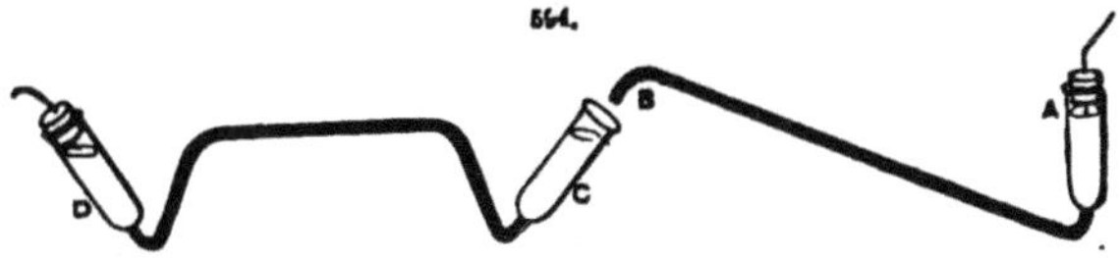

tion. But if the experiment is repeated with copper wires, a current will occur, the action of heat having quickly produced a film of oxide which has modified the constitution of the metal.

It follows from this that the action of heat has the effect of exciting in various bodies, and particularly in metals, the movement of fluids which characterize the electric current. If in a homogeneous conductor no effect is observed, it is because the two currents produced on each side of the heated point are equal and in contrary directions; but a difference of structure modifies the intensity of one of the currents, and the galvanometer shows the resultant produced from the two effects.

When the soldered joint of two different metals is operated upon, the direction of the current depends upon the nature of the metals associated, and it is impossible to give any precise rule upon this point. In the following list, the result of experiments made by Becquerel, the current traverses the heated joint, proceeding in consecutive order through the metals:—bismuth, platinum, lead, tin, copper, silver, zinc, iron, antimony.

The *intensity* of thermo-electric currents also depends upon the nature of the metals joined together; and each of these associations has an electromotive force peculiar to itself, which has been called *thermo-electric power*.

595.

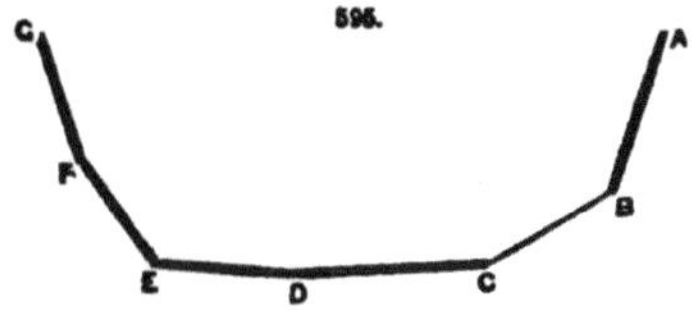

For the purpose of comparing these thermo-electric powers, Becquerel made use of a chain, A, B, C, D, E, F, G, Fig. 595, formed of several metals soldered successively one upon the end of the other. The ends of the chain being attached to a galvanometer, one of the joints was heated at a fixed temperature, 40° for instance, whilst the other joints were kept at zero.

The current produced having in each case to traverse a circle of equal resistance, its intensity may be considered as the proportional measure of the thermo-electric power of the joint, at least, from the temperature from which we operate.

It was in this manner that Becquerel obtained the following Table:—

Iron—Platinum	36·07	Copper—Platinum	8·55
Iron—Tin	31·24	Copper—Tin	3·50
Iron—Copper	27·96	Copper—Silver	2·00
Iron—Silver	26·20	Zinc—Copper	1·00

In these experiments Becquerel has proved the following fact, which is of real importance. Supposing a joint of iron and copper to be heated, we shall have a current of a certain intensity; if between the iron and copper a piece of metal is interposed, or a chain of several metals of which we heat the two end joints, we shall have exactly the same results. This is a proof that the current is really due to the difference in the propagation of heat in the metals, and not, for instance, to contact.

The intensity of the current produced by a thermo-electric couple depends on the difference of temperature of two combined joints; and, within certain limits which are variable, as well as according to the metals employed, it is sensibly proportional; but after a certain time the increase in the intensity of the current abates very considerably. Thus for a couple of iron and copper it is scarcely sensible at 300°; beyond this the intensity of the current diminishes, becomes null, and ends by changing its direction. It should be noticed that the difference of temperature is not the *only* influence on this phenomenon—the absolute temperature must also be taken into account. Thus the current has not the same intensity, one of the joints being zero and the other 20°, as if the temperatures had been at 100° and 120°.

By joining together a number of thermo-electric couples and heating simultaneously the equal joints, we obtain a Thermo-electric battery. Fig. 596 represents an apparatus of this kind invented by Pouillet. It is composed of cylinders of bismuth bent at their extremities, and connected one with the other by means of plates of copper, also bent and soldered to the bismuth. If we plunge every second joint into any cold body—melting ice, for instance—whilst the remaining joints are heated at a fixed temperature, the battery produces a current which can be collected on a conducting-wire as in an ordinary battery.

596.

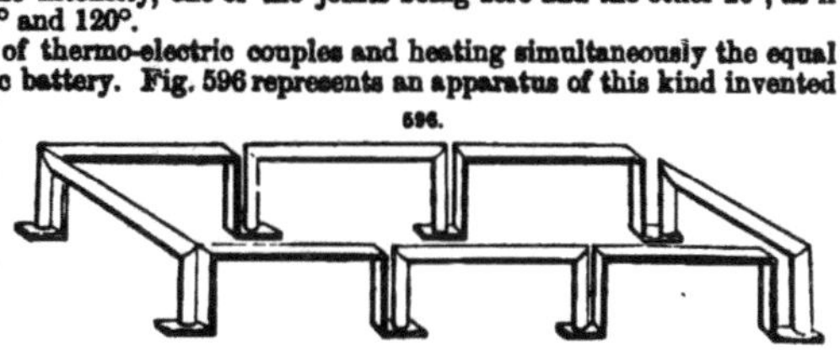

Viewed as agents for the production of the electric current, thermo-electric batteries have not as yet given very useful results as thermo-electric currents weaken rapidly, in circuits where there is little resistance, through the conductibility of the mediums in which they are generated. Thus 120 couples at least of iron and platinum are necessary to produce an appreciable decomposition of water.

Edmond Becquerel has recently introduced, however, a thermo-electric couple of exceptional power: it is formed by connecting a plate of German silver N, Fig. 597, and artificial sulphate of copper M, the positive pole corresponding to the sulphate of copper.

597.

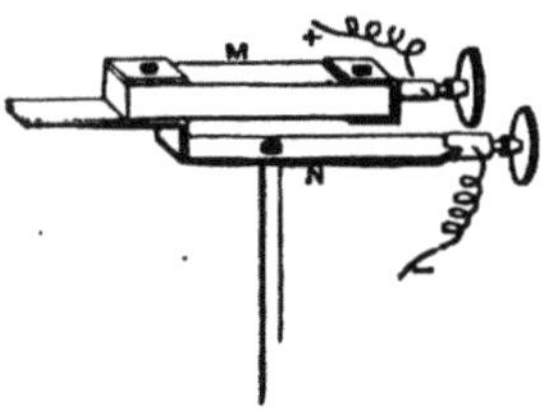

598.

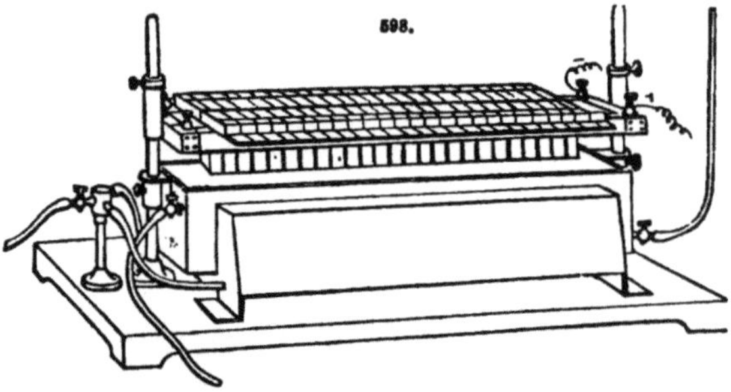

Fig. 598 shows a battery constructed upon this plan. The temperature of the joints is raised by means of common gas; and with thirty or forty elements it is capable of sensibly decomposing water, of maintaining an electro-magnet with a long coil, and of working a telegraphic apparatus.

Until lately, thermo-electric currents have only been made use of to measure temperatures under special conditions. The thermo-electric tongs invented by Peltier are, however, useful for indicating the temperature of any limited space. They consist of two thermo-electric couples, Fig. 599, of bismuth and of antimony, *a b* and *a' b'*; the bismuth of one element being united by a wire to the antimony of the other, and the circle completed by a galvanometer. It results from this arrangement that if the space comprised between the two joints *c* and *c'* becomes at all warm, a current from the bismuth to the antimony will be produced in the two couples. These two currents act, besides, upon the magnet in the same direction; and as the bulk of the instrument is small, and its calorific capacity feeble, its sensibility is very great.

599.

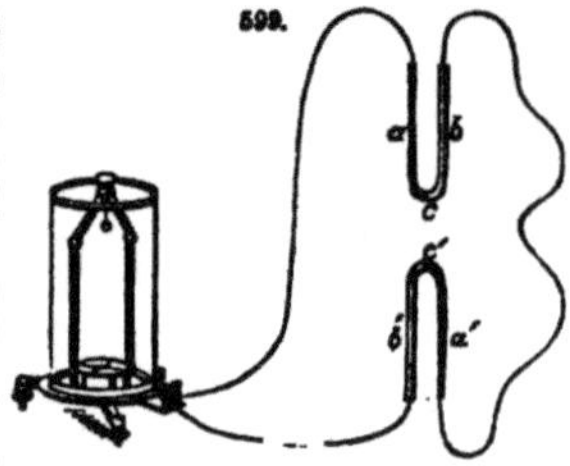

It is with a species of very fine thermo-electric plummet that physicists are able, in certain cases, to penetrate without injury those organs which they wish to have access to in order to study the temperature of organized beings, as to the variations experienced under particular circumstances.

We conclude by mentioning the apparatus employed by Becquerel to measure the temperature of the air. It consists of two thermo-electric tongs, one being placed out of doors and the other in the laboratory. A galvanometer placed in the circuit indicates zero when the two tongs are at the same temperature. It is sufficient, then, when the needle is deflected, to heat or refrigerate the second tongs in such a manner as to bring it back to zero, to tell the temperature of the air. See Boring and Blasting. Telegraphy.

BATTERY, employed to crush auriferous rock. Fr., *Moulin à pilon;* Ger., *Stampfmühle.*

The stamping mill or battery consists of a series of heavy pestles working in a rectangular mortar, each of which is alternately lifted by means of a *cam*, and subsequently let fall with its full weight upon the ore to be operated on, and of which, after being previously reduced to fragments of proper dimensions, a constant supply is introduced into the mortar, or battery-box.

When quartz-mining was first practised in California, the lifters or stems of the pestles employed were made of wood, furnished with cast-iron heads, attached by means of a wrought-iron shank driven into the lifter, and secured by two strong rectangular bands of flat iron. In most mining districts in which these wooden stems are used, the lifting of the pestle is effected by a large wooden or cast-iron drum, around the periphery of which cams are arranged in a spiral form, which, coming in contact with tongues, or tappets, fixed in the lifters, they are raised to a certain height, and, being suddenly released by the continuous motion of the axle, fall with their whole weight on whatever may happen to be beneath them. In California, however, another arrangement is employed for imparting motion to the pestles or stampers of a battery with wooden stems. Instead of a large cylindrical axle, a wrought-iron shaft is made use of, and on this are keyed a series of long curved cams, which enter mortice or slot holes, in the several stems, and cause them to be alternately lifted and released, precisely as in the case of the ordinary stamping-mill, provided with tappets and a drum-axle. When wooden stems are made use of, they are usually about six inches square, and cut out of ash, or some other hard wood, having a straight grain. These wooden stems with square heads have, however, been almost universally superseded by the rotary stamp, with a round stem of iron, to which a circular motion is given by the friction of the cam in lifting, and which, being continued up to the moment of its release, is prolonged during its descent, thus imparting a grinding action to the cylindrical head at the moment of its coming in contact with the rock to be broken.

The rotary stamp is said to be more efficient than the rectangular one, and to grind a larger quantity of rock in a given time; but however this may be, it is certain that the faces of the heads wear more evenly, and that a rotating battery requires less frequent repairs, than one made on the old principle. The battery-box is generally composed of one solid casting, and usually receives either four or five stampers; when additional reducing power is required, other similar boxes are placed on the same line. In most instances such batteries are arranged in sets of five stampers in each mortar, two of which are placed side by side in the same framing, ten stampers being thus set in motion by one shaft. Two five-stamp batteries, of a construction frequently employed, are represented, Fig. 600 being a back elevation, and Fig. 601 a transverse section; the iron rods A are the stamp stems, B the shaft, and C the cams. This shaft is provided at one end with a large pulley D, which is generally constructed of either kiln-dried wood on arms, inserted in a cast-iron boss, and then turned off in place, or is built solid of well-seasoned planks on a bored boss, and, as in the other case, turned, after being keyed to the shaft. When several of these batteries are arranged in one house, the motive power is communicated, by means of a broad belt, to the intermediate shaft B′, which is fitted with pulleys corresponding to those on the shafts B, with which they are severally connected by belts. These belts, which are manufactured out of a combination of canvas and india-rubber, are, from the first motion to the intermediate shaft, sometimes as much as 2 ft. in width. The belts from the second motion to the shaft on which the cams are keyed, are made of a thinner material, and are from 1 ft. to 14 in. wide. The lift of the stampers varies from 9 to 12 in., but 10 in. may be considered as about the average; and their weight, including the iron stem, varies from 550 to 900 lbs. The order in which the several stampers, included in one box, strike their blows, in a five-stamp battery, is not always the same in all establishments, but in most instances the first blow is struck by the central stamp. This is followed by the outside one to the right, then by the second to the left, afterwards by the second to the right, and finally by the stamper on the extreme left of the series. The number of blows struck by each stamper is from sixty to eighty per minute. The first portion of the stamper *a* is sometimes cast on to the stem, but more frequently it is fastened by wedges, and has a round aperture, in which is inserted the spill of the shoe *a′* firmly driven in or fastened by dry wooden wedges, which expand on coming in contact with water, and hold it securely in its place. The battery-box is either of iron, cast in one piece, or its bottom alone may be of cast iron, and its sides of wood; in which case the lower portion of it, together with the inside of the feed-hopper, must be lined with sheet iron, $\frac{1}{8}$ in. in thickness, fastened by $\frac{5}{8}$th bolts. Immediately under each of the stampers is placed a short cylinder of cast iron *a″*, which is retained in its position, either by fitting into a circular bedding, in which it may be keyed by wooden wedges, or it is provided with a square flange, which, coming in contact with those of the other dies and the sides of the box, act as distance-pieces, by which it is kept in its proper position. These, and the shoes of the stampers, are, when worn out, readily replaced by new, a considerable economy of time and money being the result; the parts worn out are merely coarse castings of chilled iron, without any kind of fitting. The hole *x* is for the purpose of forcing in a drift above the spill of the shoe, and thus removing it when a new one is required, whilst the hole *x′* is employed in the same way for getting off the boss from the stem. In Grass Valley, and some of the other more important mining districts, the boxes E are, almost without exception, composed of single iron castings; but in localities where amalgamation is conducted in the battery itself, the sides and ends are sometimes of wood, the bed alone being made of iron; and when this method of construction is adopted, two plates of amalgamated copper, $\frac{1}{8}$th of an inch in thickness, are often bolted at *b* on either side of the row of stampers. The rock to be crushed is introduced by a shovel at *c*, and a plate of perforated sheet iron, fastened either in a wooden frame, or retained in its place between the two rectangular iron bands, tightened by cotters, is introduced before the opening *d*. The battery-bed, whether entirely of iron, or consisting only of an iron bottom, with lined wooden sides, is firmly bolted to a block of wood, at least 2 ft. square, and of which the dimensions, when very heavy stampers are employed, are even much greater. This either forms, as in the drawing,

600

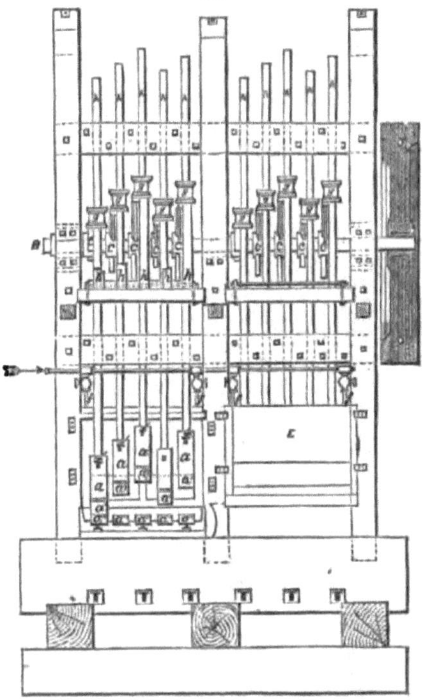

a portion of the general framing of the arrangement, or is now more frequently, to prevent jarring, made quite independent of it. It is, however, essential that this portion of the structure should be well bedded on a solid foundation, and, if possible, rest directly on the bed rock. Occasionally

601.

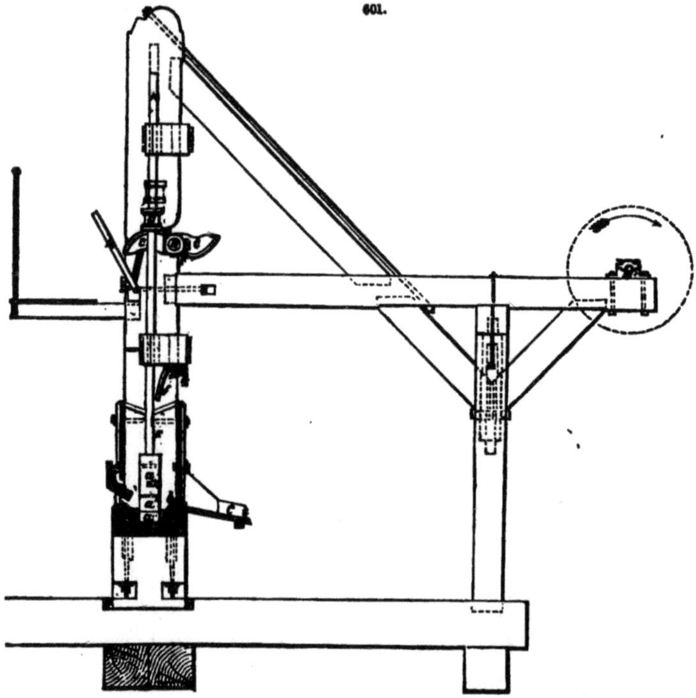

Reference.—*A*, Stamp lifters or stems. *a*, Stamp-heads. *a'*, Stamp-shoes. *a''*, Stamp-dies. *B*, Cam-shaft. *B'*, Second-motion shaft. *b*, Amalgamated copper plates. *C*, Cams. *c*, Feed-hopper. *D*, Driving-pulley. *d*, Grating. *E*, Battery boxes or coffers. *e*, *f*, Water-pipes. *G*, Blanket-board. *g*, Bosses by which stampers are lifted. *h*, Props for stampers. *x*, *x'*, Drift-holes.

quartz is crushed dry, but much more frequently water is admitted, and for this purpose a gas-pipe *e* affords the necessary supply, which enters the boxes through the branch pipes *f*, fitted with cocks for regulating the quantity introduced. The studs *g*, against which the cams come in contact, and by which the stampers are raised, are fitted on the iron stems by means of keys, which admit of their positions being readily shifted, when rendered necessary by the wearing of the shoes.

The props *h* shown in one of the batteries, but omitted in the other to avoid complication, are used for keeping up the stampers, either when the battery-box is being cleaned, or when a portion of the machinery is thrown out of action for the purpose of repair; to do this they are successively pushed forward, so as to catch beneath the several bosses *g*, when lifted by the cams to their full height. When not so employed, the props are allowed to fall back out of the way of the stems, as shown in the sectional drawing. The size of the apertures in the gratings or sieves at *d*, differ in accordance with the fineness of the gold contained in the rock treated, and is also, to a certain extent, varied in conformity with the particular views of the superintendent of the mill on that subject; but it is evident that with very small apertures the amount of quartz crushed, all other conditions being equal, will be less considerable than when a coarse screen is employed. The size of grating commonly made use of in some of the best mills in the Grass Valley district is shown, Fig. 602, which is known as No. 8.

In order to combine strength with the largest possible open surface, the apertures are sometimes made of an oblong form, and arranged as in Fig. 603.

In some establishments these gratings are fixed perpendicularly, as seen in Figs. 600, 601; but more generally they are slightly inclined outward, and this arrangement is evidently attended with certain advantages. When the grating is placed perpendicularly, a particle of quartz or other pulverized matter, splashed against the screen by the fall of the stampers, can only pass through it in case of being projected directly through one of the openings; and should it strike against a portion of the solid plate between the holes, will run down with the water on the inside, and again settle in the battery-box. If, on the contrary, the grating be placed at a considerable inclination outward, a particle of pulverized rock, which has not been projected immediately through the grate, may, on running back with the water over its inner surface, pass through one of the apertures and escape into the trough on the outside of the battery.

602.

Stamp Grate, with Round Holes.
(Full Size of Apertures.)

603.

Stamp Grate, with Oblong Holes.
(Full Size of Apertures.)

In all machines of this description, it is of importance that each particle of the rock operated upon should escape from the action of the stampers as soon as it has become sufficiently reduced in size, and with this view the grate-surface is in the Californian mills extended as much as possible, being generally made of nearly the full length of the battery-box. With a view of supporting the grating, and protecting it against injury from the impact of the water dashed against it by the falling stampers, the sheet-iron plate is externally strengthened by the application against it of some thin iron bars.

When the high cast-iron mortar is made use of, which is now generally the case, it has the form represented, Figs. 604 and 605, which the first is a transverse section, and the second a front elevation; the dies are fitted on the bottom A, and the quartz fed through the opening B, whilst the screens are fastened, by nails or screws, to a frame which is firmly secured in grooves provided for its reception at the ends of the mortar, and by two lugs at the bottom of the opening C.

604. 605.

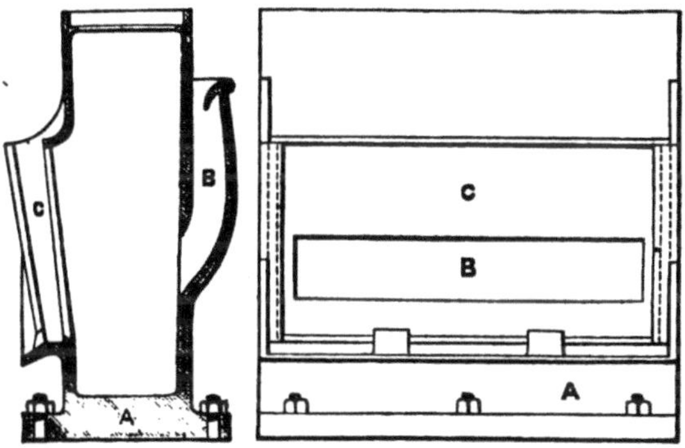

Iron Battery-box.

606.

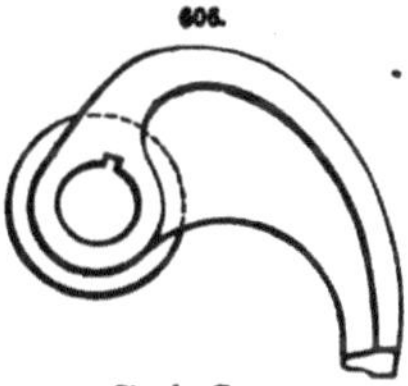

Single Cam.

In some cases, instead of employing a double cam, as seen in Fig. 601, a single one is made use of. This has generally the form shown, Fig. 606, and possesses the advantage of allowing the axle to be placed nearer the stamp stem than it can with any other cam, and also that by its use a greater number of blows can be struck per minute without danger of breakage.

The auriferous material having been reduced to the state of a finely-divided powder, it becomes necessary to provide means for the concentration and separation of the gold, which is more or less perfectly effected by an almost infinite number of different contrivances, varying slightly in their details in almost every establishment that may be visited. However much the processes employed may differ in this respect, only two decidedly distinct systems are now practically in use in California, namely, amalgamation in the battery, and crushing without the use of mercury, amalgamation being subsequently effected by means of appliances specially designed for that purpose.

Amalgamation in the Battery.—When this method is adopted, the batteries are often provided with amalgamated copper plates *b*, Fig. 601, about 5 in. in width, extending the whole length of the box; one on the feed side and the other at the discharge, the latter being protected by the sheet-iron lining of the feed-hopper, and each having an inclination of from 40° to 45° towards the stampers.

When these are not employed, spaces for the accumulation of amalgam are allowed between the dies and the sides of the box, and vertical iron bars are placed inside the gratings, between which the hard amalgam is found to collect. The copper plates are covered with mercury, by means of a rag dipped in dilute nitric acid, with which quicksilver is rubbed over the surfaces to be coated, in the same way as on those used in ordinary sluices. Quicksilver is also sprinkled into the boxes, by the feeder, at intervals of about an hour, and in quantities varying with the richness of the rock operated on. One ounce of gold requires for its collection about an ounce of mercury;

but when the gold is in a finely-divided state, the addition of another quarter of an ounce of this metal is thought advantageous. The proper proportion is, however, readily ascertained by closely watching the discharge. If any particles of amalgam which may pass through be hard and dry, a little more quicksilver must be introduced; but if, on the contrary, they be soft and pasty, or globules of mercury make their appearance, the supply in the battery must be diminished.

When the proportion of mercury has been properly adjusted, the amalgamation of the gold is very completely effected, except in cases in which the ores contain large quantities of lead or antimony, and have been previously burned for the purpose of expelling their more volatile constituents, by which treatment the particles of gold often become coated in such a way as to interfere with their combination with mercury. When the proper proportion of quicksilver has been regularly introduced, and the rock contains coarse gold, from 60 to 80 per cent. of the gold saved is caught in the battery; but when, as in the case of some of the ores of Nevada, the gold is in a very finely-divided state, and is associated with ores of silver and other sulphides, the results obtained are less satisfactory. The alloy resulting from the treatment of such ores contains silver, and in some cases affords from 300 to 400 thousandths only of gold, often producing a spongy amalgam of a dark colour, made up of an aggregation of numerous finely-divided particles. Küstel is of opinion that this effect is partially due to the presence of manganese, but it appears difficult to understand how this substance should influence, to any important extent, the combination of gold and mercury. This amalgam is exceedingly light, and is, therefore, difficult to collect, either by riffles, copper plates, blankets, or any of the other appliances commonly employed for the purpose.

When, therefore, ores contain much lead or antimony, amalgamation in the battery is not to be recommended, since this spongy amalgam is more difficult to retain than the most finely-divided gold, and a large proportion of it floats off over the blankets, riffles, or copper plates, which may be arranged for the purpose of arresting its progress. There is, besides, no evidence that battery amalgamation possesses, under any circumstances, a decided advantage—for gold ores not associated with sulphide of silver—over stamping without the use of mercury, and in some of the most productive gold districts it is seldom resorted to.

In order to collect the particles of gold and amalgam escaping from the battery-box, various ingenious contrivances are employed; but as these arrangements differ but little in their details, whether mercury be employed in the battery, or the amalgamation entirely effected after the escape of the pulverized material through the screens, we will proceed to describe the system generally in use in the northern quartz mines, in which the various arrangements are of the most improved description.

This article, with some slight alterations being made, is taken from J. A. Phillips' excellent work, 'The Mining and Metallurgy of Gold and Silver.'

BATTERY, SHIP'S. FR., *Batterie flottante;* GER., *Schwimmende Batterie;* ITAL., *Batteria galleggiante;* SPAN., *Bateria.*

Any place where cannon or mortars are mounted, for attacking an enemy or for battering a fortification, is termed a battery. This term also signifies a body of cannon for use, taken collectively; as, to discharge the whole battery at once; a ship's battery.

Cannon were introduced on board ships soon after these weapons came into general use, and they were at first applied on the non-recoil system in wooden immovable carriages; but as the size increased, it was found indispensable to allow of, but to absorb, the recoil, and to provide means of training or pointing the gun irrespective of the movements of the ship. Thus the trunnions of the gun for elevation and depression when the level was changed by the "heel" or inclination of the ship, the "trucks" or wheels on which the carriage and gun moved backwards when fired or were run "up" or "out" when loaded after firing, became necessary fittings. Then there was added, still further to meet more onerous conditions, "breechin" ropes to limit the recoil, and tackles to run the gun out. The battery was that part of the ship where the heaviest artillery was carried; smaller pieces were distributed about the poop and forecastle; and the ship was divided into two strong sections, called close quarters, forward and aft, where, in the event of boarding being successful, the crew could retire and fight, as in a castle or tower, although the enemy occupied fully the midship deck.

Modern science has recognized two broad divisions of the battery—broadside and turreted guns. In the turret system a few guns of great power are carried in the centre line of the ship, and are so fitted as to command a fire all round, where there are, as in a raft, no impediments from masts, ropes, or sails, and not much lateral motion, as pitching or rolling.

Cowper Coles, in Europe, and J. Ericsson, in America, are the chief exponents of this system. There is a subdivision of turrets into fixed and revolving. The fixed turret approaches nearly to the box-battery; the revolving turret approaches most to the single pivot-gun. Examples of each are given. See ORDNANCE. With a given tonnage, more guns of equal power can be carried on the broadside than on the turret system, since that necessitates two separate armours—the one for the gun, the other for the ship; and it is not a sufficient answer that less of the ship's side has to be armoured, and therefore less weight need be carried; since, if the same number of guns only are insisted on, no more armour is necessary in the one case than the other. In the latest development of the two principles, from twelve to sixteen guns of equal weight and power can be carried in the broadside-ship, as against four to six in the turret, on the same tonnage and at the same expense. Added to this, there is the fact that no turret-ship has yet been devised that was equal as a habitation for seamen, and therefore, in seaworthy qualities, to the broadside-ships. For coast and river defence the turret ships or rafts are eminently useful; for keeping the sea in all weathers they are utterly inefficient.

The forms of carriage for the batteries of ships have undergone a considerable change within a few years, mainly consequent on the introduction of much heavier pieces of ordnance up to 20 tons, where previously 5-ton guns were the exception, and the fact that armour outside the ship renders it possible to apply more complicated and delicate mechanism without the fear of having it too much exposed to the effects of shot and shell.

Simple as the invention of a suitable gun-carriage for heavy naval ordnance may appear, yet, in attempting this apparently easy task, before R. A. Scott succeeded in accomplishing it, many ingenious inventors failed. Captain Scott found that it was of much importance to keep heavy guns as close down to the deck as possible, to give more effect to his compressors, and not, as Scott and others suppose, that the shock is greatly diminished by the deck by having the gun so placed. The principles upon which the compressors are constructed, and those by which they are operated, were first brought into practical operation by J. Ericsson (see GUN-CARRIAGE). However, at present it is only necessary for us to show the value of Scott's gun-carriage in the formation of a ship's battery.

Fig. 607 is an elevation, and Fig. 608 is a deck-plan, of one of those carriages, constructed for one of the 10-in. broadside-guns of H.M.S. 'Hercules.' The carriage first proposed by Scott, which we gave in our specimen Part, differs in arrangement, but not in principle, from the carriage shown in Figs. 607, 608.

607.

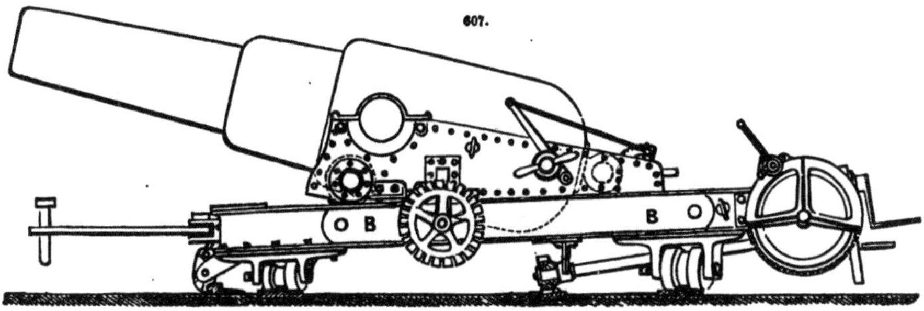

608.

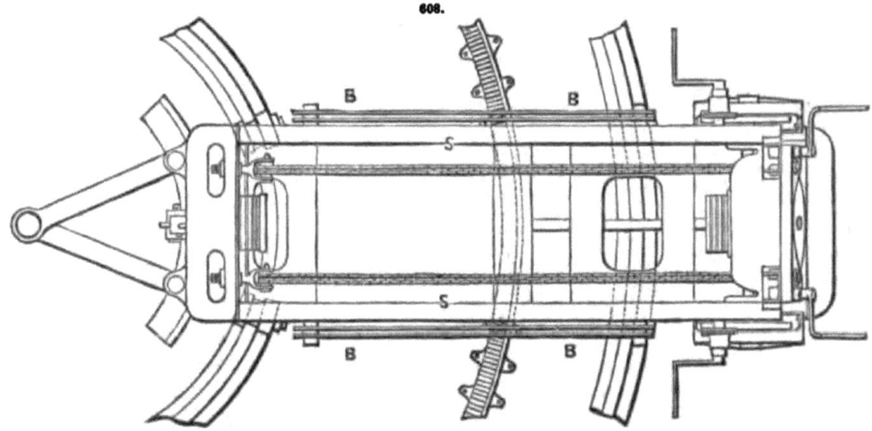

The sides S, S, Fig. 608, of the slides are of -section iron, a portion of which is hollow, and sometimes filled in with wood, which projects slightly beyond the top and bottom flanges. On the outside of the slide on each side there are supported by pins a pair of iron bars B, B, Figs. 607, 608; these bars are placed at a short distance apart, and grasped by the compressors at the same time as the sides of the slides are grasped: the parts secured by the compressors are shown in section, Fig. 609.

609.

Although Scott has succeeded in effecting the object he had in view, yet it does not appear that he is aware that he, or rather Ericsson, has introduced a mechanical principle entirely new, upon which shocks and concussions may be given without returning any great amount of rebound. Each part of his invention shows much practical skill and ingenuity; yet the whole would fail but for the symmetrical and uniform grips taken by his compressors. In fact, he can secure composite beams of wood or iron so that their vibrations become as regular as the strings of a violin: the concussions of the vibrating waves of the compound beams B S B destroy a force which in other cases produce rebounds. The friction of which Scott speaks has not the effect which he attributes to it. If the beams B S B were made of iron, they would not answer the purpose as well as wood: the wood receives all the force, and passes it forward. Wood does not offer a greater resistance than iron, but the wood passes the force applied over the parts equally; whereas the iron passes the force with a rebound or shock. The one receives it, and, to use an ordinary phrase, swallows it, and passes it more leisurely to other parts; iron would receive the force, and return it suddenly, which might cause mischief.

It should not be forgotten that a vessel of 30 tons can easily carry, and may with impunity make use of, a gun of 5 tons weight, since the vessel in such a case becomes the gun-carriage floating in water, and capable of absorbing the recoil, as a larger vessel will not do. The 'Staunch' is an example of this class of gunboat, and in her the gun, when not in use, is lowered, so as not unnecessarily to keep the centre of gravity high when not fighting. Any modification of the Moncrieff principle applied to gunboats or even vessels of larger size, would supersede turrets by the adoption of this alternate raising and lowering of the gun itself, since no armour is needed for the protection of the gun during its short exposure in pointing and firing. The broadside battery has in no great degree changed since the system was first introduced, unless by the providing more and more efficient bow and stern fire, or, as it is called, *all-round fire*, by splaying the ports, by retiring them or placing them in indents (this system having the same objection as deep embrasures, in that the enemy's shot are guided into the ship), or again, and most successfully, by better means of training on axes or turn-tables, and by muzzle-pivoting. In great naval battles, the broadside has always a marked advantage, since no such action has been fought without a necessity for engaging on both sides; and here number of guns must tell, and the turret system, which admits of but few guns, must necessarily be at fault.

A battery is said to be concentrated when a certain number of the guns are so trained and laid, as that, on being simultaneously fired, their shot will strike at nearly the same moment in nearly the same spot at a previously determined range.

This is done by measuring the distances between the centres of each port; and, after deciding the desired range to be, say 200 yards, calculating each angle of training, and inserting a mark on the deck. Now, at this range there will be no elevation. Each gun is trained by the nail on the deck; and a good marksman being posted at the gun, on which the whole battery or a part of it is concentrated, all the guns are fired at his word of command, and their shot must inevitably strike wherever his does. As the range is a fixed quantity, and forms one side of the triangle; as the angle of training of the gun chosen to concentrate on is either 90° with the line of keel for the midship gun, or about 45° for a bow or stern gun; all the other intermediate angles are easily calculated from these by the diminution of base due to the distances between the centres of gun-ports. 800, 600, and 200 yards are the usual distances for concentration

610.

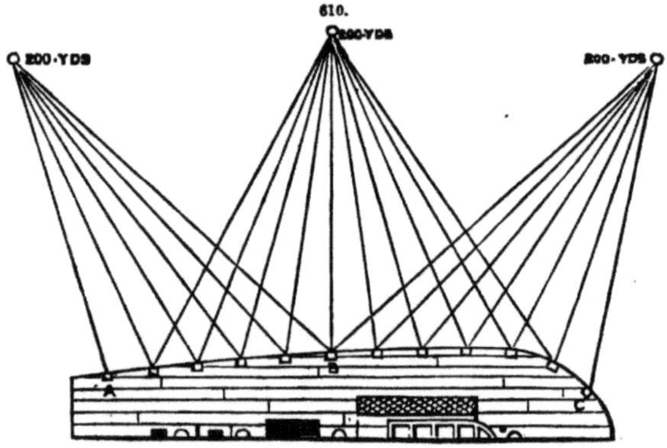

Guns concentrated on A, the after-gun; B, the midship; and C, the foremost gun.

The general arrangement and details adopted by Capt. Coles, R.N., in the construction of the cupola or turret batteries on board the 'Royal Sovereign' are shown in Figs. 611, 612.

The 'Royal Sovereign' carries four turrets, mounting in all five 300-lb. Armstrong guns, two in the forward and one in each of the after turrets. We take the forward or twin-gun turret, which is 22 ft. 9 in. in diameter, and 18 ft. 9 in. outside, Fig. 611. First, we have a massive framework of wood constructed on the main deck, and supported by the deck-beams and wrought-iron columns. The centre is formed of two large rectangular blocks of English oak, making a square of 6 ft. 3 in. × 2 ft. 6 in. deep, with a round hole in the centre 26 in. in diameter. Immediately under this is placed two balks of English oak, 18 in. × 15 in. × 30 ft. in length, running fore and aft, and bolted down to the deck-beams; on this, segments of English oak are placed, cut to an inner radius of 9 ft., and forming a ring of 9 ft. 6 in. outside diameter, which is firmly bolted to the fore-and-aft beams just mentioned; round the outside of this ring three bands of American oak are bent, each 12 in. deep × 3 in. in thickness, and bolted to the segment forming the inner ring by 1-in. bolts. Six arms or spokes radiating from the centre, each 18 in. × 12 in., like the centre and inner ring, are made of English oak; on this substantial and massive framework is bolted a turned cast-iron roller-path. In the centre is a hollow tube of wrought iron, 2 ft. 2 in. outside diameter, 3 in. in thickness, and 7 ft. 3 in. in length, forming the pivot on which the turret revolves, and acts as a safe communication with the magazine below. A large casting, which forms the centre, round which a live wrought-iron ring, lined with brass, revolves, is placed round this pivot, and sits upon the wooden framework bolted through to a somewhat similar casting below, which is supported by a forked wrought-iron column resting on the keelson. We are now come to the turn-table or platform on which the turret is built. It is a large disc of woodwork, 24 ft. 8 in. in diameter, and 12 in. in thickness, built of oak slabs, 14 in. × 6 in., bolted together, the top layer being placed at right angles to the bottom one. Near the outside the thickness is

611.

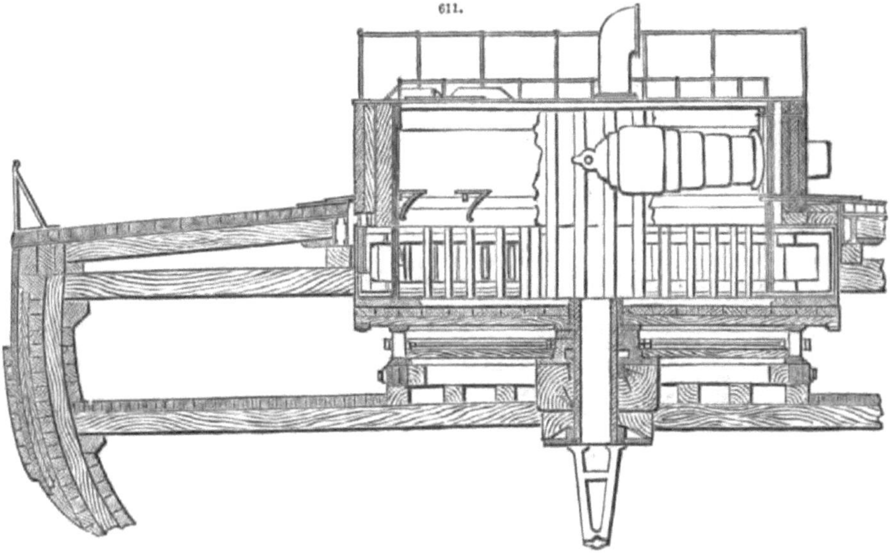

increased 4 in. by a circle of oak, 3 ft. 9 in. in width × 6 in. in thickness, let into the other portion 2 in. On the under-side of this platform is fixed the cast-iron upper rolling-path, fastened by 1-in. bolts, which pass through the frame to rings of wrought iron, 9 in. × ⅝ in., let into the top, to which they are secured by nuts. In the centre of this table is fixed a large kind of angle-iron ring of cast iron, with the flange on the bottom side, and bolted through in the same manner as the path. This casting is bored out, and a brass brush let in, ⅝ in. in thickness × 15 in. in depth, which forms the moving rubber surface round the axes. Between this casting and the lower one are arranged twelve brass conical rollers, 5½ in. in diameter at their largest part, and 5½ in. wide, supporting the centre weight of the turret. These rollers are placed in a live ring of brass; the pins round which they revolve are 1¼ in. in diameter, and screwed into the inner part of the brass ring with a jamb-nut screwed up to prevent the pin turning. A ¼-in. washer is then put on, and segments of wrought iron, 4 in. × ⅝ in., are fastened over the whole thing by 1-in. square-headed wood screws; the outside nuts of the roller-pins are tightened against these segments. The centre line through the inner rollers is 5 in. above the centre line through the outer ones, which are thirty-six in number, made of cast iron and turned conical, the largest diameter being 18 in. × 9 in. wide; they are cast H sections, the boss is bored, and a brass brush fitted. The framework in which these wheels revolve is made of an inner and outer ring of wrought iron, each 6 in. × ⅝ in. The inside diameter of the inner ring is 19 ft. 2 in., and the outside diameter of the outer ring is 21 ft. 3½ in. Like the outside ring of the centre rolling-frame, these rings are made in short segments with joint-plates on the inside, each fastened by four ⅝-in. bolts, the two outside ones being long enough to pass through both rings, with a nut at either end; round these long bolts is a piece of 1½ in. gun-barrel tube, which acts as a stay or distance-piece, to which the rings are screwed home. There are thirty-six segments to each ring. The radius-rods are 2 in. in diameter,

612.

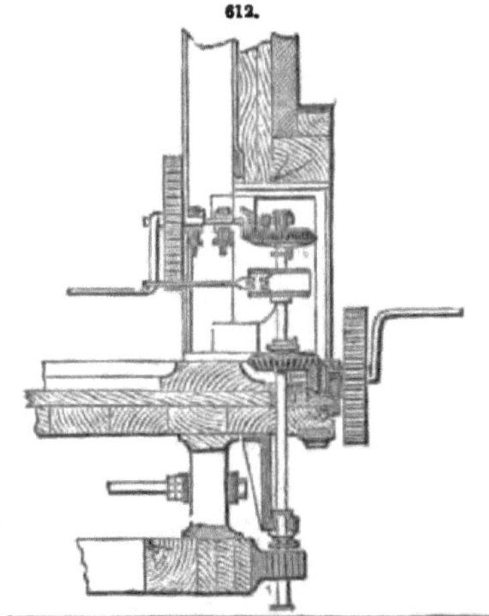

increased to 2½ in. at the ends. The inner end is screwed into the live ring before described with a jamb-nut on the outside; the opposite end passes through the roller and large rings, with two jamb-nuts on the inside, and one nut, with a pin through it, on the outside.

On this turn-table is built the turret, a little eccentric to the platform, that is to say, the centre of the turret is 9 in. from the centre of the platform, and is formed of thirty upright frames, made of ┬-irons, angle-irons, and plates. The ┬-irons which form the inside circle are 6 in. × 10 in. × ½ in. The plates on the outside are 10 in. × ½ in., with an angle-iron riveted on either side, 3 in. × 3 in. × ½ in. The ┬-irons extend from the platform to the top of the turret, a height of 9 ft. 6 in., whereas the ½-in. plates and angle-irons only extend 3 ft. 7 in. in height, and are then brought round to form the support for the woodwork and armour-plating. Between the vertical ┬-irons, and beginning at 3 ft. 7 in. from the platform, are nicely-fitted balks of teak, 18 in. × 10 in. × 6 ft. in height; round this, and let in flush with the outside of the teak, are systems of diagonal bracing or trellis-work, with one hoop top and bottom, 6 in. × ¾ in., which is also the scantling of the diagonal bars. Over this is built another ring of teak balks, 18 in. × 10 in. × 6 ft., and on the outside of this is the armour-plating, consisting of twenty plates, 5 ft. 4 in. × 3 ft. 6 in. × 5½ in. in thickness, secured to the framework of wood and iron by four 2-in. countersunk galvanized bolts to each plate, passing through to the inner lining, consisting of ½-in. plate-irons. In addition to this lining there are two wrought-iron rings in the inside, through which the bolts pass, and to which they are secured by an ordinary nut. The upper hoop is 14 in. deep × 2 in. in thickness, and the lower one 6 in. wide × ¾ in. in thickness. There is a double thickness of armour-plating, 12 ft. long, on the port-hole side of the turret, the inner plate being 4½ in. in thickness, making a total thickness of 10 in. solid plating. The roof is formed of wrought-iron rolled beams, something like a double-headed rail, except that the top flange is an inch wider than the bottom one, the top being 3¾ in., and the bottom 2¾ in. These beams are 4½ in. deep, weighing 15 lbs. to the foot-run, or 45 lbs. to the yard. They are placed 2 in. apart, and pass through the woodwork resting on the armour-plates; these are covered with 1-in. wrought-iron plates, bolted to them. There are also side-laps about every 3 ft., which are bolted to the armour-plates. There is an opening above each gun, 20 in. in diameter and 9 in. deep, packed round with wood, covered with plating; these are merely light-holes. There are also two slits over each gun, 4 ft. long by about 8 in. wide, for the "captain" of the gun to take his aim through. There are two cowl-pipes for ventilating each turret. The short standards with hand-rails are used for stowing the hammocks in, while the long ones merely serve as a hand-rail for the look-out. The port-holes are 20 in. × 3 ft., half round top and bottom, and lined in the inside with iron, ¾ in. in thickness.

We now come to the turning-gear, which consist of four winch-handles and gearing, working two vertical shafts, one on each side of the cupola, 2½ in. in diameter. Two of these handles work inside the turret, the other two on the outside. As they are all four alike, we need describe only one, which consists of a common winch-handle, Fig. 612, 15 in. radius, attached to a spur-pinion, 10 in. in diameter, 1½ in. pitch of teeth, 3¾ in. breadth of same, twenty-one in number. This pinion works into a spur-wheel, 30·4 in. in diameter, 1½ in. pitch of teeth, 3¾ in. breadth of same, sixty-three in number. To this is attached a bevil-pinion, 6 in. × 1¾ in. × 4 in. × 11 teeth, working into a bevil-wheel, 18 in. × 1¾ in. × 4 in. × 33 teeth. This wheel gives motion to the upright shaft, to the end of which is fixed a spur-pinion, 10 in. × 2½ in. × 6 in. × 14 teeth, working into the large wheel bolted to the framework, the size of which is 22 ft. × 2½ in. × 6 in. × 371 teeth.

The mechanical advantages gained by the gearing up to the periphery of the pinion which works into the rack = 27 to 1.

If we suppose the whole weight of the cupola and guns, 140 tons, to be concentrated on a circle of 15 ft. in diameter, we have: $\frac{30 \times 30 \cdot 4 \times 18 \times 264}{10 \times 6 \times 10 \times 180} = 40 \cdot 1$ to 1.

Eight men are the full complement for working the gearing, four inside and four outside the turret, the speed = 1 revolution a minute. Four men can work it, but, of course, they take double the time.

In addition to the four winches just described, there is what is termed the deck turning-gear, consisting of two winch-handles, each 15 in. radius, working a bevil-pinion 11·1 in. × 1¾ in. × 5 in. × 20 teeth, working into a bevil-wheel, 32·25 in. × 1¾ in. × 5 in. × 58 teeth. At the other end of the shaft to which this wheel is attached is a face-pinion, 10 in. × 2½ × 6 in. × 14 teeth, working into face-rack on turret, 24 ft. 6 in. × 2½ in. × 6 in. × 412 teeth, the mechanical advantage gained by this gearing from the winch-handle up to the periphery of the pinion which works into the rack = 8·7 to 1. But assuming, as before, that the weight is concentrated in a circle 15 ft. diameter, we have—

$$\frac{30 \times 32 \cdot 25 \times 294}{11 \cdot 1 \times 10 \times 180} = 14 \cdot 2 \text{ to } 1.$$

There are three small crab-winches attached to the revolving platform, two of which are used for working the guns, and one for lifting the shot.

There is a clear space of 3 in. all round the turret and the well or hole in the deck, and this space is covered by a leather flap or ring attached to the turret, and bearing on a brass ring fixed upon the deck; this leather flap is weighted with a strip of iron round the outside, in order that it may fit close to the deck.

The edge of the hole in the deck, within which the turret is placed, is strengthened by a wrought-iron well-ring, weighing about 1½ ton, and formed of 1-in. plate of wrought-iron, 2 ft. deep, with angle-irons on the outside, top and bottom. From this ring radiate iron beams, some leading transversely to the sides of the vessel, and the others to the deck-beams, to which they are firmly attached. On these beams are laid the 1-in. plates forming the deck, and upon these is placed around the turret a "glacis-plate," consisting of a ring 30 in. wide, and tapering in thickness from 3 in. at the inner to nothing at its outer edge.

Over the glacis-plate comes the deck, tapering in the opposite direction from 6 in. to nothing, and upon this is laid round the turret a ring of $\frac{3}{4}$-in. stamped wrought-iron plates, 2 ft. 6 in. wide, which serves to protect the deck from being scorched by the discharge of the gun.

The deck of the vessel rises at an angle of 5° from the sides towards the centre, and is formed of 1-in. wrought-iron plates, covered down the centre for a width of 26 ft. with 8-in. teak planking; the remainder of the surface is covered with 6-in. planking. The deck is carried by wrought-iron rolled beams placed 12 in. apart between the main wooden deck-beams, and the 1-in. plating is laid on in strips about 12 ft. long by 2 ft. 6 in. wide, joined together by $\frac{3}{4}$-in. rivets, and covering strips 4 in. × 1 in. The 1-in. plating is doubled round the openings for the turrets, hatchways, and funnel; and a double thickness is also carried, fore and aft, between the turrets and the sides of the ship. The armour-plates are bedded upon crossed diagonal planking, which was added to the original sides of the vessel on her conversion from a three-decker, making the total thickness of the timber-backing 3 ft., as shown in Fig. 611.

It may be necessary to add that the turrets are now made concentric in place of eccentric to the platform; that, in lieu of the small winches before mentioned, they put a much larger one, with an endless chain, to work the guns. See a paper by J. Baily, published in the 'Artizan,' 1868.

The ship turret-battery of Ericsson is shown in Fig. 613; the pilot-house is stationary and above the rotating turret; this arrangement enables the commander to direct the helmsman, who

613.

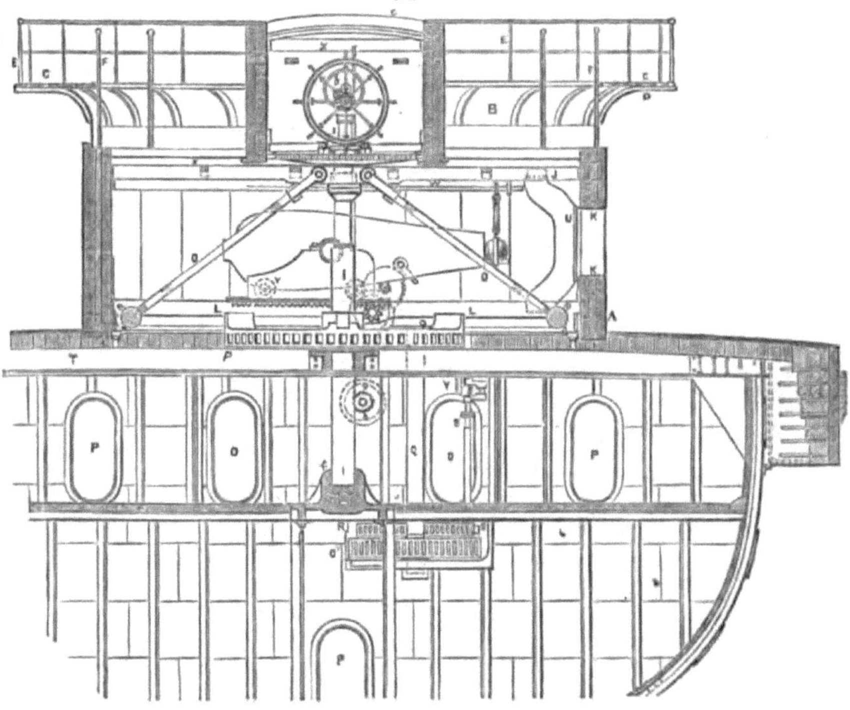

is near him, and the gunner below, while he, the commander, is observing the enemy. The turret A is 24 ft. inside diameter, 9 ft. 6 in. high, and 15 in. thick, composed of two separate cylinders formed of plates 1 in. thick, lapped and riveted together. The outer cylinder, composed of six plates, was built on a staging above the inner one of four plates, and after completion was slipped over it. The annular space of $5\frac{1}{2}$ in. between the cylinders is filled with segmental slabs, 5 in. thick, made of the best malleable iron. These slabs were made only $11\frac{1}{2}$ in. wide, in order to save time. B is an extension attached to the top of the turret, composed of plate $\frac{1}{2}$ in. thick, bent outward in the form of a trumpet, in order to throw off the sea in bad weather; C is a wooden grating extending round the turret extension, supported by and bolted to brackets D, at intervals of 3 ft.; E, stanchions for supporting a rope-rail carried round the wooden grating; F, stanchions for sustaining an awning in fine weather; G, the pilot-house, 8 ft. inside diameter, 7 ft. high, and 12 in. thick, formed of two separate cylinders, each composed of six plates 1 in. thick, lapped and riveted. After completion, the larger cylinder was forced over the smaller one. The roof is composed of three plates, 1 in. thick, covered with 8 in. thickness of wood, and an outer plating 1 in. thick. The roof is inserted below the top of the cylinders, and is thus protected from shot; both cylinders are pierced with eight elongated sight-holes. The weight of the pilot-house is supported by a broad wrought-iron cross-piece H, secured to the circumference of the house by angle-irons, this cross-piece resting on a collar near the upper end of the stationary vertical pillar I, round which the turret turns. The floor of the pilot-house consists of wooden gratings provided with grated hatches moving on hinges. J is the upper turret-beam of wrought iron, 11 in. deep, 8 in. thick in the middle, sustaining rafters which support a series of bars K, 4 in. deep, 3 in. thick, placed 2½ in. apart, on the top of which are placed perforated plates 1 in. thick covering the entire turret. L, L, are the gun-slides, four in number, 10 in. deep by 4 in. thick, of wrought iron, the ends of which rest on plate-rings secured to the turret, their middle resting on the lower turret-beam M. N is a cross-piece of wrought iron, suspended under and let into the central pair of gun-slides, and into which are tapped the diagonal braces O. By means of these braces O, the entire weight of the turret may be suspended on the collar *n'* of the turret-shaft. P is a spur-wheel bolted to the under-side of the gun-slides and lower turret-beam, and worked by a pinion Q on the vertical shaft Q', at the lower end of which is the spur-wheel Q'', driven by a pinion on the axle R, and on which is fixed the spur-wheel R'. This spur-wheel is in turn worked by the pinion S, on the lower end of the shaft S', which is provided with a crank S'', and turned by an engine consisting of two cylinders 12 in. in diameter and 16 in. stroke, placed at right angles, and bolted to the under-side of the deck-beams T. U is a port-stopper for preventing shot from entering the turret, composed of a massive block of wrought iron bent in the form of a crank, provided with bearings, in which it may be turned into such a position as to admit of the gun being rolled out, while, when turned into another position, it closes the port. V is the gun-carriage; V' a rack on the under-side of the same; V'' a pinion for moving the gun on the slides, and by means of which, combined with a friction-coupling, the recoil is checked; W is a radial sliding-bar for passing the shot into the muzzle of the gun without handling; X is the steering-wheel, and Y the double barrel for the steering-chains Y', which run upon rollers under the deck-beams. *a* is a bar of steel, 2½ in. square, inserted in a groove formed on the port side of the turret-shaft, provided with cogs on the opposite sides, by raising or lowering which, by means of a train of wheels in the steering-box *b*, the chain-barrel is worked through the pinion *c*; *e*, fore and aft bulkheads to support the weight of the turret and turret-shaft, strengthened with angle-irons, 6 in. by 4½ in., placed 18 in. apart, upon which rests the cast-iron saddle *f*; *g*, a key for regulating the height of the turret-shaft, above and below which are plates *g'* and *g''* of composition metal to prevent cutting; *h* is a cast-iron bearing for giving lateral support to the turret-shaft, into both of which square keys are let in, to prevent the shaft from turning; *i* is a wrought-iron plate, polished on the top side, secured to the deck; a corresponding plate of composition metal, with a projection on the inside, being fitted under the base of the turret, with which it revolves. This composition plate does not extend under the outside cylinder of the turret, which is supported by being bolted to the base ring *k*, as shown in the engraving: underneath this part of the turret the space is filled with oakum or similar material. Scupper-holes *l* are provided for carrying off any water which may enter the turret between the base-plates in a sea-way. Plates *m* are riveted to the four inside courses for sustaining the upward pressure of the gun-slides when the diagonal braces are screwed up, composition rings *n* being fitted between the upper turret-beam and the collar *n'* of the turret-shaft, nearly the entire weight resting upon this ring when the turret-shaft is fully keyed up. Doors *o* in the upper transverse turret-bulkhead afford communication between the berth-deck and the turret-chamber, the doors *p* forming communication to the after-part of the ship, and doors *q* in the lower bulkheads to the boiler-room and coal-bunkers.

Of the Power of Symmetrical Turret-fire, taken from 'Text-book to the Turret and Tripod Systems of Ship's Battery,' Cowper P. Coles, from plans by E. Pellew Halsted. Figs. 614, 615.

Fig. 614 is a deck-plan for a 7-turret first-rate vessel of war; and Fig. 615 is a midship section of a 7-turret first-rate vessel of war.

Mode of Armament.—Seven turrets or cupolas, with two guns each, so arranged that—1, the fire of four guns can be delivered in line of keel ahead and astern; 2, the central turrets, and very largely the deck itself, are protected from all raking fire; 3, the deck can be swept fore and aft, to prevent possibility of boarding.

COMMAND OF TURRETS: DETAILS FOR REFERENCE TO FIG. 614.

	With both Guns.	*Additional with Single Guns.*
1.	From 85° abaft port beam, to 50° abaft starboard ditto = 315°, or 28 points.	2° aft on port side; 6° on starboard ditto = 8°, or ¾ of a point.
2.	From 50° afore, to 61° abaft port beam = 111°. Then, from 2° on port side of keel forward, to 81° abaft starboard beam = 173°. Total 284°, or 25¼ points, nearly.	6° forward, and 6° aft on both sides = 24°, or 2¼ points, nearly.
3.	From 75° afore, to 71° abaft port beam = 146°. Then, from 60° afore, to 71° abaft starboard beam = 131°. Total 277°, or 24¾ points, nearly.	2° forward, and 6° aft on port side; and 7° forward, and 6° aft on starboard side = 21°, or 2 points, nearly.
4.	From 71° afore, to 71° abaft port beam = 142°. Then, from 71° afore, to 71° abaft starboard beam = 142°. Total 284°, or 25¼ points, nearly.	6° forward and aft on both sides = 24°, or 2¼ points, nearly.

5.

From 71° afore, to 60° abaft port beam = 131°. Then, from 71° afore, to 75 abaft starboard beam = 146°. Total 277°, or 24½ points, nearly.

6° forward, and 7° aft on port side; and 6° forward, and 2° aft on starboard side = 21°, or 2 points, nearly.

6.

From 81° afore port beam, to 2° on starboard side of keel aft = 173°. Then, from 61° afore, to 50° abaft starboard beam = 111°. Total 284°, or 25¼ points, nearly.

6° forward, and 6° aft on both sides = 24°, or 2¼ points, nearly.

7.

From 50° afore port beam, to 85° afore starboard beam = 315°, or 28 points.

6° forward on port side, 2° on starboard side = 8°, or ¾ of a point.

SUMMARY.

With both Guns.			Additional with Single Guns.		
No. of Turret.	Degrees.	Points.	No. of Turret.	Degrees.	Points.
1	315	28	1	8	¾
2	284	25¼	2	24	2⅛
3	277	24½	3	21	2
4	284	25¼	4	24	2⅛
5	277	24½	5	21	2
6	284	25¼	6	24	2⅛
7	315	28	7	8	¾
Mean Command ..	290° 51′	25⅞	Additional Mean Command ..	18° 35′	1⅛

This power of mean command exists when the mutual obstruction unavoidably presented by the turrets themselves is alone taken into account. But even if 50° be subtracted from it as a mean deduction for the further unavoidable obstructions presented by masts, hatchways, and the like, the remaining 240°, or more than 21 points of the compass, as a mean command for each turret, gives, says Halsted, a training-power unapproachable by any other system. The shaded spaces between the turrets, however, show the neutral surfaces, *untraversed by any fire*, to be sufficiently extensive to accommodate many of such further obstructions without any sacrifice of training whatever.

Concentration and Direction of Fire.—The whole fourteen guns, Fig. 614, can concentrate on points in direct line on either beam, 100 ft. distant from the guns of the central turret, No. 4, and can train from thence against ships or batteries, throughout an arc of 50° afore and abaft.

The four guns of 1, 2, and of 7, 6, can simultaneously concentrate on points, in direct line of keel forward and aft, 100 ft. distant from stem and stern. If engaged only forward, or only aft, this fourfold line-of-keel fire can be supplemented; if forward, by the alternate single guns of 6, 7; and, if aft, by those of 1, 2: all four of which command a line of fire of 87°, or only 3° from line of keel, forward and aft, on either side respectively. Whether chasing or chased, a fire of six guns out of fourteen, practically in line of keel, can thus be maintained.

To illustrate the bow and quarter fire:—If a radius of 300 ft. be struck from the centre of the middle cupola, it will describe two arcs practically equidistant from both bows and both quarters. And if an angle of 17°, or 1½ point, be measured from the same centre and extended on each side of the keel forward and aft, it will fix two points in each arc, 100 ft. distant, practically, from the nearest part of each bow and each quarter of the ship, as shown in plate. While the four guns of 1, 2, are still engaged in line of keel forward, six other guns can concentrate on the above point on the port bow; namely, the two guns each from 3, 6, and the single port guns from 4, 5, respectively. So, at the same time, the four remaining guns can concentrate on the other like point on the starboard bow; namely, the two guns of 7, and the single starboard guns of 4, 5; which two turrets can thus ply their single guns alternately on each of the above bow-points. But it is further obvious, on reference to Fig. 613, that Turret 1 can concentrate, or alternate, the fire of both its guns, as required, against either the starboard or port bow-point; while Turret 2 equally commands the starboard point; and the defence of the ship, in single or in general action, can thus be maintained with her whole fourteen guns—*as with an end-on broadside*—throughout an arc of 3 points, or only 17° divergent on each side of her line of keel forward. And similarly complete and symmetrical is her means of defence aft. Both guns of Turret 1, with the starboard single ones of 3, 4, can concentrate a fourfold fire on the point on the port quarter; while a sixfold fire pours on the point on the starboard quarter from the two guns each of 2, 5, backed by the single port guns of 3, 4; the whole four guns of 7, 6, being still engaged, if needs be, in line of keel aft, or both those of 7 firing, either upon the port or starboard quarter-point, and both those of 6 pouring their fire upon the port point.

The power of concentration, says Halsted, is as simple as it is perfect. It exists from the moment the turrets are "clear for action," without any complicated preparation or combination between them, and is carried out by simple direction for all or any turrets to direct their fire on any specified point of an enemy's hull within their command of training.

Securing the Turrets.—All means of securing the turrets in stress of weather are purposely omitted from the sectional design, Fig. 615:—1, in order not to complicate the drawing itself; 2, in order to keep the subject free for special study when actual construction shall be undertaken. But the following means have been fully considered, and are regarded by ablest mechanicians as ample for every practical purpose, though not incapable, of course, of further improvement.

To prevent all Lateral Motion.—Four or more powerful horizontal clamps strongly secured to under-side of upper deck, working with competent screws, and clasping or unclasping from opposite points the external circumference of turret.

To prevent all Vertical Motion.—Four or more powerful vertical or oblique chains or rods, hinged, as shown in Fig. 613, or otherwise removable, strongly binding down the turret to the stronger parts of the main deck around the turret-bed.

Security in Action.—The flanges, with the angle of cone given to the twenty-four steel rollers, and their steel roller-path—upon which, when in action, the turret revolves—supply ample means for the true and safe working of the turret in every state of weather when its guns can at all be used. But these means are further assisted by centering the floor of the turret down to the central pivot of the working shaft on the lower deck, by a system of diagonal trusses which connect the two: this arrangement fully providing against any other motion by the turret than its intended horizontal rotation.

On Upper Deck.—A clearance of 3 in. surrounds the turret on the upper deck, rendering highly improbable any permanent choking of its action by any effects of shot. But the opening on that deck is surrounded by a strong box-girder, furnished with competent rollers, by which to meet any possible circumstances making it desirable for the turret to be supported at that point. A flexible flange or valve, as in the American monitors, keeps that opening water-tight, and any possible leakage will find its ready exit to the sea through the main-deck scuppers.

The following main details are, however, freely stated:—

Construction.—Hull.—The construction of the hull is on the same principle of combined longitudinal and transverse frames as in the 'Achilles,' 'Agincourt,' &c., and other British iron-clads of war; with an inner and outer skin-plating, as introduced by Mr. Scott Russell in the 'Great Eastern' steam-ship.

The extreme ends are made cellular and water-tight, with the same preparation for ramming as in the above-named ships. The whole of the intermediate space between the inner and outer skin-platings is subdivided into sectional water-tight compartments, arranged to be used for water-ballast, to compensate, when necessary, for consumption of fuel, and so on, and for maintaining the ship at her proper trim.

Defensive Powers.—Armour and backing has been considered in its compound character, and not simply in reference to the outer armour-plates. Throughout all classes and rates, as the guns are the same, so the armour and backing are of one character and thickness. The outer armour is 6 in. thick; next, teak 11 in. thick; next, 9 in. depth of Mr. John Hughes's hollow metal backing of ⅝-in. iron; the bars in contact, running longitudinally from end to end of the

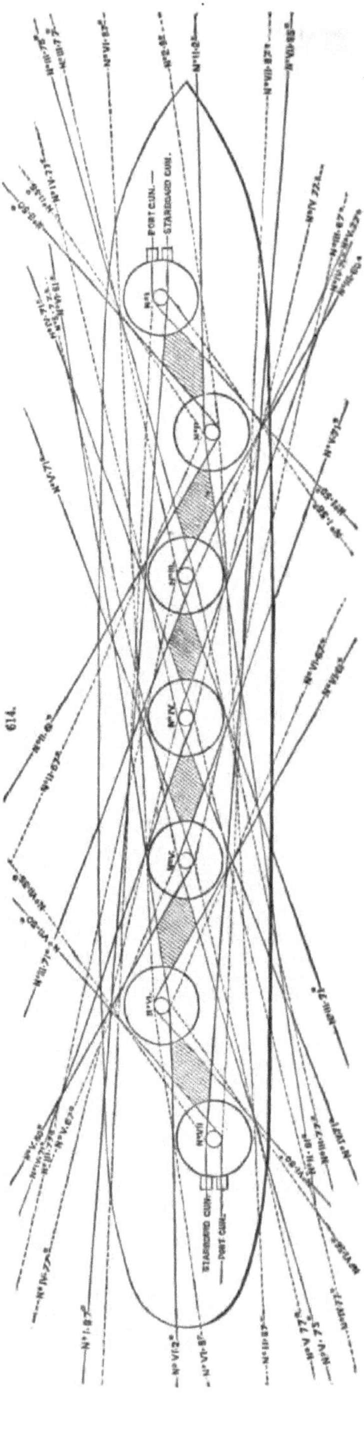

614.

ship, and securely riveted to the $\frac{3}{4}$-in. skin, as well as to the frames; thus combining with, and giving great strength and rigidity to, the whole structure. As in other irod-clads, the armour and teak-backing taper towards the ends, but not the Hughes' backing. The weight per square foot of the combined resisting media is, maximum 533 lbs., minimum 444 lbs., mean 488$\frac{1}{2}$ lbs. The resistance of the metal-backing is estimated as equal to 3-in. plating.

615.

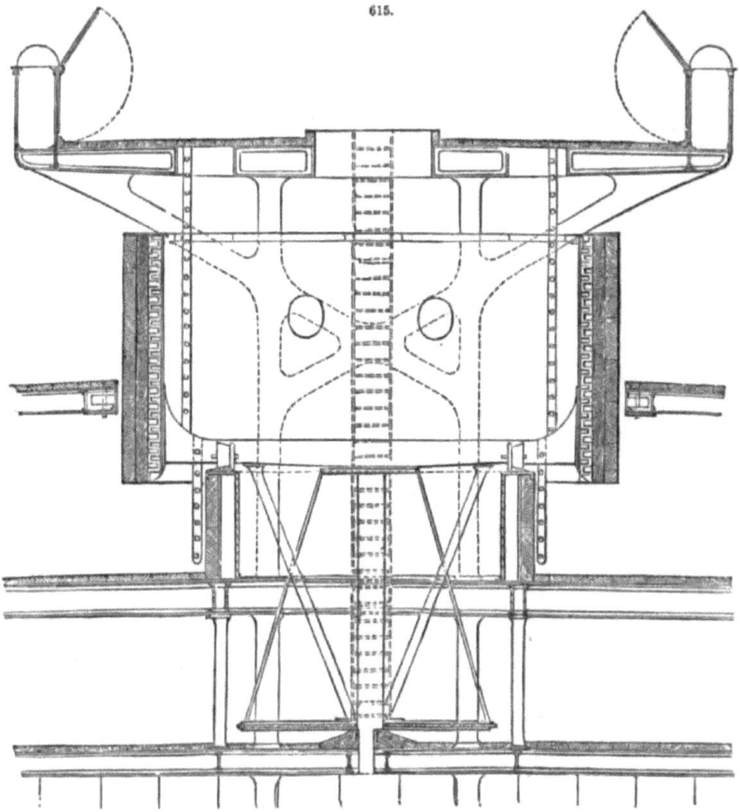

Main Deck, in all classes, is level with the top of armour, and laid with 1$\frac{1}{2}$-in. iron, covered with 6 in. of teak. The height out of water is, in ships of the line, 5 ft.; frigates, 4 ft.; corvettes, 3 ft. The only ship armour above main-deck level is that around the rudder and its working gear. The main-deck hatchways are limited to such as are requisite for provisioning, and so on; the turrets themselves constituting capacious hatchways for all other purposes. As with the upper decks of the American monitors, all main-deck hatchways are fitted with high coamings, and 1$\frac{1}{2}$-in. water-tight iron hatches, hinged, and always in place. This admits of the between-decks being flooded without the water finding its way below. Provision is made for getting rid of the water, in such case, by a large water-port beneath each main-deck gun-port on the broadside. The sides of the main deck are of $\frac{3}{4}$-plates unprotected, the armour being limited to the vital body of the ships, and all connected with the turrets. These sides are consequently liable to be riddled with shot-holes, through which the water might afterwards flood the main deck in heavy weather. In describing the turret, it will be shown that in such event complete access, as before, may still be maintained with the lower deck and all below it, even with 3 feet water over the main deck. The lower decks throughout are laid with $\frac{1}{2}$-in. plates and 4-in. teak over; the upper decks with $\frac{1}{2}$-in. plates and 4-in. teak. The lower decks are 7 ft. beneath the beams; the main decks 8 ft.; except in the corvettes and ocean-despatch, where both are 7 ft.

Spar Deck.—This most important feature in aid of the whole undertaking is adopted from the rudimental spar decks, connecting turret with turret, in the American monitors. In its application here, as a war arrangement, the leading idea has been this,—the upper deck proper being regarded as if it were the main deck of an ordinary frigate. In such case the beams of the overhead deck would ordinarily receive end support from the walls or sides of the ship, as pierced with ports for broadside-fire. But these beams would receive central supports also from two rows of stanchions, not fixed, but hinged, and removable as required. Now, the support of the spar deck represents, as it were, those two rows of central movable stanchions, converted into a system of fixed central support; the main-deck sides or walls of the ship being then altogether removed, so as to give free scope to a central, rotary, all-round turret-fire, substituted for that of the broadside. But the strength of that central system of trussed diagonal support for the spar deck enables it to fulfil also every office of a complete working upper deck, for boats, capstans, rigging, working ship, and the like, whether the turret guns beneath it be silent or in action. In all bad weather, and especially when steaming head to wind and sea, it thus constitutes a practical freeboard of ample height to secure all comfort and safety. The spar deck is constructed as a girder. Its lower convex skin is of $\frac{3}{16}$ steel plates, laid perfectly smooth; its upper convex is of $\frac{1}{2}$-in. iron, covered with 3-in. teak. Its edge is stiffened and strengthened by a continuous box-girder of $\frac{1}{2}$-in. plates, 4 ft. deep and 2 ft. wide, constituting the hammock-netting.

Resistance of Spar Deck to Explosion of Turret Guns below.—As two features of great novelty and importance are here involved, it is felt necessary thus to notice this subject. The question is obviously one of due strength. An experience of over three years, says Halsted, writing in 1867, has told us that decks of wood, with or without an underlay of iron plating, suffer no injury, beyond being blackened, by charges varying from 35 lbs. to 45 lbs. fired even less than one foot above their surface: and the nearest point of the under-side of any of these spar decks to the bore of the gun beneath it is 5 ft. 10 in. Again, long experience has determined that the angle of so-called explosive force diverges 45° from the circumference of the bore, in equal effect all round, but diminishing in effect directly as the distance. Thus the point where the explosion from the nearest muzzle will impinge on the nearest under-side of the spar deck above it will be increased to 8 ft. 9 in., presenting a smooth surface of steel. And if its strength be but equally resistant as that layer of a deck of wood, 1 ft. 4 in. distant, as in the 'Royal Sovereign,' which lies nearest to a muzzle above it, and suffices to resist all injury; it is clear no apprehension for these spar decks need be entertained. But if indeed it be found less than that wooden standard of sufficiency, then a further layer of iron plate, even to an inch in thickness, over the arc of impingement is easily added.

Figs. 613 and 615 illustrate other details of construction not here described.

Sea-going turret-ships, by Admiral Paris, Figs. 616, 617, 618, 619; taken from the 'Artizan' for April, 1869.

The subject of sea-going turret vessels, which is one of vast importance to this country, has just now acquired still greater interest from the fact lately disclosed by Mr. Childers, the present chief of the Admiralty, that our Government have at last determined to entirely leave off building wall-sided iron-clads, and construct only turret ships.

Few, if any, authorities upon this subject stand so deservedly high as Admiral Paris, who has devoted so many years to the study of its merits, and we have therefore given the following explanation of his views as communicated by him.

Since the first appearance of the 'Gloire' as a sea-going iron-clad, but few changes have arisen in the general features and form of vessels of this type. The only alteration of a notable character consists in the adoption of a *central battery*, which system has been rendered necessary, because, by reason of the increasing penetrative power of shot, it has been found practically impossible to carry armour of sufficient thickness over the whole length of wall-sided ships to afford them ample protection. One consequence of the introduction of *central batteries* has been a reduction in the number of guns carried by a vessel; thus ships of the magnitude of the 'Bellerophon' and 'Hercules' now carry only four or five guns on each side.

It is evident, however, that these guns will be more effectively used in a turret than in a central battery; and this is now beyond dispute, the doubts militating against the turret system having been cleared away by the trials of monitors in active warfare, and by their sea voyages. Also, as all iron-clads have been built of the same shape, that is to say, with vertical sides and rounded bottom, they commonly roll heavily, shipping water through their ports and over their bulwarks, and preventing the use of those guns which in the old ships of war would have probably been brought into action: to this defect must be added the danger arising from the rolling of the vessel, which tends to expose the weak or unarmoured parts of the vessel every few seconds. Hence at the present time it appears that, in the ordinary system of wall-sided iron-clads, the limited range of guns, the low height of their elevation above the surface of the water, and the rolling of the ships, are defects gravely impairing their other qualifications, which have been so greatly enhanced by the skill of many of our modern engineers.

There appeared, however, in the United States of America, a peculiar description of vessel, which was only designed for navigating rivers, but was found to possess a remarkable stability in a sea-way, only rolling about one-third as much as vessels of the ordinary build. The natural conclusion from this almost unexpected result is, let us adopt similar shaped vessels; or, in other words, let us build monitors. But if this form were adopted throughout the entire navy of any country, she would be unable to carry on a war in open sea in bad weather, as it is acknowledged to be unsuitable for a heavy sea. The problem, therefore, is how to obtain the advantages of stability and extensive range of fire as possessed by the American monitors, in vessels capable of navigating the ocean in all weathers. This problem Admiral Paris endeavoured to solve, after having had the advantage of studying the various models in the Paris Exposition, and having reliable information respecting the monitors, together with long experience at sea. To design a vessel which should be as free from rolling as a monitor; as seaworthy, and with as good accommodation, as vessels of the usual form; and carrying her guns a sufficient height out of the water, with an all-round fire, was his object.

As regards stability, or freedom from rolling, it is well known that a boat-hook floating in the water, and kept right by the weight of the iron end, is least affected by the waves. Again, a

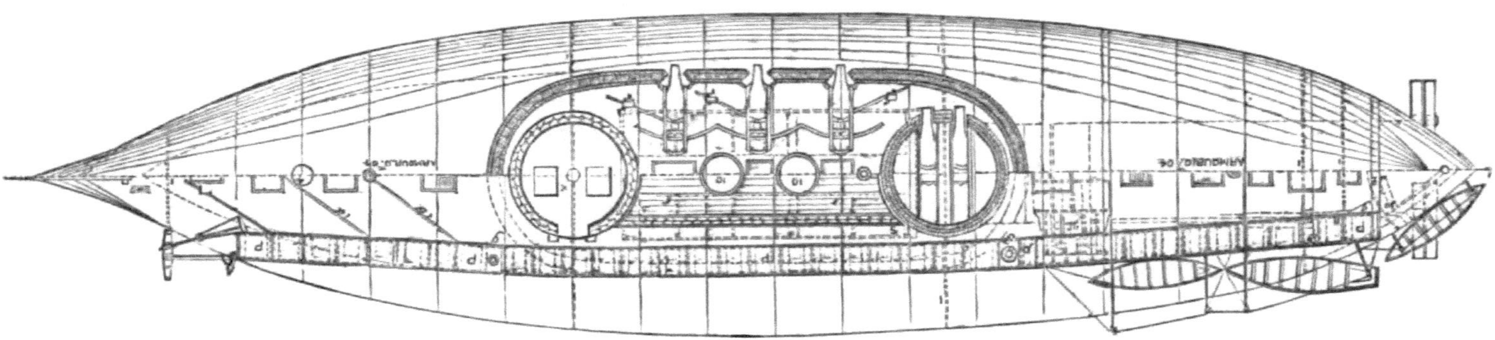

617.

616.

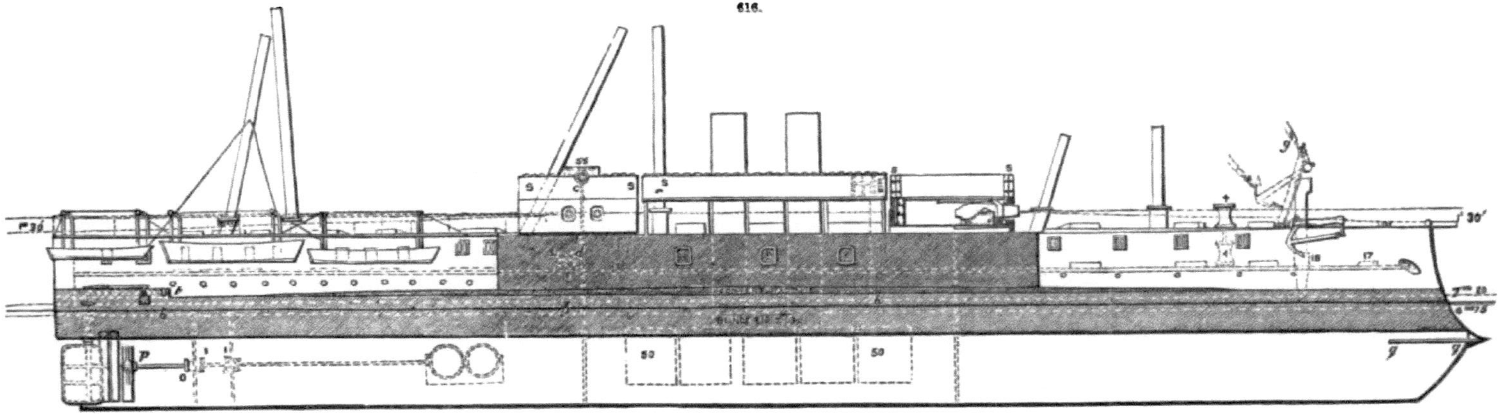

plank weighted on the bottom side will not roll; while a body in the form of a sphere, or a cylinder, has no stability of its own, and requires a large amount of ballast to keep it upright. The contrast between the motion of different shapes of buoys in a heavy sea, is a simple illustration of varying stability. A body of almost equal specific gravity to, and which consequently has its upper side level with, the water, such as a boat filled with water, or a piece of ice, moves with the waves, but only so far. The first of these forms would, of course, be impossible, as the vessel would be impracticably deep; besides which, a complete immobility is undesirable. The merits of the circular form are about to be proved by the Winan's yacht, and presents many difficulties; while the monitors have proved, on a large scale, the influence of waves upon a body of a specific gravity nearly equal to water.

618.

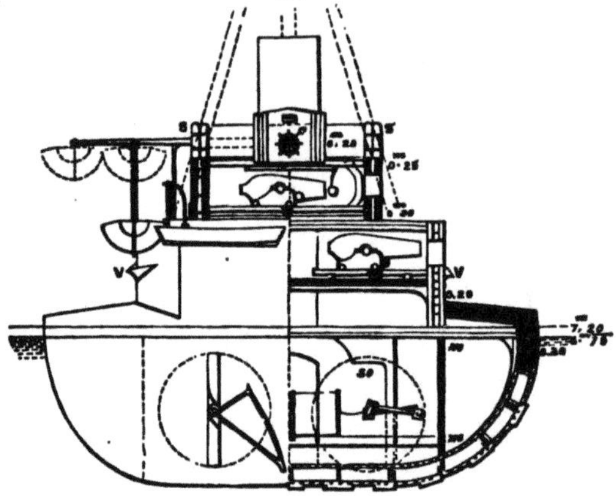

619.

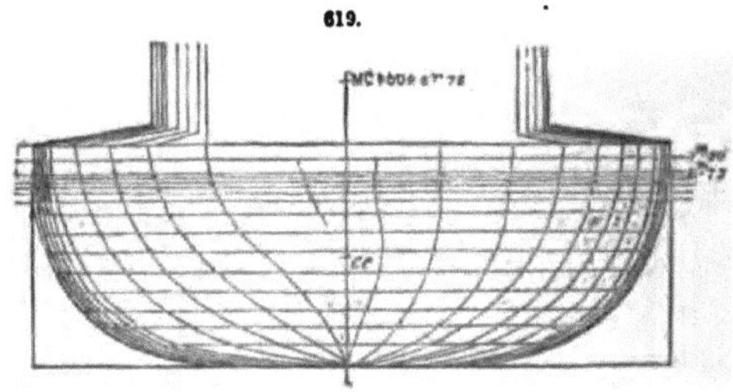

These three illustrations, applied to the usual construction of a vessel, which may be described as a semi-cylinder, with a parallelopipedon over it, show that it is the form of maximum rolling, especially if weight is fixed outside to add to the inertia. It is well known that a raft rolls less than a boat; while a boat-hook, floating close to them, is quite motionless, for which reason it has been employed as a wave-tracer for showing the rolling of ships and the motion of waves, and has been found to act very perfectly. It is, moreover, necessary to rely upon the well-established fact of the non-rolling of the monitors, as the excessive motion of iron-clad vessels have fully proved our ignorance of the laws which govern the motion of ships, because if any theory upon this point really existed, it would be unpardonable not to have applied it.

For this reason, when it was first proposed to construct iron-clad vessels, Admiral Paris proposed that a model should be built—say about 130 ft. long—with iron frames and wood planking fastened with screw bolts, thus making a composite vessel. With such a vessel experiments might have been tried with weights in every possible position; and after having experimented sufficiently, the boat might be hauled ashore, the planks removed, and the frames put in the furnace and afterwards bent, so as to form a vessel of some different shape upon which to try similar experiments. By these means a considerable amount of practical knowledge could have been obtained as to the best form of vessel for war purposes.

It has been already mentioned, the monitor is the best form of vessel for stability, and is also perfectly adapted to the smooth surface of rivers, where the ship's company have frequent opportunities to breathe pure air ashore, though they cannot be considered as adapted for long voyages at sea. The first monitor was sunk in weather which permitted another ship to use her boats; while the 'Weehawkee' and the 'Affondatore' suffered the same fate, in consequence of opening the hatches to breathe a little air after a battle.

The point, therefore, is to attain as near as possible to the non-rolling quality of the monitors, and avoid their unseaworthiness, which the Admiral endeavoured to accomplish in the following manner. It will be seen from Figs. 616, 617, 618, 619, that the midship section of the proposed vessels are formed of parts of circles for the sides with a flat bottom, the sides rising but a short distance above the water-line, and covered with a deck for a certain distance inwards: when the sides again rise to a sufficient height to admit of ports being placed well above water, this portion being long and narrow, somewhat similar to a river steamer in shape. Upon reference to the cross-sections, Figs. 617, 618, it will at once be observed that, with the exception of the comparatively narrow raised portion, the vessel is similar in section to a monitor, and consequently possesses the similar properties of stability. This element being the most important point, it was necessary to study it carefully, and, therefore, Paris calculated the positions of centre of *carène* (that portion of a vessel which is below the water-line, when floating in any position, whether upright or inclined) and the metacentre, both when the vessel was upright and also when heeled over at various angles up to 20°, from which the following Table was obtained.

Inclination.	Draught to lowest point of immersion.	Displacement of *carène*, P.	Distance of centre of *carène*: To water-line.	Distance of centre of *carène*: To longitudinal plan.	Height of metacentre over centre of *carène*.	Height of centre of gravity to water-line, + over − under.	Height of centre of gravity over centre of *carène*, a.	Value of $\rho - a$.	Moment of stability, M.	Momentum of sails, M'.	Value of $\frac{M}{M'}$.
	m.	tons.	m.	m.	m.			m.			
Vessel upright	6·750	8368	2·875	0·0	5·85	− 0·58	2·30	3·55	29716	72766	0·408
Inclined 5° ..	6·765	"	2·90	0·42	4·05	− 0·60	2·31	1·74	14560	..	0·209
" 10° ..	6·795	"	2·95	0·73	3·40	− 0·62	2·335	1·065	8912	..	0·106
" 15° ..	6·840	"	3·03	1·02	3·30	− 0·65	2·375	0·925	7740	..	0·092
" 20° ..	6·92	"	3·12	1·29	3·48	− 0·70	3·415	1·085	9079	..	0·107

Vessels of this form are notable clippers; but, for the sake of comparison, we will take a spread of canvas equal to that of an old three-decker with 3000 square mètres, the centre of effort being $26^m \cdot 20$ above the water-line. The Admiral then compared the moment of stability at various inclinations with that of vessels built upon the old system, and given in a Table in a most instructive work by M. Fréminville, sub-director of the school of Génie Maritime, entitled 'Guide du Marine,' which is as follows:—

Names of Ships.	Value of $\frac{M}{M'}$ for the ships indicated.	Ratio of the stability of the proposed ship, Fig. 616, to that of each of the vessels named below, for inclinations of: Normal.	5 deg.	10 deg.	15 deg.	20 deg.
		times.	times.	times.	times.	times.
'Breslau,' 100 guns, with full complement ..	0·082	5	2·55	1·29	1·12	1·30
Stores having been used	0·059	6·9	3·54	1·8	1·55	1·84
'Tage,' 90 guns, with full complement	0·096	4·25	2·37	1·11	0·95	1·12
Stores having been used	0·078	5·1	2·69	1·36	1·18	1·37
'Breslau,' with full complement	0·076	5·37	2·75	1·40	1·21	1·41
'Alceste,' 52 guns, with full complement	0·067	6·09	3·12	1·58	1·37	1·60
Stores having been used	0·053	7·59	3·94	2·00	1·73	2·00
'Jeanne d'Arc,' 44 guns, with full complement ..	0·063	6·47	3·31	1·68	1·46	1·70
Stores having been used	0·051	8·00	4·10	2·08	1·80	2·10
'Eurydice,' 30 guns, with full complement ..	0·077	5·30	2·71	1·37	1·195	1·39
'Obligado,' 10 guns, with full complement ..	0·065	6·27	3·21	1·63	1·41	1·64
Stores having been used	0·059	6·91	3·37	1·80	1·56	1·88

From these figures it will be seen that the stability of the proposed ship, Figs. 616, 617, 618, 619, which has been calculated for a light cargo, or at least for a mean one, is eight times that of some of the old ships after the stores were used; and if compared with the 'Tage,' with full complement of stores and provisions, the ratio is 4·25 times, the position of 616 being normal. But it is to be observed that at 5° the ship, Fig. 616, has 5 per cent. less stability compared with the 'Tage' with full complement, but 1·18 more after the stores have been used. It has also to be remarked that though the stability is a minimum at the angle of 15°, it rises afterwards, and at 20° is much higher.

Admiral Paris is aware that the comparison with the ships mentioned is not so fair as would be one with old ships having "tumbling home" sides, such as the 'Royal Louis,' and so forth, designed by M. Ollivier in 1750, and such as that proposed by the Admiral formerly. These have a *carène* like other ships—that is to say, 88^m long, $18^m \cdot 20$ beam, and 8^m draught but the upper deck would not be proportionately so wide as the more modern vessels, so that the metacentre would be motionless, as in the old class of ships, as shown in the work entitled 'L'Art Naval à Exposition de 1867.' The plan first proposed, and here alluded to, was but a first step towards the present arrangement.

It should here be observed that all the calculations relating to the present proposed form of turret-ships have been made in reference to the largest, shown in Figs. 616, 617, 618, 619, and proportionately reduced for others, of which the lines and proportions are relatively the same.

It may be of advantage to those engaged in the examination of such questions as that under present consideration to be made acquainted with the method pursued in the calculations to economize time and labour.

The calculation of the centre of *carène* for a ship when inclined is exceedingly tedious, wherefore the writer employed a term called the *mean section*, which is obtained by taking the arithmetical mean of the ordinates to each water-line, and setting off the height as obtained in this manner. Then is the centre of *carène* situate in the centre of this figure, as may be proved by arithmetically calculating a centre of *carène* according to Stirling's differential method, as employed by the Swedish constructor Chapman, the French naval architect Clairbois, and now adopted by naval architects generally.

Figs. 620, 621, 622, 623, represent one of Captain Cowper P. Coles' turret-ships lately designed: this is a vessel of 4272 tons, or about the same as the 'Bellerophon,' but with low freeboard from 3 ft. to 4 ft. out of the water. Her length between perpendiculars would be 320 ft.; beam, 53 ft.; two turrets, with guns 12 ft. above the water and 11 ft. horizontally from the sea. Between the turrets a deck-house is built, of the ordinary scantling of an iron ship, extending towards the vessel's sides at an angle that allows the turrets to cross their fire at 68° on the beam, and commanding an uninterrupted fire around bow and stern. This deck-house may be one, two, or three stories high, according to the accommodation required—in the present instance it is two stories high, surmounted by a hurricane deck 22 ft. out of the water. In the upper or second story, *b, b, b*, good accommodation is provided for the captain, the dining cabin being 24 ft. by 13 ft., with a suite of cabins in proportion, a commander's cabin, a wardroom, and four other cabins for officers, office, galley for the officers, a sick bay, baths, and w.cs. for both officers and men. The cabins and apartments on this deck would receive light and air from ports 2 ft. square, and 18 ft. out of the water, besides ventilation from the hatchways communicating with the hurricane deck. The lower or first story, *c, c, c*, would have mess-tables for 300 men, her fighting complement, a galley, midshipmen's, engineers', and warrant-officers' berths, besides four cabins for officers. This deck would receive air and light by means of ports 18 in. square, 11 ft. 6 in. out of the water, beside hatchways. All the hatchways on this deck leading below would have the means of being closed and kept water-tight the same as in monitors. The remaining nine cabins for officers would be on the lower deck, which corresponds with the 'Captain's' and 'Royal Sovereign's,' receiving light from deck-lights, and ventilation by tubes communicating direct with the outer air through the deck-house; there would also be provision made on this deck for messing 300 extra men, should it be required.

From the central position of the deck-house, as well as the horizontal and vertical distance from the water to the ports, and openings in it, free ventilation will be ensured in all weathers, whilst the comfort and accommodation will be second to no vessel in Her Majesty's navy.

From the form of this vessel above water, and the little resistance shown to wind or sea, lighter anchors and cables might be used than usual; but in this case we will suppose them to be the same weight as the '*Captain's*.' The cables would be led through fairleads at the edge of the bow along the upper deck, and through hawse-pipes, 8 ft. above the water, into the lower story of deck-house, where they would be worked by two steam capstans, one on each side, as in the 'Royal Sovereign.' The bower-anchors at sea would be stowed upon the deck without interfering with the firing or depression of the turret guns, and the spare anchors against the deck-house. The anchors can be worked on the bow or stern with equal facility. She would have the usual complement of boats, namely, two 42-ft. launches, two 32-ft. steam cutters, two 28-ft. life-boats, a captain's galley, and an officers' gig, which would be hoisted up by davits to the deck-house, and so arranged that at any moment they can be turned inboard when firing at sea.

Opinions will differ upon the question of masts; but, after considering the report of our iron-clad squadron, as well as other practical evidence, from which it appears that our ships as now masted are incapable as a fleet of manœuvring under sail without the aid of steam, the balance of advantage is, in Captain Coles' opinion, against such useless masts and sails, and in favour of fuel being substituted for the weight thus gained. Masts may be desirable to assist in decreasing the rolling propensities of high freeboard ships; but for vessels of low freeboard, with their decreased rolling, the Captain maintains that we can afford at once to sweep away the masts, gear and rigging, as antagonists to the steaming and fighting powers of war ships, and can at any time resort to temporary or small ones for fore and aft sails, such as 'Royal Sovereign' and 'Prince Albert' have, should it be considered desirable. In the stowage of fuel, weight and space have to be considered. In this vessel every 100 tons would make a difference of about 3 in. draught of water. In an armoured vessel the saving in weight, inherent to low freeboard, and the absence of masts gets rid of the greatest part of the difficulty under the head of weight; and the author of the design finds that, after giving ample accommodation, as before explained, for a crew of 300 men, and extra accommodation for 300 more men, making in all 600 men besides the officers, she would stow upwards of 1000 tons of coals, and even more, at the sacrifice of 3 in. immersion for every 100 tons.

The armament would consist of two turrets, with 13-in. armour, carrying four of the heaviest guns that can be procured, say 600-pounders, 12 ft. above the water; and two pivot-guns, if thought expedient, on the hurricane deck, 22 ft. out of the water: the latter may be found advantageous for firing down upon a ship's decks; and it is believed a vigorous fire kept up by rockets and breech-loading rifles, from the position and height of this hurricane deck, would have a great effect upon an enemy's upper deck and open ports. Leaving the pivot-guns out of the question, the fighting powers of this ship's turrets would remain as follows: 88° of the circle are commanded by two 600-pounders, throwing a broadside of 1200 lbs., and the remaining 272° by four 600-pounders, throwing 2400 lbs.; and it will be observed that this ship can engage end-on with two 600-pounders, and at 22° from the line of keel, or only two points on the bow, with *four* 600-pounders, which would be her strongest point of attack; the importance of this is much increased when it is considered that all well-designed iron-clads will endeavour to present their bows to the enemy, when this ship would show the deflecting surface of a sharp protected bow, not more than from 3 ft. to 4 ft. out of the water, and a circular turret 9 ft. high above it; whilst in broadside-ships of approved model, a bulkhead some 20 ft. above the water is placed across the ship to protect their guns, which, when fighting end-on, presents a large and weak target at right angles to the enemy's raking fire, which might place their battery *hors de combat* before a chance of using their broadside-guns occurred.

In comparing the defensive powers of Captain Coles' ship with the high and weaker sides of broadside-ships, and of the 'Monarch,' when we say that her sides and turrets would be protected with 13 in. of iron, we give but a small idea of her great superiority, in some respects, over any

Profile.

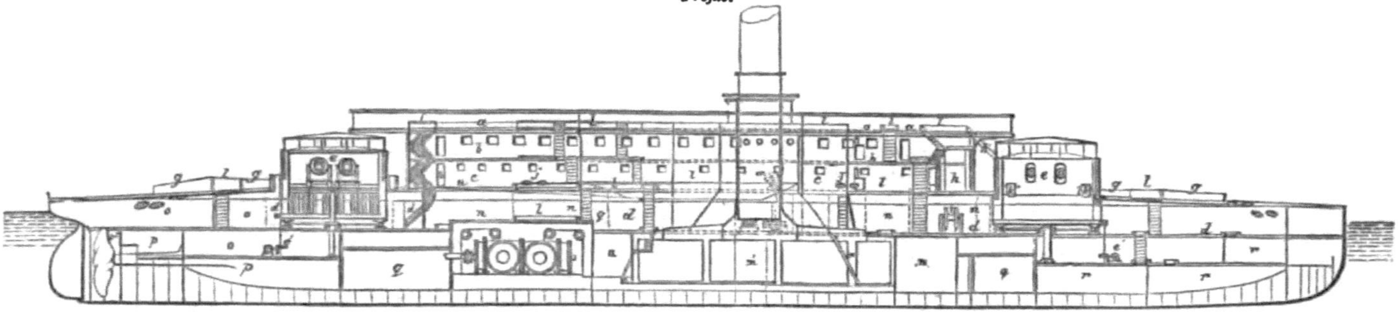

321.

Hurricane Deck.

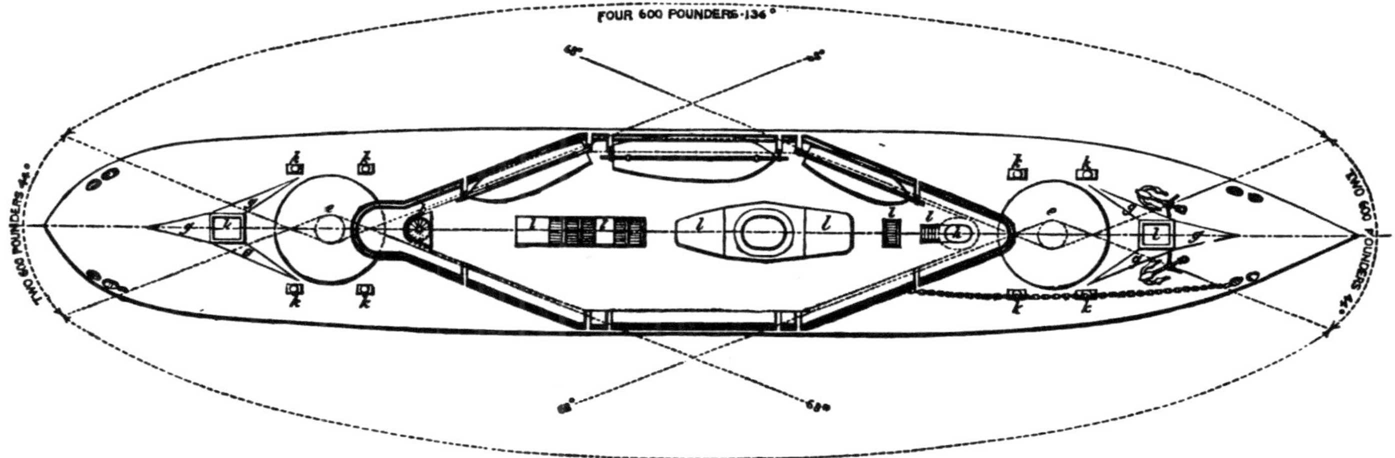

Reference.—*a, a, a,* Hurricane deck. *b, b, b,* Upper story, with accommodation for the captain and wardroom officers. *c, c, c,* Lower story, with accommodation for ship's company and junior officers. *d, d, d,* Lower deck. *e, e,* Turrets, with four 600-pounders. *e′, e′,* Engines for working turrets. *f, f,* Alternative of fixed breastworks round turret. See Figs. 622 and 623. *g, g, g,* Breakwaters. *h,* Pilot-house. *h′,* Slot in pilot-house for sight. *i,* Wheel. *j, j,* Hawse-pipes. *k, k,* Bitts. *l, l, l, l,* Ventilation and ladderways. *m, m,* Steam capstans and engines. *m′,* Chain lockers. *n,* Coals. *o, o, o,* Bread, slops, and stores. *p,* Provisions and spirits. *q, q,* Magazines. *r, r,* Stores, water, &c. *s, s, s,* Captain's apartments. *u,* Commander's cabin. *v, v, v,* Officers' cabins.

high freeboard vessel yet built, for her lowness of freeboard will give her antagonists but a sorry chance of hitting her, whilst her steadier platform will reduce the chances of her being hit below the water-line to the minimum. Lowness of freeboard may be said to facilitate being boarded; but, supposing an enemy to get upon the turret deck, he would have to take possession of the deck-house, which, from its position and height, as well as being enfiladed by the turrets, would be a matter of great difficulty.

622.

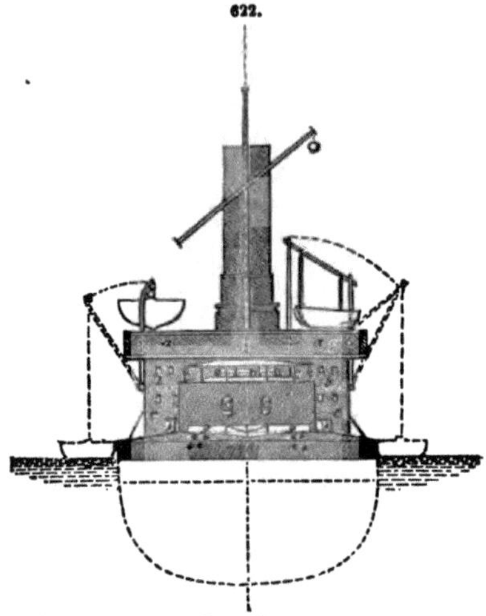

Bow View.

Twin screws and no masts, in combination with the peculiar form of the vessel above water, showing so little resistance at her ends to the wind and sea, it is believed will assist in ensuring both speed and handiness to the fullest extent, as well as her absence of motion as compared with our present iron-clads, whose great loss of speed at sea may be attributed in a great measure to their floundering propensities. Her engines would be of 900 horse-power, and capable of propelling her at the rate of 14 knots. The principle of a deck-house can be modified to suit vessels of all sizes, carrying from one turret upwards.

A one-turret ship would have the deck-house extending within about 10 ft. of the stern, balancing the turret by the other weights in the ship. A three-turret ship would have the third turret mounted on the hurricane deck before the funnel, bringing four guns right ahead. A four-turret ship would have the second and third turrets mounted at each end of the hurricane deck, bringing four guns right ahead and astern. It might be thought advisable in some instances to make the upper turrets of lighter iron than the lower ones, merely covering the guns mounted in pairs on turn-tables by a light iron turret, protecting their crews from rifles, grape, and canister.

623.

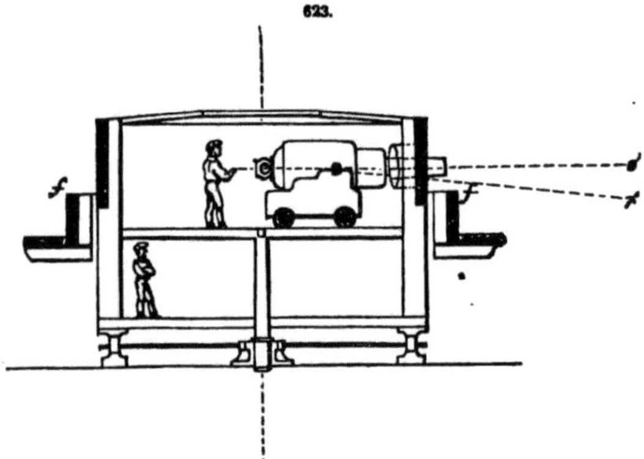

Fixed Breastwork round Turret and Two-story Turret.

BAY. Fr., *Baie (de porte, de fenêtre)*; Ger., *Offnung für Thür oder Fenster*; Ital., *Apertura*; Span., *Abertura en una pared.*

In builders' work, the space or extent embraced by one mode of construction, as a *bay of joists*, which is the surface covered by the joisting filled in between any two binding-joists; or a *bay of roofing*, which is the part filled in with common rafters between a pair of principal rafters.

The openings between the supports of a *bridge* are also called *bays*.

BAY OF JOISTS. Fr., *Panneau, Assemblage de Charpente*; Ger., *Füllung, Fachwerk*; Ital., *Commettitura*; Span., *Luz de traviesas.*

See Bay.

BAY OF ROOFING. Fr., *Travée de comble*; Ger., *Dachstuhlfette, Dachlatten*; Ital., *Scompartimento del tetto*; Span., *Luz de cobertizo.*

See Bay.

BEAD. Fr., *Baguette*; Ger., *Riefe oder Stab*; It., *Astragalo*; Sp., *Astragalo, Tondino*; *Junquillo.*

A moulding on wood or other material, generally forming, in section, a part of a circle, is designated a bead, which when on wood it is said to be *struck* on if formed with a plane which cuts the wood into the shape required.

A bead on the edge of a board, as shown in Fig. 624, is called a *nosing;* if on the face, but flush with the surface, and with one *quirk* only, as in Fig. 625, it is called a *quirk-bead.* But if with two quirks, Fig. 626, it is called a double quirk-bead; and if struck on the angle of a piece of work, Fig. 627, it is called a *return-bead.*

624. 625. 626. 627. 628. 629.

A bead projecting beyond the surface, Figs. 628, 629, is called a *cock-bead,* and sometimes a *cocked bead.*

Bead and Batten Work.—An expression used by carpenters to denote a rough style of work composed of battens with a bead run along the edge in the direction of the grain of the wood.

Bead-butt.—A term in joining applied to work framed in panels, when the latter are flush with the framing and a bead is struck or run on two sides only of the panel in the direction of the grain of the wood. The ends of the bead are made to stop or butt against the rails. See Door.

Bead-flush.—A term applied to framed work, which differs from the last, in having the bead run on the framing instead of on the panel, and in being all round, instead of on two sides only.

BEAM. Fr., *Poutre;* Ger., *Balken, Träger;* Ital., *Trave;* Span., *Viga, Tirante.*

A beam is a piece of timber or metal, or both combined, used for sustaining a weight, or counteracting forces by tension and compression in the direction of its length, and by widening and narrowing in the directions of the breadth and thickness.

The word beam in builders' work is most frequently subjoined to another word, used adjectively or in apposition, to show its use or form, as *binding-beam, box-beam, built-beam, camber-beam, collar-beam, hammer-beam, dragon-beam, straining-beam, tie-beam, trellis-beam, trussed-beam,* and so on. See Strength of Materials.

BEAM-FILLING. Fr., *Maçonnerie de remplissage au niveau des poutres;* Ger., *Mauerwerk zwischen der Balkenlage;* Ital., *Riempimento;* Span., *Relleno.*

Masonry, brickwork, or concrete filled in from the level of the under-edges of the beams to that of their upper edges, is known as *beam-filling.* Beam-filling occurs either between joists or floor-beams, or in filling up the triangular space between the top of the wall-plate of the roof and the lower edges of the rafters, or even to the under-surface of the boarding or lath for slates, tiles, or thatching. This operation is necessary in garret-rooms, where the walls form sides of apartments, and where the tie-beams are placed above the bottom of the rafters.

BEARER. Fr., *Lierne;* Ger., *Bindesparren;* Ital., *Sostegno, Portante;* Span., *Cadena.*

Generally, this term is applied to any member of a structure which has to support a weight above it, as the joists in a flat roof, or the short pieces nailed to the rafters to support the gutter-board in a roof-valley, Fig. 630.

630.

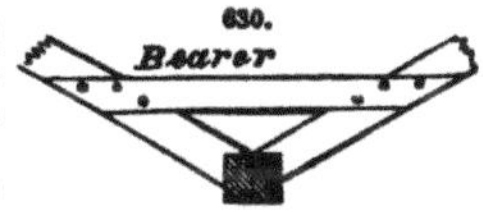

The term may also be applied to any beam, whether of wood or iron, placed horizontally, which has to support a weight above it, as to a breesummer, which is only a particular application of a bearer.

BEARING. Fr., *Collet;* Ger., *Lager;* Ital., *Colletto;* Span., *Punto de apoyo.*

Bearing of a girder or beam is the portion of it which rests on the supports.

Timbers or lintels let into a wall have usually a bearing of 9 in. at each end; stone steps should also have a bearing of 9 in.

The bearing of joists or other beams supported at both ends is regulated by the resistance of the material to crushing. A bearing of 4½ in. on the sleepers or wall-plates is usually considered sufficient in ordinary dwelling-houses.

Girders and beams are said to have a solid bearing when supported throughout their whole length. But in the case of sills, and so forth, when not so supported, they are said to have a false bearing.

The *bearing-distance* is the unsupported part between the bearings.

BEARING OUT.

This expression, generally applicable to new work, is used by painters to imply that a third, fourth, or fifth coat of paint has been so fully and evenly laid upon the previous ones, that the original colour of the body, the knotting, the priming, and the under-coats, cover well, and give no sign of disfiguring the glossy surface of the finishing coat.

BED. Fr., *Semelle;* Ger., *Tragseite;* Ital., *Corso.*

In masonry and brickwork is the upper and under side of a stone or brick. In arch-stones the beds are the joints which radiate towards the centre. In slates the under-side is the bed.

Stones are usually specified to be laid on their quarry-bed, that is, the same relative position they had before they were quarried.

In builders' measurements the term beds and joints is used to designate all the parts of a stone covered in the work, in contradistinction to the face. The term also includes the back, if it be measured.

BEDDING TIMBERS, the process of. Fr., *Poser les racinals;* Ger., *Legen der Grundhölzer;* Ital., *Fare il letto alle travi.*

Bedding Timbers.—Laying them on a wall or otherwise on a bed of mortar, cement, or putty, so that the bearing may be solid or uniform throughout. Wall-plates and sleepers are bedded in this way. Care should be taken when bedding timbers to allow a free circulation of air around them, otherwise they are liable to rapid decay, particularly if in contact with lime-mortar in a moist state, as the hydrate of lime in the mortar abstracts the carbonic acid gas from the wood, which hastens its decay.

Bedding Stone.—See SETTING.

BED-PLATE. FR., *Plaque de fondation;* GER., *Grundplatte;* ITAL., *Piastra di fondazione;* SPAN., *Plancha de cimiento, carrera.*

A *bed-plate* or *bed-piece* is the principal or foundation framing or piece, in machinery, by which the other parts are supported and held in place; the bed. It is also called the *base-plate* and the *sole-plate.*

BEETLE. FR., *Maillet;* GER., *Holzschlegel;* ITAL., *Mazzuolo;* SPAN., *Maleta.*

Beetle is a name given to a heavy mallet or wooden hammer, used to drive wedges, beat pavements, and the like. The term beetle is also applied to a machine used to produce figured fabrics by pressure from corrugated or indented rollers.

BELL. FR., *Cloche;* GER., *Glocke;* ITAL., *Campana;* SPAN., *Campana.*

631.

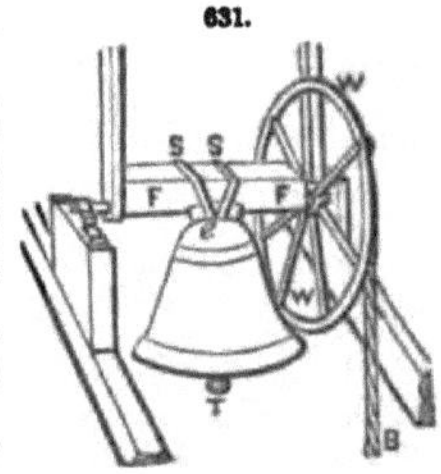

A bell is a hollow metallic vessel, which gives forth a clear, musical, ringing sound on being properly struck. The most common form of bell is shown in Fig. 631. In this form it is expanded at the lower part, is furnished at the top with an ear *e*, for the purpose of suspension, and has within it a tongue or clapper T, by the blow of which the sound is produced; S S, straps which secure the bell to the rock-shaft F, F; B is the bell-rope, and W W the vibrating wheel.

Another form, especially of small bells, is that of a hollow body of metal perforated, and containing a loose solid ball, to make a sound when it is shaken.

In the formulas, Table I., D = the diameter of the bell at the mouth, in inches; d = the diameter of the bell at the crown; h = height of the bell from the mouth to the crown; S = the thickness of the sound-bow, in inches; W = weight of the bell in lbs. avoirdupois (7000 grains = 1 lb. avoirdupois, and 5760 grains = 1 lb. Troy); n = the number of vibrations a second, corresponding with the key-note of the bell, see Table II.; k = the coefficient, expressing the relative thickness of the sound-bow to the diameter of the bell—it varies from ·07 to ·08. In peals of bells the sound-bow is generally put, S = ·08 × D for the triple; S = ·07 × D for the tenor; and the intermediate bells in the peal, proportions lying between these for the respective sound-bows.

TABLE I.

$W = 0 \cdot 25\, D^2 S$ .. .. [1]	$D = 2 \sqrt{\frac{W}{S}}$ .. .. [7]	$S = \frac{n D^2}{58000}$ .. .. .. [12]
$W = \frac{D^4 n}{232000}$ .. .. [2]	$D = 240 \cdot 83 \sqrt{\frac{S}{n}}$ .. [8]	$S = \frac{4\, W}{D^2}$ .. .. .. [13]
$W = 0 \cdot 25\, D^3 k$ .. .. [3]	$D = 21 \cdot 947 \sqrt[4]{\frac{W}{n}}$ .. [9]	$S = k D$ .. .. [14]
$n = 58000 \frac{S}{D^2}$ .. .. [4]	$D = \sqrt[3]{\frac{4\, W}{k}}$ .. .. [10]	$k = \frac{S}{D}$ .. .. [15]
$n = 232000 \frac{W}{D^4}$ .. .. [5]	$D = 58000 \frac{k}{n}$ .. .. [11]	$k = \frac{4\, W}{D^3}$ .. .. .. [16]
$n = 58000 \frac{k}{D}$ .. .. [6]		

$$W = D d S (0 \cdot 5 - 0 \cdot 0002816\, d) + \cdot 00375\, h\, d^2 S \quad .. \quad .. \quad [17]$$

Example 1, to illustrate [1], Table I.—Required the weight (W) of a bell, the diameter D = 60 in., and the thickness of the sound-bow S = 4·8 in.

$$[1]. \quad \cdot 25\, D^2 S = W = \cdot 25 \times 60^2 \times 4 \cdot 8 = 4320 \text{ lbs.}$$

Example 2.—A bell of 2636·4 lbs. (W) is to be constructed with a sharp note, putting for the sound-bow K = ·075; what is the diameter (D) of the bell?

$$\text{From Table I., [10], } D = \sqrt[3]{\frac{4\, W}{k}} = \sqrt[3]{\frac{4 \times 2636 \cdot 4}{\cdot 075}} = 52 \text{ in.}$$

Example 3.—Required the diameter (D) of a bell with the key-note D♯ in the first octave above zero, the bell to be light with a full note.

In this case $k = \cdot 07$, and $n = 152 \cdot 25$. Table II.

$$D, \text{ the diameter} = 58000 \frac{k}{n} = \frac{58000 \times \cdot 07}{152 \cdot 25} = 26\tfrac{2}{3} \text{ in. [11] Table I.}$$

Example 4.—Required the key-note of a bell with D = 44 in., and S = 3·52 in.

$$[4]. \quad n = 58000 \times \frac{3 \cdot 52}{(44)^2} = 58000 \frac{S}{D^2} = 105 \cdot 45.$$

In Table II. the nearest number 105·45, in the first octave below zero is 107·63, which answers to the key-note A.

TABLE II.—VIBRATIONS A SECOND = n.

Key-Note.	Bass.			Zero.	Treble.		
	3rd Octave.	2nd Octave.	1st Octave.		1st Octave.	2nd Octave.	3rd Octave.
C	16·000	32·000	64·000		128·00	256·00	512·00
C♯	16·947	33·895	67·790		135·58	271·00	542·32
D	17·960	35·920	71·840		143·68	287·36	574·72
D♯	19·027	38·055	76·110		152·22	304·44	608·88
E	20·159	40·318	80·636		161·27	322·54	645·09
F	21·357	42·715	85·430		170·86	341·72	683·44
F♯	22·627	44·255	90·510		181·02	362·04	724·08
G	23·972	47·945	95·890		191·78	383·56	767·12
G♯	25·398	50·797	101·59		203·19	406·37	812·75
A	26·908	53·817	107·63		215·27	430·53	861·07
A♯	28·508	57·017	114·03		228·07	456·13	912·27
B	30·204	60·409	120·82		241·63	483·27	966·54
C	32·000	64·000	128·00		256·00	512·00	1024·0

Example 5.—A bell has to be cast with the key-note C in first octave below zero; required the diameter D, when the weight = 6561 lbs.

We find in this case $n = 64$, Table II.

$$\therefore \text{ [9], Table I., } D = 21\cdot947\sqrt[4]{\frac{6561}{64}} = 69\cdot84 \text{ in.}$$

Example 6.—What is the thickness (S) of the sound-bow of the bell in Example 5, D = 69·84; $n = 64$?

$$\text{From [12], Table I., } S = \frac{n\,D^2}{58000} = \frac{64 \times (69\cdot84)^2}{58000} = 5\cdot38 \text{ in.}$$

TABLE III.

Abscissa x.	Ordinate y.	Thickness of Metal.			
		S = 1.	S = ·07 D.	S = ·075 D.	S = ·08 D.
1	0·4142	1	·700	·750	·800
1½	0·686	·800	·560	·600	·640
2	0·867	·653	·459	·490	·522
2½	0·974	·547	·382	·410	·437
3	1·025	·474	·331	·355	·379
3½	1·030	·423	·295	·317	·338
4	1·000	·380	·266	·285	·304
4½	0·955	·351	·245	·263	·281
5	0·875	·327	·228	·245	·261
5½	0·775	·301	·211	·226	·241
6	0·665	·291	·203	·218	·233
6½	0·530	·286	·200	·214	·228
7	0·390	·279	·195	·209	·223
7½	0·235	·272	·190	·204	·217
8	0·075	·267	·186	·200	·213
8·74	0·780	·333	·233	·250	·266

Example 7.—If D = 50 in., $d = 27$ in., $h = 36$, and S = 4·, what is the weight (W) of the bell?

$$\text{From [17], Table I., } W = D\,d\,S\,(\cdot5 - \cdot0002816\,d) + \cdot00375\,h\,d^2\,S = 50 \times 27 \times 4\,(4\cdot9923968) + 393\cdot66 = 3092\cdot13936 \text{ lbs.}$$

When a bell is to be constructed, we generally have the weight or key-note given, the diameter and sound-bow are calculated by the preceding formulas and examples, and we may then proceed with the construction, shown in Figs. 632, 633, 634.

The diameter of the bell at the mouth is divided into 10 equal parts, called strokes, which then is the scale and measurement for the construction.

Shrinkage to be allowed for $\frac{1}{8}$th of an inch to the foot.

The section of a bell is generally laid out on a piece of board represented by the lines a, b, c, d, which then is cut out and used for turning up the mould for the bell. The board should be about 11 strokes long, and 2·5 strokes wide. Through the centre of the board draw the line p, q, parallel to b, c; bisect the line p, q, and set four (4) strokes from the bisecting point towards each end; divide the strokes into halves, and number them as shown on the accompanying drawing, Fig. 632. Through each division draw lines at right angles to p, q, set off the corresponding ordinates y expressed in strokes, Table III., and join them by a curve-line, which then will be the centre of thickness of metal in the bell.

632. 635.

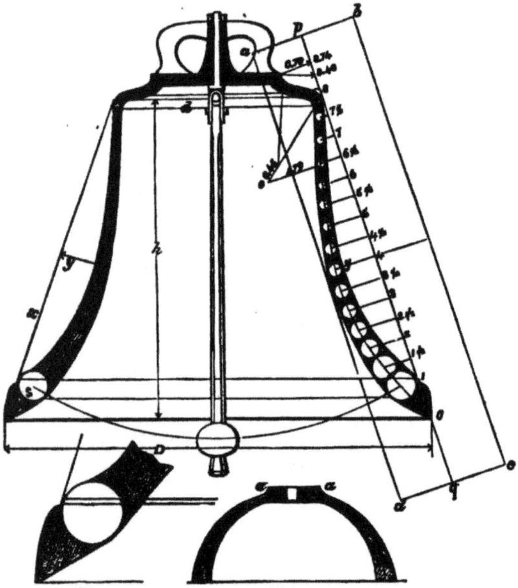

634. 633.

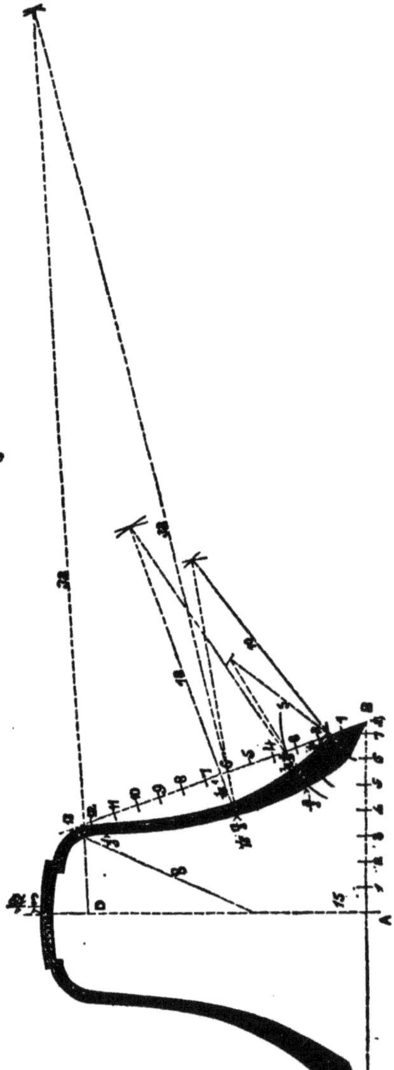

At the end of the first ordinate, as a centre, draw a circle with a diameter equal to the desired thickness of the sound-bow, which should be from 0·7 to 0·8 stroke. At every succeeding ordinate draw a circle with the diameter noted in Table III.: for instance, if the thickness of the sound-bow is $4\frac{1}{2}$ in., then the thickness of metal or diameter of the circle at the third ordinate will be $4 \cdot 5 \times 0 \cdot 474 = 2 \cdot 133$ in.; but if the sound-bow is 0·7, 0·75, or 0·8 stroke, the thickness of metal at the third ordinate will be 0·331, 0·355, or 0·379 stroke. When all the thicknesses are thus drawn, draw the two lines tangenting the circles on each side of the centre line of the metal.

From 0 to 1 make a moulding of 0·1 stroke thick over the line, as shown in Fig. 633. Prolong the $6\frac{1}{2}$ ordinate, and set off 1·79 stroke to e, which then is the centre for the curve on the top; draw the arc through the centre of the small circle at the eighth ordinate; join e, 8, set off from e, 0·46 stroke to the centre for the inside curve at the top.

Thickness of metal of the top should be 0·3 the sound-bow at 8, and 0·333 at r. Draw the ordinate at 8·74, set off 0·78 to r, join r and the abscissa 8·48, and prolong the line through r; then finish the drawing, as shown in Fig. 632.

When the board is cut out and ready for turning the mould, it must be carefully set, so that the outside diameter of the crown will be half the diameter of the mouth of the bell.

This form of bells gives the greatest possible gravity of tone with the least possible quantity of metal. Bells can be made almost in any form without seriously affecting the quality of tone; but the thickness of metal should always be in proportion as the square of the diameter, taken at the centre of the metal, as in Fig. 634.

TABLE IV.—PROPORTIONS OF A PEAL OF EIGHT BELLS.

Bells.	Key-Note.	n.	k.	S. in.	D. in.	W. lbs.	Clapper.
							lbs.
Tenor ..	D	71·84	0·070	3·95	56·5	3156	63
2nd ..	E	80·64	0·071	3·62	51·1	2366	48·6
3rd ..	F♯	90·51	0·072	3·32	46·1	1765	37·2
4th ..	G	95·89	0·073	3·22	44·2	1575	34·1
5th ..	A	107·63	0·075	3·08	40·5	1262	28·1
6th ..	B	120·82	0·077	2·85	37·0	976	22·4
7th ..	C♯	135·58	0·079	2·67	33·8	763	18·2
Triple ..	D	143·68	0·080	2·58	32·3	673	16·8

Clapper.—The weight of the clapper should be from $\frac{1}{40}$th to $\frac{1}{50}$th the weight of the bell; the smaller bells take the largest clappers.

The tracing of bells rests upon a fixed basis, called the *Bell-scale*, or *Jacob's Staff*, the result of long experience, and handed down from generation to generation among founders. It depends upon certain proportions which, like the modules in architecture, serve to regulate and to harmonize the different parts of the bell. The *bow*, or, in other terms, the thickest part of the bell, constitutes the principle of all the measurements.

The following Table, which we borrow from M. Guettier's work on Casting, gives the diameter of bells, and the thickness of the bow, from the weight of 3 kilos. to 12,000 kilos. It is nothing more than a scale, presented under another form and in metrical measure.

TABLE V.

Weight of Bells.	Thickness of Bow.	Large Diameter.	Weight of Bells.	Thickness of Bow.	Large Diameter.	Weight of Bells.	Thickness of Bow.	Large Diameter.
3	·008	·120	200	·047	·705	3500	·123	1·845
4	·011	·165	250	·050	·750	4000	·128	1·920
5	·013	·185	300	·055	·825	4500	·134	2 010
6	·015	·225	350	·058	·870	5000	·137	2·055
10	·019	·285	400	·060	·900	5500	·141	2·115
15	·021	·315	450	·063	·945	6000	·146	2·190
20	·022	·330	500	·065	·975	6500	·150	2·250
25	·023	·345	600	·068	1·020	7000	·154	2·310
30	·025	·375	750	·074	1·110	7500	·158	2·370
35	·027	·405	1000	·081	1·215	8000	·160	2·400
40	·028	·420	1250	·087	1·305	8500	·164	2·460
45	·029	·435	1500	·093	1·395	9000	·168	2·520
50	·030	·450	1750	·098	1·470	9500	·170	2·550
75	·034	·510	2000	·103	1·545	10000	·173	2·595
100	·037	·555	2250	·108	1·620	11000	·181	2·715
125	·040	·600	2500	·110	1·650	12000	·190	2·850
150	·043	·645	2750	·114	1·710			
175	·045	·675	3000	·117	1·755			

Several methods are employed for tracing bells. The one mostly used in France gives 15 thicknesses of the bow to the diameter, 7½ to the diameter of the crown, 12 to the line joining the lower ridge of the bell and the base of the crown, and finally 32 to the great radius serving to trace the profile of the bell proper. Fig. 635, where each line of construction has its dimension marked—the thickness of the bow being taken as unity—will be sufficient to show how the process is carried out.

BELL-CRANK. FR., *Levier brisé*; GER., *Winkelhebel*; ITAL., *Zanca*; SPAN., *Cigüeña*.

An iron or brass lever in the shape of a quadrant of a circle, attached to an iron holdfast which is driven into a wall, receives the name of bell-crank, because it is used to connect bell-wires at the angles or corners of a room. Any rectangular lever, Fig. 636, by which the direction of motion is changed through an angle of 90°, is termed a bell-crank.

636.

BELL-HANGING, DOMESTIC. FR., *Pose des sonnettes*; GER., *Befestigen der Glockenzüge in den Wohnhäusern*; ITAL., *Mettere i campanelli*.

The art of domestic bell-hanging is quite modern. It is believed not to have been in practice much before the present century. Within the writer's recollection it was usual, in even the best houses, to expose the wires to view along the walls and ceilings, in the angles of which they were fixed, sometimes to the great disfigurement of the room. Within late years the "secret system" of bell-hanging has been introduced, which consists in carrying the wire and cranks in tubes and boxes concealed by the finishings of the walls. The tubes are usually of tinned iron or zinc; but they ought to be either of brass or strong galvanized iron. Zinc is not to be depended upon. In some places it will moulder away. If not soldered, it opens, and the wires work into the joinings of the tube, which stops their movement.

The proper time to commence bell-hanging is when the work is ready for lathing; but it should not be delayed after the rough-cast plastering has commenced. If the work be performed at this period, it enables the bell-hanger to see his way more clearly, and prevents much cutting away of the plasterers' work afterwards.

The bell-board is usually placed in some conspicuous place, where the bells can be both seen and heard by the attendant. It should be painted white, and each bell should be designated by a letter or number painted on the board.

BELL-METAL. Fr., *Métal des cloches*; Ger., *Glockenmetall*; Ital., *Lega delle campane*; Span., *Metal campanil.*

A good average bell-composition is 75 copper, 25 tin; 30 of tin to 100 copper is also a good proportion. Large bells are cast of 80 copper, 6 zinc, 10 tin, 4 lead. A very fine large bell consisted of 71 copper, 26 tin, 2 zinc, and 1 iron. No definite ratios, however, between the metals of which *bell-metal* is composed have as yet been established. See Alloys. Copper. Tin.

BELLOWS. Fr., *Soufflet*; Ger., *Blasebalg*; Ital., *Mantice*; Span., *Fuelle.*

A *bellows* is an instrument, utensil, or machine for propelling air through a tube, for various purposes, as blowing fires, filling the pipes of an organ with wind, and so on. The common bellows is formed of two boards, Fig. 637, with a skin of leather *l* nailed to their edges and hanging loosely between them; thus forming a sort of chamber, which is capable of being enlarged or contracted at pleasure. To the lower board is fixed a metal nozzle *n*, communicating with the wind-chamber; and this board is also furnished with a *clack-valve*, *c*, for the admission of air.

The blast obtained by means of the common bellows is intermittent; and if a continued blast is required, a bellows with a double chamber and an additional valve is necessary. The long shape forge-bellows, Fig. 638, is a bellows of this description, having an extra chamber *a* and an extra valve *b*. The centre-board *d* should be a fixture; the nozzle connected with the upper chamber; the upper board loaded or made of heavy materials, and the lower board moved when in use. The two valves should open alternately.

637.

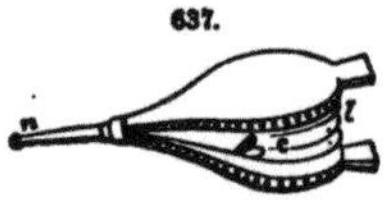

638.

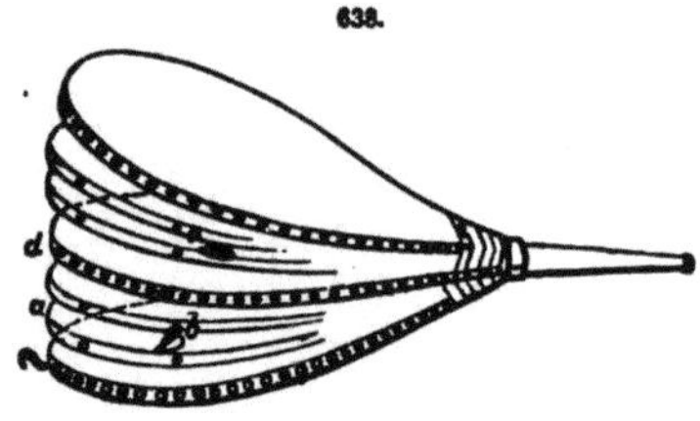

639.

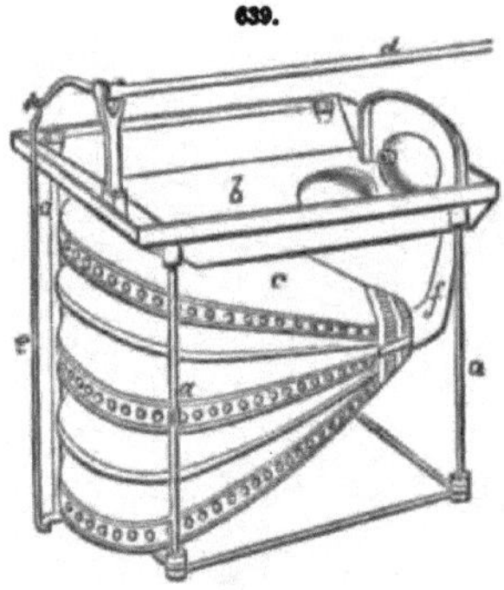

640.

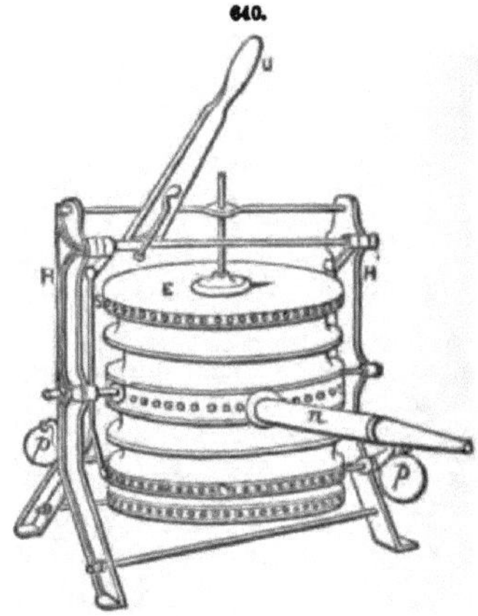

641.

Fig. 639 shows the application of a double-chambered bellows to a portable forge. *a*, *a* are legs supporting the hearth *b*; *c*, the bellows fastened by the centre-board to the legs *a*, and worked by the handle *d* inserted in a hook at the top of the lever *e*. By depressing the handle *d*, the lever *e* lifts the bottom flap of the bellows, which falls by its own weight. The collapsing of the upper flap forces the blast through the pipe *f* to the fire.

Double-blast bellows are frequently made of a circular shape. Fig. 640 shows this arrangement. H is the frame supporting the bellows E, which is worked by the handle U communicating with the lever S. *p*, *p* are weights attached to the lower board *o*, and *n* the nozzle fixed to the centre-board. An arrangement for obtaining a continuous blast of air by the weight and easy movement of a person standing on two bellows, and resting his weight alternately on them, invented by Henry Neumeyer, is shown in Fig. 641. It consists in constructing two bellows B B, connected by a rope C, and fastened to a centre-board A. These bellows rising and falling alternately, by means of valves properly arranged keep the wind-chamber F filled.

BELL-TRAP. FR., *Pommelle de puisard;* GER., *Rost im Senkloch;* ITA , *Chiavica a tenuta d'aria.*

A small *stench-trap*, from 3 to 6 in. diam., Fig. 642, usually fixed over the waste-pipe of a sink or other inlet to a drain. The foul air is prevented from rising by an inverted cup or bell, the lips of which dip into a chamber filled with water surrounding the top of the pipe.

642.

The grating to which the bell is attached should never be fastened down, as the opening between the lip and side of the pipe frequently becomes choked, and it is desirable to have the means of freeing it.

The bell should be made to dip deep into the water, to prevent the foul air escaping. Imperfectly-constructed bell-traps, by permitting a communication from the sewer, and so contaminating the air of the dwelling, is the cause of more unhealthiness to the occupants than is commonly supposed. See DRAINAGE and STENCH-TRAPS.

BELTS, TRANSMISSION OF MOTION BY MEANS OF. FR., *Courroie;* GER., *Riemen;* ITAL., *Coreggia;* SPAN., *Cintura, Correa.*

There are two theories upon which the transmission of motion, by means of endless belts or cords, are founded. The first, that of M. Prony, relative to the sliding of a cord or belt upon the surface of a drum; the second, that of Poncelet, refers to the variation of tension in the two parts of the belt or cord employed in these transmissions.

Morin proved, by special experiments, the consequences of these two theories; we give a succinct account of the results of these researches.

In explaining the first of these theories, with respect to the slipping of belts upon cylinders, let us consider a belt or cord enveloping a portion of the surface of a cylinder, and acted upon at one end by a power P, and upon the other by a resistance Q, Fig. 643. It is clear that, to produce slipping of the belt, the power P should be equal to the resistance Q, increased by the resistance opposed by the friction of the cord upon the surface of the cylinder. Let us seek to determine this friction.

643.

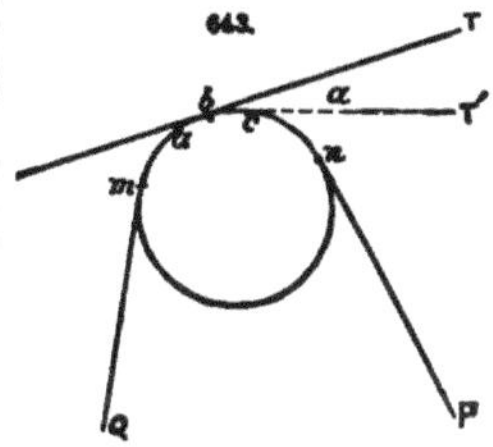

For this purpose, we consider the two consecutive elements $a\,b$ and $b\,c$ of the belt, and call T the tension of the cord in the element $a\,b$; T′ the tension of the cord in the element $b\,c$. It is evident that the tension T′ exceeds the tension T by an infinitely small quantity t, which is precisely the measure of the resistance opposed by the friction; we have then $T' = T + t$; and passing from one element to the other, from the point n of contact of the direction n P, where $T = P$, to the point m of contact of the direction m Q, where $T = Q$, the sum of all the increments of tension produced by the friction at the moment of slipping, will give the total tension.

The friction or elementary increase of tension t, from the element $a\,b$ to the element $b\,c$, is produced by the pressure resulting from the component of tension T′, normal to the surface, which is T sin. α, calling α the infinitely small angle at the intersection of the two elements $a\,b$ and $b\,c$, or simply $T\alpha$, since T differs by an infinitely small quantity from T′, and the sine α from α; we have then $t = f.T\alpha = Tf\,\frac{S}{R}$, f being the ratio of the friction to the pressure.

The sum of all these increments of tension, taken from the point m, where $T = Q$, to the point n, where $T = P$, leads, according to analysis which we will presently discuss, to the formula

$$\log. P = \log. Q + 0{\cdot}434 f\,\frac{S}{R}, \text{ or } P = Q \times 2{\cdot}718^{f\frac{S}{R}} = Q\,e^{f\frac{S}{R}},$$

S being the total length embraced by the cord, and R the radius of the circle.

We see by this expression that the tension of the motive power increases from $P = Q$, answering to $S = 0$, proportionally to the opening of the angle $\frac{S}{R}$, embraced by the belt, and not to the absolute extent of the arc; which shows, from theoretic considerations, that for an increase of the friction of slipping of cords or belts, it is not essential to enlarge the diameter of the cylinder, but that the proportional part of the circumference to be enclosed should be increased.

The preceding formula relates to the case where the power P is to overcome the resistance Q, and consequently, besides this, to surmount the friction of the cord or belt upon the drum. When, however, as is frequently the case, the force P is to yield to the force or weight Q, for moderating its action, or resisting it altogether, as, for example, in the lowering of goods, the friction acts in favour of the force P, and we have

$$\log. P = \log. Q - 0{\cdot}434 f.\,\frac{S}{R}, \text{ or } P = \frac{Q}{2{\cdot}718^{f\frac{S}{R}}}.$$

Such are the relations which theory indicates between the forces P and Q, the arc of contact, the radius of the drum, and the coefficient of friction. It remains to determine by experiment the correctness of these relations.

Experiments upon the Slipping of Cords and of Belts upon the Surface of Wooden Drums, and of Cast-iron Pulleys.—For this purpose Morin made use of three wooden drums, with diameters of 2·741 ft., 1·338 ft., and 0·328 ft., placing them horizontally in a fixed position, so that they could not turn, and over them was passed a belt of black curried leather, nearly new, but having acquired a certain pliability from previous use. Its breadth was 0·164 ft., and thickness 0·173 ft.; its rigidity seemed so feeble that Morin found himself justified in neglecting it in its ratio to the friction of slipping upon the surface of the drum.

The two strips of the belt hung vertically in equal portions on each side of the drum, and to each of them was attached a scale to receive the weights. The belt weighed 5·06 lbs., each scale 0·5 lb.; consequently, the weight of each strip, of equal length, was, with its plate, 3·03 lbs. The arc embraced was equal to the semi-circumference. At first, equal weights were put in the scales, then gradually was added to one of them the weights necessary to make the belt slide upon the drum.

We see from this, that the tension Q of the ascending strip was equal to 3·03 lbs. plus the weight contained in the corresponding scale, and that the tension P of the descending strip was equal to Q increased by the weight added, over and above the primitive load.

This established, the preceding formula becomes

$$\log. P = \log. Q + 0{\cdot}434 f . \frac{S}{R}, = \log. Q + 0{\cdot}434 f \times 3{\cdot}1416,$$

whence we deduce

$$f = \frac{\log. P - \log. Q}{0{\cdot}434 \times 3{\cdot}1416} = \frac{\log. P - \log. Q}{1{\cdot}363}.$$

By introducing in this formula the values of P and Q furnished by experiments, we are enabled to calculate the different values of the ratio f of the friction to the pressure, and to be assured that they confirm the theoretic consequences which we have unfolded.

The two following Tables contain the results of the experiments:—

Experiments upon the Friction of Belts upon Wood Drums.

Width of Belt.	Condition of the Belt.	Diameter of Drum.	Length of Arc embraced.	Tension of the Part. Rising, Q.	Falling, P.	Ratio of Friction to Pressure, f.
ft.		ft.	ft.	lbs.	lbs.	
0·164	Dry, somewhat oily	2·741	4·306	14·060	66·992	0·497
				14·060	64·786	0·486
				14·060	64·786	0·492
				36·114	167·341	0·488
				36·114	153·336	0·460
				36·114	151·461	0·458
				25·087	111·102	0·473
				25·087	95·603	0·426
					Mean ..	0·472
0·164	Dry, somewhat oily	1·338	2·099	14·060	63·683	0·472
				14·060	69·197	0·458
				14·060	63·242	0·507
				36·114	140·875	0·479
				36·114	140·875	0·433
					Mean ..	0·462
0·164	Dry, somewhat oily	0·328	0·514	14·060	73·608	0·526
				14·060	75·813	0·541
				25·087	91·252	0·411
				25·087	98·975	0·438
				25·087	94·560	0·422
				36·114	161·827	0·477
				36·114	168·576	0·490
					Mean ..	0·472
0·091	Very dry and rough.	2·741	4·306	11·911	71·458	0·570
				11·911	72·560	0·575
				22·938	114·465	0·512
				22·938	104·541	0·483
				33·965	137·622	0·446
				33·965	136·519	0·443
					Mean ..	·504
					General Mean	0·477

EXPERIMENTS UPON THE FRICTION OF BELTS OF CURRIED LEATHER UPON CAST-IRON PULLEYS.

Breadth of Belt.	State of the Belt.	Diameter of the Pulley.	Arc embraced S.	Tension of Strip. Ascending Q.	Tension of Strip. Descending P.	Ratio of Friction to Pressure f.	Remarks.
ft.		ft.	ft.	lbs.	lbs.		
0·164	Dry, a little unctuous	2·000	3·143	14·060	29·719	0·338	This belt was old, having been used a long time in a spinning-mill. The pulley was not turned.
				14·060	36·996	0·308	
				14·060	34·791	0·288	
				25·087	64·566	0·301	
				36·114	89·047	0·282	
				36·114	82·436	0·262	
					Mean..	0·279	
0·164	Dry, a little unctuous	2·000	3·143	14·060	35·286	0·300	This belt was new. The pulley was not turned.
				25·087	61·478	0·285	
				25·087	57·067	0·271	
				36·114	80·224	0·254	
				36·114	80·224	0·254	
				58·170	160·724	0·323	
					Mean..	0·281	
0·164	Dry, a little unctuous	0·361	0·566	14·060	31·704	0·259	The pulley was turned; its width was only ·098 ft., and so reduced the slipping part of the belt to ·098 ft.
				14·060	40·525	0·336	
				25·087	59·273	0·273	
				25·087	68·095	0·318	
				36·114	81·328	0·259	
				36·114	81·328	0·259	
					Mean..	0·284	
0·164	Moistened with water	2·000	3·143	25·087	68·095	0·317	
				14·060	43·834	0·361	
				14·060	43·834	0·361	
				36·114	114·410	0·366	
				36·114	127·643	0·401	
				47·142	199·321	0·458	
					Mean..	0·377	

We see by the results of these experiments, in which the arc of contact varied in the ratio of 8·3 to 1 nearly, and where the tension has reached very nearly the limits assigned to the belts of machinery, that the value of the ratio f, of friction to the pressure, remained very nearly constant.

The three first series of the first Table fully confirm the theoretic considerations. The fourth series relates to a belt quite new, and very stiff, and to this we attribute the small increase presented by it in the mean value. This belt having, moreover, only a width of ·091 ft., or about the half of the preceding, we see that this last series confirms, as to belts, the law of the independence of surface.

In the experiments of the second Table, the extent of arc embraced varied in the ratio of 6 to 1, the breadth of the belt pressed against the pulley in that of 2 to 1, the tension from 1 to 3 and from 1 to 6, and still the value of the ratio f, of friction to the pressure, remained sensibly constant, and equal in the mean, for the dry belt and dry pulleys, $f = 0·282$. When the pulley was moistened with water we had $f = 0·377$.

Conclusions.—In considering the results of these two series of experiments upon the friction of belts upon wooden drums and cast-iron pulleys, we see that we are justified in admitting that the ratio of the resistance to the pressure is:—

1st. Independent of the width of the belt and of the developed length of the arc embraced, or of the diameters of the drums, or, what amounts to the same, are independent of the surface of contact.

2nd. Proportional to the angle subtended by the belt at the surface of the drum.

3rd. Proportional to the logarithm of the ratio of the tension of the strips, and expressed by the formula

$$f = \frac{\log.\left(\frac{P}{Q}\right)}{1·363}.$$

Experiments upon the Variation of the Tension of Endless Cords or Belts used in Transmitting Motion.—We pass now to an experimental proof of the theory given by M. Poncelet, upon the transmission of motion by endless cords or belts, and will first give a description of its nature.

When a cord or belt surrounds two pulleys or drums, between which it is designed to maintain a conjoint motion, care is taken to give it a sufficient tension, which is usually determined by trial, but which it would be best to calculate, as we shall see hereafter. The *primitive* tension is, at the commencement, the same for both parts of the belt; and this equality, established in repose, is only destroyed by the friction of the axles, which may act in either direction, according to that of the motion of the pulleys.

Let us examine how this motion is transmitted in such a system. Let C be the motive drum; C′ the driven drum; T_1 the primitive tension common to the parts A A′ and B B′ of the belt, from the moment when the drum C begins to turn until it commences to turn the drum C′.

644.

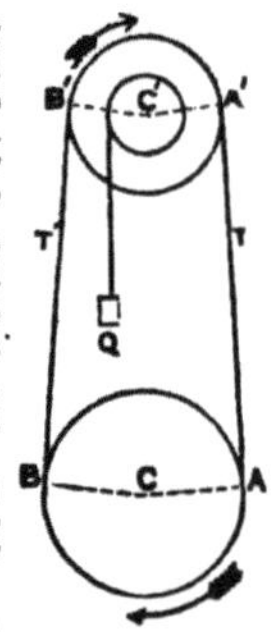

The point A of primitive contact of the part A A′ advances, in separating from the point A′, in the direction of the arrow; the strip A A′ is stretched, and its tension increased by a quantity proportional to this elongation, according to a general law proved by experiment upon traction.—(See Lessons upon 'Résistance des Matériaux.') At the same time, the point B of contact of the part B B′ approaches by the same quantity towards the point B′, so that the portion B B′ is diminished by a quantity equal to the increase of that of A A′. If, then, we call T the tension of the driving portion A A′, at the instant of its being put in motion, T′ the tension of the driven part B B′, t the quantity by which the primitive tension T_1 is increased in the portion A A′, and diminished in the part B B′, we shall have $T = T_1 + t$, and $T' = T_1 - t$, and consequently $T + T' = 2\,T_1$. Then, at any instant, the sum of the two tensions T and T′ is constant and double the primitive tension.

Now it is evident that in respect to the driven drum C′ the motive power is the tension T, and that the tension T′ acts as a resistance with the same lever arm, so that the motion is only produced and maintained by the excess $T - T'$ of the first over the second of these tensions.

If the machine is, for example, designed to raise a weight Q acting at the circumference of an axle with a radius R′, it is easy to see, according to the theory of moments, that at any instant of a uniform motion of the machine we must have the relation

$$(T - T')\,R = Q\,R' + f\,N\,r,$$

N being the pressure upon the journals, and r their radius.

The pressure is easily determined; for calling a the angle formed by the directions A A′ and B B′ of the belts with the line of the centres C C′, M the weight of the drum, we see immediately that $N = \sqrt{[M + Q + (T - T')\sin. a]^2 + (T + T')\cos.^2 a}$, an expression which, according to the algebraic theorem of M. Poncelet, has for its value a fraction equal to $\frac{1}{25}$ nearly, when the first term under the radical is greater than the second, $N = 0{\cdot}96\,[M + Q + (T - T')\sin. a] + 0{\cdot}4\,(T + T')\cos. a$. This value of N being introduced into the formula for equality of moments, we have a relation containing only the values of the resistance Q and of the tensions. But as it may be somewhat complicated for application, observing that in most cases the influence of the tensions T and T′ upon the frictions will be so small that it may be neglected, at least in a first approximation, we proceed as follows:

First, neglecting the influence of the tensions upon the friction, we have simply, in the actual case, $N = M + Q$, and consequently $(T - T')\,R = Q\,R' + f\,(M + Q)\,r$, whence we deduce

$$T - T' = \frac{Q\,(R' + f\,r) + f\,.\,M\,r}{R} = Q,$$

which furnishes a first value for the difference of tensions, which is the motive power of the apparatus.

But this is not sufficient to make known these tensions, and it is necessary to determine the primitive tension T′, so that in no case the belt may slip.

According to the theory of M. Prony, we have, at the instant of slipping, between the tension, T and T′ the relation $T = T' \times 2{\cdot}718 + f\,\frac{S}{R}\,T' = K\,T'$, the number K being a quantity depending upon the nature and condition of the surfaces of contact, as well as upon the angle $\frac{S}{R}$ embraced by the belts upon the drum C′. These quantities are known, and we may in each case calculate the value of K by this formula, or take it from the following Table, which answers to nearly all the cases in practice.

By means of this Table, we shall have then the value of $T = K\,T'$, and consequently $T - T' = (K - 1)\,T' = Q$, Q representing the greatest value which the difference of tensions should attain to overcome the useful and passive resistances.

From this relation we may derive the smallest tension to be allowed to the driven portion of the belt to prevent its slipping: we thus have $T' = \frac{Q}{K - 1}$.

We should increase this value by $\frac{1}{10}$ at least, to free it from all hazard of accidental circumstances, and to restore the account of the influence of the tensions upon the friction, which was neglected. This established, we have $T = Q + \frac{Q}{K - 1}$, and consequently

$$T_1 = \frac{T + T'}{2} = \frac{1}{2}\,\frac{K + 1}{K - 1}\,Q.$$

All the circumstances of the transmission of motion will then be determined.

If these first values of T, T', and T_1 are not considered as sufficiently correct, we may obtain a nearer approximation by introducing them in the value of the pressure N, and thus deduce a more exact value of Q, which will serve to calculate anew T', then T and T_1.

Ratio of the Arc embraced to the Circumference.	Value of the Ratio K.					
	New Belts upon Wooden Drums.	Belts in usual Condition.		Moistened Belts upon Cast-iron Pulleys.	Cords upon Wooden Drums or Axles.	
		Upon Wooden Drums.	Upon Cast-iron Pulleys.		Rough.	Smooth.
0·20	1·87	1·80	1·42	1·61	1·87	1·51
0·30	2·57	2·43	1·69	2·05	2·57	1·86
0·40	3·51	3·26	2·02	2·60	3·51	2·29
0·50	4·81	4·38	2·41	3·30	4·81	2·82
0·60	6·59	5·88	2·87	4·19	6·58	3·47
0·70	9·00	7·90	3·43	5·32	9·01	4·27
0·80	12·34	10·62	4·09	6·75	12·34	5·25
0·90	16·90	14·27	4·87	8·57	16·90	6·46
1·00	23·14	19·16	5·81	10·89	23·90	7·95
1·50	..	..	..	..	111·31	22·42
2·00	..	..	..	..	535·47	63·23
2·50	..	..	..	..	2574·80	178·52

Experiments upon the Variations of Tensions of Endless Belts employed for the Transmission of Motion.—To verify by experiment the exactness of these considerations, Morin placed vertically above the axis of a hydraulic wheel, and of a pulley mounted upon its axle, a cylindrical oak drum, 2·74 ft. in diameter, and whose axis was 9·84 ft. from that of the wheel. Around this drum A'B', and the pulley A B, was passed a belt which, instead of being in one piece, was in two parts, joined at each end by a dynamometer, with a plate and style, of a force of 441 lbs. Moreover, these *dynamometers* were easily secured in positions such that that of the descending portion of the belt was near the upper drum, and that of the ascending near the lower drum. Thus the belt could be moved over a space of 6·56 ft. without the risk of the instruments being involved with the drums.

645.

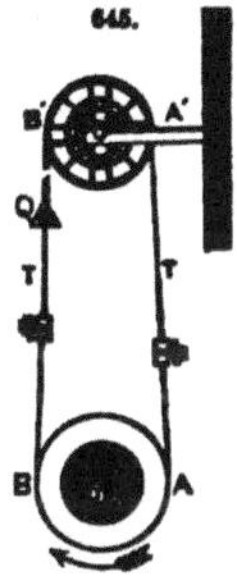

A thread wound several times around the circumference of one of the grooves of the plate of each of the dynamometers, and attached by the other end to a fixed point, caused the plate to turn when the apparatus was in motion, and the paper with which the plate was covered received thus the trace of the style of the dynamometer.

The belt being passed over the two drums, the tensions of the parts were varied at will in either direction, by suspending at the circumference of the upper drum a plate Q charged with weights. As to the primitive tension, it was increased by bringing nearer together the ends of the belt, or in diminishing its length before the experiment.

The apparatus being thus prepared for observations, before loading the plate Q, we traced the circles of flexure of each of the dynamometers, so as to have the tensions of the belt at rest, and to obtain by their sum the double of the primitive tension T_1. We may conceive that these two tensions can never be quite equal; but that is not important, inasmuch as we have to deal only with their sum.

This obtained, we load the plate with a weight which, being suspended upon the circumference by a cord of a diameter equal to the thickness of the belt, has the same lever arm as the tensions. That part of the belt opposed to this weight is stretched, and the part on the same side is slackened, and we trace the new curves of the flexure of the dynamometers.

For the same primitive tension we may make a series of experiments up to the motive weight, under the action of which the belt slides upon either drum.

In these experiments facilities were afforded for allowing the two drums to turn a certain amount under the action of the tensions, so that we could realize the three cases in practice, to wit, that of the variation of tensions before motion was produced, that of the variation during motion, and finally, that of the slipping.

The belt used in these experiments was very pliable, soft, and little liable to be polished in slipping. In calculating the ratio of the friction to the pressure for this belt, by means of experiments 3, 13, and 19, we find respectively $f = 0·578$, $f = 0·596$, and $f = 0·544$, the mean being $f = 0·573$.

Remarks upon the Results contained in the following Table.—We see that the first line of each series corresponds to the case where there was no additional weight, and where each portion of the belt took the primitive pressure corresponding to the distance apart of the axes. As the weight suspended from the drum was increased, the tension of one of the strips was increased, and that of the other was diminished; but so that their sum remained constant, as is shown by the fifth column of the Table.

These results, which completely confirm the theory of M. Poncelet, being relative to tensions whose sum reaches 198 lbs. and more, where the greatest rise as high as 169 lbs., and the smallest fall as low as 11 lbs., comprise nearly all the cases in practice, and show that this theory may with safety be applied to the calculation of transmission of motion by belts.

EXPERIMENTS UPON THE VARIATION OF THE TENSIONS OF ENDLESS BELTS EMPLOYED IN TRANSMITTING MOTION BY PULLEYS OR DRUMS.

Number of Experiment.	Weight suspended at the Circumference.	Tension of the Part.		Sum of the Tensions, $T + T' = T_1$.	Remarks.
		Rising or stretched, T.	Descending or slackened, T'.		
	lbs.	lbs.	lbs.	lbs.	
1	0·00	38·57	32·84	71·41	
2	44·61	60·07	12·84	72·91	
3	59·55	63·14	10·19	73·33	The belt slipped.
4	0·00	64·86	57·41	122·27	
5	22·56	75·09	46·82	121·90	
6	44·61	84·51	36·26	120·77	The dynamometers moved about 3·28 ft.
7	66·67	97·96	24·17	122·13	
8	97·54	109·92	20·76	130·67	
9	0·00	73·73	62·32	136·05	Ditto ditto.
10	55·64	99·00	41·53	140·53	
11	110·78	117·77	20·38	138·15	
12	0·00	66·91	57·94	124·85	
13	115·19	103·78	15·86	119·64	The belt slipped.
14	0·00	107·53	98·91	206·44	
15	55·64	130·05	70·26	200·31	
16	110·78	157·02	47·57	204·59	
17	0·00	97·24	88·75	185·99	
18	110·78	154·29	40·78	195·07	
19	174·23	170·67	43·42	214·09	Ditto ditto.
20	0·00	86·72	71·34	158·06	
21	88·72	134·84	44·17	179·01*	

* Besides the load Q, there was suspended to the main circumference of the floats of the wheel, at 6·05 ft. from the axis, a weight of 22·56 lbs., which broke the equilibrium.

In conclusion, we would add that belts designed for continuous service may be made to bear a tension of 0·551 lb. per ·0000107 sq. ft., or ·00155 sq. in. of section, which enables us to determine their breadth according tc the thickness.

We give, from the 'Journal of the Franklin Institute' (1868), an account of the experiments and comparisons of Robert Briggs and H. R. Towne, relative to the transmission of force by belts and pulleys. The results so independently obtained by these investigators will, we have no doubt, be useful to those engaged in the construction and working of machinery.

R. Briggs observes:—"There are few mechanical engineers who have not been frequently in want of tabular information or readily applicable formulas, upon which they could place reliance, giving the power which, under given conditions and velocity, is transmitted by belts without unusual strain or wear. The formula of the belt or brake is well known and simple; and it is only necessary to acknowledge and adopt a value for the coefficient of friction (or of adhesion, which is perhaps the better term), to allow this formula to be applied in daily use. And this coefficient of friction has been carefully established by the experiments of General Morin and M. Prony, and has been made available to English and American engineers, by the translation of Bennett. It must be remarked that there are some mistakes in the text of Bennett's translation, which will lead to serious errors, unless read by a careful investigator.

"With every point needed, therefore, at the command of the engineer, it is somewhat surprising that a more extensive publication and general use of the data has not followed.

"But notwithstanding the existence of this correct mathematical and experimental information, the numerous tables which have been given by mechanical engineers appear to have had only that kind of practical basis which has come from guessing that an engine or a machine, either the driving or the driven, with a belt of given width, was producing or requiring some quantity of power, which might be expressed in terms generally without any stated arc of contact."

The terms referred to are vulgarly called foot-pounds, but should be nominated *units of work*. See PRINCIPLE OF WORK.

"Three rules given by practical mechanics vary so much, as to give as bases for estimate (without regard to arc of contact) 0·76 horse-power, 0·93 horse-power, and 1·75 horse-power, respectively, for the power of a belt 1 in. wide running 1000 ft. per minute.

"It was the requirement to know the exact useful effect of a novel disposition involving an unusual small arc of contact of the belt upon the pulley, where much embarrassment would result if the application proved itself unsatisfactory, that led to the present inquiry. As the writer was not able to give the time demanded for making such experiments as would establish the practical coefficient of adhesion, he, Briggs, suggested what was desired to H. R. Towne; and the numerous experiments, of which he gives the accompanying report, are the result of the labour and care of H. R. Towne.

"It was not until after the experiments were completed," says H. R. Towne, "that either he or R. Briggs knew of the publication of M. Prony or General Morin, although Bennett's translation rested upon the shelves of the writer's library; but, aside from the gratification which we feel at the corroboration, we think the reader of this article will be pleased to know that our data is founded upon the ordinary pulleys and belts of the workshop, and our experiments were not impaired by any niceties which common workmen would not apply.

"Even the crudeness of our experimental apparatus, and the general not over-exact method adopted, will serve to demonstrate to the minds of practical men the possibility of relying upon figures which have been established so nearly in accordance with the customs of the workshop."

We have before shown that $\frac{T_1}{T_2} = \epsilon^{f\frac{l}{r}}$, or, $\frac{Q}{P} = \epsilon^{f\frac{l}{r}}$, where ϵ is the base of hyperbolic logarithms; T_1 = the tension of the belt on the tight side; T_2 = the tension of the belt on the loose side; f = the coefficient of friction; r = radius of pulley; and l = the length of the arc of contact. We can further transform this equation, by substituting the ratio of the angle in degrees for the length of contact on the arc, compared to the radius, $\epsilon = 2\cdot718281828$.

Thus $\frac{2r\pi}{360}$ = arc of 1°, let $l = a\left(\frac{2r\pi}{360}\right) \therefore \frac{T_1}{T_2} = \epsilon^{f\frac{2\pi a}{360}}$, and taking the numerical values of ϵ, π, and dividing out the 360, $\therefore \frac{T_1}{T_2} = 2\cdot718^{0\cdot017456 f a}$

$$\therefore \log.\left(\frac{T_1}{T_2}\right) = 0\cdot4343\ (0\cdot017456 f a), \qquad [1]$$

$$\therefore \log. T_1 - \log. T_2 = 0\cdot00758 f a$$

$$\therefore \frac{T_1}{T_2} = 10.^{0\cdot00758 f a}, \qquad [2]$$

$$\therefore f = \frac{\log.\left(\frac{T_1}{T_2}\right)}{0\cdot00758\, a} \qquad [3]$$

As we assumed, $P = T_1 - T_2$. $\therefore T_2 = T_1 - P$, which inserting in equation [2]

$$\therefore \frac{T_1}{T_1 - P} = 10^{0\cdot00758 f a}$$

$$\therefore P\left(10^{0\cdot00758 f a}\right) = T_1\left(10^{0\cdot00758 f a}\right) - T_1$$

$$\therefore P = T_1\left(1 - 10^{-0\cdot00758 f a}\right) \qquad [4]$$

The third equation is the one to which we would now call attention. By it, for any given values for the ratio $\frac{T_1}{T_2}$, we can determine the coefficient of friction, when, by experiment, we have fixed the greatest difference of the two strains without slipping on a pulley with a given arc (measured by a) of contact.

We would here, says the experimenter, make a very important observation, which forms the key of the whole system of transmission of force by belts. In practice, *all belts are worked at the maximum coefficient of friction*. A belt may, when new or newly tightened, work under heavy strain, and with a small coefficient of friction called into action; but in process of time it becomes loose, and it is never tightened again until the effort to perform its task is greater than the value of the coefficient with a given tension of belt and the belt-slips. We, says Briggs, run our belts as slack as possible, so long as they continue to drive.

It has been shown that the value of $T_1 + T_2$, or the sum of the strains upon the two sides of a belt (loose and tight), is a constant quantity—that is, when a belt is performing work it will become loose on the one side to the exact amount that it is strained on the other, and when at rest, not transmitting force, the tensions will become equal, and their sum be the same as before. It is manifest that the limit of the strength of a belt is found in the maximum tension T_1 and that this strength being known, the effective pull (P) is further limited with any given arc of contact by the value f, of the coefficient of friction.

The discussion has so far been limited to the pull exerted by a belt; when we would include the power which belts will transmit, we have only to multiply the pull by some given or assumed velocity, to transform our equations into work performed.

By means of the third equation, we will now deduce a value for the coefficient of friction as given by the experiments.

All the experiments were with the arc of contact = 180° = a, which, substituting $f = \frac{\log.\frac{T_1}{T_2}}{1\cdot3644}$, and the result of 168 separate experiments has given, under tensions of T_1 from 7 to 110 lbs. to an inch of width of belt, $\frac{T_1}{T_2} = 6\cdot294$. $\therefore f = \frac{\log. 6\cdot294}{1\cdot3644} = 0\cdot5833$. Bennett's Morin, page 306, ¶ 253, gives $f = 0\cdot573$.

In this case T_1 has in all cases been so much in excess of T_2, as to slip the belt at a defined, slow, but not accelerating, motion.

From an examination of the report of the experiments, we think the reader will coincide with our conclusion that $\frac{8}{10}$ of this value of $\frac{T_1}{T_2}$ can be taken as a suitable basis for the working friction or adhesion which will cover the contingencies of condition of the atmosphere as regards temperature and moisture; or $\frac{T_1}{T_2} = 3 \cdot 7764$ (maximum practical value) $\therefore f = \frac{\log. 3 \cdot 776}{1 \cdot 3644} = 0 \cdot 42292.$

It should be noted that the experiments were made without any appreciable velocity of belt, and throughout this paper no regard has been paid to the effect of velocity or of the dimensions of the pulleys upon the value of the coefficient of friction.

For pulleys less than 12 in. diameter (with the belts of the ordinary thickness of about $\frac{3}{16}$ths in.), and for velocities exceeding about 1000 ft. a minute, allowance must be made for the rigidity of the belt in the one case, and for the *interposition of air between the pulley and the belt* in the other. At high speeds, say 3000 ft. velocity of belt a minute, the want of contact can be seen, sometimes, to the extent of one-third the arc encompassed by the belt. Briggs has proposed to place a deflector or stripper near the belt, to take off the stratum of air moving with it, but has never tried the experiment, although he has little doubt of its giving some advantage.

The experiments further show that 200 lbs. to an inch of width of belt is the maximum strength of the weakest part—that is, of the lace-holes. Taking this, with a factor of safety at one-third, we have the working strength of the belt, or the practical value for $T_1 = 66\frac{2}{3}$ lbs. Bennett's Morin, page 306, ¶ 253, gives 55·1 lbs. the inch of width as admissible. In the case when belts are spliced instead of laced, a great increase of strength has been shown, the experiments giving 380 lbs. to an inch of width, or 125 lbs. safe working strength.

If we insert these values of f and T_1 in [4],

$$\therefore P = 66\tfrac{2}{3}\left(1 - 10^{-0 \cdot 0078 \times 0 \cdot 42292\, \alpha}\right)$$

$$\therefore P = 66\tfrac{2}{3}\left(1 - 10^{-0 \cdot 003206\, \alpha}\right) \qquad [5]$$

This equation [5] is the really important one in practice, and by means of logarithms can be solved for any values of $\alpha°$ readily; but as some of those who may wish to use it may not be at once prepared to use the logarithmic notation, from want of use or practice, we give an example. Suppose we take an angle of 90°, the negative exponent then becomes $-0 \cdot 003206 \times 90 = -0 \cdot 28854$; subtracting this from 1, we have $-1 \cdot 71146$. This term thus becomes $10^{-1 \cdot 71146}$. Now this expression is only the notation for anti-logarithm $-1 \cdot 71146$, or in words the number for which $-1 \cdot 71146$ is the logarithm. Logarithmic tables give this number $= 0 \cdot 51505$, and the equation

$$P = 66\tfrac{2}{3}\left(1 - 10^{-0 \cdot 003206 \times 90}\right) = 66\tfrac{2}{3}\left(1 - 10^{-1 \cdot 71146}\right) = 66\tfrac{2}{3}\,(1 - 0 \cdot 51505) = 66\tfrac{2}{3} \times 0 \cdot 48495.$$

$$\therefore P = 32 \cdot 33.$$

The largest possible angle for an open belt, without a carrier or tightener, is 180°, as upon either the driving or the driven pulley this cannot be exceeded; but for crossed, or carried, or tightened belts, the angle may be as large as 270°.

Briggs and Towne give the following Table of results for different arcs of contact (corresponding to $\alpha°$) within the usual limits of practice.

TABLE I.—STRAIN TRANSMITTED BY BELTS OF ONE INCH WIDTH UPON PULLEYS WHEN THE ARCS OF CONTACT VARY AS THE ANGLES OF

90°	100°	110°	120°	135°	150°	180°	210°	240°	270°
lbs.	lbs.	lbs.	lbs.	lbs.	lbs.	lbs.	lbs.	lbs.	lbs.
32·33	34·80	37·07	39·18	42·06	44·64	49·01	52·52	55·33	57·58

If we suppose the pulley to be 1 ft. in diameter, and to run some number, N, of revolutions a minute, we have the power transmitted $= N \pi P$.

And we give the following Table for different arcs of contact (corresponding to $\alpha°$) within the usual limits of practice.

TABLE II.—POWER TRANSMITTED BY BELTS ON PULLEYS ONE FOOT IN DIAMETER ONE REVOLUTION A MINUTE. ARCS OF CONTACT OF BELTS UPON PULLEYS CORRESPONDING TO THE ANGLES:

Inches of Width of Belt.	90°	100°	110°	120°	135°	150°	180°	210°	240°	270°
	foot-lbs.	foot-lbs.	foot-lbs.	foot-lbs.	foot-lbs.	foot-lbs.	foot-lbs.	foot-lbs.	foot-lbs.	foot-lbs.
1	102	109	116	123	132	140	154	165	174	181
2	203	219	233	246	264	280	308	330	348	361
3	305	328	349	369	396	420	462	495	521	542
4	406	437	466	492	528	560	616	660	695	723
5	508	547	582	615	660	701	770	825	869	904
6	609	656	699	738	792	841	924	990	1043	1084
7	711	766	815	861	924	982	1078	1155	1217	1265
8	813	875	932	985	1056	1122	1232	1320	1391	1446
9	914	984	1048	1108	1188	1262	1386	1485	1564	1626
10	1016	1094	1165	1231	1321	1402	1540	1650	1738	1807

The application of Table II. to any given cases of known angle of the arc of contact, width of belt in inches, diameter of pulley in feet, and number of revolutions, is simply to take the figures from the Table for the first two, and multiply by the two succeeding conditions, to obtain the foot-pounds of power transmitted.

These experimenters take the following examples:—1st. Schenck (of New York) found an 18-in. wide belt running 2000 ft. a minute, the pulleys being 16 ft. to 6 ft., would give 40 horse-power, with ample margin (one-fourth).

If we take the distance from centre of the 16-ft. pulley to that of the 6-ft. to be 25 ft. (about the usual way of placing the fly-wheel pulley of an engine in regard to the main line of shafting), we have the arc of contact subtending about 153°. From Table I. the strain transmissible is 45·1 lbs. × 18 × 2000 = 1,623,600 foot-pounds or units of work = 49·2 HP.

2nd. William B. Le Van (of Philadelphia) found by indicator that an 18-in. wide belt running 1800 revolutions a minute, the pulleys being 16 ft. and 5 ft., respectively, transmitted 43 HP., with maximum power transmissible unknown. If we take the centre's distance, as before, at 25 ft., we have the arc of contact subtending about 150°.

From Table I. we derive 44·64 lbs. as the strain transmissible × 18 × 1800 = 1,446,336 foot-pounds = 43·83 HP. The same authority found by indicator that a 7-in. wide belt over two 2-ft. 6-in. pulleys, 11 ft. centre to centre (horizontal), moving 942 ft. a minute, gave 8 HP. From Table I., for 180° angle, we take 49·01 × 7 × 942 = 323,172 foot-pounds = 9·79 HP. This belt was stated to be very tight.

3rd. A. Alexander ('Engineer,' March 30, 1860) gives a rule that a 1-in. belt will, at 1000 ft. velocity, transmit 1⅔ HP.

If we take the contact at 180° from Table I., 49·01 × 1000 = 49,000 foot-pounds, we have only 1½ HP.

4th. William Barbour (same journal, March 23, 1860) gives as the power a 1-in. belt will transmit with 1000 ft. velocity = 0·927 HP., when we derive with 180° angle from our Tables = 1½ HP.

5th. A. B. Ex (same journal, April 6, 1860) gives a rule

$$\frac{\text{diameter in inches} \times \text{revolutions a minute} \times \text{breadth in inches}}{5000} = \text{N HP},$$

ratio of pulleys not to exceed 5 to 1. Changing this rule to

$$\frac{\text{diameter in feet} \times \text{revolutions a minute} \times \text{width in inches}}{5000 \div 12 \div 33{,}000} = \text{N foot-pounds}.$$

$$\therefore \frac{\text{Diameter in feet} \times \text{revolutions a minute} \times \text{width in inches}}{0{\cdot}01263} = \text{N foot-pounds}.$$

$$\therefore 79{\cdot}2 \times \text{diameter in feet} \times \text{revolutions a minute} \times \text{width in inches} = \text{N foot-pounds}.$$

From Table II. the angle of 120° gives 123 in place of 79·2, and it would appear this authority adopts about ⅔ the effect we take.

6th. W. Fairbairn gives ('Mills and Mill Work,' Part II., page 4) a table of approximate width of leather straps in inches necessary to transmit any number of horse-power; the velocity of the belt being taken at 25 to 30 ft. a second (1500 to 1800 a minute), 1-foot pulley, 3·6 in. wide, gives 1 HP.

Assume 1650 ft. a minute, contact 180°, we have from Table I., 1650 × 49·01 × 3·6 × 1 = 29,112 foot-pounds = 0·87 HP.

7th. Rankine gives ('Rules and Tables,' page 241) 0·15 as the coefficient of friction, probably applicable to the adhesion of belts on pulleys to be used with his formulas in estimating the power transmitted. Neither experiments nor practice give so small a coefficient as this.

We, says Briggs, could multiply authorities on these points, but think the corroboration of those we quote with our Tables sufficient to establish our experimental and estimated coefficient of friction, $f = 0{\cdot}423$, as a proper practical basis.

These experimenters give the two following cases, not only to show the application of the formula 5, but as matters of some interest.

In the construction of one of the forms of centrifugal machines for removing water from saturated substances, the main or basket spindle is driven by cone-formed pulleys, one of which, being covered with leather, impels the other by simple contact.

In the particular instance taken, the iron pulley on the spindle was 6 in. largest diameter, and the leather-covered driving-pulley was 12 in. largest diameter; the length of cones on the face was 4 in., this last dimension corresponding to width of belt in other cases. By covering the leathered pulley with red-lead, we were able to procure an *impression* on the iron pulley, showing the width of the surfaces of contact when the pulleys were compressed together with the force generally applied when the machine was at work. This width was, at the largest diameters, almost exactly ¼ in. From the nature of the two convex surfaces compressing the leather between them, the actual surface of efficient contact cannot be taken at over half this width. (The slight error in estimating this contact as straight lines in place of circular arcs may be neglected.) This gives the angle subtended by the arc of contact on the iron pulley = 2½°, taking equation [5]. $P = 66\tfrac{2}{3}\left(1 - 10^{-0{\cdot}003206\,a^{\circ}}\right) = 66\tfrac{2}{3}\left(1 - 10^{-0{\cdot}008015}\right) = 66\tfrac{2}{3}\left(1 - 10^{-1{\cdot}991985}\right) = 66\tfrac{2}{3}\,(1 - 0{\cdot}98171) = 66\tfrac{2}{3}\,(0{\cdot}01829) = 1{\cdot}3717.$

Now, the average diameter of the iron pulley in the middle of its 4-in. face is 4·708 in. = 0·3923 ft., with a circumference of 1·2326 ft., and it is usual to run, at the least velocity, 1000 revolutions a minute; whence the power given by these pulleys = 1·3717 lb. × 4 in. × 1·2326 ft.

× 1000 revolutions = 6757 foot-pounds = $\frac{1}{5}$ HP. As the work performed by one of these centrifugal machines is the acquirement of velocity under the resistance of the friction of the machine and of the air, and the work of expelling the moisture is so insignificant in comparison that it may be neglected in estimating, it can be taken as probable that the real power demanded to keep the machine in motion is very nearly that given by calculation. It should be stated that the basket belonging to this particular machine is 29 in. diameter and 12 in. deep.

The second special case we instance at present consists in a proposed arrangement for driving a fan which had previously been found to demand an 8-in. belt on a 10-in. pulley to run it 1275 revolutions a minute. (The arc of contact here was 162°, so that the apparent power with a very tight belt was 37½ HP.: but about ¼ of this was defective adhesion from running a rigid belt over so small a pulley.) It was thought desirable to avoid the fast-running countershafts, and drive this direct from an engine-pulley fly-wheel, by impingement, so to speak, of the belt on its tight side between the fan-pulley and another larger carrier-pulley, against a portion of the periphery of the fly-wheel.

If we suppose the force demanded measured on the fan-pulley, as before, to be 40 HP. = 1,320,000, and the fan-pulley to be 10 in. diameter × 16 in. wide, and to run 1250 revolutions a minute; $\therefore \frac{1,320,000}{1250 \times \frac{10}{12} \times 16 \times \pi} = 25\cdot2$ as the pull, P, on each inch of width of the belt as it comes from the 10-in. pulley. By substituting this value for P in equation [5], and then reducing the equation to find the value for a°, we have $a^\circ = 65^\circ$, which is the angle of contact demanded to give the necessary adhesion.

It will be noticed that this angle is independent of the diameter of the fly-wheel pulley, it being only requisite that that diameter should be such as with the given or assumed number of revolutions will produce the given velocity. In the case taken for example, the fly-wheel pulley was 16 ft. diameter × 16 in wide, with 70 revolutions a minute velocity.

As we have before remarked, the sum of the two tensions on the belt is constant, whether the belt is performing work or not; that is, $S = T_2 + T_1$; but $P = T_2 - T_1$. $\therefore T_2 = T_1 - P$. $\therefore S = 2T_1 - P$.

As we assumed in equation [5] T to equal 66⅔ lbs., we can substitute the value of P as in Table I. in the equation, $S = 2\,(66\tfrac{2}{3}) - P = 133\tfrac{1}{3} - P$, from which it is evident that the sum of the tensions will vary with P or with the angle of contact. It is evident, also, that the load upon the shaft proceeding from the tensions T_2 and T_1 will be the resultant of whatever angle the belt makes with a line joining the centres of the two pulleys, or as the cosine of that angle.

By constructing on paper a pair of pulleys, it will be readily discerned that the angle in question for small pulleys $= 90^\circ - \frac{a}{2}$, and for large and crossed ones, $= \frac{a}{2} - 90^\circ$, we can consequently form the following Table:—

TABLE III.—STRENGTH OF LACING OF JOINT 66⅔ LBS. THE INCH WIDE,

Showing, first, the sum of tensions on both sides of a belt to each inch of width, whether in motion or at rest, when strained to transmit the maximum quantity of power in general practice; and showing, second, the load carried by the shafts and supported constantly by the journals the inch of width of belt, when the arcs of contact vary as the angles of

	90°	100°	110°	120°	135°	150°	180°	210°	240°	270°
	lbs.	lbs.	lbs.	lbs.	lbs.	lbs.	lbs.	lbs.	lbs.	lbs.
1st,	101·	98·53	96·26	94·15	91·27	88·69	84·32	80·81	78·	75·75
2nd,	71·42	75·47	78·85	81·53	84·32	85·67	84·32	78·05	67·59	53·56

When machinery is driven by gearing, the shafts only carry the running wheels and the weight, and when the machines are thrown on, the friction of the lines increases with the work; but with belts and pulleys the load on the line and its frictional resistance is constant, whether the machinery works or lies idle.

It is not proper to assume that the load produced by the belt on the shaft is exactly that given by the second line in Table III.; but we can be safe in taking those weights as rarely exceeded, because belts begin to fail when they are; and as rarely much less, because few of our machines are not worked up to their belt of capacity.

The advantages shown by the figures on all the Tables, but especially on the last, in those arcs of contact over 180° where crossed belts are used, have the substantial ground of practice, although many mechanics are unaware of the facts. R. Briggs instances a case of several heavy grindstones having from main to counter lines 8-in. crossed belts on pulleys, 3 ft. diameter, running 120 revolutions, only 8 ft. centre to centre, where belts have already lasted, day and night use, three-and-a-half years. For the same purpose, 6-in. open belts were formerly used, with an average duration of a few weeks only.

Another use of a crossed belt is for long belts, the crossing effectually preventing those waves which generally impair, if they do not destroy, such belts when open.

We give a tabular record of the experiments of H. R. Towne, which will repay examination, as exhibiting several interesting and instructive facts connected with the efficiency of leather belts.

These experiments were made with leather belts of 3 and 6 in. width, and of the usual thickness—about $\frac{3}{16}$ths of an inch. The pulleys used were respectively of 12, 23¼, and 41 in. diameter, and were in each case fast upon their shafts. They were the ordinary cast-iron pulleys, turned on the face, and, having already been in use for some years, were fair representatives of the pulleys usually found in practice.

Experiments were made first with a perfectly new belt, then with one partially used and in the best working condition, and finally with an old one, one which had been so long in use as to have deteriorated considerably, although not yet entirely worn out. The adhesion of the belts to the pulleys was not in any way influenced by the use of unguents or by wetting them: the new ones, when used, were just in the condition in which they were purchased; the others in the usual working condition of belts as found in machine-shops and factories—that is, they had been well greased, and were soft and pliable.

The manner in which the experiments were made was as follows:—The belt being suspended over the pulley, in the middle of its length, weights were attached to *one* side of the belt, and increased until the latter slipped freely over the pulley; the final, or *slipping*, weight was then recorded. Next, 5 lbs. were suspended on *each* side of the belt, and the additional weight required upon *one* side to produce slipping ascertained as before, and recorded. This operation was repeated with 10, 20, 30, 40, and 50 lbs., successively, suspended upon both sides of the belt. In the Tables, these weights, *plus* half the total weight of the belt, are given as the "equalizing weights" (T_2 in the formulas); and the *additional* weight required upon one side to produce slipping, is given under the head of "unbalanced weights;" this latter, *plus* the equalizing weight, gives the total tension on the loaded side of the belt (T_1 in the formulas).

The belt, in slipping over the pulley, moved at the rate of about 200 ft. a minute, and with a constant, rather than increasing, velocity; or, in other words, the final weight was such as to cause the belt to slip smoothly over the pulley, but not sufficient to entirely overcome the friction tending to keep the belt in a state of rest. In this case, that is, with an excessive weight, the velocity of the belt would have approximated to that of a falling body; while in the experiments its velocity was much slower, and was nearly constant, the friction acting precisely as a brake. By being careful that the final weight was such as to produce about the same velocity of the slipping belt in all of the experiments, reliable results were obtained.

It became necessary to make use of a weight such as would produce the positive motion of the belt described above, as it was found impossible to obtain any uniformity in the results when the attempt was made to ascertain the minimum weight which would cause the belt to slip. With much smaller weights *some* slipping took place, but it was almost inappreciable, and could only be noticed after the weight had hung for some minutes, and was due very probably to the imperceptible jarring of the building. After essaying for some time to conduct the experiments in this way, and obtaining only conflicting and unsatisfactory results, the attempt was abandoned, and the experiments made as first described.

In this way, as may be seen, results were obtained which compare together very favourably, and which contain only such discrepancies as will always be manifest in experiments of the kind. It is only by making a great number of trials and averaging their results, that reliable data can be obtained.

The value of the coefficient of friction which we deduce from our experiments is the mean of no less than 168 distinct trials.

It will be noticed, however, that the coefficient employed in the formulas is but *six-tenths* of the full value of that deduced from the experiments, the latter being 0·5853 and the former 0·4229. This reduction was made, after careful consideration, to compensate for the excess of weight employed in the experiments over that which would just produce slipping of the belt, and may be regarded as safe and reliable in practice.

A note is made, over the record of each trial, as to the condition of the weather at the time of making it—whether dry, damp, or wet; and it will be noticed that the adhesion of the belts to the pulleys was much affected by the amount of moisture in the atmosphere. It is to be regretted that this contingency was not provided for, and a careful record of the condition of the atmosphere kept by means of an *hygrometer*. The experiments indicate clearly, however, that the adhesion of the old and the partially-used belts was much increased in damp weather, and that they were then in their maximum state of efficiency. With the new belts, the indications are not so positive; but their efficiency seems to have been greatest when the atmosphere was in a dry condition.

Experiments were also made upon the tensile strength of belts, with the following results:—The weakest parts of an ordinary belt are the ends through which the lacing-holes are punched, and the belt is usually weaker here than the lacing itself. The next weakest points are the *splices* of the several pieces of leather which compose the belt, and which are here perforated by the holes for the copper rivets. The strengths of the new and the partially-used belts were found to be almost identical. The average of the trials is as follows:—

3-in. belts broke through the lace-holes with	629 lbs.
" " rivet " "	1146 lbs.
" " solid part "	2025 lbs.

These give as the strength to the *inch of width*:—

When the rupture is through the lace-holes	210 lbs.
" " rivet "	382 lbs.
" " solid part	675 lbs.

The thickness being $\frac{7}{32}$ in. (= ·219), we have as the tensile strength of the leather 3086 lbs. a sq. in.

From the above we see that 200 lbs. an inch of width is the ultimate resistance to tearing that we can expect from ordinary belts.

The experiments herein described are strikingly corroborative of those already on record, and this gives increased assurance of their reliability; and, although there is nothing novel either in them or in their results, it is hoped that they will prove of interest, and that an examination of them will lead to confidence in the formulas which are based upon them.

Three-Inch New Belt.

On 12-inch Pulley (between Seams).						On 12-inch Pulley (on Seams).					
Equalising Weights, or Initial Tension on each side of Belt $= T_2$.	Unbalanced Weight. Trial No. 1.	Unbalanced Weight. Trial No. 2.	Unbalanced Weight. Average of Trials.	Total Weight on loaded side of Belt when Belt slipped $= T_1$.	$\frac{T_1}{T_2}$	Equalising Weights, or Initial Tension on each side of Belt $= T_2$.	Unbalanced Weight. Trial No. 1.	Unbalanced Weight. Trial No. 2.	Unbalanced Weight. Average of Trials.	Total Weight on loaded side of Belt when Belt slipped $= T_1$.	$\frac{T_1}{T_2}$
	damp.	dry.					damp.	damp.			
2·94	17	20	18·5	21·44	7·29	2·94	14	14	14·0	16·94	5·76
7·94	43	58	50·5	58·44	7·36	7·94	30	32	31·0	38·94	4·90
12·94	70	86	78·0	90·94	7·03	12·94	48	51	49·5	62·44	4·83
22·94	108	167	137·5	160·44	6·99	22·94	92	87	89·5	112·44	4·90
32·94	124	226	175·0	207·94	6·31	32·94	107	110	108·5	141·44	4·29
42·94	159	311	235·0	277·94	6·47	42·94	134	138	136·0	178·94	4·17
52·94	210	343	276·5	329·44	6·22	52·94	191	188	189·5	242·44	4·58
				Mean.	6·810					Mean.	4·775

Three-Inch New Belt.

On 23⅝-inch Pulley.						On 41-inch Pulley.					
Equalising Weights, or Initial Tension on each side of Belt $= T_2$.	Unbalanced Weight. Trial No. 1.	Unbalanced Weight. Trial No. 2.	Unbalanced Weight. Average of Trials.	Total Weight on loaded side of Belt when Belt slipped $= T_1$.	$\frac{T_1}{T_2}$	Equalising Weights, or Initial Tension on each side of Belt $= T_2$.	Unbalanced Weight. Trial No. 1.	Unbalanced Weight. Trial No. 2.	Unbalanced Weight. Average of Trials.	Total Weight on loaded side of Belt when Belt slipped $= T_1$.	$\frac{T_1}{T_2}$
	damp.	damp.					dry.	dry.			
2·94	12	16	14·0	16·94	5·76	2·94	13	13	13·0	15·94	5·42
7·94	34	44	39·0	46·94	5·91	7·94	38	39	38·5	46·44	5·85
12·94	54	77	65·5	78·44	6·06	12·94	55	61	58·0	70·94	5·48
22·94	110	135	122·5	145·44	6·34	22·94	..	..	..	..	..
32·94	161	191	176·0	208·94	6·34	32·94	188	176	182·0	214·94	6·53
42·94	207	234	220·5	263·44	6·14	42·94	..	..	..	..	..
52·94	277	282	279·5	332·44	6·28	52·94	289	295	292·0	344·94	6·52
				Mean.	6·118					Mean.	5·960

Three-Inch Belt.

Partially used and in good order—on 12-inch Pulley.								
Equalising Weights, or Initial Tension on each side of Belt $= T_2$.	Unbalanced Weight. Trial No. 1.	Unbalanced Weight. Trial No. 2.	Unbalanced Weight. Trial No. 3.	Unbalanced Weight. Trial No. 4.	Unbalanced Weight. Trial No. 5.	Unbalanced Weight. Average of Trials.	Total Weight on loaded side of Belt when Belt slipped $= T_1$.	$\frac{T_1}{T_2}$
	damp.	damp.	wet.	dry.	dry.			
3·2	13	14	26	12	11	15·5	18·7	5·84
8·2	34	36	53	32	32	37·4	45·6	5·56
13·2	57	63	77	50	50	59·4	72·6	5·50
23·2	102	109	149	90	99	109·8	133·0	5·73
33·2	123	128	226	144	144	153·0	186·2	5·61
43·2	174	185	299	183	..	210·2	253·4	5·87
53·2	226	265	338	249	..	268·5	322·7	6·07
							Mean ..	5·754

THREE-INCH BELT—PARTIALLY USED AND IN GOOD ORDER.

On 41-inch Pulley.						On 23½-inch Pulley.						
Equalising Weights, or Initial Tension on each side of Belt $= T_2$.	Un-balanced Weight. Trial No. 1.	Un-balanced Weight. Trial No. 2.	Un-balanced Weight. Average of Trials.	Total Weight on loaded side of Belt when Belt slipped $= T_2$.	$\frac{T_1}{T_2}$	Equalising Weights, or Initial Tension on each side of Belt $= T_2$.	Un-balanced Weight. Trial No. 1.	Un-balanced Weight. Trial No. 2.	Un-balanced Weight. Trial No. 3.	Un-balanced Weight. Average of Trials.	Total Weight on loaded side of Belt when Belt slipped $= T_1$.	$\frac{T_1}{T_2}$
	dry.	dry.					damp.	damp.	dry.			
3·2	26	25	25·5	28·7	8·97	3·2	23	22	23	22·6	25·8	8·06
8·2	59	55	57·0	65·2	7·95	8·2	53	54	53	53·3	61·5	7·50
13·2	86	82	84·0	97·2	7·36	13·2	86	89	82	85·6	98·8	7·48
23·2	..	..	..	..	..	23·2	..	..	141	..	..	..
33·2	188	194	190·5	223·7	6·74	33·2	194	210	181	195·0	228·2	6·87
43·2	..	..	..	..	..	43·2	..	..	223	..	..	..
53·2	306	323	314·5	367·7	6·91	53·2	285	316	272	291·0	344·2	6·47
				Mean .	7·586						Mean .	7·276

THREE-INCH OLD BELT.

On 12-inch Pulley.						On 23½-inch Pulley.					
Equalising Weights, or Initial Tension on each side of Belt $= T_2$.	Un-balanced Weight. Trial No. 1.	Un-balanced Weight. Trial No. 2.	Un-balanced Weight. Average of Trials.	Total Weight on loaded side of Belt when Belt slipped $= T_1$.	$\frac{T_1}{T_2}$	Equalising Weights, or Initial Tension on each side of Belt $= T_2$.	Un-balanced Weight. Trial No. 1.	Un-balanced Weight. Trial No. 2.	Un-balanced Weight. Average of Trials.	Total Weight on loaded side of Belt when Belt slipped $= T_1$.	$\frac{T_1}{T_2}$
	damp.	damp.					damp.	wet.			
2·75	12	13	12·5	15·25	5·55	2·75	16	20	18·0	20·75	7·55
7·75	35	39	37·0	44·75	5·77	7·75	48	70	59·0	66·75	8·61
12·75	61	66	63·5	76·25	5·98	12·75	72	106	89·0	101·75	7·98
22·75	102	109	105·5	128·25	5·64	22·75	157	174	165·5	188·25	8·27
32·75	142	158	150·0	182·75	5·58	32·75	220	232	226·0	258·75	7·90
42·75	203	220	211·5	254·25	5·95	42·75	256	276	271·0	313·75	7·34
52·75	259	273	266·0	318·75	6·04	52·75	301	311	306·0	358·75	6·80
				Mean .	5·787					Mean .	7·778

SIX-INCH NEW BELT.

On 23½-inch Pulley.						On 41-inch Pulley.					
Equalising Weights, or Initial Tension on each side of Belt $= T_2$.	Un-balanced Weight. Trial No. 1.	Un-balanced Weight. Trial No. 2.	Un-balanced Weight. Average of Trials.	Total Weight on loaded side of Belt when Belt slipped $= T_1$.	$\frac{T_1}{T_2}$	Equalising Weights, or Initial Tension on each side of Belt $= T_2$.	Un-balanced Weight. Trial No. 1.	Un-balanced Weight. Trial No. 2.	Un-balanced Weight. Average of Trials.	Total Weight on loaded side of Belt when Belt slipped $= T_1$.	$\frac{T_1}{T_2}$
	dry.	damp.									
6·75	30	36	33·0	39·75	5·89	6·75	32	29	30·5	37·25	5·52
11·75	53	63	58·0	67·75	5·77	11·75	53	50	51·5	73·25	6·23
16·75	78	81	79·5	96·25	5·75	16·75	72	72	72·0	88·75	5·30
26·75	125	135	130·0	156·75	5·86	26·75	118	115	116·5	143·25	5·36
36·75	161	177	169·0	205·75	5·60	36·75	159	176	167·5	204·25	5·55
46·75	216	223	219·5	266·25	5·70	46·75	209	236	222·5	269·25	5·76
56·75	265	269	267·0	323·75	5·70	56·75	265	275	270·0	326·75	5·76
				Mean .	5·752					Mean .	5·640

Belts and drums constitute the most convenient method of transmitting rotary motion from one shaft to another. Of late years their use has enormously increased, on account of their easy application in almost every case where power is required to be transferred. Compared with cog-wheel gearing, they possess several advantages, namely, the driving and driven shafts may be at a considerable distance from each other, and need not be parallel or in the same plane.

Belts and drums form very effective friction-couplings. If a machine driven by a belt becomes accidentally overloaded, the belt slips upon the drum, and a break-down is generally prevented.

By the introduction of fast and loose pulleys, the driven shaft can be set in motion or stopped with perfect safety, whilst the driving shaft is running at full speed. In cotton mills and other factories, where a number of independent machines are driven from a single shaft, this contrivance is of great value.

The motion of belts and drums is much smoother than that of gearing, and they can be readily applied to machines which require a high velocity, where ordinary gearing would be quite inadmissible.

Material and Mode of Manufacture.—Leather is generally used for the manufacture of belting. Other materials have been tried, such as india-rubber, gutta-percha, woven hemp, &c., but leather has been found to be the most reliable and economical in wear.

The best description of leather for the purpose is English ox-hide tanned with oak bark by the slow old-fashioned process, and dressed in such a way as to retain firmness and toughness, without harshness and rigidity.

646.

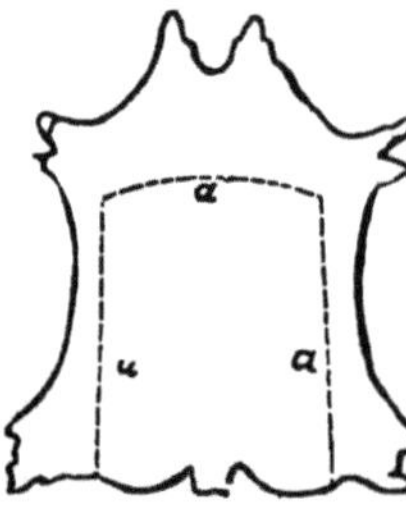

The prime part of the hide only, called the *butt*, should be used, the other portions being comparatively loose, and only fit for inferior purposes.

The dotted lines *a*, *a*, *a*, in Fig. 646 show the shape and size of the *butt*.

Butts are generally cut out of hides in the preliminary preparing process, and tanned by themselves. The time required for tanning best leather varies from twelve to eighteen months, according to the thickness. When the tanning is thoroughly completed, the butt is curried or dressed. During this process it must be stretched. This is effected by machinery, the leather being allowed to dry in its extended state. Strap-butts of best leather can be permanently extended from 4 to 5 in. For light work, belts of single substance are sufficient, the strips of leather being joined together by feather-edged splices, first cemented and then sewn. Single belting varies in thickness from $\frac{3}{16}$ to $\frac{1}{4}$ in. For heavy work, double and sometimes treble layers of leather are required, cemented and sewn through their entire length. The material used for the sewing is either strong well-waxed hemp, or thin strips of hide prepared with alum. The latter is generally used in the North of England; but its advantage over good waxed hemp is doubtful. The thickness of double belting is from $\frac{5}{16}$ to $\frac{7}{16}$ in.

An improvement in the ordinary double belt, shown in Figs. 647, 648, has been introduced by Messrs. Hepburn and Sons, of Southwark, who have given much attention to this branch of leather manufacture.

647.

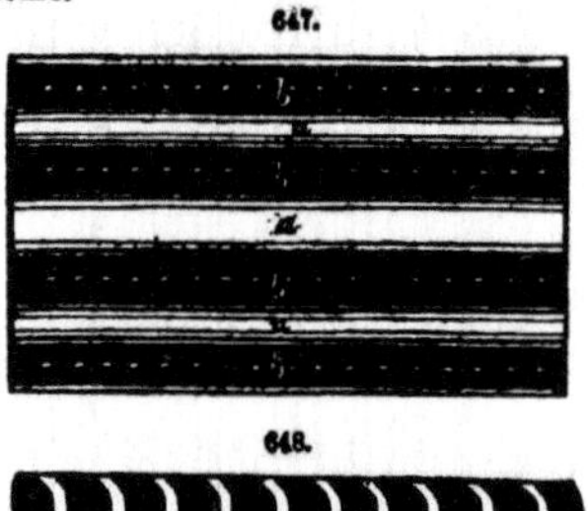

649.

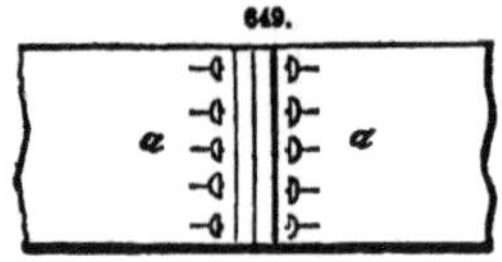

650.

648.

651.

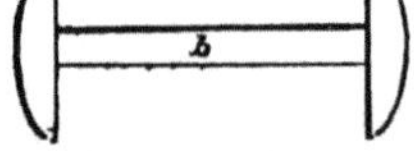

It consists in the substitution of a strip of prepared untanned hide for the outer layer of the belt, corrugated, as marked by *a a a* in Fig. 647; the inner one next the drum being made of tanned leather as usual, and not corrugated. This combination, or *composite belt* as it is called, has much greater strength than a belt composed of tanned leather only, and is not nearly so liable to stretch in working.

The sewing is also different. Instead of hand-labour being employed, these belts are sewn, or rather riveted, by machinery, and copper or malleable iron wire is used in place of waxed hemp, as shown in Figs. 647, 648.

Fig. 648 represents a longitudinal section of the belt through one of the lines of sewing, and shows the way in which each stitch or rivet is clenched and turned in.

This metallic sewing is also applied to double belting made entirely of leather, and has been found to work well, and to be more durable than ordinary hand-sewing.

Fastenings of Belts.—Belts are usually laced together by thin strips of hide prepared with alum; from 12 to 18 in. is allowed for lap; and rows of holes at equal distances are punched in each end of the belt. A belt-fastener, which is extremely simple and effective, and requires no lap, is given full size in Fig. 651. It is made of tough metal, of various sizes to suit different thicknesses of belting; *a*, *a*, Figs. 649, 650, shows how this fastener is applied.

Various Modes of Driving Drums.—Fig. 652 represents the direct method, where the shafts of the drums are parallel and in the same plane. A is the driving drum, revolving in the direction of the arrow; B is the driven drum. C is called the *leading* side of the belt; D the *following* side.

Belts are more effective when the leading side is underneath, as in Figs. 654, 655.

652.

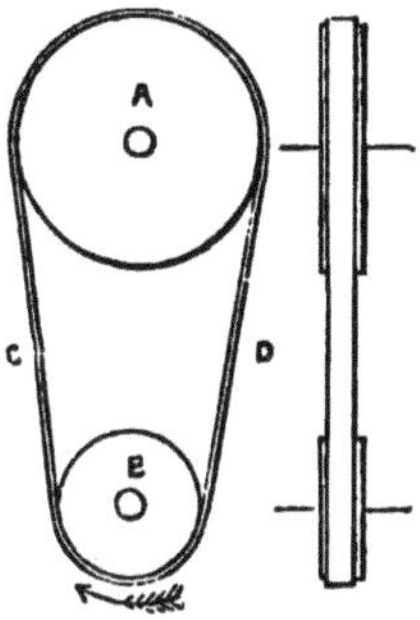

653.

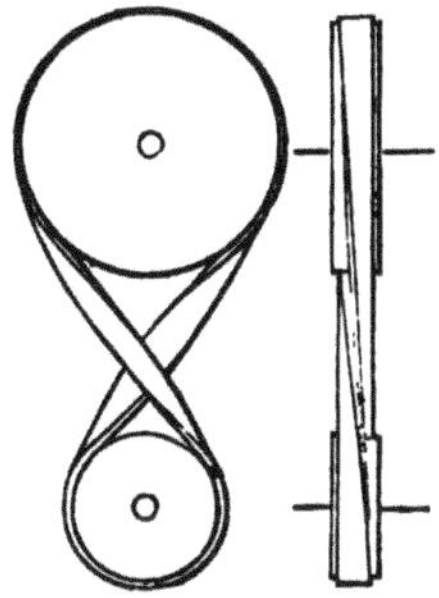

654.

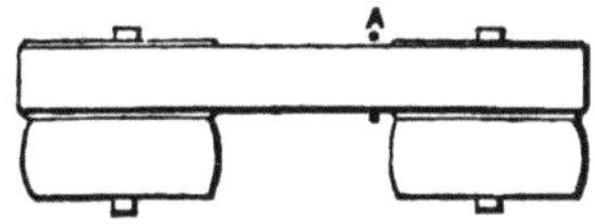

655.

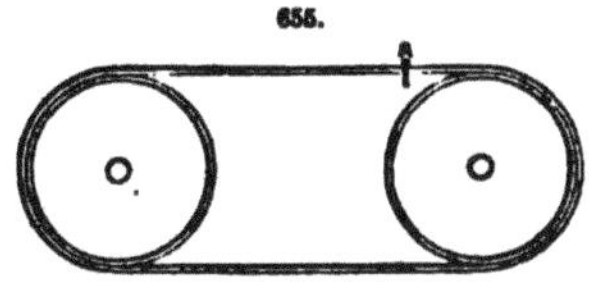

656.

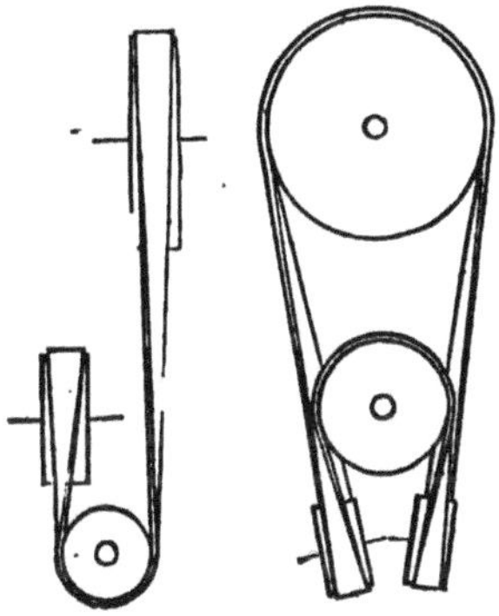

657.

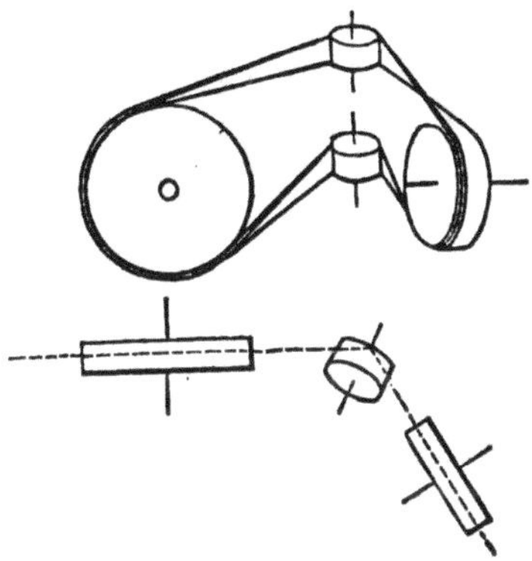

658.

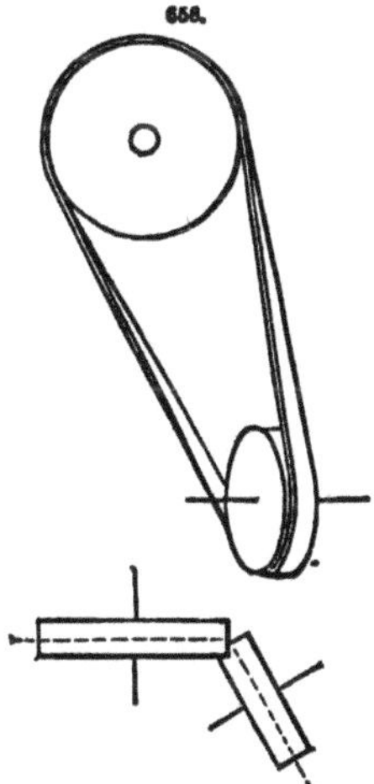

659.

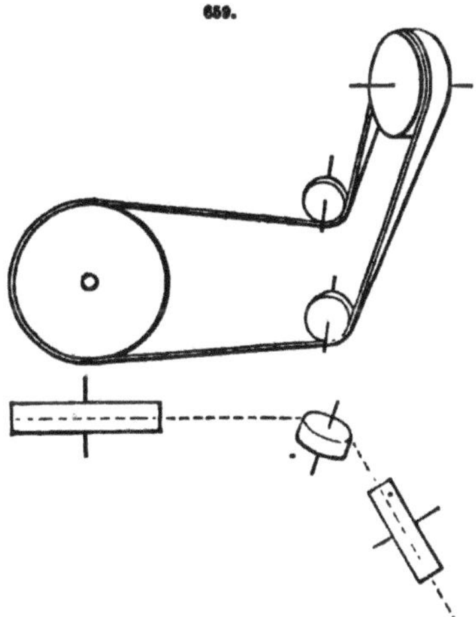

The drum should be rather wider than the belt, and slightly rounded on the face; $\frac{3}{16}$ths of an inch to the foot of width is sufficient convexity, except in the case of small high-speed pulleys, when it should be from $\frac{1}{4}$ to $\frac{1}{2}$ in.

Fig. 653 represents the crossed belt, which has the effect of making the drums revolve in opposite directions.

Figs. 654, 655, represent the fast and loose pulley.

The fork for shifting the belt from the fast to the loose pulley must be placed at A, near the pulleys, and act on the following part of the belt.

Fig. 656 is a case where the shafts are parallel, but the drums not in the same plane; here it is necessary to lead the belt over guide-pulleys.

Fig. 657 is a similar case, where the shafts are in the same plane, but not parallel. In Fig. 658 the shafts are neither in the same plane, nor parallel; yet in this case no guide-pulleys are needed. Fig. 659 is a case where guide-pulleys are required.

Fig. 660 represents the most general way of varying the velocity-ratio of two shafts. The proportion of the respective diameters of each pair of pulleys is such that the sum of the two is always constant, and, therefore, the same length of belt will do for all the speeds.

660.

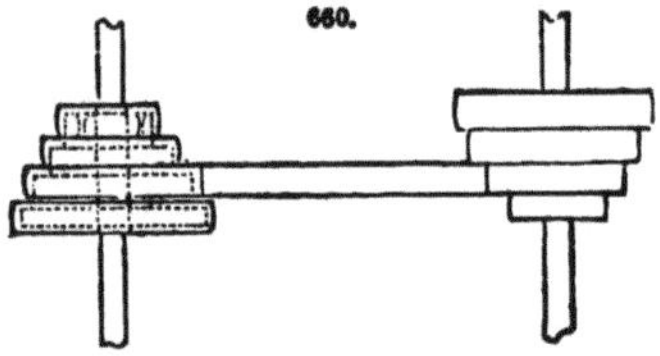

662.

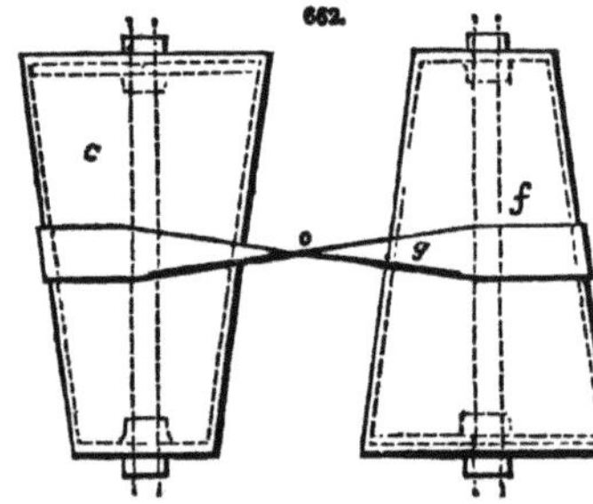

661.

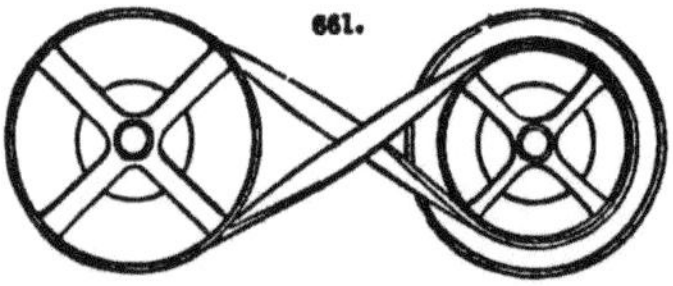

Figs. 661, 662, is a mode of varying the speed whilst the machinery is in motion. Two conical drums are placed in a reversed way to each other, so that the sum of any two corresponding diameters is always a constant quantity. The belt is made to move laterally along the surface of the cones, by certain contrivances; and thus a gradual increase or decrease of speed in the driven drum is brought about, on account of the constantly varying proportions of the two diameters, while the *angular velocity* of the driving drum remains constant.

Tension-rollers.—In order that the natural tension of the belts shall remain constant, and not exceed, though equalling, the value calculated, it is requisite to use *tension-rollers.*

The weight q of these rollers is found by the approximate expression $q = \frac{2\,T_1 \cos. a}{\cos. b}$, wherein a is half the obtuse angle A D B formed by the two branches of the belt upon which the weight rests, and may be assumed *a priori;* and b the angle between the line A B and the horizontal line A C, Fig. 663. That is angle B A C = b.

This formula is reduced to the following rule:—

To calculate the weight of a tension-roller, capable of producing its pressure upon the two branches of a belt, a given normal tension, as in Fig. 663.

663.

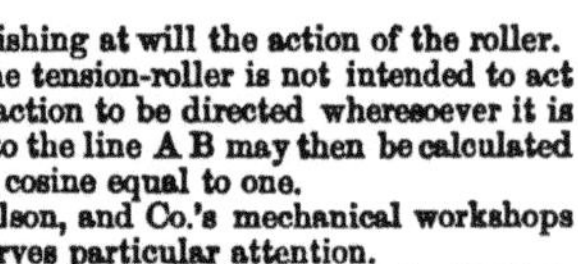

Multiply the given normal tension by twice the cosine of half the obtuse angle formed by the two branches of the belt, and divide the product by the cosine of the angle formed by the common tangent of the two drums with the horizontal line A C.

In fixing the belt, care must be taken to give it such a length that, when in repose, it shall only have a minimum flexure, and then the tension T will very nearly equal the value assigned to it by our previous developments.

Means must, of course, be reserved of increasing or diminishing at will the action of the roller.

If, owing to the particular arrangement of the drums, the tension-roller is not intended to act vertically, a suitable combination of levers may enable its action to be directed wheresoever it is necessary. The effort it exerts upon the belt perpendicularly to the line A B may then be calculated by the above rule, in supposing the angle b to be nul and its cosine equal to one.

The general arrangement of the belting of Nasmyth, Wilson, and Co.'s mechanical workshops at Patricroft, near Manchester, shown in Figs. 664, 665, deserves particular attention.

The shop consists of two spans of 38 ft. and 48 ft. respectively, each having a length of 102 ft. It is lighted from the top by means of skylights, as indicated in Fig. 664. The smaller shop derives additional light from side-windows, close to which a row of small lathes is placed. The shafting A, A, is driven from a pair of vertical engines, fitted with balanced slide-valves, and having a pair of large pulleys fixed to their crank-shaft B, B, Fig. 665, in lieu of a fly-wheel. The straps from these pulleys are carried in opposite directions, so as to balance the strains produced by them; and one strap C, C, supplies power to the machinery of each span of the shop. The driving-shaft D, D, on the top of each span transmits power to two parallel lengths of shafting, one at each side of the two buildings; and for these also the belts balance their strains upon the

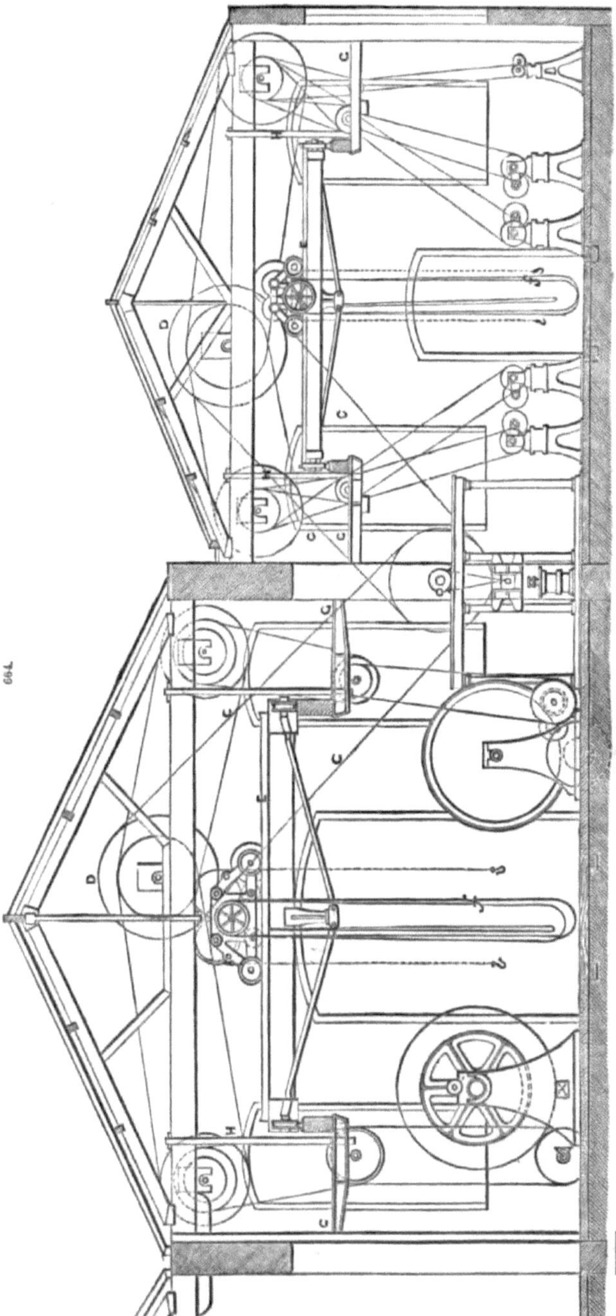

664.

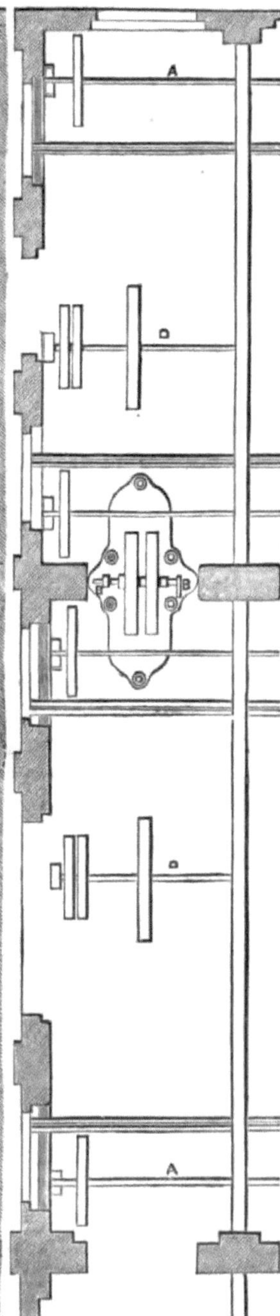

665.

driving-shaft by being placed symmetrically in opposite directions. The machines are placed in parallel rows lengthways at both sides of the buildings, so as to have a clear wide space throughout the entire length of the shop for internal locomotion. By placing the smaller lathes in double rows in close proximity, and with their back-gearing against each other, it has been possible to put an unusual number of these tools into a limited space, still allowing free access to each machine, and giving a clear space in the centre for a passage. The workmen stand face to face, having the machines between them, and there is sufficient room for all their operations.

Each span is traversed by a travelling-crane E of suitable power. The cranes are capable of reaching every tool in the shop, and are worked by hand from below by means of endless cords *f*, *f*, passing round grooved pulleys attached to the gearing. The movements are in three directions; and, by provision of double-winding drums for the lifting-chains, it is possible to reach the extreme ends of the shop on either side. To facilitate the passage of the travellers and their chains, it was found necessary to avoid the use of vertical columns for supporting the longitudinal timbers which carry the crane-rails, and another mode of construction was substituted. The crane-rails are supported on horizontal cast-iron brackets G, G, secured at one end to the side columns or wall-standards, and having their free ends suspended from the cross-beams of the roof, as shown at H, H, Fig. 664. This mode of supporting the travellers is found to answer well, and it forms an element of great convenience with regard to the movement of heavy masses within the shop. The cast-iron brackets, at the same time, serve for the attachment of the hangers or bearings for the shafting, for which purpose wooden planks are fixed to them throughout the entire length of the shop in lines parallel to the side walls.

We append, with some slight alterations, an article on the driving-belt, by Edward Sang. This article, taken from the 'Practical Mechanic's Journal' (1866), will render clear the methods of investigation employed by Prony and Poncelet.

The first problem which presents itself when we consider the arrangement of the driving-belt, is to compute the length of the belt when the diameters of the wheels and the distance between their centres are known. If the machinery be in position, it is a very easy matter to measure the length of the required belt; and this direct measurement is even to be preferred to calculation, because ultimately the stretch of the belt has to be suited to the desired strain by trials. The converse problem, "Having given the length of the belt, to compute the diameters of the wheels," is much more important, because the process of repeated trial would be both tedious and expensive.

This converse problem arises when we have to design a set of speed-cones, which may, with one belt, give a variety of speeds. The solution of this converse problem is attended with considerable difficulty: it can only be accomplished, for practical purposes, by help of that modification of the method of trial and error which consists in tabulating a regular series of computed results.

In the 'Edinburgh Philosophical Journal' for April, 1831, E. Sang gave a table for this purpose, which had hardly been printed when an obvious improvement suggested itself. He reconstructed the table with this improvement, and thus describes the manner in which it is to be used.

In the first place, it is to be observed that there are *three* principal dimensions which have to be taken into consideration: these are the diameter of the wheel, the diameter of the pulley, and the distance between the centres. Now, it would be impossible to make a table of triple entry in which these three dimensions should enter as arguments, and the corresponding length of the belt as a result.

It is necessary, for the sake of abridgment, to assume one of these as constant; and that one which answers best is the distance between the centres. In the subjoined table this distance is announced to be *unit*.

666.

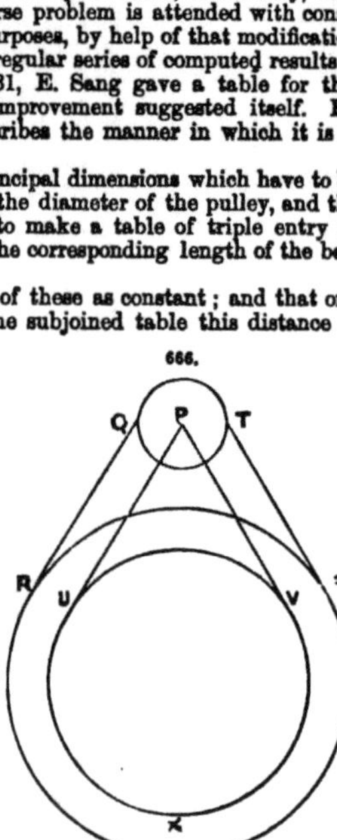

In the second place, if P represent the centre of the pulley, W that of the wheel, and Q R S T the belt passing over them, the inclination of the free parts, Q R and S T, to the line of centres depends only on the difference between the two diameters; so that if another pair were placed at the same centres, and having their diameters each 1 in., or any other quantity, more than those of the former pair, the inclination of the free parts of the band would be unchanged.

Thirdly, if we describe a circle round the centre, W, with a radius equal to the difference between the two radii, and draw from P the lines P U, P V to touch it, the entire line, P U X V P, is less than the length of the band, Q R S T, by the circumference of the pulley. Hence for each difference between the diameters, or for each inclination of the free part of the belt, there is a corresponding excess of the length of the belt above the circumference of the pulley, and also a corresponding excess of the same length of belt above the circumference of the wheel. By attention to these matters we can make our table one of single entry.

In the former table, Sang made the difference between the diameters the argument, and placed opposite to each difference the corresponding excess of the belt above the two circumferences. This arrangement made it necessary to multiply and to divide by 3·1415926. By the arrangement of the present table these multiplications and divisions are avoided.

If we compute the diameter of a wheel round which the belt would just go, and call the diameter of this wheel the *belt-diameter*; then, for each inclination, the excesses of the belt-diameter above the diameters of the wheel and pulley are determined. These excesses are entered in the table. Furthermore, Sang has made the inclination of the free part of the band the primary argument. In this way the arrangement of the table is as follows:—

Column 1 contains the inclinations of the free part of the belt to the line of centres, given for each half-degree of the centesimal system, that is, for each multiple of 27′ of the ordinary division; the values being given in ancient degrees and minutes, as well as in decimal parts of the right angle.

Column 2 contains the corresponding differences, W − P, between W, the diameter of the wheel, and P, the diameter of the pulley; the differences of these values are also given for the purpose of interpolation.

Column 3 contains the corresponding excesses, B − P, of the belt-diameter, B, above the diameter of the pulley, with the differences.

Column 4 contains the values of the excesses, B − W, of the same belt-diameter above the diameter of the wheel, also with their differences; the whole being given in decimal parts of the distance between the centres.

TABLE OF THE RELATIVE DIMENSIONS OF BELTS AND PULLEYS.

Inclination.			W−P.		B−P.		B−W.	
°	′	Decim.	Value.	Diff.	Value.	Diff.	Value.	Diff.
0	00	·0000	0·00000	1571	0·63662	787	·63662	783
0	27	·0050	0·01571	1570	0·64449	792	·62879	780
0	54	·0100	0·03141	1571	0·65241	795	·62099	775
1	21	·0150	0·04712	1570	0·66036	798	·61324	772
1	48	·0200	0·06282	1570	0·66834	803	·60552	767
2	15	·0250	0·07852	1569	0·67637	806	·59785	763
2	42	·0300	0·09421	1569	0·68443	810	·59022	759
3	09	·0350	0·10990	1568	0·69253	814	·58263	755
3	36	·0400	0·12558	1567	0·70067	816	·57508	750
4	03	·0450	0·14125	1567	0·70883	821	·56758	746
4	30	·0500	0·15692	1565	0·71704	824	·56012	742
4	57	·0550	0·17257	1565	0·72528	827	·55270	737
5	24	·0600	0·18822	1563	0·73355	830	·54533	733
5	51	·0650	0·20385	1562	0·74185	834	·53800	728
6	18	·0700	0·21947	1560	0·75019	837	·53072	724
6	45	·0750	0·23507	1560	0·75856	840	·52348	719
7	12	·0800	0·25067	1557	0·76696	843	·51629	714
7	39	·0850	0·26624	1556	0·77539	846	·50915	710
8	06	·0900	0·28180	1554	0·79385	849	·50205	705
8	33	·0950	0·29734	1553	0·79234	852	·49500	701
9	00	·1000	0·31287	1550	0·80086	855	·48799	696
9	27	·1050	0.32837	1549	0 80941	857	·48103	691
9	54	·1100	0·34386	1546	0·81798	860	·47412	686
10	21	·1150	0·35932	1544	0·82658	863	·46726	681
10	48	·1200	0·37476	1542	0·83521	865	·46045	677
11	15	·1250	0·39018	1539	0·84386	868	·45368	671
11	42	·1300	0·40557	1537	0·85254	870	·44697	667
12	09	·1350	0·42094	1535	0·86124	873	·44030	662
12	36	·1400	0·43629	1531	0·86997	875	·43368	656
13	03	·1450	0·45160	1529	0·87872	877	·42712	652
13	30	·1500	0·46689	1526	0·88749	880	·42060	646
13	57	·1550	0·48215	1523	0·89629	881	·41414	042
14	24	·1600	0·49738	1520	0·90510	883	·40772	636
14	51	·1650	0·51258	1517	0·91393	886	·40136	632
15	18	·1700	0·52775	1513	0·92279	887	·39504	626
15	45	·1750	0·54288	1510	0·93166	889	·38878	621
16	12	·1800	0·55798	1507	0·94055	891	·38257	616
16	39	·1850	0·57305	1503	0·94946	892	·37641	611
17	06	·1900	0·58808	1500	0·95838	895	·37030	605
17	33	·1950	0·60308	1495	0·96733	895	·36425	600
18	00	·2000	0·61803	1492	0·97628	897	·35825	595
18	27	·2050	0·63295	1488	0·98525	899	·35230	590
18	54	·2100	0·64783	1485	0·99424	899	·34640	584
19	21	·2150	0·66268	1480	1·00323	901	·34056	579
19	48	·2200	0·67748	1475	1·01224	902	·33477	574

TABLE OF THE RELATIVE DIMENSIONS OF BELTS AND PULLEYS—*continued.*

Inclination.			W–P.		B–P.		B–W.	
°	′	Decim.	Value.	Diff.	Value.	Diff.	Value.	Diff.
20	15	·2250	0·69223	1472	1·02126	904	·32903	568
20	42	·2300	0·70695	1467	1·03030	904	·32335	563
21	09	·2350	0·72162	1463	1·03934	905	·31772	558
21	36	·2400	0·73625	1458	1·04839	906	·31214	552
22	03	·2450	0·75083	1454	1·05745	906	·30662	547
22	30	·2500	0·76537	1449	1·06651	908	·30115	542
22	57	·2550	0·77986	1444	1·07559	908	·29573	536
23	24	·2600	0·79430	1439	1·08467	908	·29037	531
23	51	·2650	0·80869	1434	1·09375	909	·28506	525
24	18	·2700	0·82303	1429	1·10284	909	·27981	520
24	45	·2750	0·83732	1424	1·11193	910	·27461	514
25	12	·2800	0·85156	1419	1·12103	910	·26947	509
25	39	·2850	0·86575	1413	1·13013	909	·26438	503
26	06	·2900	0·87988	1408	1·13922	910	·25935	499
26	33	·2950	0·89396	1402	1·14832	910	·25436	492
27	00	·3000	0·90798	1397	1·15742	910	·24944	487
27	27	·3050	0·92195	1391	1·16652	909	·24457	482
27	54	·3100	0·93586	1385	1·17561	909	·23975	476
28	21	·3150	0·94971	1380	1·18470	909	·23499	471
28	48	·3200	0·96351	1373	1·19379	908	·23028	465
29	15	·3250	0·97724	1368	1·20287	908	·22563	460
29	42	·3300	0·99092	1361	1·21195	907	·22103	454
30	09	·3350	1·00453	1355	1·22102	906	·21649	449
30	36	·3400	1·01808	1349	1·23008	906	·21200	444
31	03	·3450	1·03157	1343	1·23914	904	·20756	438
31	30	·3500	1·04500	1336	1·24818	904	·20318	432
31	57	·3550	1·05836	1329	1·25722	902	·19886	427
32	24	·3600	1·07165	1323	1·26624	901	·19459	422
32	51	·3650	1·08488	1317	1·27525	900	·19037	416
33	18	·3700	1·09805	1309	1·28425	899	·18621	411
33	45	·3750	1·11114	1303	1·29324	897	·18210	406
34	12	·3800	1·12417	1295	1·30221	896	·17804	400
34	39	·3850	1·13712	1289	1·31117	894	·17404	394
35	06	·3900	1·15001	1282	1·32011	892	·17010	390
35	33	·3950	1·16283	1274	1·32903	891	·16620	384
36	00	·4000	1·17557	1267	1·33794	888	·16236	378
36	27	·4050	1·18824	1260	1·34682	887	·15858	373
36	54	·4100	1·20084	1252	1·35569	884	·15485	368
37	21	·4150	1·21336	1245	1·36453	883	·15117	363
37	48	·4200	1·22581	1238	1·37336	880	·14754	357
38	15	·4250	1·23819	1230	1·38216	877	·14397	352
38	42	·4300	1·25049	1222	1·39093	876	·14045	347
39	09	·4350	1·26271	1214	1·39969	872	·13698	341
39	36	·4400	1·27485	1206	1·40841	871	·13357	337
40	03	·4450	1·28691	1199	1·41712	867	·13020	331
40	30	·4500	1·29890	1190	1·42579	864	·12689	326
40	57	·4550	1·31080	1182	1·43443	862	·12363	320
41	24	·4600	1·32262	1175	1·44305	859	·12043	316
41	51	·4650	1·33437	1166	1·45164	855	·11727	310
42	18	·4700	1·34603	1157	1·46019	853	·11417	306
42	45	·4750	1·35760	1149	1·46872	849	·11111	300
43	12	·4800	1·36909	1141	1·47721	845	·10811	295
43	39	·4850	1·38050	1133	1·48566	842	·10516	290
44	06	·4900	1·39183	1123	1·49408	839	·10226	285
44	33	·4950	1·40306	1115	1·50247	835	·09941	281

TABLE OF THE RELATIVE DIMENSIONS OF BELTS AND PULLEYS—*continued.*

Inclination.			W−P.		B−P.		B−W.	
°	′	Decim.	Value.	Diff.	Value.	Diff.	Value.	Diff.
45	00	·5000	1·41421	1107	1·51082	831	·09660	275
45	27	·5050	1·42528	1097	1·51913	827	·09385	270
45	54	·5100	1·43625	1089	1·52740	824	·09115	265
46	21	·5150	1·44714	1080	1·53564	819	·08850	261
46	48	·5200	1·45794	1070	1·54383	815	·08589	256
47	15	·5250	1·46864	1062	1·55198	811	·08333	250
47	42	·5300	1·47926	1053	1·56009	806	·08083	246
48	09	·5350	1·48979	1043	1·56815	803	·07837	242
48	36	·5400	1·50022	1034	1·57618	797	·07595	236
49	03	·5450	1·51056	1025	1·58415	793	·07359	232
49	30	·5500	1·52081	1016	1·59208	788	·07127	227
49	57	·5550	1·53097	1006	1·59996	784	·06900	223
50	24	·5600	1·54103	996	1·60780	778	·06677	218
50	51	·5650	1·55099	987	1·61558	774	·06459	213
51	18	·5700	1·56086	977	1·62332	768	·06246	209
51	45	·5750	1·57063	968	1·63100	763	·06037	205
52	12	·5800	1·58031	958	1·63863	758	·05832	200
52	39	·5850	1·58989	948	1·64621	753	·05632	195
53	06	·5900	1·59937	938	1·65374	747	·05437	191
53	33	·5950	1·60875	928	1·66121	741	·05246	187
54	00	·6000	1·61803	919	1·66862	736	·05059	183
54	27	·6050	1·62722	908	1·67598	730	·04876	178
54	54	·6100	1·63630	898	1·68328	724	·04698	174
55	21	·6150	1·64528	888	1·69052	718	·04524	170
55	48	·6200	1·65416	878	1·69770	713	·04354	165
56	15	·6250	1·66294	867	1·70483	705	·04189	162
56	42	·6300	1·67161	858	1·71188	700	·04027	158
57	09	·6350	1·68019	847	1·71888	694	·03869	153
57	36	·6400	1·68866	836	1·72582	687	·03716	150
58	03	·6450	1·69702	826	1·73269	680	·03566	145
58	30	·6500	1·70528	815	1·73949	674	·03421	142
58	57	·6550	1·71343	805	1·74623	667	·03279	138
59	24	·6600	1·72148	795	1·75290	660	·03141	134
59	51	·6650	1·72943	783	1·75950	653	·03007	130
60	18	·6700	1·73726	773	1·76603	647	·02877	127
60	45	·6750	1·74499	762	1·77250	639	·02750	122
61	12	·6800	1·75261	752	1·77889	632	·02628	120
61	39	·6850	1·76013	740	1·78521	625	·02508	115
62	06	·6900	1·76753	730	1·79146	617	·02393	113
62	33	·6950	1·77483	718	1·79763	610	·02280	108
63	00	·7000	1·78201	708	1·80373	602	·02172	106
63	27	·7050	1·78909	697	1·80975	595	·02066	101
63	54	·7100	1·79606	685	1·81570	587	·01965	99
64	21	·7150	1·80291	674	1·82157	579	·01866	95
64	48	·7200	1·80965	664	1·82736	571	·01771	92
65	15	·7250	1·81629	652	1·83307	564	·01679	89
65	42	·7300	1·82281	640	1·83871	555	·01590	86
66	09	·7350	1·82921	630	1·84426	547	·01504	82
66	36	·7400	1·83551	618	1·84973	538	·01422	80
67	03	·7450	1·84169	607	1·85511	530	·01342	77
67	30	·7500	1·84776	595	1·86041	522	·01265	73
67	57	·7550	1·85371	584	1·86563	513	·01192	71
68	24	·7600	1·85955	573	1·87076	505	·01121	68
68	51	·7650	1·86528	561	1·87581	496	·01053	65
69	18	·7700	1·87089	549	1·88077	486	·00988	63

Table of the Relative Dimensions of Belts and Pulleys—*continued.*

Inclination.			W – P.		B – P.		B – W.	
°	′	Decim.	Value.	Diff.	Value.	Diff.	Value.	Diff.
69	45	·7750	1·87638	538	1·88563	478	·00925	60
70	12	·7800	1·88176	526	1·89041	470	·00865	57
70	39	·7850	1·88702	515	1·89511	459	·00808	55
71	06	·7900	1·89217	503	1·89970	451	·00753	52
71	33	·7950	1·89720	491	1·90421	442	·00701	49
72	00	·8000	1·90211	480	1·90863	432	·00652	48
72	27	·8050	1·90691	468	1·91295	423	·00604	45
72	54	·8100	1·91159	456	1·91718	413	·00559	43
73	21	·8150	1·91615	444	1·92131	404	·00516	40
73	48	·8200	1·92059	432	1·92535	394	·00476	39
74	15	·8250	1·92491	420	1·92929	384	·00437	36
74	42	·8300	1·92911	409	1·93313	374	·00401	34
75	09	·8350	1·93320	397	1·93687	364	·00367	32
75	36	·8400	1·93717	384	1·94051	355	·00335	31
76	03	·8450	1·94101	373	1·94406	344	·00304	28
76	30	·8500	1·94474	361	1·9475[illegible]	334	·00276	27
76	57	·8550	1·94835	348	1·95084	324	·00249	24
77	24	·8600	1·95183	337	1·95408	313	·00225	24
77	51	·8650	1·95520	325	1·95721	303	·00201	21
78	18	·8700	1·95845	312	1·96024	293	·00180	20
78	45	·8750	1·96157	300	1·96317	282	·00160	18
79	12	·8800	1·96457	289	1·96599	271	·00142	17
79	39	·8850	1·96746	276	1·96870	261	·00125	16
80	06	·8900	1·97022	264	1·97131	250	·00109	14
80	33	·8950	0·97286	252	1·97381	239	·00095	13
81	00	·9000	1·97538	239	1·97620	228	·00082	12
81	27	·9050	1·97777	228	1·97848	217	·00070	10
81	54	·9100	1·98005	215	1·98065	205	·00060	10
82	21	·9150	1·98220	203	1·98270	195	·00050	8
82	48	·9200	1·98423	191	1·98465	183	·00042	7
83	15	·9250	1·98614	178	1·98648	172	·00035	7
83	42	·9300	1·98792	166	1·98820	161	·00028	5
84	09	·9350	1·98958	154	1·98981	149	·00023	5
84	36	·9400	1·99112	142	1·99130	138	·00018	4
85	03	·9450	1·99254	129	1·99268	126	·00014	4
85	30	·9500	1·99383	118	1·99394	114	·00010	3
85	57	·9550	1·99501	104	1·99508	103	·00007	2
86	24	·9600	1·99605	93	1·99611	90	·00005	1
86	51	·9650	1·99698	80	1·99701	79	·00004	2
87	18	·9700	1·99778	68	1·99780	67	·00002	1
87	45	·9750	1·99846	55	1·99847	55	·00001	0
88	12	·9800	1·99901	43	1·99902	43	·00001	1
88	39	·9850	1·99944	31	1·99945	30	·00000	0
89	06	·9900	1·99975	19	1·99975	19	·00000	0
89	33	·9950	1·99994	6	1·99994	6	·00000	0
90	00	1·0000	2·00000	..	2·00000	..	·00000	..

In order to make the use of this table clear, we give an example:—

Let the distance between the centres be 26 in., the first diameter on the wheel 27 in., the first on the pulley 3·65, and the fourth on the pulley 7·80, it is required to compute the fourth diameter on the wheel.

The first business is to reduce these dimensions to decimal parts of the distance between the centres; this is done by dividing them all by 26; the results are $W_1 = 1·03846$; $P_1 = ·14038$; $P_4 = ·30000$. From these we find $W_1 - P_1 = ·89808$, and we have to seek, by help of the table, the corresponding value of $B - P_1$. Now, the nearest number to the above ·89808 which we find in the table is ·89396, which corresponds to $B - P = 1·14832$; we have, therefore, to make a correction by interpolating by means of the proportion 1402 : 910 :: 412 : 267; wherefore the value of $B - P_1$ comes out 1·15099, and adding to this the value of P_1 we have the belt-diameter $B = 1·29137$. (If it were worth while, we might convert this into inches, obtaining 33·5756 in.

for the diameter of the wheel round which the belt would just fit, and giving 105·48 in. for the length of the belt itself.)

Subtracting the fourth pulley diameter, namely, ·30000, from this, we find $B - P_4 =$ ·99137. Seeking in the proper column for this number, we find ·98525, differing from what is wanted by 612. The corresponding tabular value of $W - P$ is ·63295, which has to be corrected by interpolation, thus, 899 : 1488 :: 612 : 1013, giving the true value $W_4 - P_4 =$ ·64308: wherefore $W_4 =$ ·94308, or in inches 24·52, which is the dimension sought for. The calculation may be arranged thus:—

W_1	=	1·03846	
P_1	=	·14038	
$W_1 - P_1$	=	·89808	
Table		·89396	
Tabular Error		412	; 1402 : 910 :: 412 : 267
Table		1·14832	
Correction	+	267	
$B - P_1$	=	1·15099	
P_1	=	·14038	
B	=	1·29137	
P_4	=	·30000	
$B - P_4$	=	·99137	
Table		·98525	
Error		612	; 899 : − 1488 :: 612 : 1013
Table		·63295	
Correction	+	1013	
$W_4 - P_4$	=	·64308	
P_4	=	·30000	
W_4	=	·94308 = 24·520 in.	

To this it may be necessary to add an example in which the diameter of the pulley is wanted:—

Let the distance between the centres be 27·5 in., the lesser diameter on the pulley 3·5, the greater diameter on the wheel 27, and the lesser diameter on the wheel 24; and let us compute the corresponding greater diameter on the pulley.

Dividing all by 27·5, we obtain $P_1 =$ ·12727, $W_1 =$ ·98182, $W_2 =$ ·87273; whence

W_1	=	·98182	
P_1	=	·12727	
$W_1 - P_1$	=	·85455	
Table		·85156	
Error		299	; 1419 : − 509 :: 299 : − 107
Table		·26947	
Correction	−	107	
$B - W_1$	=	·26840	
W_1	=	·98182	
B	=	1·25022	
W_2	=	·87273	
$B - W_2$	=	·37749	
Table		·38257	
Error	−	508	; − 616 : 1507 :: − 508 : 1243
Table		·55798	
Correction	+	1243	
$W_2 - P_2$	=	·57041	
W_2	=	·87273	
P_2	=	·30232 = 8·314 in.	

If a set of pulleys and wheels be suited to each other, so that one band may go round each pair, and if another set be made by augmenting all the diameters by one and the same quantity, then this second set also will have the property of being suited to one band. Hence, if a pair of driving cones be suited to a belt of one thickness, they will also suit a belt of any other thickness.

When we have to do with wheels to be driven by means of round bands, and in which the edges are grooved, it is convenient to make the calculations as for the diameters of the cylindric surfaces A A, into which the grooves are to be cut, and we must take care, after these cylindric fillets are dressed, to cut all the grooves to one gauge. This being done, and a band of any thickness being fitted to one pair of grooves, the same band will fit the remaining pairs.

667.

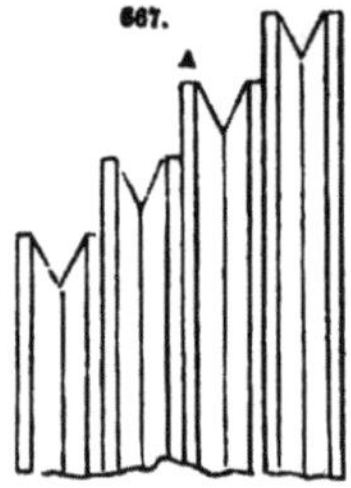

When the belts are to be crossed, there is no trouble in computing the diameters; for if the sum of the two diameters be kept the same, the length of the belt remains unchanged. In order to compute the length of a crossed belt, we have only to enter the column titled W − P with the sum of the diameters instead of their difference, and then the corresponding value in the column B − P is the belt diameter. Such a computation is very seldom needed, but merely for the sake of completeness an example is added.

The distance between the centres being 30 in., and the diameters of the wheel and pulley being 43 in. and 7 in. respectively, the length of the crossed belt is wanted.

Here the sum of the diameters is 50 in.: this divided by 30 gives W + P = 1·66667: hence

W + P		1·66667	
Table		1·66294	
Error		373	; 867 : 705 :: 373 : 303
Table		1·70483	
Correction	+	303	
B	=	1·70786	= 51·236 in.

whence the length of the crossed belt is 161·05 in.

When a belt is passed over the circumferences of two pulleys, so as to connect the motion of the one with that of the other, it seems as if the velocities of these two circumferences ought to be alike; and that difference which is observed whenever a belt is used for communicating determinate velocities, has been attributed to the slipping of the belt, or to that imaginary cause, *the imperfection of machinery*. The actions of machines, however, are governed by laws as exact and invariable as those which regulate the motions of the planets; and what we call the imperfections of those actions are only evidences of our ignorance of, or our inattention to, those properties of matter from which they inevitably follow. These accommodations, as we may call them, are essential to the comfortable action of machines, and bring within our reach the use of contrivances which would otherwise demand unattainable perfection in workmanship. An investigation of the action of the driving-belt shows that its supposed imperfection is an accompaniment essential to this mode of propulsion, and governed, like all other natural phenomena, by precise laws.

To place the nature of the action in a clear light, let Fig. 668 represent a pair of wheels connected by a belt, A being the centre of the driving, B that of the driven wheel, and C D E F G H the belt moving in the direction of the letters.

The apparatus being brought to rest, and the tensions on the two open parts, C D and F G, of the belt being alike, let us attempt to put it in motion by applying pressure to the wheel A. The wheel B will not begin to move until A has first turned a little round, so as, by augmenting the tension on F G, and relaxing that on C D, to create a difference of tension sufficient to overcome the resistance offered to the motion of the wheel B. But this difference of tension can only be

668.

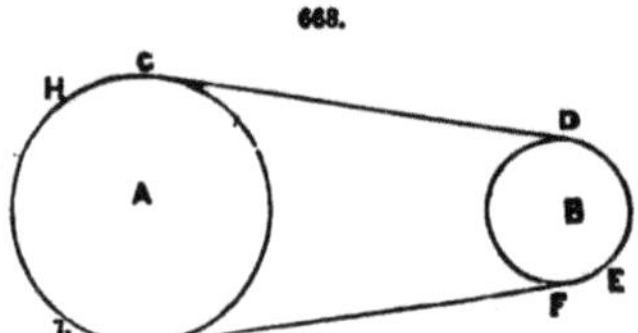

669

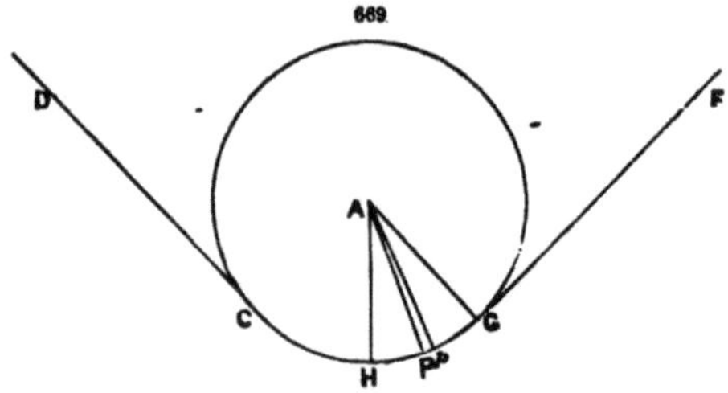

created by the adhesion of the belt G C H to the surface of the wheel A, and can only be permitted to exist by the similar adhesion of D E F to the surface of the wheel B. And again, these frictions necessarily depend on the pressure which the belts exert upon the surface of each wheel; so that the first branch of our inquiry must be into the law of the pressure of a belt against the curved surface to which it is applied.

Let then F G H C D, Fig. 669, represent a belt which, while kept to a uniform tension, has been partially wound on the circumference of the wheel A. Our business is to discover the pressure which the belt exerts against each portion of the arc G C H.

Having bisected the arc of contact in H, assume a small element Pp of the half H G; the pressure on this minute portion of the circumference of the wheel, necessarily directed to the centre A, may be decomposed into two portions, one parallel to and the other perpendicular to H A. That part which is perpendicular to H A is balanced by a similar pressure obtained from the corresponding element of the arc H C, and so is eliminated; but that part which is parallel to H A is augmented by the analogous pressure on the other side, and the sum or integral of all such pressures must make up the resultant of the tensions of the two free parts of the belt. If we put r for the radius of the wheel, a for the angle H A P, and $d\,a$ for the minute increment P A p ($d\,a$ is termed the differential of a very small angle, not $d \times a$; this is the very clumsy notation of the differential calculus), also p for the pressure of the belt against a linear unit of the circumference, $p\,r\,d\,a$ must represent the pressure of the belt against the arc Pp, of which the length is $r\,d\,a$.

This pressure, $p\,r\,d\,a$, acting in the direction P A, gives, when estimated in the direction H A, the value $p\,r$, cos. a, $d\,a$. The integral of this for the angle a is $p\,r$ sin. a, and therefore if A represent the angle H A G, $2\,p\,r$ sin. A is the value of the whole pressure of the belt against the surface G H C, estimated in the direction H A. But if s be put for the uniform strain on the belt, $2\,s$ sin. A is the value of the resultant of the two tensions G F and C D, wherefore $p\,r = s$, because $2\,p\,r$ sin. A $= 2\,s$ sin. A; that is to say—

The pressure of the belt against a portion of the circumference equal in length to the radius is just equal to the strain on the belt.

Or, as it may otherwise be stated, *The pressure against a linear unit of the circumference is equal to the tension of the belt divided by the number of the linear units in the radius of curvature, since* $p = \frac{s}{r}$.

Before proceeding to apply this principle to the investigation of friction, it is necessary to point out one or two important corollaries:—

When a cord, supposed to be of uniform tension, is wound round any object which is not cylindric, the pressure is greatest upon those parts which are most salient, since the radius of curvature there is the least, and therefore the tendency is to bring the object nearer to the cylindric form.

Again, if a belt be passed over two wheels of different diameters, the pressure per linear inch is greater on the circumference of the smaller wheel, and hence the tear and wear both of the belt and of the surface must be greater there: and since the parts of the smaller wheel come more frequently round to be acted on than do those of the larger, it follows that the abrasion on the lesser wheel is greater than that on the larger, roughly, in the inverse ratio of the squares of their diameters. This is in accordance with the fact, well known to turners, that the pulley-groove wears away many times faster than the groove in the fly-wheel.

When a guide-pulley is used to deflect or to tighten a band, the pressure upon it is inversely proportional to the radius of the pulley; and the injury done to it must, on account of the greater flexure, be in a still higher ratio: hence the importance of having such pulleys made as large as circumstances will allow. It is to be remarked that the angle of deflection has nothing to do with this action, which remains the same however small or however great the arc of contact may be; the amount of deflection, however, has to do with the friction on the axis of the pulley, and cannot be overlooked in an estimate of the general working of the machinery.

When the radius of curvature is very small, the injury done to the surface of the belt by what may be called a mere contact, becomes very great, being inversely as the radius of curvature and directly as the tension: hence the ease with which a tight cord is cut. The edge of the knife has its radius of curvature excessively small, and its penetrating power is enormous.

In this investigation we have assumed the belt to be uniformly tense, and the surface to be uniformly curved throughout. To accommodate the result to those cases in which the tension and the curvature are variable, we must restrict the formula to an infinitesimal portion of the arc of contact. Thus, if a be the inclination of the normal to a fixed line, r the radius of curvature of the arc, and therefore $r\,d\,a$ the length of an element of that arc, we shall have $p\,r\,d\,a = s\,d\,a$, so that the pressure of the belt against a small portion of the curved surface is proportional to its angular extent and to the tension of the belt, without reference to the radius of curvature. Supposing the tension to be alike, the pressure of the belt on one degree of the circumference of the pulley is just equal to its pressure on one degree of the circumference of the wheel.

When the tension of the belt is not uniform, there must be a tendency to slip from those parts where the tension is small towards those where the tension is great, and this tendency can only be counteracted by the friction on the intermediate surface. So long as this friction exceeds the difference between the two tensions there can be no change; but whenever the difference of tension exceeds the friction on the intermediate arc of contact the belt must slip. Thus, if the belt at G, Fig. 670, be strained to a tension S, while at P the tension is only s, there can be no slipping when the friction on the intermediate arc G P is greater than S $- s$; and therefore, if the belt be actually slipping, or on the point of slipping, the friction on G P must be exactly equal to the difference between the strains S and s.

670.

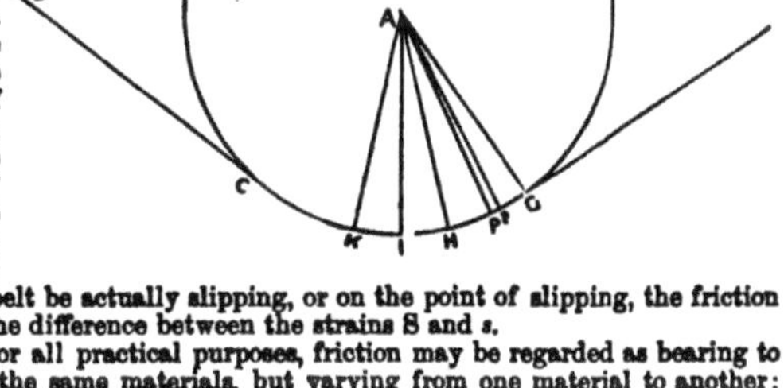

Experiment has shown that, for all practical purposes, friction may be regarded as bearing to the pressure a ratio constant for the same materials, but varying from one material to another;

and this ratio, represented by what is called the coefficient of friction, has been determined for a great variety of substances. Let us put f for this coefficient, so that, p being the pressure, fp may be the corresponding friction.

Take now Pp, a minute increment to the arc CP; the difference between the tensions at P and p, when the belt is slipping, must be equal to the friction on the intermediate arc Pp. Now, although in strictness the tension augments from P to p, yet when Pp is supposed to be infinitesimally minute, we may neglect this variation, and assume that the pressure upon the arc Pp is $s\,da$, and that, therefore, the friction upon that arc is $fs\,da$. In this way the increment of the tension from P to p becomes $ds = fs\,da$, whence $\frac{ds}{s} = f\,da$.

It is well known that the quotient of the differential of a variable by the variable itself, is the differential of the hyperbolic logarithm of that variable; wherefore, passing from differentials to their integrals, we have $fa = \log._{e} s +$ constant.

If then A and a be the inclinations of two normals to a fixed line, S and s the tensions of the belt at their extremities, we must have $f\text{A} = \log._{e}\text{S} +$ constant, $fa = \log._{e} s +$ constant, and subtracting, in order to eliminate the intermediate constant introduced by the integration, we get

$$f(\text{A} - a) = \log._{e} \frac{\text{S}}{s}.$$

It thus appears that, when a belt is slipping, the difference between the logarithms of the tensions at two points is proportional to the angular interval between those points, without any regard to the radius of curvature, and that, therefore, this law holds good of unequably as well as of uniformly curved surfaces.

If we measure off a series of equal angles, GAP, PAH, HAI, IAK, and so on, the logarithms of the tensions at the points G, P, H, I, K, and so on, must be in arithmetical progression, and the tensions themselves in continued proportion: this is also evident when we put the above equation under the form $\frac{\text{S}}{s} = \epsilon^{f(\text{A}-a)}$, in which ϵ is the basis of the hyperbolic system of logarithms.

In order to prepare the above formula for actual use, we convert the neperian into common logarithms by multiplying each side of the equation by M = ·43429 44819, and so on, the modulus of common logarithms, and obtain $\text{M}f(\text{A} - a) = \log. \frac{\text{S}}{s}$.

But here again the angles A and a are represented by arcs measured in parts of the radius, whereas it is customary to estimate them in degrees. Now, for half a turn A − a is π = 3·14159 26536, wherefore for 1° of the ancient (or common) division the difference of logarithms must be $\text{M}f\frac{\pi}{180}$, or for one degree of the modern division $\text{M}f\frac{\pi}{200}$. Wherefore if a° be the number of ancient, a^{c} the number of centesimal degrees over which the belt touches,

$$\log. \frac{\text{S}}{s} = fa^{\circ} \times \cdot 00757\ 98686$$
$$= fa^{c} \times \cdot 00682\ 18818$$

will give the ratio of the tensions at its two extremities when the belt is slipping. For half a turn the difference of the logarithms is $f \times$ 1·36437 63538; and thus, if the friction between the two surfaces were equal to the pressure, the ratio between the tensions of the two ends of a belt bent half round any cylindroid would need to be 23 : 1 before the belt would begin to slip.

When we know the coefficient of friction and the angle of contact, we can compute, by the above formula, the ratio of the tensions; conversely, if we have obtained the ratio of the tensions by experiment, we can compute the coefficient of friction; or, lastly, having given the coefficient of friction, we can calculate the angular extent which may cause a given ratio between the tensions.

If, in the material used, the friction were one-fourth part of the pressure, and if a belt were thrown over a fixed pulley, as shown in Fig. 671, the logarithm of the ratio of the tensions would be 0·3410941, and the ratio itself would be 2·19328 : 1; so that if a weight of 2·19 lbs. were hung on at Q, a weight of 1 lb. at R would be sufficient to prevent it from falling, whereas it would take 4·81 lbs. at R to draw Q up. If the belt were thrown round so as to make the contact over three half-turns, the ratio of the strains would be the cube of the preceding, or 10·5572, so that 1 lb. at R would prevent the fall of 10 lbs. at Q.

671.

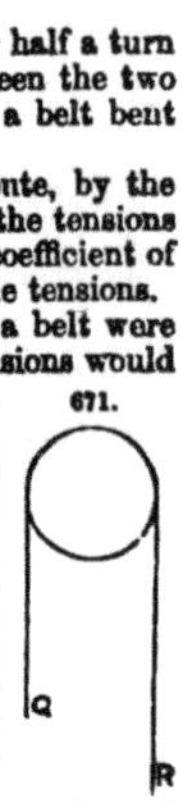

Hence the advantage which a sailor obtains by casting a line round a spar. With one turn he is able to resist a strain, say, four times that which he can exert directly; with another turn he squares the ratio and obtains an advantage of sixteen times; with a third turn an advantage of sixty-four times. In this way he is able to command the strain on the cable, so as by paying it out to stem gradually the motion of the ship. Hence also it is that one man pulling at the loose end of a rope can prevent it from slipping on a windlass at which several men are working.

Another example of the application of this law is as follows:—Having tied a weight to the end of a string, let us give the string a few turns round the smooth axis of a fly-wheel, and then pull at the free end. If the weight be, say, 1 lb., we may exert a strain of perhaps 60 lbs. without causing the cord to slip upon the axis: and by steadily continuing to pull, we put the fly-wheel in motion. But when we cease to pull, the force which the fly-wheel has acquired carries it forwards so as to raise the weight and relieve the tension of that end of the cord by which we hold. As soon as this tension is reduced to the sixtieth part of a pound, the cord slips, and, if we let back the string, the weight falls. In this way, by alternately pulling and letting down, we generate motion only in one direction, since the resistance downwards is but a small fraction of the pressure exerted upwards.

This rapid accumulation of resistance explains the actions of ties, splices, and knots of all kinds: the manner in which it takes place is, on that account, deserving of careful study.

Having now considered the elementary principles on which the action of the driving-belt is founded, E. Sang proceeds to the proper subject of his essay.

Let us suppose that, while a definite resistance is offered to the motion of the wheel B, the apparatus represented in Fig. 668 is actually turning, in consequence of a force communicated through the wheel A. The free part, G F, of the belt has been strained, and the part C D relaxed, until the difference between the two strains has become equal to the offered resistance. Thus at G the belt is lapped upon the driver in a distended state; but it cannot remain in this state until it reach C, for we have seen that any difference in tension between two parts of the belt can only exist in consequence of the friction on the intermediate surface. If the angular distance C H correspond to the ratio of the two tensions, the parts of the belt, as soon as they reach H, must begin to contract, and so fall behind the surface of the wheel in its onward progress; and this contraction must go on until each part arrive at C in that particular state of distension which is due to the strain on C D.

The belt, in its now contracted state, is applied at D to the surface of the pulley B, and proceeds along with that surface until it reaches a point E, having the angular distance E B F equal to H A C; there it begins to be distended, and it reaches F in a state to exert the tension on F G.

Thus it appears that in the ordinary and perfect action of a belt there is necessarily a slipping over the surfaces of the wheel and of the pulley; a slipping which is evinced by the polish and tear and wear of those surfaces. The ordinary notion that the velocity of the wheel should be just that of the belt is thus seen to be altogether erroneous, and the idea of communicating a determinate velocity by means of a belt to be quite a fallacy.

If, on account of increased resistance to the motion of B, the ratio of the tensions on F G and C D be augmented, the extent of the arcs of slipping must increase also; and as soon as the ratio becomes what is due to the arc D F, the belt will be drawn over the pulley without carrying it along, and then the belt slips in the ordinary acceptation. Now the angle of contact is greater on the larger wheel, and therefore we may be prepared to expect that the belt should slip on the pulley. With crossed belts the angles of contact are nearly alike on both surfaces, and minor circumstances are then sufficient to determine on which of the two the slipping will happen. Hence it is useful to arrange the machinery so that the curvature of the slack part of the belt may augment the angle of contact; in general, it is best to have the upper free part slack.

The remedy for the slipping is to augment the general tension of the belt either by increasing the distance between the centres or by shortening the belt; in some cases by using deflecting pulleys. These remedies would have been ineffectual if the angular distance C H had been proportional to the difference between the two tensions instead of to the difference between their logarithms; but as it is, when we augment equally both strains we diminish the ratio between them, and thus lessen the distance over which the slipping takes place, so as to bring it within F D.

In practice it is clearly advantageous to have the belt as slack as the nature of the work will permit, in order to have no avoidable strain and friction upon the axes. This arrangement is readily made by trial, and must, indeed, be repeatedly performed in consequence of the permanent elongation which takes place in new belts; nevertheless, in order to leave no essential part of the theory untouched, we propose to compute the actual quantity of belt which must be used to overcome a stated resistance.

From the dimensions of the machinery we can easily ascertain the angle D B F, and from that, if the coefficient of friction be known, the utmost ratio that can be allowed between the strains on the two free parts of the belt. The nature of the work to be done gives us the difference between those strains, and thus, by the solution of a very easy equation, we can compute the strains themselves.

If d be the required difference, and if ρ be the ratio of the two strains, we must have

$$S = \frac{\rho d}{\rho - 1}; \; s = \frac{d}{\rho - 1},$$

where, according to what has been already shown, $\rho = e^{f(DBF)}$.

Having now ascertained the values of the strains to which the two parts of the belt must be subjected, we can discover how much belt must be used by attending to the law of its extension. Having suspended a piece of it by one end, and measured the distance between two marks made on it, we attach a known weight to the lower end, and measure the increased distance between the same marks. By this experiment we discover the law of distension of all belts of the same material. Let x be the extension of one unit in length when its own weight is hung on; then if w be the weight per foot of another belt of the same material, l its length when unstrained, and s the strain to which it is subjected, its length on being strained becomes $l\left(1 + \frac{x}{w} s\right)$.

For ordinary practice it is sufficient to suppose that the whole length of the belt is subjected to the average of the two strains, S and s, so that if Q be the quantity of belt, that is, the length measured when there is no tension, and L the true length of the line C D E F G H,

$$L = Q\left(1 + x\frac{S + s}{2w}\right); \; Q = \frac{2wL}{2w + x(S + s)}.$$

When the material is not very extensile, the quantity of extension, or the distance which the two ends should want of meeting before they are joined together, is given to a sufficient degree of precision by the formula, $L - Q = L\,\frac{x}{w}\,\frac{S + s}{2}$, from which we see that the distension of the band is nearly proportional to the strain on the axes.

This computation is, however, only approximate. If we wish to obtain a true result we must look more narrowly into the conditions of the problem; and although, for practical purposes, this be a mere refinement of exactitude, it is yet useful in leading to a just perception of the principles involved.

Let C be the distance between the centres, R and r the radii of the two wheels, and i the inclination of each free part of the belt to the straight line joining the centres: then in the case of the plain belt we have

$$\begin{aligned} R - r &= C \sin. i, \\ CD &= a \cos. i, \\ DF &= r(\pi - 2i), \\ GC &= R(\pi + 2i), \end{aligned}$$

so that the entire length of the belt, taking no account of the deviation from straightness caused by its own weight, is $L = 2 C \cos. i + \pi (R + r) + 2 i (R - r)$.

The belt is in various states of tension at different places. From C to D there is the uniform tension s; from D the tension gradually increases to F, where it becomes equal to S; from F to G the tension remains constant, nor does it begin to change until we reach the point h, so taken that the angular distance C h is equal to D F; at that point the tension begins to decrease, until it is reduced to S at the point C. These are the phases when the belt is plain and when the larger wheel drives.

Putting k for $\frac{x}{w}$, that is, for the extensibility of the actual belt, the quantities on the two free parts C D and F G are $\frac{a \cos. i}{1 + ks}$ and $\frac{a \cos. i}{1 + kS}$; the length of the arc G h is L, 4 R i, and since the tension on it is S, when the larger wheel drives, the quantity of belt is $\frac{4 R i}{1 + KS}$.

There remain yet to be computed the quantities of belt on the two arcs D F and $h c$, over which the tension is variable. Since these are similar arcs of different circles, and since the gradation of strain is alike in both, the quantities of belt on them must be proportional to their radii, and so one investigation is enough.

Let, then, θ be an arc over which the tension varies, and let $d\theta$ be an element of that arc, the radius being unit; then if s be the strain at the beginning, $s \epsilon^{f\theta}$ is the strain at the end of the arc, and the quantity of belt applied to $d\theta$ is $dq = d\theta 1 + k\epsilon^{f\theta} s)^{-1}$.

The integral of this gives the quantity of belt on the whole arc s to be

$$q = \theta + \frac{1}{f} \log._\epsilon \frac{1 + ks}{1 + ksc^{f\theta}}.$$

To adapt this general formula to our present case, we must put $\theta = \pi - 2i$, and observe that then $s\epsilon^{f\theta}$ becomes the major strain S, so that $(R + r)\left\{\pi - 2i + \frac{1}{f} \log._\epsilon \frac{1 + ks}{1 + kS}\right\}$ is the quantity of belt covering these two arcs. Hence the length of the entire belt in its unstretched state must, in order to give the requisite strain, be

$$Q = a \cos. i \left(\frac{1}{1 + ks} + \frac{1}{1 + kS}\right) + \frac{4 i R}{1 + kS} + (R + r)\left(\pi - 2i - \frac{1}{f} \log._\epsilon \frac{1 + kS}{1 + ks}\right),$$

and the amount of contraction when the belt is taken off the wheels should be

$$L - Q = a \cos. i \left(\frac{ks}{1 + ks} + \frac{kS}{1 + ks}\right) + 4 i R \frac{kS}{1 + kS} + \frac{R + r}{f} \log._\epsilon \frac{1 + kS}{1 + ks}.$$

When the smaller wheel leads the larger, the tension on the arc G h is only s, and the above formula becomes

$$L - Q^1 = a \cos. i \left(\frac{ks}{1 + ks} + \frac{kS}{1 + kS}\right) + 4 i R \frac{ks}{1 + ks} + \frac{1}{f} (R + r) \log._\epsilon \frac{1 + kS}{1 + ks}.$$

And, when the belt is crossed, we have $a \sin. i' = R + r$, and the angles of contact, $\pi + 2i'$ each, the point, h, merging into G; hence the entire length in that case is

$$L' = 2 a \cos. i' + (R + r)(\pi + 2i'),$$

and the amount of distension, $L' - Q'' = \cos. i' \left(\frac{ks}{1 + ks} + \frac{kS}{1 + kS}\right) + (R + r) \frac{1}{f} \log._\epsilon \frac{1 + kS}{1k + s}$.

If we develop the above values of the contraction in series, according to the powers of k, and omit the second and higher powers, which are always exceedingly small fractions, we obtain the approximations,

$$L - Q = k \left\{a \cos. i (S + s) + 4 R i S + \frac{1}{f}(R + r)(S - s)\right\}$$

$$L - Q' = k \left\{a \cos. i (S + s) + 4 R i s + \frac{1}{f}(R + r)(S - s)\right\}$$

$$L' - Q'' = k \left\{a \cos. i' (S + s) + 4 R i s + \frac{1}{f}(R + r)(S - s)\right\}$$

The last term of each of these expressions depends on $S - s$, that is, on the work to be done, while the first terms depend on the manner in which that work is accomplished.

Throughout the whole of these inquiries, the superiority of the crossed belt is apparent. Thus, in plain belts the ratio of S to s is much less than in crossed belts, and therefore the general tension and consequent friction must be greater.

Having now examined what may be called the statical part of our subject, and ascertained the relation between the resistance to be overcome and the length of the belt, Sang proceeds to consider those phenomena which attend the continuous motion of the apparatus.

It at once follows, from the continuity of the belt, that if the resistance remain unchanged, equal quantities must pass over with the circumference of both wheels. Now it is wrapped upon the driving in a more distended state than upon the driven wheel, so that the linear velocity of the latter must be less than that of the former. If B be the entire *quantity* of belt which has passed a given point, the distances traversed by the driving and driven wheels must have been $B(1 + kS)$ and $B(1 + ks)$ respectively. Now the force, or quantity of work done, is the product of the resistance by the distance passed over; so that the force given out by the driver must have been $B(S - s)(1 + kS)$, while that communicated to the pulley has only been $B(S - s)(1 + ks)$. Hence a belt does not transmit the whole of the force delivered to it. The origin and nature of this loss have not, so far as I am aware, ever been adverted to, and on this account I shall examine it the more minutely.

Subtracting the second expression from the first, and replacing for k its equivalent $\frac{x}{w}$, we have

loss of force $= B\,\frac{x}{w}(S - s)^2$.

From this it appears that the loss of force is directly proportional to the extensibility of the material of the belt, and inversely to the weight of a given length of it: and hence the advantage of using a material not easily stretched, and also, with a limitation to be hereafter noticed, heavy rather than light belts.

Again, if we have to create a given tension, we may, by augmenting the diameters of both wheels proportionally, reduce $S - s$; for example, with double diameters we should have $S - s$ halved, and its square quartered; but then the belt would pass with double speed: so that, taking all into account, the loss of force would be halved, provided the strength of the belt be not changed. Hence we have these maxims:—

The loss of force is directly proportional to the specific extensibility of the material, inversely proportional to the weight of one foot of the belt, and also inversely proportional to the diameters of the wheels when their ratio and the work to be done are fixed.

The only absorbent of this lost force is the friction of the belt upon the surfaces of the two wheels. In order to compute the amount of this friction, let σ be the strain at any part of the arc over which the slipping takes place, and $\frac{B}{\tau}$ the quantitative velocity of the belt, then $\frac{B}{\tau}(1 + k\sigma)$ is the linear velocity of that part of the belt, while $\frac{B}{\tau}(1 + kS)$ is the velocity of the surface of the wheel; so that the rate of slipping must be $\frac{B}{\tau}k(S - \sigma)$. Now the friction on an element of the arc is necessarily $d\sigma$, so that $\frac{B}{\tau}k(S - \sigma)\,d\sigma$ is the rate at which force is absorbed by friction upon this element. Integrating from $\sigma = S$ to $\sigma = s$, and extending the computation over the whole time, the force lost by friction on the wheel is $\frac{1}{2}Bk(S - s)^2$: that consumed by friction on the pulley is also $\frac{1}{2}Bk(S - s)^2$; and thus the total loss is $Bk(S - s)^2$, precisely what we found by comparing the velocities of the two wheels. We may observe here that the same result would have been found although any other law of friction or of extensibility had been assumed; and it is also worthy of notice that this force has been distributed equally between the two wheels, irrespective of their diameters.

The amount of force delivered to the driver is $(S - s)(1 + ks)B$, but the whole of this is not communicated to the free parts of the belt, a portion $(S - s)\frac{1}{2}Bk$ being previously absorbed by friction on the surface of the driver. The force received by the free parts of the belt, and by them transmitted to the pulley, is thus, $B(S - s)\left\{1 + k\frac{S + s}{2}\right\}$. Of this, again, a second portion $\frac{1}{2}Bk(S - s)$ is intercepted by the friction on the surface, and only $B(S - s)(1 + ks)$ communicated to the shaft of the pulley.

In the above formula, for the force actually transmitted along the free parts of the belt, we may observe that the last factor is, for all practical purposes, a constant quantity for a given belt when once put on; for although the apparatus may encounter a variety of resistances, $S - s$, the sum of the two tensions, S and s, hardly varies.

Now the distances passed over by the circumferences of the two wheels are susceptible of easy measurement; let us suppose that, in the time τ, these distances have been found to be D and d; then we shall have $D = B(1 + kS)$; $d = B(1 + ks)$; whence

$$S - s = \frac{D - d}{Bk};\ B\left(1 + k\frac{S + s}{2}\right) = \frac{D + d}{2};$$

so that the force transmitted through the free parts of the belt is $\frac{D - d}{k} \cdot \frac{D + d}{2B}$.

In perfect strictness this formula is only true if the tensions remain uniform during the whole time, for the last factor $\frac{D + d}{2B}$ varies slightly with the resistance to be overcome. In practice, however, this factor scarcely differs from $\frac{L}{Q}$, and the error of substituting this latter for it is contained within very narrow and ascertainable limits; hence we may assume, for all practical purposes, the formula

$$\text{Transmitted force} = (D - d)\,\frac{L}{Q}\cdot\frac{w}{x},$$

as giving the amount of force transmitted during the time over which the measurements extend.

The entire loss of force in the same time is $2\,\frac{(D-d)^2}{D+d}\,\frac{L}{Q}\,\frac{w}{x}$, one-half being lost at each surface.

Knowing, then, the specific extensibility of the material which is denoted by x, and w the weight of a linear unit of the belt used, we are prepared to register the amount of force actually transmitted by it. For this purpose, we bring the edges of two light pulleys to bear gently upon the two free parts of the belt, so as to register the distances through which they move, and then the differences between these distances is proportional to the transmitted force.

Even although the belt be not quite uniform in thickness, this method of registration still answers, particularly if the values of x and w be determined by experiments made on the whole length of the actual belt.

Hitherto we have reasoned as if the belt had no thickness, as if the surfaces of the pulleys were truly cylindric, and as if the material were perfectly elastic. We have yet to take all these circumstances into account.

For the purpose of obviating the tendency which a flat belt has to run off the pulleys, the surfaces of these are made convex in the direction of the axis. At first thought it would appear that such a form would increase the tendency to run off, and that a grooved or hollow contour would rather be needed; but a little reflection clears the matter up. When a band of rope or of ox-gut is used, the wheels are made with deep grooves; in these cases the roundness of the band allows it to roll, so to speak, down the declivity of the groove whenever the inaccurate position of the wheels has tended to lead it out of its proper place. The spiral motion resulting from this action is familiar to turners. Since the pressure upon the sides of the groove is augmented in proportion to the cosecant of the half angle, the friction is augmented in the same proportion, and hence deep angular grooves give a better hold to the band, and enable the lathe to run more lightly than round-bottomed grooves do.

But when a flat belt is used, this rolling cannot take place, and the adhesion of the surfaces causes an action which may be best explained by help of a simple mechanical illustration.

Having provided a slightly tapered cone A B C D, let a belt E F be attached to its surface in a direction perpendicular to the axis; and then let the cone be turned gently round while the tape E F, with a strain upon it, is obliged to pass through a fixed aperture at F. That side of the belt which is toward the larger end of the cone is more rapidly taken up than the other side, and the belt, in consequence, gradually inclining from the perpendicular, runs up toward the thicker end of the cone until the obliquity of the strain directed to F causes it to slide.

672.

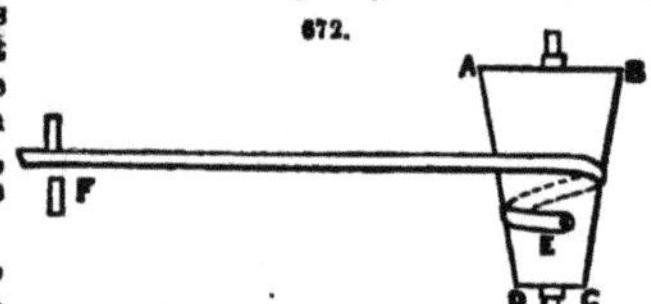

If the surfaces of the wheels were truly cylindric, and exactly opposite to each other on parallel shafts, and if the belt were perfectly straight and uniform, there would be no tendency to move to the one side or to the other. But belts, and especially those of leather, are liable to irregularities, and their joints are with difficulty made straight, so that the two edges are not of equal length; besides, we are unable to arrange machinery with perfect accuracy; hence arises a tendency of the belt to move aside: this tendency is corrected by the rounding of the peripheries. The effect of this rounding is easily seen. Whenever the belt has chanced to go to one side, that edge which is nearer the middle of the pulley is taken up faster than the other edge, and so the belt is soon brought into its place again. It may be here observed that this very provision which keeps the belt in its place so long as the machinery is in proper action, tends to throw it off whenever the resistance becomes so great as to cause a slipping. The degree of rounding of the rims depends on the character of the belts, and on the precision with which the machinery is erected; it can, therefore, only be ascertained by experience. The rounding should be made as slight as is consistent with security, since, as we shall see immediately, every deviation from the cylindric form is accompanied by a loss of force.

In their progress round the wheels, the different parts of the belt are stretched and relaxed alternately. Now, if the material were perfectly elastic, the force expended on the distension would be reproduced on the contraction of the belt; but the imperfection of the elasticity prevents more than a portion of this force from being restored, and hence arises a loss, the character and amount of which we may attempt to examine.

There do not exist any series of experiments on the phenomena of imperfect elasticity of sufficient extent or exactitude to make known the law according to which force is lost by it. Analogy and a few experiments have led to the belief that, within the limits of ordinary practice, the force reproduced bears to that which had been expended in disturbing the natural condition of the particles a ratio constant for a given material; but observations that we have lately made have satisfied us that this law is far from being true, and that *time* has a great influence on the phenomena. For the present, it will be enough to observe that the loss of force will certainly be greater the greater the disturbance of the particles.

Now, in changing from the state of tension S to that of s, the amount of linear disturbance is proportional inversely to the weight, w, of a given length of the belt, and therefore, in respect of this part of the action also, the loss of force is the less the heavier the belt is made. So far as we have yet seen, it is preferable to use heavy belts.

When bent round the circumference of a wheel, the outer parts of the belt are distended, the inner parts relaxed; and supposing the section of the belt to be rectangular, the amount of force expended in making these changes is proportional directly to the breadth, to the square of the thickness, and inversely to the diameter of the wheel. Hence if two belts be of like strength, but the one broad and thin, the other narrow and thick, the amounts of force expended in bending them must be proportional directly to their thicknesses; and hence the advantage of using broad thin belts.

The practice of strengthening belts by riveting on an additional layer must be exceedingly objectionable; indeed, it is difficult to see how any additional strength is gained, for the outer layer must be tight when on the wheel, and slack when free; so that, in reality, the strength of only one layer can be available: the parts of the compound belt are puckered and open alternately, as evinced by the crackling noise. The proper procedure is to increase the breadth of the belt.

If, as must be the case in all driving-belts, the rim of the wheel be rounded, the loss of force by flexure is augmented; because the difference between the radii of the middle of the outer surface, and of the edge of the inner surface, is greater than if the rim had been cylindric; and hence the importance of having the rims as little rounded as possible.

If the law of imperfect elasticity were known, we could reduce these various losses to calculation; as it is, we can only take a limited and very unsatisfactory view of the subject.

The following investigation was undertaken by J. B. Francis, for the Merrimack Manufacturing Company, of Mass., U.S., for the purpose of determining the relative fitness of shafting and belting, of particular materials, for a cotton factory being erected by this company.

In factories and workshops, power is usually taken off from the lines of shafting, at many points, by pulleys and belts, by means of which the machinery is operated, as shown in Figs. 664, 665. When the machines to be driven are below the shaft, there is a transverse strain on the shaft, due to the weight of the pulley and tension of the belt, which is in addition to the transverse strain due to the weight of the shaft itself. Sometimes the power is taken off horizontally on one side, in which case the tension of the belt produces a horizontal transverse strain; and the weight of the pulley acts with the weight of the shaft to produce a vertical transverse strain. Frequently the machinery to be driven is placed above the floor to which the shaft is hung in the story below; in this case the transverse strain produced by the tension of the belt is in the opposite direction to that produced by the weight of the pulley and shaft. Sometimes power is taken off in all these directions from the part of a shaft between two adjacent bearings. To transmit the same power, the necessary tension of a belt diminishes in proportion to its velocity; consequently, with pulleys of the same diameter, the transverse strain will diminish in the same ratio as the velocity of the shaft increases. In cotton and woollen factories with wooden floors the bearings are usually hung on the beams, which are usually about eight feet apart; and a minimum size of shafting is adopted for the different classes of machinery, which has been determined by experience as the least that will withstand the transverse strain. This minimum is adopted independently of the size required to withstand the torsional strain due to the power transmitted; if this requires a larger diameter than the minimum, the larger diameter is, of course, adopted. In some of the large cotton factories in this neighbourhood, in which the bearings are about 8 ft. apart, a minimum diameter of $1\frac{7}{8}$ in. was formerly adopted for the lines of shafting driving looms. In some mills this is still retained, in others $2\frac{1}{8}$ in. and $2\frac{3}{16}$ in. have been substituted. In the same mills the minimum size of shafts driving spinning machinery is from $2\frac{1}{8}$ to $2\frac{11}{16}$ in. In very long lines of small shafting fly-wheels are put on at intervals, to diminish the vibratory action due to the irregularities in the torsional strain.

We can deduce from formula [1], which will be presently established, the *breaking power*, or, in other words, the power which, being transmitted by a shaft, will produce a torsional strain upon it equal to its total resistance to that force.

Put p = the breaking power, in horse-powers of 33,000 foot-pounds.
N = the number of revolutions of the shaft per minute.

$$p = \frac{2\pi \mathrm{R\,N\,W}}{12 \times 33000},$$

from which we deduce,

$$\mathrm{W\,R} = \frac{12 \times 33000\,p}{2\pi \mathrm{N}}.$$

Substituting this value in [1], we find,

$$p = \frac{\pi^2 \mathrm{N}\, d^3\, \mathrm{T}}{8 \times 33000 \times 12} = 0\cdot000003115\,\mathrm{N}\,d^3\,\mathrm{T}. \qquad [8]$$

Substituting the values of T, adopted above for iron and steel, we have

$$\text{For wrought iron, } p = 0\cdot1558\,\mathrm{N}\,d^3, \qquad [9]$$
$$\text{„ steel, } p = 0\cdot2492\,\mathrm{N}\,d^3, \qquad [10]$$
$$\text{„ cast iron, } p = 0\cdot0935\,\mathrm{N}\,d^3. \qquad [11]$$

A formula for the wrought-iron shafts of prime movers and other shafts of the same material, subject to the action of gears, which Francis adopted in numerous cases in practice during the last twenty years, and found to give an ample margin of strength, is

$$d = \sqrt[3]{\frac{100\,\mathrm{P}}{\mathrm{N}}}, \qquad [12]$$

in which P = the power transmitted, and from which we deduce

$$P = 0\cdot01\,N\,d^3. \qquad [13]$$

For simply transmitting power, the formula used is

$$d = \sqrt[3]{\frac{50\,P}{N}}, \qquad [14]$$

from which we deduce

$$P = 0\cdot02\,N\,d^3. \qquad [15]$$

The following Table gives the power which can be safely carried by shafts making 100 revolutions per minute. The power which can be carried by the same shafts at any other velocity may be found by the following simple rule:—

Multiply the power given in the Table by the number of revolutions made by the shaft a minute; divide the product by 100; *the quotient will be the power which can be safely carried.*

Diameter in Inches.	Horse-power which can be safely carried by Shafts for Prime Movers and Gears, well supported by bearings, and making 100 revolutions a minute; if of			Horse-power which can be safely transmitted by Shafts making 100 revolutions a minute, in which the Transverse Strain, if any, need not be considered; if of		
	Wrought Iron, computed by Formula [13]	Steel, computed by Formula [16].	Cast Iron, computed by Formula [20].	Wrought Iron, computed by Formula [15].	Steel, computed by Formula [18].	Cast Iron, computed by Formula [22].
1·00	1·00	1·60	0·60	2·00	3·20	1·20
1·25	1·95	3·12	1·17	3·90	6·24	2·34
1·50	3·37	5·39	2·03	6·74	10·78	4·06
1·75	5·36	8·58	3·22	10·72	17·16	6·44
2·00	8·00	12·80	4·80	16·00	25·60	9·60
2·25	11·39	18·22	6·83	22·78	36·44	13·66
2·50	15·62	24·99	9·37	31·24	49·98	18·74
2·75	20·80	33·28	12·48	41·60	66·56	24·96
3·00	27·00	43·20	16·20	54·00	86·40	32·40
3·25	34·33	54·93	20·60	68·66	109·86	41·20
3·50	42·87	68·59	25·72	85·74	137·18	51·44
3·75	52·73	84·37	31·64	105·46	168·74	63·28
4·00	64·00	102·40	38·40	128·00	204·80	76·80
4·25	76·77	122·83	46·06	153·54	245·66	92·12
4·50	91·12	145·79	54·67	182·24	291·58	109·34
4·75	107·17	171·47	64·30	214·34	342·94	128·60
5 00	125·00	200·00	75·00	250·00	400·00	150·00
5·25	144·70	231·52	86·82	289·40	463·04	173·64
5·50	166·37	266·19	99·82	332·74	532·38	199·64
5·75	190·11	304·18	114·06	380·22	608·36	228·12
6·00	216·00	345·60	129·60	432·00	691·20	259·20
6·25	244·14	390·62	146·49	488·28	781·24	292·98
6·50	274·62	439·39	164·78	549·24	878·78	329·56
6·75	307·55	492·08	184·53	615·10	984·16	369·06
7·00	343·00	548·80	205·80	686·00	1097·60	411·60
7·25	381·08	609·73	228·65	762·16	1219·46	457·30
7·50	421·87	674·99	253·13	843·74	1349·98	506·26
7·75	465·48	744·77	279·29	930·96	1489·54	558·58
8·00	512·00	819·20	307·20	1024·00	1638·40	614·40
8·25	561·52	898·43	336·91	1123·04	1796·86	673·82
8·50	614·12	982·59	368·47	1228·24	1965·18	736·94
8·75	669·92	1071·87	401·95	1339·84	2143·74	803·90
9·00	729·00	1166·40	437·40	1458·00	2332·80	874·80
9·25	791·45	1266·32	474·87	1582·90	2532·64	949·74
9·50	857·37	1371·79	514·43	1714·74	2743·58	1028·86
9·75	926·86	1482·98	556·12	1853·72	2965·96	1112·24
10·00	1000·00	1600·00	600·00	2000·00	3200·00	1200·00

Comparing formula [9] with [12] and [13], and also with [14] and [15], it will be seen that the formulas [12] and [13], used for shafts for prime movers, give a strength 15·58 times the breaking power; and the formulas [14] and [15], for shafts simply transmitting power, give a strength 7·79 times the breaking power.

In applying the rules for the strength of materials to constructions in which there is no movement, it is usual to make the computed strength from three to five times the breaking strain. Bodies in rapid motion, however, usually require a greater margin of strength, in order to provide for the tendency to vibration. In cases where shafting for simply transmitting power is very accurately finished, and firmly supported by bearings at short intervals, an excess of strength two-thirds of that given by formulas [14], [19], and [23], will undoubtedly suffice. In ordinary cases, however, the strength given by these formulas should be adopted.

It must be understood that the shafts to which formulas [12] and [13] are applied, are supported by bearings sufficiently near to each other to guard against the transverse strain caused by the prime mover or gear.

To find formulas for steel shafts of the same strength as those for wrought iron, we have for prime movers $p = 15\cdot58\,\mathrm{P}$: substituting this value of p in [10], we have

$$\mathrm{P} = 0\cdot016\,\mathrm{N}\,d^3, \qquad [16]$$

from which we deduce

$$d = \sqrt[3]{\frac{62\cdot5\,\mathrm{P}}{\mathrm{N}}}. \qquad [17]$$

Similarly, we find for steel shafts for simply transmitting power,

$$\mathrm{P} = 0\cdot032\,\mathrm{N}\,d^3, \qquad [18]$$

$$\text{and} \quad d = \sqrt[3]{\frac{31\cdot25\,\mathrm{P}}{\mathrm{N}}}. \qquad [19]$$

Similarly, for cast iron, we find for prime movers,

$$\mathrm{P} = 0\cdot006\,\mathrm{N}\,d^3, \qquad [20]$$

$$d = \sqrt[3]{\frac{167\,\mathrm{P}}{\mathrm{N}}}. \qquad [21]$$

For simply transmitting power,

$$\mathrm{P} = 0\cdot012\,\mathrm{N}\,d^3, \qquad [22]$$

$$d = \sqrt[3]{\frac{83\,\mathrm{P}}{\mathrm{N}}}. \qquad [23]$$

Comparing formulas [14] and [19], it will be seen that the diameters of shafts of wrought iron and steel to transmit the same power are in the ratio of the cube root of 50 to the cube root of 31·25, or as 1 to 0·855. The weights of the shafts will be as the squares of the diameters, or as 1 to 0·731. The power required to overcome the friction of the shafts in their bearings, assuming that the coefficient of friction is the same for wrought iron and steel, will be as the products of the weights into the velocities of the rubbing surfaces. The number of revolutions in a given time being the same in both, the velocities of the rubbing surfaces will be as the diameters, and the weights will be as the squares of the diameters; the power required to overcome the friction will therefore be as the cubes of the diameters, or as 1 to 0·625. That is to say, the power which must be expended to overcome the friction of a steel shaft is five-eighths of that required to overcome the friction of a wrought-iron shaft of equal strength.

The superiority of steel to resist transversal strain is much less than to resist torsional strain. The relative diameters of wrought-iron and steel shafts, to resist equal transverse strains, exclusive of their own weights, are inversely as the fourth roots of the respective values of E, or as $\left(\frac{1}{3500000}\right)^{\frac{1}{4}}$ to $\left(\frac{1}{3800000}\right)^{\frac{1}{4}}$, or as 1 to 0·98. That is to say, steel shafts, to offer the same resistance to external transverse strains, may be 2 per cent. less in diameter than wrought-iron shafts. The weights of such steel shafts will be about 4 per cent. less than the weights of wrought-iron shafts of equal stiffness; and the power required to overcome the friction of the bearings will be about 6 per cent. less.

The constant expressing the resistance of cylindrical bars to torsion, is deduced from Navier's formula (see 'Résumé des Leçons sur l'Application de la Mécanique'),

$$\mathrm{T} = \frac{16\,\mathrm{W}\,\mathrm{R}}{\pi\,d^3}, \qquad [1]$$

in which,

T = a constant for the same material.

W = the weight, in pounds, which, if applied at the distance R, in inches, from the axis, will just fracture the bar.

π = the ratio of the circumference of a circle to its diameter.

d = the diameter, in inches, of the bar at the place of fracture.

The bars subjected to torsion were finished in the form of the following diagram; the ends being 2 in. square, and the middle turned down to a diameter of $\frac{7}{8}$ in., in order to ensure the fracture taking place in that part of the bar.

673.

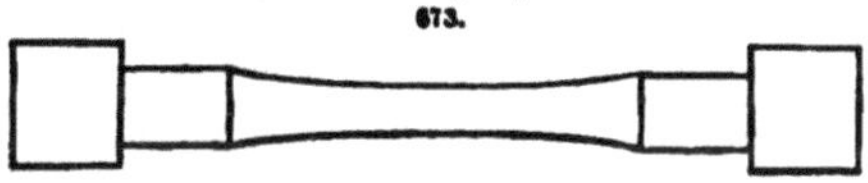

The weight producing the torsion was applied at the end of a lever, of the effective length of 35·975 in., fitted to the square boss at one end of the bar. The tendency of the bar to revolve under the action of the weight, was controlled by a worm-wheel about 15 in. in diameter and 138 teeth, fitted to the square boss at the other end of the bar. This wheel could be moved through any arc by means of a worm. As the bar became twisted by the torsional strain, the worm-wheel was moved through an arc sufficient to bring the lever to an horizontal position.

A graduated circle on one face of the worm-wheel furnished the means of measuring the arc of torsion.

The effective weight of the lever and scale at 35·975 in. from the axis, where the scale was hung on a knife-edge, was 48·5 lbs., and was the least effective weight which could be applied to produce torsion.

EXPERIMENTS ON TORSION.

Description of the Bar.	Mean Diameter of the reduced part of the Bar, in inches.	Arc of Torsion just before Fracture.	Weight producing Fracture, in pounds.	Mean Temperature of the Air.	Value of T.
English refined wrought iron, from a bar 2 in. in diameter, marked A, 13	0·750	416°·8	113·17	58°·8	49,148
Same, marked 13	0·750	596·0	125·69	66·0	54,585
Wrought iron, from the Pembroke Iron Works, Maine, marked 14	0·753	641·3	143·72	62·3	61,673
Decarbonized steel, from the Farist Steel Company, Windsor Locks, Conn., from a bar 2 in. square, marked B, 6	0·752	390·5	192·48	70·2	82,926
Spindle steel, from the same, from a bar 2 in. square, marked A, 5	0·750	284·3	235·17	68·3	102,131
Steel, from the Nashua Iron Company, Nashua, N.H., from a bar 2 in. square, marked 2	0·751	611·3	198·73	65·5	85,961
Same, marked *d*, 2	0·752	557·0	203·23	63·7	87,557
Steel, from same, from 1½ in. octagonal bar, marked 4	0·752	475·0	221·0	67·5	95,213
Same, marked 3	0·751	508·3	217·25	61·2	93,972
Steel, from the works of Hussey, Wells, and Co., Pittsburgh, from a bar 2 in. square, marked E, 1 ..	0·751	308·0	202·66	63·6	87,661
Same, marked 1	0·748	297·3	196·50	68·0	86,023
Bessemer steel, from the works of Messrs. Winslow and Griswold, Troy, New York, from a bar 2 in. square, marked 16	0·748	215·5	181·97	66·0	79,662
Same, marked 16 x	0·748	268·5	174·50	67·0	76,392

EXPERIMENTS ON DEFLECTION.

Description of the Bar.	Diameter of Bar at the middle, in inches.	Deflection, in inches.	Mean Temperature of the Air.	Value of E.
Spindle steel, from the Farist Steel Company, Windsor Locks, Conn., from a bar $1\frac{1}{16}$ in. in diameter, marked A, 7	0·995	0·2330	48°·0	3,853,590
Same, marked A x 7	0·977	0·2315	53·8	3,847,530
Decarbonized steel, extra, from the Farist Steel Company, from a bar $1\frac{1}{16}$ in. in diameter, marked A A x	0·993	0·2310	53·0	3,918,360
Same, marked A A 8	0·995	0·2327	53·7	3,858,557
Decarbonized steel, from the Farist Steel Company, from a bar $1\frac{1}{16}$ in. in diameter, marked 9 x B ..	0·992	0·2330	54·2	3,900,420
Same, marked 9 B	0·995	0·2307	53·3	3,892,008
Steel, from the works of Hussey, Wells, and Co., Pittsburgh, from a bar $1\frac{1}{16}$ in. in diameter, marked 15	0·998	0·2337	52·2	3,796,060
Same, marked 15 x	0·996	0·2337	49·8	3,826,641
Bessemer steel, from the works of Messrs. Winslow and Griswold, Troy, New York, from a bar $1\frac{1}{16}$ in. in diameter, marked 17 x	1·000	0·2330	49·4	3,777,095
Same, marked 17	1·000	0·2315	52·0	3,801,566

The experiments on deflection were made on round bars turned to a diameter of about 1 in. The distance between the points of support was 48 in. Observations were made of the deflections produced by a weight of 150 lbs. suspended at the middle point between the supports. This weight was not sufficient to cause any sensible set in the bar after the weight was removed; and no sensible increase in the deflection was produced by allowing the weight to remain suspended on the bar for several days.

The constant E for deflection has been computed by Navier's formula,

$$s = \frac{l^3 W}{6 \pi d^4 E}, \qquad [2]$$

in which

l = the distance between the points of support, in inches.
W = the weight at the middle point between the supports, in pounds.
π = the ratio of the circumference of a circle to its diameter.
d = the diameter of the bar, in inches.
s = the deflection at the middle point between the supports, in inches.

Several specimens of the steel have been tested for tensile strength, at the works of the South Boston Iron Company, by F. Alger, in the apparatus designed by Major W. Wade, for testing metals for cannon, a description of which may be found in 'Reports of Experiments on the Strength and other Properties of Metals for Cannon,' published in 1854, by authority of the Secretary of War, U.S.

EXPERIMENTS ON TENSILE STRENGTH.

Description of the Specimen.	Diameter at the place of Fracture, in inches.	Weight producing Fracture, in pounds.	Tensile Strength a square inch, in pounds.	Specific Gravity.
Spindle steel, from the Farist Steel Company, Windsor Locks, Conn., marked A 10 A 1	0·597	40,800	145,754	7·8401
Same, marked A 10 A 2	0·598	39,500	140,639	7·8287
Decarbonized steel, from the same, marked B 11 B 1	0·596	34,500	123,662	7·8583
Same, marked B 11 B 2	0·597	35,200	125,750	7·8514
Decarbonized steel, extra, from the same, marked A A x 1	0·600	30,500	107,862	7·8417
Same, marked A A x 2	0·600	30,900	109,271	7·8579
Same, marked A A x; ends upset in order to form the specimen	0·600	30,800	108,901	7·8484
Same as next preceding specimen, marked A A x 2	0·600	29,700	105,053	7·8534
Steel, from the works of Hussey, Wells, and Co., Pittsburgh, marked c 12, 1	0·594	40,400	145,790	7·8530
Same, marked c 12, 2	0·594	40,200	145,070	7·8496

From the many experiments on the fracture of iron and steel by torsion, Francis deduced the following values of T; using the above formula for cylindrical bars, and Navier's formula,

$$T = \frac{3\sqrt{2}\,W\,R}{b^3} \qquad [3]$$

for square bars, in which b = the side of the square in inches, and W and R the weight in pounds, producing fracture, and the distance from the axis in inches at which it is applied.

Experiments by Rennie, given in the 'Philosophical Transactions of the Royal Society' for 1818.

Bar of English wrought iron, 0·25 in. square	T = 65,982
„ Swedish wrought iron, 0·25 in. square	T = 61,909
„ shear steel, 0·25 in. square	T = 111,191
Average of 3 bars of iron cast horizontally, 0·25 in. square ..	T = 64,776

Experiments given in the Fifth Edition of 'Haswell's Engineers' and Mechanics' Pocket-Book.'

Bar of Ulster Iron Company's wrought iron, 1 in. diameter ..	T = 87,090
„ Swedish wrought iron, 1 in. diameter	T = 93,965

Experiments made at the Royal Gun Factories, Woolwich, England, on many Varieties of Cast Iron. Parliamentary Document, July 30, 1858.

Experiments are given on fifty-one varieties of British cast iron, besides several varieties from other countries. Francis selected the experiments on four varieties of British iron, namely, the strongest, two of medium strength, and the weakest; each result being deduced from a mean of several experiments on bars about 1·8 in. in diameter.

From West Hallam Iron Works, Ilkeston	T = 38,217
„ Netherton Iron Works	T = 34,490
„ Butterley „	T = 33,949
„ Hæmatite Iron Company	T = 22,132

Experiments made at the Fort Pitt Foundry, in 1846, *on bars of different forms and dimensions, of Common Foundry Iron, given in 'Reports of Experiments,' &c., above cited.*

Bar about 1 in. square	T = 36,846
„ 1·415 in. square	T = 34,443
„ about 1·749 in. square	T = 42,821
„ 1·135 in. in diameter	T = 37,445
„ 1·595 „ „	T = 42,047
„ 1·955 „ „	T = 38,851

Experiments made at the West Point Foundry, in 1851, on Greenwood Iron of different grades, mixtures, and fusions, given in 'Reports of Experiments,' above cited.

Mean deduced from eighteen experiments on bars about 1·9 in. diameter	T.= 44,957
The value of E, for wrought iron, we have previously deduced from English experiments, and tested by a single experiment on a shaft 2 in. in diameter and about 180 in. between bearings. From these experiments we find	E = 3,492,539

There being such great irregularities in the values of T, it will not be safe, in practice, to take a mean value, but one near the lowest value.

The values for wrought iron vary from 49,148 to 93,965. For safety, we take for wrought iron	T = 50,000
The values for steel vary from 76,392 to 111,191. For safety, we take for steel	T = 80,000
The values for cast iron vary from 22,132 to 64,776. For safety, we take for cast iron	T = 30,000
We also take for wrought iron	E = 3,500,000
And for untempered steel	E = 3,800,000

Shafts for transmitting power are subject to two forces, namely, transverse strain and torsion. In shafts of wrought iron or steel, in which the bearings are not very near to each other, a transverse strain, too small to cause fracture, will produce sensible deflection; if this is too great, it will produce sensible irregularities in the motion, and tend towards the rapid destruction of the shaft and its bearings. This limits the distance between the bearings, as the weight of the shaft itself will produce an inadmissible amount of deflection whenever this distance exceeds a certain amount, which varies with the material and diameter of the shaft.

The deflection of a cylindrical shaft from its own weight, supported at each end, but disconnected from other shafts, is given by the formula [4], which is deduced from Navier's formula for the deflection of a cylindrical bar. See 'Journal of the Franklin Institute' for February, 1862.

$$\delta = 0\cdot007318 \, \frac{l^4}{d^2 E}. \qquad [4]$$

If the several parts are so connected as to be equivalent to one continuous shaft, it will correspond to the case of a beam fixed at both ends, for which case Barlow gives δ equal to two-thirds of its value in the case of a beam supported at both ends, given by formula [4]. Navier, taking into account the effect of the deflection in the adjacent divisions, finds δ equal to one-fourth of its value by formula [4]. In order to decide which of these authorities to follow, Francis appealed to experiment.

Experiment 1.—A bar of wrought iron purchased as "English refined," 12 ft. 2$\frac{3}{8}$ in. long, 0·367 in. deep, 1·535 in. wide, was supported at four equidistant points, 4 ft. apart. When loaded at the middle points of each division with 52 lbs., the deflection in the middle division was 0·069 in., and the mean deflection in the other two divisions was 0·371 in. The weight on the middle division was then increased until the deflection was alike, namely, 0·281 in. in each division; the weight being 82·84 lbs. in the middle division, and 52·00 lbs. in each of the other divisions. 4 ft. was then cut off of each end of the bar, when the deflection, with 82·84 lbs. on the middle division, was 1·102 in.

Experiment 2.—A bar of iron of the same quality and length as in Experiment 1, 0·551 in. square, was laid on the same supports. When loaded at the middle points of each division with 52 lbs., the deflection in the middle division was 0·058 in., and the mean deflection in the other two divisions was 0·314 in. The weight on the middle division was then increased until the deflection was 0·241 in. in each division; the weight being 82·84 lbs. in the middle division, and 52·00 lbs. in each of the other divisions. 4 ft. was then cut off of each end of the bar, when the deflection, with 82·84 lbs. on the middle division, was 0·984 in.

In the case in which the deflections were alike in the three divisions, the middle division corresponds to the case of a continuous shaft supported by numerous equidistant bearings; and the case where the bar was reduced in length, corresponds to that in formula [4]. Comparing the deflections in the two cases in the above experiments, we find by Experiment 1 that the ratio of the deflection of the shaft, simply supported at each end, to that of the continuous shaft, is as 1 to 0·255. In Experiment 2, the corresponding ratio is as 1 to 0·245; the mean of the two experiments giving a ratio of 1 to 0·25, which agrees with Navier, and we must adopt for the deflection of a continuous shaft, from its own weight, the formula

$$\delta = \tfrac{1}{4} \times 0\cdot007318 \, \frac{l^4}{d^2 E}. \qquad [5]$$

These experiments indicate the effect of connecting the chords of truss-bridges over the piers. Assuming that in a bridge of not less than three equal spans, the top and bottom chords have equal resisting powers, and the whole length of the bridge is uniformly loaded, if the chords are continuous throughout the whole length of the bridge, the deflection of any span, except the end spans, will be one quarter of the amount that it would be if the chords were disconnected at the piers.

The greatest admissible value of δ in proportion to the length must be determined by experience. Tredgold assumes that, for cast iron, it might be 0·01 in. for each foot in length, or $\frac{1}{1200}$ part of the length, *whatever may be the diameter;* but the transverse strain to produce this deflection is a greater fraction of the transverse strain that will produce fracture in a large shaft than in a small one. The maximum strains of extension and compression in a shaft, for the same deflection, are in proportion to the diameter, while the deflection itself, from the weight of the shaft, is inversely as the square of the diameter; consequently, the deflection, to produce the same maximum strains, must be inversely as the diameter.

Adopting this principle, and the assumption that a shaft of wrought iron or untempered steel 2 in. in diameter may deflect from its own weight 0·01 in. a foot in length between the bearings, we may determine the greatest admissible distances between the bearings of shafts of other diameters, as follows:—

The greatest admissible deflection for any diameter d, is

$$\delta = \frac{2\,l}{1200\,d} = 0{\cdot}00167\,\frac{l}{d}. \qquad [6]$$

Substituting this value of δ in [5] and reducing, we have

$$l = \sqrt[3]{0{\cdot}9128\,d\,\mathrm{E}}. \qquad [7]$$

Table of the Greatest Admissible Distances between the Bearings of Continuous Shafts, subject to no Transverse Strain except from their own Weights; computed by Formula [7].

Diameter of Shaft, in inches.	Distance between Bearings, in feet.		Diameter of Shaft, in inches.	Distance between Bearings, in feet.	
	If of Wrought Iron.	If of Steel.		If of Wrought Iron.	If of Steel.
1	12·27	12·61	7	23·48	24·13
2	15·46	15·89	8	24·55	25·23
3	17·70	18·19	9	25·53	26·24
4	19·48	20·02	10	26·44	27·18
5	20·99	21·57	11	27·30	28·05
6	22·30	22·92	12	28·10	28·88

In practice, long shafts are scarcely ever entirely free from transverse strains. However, in the parts of long lines which have no pulleys or gears, with the couplings near the bearings, the interval between the bearings may approach the distances given in the table. Near the extremities of a line, the distances between the bearings should be less than those given in the table. The last space should not exceed 60 per cent. of the distance there given, the deflection in that space being much greater than in other parts of the line. In shafts moving with high velocities, it will usually be necessary to shorten the distances between the bearings, as given in the table, in order to obtain sufficient bearing-surface to prevent heating.

BELTING, Chain. Fr., *Ceinture de chaine;* Ger., *Kettenseil.*

Clissold's chain belting, Figs. 674, 675, 676, 677. This belting is composed of iron links of a peculiar shape, an enlarged cross-section of which is shown in Fig. 677. The links are coupled by

674.

675.

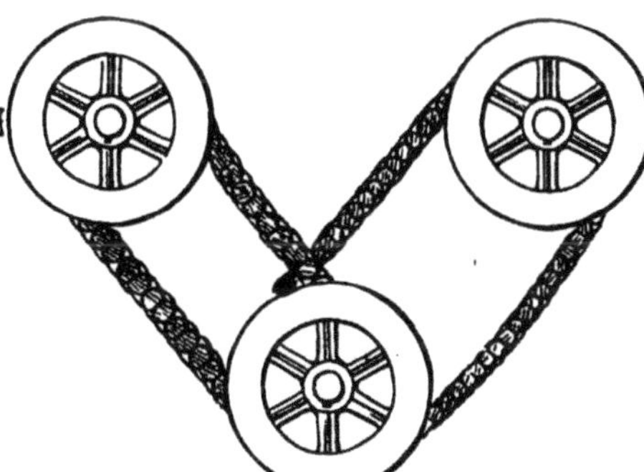

676.

677.

pins, as shown in Figs. 674, 675, 676; each alternate is formed with sockets, in which pieces of hard wood, bevelled at the ends to fit the pulleys, are inserted. It may be observed from the detached section, Fig. 677, that the shape of the groove in the pulleys is such that the belt is clipped between the sides, and a firm hold is obtained. Chain-belts of this kind have now been in use for several years, some of them running at a speed of 1700 ft. a minute; they have been found durable, whilst they work smoothly and without slip.

John Fielden's cast link-chain is shown in Figs. 678, 679. This chain shows that John Fielden has much ingenuity; it is the only faultless cast link-chain which has fallen under our notice.

The chain-wheel round which the chain, Figs. 678, 679, may be moved, is given in plan and section, Fig. 682. It has been found that the chain works best when there is a flat web passing from tooth to tooth, as shown in the plan, Fig. 682. Many of these chains, made of malleable cast iron, are successfully used in the woollen-mills of Lancashire and Yorkshire.

678.

679.

680.

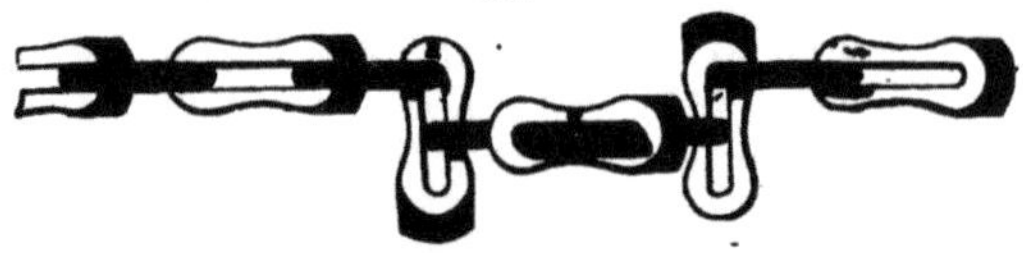

681.

682.

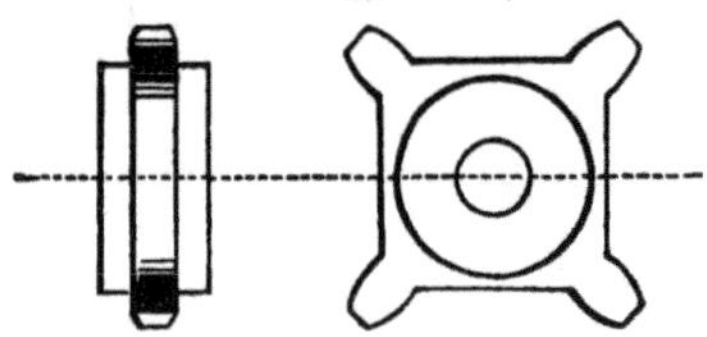

The Fielden chain, made of cast brass, is much in use in dye-works. One of the chief peculiarities of this chain is, that it cannot come uncoupled, no matter whether the chain be slack or tight when at work; and yet, a workman once shown how, may uncouple or recouple it in a very short time.

The chain is easily coupled together by commencing at one end, and keeping all the links with lip in one direction when on the plates, as shown in Figs. 678, 679.

For the method of uncoupling, see Figs. 680, 681.

Set the links 2 and the links 5 at right angles to the plates 3 and 6. Place the ends of plates 3 and 6 together, as shown in Fig. 680. It will now be found that the lip of links 4 can be pushed forward into the hollow side of links 5.

The links 4 will then open over the ears of plate 6, so that plate 3 may be removed away. See Fig. 681.

The links 4 may now be put back into their respective places upon plate 6, then turned round and be easily removed from plate 6.

BENCH. Fr., *Établi;* Ger., *Hobelbank;* Ital., *Banco da falegname;* Span., *Banco.*

A bench is a table on which carpenters, joiners, and others prepare their work. It is usually from 10 ft. to 14 ft. long, 2 ft. 3 in. wide, and 2 ft. 8 in. high.

The carpenter's bench is furnished with a screw-board A, Fig. 683, which holds the bench-vice B and the bench-pin C. The vice consists of a cheek, having a screw E working into a nut

683.

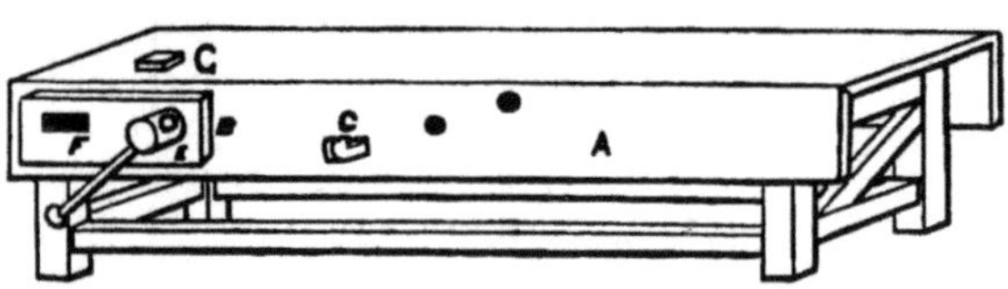

fixed at the back of the screw-board, and a guide F. The bench-pin C is made to fit tightly into holes placed at different elevations in the screw-board, its use being to assist the bench-vice in retaining the board whose edges are about to be planed, or, as it is technically termed, *shot.*

On the top of the bench is the bench-stop G, which is a piece of iron made with teeth to catch in the end of the piece of wood to be worked, and prevent it from being pushed forward by the force of the plane.

Bench-hook.—A movable pin, passing through a mortice in the top of the bench, for preventing the stuff from sliding while being wrought by the plane.

BENCH-MARKS. Fr., *Repère;* Ger., *Merkzeichen;* Ital., *Punto di paragone;* Span., *Cotas de referencia.*

See Railway Engineering.

BEND. FR., *Tuyau coudé*; GER., *Knierohr*; ITAL., *Gomito*.

A piece of curved pipe connecting two straight portions is designated a *bend*.

Earthenware or stoneware bends are usually double the price of a straight piece of the same length. The price of cast-iron bends is also increased, owing to the pattern being more costly, and, in most cases, having to be specially made.

Wrought-iron pipes can be bent while cold by filling them with lead, and afterwards making it out by heating the pipe.

BERM. FR., *Berme*; GER., *Berm, Wallabsatz*; ITAL., *Banchina*; SPAN., *Berma*.

See FORTIFICATION.

BETON. FR., *Bêton*; GER., *Grundmörtel*; ITAL., *Calcestruzzo*; SPAN., *Hormigon*.

See CONCRETE.

BEVEL OR BEVIL. FR., *Angle qui n'est pas droit*; GER., *Spitzer oder stumpfer Winkel*; ITAL., *Smusso*; SPAN., *Chaflan*.

A bevel in masonry or brickwork is a sloped or canted plane surface.

Any angle except one of 90° is called a *bevel-angle*. See HAND-TOOLS.

BINDERS, OR BINDING-JOISTS. FR., *Travon, chevesteau*: GER., *Unterzug, Mittelbalken*, SPAN., *Traviesas, tirantes*.

See JOISTS.

684.

BIRD'S MOUTH. FR., *Joint en biseau, About en Geule*; GER., *Keilsteg*; ITAL., *Commettitura abecco d'uccello*; SPAN., *Picolete*.

Bird's Mouth.—A notch cut at the end of a piece of timber, as Fig. 684.

In bricklaying, a notch cut in a brick to adapt it to any internal angle less than 90° is a bird's mouth.

BISCUIT MACHINE. FR., *Machine à faire le biscuit*; GER., *Bisquit Maschine*; ITAL., *Macchina da biscotti*.

See BREAD-MAKING MACHINERY.

BISMUTH. FR., *Bismuth*; GER., *Wismuth*; ITAL., *Bismuto*; SPAN., *Bismuto*.

Bismuth is a rare metal, but its distinguished qualities are that it is very fusible, and causes other metals to become fusible also. Like *antimony*, it is very brittle, and of a brilliant lustre; its colour is white, tending to flesh-colour. It melts when pure at 480°; it may be distilled in a close vessel, and then crystallizes in laminæ. Water being put = 1, its specific gravity is 9·83, which may be increased to 9·98 by hammering. Bismuth is peculiarly suitable for castings, as it expands in the act of cooling, which renders it peculiarly suitable for castings.

Ores of Bismuth.—There are many minerals which contain bismuth, but they do not often occur in such quantities as to make the extraction of the metal profitable. The metal is not very valuable, and, notwithstanding its scarcity, it is sold at a low price. It occurs native, and is then easily obtained. Native bismuth is found in Monroe, Ct., where it is associated with wolfram, galena, blende, and quartz; also in Chesterfield, South Carolina; and, of course, in many localities of other parts of the world. Sulphuret of bismuth occurs at Haddam, Conn. The carbonate is found in the gold district of Chesterfield, South Carolina; and the sulphuret and lead and copper, at Lubec lead mines, in Maine. Telluric bismuth exists in the gold regions of Virginia and North Carolina, U.S. All the metal in market is obtained almost exclusively from cobalt-speise, at the smalt works of Germany. This residuum, from which also nickel is extracted, contains on the average 7 per cent. of bismuth.

685.

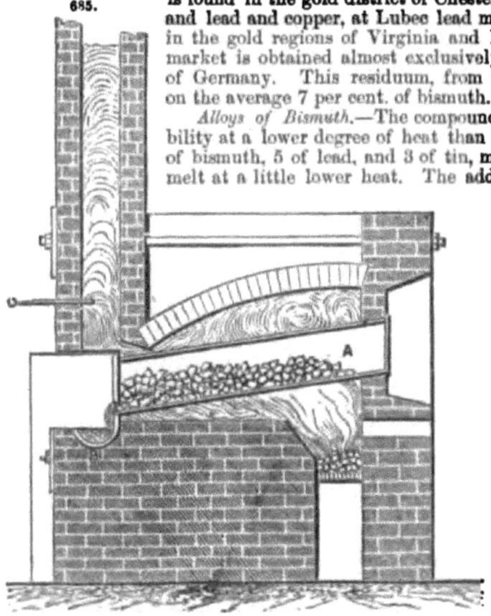

Alloys of Bismuth.—The compounds of bismuth are distinguished by fusibility at a lower degree of heat than those of most other metals. Eight parts of bismuth, 5 of lead, and 3 of tin, melt at 202°. Two bismuth, 1 lead, 1 tin, melt at a little lower heat. The addition of mercury increases the fusibility of these alloys. One bismuth, 2 tin, 1 lead, is soft solder for pewter. Clichés for stereotypes are composed of 3 lead, 2 tin, 5 bismuth: this alloy melts at 199°. 45·5 bismuth, 28·5 lead, 17 tin, and 9 mercury, is an alloy for plugging teeth; it fuses at 149°. An amalgam of 20 bismuth and 80 mercury is used for silvering the interior of glass globes. Like antimony, bismuth forms an alloy readily with the alkaline metals. Its affinity for arsenic is very weak, like that of phosphorus; both of these substances may be evaporated from the hot metal almost entirely. All its compounds with precious metals are very brittle. Bismuth has been proposed instead of lead for refining silver; but the experiments performed with it were not satisfactory. A compound of tin and bismuth is stronger, harder, and more sonorous than pure tin; and for these reasons it is added to pewter. An alloy of equal parts of lead and bismuth is heavier than the mean density of the two metals, it being 10·709.

Uses.—Bismuth is scarcely used alone; it is chiefly employed for imparting fusibility to alloys. Besides the above-mentioned applications, it is used in the alloys of which safety-plates and plugs in steam-boilers are made. Its oxides are used as cosmetics; also as paints, and printing colours.

Manufacture.—The operation of smelting bismuth is extremely simple; the metal having but a weak affinity for other substances is obtained by simply heating its ore. The cut, Fig. 685, shows a modern liquation furnace, by which the metal is obtained. A is a cast-iron retort, at the highest part of which the crude ore is charged; B shows a cast-iron bowl into which the metal flows. About half a hundredweight of broken ore is charged in each retort, of which there are four in a furnace side by side. This quantity nearly half fills a retort, so that the upper part of it is empty. The lower end of it is closed with a clay plate, or slab, provided with an aperture for the discharge of the melted metal. The pipes, when properly ignited, soon cause the metal to flow into the dish B, which contains some charcoal-dust. By applying a brisk fire and some stirring to the ore, all the metal contained in it is obtained within half an hour. The residuum of the ore is now scraped out of the retort into a trough with water, and the pipes are filled afresh. About a ton of ore is smelted in a day of eight hours. The metal is remelted, cast into iron moulds in the form of ingots, and is now ready for the market.

The metal thus obtained is not pure but it may be purified by remelting in a flat earthen, or rather a bone ash-dish, at a low heat, removing the dross as it appears on the surface of the metal. It is advisable to melt the metal thus obtained in a purer form in a blacklead pot, and then cast it into the mould for ingots. Bismuth cannot be freed from silver by these means, in consequence of which the article of commerce always contains some of that metal. The annual production of this metal amounts to nearly 19,000 lbs.

BITS. Fr., *Bittes;* Ger., *Grosse Beting;* Ital., *Bitte.*

See Augers, Braces, and Bits.

BLASTING. Fr., *Pétardement;* Ger., *Sprengen;* Ital., *Minare.*

See Boring and Blasting.

BLAST FURNACE. Fr., *Fourneau à courant d'air forcé;* Ger., *Schachtofen;* Ital., *Forno ad aria forzata.*

Furnaces are classified as *wind* or *air furnaces* when the fire is urged only by the natural draught; as *blast furnaces* when the fire is urged by the injection artificially of a forcible current of air; and as *reverberatory furnaces* when the flame of the fire, in passing to the chimney, is thrown down by a low arched roof upon the materials operated upon.

In general terms, a *furnace* is an enclosed place where a hot fire is maintained, as for melting ores or metals, for warming a house, for baking bread or pottery, or for other useful purposes; as an iron *furnace;* a hot-air *furnace;* a glass *furnace;* an engine *furnace,* and the like. See Furnaces.

The right construction and suitable arrangement of blast furnaces for either hot or cold blast are of considerable importance in the smelting and manufacturing of iron.

Smelting is an operation which is performed in the blast furnace, as it is termed, because of its size and auxiliaries. In it the separation of the metal from the ore depends on the presence of heat, carbon, and the condition that the metal is heavier than the oxidized substances which form the slag. Blast furnaces are used exclusively in America, for smelting fluid iron, and mostly grey iron. In some parts of Europe a lump of solid iron is formed in the hearth of the furnace. But this is an expensive way of smelting iron, and not proper for imitation.

686.

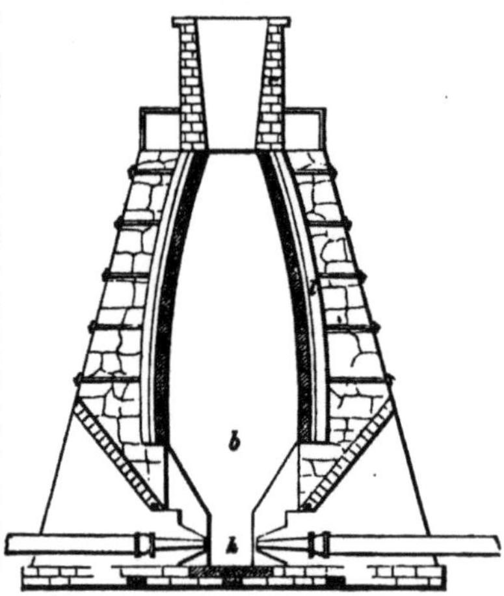

Fig. 686 shows a vertical section of a modern blast furnace. These furnaces are from 25 to 50 ft. high. In almost all instances, the bulk of the mason-work is constructed of rough stones. Sandstone is preferable, but any kind may be used except limestone. The furnace itself forms a pyramidal mass of masonry, commonly as wide at the base as the height from the floor to its mouth. The interior of the furnace is formed of fire-proof material, the lower parts of sandstone, and the upper of fire-brick. The lower part, marked *h*, forms the crucible, or hearth, at which is the strongest heat, and where that part of the ore which has not been smelted in higher parts of the furnace is melted. This part is most commonly square, its sides are from 20 in. to 6 ft. wide, and it is never less than 5 ft., often 8 ft. high. The stones of which they are built in America are exclusively sandstones, while in Europe we find them constructed, not only of this material, but also of granite, gneiss, and even limestones; the latter, however, are becoming rare. Above the hearth *h*, the furnace widens rapidly and forms a gentle slope, *b* the boshes, where the furnace is gradually converted from a square to a round form. At the top, or widest part of the boshes, which varies from 8 to 18 ft. in diameter, the horizontal section of the interior of a furnace is a perfect circle, which is continued up to its mouth. This round part of the furnace is most generally formed of fire-brick, but in some instances of sandstone or shale. It has the form of an inverted cone, in which the sides are more or less curvde. This part of the furnace, marked *i*, is termed the in-wall or lining. All those parts below the lining are solid stones, and closely joined to the rough walls. The lining itself is not close to the rough wall; there is a space between marked *l*, from 6 to 8 in. wide, filled with broken stones, or broken furnace-slags; these are loose, so as to admit of an independent motion of the in-wall, which is for these reasons made of fire-brick.

Rough stones expand and contract more than fire-brick, and are more liable to fractures; and as injury to the in-wall may cause serious losses, the safest plan is to use good fire-brick for its construction. The bricks are generally moulded in the proper manner for forming a circle, and are from 13 to 18 in. in length, which size decides the thickness of the in-wall. The in-wall rests on the rough wall of the stack, and is in many instances supported by heavy cast-iron beams, which form, in the meantime, the tuyere arches. The mouth of the furnace is, in some instances, very narrow, in others wide; this depends on the size of the furnace, kind of ore, fuel, blast, and management. The diameter of this throat varies from 20 in. to more than 10 ft. In the majority of cases the mouth is provided with a cast-iron cylinder, which forms the throat. This cylinder receives the cold material, and is thus prevented from melting, or from injury. The top of the furnace is generally crowned with a chimney, *c*, as wide, or somewhat wider, than the mouth of the furnace; it is provided with one door at small furnaces, and with several at large furnaces. Through these doors the smelting materials are charged.

At the lower part of the furnace may be seen arches, or recesses in the masonry of the stack. These are formed by dividing the basis of the furnace into four piers, as shown in Fig. 687, and are called side arches, H H, and back-arch, F, and work-arch, G. These recesses are generally covered by semicircular brick arches; in few instances they are formed of cast-iron beams. The arches are from 8 to 16 ft. wide, according to the size of the furnace. At large furnaces, a communication between these arches is effected by a gangway, I I I I, piercing the piers. The front or work arch, often called tymp-arch, shows that the crucible is open here; the discharge of the metal and slag is prevented by the dam-stone K, which is of a triangular section, bedded in fire-clay

687.

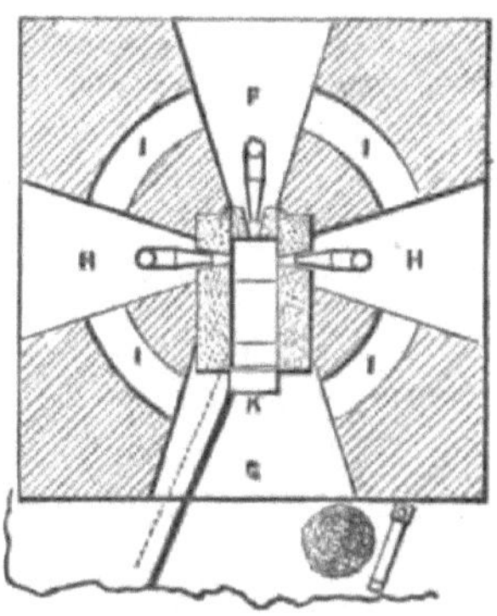

688.

upon the bottom stone. L is the tymp-stone, Fig. 688; it forms, by being raised from 15 to 24 in. above the bottom stone, the aperture for the discharge of the smelted matter, and affords ample space for the removal of any obstructions which may happen to be formed in the hearth. There are some peculiarities in the relative position of dam-stone and tymp, which we shall point out hereafter. The tymp as well as dam-stone are covered with a heavy cast-iron plate, to prevent their being injured by charging heat. From the top of the dam a gentle slope is formed for the discharge of slag which floats continually from the furnace. At the base of the dam-stone a small aperture—the tap-hole—is formed by cutting a part off from this stone before it is bedded.

In Fig. 688 an enlarged view of the hearth and boshes is represented, which presents all their parts more distinctly. It shows the principal joints of the hearthstones, and the manner in which the boshes are formed. These are, in small furnaces, constructed of a mixture of clay and sand, and in large charcoal, anthracite, and coke furnaces, either of fire-brick or of sandstones.

In the plan here represented the furnace is provided with three tuyeres, T. The blast pipes are conducted from the blast machine under the bottom stone of the hearth, and branches from it are led to the tuyeres. Small charcoal furnaces, which smelt from 20 to 25 tons of metal a week, work by one tuyere from one of the side arches. Large charcoal furnaces are worked by two tuyeres on the opposite sides; while anthracite or coke furnaces generally have three, and some five or six tuyeres. Conducting the blast pipes under ground, it has advantages in respect to saving room, but it causes vexation in case any accidents happen to them, which often occur by using hot blast. It affords, however, in the meantime, the security of a dry bottom stone, which is of great value at any furnace. If the blast pipes are thus conducted under the hearth, they should be placed in a spacious channel, so that necessary repairs may be effected at any time. The bottom stone is laid upon a strong cast-iron plate which covers this channel.

The rough walls of a furnace may be erected with little lime mortar in the joints; in fact, roughly-laid stones appear to form the best stacks. When the masonry is well joined, and filled close with mortar, a system of air-channels is required to facilitate the escape of moisture which adheres tenaciously to any masonry. In all instances, a stack may be erected of hewn stones, bricks, or ore, roughly put together; but a well-arranged system of iron binders is required to prevent a separation of the mason-work in consequence of the ever-active expansion and contraction of the building materials. The mode of affixing these iron binders, or ties, is indicated in the various drawings; and more particularly in Fig. 689, which presents a horizontal section through the widest part of the boshes. The particulars respecting the arrangement of these ties are subject to the discretion of the builder; but we may remark that there never can be too many binders in a stack. A large number of light bars is preferable to a small number of heavy ones. These binders are wrought iron, generally square bars from $1\frac{1}{2}$ to 2 in., provided with keys at both ends, in preference to screws and nuts, which are not often used. Each end of a binder is also provided with a large cast-iron washer, which covers the channel as well as the stones nearest to the binder. As we have said, the form of these binders is generally that of a square bar; but a flat form of bars is preferable. These binders are located in spacious channels, so that they may be taken out and mended in case any of them break.

689.

The furnace represented, Fig. 686, is located near the side of a steep hill. The hill and furnace are then connected by a bridge, constructed of wood, or in some instances of stones, or bricks. Upon this bridge a light building, the bridge-house, is erected, which serves as a store-house for fuel, ore, and flux, sufficient to feed the furnace for one or two days. Dry stock is thus protected against rain or snow. At large furnaces, no such use is made of the bridge-house, because it would require to be of too large dimensions. When a furnace is erected on a level place, or when no advantages can be obtained by locating the stack near a hill, which is decided by the mode of supplying the ore and coal, these materials are hoisted by machinery into a tower. Wheel-barrows, which contain the smelting materials, are pushed upon platforms and are raised by chains to the top of the furnace. In Fig. 690 such an arrangement is represented. The tower N is

690.

generally erected of strong timbers, and its top connected with that of the furnace by means of a wooden bridge. A platform is made, which forms in the meantime a cassoon, for the reception of so much water as will balance the load of ore, or coal. Two such platforms are suspended on a strong chain, over a rope-barrel, and when the lower platform is loaded, a current of water is conducted by means of leather hose, into the box, or cassoon, which forms the upper platform. When the amount of water, which flows from a reservoir placed at the top, together with the empty barrows, is heavier than the loaded platform below, the water is shut off, and the loaded platform ascends, while the empty one descends. When the cassoon with water arrives beneath, the upper platform is locked, and the water below is discharged by a self-acting valve, into a drain below the level of the floor. The rope-barrel is provided with a strong brake by which to arrest the machinery, in case an accident happens to any part of it.

This machinery for hoisting is convenient, inasmuch as the power to set it in motion is easily applied, and always at the command of the workmen, provided the cistern is always supplied with water. At the furnace here represented the hot-blast apparatus is placed at the top. The cold blast is conducted upward, and the hot air down to the tuyere. Under this arrangement considerable pressure in the blast is lost, which may be in some measure modified by employing wide pipes. At most furnaces which have been recently erected, both hot-blast apparatus and steam-boilers, which supply the blast-engine with steam, are located on the top of the furnace. For these reasons the area at the top is enlarged. The furnaces are thus made more massive, consepuently there is less loss of heat by radiation from the furnace, and room for a large mouth. The hot-blast apparatus is, in these instances, located behind the steam-boilers; it receives the waste heat when it has passed the boilers. In some instances the top flame is divided, and partly led under the boilers and partly into the hot-air stove.

Whatever may be the dimensions of a furnace, or whatever kind of fuel or ore is used, the work is more or less modified by local circumstances. When a furnace is newly built, or has been out of blast, or has a new hearth put in, a slight fire is at first kindled at the bottom while the dam-stone is wanting. In order to protect the hearthstone against the immediate contact of a strong heat, it is lined with common bricks, which prevents the flying of these stones. The aperture formed by the tymp, bottom stone, and side stones, is walled up by common brick, and only a few small apertures admit of air for combustion. The hearth and stack are thus slowly dried, which may require from three to ten days. When the hearth has been for some days thoroughly warm, the brick lining is removed, and it is filled to the top of the boshes with either charcoal, anthracite, or coke, whichever may be the combustible used for smelting. The tymp is open, in case the hearth is warm and dry; but when any doubt exists as to its being dry, ashes or sand is thrown on the coal in the tymp to prevent a rapid fire. In order to remove clinkers which may be formed in the hearth, it is cleaned every twelve or twenty-four hours; and when the heat is strong, or an early starting of the furnace is contemplated, a grate is formed by means of ringers—long iron bars—as shown in Fig. 691. Thus a strong draught is produced, a rapid combustion ensues, and the heat is augmented. If these bars are withdrawn after fifteen minutes or half an hour's time, the hot coal, descending on the clean hearth-stones, will heat them thoroughly, and prepare them as well as the bottom stone for the reception of hot metal. One day, sometimes two or three days' heat, which time may be shortened by the repeated formation of grates, will prepare the hearth; the fuel has been, all this time, held as high as the boshes.

691.

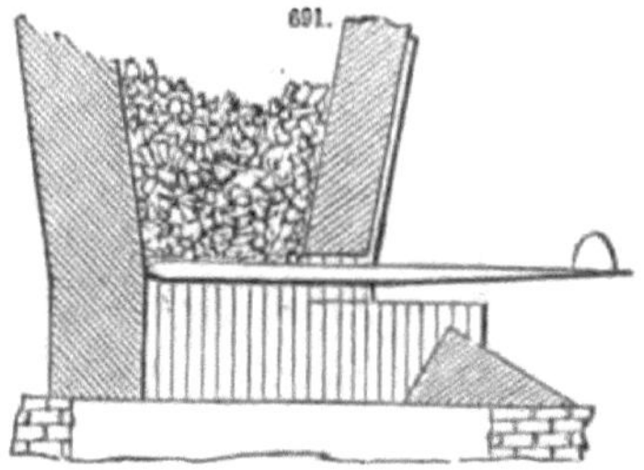

When thus far heated the furnace may be charged with ore. In small charcoal furnaces the coal is generally filled higher than the boshes; but in large ones, and those which are thoroughly heated, there is no need of having much fuel. The furnace is now regularly charged alternately with ore and coal; the ore amounts to only half of a full charge, but the measure of coal is always the same. These charges are not made in rapid succession, but the flame is allowed to become visible on the top of the last charge before another is filled. The furnace is thus slowly fed by alternate charges of ore and coal; and, in order to facilitate the ascent of the gases, coarse coal is selected; or, when charcoal is used, brands or wood are mixed with the coal. When full to the very top, the furnace is ready to receive blast, and not sooner. Some founders usually let on blast before a furnace is quite full. This is imprudent; it causes disorder from the start. When the furnace is thus filled, the ore is drawn down by repeated gratings, which are so managed that the bottom is properly heated. When the first signs of slag or iron appear at the tymp or the tuyeres, the bottom is once more cleared of all adhering cinder, the dam-stone put in its place, and the dam plate bedded in clay upon it. A large coal or coke, or, what is better still, a mixture of fine damp coal and a little clay, is filled into the tap-hole; a stopper, or at first only heavy dust, is filled under the tymp, and the blast put on. At the first, only weak blast is used; in fact, for the next two or three weeks the furnace does not receive full blast, in order not to injure the new hearth and in-walls by a too sudden and strong heat. A furnace is stouter with about half the pressure which it will take, and that gradually increased in the course of some weeks. A few hours' blast will raise the fluid cinder to the top of the dam-stone; the blast may now be stopped for a few minutes, the hearth tried by a light bar as to cleanliness, and if found free from clinkers or cold slag, a light stopper is formed of clay and coal-dust, the tymp shut, and the blast let in again. The melted iron accumulates now at the bottom of the hearth, and the slag runs over the dam-plate and is carried away. The furnace, of course, is perpetually filled with coal and ore, so that the materials are always level with the top. It must be a standard rule never to blow a furnace by low stock, no matter how it works: it must be full. In cases of imminent danger of chilling, a sinking of charges is excusable, but only to be refilled by dead charges; that is, coal without ore.

692.

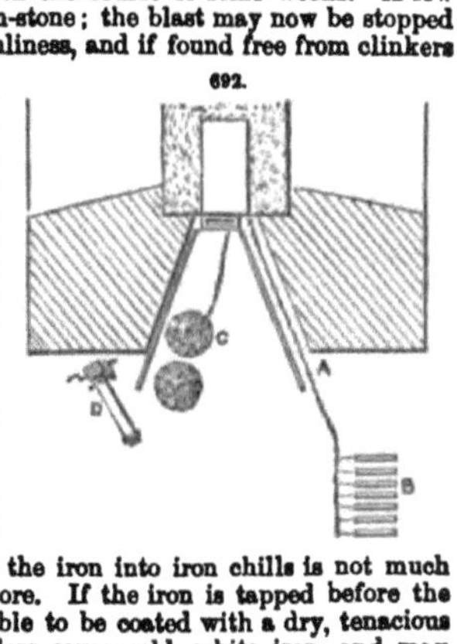

It will require, according to the kind of furnace, from twelve to twenty-four hours to fill the hearth with iron. If possible, the iron ought to come near the top of the dam, before the tap-hole is opened for the first time. The tap-hole is generally at the right-hand side in the tymp-arch, near A, Fig. 692. A channel,—run,—dug in moulding sand, conducts the iron to the pig-bed, B, where the pigs are previously moulded into sand or coal-dust, or dust of anthracite or coke, by means of wooden patterns. Running the iron into iron chills is not much practised. It is confined to only a few furnaces near Baltimore. If the iron is tapped before the pool is quite full, the hearthstones below the tuyeres are liable to be coated with a dry, tenacious cinder, which may cause serious vexation. Such dry cinders cause cold, white iron, and may occasion the freezing of the iron in the bottom. When the iron is thus tapped from the furnace, the blast is slaked, or stopped, the clinkers and cold cinder removed, and a fresh stopper of clay and sand placed under the tymp, and the blast put on again.

The first iron made should always be grey iron; for these reasons the ore charges are light. An increase of ore must be made gradually and very slowly, proceeding with the greatest caution as to the increase of burden. White or mottled iron, in the first week of a blast, is an indication of scaffolding the furnace. The fluid cinder thus formed is liable to adhere to the in-walls and cause troublesome concretions. When grey iron is smelted the cinder is not very fluid, and may descend into the crucible before it becomes sticky, where the heat is strong enough to remove it at any time.

The tymp-arch is divided into two parts, as shown in Fig. 692. The run for the fluid iron is as much lower than the part C in the middle as the height of the dam-stone. C forms somewhat of a slope, falling from the dam-stone gently. At the left-hand side are two cavities, into which the cinder runs alternately. About a ton of it is necessary to fill such a cavity with slag. A piece of pig iron, or any other iron, is set vertically into the centre of the empty cavity, the cinder flows around it, by which it is firmly held, and, when the mass is nearly cold, it may be lifted by means of a crane located at D. It is deposited on a horse-cart, and carried away. The slope C is separated from the run A, and from the slag-trough, by cast-iron plates, set so close to both sides as to afford only sufficient room for either the iron or the slag to be removed. The room, or slope, thus formed, is necessary for the furnace-men to stand upon and work the furnace. In order to make this space as large as possible, the tymp-arch is considerably larger than the tuyere-arch.

Having thus far given a general description of a blast furnace, its construction and mode of operation, we will now take notice of some of its most important particulars. To commence with the bottom stone. This part of the furnace should be particularly dry, and, if convenient, even warm A cold or damp bottom causes white iron and waste of fuel. In some parts, particularly in Sweden, the bottom is purposely kept cool; but not so in this country. It happens at some old furnaces that the foundation is not perfectly dry, for want of drains; but furnaces recently erected are well provided with means for the removal of moisture. Some kinds of ore, but chiefly the quality of iron smelted, afford the reasons for having a cold bottom stone; considering, however, the greater use of fuel incident to it, the advantages are in favour of a dry and warm bottom. The leading form should be the one represented in Fig. 693, for the foundation of a furnace. A spacious archway crosses under the furnace between the pillars, so that a man may enter and examine it. Any moisture which happens to penetrate from above, which is often the case at hot-blast furnaces, thus subsides quickly, and cannot do much harm. In the meantime, it affords an opportunity of correcting the discharge of water, in case there is any obstruction. A damp bottom stone is not only the cause of waste of fuel, but it produces vexatious concretions of cinder below and around the tuyere, which cause much trouble to the founder. The bottom stone should be in one piece, if possible, but there is not much harm done if it is spliced, provided the joint is close, and the stone safely bedded. It should be a hard, well-dried sandstone, with a uniform grain.

693.

694.

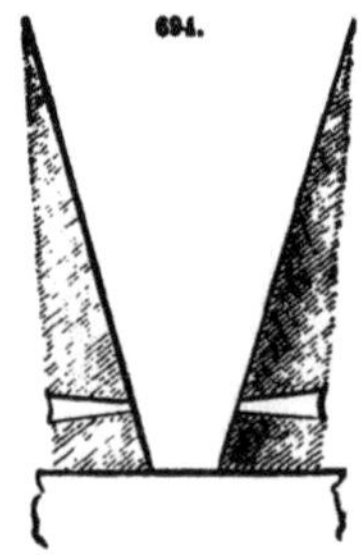

The plan of the hearth is a square, and seldom round or elliptic; the dimensions of the hearth depend entirely on circumstances. A furnace in western New York—Siscoe furnace—which melts a mixture of magnetic ore and hematites, principally the former, is 2 ft. 10 in. wide, 18 in. high below the tuyere, and 20 in. above the tuyere, where the boshes commence. Such a low hearth is suitable for magnetic ore, spathic ore, and some specular ores, but it would not work well with hematites; for the latter kind of ore requires a higher hearth above the tuyere. The charcoal furnaces of Pennsylvania, U.S., are not often less than 4 ft. high above the tuyere. Ores which melt easily, or which are porous and form grey iron, ought to be smelted in a high hearth. The height of a hearth is regulated by the ore, but the size of it at the tuyere is determined by the fuel. A hearth for anthracite or coke is not higher than a charcoal hearth. For ores which melt with difficulty, a low hearth—in fact, one where the boshes commence at the tuyere, as shown in Fig. 694, is considered profitable, and for porous hematites it may reach 1·5 ft. above the tuyere. The space below the tuyere is generally plumb; above it, the batter is from $\frac{1}{24}$ to $\frac{1}{6}$; that is, one half-inch to the foot for very mild ores, and 2 in. to the foot for refractory ones. A high crucible has always more taper than a low one; and one for rich or refractory ores more than one for impure and fusible ores. When forge-pig is smelted, the hearth is lower or more tapered than for grey or foundry pig. The hearth should be wider, and have more batter, when much than when only a little iron is to be smelted. The height and batter of a hearth are in fact not of so much importance as is commonly supposed. It is expensive to smelt grey iron in a low or much-tapered hearth, and it is expensive to manufacture forge-pig in a high hearth. A high hearth always causes inferior forge-iron. The high crucible saves fuel and ore, but works slow If we assume that a furnace without a hearth, where the tuyere is only 8 or 10 in. above the bottom, and the batter of the boshes drawn down to the tuyere, as shown in Fig. 694, and also that a furnace of this construction produces the best forge-iron—then regarding this as one extreme of the forms of a hearth, and considering the other extreme to be a hearth 6 ft. high, and only ½-in. batter to the foot, and assign to this the capacity of producing the best foundry-pig iron—we shall have the intermediate forms nearly in the following order for ores. Starting with no hearth, or the lowest hearth, low pressure in blast, forge-pig, and much iron, the ores which may be smelted to advantage are as follows: raw sparry carbonates, raw magnetic ore, silicates and forge-cinder, raw argillaceous ore, crystallized peroxide, specular ore, compact peroxide, red oxides, roasted carbonates, roasted magnetic ore, roasted argillaceous ore, raw hematites, roasted hematites, pure bog ores, and bog ores which contain much phosphate. The series of ore inverted will work in a high hearth, strong blast, foundry-pig, and smelt slowly. Bog ore may be smelted in a low hearth, but not to advantage. As the ores are generally impure, a great deal of iron is lost in the slags, and consequently much coal is used; the yield is bad, and however good the iron may be in the forge, it is of no use in the foundry. If, on the contrary, we smelt refractory ore, commencing with the series, in a high hearth, the yield is poor, much coal is used, the iron never good for the forge, and not useful in the foundry. We thus see how much the form of a hearth is dependent upon a variety of circumstances, which must be considered in its construction. If we erect a cylindrical high hearth for smelting magnetic ore, and intend to smelt good forge-pig, and much of it, we certainly fail in the attempt. And if we desire to produce foundry-pig, of bog ores, in a furnace without a hearth, we shall find the iron very poor, weak, and hard, consuming much coal and ore in its manufacture, and not suitable at all to be worked in the forge. By considering these facts, we distinguish easily the correct form of hearth for certain kinds of ore, as well as quality and yield of iron.

One side of a horizontal section of the hearth, or the distance between the tuyeres, is never less than 18 in., and not larger than 8 ft. Round or oval sections of crucibles are not often used, and we shall not allude to them. The true measure of a hearth is the contents of the area, for which we assume one side of a square. These dimensions are somewhat controlled by the nature of the ore, but depend chiefly on the quality and kind of fuel, on the quantity and kind of iron to be smelted, on the pressure of the blast, and on the number of tuyeres. A hearth of only 18 in. square at the tuyere, which is worked by one tuyere, will make very little iron—2 or 2½ tons in twenty-four hours. These dimensions are only suitable for working by weak blast, with ½-lb. pressure, and charcoal. In fact, all dimensions below 30 in. are for charcoal only. A hearth of 24 inches may produce from 3 to 5 tons per diem, with ⅝-lb. blast and one tuyere; two tuyeres may bring the yield to 6 tons a day. A hearth of 30 in. may be worked by three tuyeres, ⅝ to 1 lb. pressure, and produce from 6 to 10 tons of metal in twenty-four hours. The ore has much influence on the yield. A hearth of at least 30 in., and from that to 4 ft. in width, is used for smelting by coke, the yield of the furnace being in ratio to the size and amount of blast; it varies from 10 tons per diem to 16 or 17 tons. We find in anthracite furnaces—the largest hearths in them—the distance between the opposite tuyeres is not less than 40 in., and sometimes it is as wide as 6 ft. An old hearth is frequently found to work well with a width of 8 ft. The yield in these furnaces varies from 10 tons per diem to 30 tons, according to size, ore, pressure of blast, and number of tuyeres. Large-sized hearths are generally of a round form.

Pressure of Blast.—The density of blast depends strictly on the quality of fuel. It has been observed that soft charcoal works best with ½ to ⅝ lb. pressure to the sq. in.; hard charcoal, with ⅝ to 1 lb. pressure. The best wood charcoal will not bear more than this density. Raw bituminous coal, or coke, is worked to advantage with 2½ lbs. to 4 lbs. pressure, and anthracite should have at least 4 lbs. We have no evidence that more density is injurious to the operation with anthracite. When less pressure than this is at our disposal, either from want of power or an imperfect blast-machine, the width of the hearth should be reduced, to produce the necessary force of current in the fuel. When hot blast is used, the densities are as above stated; but with cold blast they may be considerably increased. As the effects of hot blast may be in some measure produced by higher densities, the best results must, as a matter of course, be obtained when pressure and temperature are so regulated as to work the ore with the smallest amount of fuel. We are not informed what density of cold blast anthracite coal will bear; but we know that strong coke will bear 6 lbs., hard charcoal 1 to 1½ lb., and soft charcoal to ⅝ and 1 lb.

695.

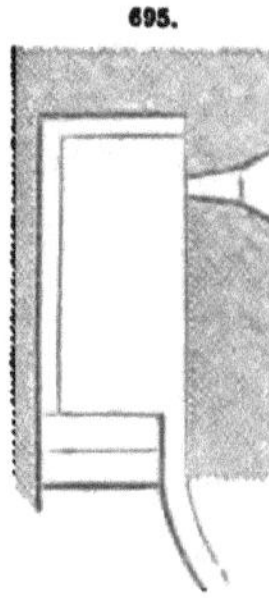

Number of Tuyeres.—The number and size of tuyeres depend on the size of hearth, the quantity of iron to be made, and whether hot or cold blast is used. In a small furnace, where charcoal is used, and the production is limited, but one tuyere is used; and this is applied at one side of the hearth, as shown in Fig. 695, and at the side of the tap-hole. It is a remarkable fact, that all attempts have proved futile to work a furnace by placing the tuyere in the back stone, opposite the *tymp*. This appears to be its natural position, but in practice it does not prove so. Good coal, fusible ore, strong blast, and a well-sized hearth, will produce a large quantity of iron with one tuyere. Some furnaces smelt as much as 7 tons per diem by these means. There are great advantages in working one tuyere, particularly with refractory ores and cold blast. All clayish and siliceous ores work better with a single one. When a second tuyere is used, it is placed opposite the one shown above. At charcoal furnaces we do not often find a third tuyere. At coke and anthracite furnaces we find at least two tuyeres, and in most cases three; and sometimes as many as five or six. Then the section of the hearth is round, and the tuyeres are placed as shown in Fig. 696. This arrangement is well adapted to work by hot blast, but troublesome in using cold blast, on account of the cooling influence of the many apertures. Blast, strongly heated so as to prevent chills at the tuyeres, works admirably well by such an arrangement. A wide hearth and hot blast will admit of the use of more tuyeres than a narrow hearth and cold blast.

696.

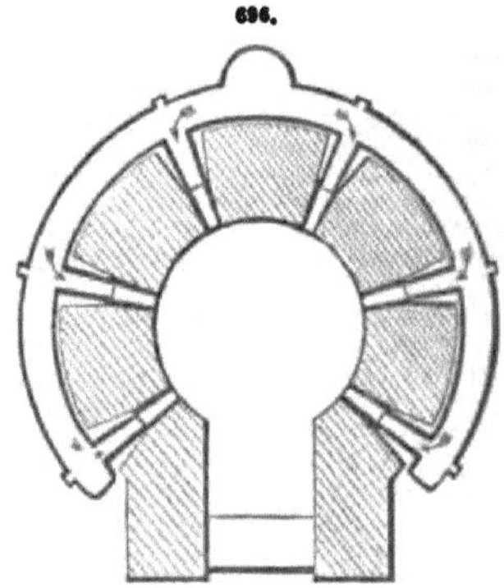

Size of Boshes.—That part of the furnace commencing at the top of the crucible, which forms a slope more or less steep, the form of which varies very much, will be easily understood after the preceding remarks. The width of boshes, which means the largest diameter of the interior furnace, depends in some measure on the fuel, but chiefly on the quantity of blast which is brought to bear upon the fuel. The diameter, or rather the slope of the boshes, depends also on the kind of ore which is smelted. We may reasonably assume that this slope is intended to perform a certain service, and that can be no other than gradually to diminish the force of the current of blast. As has been demonstrated in previous pages, the current of blast carries along with it particles of carbon, which may be either dissolved in the gases or not. They will be deposited where the current or temperature is too weak to hold them longer in suspension. This fine carbon is absorbed by the porous ore. The size of the boshes must be, therefore, in ratio to the quantity of blast and the kind of fuel; assuming that both current and temperature are at the greatest diameter, so far diminished as to deposit the particles of carbon. If the boshes are too narrow for a certain quantity of blast, the point of depositing carbon is carried higher and the smelting of the ore commences where it is liable to deposit refractory slag on the slope, causing scaffolding. If the diameter is too large, the ores are carbonized too low, and the slightest alteration of heat must inevitably deposit partially melted ore in the widest part of the boshes, causing concretions. In cases of doubt, it is preferable to have the boshes too narrow, rather than too wide; but we must be aware that nothing has more influence upon the quantity of metal smelted than the dimension of the boshes. But if the furnace cannot be supplied with sufficient blast, it is very vexatious to have the boshes too wide. The extreme diameters in use are from 7 to 18 ft. Charcoal furnaces will bear 9½, and in some instances 10 ft. of width; but the latter is rather a large size. Coke furnaces are not often less than 12 ft., and do not work well if larger than 15 ft. Anthracite appears to afford a wide range; we find furnaces of 10 ft. boshes, and also of 18 ft., or nearly four times larger. As the quantity of blast is in proportion to the fuel, and that in some measure controlled by the quantity of metal made, we find that the production of a furnace is nearly in proportion to the size of the boshes; still, this rule is not so perfect as to admit of correct deductions. The kind of ore and quality of iron smelted exert almost as much effect on the yield of a furnace as the size of boshes. Fusible, well-fluxed ore furnishes more iron, and a larger quantity of forge than of foundry iron may be made in the same time by boshes of the same size. When the diameter depends on the quantity of blast, the slope of the boshes is regulated by the ore and the quality of iron smelted. A slope of 50° is commonly adopted in small furnaces where fusible bog ores are smelted; even 45° are not considered too flat. Raw ores, of the primitive formation, are smelted in slopes of from 70° to 75°, as shown in Fig. 694. Between these two extremes we observe many varieties of slopes. Close, compact ores, which do not form grey iron, are smelted in steep boshes; and ores which are inclined to produce a grey fusible iron may be smelted in flat ones. Foundry iron is better when made in flat boshes, and forge when made in steep ones. The yield of a furnace is greater in the latter than in the former. The use of fuel is also greater in steep than in flat boshes. This depends, however, so much on the form and composition of the ore, that in these respects little influence is exerted by the slope of boshes.

That part of the furnace commencing at the widest part of the boshes and extending to the top is always of a conical form, with straight, or more or less curved sides. By examining the use of this part of the furnace, we arrive at its correct form. In practice, we find it such as is represented in Fig. 697. We shall not consider the advantages or disadvantages of these forms of in-walls, but proceed to define the use of this part of the furnace. Assuming that the operation of reviving and melting the metal and the slag is carried on in the lower part of the furnace, from the largest diameter downwards—which is not always true, as we shall see hereafter—we discern the use of the space enclosed by the in-wall. Nothing is performed in it except the evaporation of water, and of gases from the ore and coal. In reducing the ore, these substances should not be present. Water, as well as hydrogen, free oxygen, or nitrogen, are of no use in the crucible; the first and the second are actually hurtful. The object of this space, therefore, is to evaporate water from ore and coal without causing injury to the form of these substances. A high heat, of course, will evaporate water sooner than a low one; and it will also break coal and some kinds of ore, and form dust of them. For these reasons, a high heat at the mouth of the furnace is often found to be injurious to the smelting operation. Charcoal requires at least twenty-four hours to dry it at a low

A B 697. C D

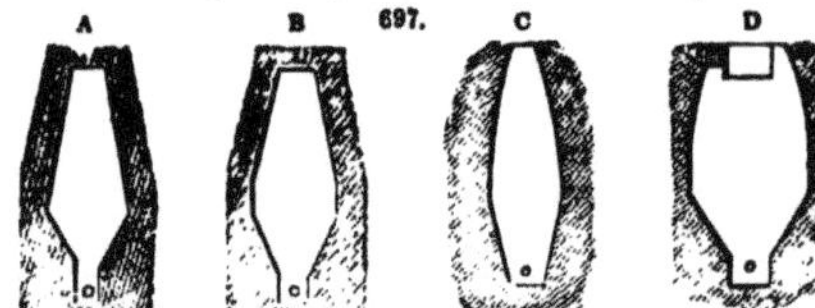

heat, and some kinds of clay or argillaceous ore three times that length of time. When coal and ore may be dried in twenty-four hours, without injury to form, the size is sufficient when the capacity of the furnace above the boshes is sufficiently large to hold ore enough to work for twenty-four hours. When 10 tons of iron are smelted in that time, and 3 tons of ore and 200 bushels of coal are required for 1 ton of iron, the furnace must hold 30 tons of ore and 2000 bushels of coal above the boshes. Generally we find the capacity somewhat greater; but there is no necessity to have more stock in the furnace, whether charcoal or anthracite. Where coke or raw bituminous coal is used, the case is different. This fuel contains always more or less hydrogen, the pressure of which is highly injurious in that part of the furnace where the iron is received. It requires a red heat and a liberal supply of air to expel hydrogen from a large body of coal, and also much time. In this case little can be done in twenty-four hours, and therefore such furnaces have a capacity for three days' stock. The form of the in-wall, if curved or straight, cannot therefore have much influence on the operation; but a gentle curve, or a cylindrical part above the boshes, is found advantageous. This will afford some play for the oscillations of blast, and prevent scaffolding on the slope of the boshes. The curved form shown in C, D, Fig. 697, affords one advantage—namely, the same capacity with less height; and for this reason the curve is advantageous. It has been observed, that beyond a certain limit there is no advantage in increasing the height of a furnace. This may be 40 ft. for charcoal and anthracite, and 50 ft. for coke or bituminous coal. A lower furnace works easier, makes better iron for the forge, and, when well arranged—that is, of sufficient capacity—does not consume more coal in proportion to the iron smelted than a high furnace. It is, therefore, desirable to operate with the least height. This can be accomplished only by curving the lining so as to gain in the capacity. If this space of the furnace serves no other purpose, we may give any form we choose, without injury to the work. It has been proved by experience that the curved lining and a low furnace work better than a straight lining and high stack. The curvature is, of course, never to extend beyond the largest diameter of the boshes.

Size of Mouth.—On this subject a great deal of controversy has arisen without any satisfactory result. It is settled that narrow tops and too wide mouths cause a waste of fuel.

If we consider the object of this aperture we shall be able to determine its size. The mouth is chiefly for charging the furnace with ore, coal, and flux, and for the escape of the waste gases. If the current of the gas is too strong at the top, a large quantity of small particles of carbon are torn loose and escape, thus causing a loss of fuel. If the throat is too narrow, the ore is apt to bake and form lumps, which pass into the centre of the furnace, and descend to the hearth unsmelted. A mouth of the proper size will cause some ore to move towards the in-walls, while the large mass will remain in the centre. When the throat is narrow, a marked difference is found in the operation of the hearth when a fresh charge is thrown in and when it is down. This is particularly the case at small, low charcoal furnaces. The oscillation in heat thus produced causes a waste of fuel. A narrow throat will work hotter than a wide one, and thus cause the flying of coal and ore, which makes dust in the boshes and forms an obstruction to blast. There would be no objection to making the mouth as wide as the boshes, were it not that by so doing the ore is thrown chiefly near the in-walls, which in melting will cause scaffolding; and if in this case the ore is charged in the centre of the furnace, the waste heat escapes chiefly at the in-walls, preventing it from becoming sufficiently dry before arriving at the hearth. These reflections lead to the conclusion that half the diameter of the boshes should be nearly the diameter of the mouth, which is confirmed by practice. At charcoal furnaces the diameter of the mouth is generally made narrower, under an apprehension that heat will be lost. At coke and anthracite furnaces, we find the throat wider than half the diameter of the boshes; it ranges between $\frac{1}{2}$ and $\frac{2}{3}$ of it. There is less danger of losing heat by a wide than a narrow mouth; the latter always consumes more fuel than the former, but it requires more attention on the part of the founder, because of its tendency to cause scaffolding. The mouth may be formed of an iron cylinder or a brick wall.

The mouth is sometimes surmounted by a chimney; this is required to protect the workmen against injury from the flame. An open mouth causes improper filling, because the men sometimes cannot get near the proper place, in consequence of the flame which is driven there by the draught or wind. It is in all instances proper to erect some protection for the fillers. A brick chimney, well provided with binders, and some apertures for charging, is all-sufficient for the purpose.

Blast furnaces have in most cases, and should have in all cases, a roof over the top of the stack, a bridge-house, and a moulding-house. The uses of these buildings are obvious—for the

698

protection of those who work the sand in the pig-bed, and the ore, coal, and furnace, from the effects of rain and snow. The whole of a blast furnace, including all these buildings, assumes then a form such as is represented, Fig. 698. All these buildings should be constructed of iron, or coated with a fire-proof paint; for any losses in consequence of a conflagration in these establishments are of a serious nature, because they cannot well be covered by insurance.

We recognize two distinct and active principles in smelting iron ore, or, in fact, any other ore. The one relates to those instances where porous ore is smelted; the other is that in which compact ores are reduced. In former pages we had occasion to allude to this subject, and remarked that carbon must be dissolved in the gases of the lower part of the furnace, or we could not account for grey iron. The carbon may be dissolved in hydrogen, or any other gas; it will be attracted by the oxygen of the ore, the pores of which it will fill if it is porous. By whatever means the carbon is dissolved, or set in motion, the effect must be in all cases the same. If ore is very porous, it will absorb much carbon, and form a black or brownish-black mass, which is more refractory than pure ore, and which melts only in the high heat of the crucible. This condition of the ore in the furnace has been proved by actual experiments. In drawing out the contents of a furnace, when in the best condition for smelting, the ore was found to be in this state. Under these circumstances, iron is smelted with the smallest amount of fuel, grey iron follows with the greatest facility, and good quality can be depended on. This mode of smelting can be practised only on porous ores, bog ores, hematites, soft red oxides, and roasted ores. It requires a large space above the boshes in order to saturate the ore with carbon, and flat boshes to concentrate the heat below them and in the centre of the furnace; it also requires a highly-charred fuel, which is free from hydrogen. It has been observed that, when it is desirable to smelt grey iron, raw fuel or imperfectly-charred fuel does not facilitate the operation. On the contrary, raw fuel leads to the formation of white iron. Grey iron, of course, is smelted by raw bituminous coal, and also by wood; but we speak here only of ordinary circumstances, in which the fact is as we state. If the medium in which the carbon is dissolved is indifferent in respect to the formation of grey iron, carburetted hydrogen ought to be particularly suitable for the formation of it. But this is not the case. Fuel which contains hydrogen forms, at best, an impure iron: and if grey, it contains but little carbon. Carburetted hydrogen certainly deposits more carbon in the ore than can by any other means be accomplished; but hydrogen reduces many substances, such as silex or lime, which are not reduced by carbon except under peculiar circumstances. When hydrogen penetrates oxide of iron at a low temperature, it forms a powder of metallic iron, which does not so readily combine with carbon as a powder of oxide of iron exposed to the same degree of heat. We discern thus very readily the means by which grey cast iron, of good quality, can be formed.

In these remarks we have alluded to one extreme—that is, the formation of grey iron from porous ore; we shall now examine the other, or the formation of white iron from compact ore. The best material to serve as an illustration is a silicate of iron. When forge-cinder is charged to a blast furnace, it cannot absorb carbon in its pores: for it is compact, and not accessible to any gas, so far as its interior is concerned. This substance will move unaltered in the furnace, to a point where the heat is strong enough to melt it. Here it is converted into a fluid cinder, or slag, and trickles down through the hot coals; these absorb oxygen from iron and other metals, and if the height of the column of hot coals is sufficient, all the iron may be reduced before the fluid slag arrives at the tuyeres. No carbon can be absorbed by the iron in this case: for it forms large globules, and is not in a condition, or not in so close contact with the metal as to combine with it. The fluidity of the metal is in this case most generally produced by substances which are nearest to the iron; and as these are phosphates, sulphates, and silex, their respective bases will combine with the metal, which, when once fluid, does not absorb carbon, but rapidly descends into the hearth. In this case we perceive that a certain height of a column of hot coal is required to reduce the oxide; if that column is too low, the iron arrives, in the form of oxide, at the tuyere, and cannot be converted into metal but by the presence of grey iron in the pool of the crucible, the carbon of which will reduce the oxide which may be in the slag. The height of the column of hot coal required will depend on the nature of the ore. If it is a very fusible silicate, such as forge-cinder, a considerable height is necessary, because it will descend rapidly and escape the action of carbon.

In thus analyzing the operation of a blast furnace, we see that, in the one case a very low column of heat, and in the other a very high one, is required. Between these two extremes we find the proper height. Very porous bog ores containing phosphorus, smelt in a narrow, almost cylindrical hearth, 5 feet in height; a red heat is hardly perceived at the boshes. Forge-cinder, by itself, requires at least a column of 25 or 30 feet in height of a red heat; and thus high the crucible ought to be, in order to obtain the necessary intensity of heat. This shows very clearly the principle involved in the construction of a furnace-hearth and the boshes. Where the latter commence, there the smelting of the ore begins, and not elsewhere. But as this rule would cause the crucible to be very high for refractory ores, it and the boshes form one general slope, which may be very high, as the ore requires no preparation in the upper parts of the furnace, and only the fuel is to be freed from moisture.

These principles are not confined to the kind of ore; fuel exerts more or less influence on the height as well as dimensions of a furnace. Hard, dry fuel, such as anthracite, requires little preparation in the furnace, and low stacks will work with it profitably. Coke requires more preparation, and charcoal most; and as a high furnace has a tendency to draw the heat up, it is found necessary to reduce the height of a charcoal furnace, in order to save fual, by reducing the column of heat, and consequently radiation of heat. In the difference of the height of heat in furnaces, or in the radiation, which is the necessary consequence, may be found the chief cause of the great difference in the consumption of fuel. This accounts for the fact that charcoal furnaces which smelt mild, fusible bog ores, will produce a ton of iron with less than a ton of coal, when anthracite furnaces use from 1·6 to 2 tons and more, and coke furnaces do not work with less than 2 tons of coke, which is equal to 3 or 4 tons of coal, for smelting the same amount of metal.

Hot Blast.—The application of hot blast at blast furnaces is general; with few exceptions at charcoal furnaces, it is done everywhere. In Figs. 699 and 700 we have represented the apparatus as it is most commonly constructed, which may be considered as its best form. In some instances the hot-blast stove is placed near the tuyeres, as shown in Fig. 701; and each tuyere has its own

699.

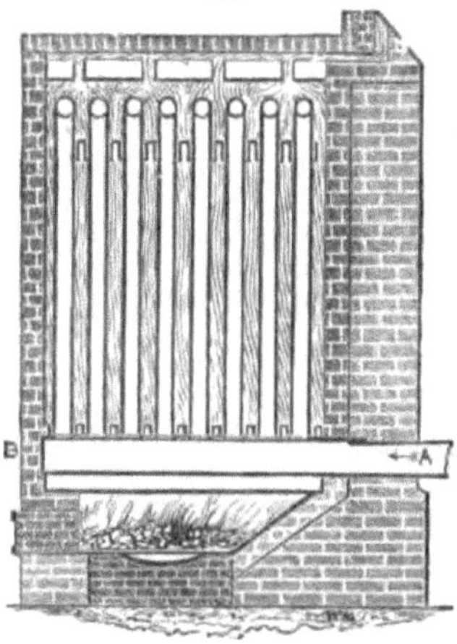

700.

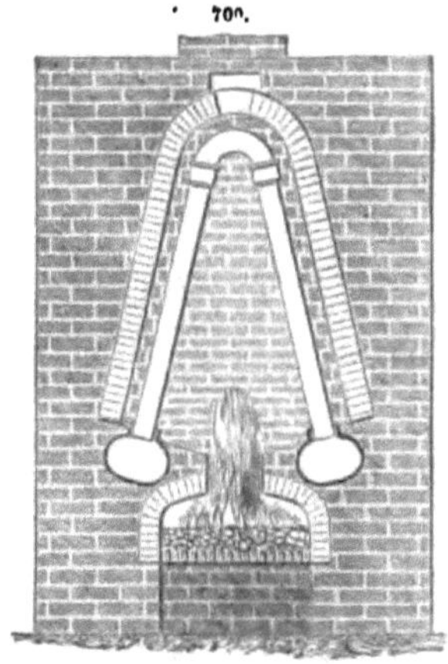

701.

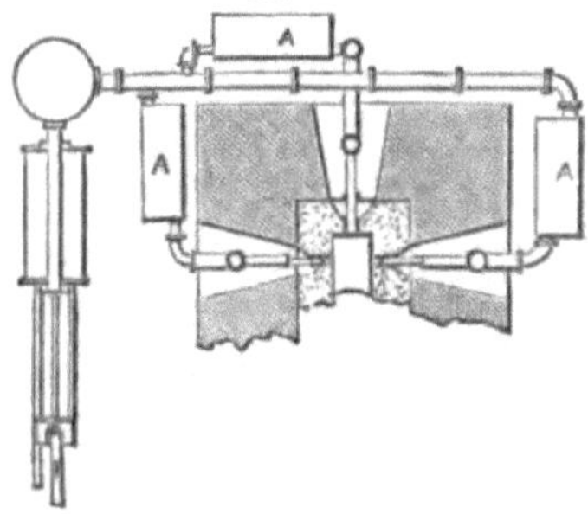

702.

stove, A A A, which enables the founder to heat the blast for one tuyere more than for the other, as its condition may require. At large furnaces it frequently happens that one tuyere does not work so hot as the other, and, in order to remedy the evil, more heat is applied at the cold one.

In general, one apparatus is placed conveniently near to all the tuyeres, and this heats the blast for all of them, however many there may be. In this case the most convenient position is behind the furnace, somewhat elevated above the tuyeres, as shown in Fig. 702. The hot-blast pipes are then above the heads of the workmen, and easily accessible for repairs.

These cases refer to the method of heating the blast with separate fuel, which is not often practised. The most common mode is to heat it at the top of the furnace, or at some distance below it, even as low as represented in the last engraving. The first instance has been represented, Fig. 690; and in the latter, the arrangement takes the form shown in Fig. 703. The waste heat is conducted from the top of the furnace, either in large iron pipes, or in channels of masonry, to that point where the hot-blast stove is located. In some instances we find the stove provided with a

703.

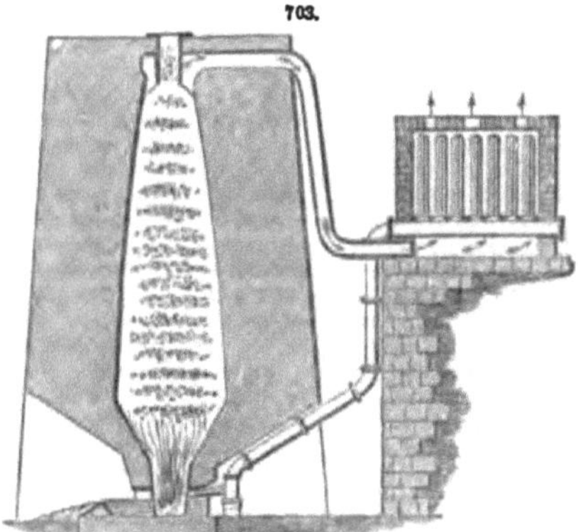

furnace, or grate, for burning coal or wood. This precaution is taken to provide heat by extra fuel, in case the waste heat from the furnace is not sufficient to heat the blast to the degree required. This does not happen at anthracite or coke furnaces, but is confined to charcoal ovens.

At some furnaces we find the hot-air pipes enclosed in wrappings, which consist of articles which are bad conductors of heat; at others, the pipes are walled in, in the rough masonry of the stack. Whatever be the mode of conducting blast, the pipes ought to be spacious, for the increased volume of the hot air, compared with it when cold, causes much loss of power, or pressure, if the pipes are too narrow.

Effect of Hot Blast.—The apparently singular effect of hot air in a smelting furnace is chiefly of a chemical nature. The heat introduced by the hot air amounts at best to $\frac{1}{6}$, and in most instances only to $\frac{1}{10}$ of that generated; and still a considerable amount of fuel is saved by it, which at charcoal furnaces amounts to $\frac{1}{4}$, at coke furnaces to $\frac{1}{3}$, and at anthracite furnaces to nearly $\frac{2}{3}$ of that used by cold blast. The immediate advantages are, the quantity of heat introduced, in case that is derived from waste heat, a small increase of temperature, and a fluid cinder, by which flux is saved. The latter effect is in consequence of the absence of the chilling effect of cold air, and a more intimate union of the ingredients. The essential effect of hot blast consists in its facilitating the union of carbon and the oxygen of the blast, by which means carbonic oxide is more readily formed, in which gas carburetted iron may descend without loss of carbon. Cold blast will produce a larger atmosphere of carbonic acid around the tuyere than hot blast, and this gas will not only absorb carbon, but oxidize silicon and iron. As the influence of hot blast on ore is of such a nature as to facilitate the revival of metals, many other substances besides iron ore are reduced, and form an alloy with the metal. To these foreign substances belong particularly those which are in close contact with the particles of iron, such as phosphorus, sulphur, and silex; calcium is often reduced from the limestone used as flux, when the blast is heated beyond a reasonable temperature. By experience it has been found that, for charcoal, a heat beyond 300°, for coke 400°, and for anthracite 500°, is of not much advantage.

The quality of iron smelted by hot blast must naturally be inclined to grey iron, because all the oxygen and other gases being perfectly saturated with carbon, there is no opportunity for the ore to escape being carbonized. But it has been observed, and must naturally be expected, that hot-blast iron is more impure than cold-blast iron. It contains, particularly, more of the basis of silex, because this substance is everywhere associated with iron ore, and is subject to reduction by carbon at a high heat in the presence of iron, and in the absence of carbonic acid. The quality of the metal is, therefore, eminently suited for use in the foundry. It is, on account of the amount of its impurities and the metallic form in which they are present, very fluid, and remains so a long time, which is the cause of its forming grey, tempered castings. Whatever may be the opinion and experience of some engineers, there cannot be any doubt that cold-blast iron with the same amount of carbon as hot-blast iron, and cast into dry moulds, is stronger than hot-blast iron, smelted from the same kind of ore. Hot-blast iron forms a superior foundry iron for small castings, but it is weak in large castings, the cause of which is obvious. The mixture of carbon, impurities, and iron, which causes its fluidity, makes it also a bad conductor of heat; it will not cool so quickly as strong and pure iron, and consequently it is not so liable to crystallization. This iron may be, therefore, a superior foundry iron for small castings; but it must be always inferior to cold-blast for heavy ones, and particularly for the forge.

The large quantity of heat lost at the top of a blast furnace has been the cause of frequent speculations to devise some plan for its use since the earliest adoption of these furnaces. It has of late led to a great deal of controversy, and occasioned much examination of the nature of these gases, as well as of those in the interior of the furnace. The subject is so far settled at present, that it is found injurious to abstract gases lower down from the top than where they consist chiefly of carbonic acid, vapours of water, a little carbonic oxide and nitrogen, and some other substances; in fact, these gases are not abstracted until they cease to be useful in the furnace. We may tap gases from the furnace lower down in the stack; but they are not of more use than those near the top, and such an operation is more or less injurious to the working of the ores. When these gases are abstracted at a height where they cease to be useful, we may term them waste heat; but if we tap lower down they cease to be waste heat; for the highly carbonized combustible gases are essential to the good effects of the furnace, as must be evident from our preceding remarks.

At a variable height, 8 ft. on an average below the top in charcoal furnaces, 8 or 10 ft. in anthracite furnaces, and 12 or 16 ft. in coke furnaces, the gases are of the same, or similar composition. They consist here chiefly of carbonic acid, nitrogen and steam, and some carbonic oxide. It is a mere matter of convenience, so far as regards the furnace, at what precise spot we abstract the gas. Below these various heights it changes suddenly in its composition. It is composed chiefly of carbonic oxide, some hydrogen, nitrogen, and moisture. These are substances which are essential to the reduction of the ore, and which ought not to be removed.

We have already shown the mode of abstracting the waste heat from the furnace. The most common method is by means of a cast-iron cylinder of 5 to 8 ft. in length, as shown Fig. 703 and in other drawings. The depth to which a cylinder is lowered has no effect upon the amount of heat obtained; this is regulated by the distance to which the heat is to be conducted. A long or deep cylinder affords more pressure; consequently the gas may be conducted farther from it. When steam-boilers, or a hot-blast stove, are at the top of a furnace, the insertion of a cylinder is not necessary; in fact, it is of no advantage in any case, for sufficient heat is given out at the top in all instances to heat steam-boilers and hot-blast stoves. In this case the arrangement is such as is shown in Fig. 704. A chimney at the end of the boilers, or at the top of the stove, produces the necessary draught. A plain cast-iron plate with a narrower mouth than that of the brick below, affords a chamber on the top of the fuel. When desired, this aperture in the iron plate may be covered by a door which is occasionally removed for charging fresh ore and coal. This plan works well enough in small charcoal furnaces; but at large furnaces, with a wide mouth, and whose boilers require a large quantity of heat, the effects are not certain. If any objection exists to the application of an iron cylinder, which may be the case when the top works very hot, an arrangement such as is represented in Fig. 705 is equally effective. It is particularly employed,

704.

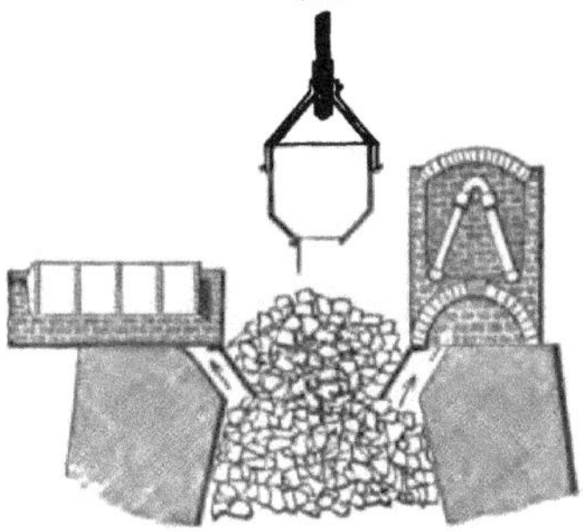

705.

and useful for burning lime, or heating blast. Over the mouth of the furnace a chimney is erected, provided with a damper on its top. Some iron doors, which are opened by pushing the wheel-barrow against them, and shut themselves when it is withdrawn, afford access to the interior for charging. By these means all the heat at the top is saved, and may be conducted to any place where it will be useful.

The amount of waste heat at a blast furnace is very large, but even when tapped low in the stack and burned with the addition of fresh atmospheric air, its temperature is so low that it cannot be employed to advantage for melting, puddling, or welding iron. At the top it produces a high red heat, and at anthracite or coke furnaces a white heat, well adapted for generating steam, heating blast, burning lime or bricks, and similar purposes. In conclusion, we insert some tables, which will be found useful for reference, explaining many things which could not be referred to in this short exposition.

Charcoal Furnace.—At a charcoal furnace the following persons are employed:—one founder, two firemen, two fillers, one gutter-man, one coal-drawer, a bank hand, and a horse, cart, and driver, and if there is a *stamping-mill*, or *battery*, a hand to attend to it. This is the smallest number of hands necessary to manage a furnace; generally there is twice that number. When ore is to be broken or roasted, flux to be broken, and similar work to be done, an additional number of hands is required. There are charcoal furnaces which consume 250 bushels of coal to a ton of iron; 200 bushels is an average in the Western States of America. In Western New York, some furnaces smelt a ton of iron, from magnetic ore, to 150 bushels of coal; and in the St. Lawrence district, where specular ore and red hematites are chiefly smelted, as low as 100 bushels of coal are used to the ton of iron. A stack in that region, which operates well, is about 30 or 33 ft. high; 7 or 8 ft. boshes, with a cylindrical part, 2 ft. high, above the boshes; mouth, 31 in., and from that to 36 in. (when an iron cylinder is used it is of the same size); width of hearth between the tuyeres, 22 in., and 32 in. at the top; height of hearth, 5½ ft.; tuyeres, 22 in. above the bottom; the in-wall a curve, as shown Fig. 677, C; such a furnace smelts from 5 to 8 tons a day.

The Siscoe furnace, on Lake Champlain, working magnetic ore, is 44·75 ft. high; 13 ft. boshes; 2 ft. 10 in. hearth, across the tuyere; hearth, 38 in. high: slope of boshes, 64°, with a cylindrical part above the slope of 3½ ft. high; mouth, 4 ft. 3 in. wide. This furnace uses about 160 bushels of coal to a ton of iron; its erection has cost about 30,000 dollars, exclusive of eight kilns for charring wood, which cost an additional sum of 10,000 dollars.

Anthracite Furnaces.—The form and dimensions of these vary exceedingly. They are not often above 33 ft. high; from 10 to 18 ft. boshes; 3½ ft. to 5½ ft. across the tuyeres; hearth, from 3 to 5 ft. in height, generally much battered; boshes, from 50° to 70°; top, 5 ft. to 9 ft. in width. A small anthracite furnace produces from 80 to 120 tons of iron, large furnaces from 180 to 200 tons, per week.

Coke Furnaces.—These are generally 50 ft. high, and as wide at the base; boshes, 15 ft.; slope, 65° to 70°; hearth, across the tuyere, 4 ft.; at top, 5 ft.; height of hearth, 6 ft., and tuyere above bottom stone, 2 ft. The cost of erection is equal to that of an anthracite furnace; iron made per week is 80 to 100 tons, using 2 tons of coke to 1 ton of iron.

The coal charges at furnaces are always of the same measure—about 15 bushels, more or less. The weight of ore is regulated according to the capacity of the coal for smelting. The number of charges in a certain time, say twelve hours, varies from 12 to 30, according to the quantity of blast injected into the furnace.

The number of blast furnaces in the United States may be estimated at 1200; of which about 100 are anthracite furnaces, 8 bituminous coal furnaces, and a similar number which use coke. The others are charcoal furnaces.

Description of the Blast Furnaces at Iron Works at Grosmont, by Hiram C. Coulthard, of Blackburn. Taken from the 'Proceedings Inst. M. E.'—In the Cleveland iron district, where the Grosmont Iron Works are situated, there were, in 1863, 63 blast furnaces in full operation, 17 furnaces not in operation, standing for repairs or other causes, and 11 furnaces in various stages of progress.

Grosmont, near the coast of Yorkshire, is situated about seven miles from the port of Whitby, 20 miles from the Durham coalfield, and about the same distance from the lime district of Pickering, whence the supply of lime is derived. Fig. 706 is a general plan of the entire works, which are adjacent to the main line of railway from Whitby to Castleton, joining the North Yorkshire and Cleveland Railway, and thus in connection with the Newcastle and Durham coal and coke districts. A siding from the main line runs into the works.

706.

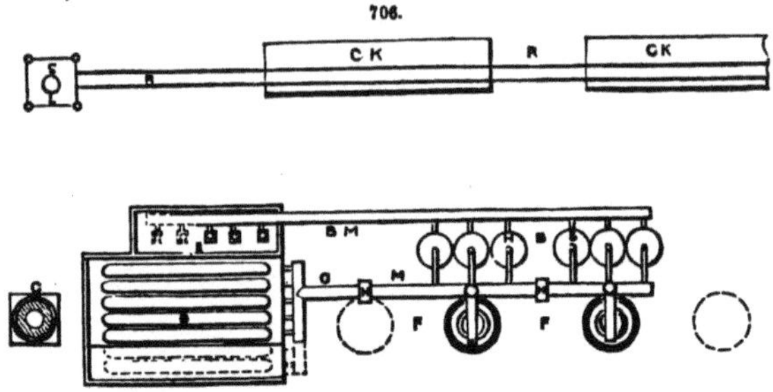

Reference.—S L, Steam Lift. C K, Calcining Kilns. R R, Railway. B, Boilers. E, Engines. B M, Blast Mains. G M, Gas Main. F F, Furnaces. H B S, Hot Blast Stoves. C, Chimney.

These blast furnaces are constructed on a very efficient and economical plan. Each furnace is capable of producing 250 tons of pig-iron a week, allowing for stoppage on Sundays. Fig. 707

707.

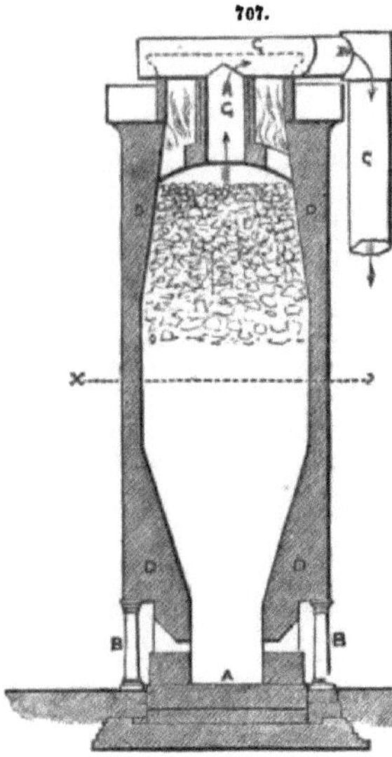

708.

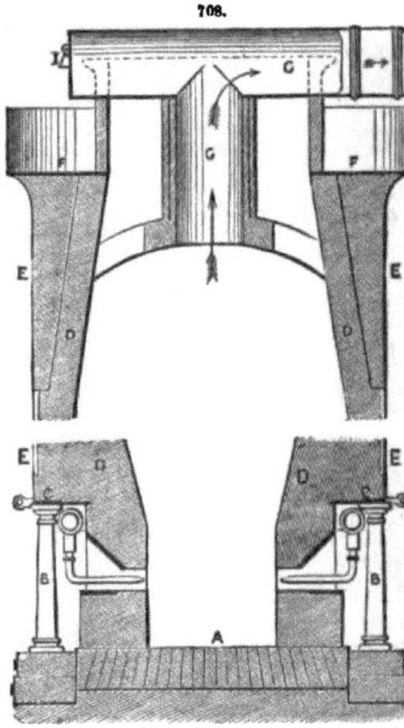

is a vertical section of one furnace, and Fig. 708 shows an enlarged vertical section of the top and bottom of the furnace. Figs. 709 to 713 are transverse sections of the furnace at the tuyeres, tapping-hole, and hearth, and through the body of the furnace.

Each furnace measures 18 ft. diameter at the boshes, and a total height of 63 ft. from groundline to level of charging-floor. The foundations were dug out to a depth of about 9 ft., to rock on one side, and hard blue clay on the other, the ground sloping in the direction of the dip of the rock. The stone foundations, both for the hearth and casing of the furnace, are shown in the vertical sections, Figs. 707, 708, and consist of ring-courses of masonry built on concrete, about 26 ft. diameter, each course being bound by a wrought-iron ring, 5 in. wide and $\frac{1}{2}$ in. thick, Fig. 708. In the interior of the uppermost ring-course is built the fire-brick hearth A, Fig. 708; the blocks of which this is formed are shown in plan and vertical section in Figs. 711, 712. These blocks are set in ground fire-clay in a moist state, special care being taken to secure a perfectly homogeneous mass, as the whole of the superstructure of the furnace and its contents, when in working order weighing about 1200 tons, rest upon this foundation. On the top course of masonry the founda-

709.

710.

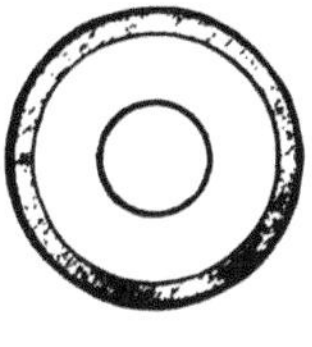

711.

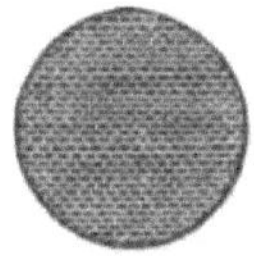

712.

713.

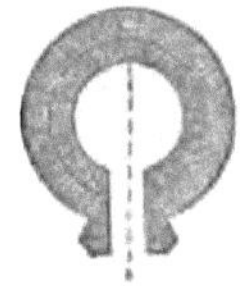

tion-plates of cast iron, 3 ft. 6 in. square and 4 in. thick, are bedded in fire-clay, to which are bolted the cast-iron columns B B, Fig. 708, 17 in. diameter, for carrying the superstructure. These columns are united at the top by a cast-iron ring or cornice C, in segments $3\frac{1}{4}$ in. thick, each segment having a semicircular snug cast on its under-side, which, when the work is joined together, fits into the top of the column B, thus binding the whole of the segments into one ring.

The entire lining of the furnace inside is of refractory fire-brick, D, Fig. 708; the furnace is cylindrical on the outside and entirely cased with wrought-iron plates E, $\frac{3}{4}$ in. thick at the bottom of the furnace, and towards the top of the furnace diminished in thickness to $\frac{5}{16}$ in. This casing weighs about 30 tons and costs about 400*l*., and is now being generally used in the place of the massive stack of masonry formerly used. There are ten cast-iron pillars B for carrying the superstructure, placed at a distance of 7 ft. apart, except where the tapping-hole is situated, where the distance is increased to 10 ft., as seen in Fig. 709. Brackets are cast on these pillars, Fig. 708, for the purpose of carrying the circular pipes that convey the blast and water round the furnaces for distribution to the various tuyeres. There are five tuyeres to each furnace, one of which is shown in longitudinal section in Fig. 714.

714.

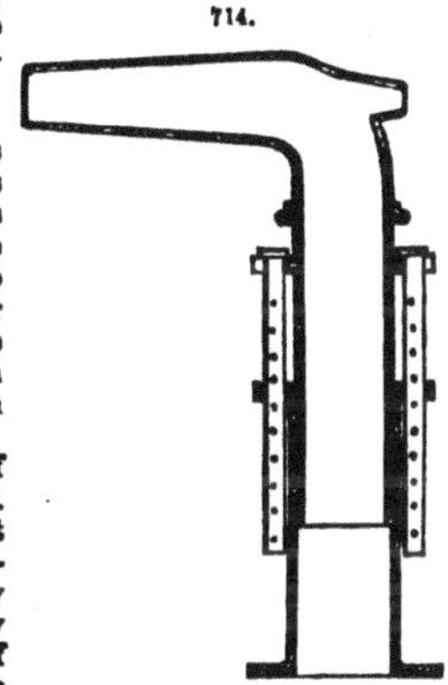

Fig. 713 shows a transverse section of furnace at X X.

At the top of the furnace a wrought-iron plate-cornice F is fixed, Fig. 708, forming the charging-floor; and the two furnaces are connected by means of two longitudinal wrought-iron girders 4 ft. and 3 ft. deep respectively, the larger one prepared to receive the wrought-iron beams that form the roadway of the incline up which the materials for smelting are drawn by means of a pair of fixed horizontal engines. These girders are united by nine intermediate cross-girders of wrought iron, and, when covered with plates, form the roadway of the charging-floor, having a screen 3 ft. 6 in. high running round for protection.

The throat of the furnace, Fig. 708, is adapted for taking off the waste gas, which is collected in a wrought-iron tube G, 5 ft. diameter, which extends down the throat of the furnace about 5 ft., and is lined inside and cased outside with refractory fire-brick 6 in. in thickness. This tube is fixed to and supported by a crown or dome, built in the throat of the furnace, of specially moulded lumps of fire-clay, supported by six buttresses built of the same material. The crown has six openings formed at the sides for charging purposes, and one opening in the centre, through which the gas passes into the tube G. There is the usual brick chimney at the top of the furnace, with wrought-iron swing-doors corresponding with the openings in the crown. The gas is conveyed from the furnace-top to the boilers and hot-blast stoves by a wrought-iron tube 5 ft. 6 in. diameter, large enough to take off the gas from two additional furnaces; and square boxes, H, Fig. 706, are fixed at intervals along the tube to allow for expansion. A flap-valve, I, Fig. 708, opening outwards for cleaning purposes, is fixed at the end of the tube over the furnace.

Figs. 715, 716, show a vertical section and sectional plan of one of the hot-blast stoves. Three of these are built to each furnace, of common brick made on the estate, lined with refractory fire-brick, and externally bound firmly together by wrought-iron hoops 4 in. wide and $\frac{3}{4}$ in. thick, placed at intervals of 3 ft. The stoves are heated by the gas being admitted at the top, J, and a small fire is kept on the grate at the bottom for the purpose of ensuring that the gas is always ignited. Four flues, K K, Fig. 716, pass away from the bottom of the stove to the main chimney-flue L, Fig. 715, which is in connection with the chimney-stack, Fig. 706, of 180 feet height. A simple disc-valve J is fixed at the top of the stove where the gas enters, to cut off the supply of gas from the stove at any time. The pipes M, through which the blast passes, consist of ten pairs to each stove, 12 in. diameter, each pair being arched at the top and united at the bottom by connecting foot-boxes, thus forming one continuous course of pipes for the blast to pass along. The blast enters on one side of the oven, and, after circulating through the pipes M, passes out at the other side into the main pipe N for the service of the tuyeres, as shown by the arrows. A stop-valve O serves to cut off the communication of each stove with the blast-main, which is 5 ft. 6 in. diameter, and thus forms also the blast reservoir. The temperature of the blast is from 600° to 700° Fahr., and the quantity blown by each engine is 6000 cub. ft. per minute, at a pressure of 3 lbs. per square inch. These hot-blast stoves have been found most effective; from the enlarged capacity of the pipes, the blast is much longer in passing through them, and consequently they are not required to be kept at such a destructive heat.

715.

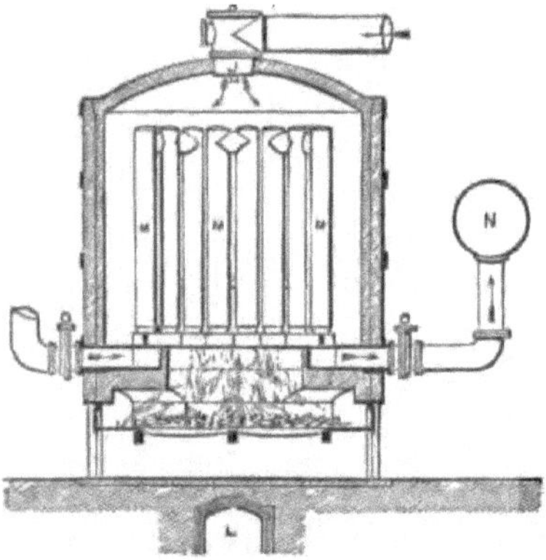

716.

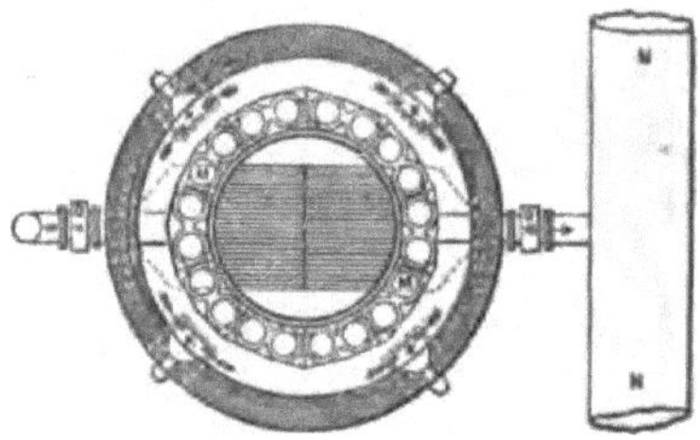

The blast is supplied by three direct-acting high-pressure engines, quick-moving, having air-cylinders 57½ in. diameter, with a stroke of 8 ft. Two engines are sufficient for the work of two furnaces, a third one being provided in case of emergency. The reason for separate engines being used is that in the case of an accident to the blowing-engine, when only one engine is used, the whole of the furnaces are thrown idle. Moreover, the cost of machinery for two furnaces is much less in these engines, taking into consideration the expensive nature of the stonework, and so on, required for the foundation of one large beam-engine. The only foundation required for these engines is about 3 ft. depth of brickwork, with a framework of timber on which to bolt the foundation-plates.

The engine-house is of brick, the roof is formed by the water-tank, which contains the water-supply for the tuyeres, pig-beds, and so on. In the engine-house is fixed a travelling crane, for the convenience of examining any portion of the engines; this is found a most useful appendage. The boilers are five in number, each 73 ft. long by 5 ft. diameter, of the plain egg-ended form, heated by the waste gas from the blast furnaces. They are suspended by means of cast-iron bridges from the top of the boiler seats, and are fed by three donkey-engines, all connected to one pipe over the boilers. The steam pressure is 60 lbs. a square inch above the atmosphere.

A steam lift is fixed in the works in the position shown S L, Fig. 706, for the purpose of raising the minerals from the line of railway to the top of the calcining *kilns*.

Description of a Method of Taking off the Waste Gases from Blast Furnaces, by Charles Cochrane.— In the 'Proceedings of the Inst. of M. E.' for 1860, Cochrane observes, there is no novelty in the fact of taking off the waste gases from a blast furnace; for many methods have been and are at present employed for accomplishing this object. Though Cochrane was unaware of any similar method, he does not desire to claim originality in that about to be described; but as there is such acknowledged diversity of opinion as to the respective merits of different plans, and great difficulty in procuring reliable information on any, it is proposed to give a description of an arrangement which has been in successful operation for some time at the Ormesby Iron Works, Middlesborough. The large waste of fuel from the mouth of a blast furnace where the escaping gases are allowed to burn away is well known, and amounts to more than 50 per cent. of the fuel burnt; hence there is considerable margin for economy, bearing in mind the large quantity of coals consumed in raising steam for generating the blast and the further quantity necessary to heat that blast to the required temperature. In fact, assuming a consumption of 300 tons of coke a week to make 200 tons of iron, about 100 tons of coal would be required to generate steam and heat the blast. Taking off the gases from one furnace under such conditions does, according to actual experiment, furnish gas equivalent to upwards of 150 tons of coal a week. This is obviously an important matter where coals are expensive.

The blast furnace is alternately charged with coke, ironstone, and limestone, in proportions depending upon the quality or "number" of iron desired. The arrangement of these materials in the furnace is generally deemed important, though it admits of considerable latitude without any appreciable alteration in the working of the furnace. Thus it does not seem to be of any importance whether the charge of coke be 12 cwt. or 24 cwt., the amount of load of ironstone and limestone being in the same proportion of 1 to 2. The chief point, if there be one, to be gained in the arrangement of the material is, to distribute it pretty equally over the furnace, not allowing all the large material to roll outwards, and the small to occupy the centre of the furnace, or *vice versâ:* for it is supposed the ascending gases will pass through the more open material of the furnace to the injury of the closer; thus the two reach the active region of reduction in different states of preparation, and the operations of the furnace are interfered with. To provide for this contingency, which is met in an open-topped furnace by filling at the sides at three, four, or even six points of the circumference of the throat, allowing the material to slide inwards 2 or 3 ft. on a sloping plate, it was considered expedient in the present instance to make the filling aperture as large as practicable: it was therefore made 6 ft. 6 in. diameter, as shown in Fig. 717, so that the material tends to arrange itself in a circle a little outside the centre, thus correcting the tendency of large material to roll outwards by causing a similar tendency to roll towards the centre also.

This point is gained by one of the simplest methods in use for closing the top of a blast furnace, where a cone is used to lower into the furnace for filling; but it is secured at the expense of the height of material in the furnace. A certain height is necessary for the efficient working of the

717.

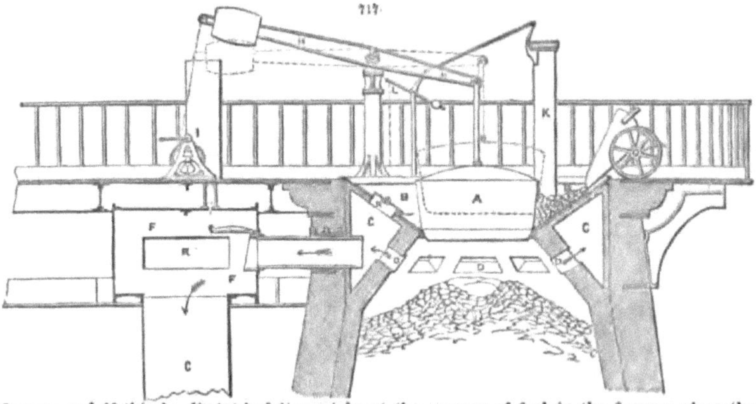

furnace, and if this be diminished it must be at the expense of fuel in the furnace, since the absorption of heat from the gases depends on the height of material through which they have to pass up: if this be diminished, the gases issuing from the throat of the furnace will escape at a higher temperature; if increased, at a lower.

But there is an important difference to consider in the conditions of a closed and an open-topped furnace, to which Cochrane is not aware that attention has hitherto been drawn; a difference which acts somewhat in favour of the open-topped furnace. The working of the furnaces themselves seems to show that an open-topped furnace is less sensitive to irregularities of moisture in the material, quantity of limestone, size of material, and so on, which can be accounted for only by the fact that the open-topped furnace has the advantage of a large amount of surplus heat, due to the combustion of the waste gases at its throat, which serves to dispel moisture and calcine the limestone, and helps to warm up the large pieces of ironstone: all of which operations in the close-topped furnace are effected only at a lower point of the furnace, thus necessitating a larger consumption of coke. With the same proportion of ironstone to limestone, it has been found to require about 10 per cent. more fuel to produce the same number or quality of iron in a close-topped than in an open-topped furnace. In the close-topped furnaces the gases pass away at a temperature of about 450° Fahr.; whilst in the open-topped a temperature of between 1000° and 2000° is generated in the throat of the furnace by their combustion.

In comparing the extra quantity of coke consumed in a close-topped blast furnace with the saving in coals for the boilers and hot-blast stoves, it is obvious that the economy to be derived by taking the gases off depends on the comparative value of coke and coal. In the Middlesborough district, where coal is expensive, it is an undoubted source of economy; where coke is very dear, however, and small coal can be obtained at a mere nominal cost for boiler and stove purposes, the use of the waste gases would possibly do little more than compensate for the outlay involved. Here, no doubt, is one source of the variety of opinion entertained in various districts as to the advantage of taking off the gas. Cochrane's experience at Middlesborough has been, that the waste gases can be taken off without affecting the quality of the iron produced, though at the expense of more fuel.

718.

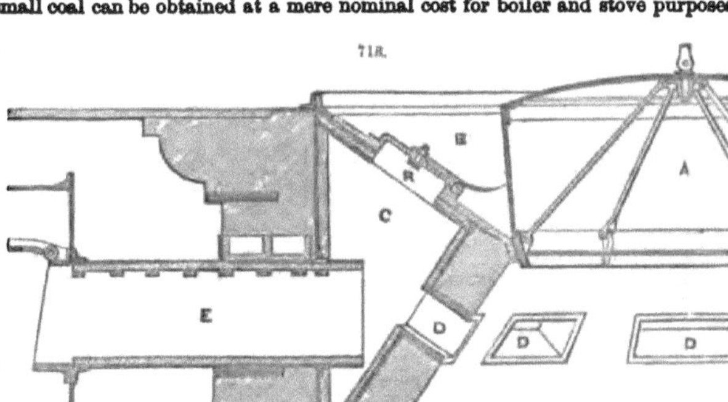

The mode of closing the furnace-top and taking off the gases at the writer's works is shown in Fig. 717. The top of the furnace is closed by a light circular wrought-iron valve A, 6 ft. 6 in. diameter, with sides tapering slightly outwards from below, as shown enlarged in Fig. 718, to admit of being easily drawn up through the materials, which are tipped at each charge into the external space B. To prevent excessive wear upon the body of the valve, shield-plates are attached at four points of its circumference, against which the material strikes as it rolls out of the barrows. An annular chamber C encircles the throat, triangular in section, into which the gas pours through the eight orifices D D from the interior of the furnace, and thence passes along the rectangular tube E into the chamber F. At the extremity of the tube E is placed an ordinary flap-valve opened by a chain, by means of which the communication between the furnace and the descending gas-main G may be closed. The valve A is partially counterpoised by the balance-weight at the other extremity of the lever H, and is opened by a winch I when the space B is sufficiently full of materials. At the time when the blast is shut off for tapping the furnace, the gas escapes direct into the atmosphere through the ventilating tube K, which is connected by levers L with the blast inlet-valve below.

Fig. 719 shows the connection between the furnace-top and the hot-blast stoves to be heated by the waste gases, which pass down the descending main G into the horizontal main M running parallel and close to the line of stoves N, from which descend smaller pipes O to each stove, as shown in Figs. 720 and 721. The supply of air for burning the gas in the stoves is admitted

721. 720. 719.

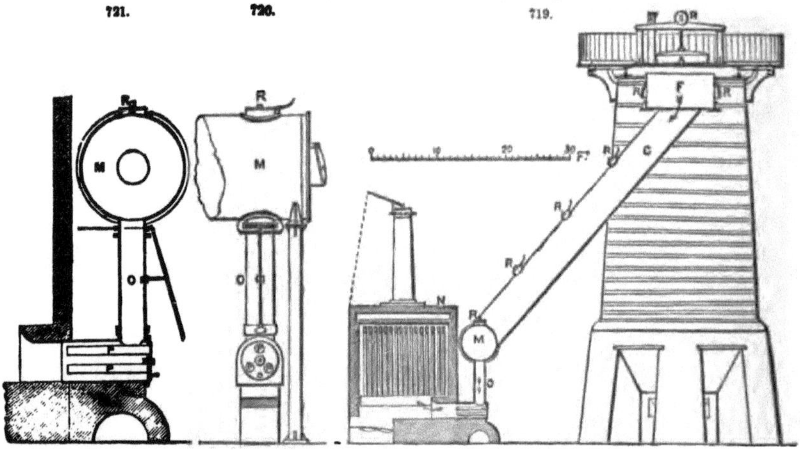

through the three tubes P, and can be regulated at pleasure by the circular slide closing the ends of the tubes, which has an aperture corresponding to each tube, and is planed on the rubbing-face, as is also the surface against which it works, in order that the slide may be sufficiently air-tight when closed. The ignition takes place where the air and gas meet, the ignited gas streaming into the stove and diffusing its heat uniformly over the interior. An important element in the working of an apparatus of this description is to provide for explosions, which must take place if a mixture of gas and air in certain proportions is ignited. To provide for this contingency, escape-valves R are placed at the ends and along the tops of the main tubes G and M; but to prevent explosions as far as possible, the ventilating tube K, Fig. 717, is used at the top of the furnace, connected with the blast-valve at the bottom, so that when the valve is closed, as at casting time, the act of closing opens the ventilating tube, and allows the gas to pass away direct into the atmosphere. The gas would otherwise be in danger of slowly mixing with air passing back through the stoves or otherwise gaining access into the tubes, and would thus give rise to an explosion. Until the ventilating tube was provided, it was necessary to lift the valve A closing the mouth of the furnace when the blast was taken off, otherwise slight explosions took place from time to time.

In the use of Durham cokes in the blast furnace, an inconvenience arises from the large deposit which takes place in the passage of the gas from the furnace and in the stoves and boilers. Under the boilers this deposit is a great objection, as it is a very bad conductor of heat, and needs to be frequently removed: in the stoves it is not so objectionable, though these need a periodical cleansing. The deposit does not arise altogether from the cokes, it is true; and it may be interesting to know its composition, which is as follows:—

Silica	18·86
Carbon	16·14
Alumina	13·87
Sulphate of lime	13·61
Lime	11·01
Protoxide of zinc	10·31
Peroxide of iron	9·01
Protoxide of manganese	2·56
Potash	2·13
Protoxide of iron	1·25
Magnesia	1·25
Chloride of sodium	0·60
	100·60

At a temperature of upwards of 3000° this mixture melts in a yellowish slag, dispelling the zinc; but there are no signs of fusion at the temperature produced by the ignition of the gas in the stoves, which must roughly approximate to that of melting iron, from the results of a few experiments made to ascertain this point: though thin pieces of cast iron were not fairly melted down, they reached the rotten temperature, which is only a few degrees below melting, and gave further signs of nearly melting by throwing off sparks when quickly withdrawn from the stoves and struck smartly against another object.

C. Cochrane says, that he has heard it asserted that the closing of the top of the furnace is the source of mischief to its working by producing a back-pressure in it. Under ordinary circumstances, with the furnace-top open, the blast enters the tuyeres at a pressure ranging from 2½ to 3 lbs. a square in. In the present close-topped furnace there are eight outlet-orifices D, Fig. 717, each 2 ft. by 1 ft., giving a total area of 16 sq. ft. for the passage of between 5000 and 6000 cub. ft. of gas per minute, raised to a temperature of 450° Fahr.; and the actual back-pressure of the gas, as measured by a water-gauge inserted into the closed top of furnace, is from ½ to ¾ in. column of water, or about $\frac{1}{40}$th or $\frac{1}{80}$th of a lb. the sq. in., an amount so trivial as compared with a pressure of from 2½ to 3 lbs. as to be unworthy of notice. Of course, if the tubes are contracted in size, a greater back-pressure will be produced; and it is quite possible that where attention has not been paid to the circumstance, the back-pressure may have interfered with the working of the furnace by preventing the blast entering so freely.

As regards economy in the wear and tear of hot-blast stoves of the ordinary construction, there can be no question the pipes last much longer when heated by gas, provided the temperature of the stove be carefully watched to prevent its rising too high; whilst the value of the same heating-surface compared with its value when coals are used is greatly increased, owing to the uniform distribution of the ignited gases throughout the stove. In the use of the gases at Cochrane's works, this economy of surface is such that two stoves heated by gas will do the work of a little more than three heated by coal fires.

On the Working and Capacity of Blast Furnaces, in the 'Proceedings of the Inst. M. E.' (1864), C. Cochrane further observes,—referring to Fig. 722, which shows the original construction of closed-top and lifting-valve for charging, the materials for the charges being filled into the exterior space B surrounding the charging-valve A, which is drawn up into the position shown by the dotted lines for allowing the materials to fall into the furnace; while the gas is taken off from the furnace-top by the passage E,—that the usual plan of closed-top adopted in blast furnaces is that represented in Fig. 723, in which it will be seen that the materials are filled in against a lowering-cone C, placed in the throat of the furnace, which on being lowered into the position shown dotted, permits their fall into the furnace. The tendency of the material in this case is to roll outwards from the charging-cone to the side of the furnace, and thence back again to the centre, as shown in the drawing.

722.

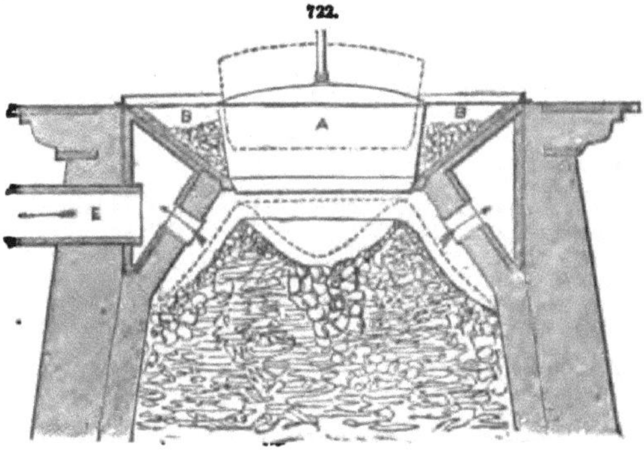

723.

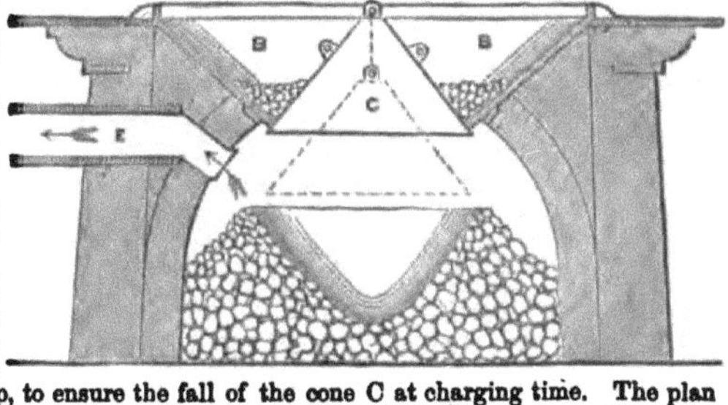

It was thought at the time of adopting the plan shown in Fig. 722, that the height of the materials carried by the same furnace would be increased, and that a corresponding economy in consumption of fuel would result, owing to the circumstance that where the plan shown in Fig. 723 is adopted, the level of the materials must always be maintained at a certain distance below the top, to ensure the fall of the cone C at charging time. The plan shown in Fig. 722 was devised with due regard, as it was thought, to the arrangement of the materials in the furnace; and it was intended that they should arrange themselves as shown by the dotted line in that drawing, part of the larger material rolling to the outside of the furnace and part to the centre.

As long as the furnace could be kept so full as to ensure the arrangement of materials shown by the dotted line in Fig. 722, there was no reason that it should not work uniformly; but the practical result was that it was found impossible to keep the furnace sufficiently full to secure the distribution of the materials in the manner intended. The level of the surface of the materials was generally below that intended, the consequence of which was that the material on falling into the furnace was shot into the centre, from whence the largest pieces rolled outwards, and the whole charge arranged itself as shown by the full line in Fig. 722. The result of this was irregular working of the furnace over a period of many months, during which an explanation of the irregularity was in vain sought for. At one time it was thought the back-pressure of the escaping gas had something to do with the irregularity; at another the cause was sought for in the difficulty of keeping the hopper-valve A of the furnace tight, and the necessity for using small material around the valve, as a kind of lute between every charge, to prevent the escape of the gas; until it occurred to C. Cochrane that the arrangement of the materials in the furnace was the sole explanation of the difficulty, and that as all the material was shot into the centre of the furnace, the small pieces would remain there, whilst the large would roll to the outside. Believing that it was of great importance, in order to secure uniform results, that there should be a uniform distribution of the heated gas from the hearth over the entire horizontal area of the furnace at each stage of its height, he considered that the effect of any small material being collected in any portion of the area would be to obstruct the passage of the gas at that part, and so prevent that portion of the material from being heated to its proper degree of temperature.

Deeming this to be the explanation of the irregularities experienced in the working of the furnace, Cochrane devised a method of distributing the material so as to prevent such a result, by the introduction of a frustum of a cone D, Fig. 724, suspended inside the throat of the furnace, which was found to be all that was necessary. The materials then arranged themselves in the desired manner, as shown in Fig. 724; and the result has since been a perfect uniformity in the working of the furnace. Where previously a yield of foundry iron from the same furnace could not be relied upon for more than about 24 hours at a time, and the annoyance was incurred of the furnace suddenly changing to white iron, the production of white iron, except when desired, is now unknown. A consideration of these facts will lead to a fair estimate of the importance of the arrangement of the materials in a blast furnace. Anything that opposes the free passage of the ascending heated gas at any part of the furnace must direct the gas into another channel, and the material thus left insufficiently acted upon finds its way into the hearth at a low temperature, and white iron is the result.

724.

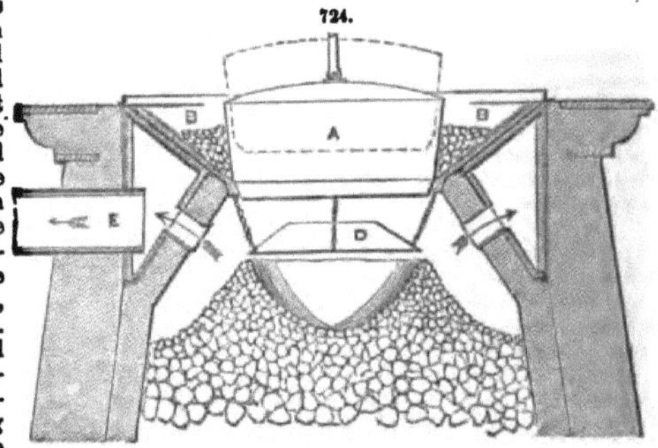

The effect produced on the distribution of material by this internal frustum of a cone is obviously similar to that of the ordinary lowering-cone when lowered, shown in Fig. 723; and the latter has now consequently been finally adopted at the Ormesby Iron Works as the permanent form of the arrangement, and is now being carried out there.

The most perfect action of a blast furnace C. Cochrane conceives to consist in the development of the highest temperature needed for the production of the required quality of iron, in a layer or stratum as little removed from the tuyeres as possible; and the gradual absorption of the heat from the ascending gas by the materials through which it passes, until it leaves the throat of the furnace at the lowest possible temperature. Anything which tends to cause a more perfect absorption of the heat developed in the hearth, or to lower the level of the region of highest temperature in the furnace, will thus be beneficial.

With regard to the absorption of the heat from the gas, it is obvious that the hotter the temperature at which the gas escapes, the more wasteful must be the effect; and theoretically the height of a furnace should be increased until the temperature of the escaping gas is reduced to that of the materials on their introduction into the furnace-top. This is the theoretical limit to the height of a blast furnace: but it must not be forgotten that the less the difference in temperature between two bodies, the less rapid is the communication of heat from the hotter to the cooler; hence for the absorption of the last few degrees of temperature from the ascending gas a much greater height of material is necessary than where the gas and the material differ more widely in temperature. Already with 50 to 60 ft. height of blast furnace in the Middlesborough district the temperature of the escaping gas does not exceed 500° to 600° Fahr.; and it is a question to be answered only by experiment, how far the gain from the heights of 70 to 75 ft. already accomplished at Middlesborough, and further heights of 10 or 20 ft. additional that are contemplated, will compensate for the extra work in raising the materials to the additional height and for the more substantial plant required. In the direction of height there is unquestionably on this account a limit which will speedily be attained; supposing the limit be not previously determined by the necessity for increased pressure of blast and by the increased difficulty in working the furnaces.

Taking off the Waste Gas from Open-topped Blast Furnaces, by George Addenbrooke.—We take the succeeding account of this method from the 'Proceedings Inst. M. E.':—Writing in 1865, Addenbrooke observes that the utilization of the waste gas from blast furnaces has now become not only an accomplished fact, but a great commercial success, and consequently an important part of furnace management. This gas, or rather mixture of gases, issues in large quantities from all the interstices between the last charge of materials in the furnace-throat; and it passes off with such rapidity as to prevent a sufficient mixture of air taking place to render it inflammable until it has risen to some little height above the top of the materials in the furnace-mouth. As soon, however, as this mixture of air takes place, a very considerable portion of the gas is consumed, in the case of the ordinary open-topped furnaces that do not utilize the waste gas. This combustion develops a great amount of heat; and the question therefore arises, how can the waste gas be made further useful, without in any way injuring the yield, the working of the furnace, or the quality of the iron made; for if any injury were occasioned in either of the above respects by taking off the waste gas, the utilization of the gas ought certainly not to be attempted. It is evident that there must always be an escape of surplus gas from the top of the materials in the furnace-throat, from the consideration that the heat in the lower part of the furnace distils off the gas from the fuel in the upper part; and this gas, not meeting with a supply of oxygen inside the furnace, passes up unconsumed to the furnace-mouth, where, upon mixing with the external air, it burns away to waste, unless taken off previously in order to be usefully burnt elsewhere.

The utilization of the waste gas has been extensively carried out in two different modes, each capable of being applied and worked in several different ways. The one mode is known as the Close-top system, and the other as the Open-top system.

Addenbrooke was of opinion that the waste gas ought to be utilized for the following reasons, namely, that a furnace would work to better yield where the gas was utilized, and with greater regularity as to the quality of iron made; and that there would be a very considerable saving in repairs to hot-blast stoves and boilers by heating these with the waste gas, together with greater regularity in the heat and pressure of the blast, because of a more even temperature being maintained under the boilers and in the stoves; while there would also be a considerable saving in wages, and the men would be made more regular in charging the furnace.

The principle upon which the waste gas is taken off in the case of the close-topped furnaces is that, by keeping the furnace-top closed, the gas must necessarily pass away through any openings which are made for its escape, and may thus be made to travel even to a distance of more than a quarter of a mile from the furnace, as is done at the Dowlais Iron Works. In the open-topped furnaces the idea is that, after the gas has done very nearly all its work in the furnace, on arriving within about 5 ft. of the top of the materials in the furnace-mouth the greater part can be drawn off from the furnace by applying a mild suction, and employed to advantage for heating purposes elsewhere; at the same time, as no considerable amount of force is used for drawing off the gas, either by the suction of a chimney or otherwise, all surplus gas generated in the furnace beyond the amount drawn off escapes at the open top of the furnace, by passing up through an average of 3½ ft. depth of charged materials above the point of taking off the gas.

The open-topped plans of taking off the waste gas may here be divided into two classes:—those taking off the gas at a less depth than 5 ft. below the top of the materials in the furnace-throat; and those taking it off below that level. In the former the gas is taken off with due regard to the effect on the yield and working of the furnace; while in the latter the utilization of the gas is made the chief object.

In order to carry out the utilization of the gas without risk of interfering with the successful working of the furnace, it is of very great consequence not to take off the whole of the gas, but to leave a certain portion always to escape at the furnace-mouth, so that it may continue the process of preparing the newly-charged materials, and begin to dry and warm them immediately upon their being charged, and also prevent any downward current of air taking place from the furnace-top. Such a downward current of air must necessarily take place frequently, where the whole of the gas is drawn off; as the chimney-power requisite for this purpose would be quite sufficient to draw down the air through the average depth of 3½ ft. of materials in the furnace-throat above the gas-openings, at any time when there was not an ample supply of gas to be drawn off. The result would then be that where the ascending gas and the descending air met in the furnace a bright flame would be produced, which taking place amongst the fuel must occasion a very serious loss, by causing combustion of the fuel before it reaches the part of the furnace where its combustion is useful; and it appears doubtful whether fuel thus once lighted would not continue smouldering the whole of the way down in the furnace. On the other hand, if the fuel is properly covered in the upper part of the furnace by a sufficient depth of materials, so as to be protected from the air, Addenbrooke doubts whether it will begin to burn till it reaches the zone of fusion, where it then changes from a mere highly-heated state to one of active combustion caused by the presence of air supplied from the tuyeres. He believes that in the fact of covering up the fuel, without ignition being allowed to take place, lies one of the chief sources of saving in the yield of the fuel; and he considers that it is this alone, in the close-topped furnace, which to a very great extent makes up for the loss of yield of fuel that must inevitably result from the use of the close-topped system with its consequent back-pressure. This saving, however, is more than counterbalanced by the fact that neither drying nor warming nor any other preparation of the materials can be carried on in the close-topped furnace except by the heat of the gas coming up from below.

Were it not for the back-pressure produced in the furnace by a closed top, this system would doubtless work to a much better yield than the open top; but the entire prevention by the closed top of any drying or warming of the materials taking place until they have descended some distance within the furnace is a serious objection, in G. Addenbrooke's opinion, to the close-topped plan; whilst, on the other hand, in a well-worked open-topped furnace the preparation of the materials begins at once upon their being charged. Moreover, there is no way of so regulating the driving of a furnace or rate of descent of the materials in the interior as that in every hour the furnace shall take the same quantity of blast; but whenever the steam-pressure happens to rise above the average, causing the engine to force more blast into the furnace, or whenever the materials happen to lie more open in the furnace, or to be drier, an increased driving of the furnace will be occasioned, which will give an increased production of gas to pass off from the furnace. As this larger quantity of gas has in the close-topped furnace to pass off through the same openings which previously carried off a smaller quantity, the result must be an increase of the back pressure in the furnace; whereas with an open top the surplus gas readily escapes at the mouth of the furnace, without producing any back-pressure inside the furnace. Another objection to the closed top is that the pressure coming from the lower part of the furnace will carry off with it from the furnace more dust than would be drawn off by suction.

The requirement in the utilization of the blast furnace waste gas appears therefore to be an open top where the fuel can be buried or covered, without its being ignited to any material extent either by the escaping gas, or, worse still, by the gas being drawn off so completely from the furnace-throat that air is drawn down also through the materials, causing a hidden fire to be raging beneath the surface of the materials while everything seems to be going on well. Addenbrooke states that he has frequently seen the gas drawn off so completely that a man could walk inside the mouth of a large furnace while the blast was on; but experience shows that such a state of the furnace-top is altogether wrong, and if ever it occurs, the main gas-valve ought at once to be closed sufficiently to ensure some gas passing off at the furnace-top, in order to begin the preparation of the materials and prevent any downward current of air.

The level at which the gas is taken off from the furnace is a most important point, as in effect the working height of the furnace nearly terminates at the level where the gas is taken off. The most satisfactory working that Addenbrooke has known of a blast furnace has been where the gas was partially and not wholly taken off, and at a depth of 4 ft. 6 in. below the top of the materials. In one case that came under his experience, one of Darby's bells was used very successfully for a considerable time, taking off the gas at a depth of 4 ft. 6 in. below the top; but when it became necessary to change it, the new bell was inserted to a depth of 5 ft. 6 in., in order to give the fillers a little more range of level. This was, however, found to work so much less satisfactorily, that it was altered back again to the original depth of 4 ft. 6 in.

The very great sensitiveness of a furnace to the least lowering of its working height is undoubtedly the cause of more than one-half of the mottled and white iron that is made where grey forge or forge-iron was expected. The change is usually caused by delay in filling, most frequently at night, when the men often neglect their work, and allow the furnace to drive, so that the surface of the materials is lowered a considerable distance; and the effect is then observed about three casts later by the production of mottled or white iron. The worst consequence of this neglect is that the furnace manager then finding white iron made, probably alters the burden at once in order to correct the fault; and, after charging it up to the next casting time with either less ore or more fuel, he probably finds this next cast all right, and therefore alters the burden back again; but in another cast or two comes the iron made from the lighter burden, rather too grey to suit the purposes it is wanted for. With a closed top there can be no check upon irregularity in filling, day or night, except by constantly watching the filling. But in an open-topped furnace, where the gas is taken off through openings not lower than 5 ft. below the top of the materials, if the filling were delayed more than half-an-hour, the greater portion of the gas would begin to escape at the furnace-mouth instead of being drawn off through the gas-openings, as the surface of the materials would be lowered nearly to the level of the gas-openings. The fillers' neglect would then be detected by an increased flame from the furnace-mouth, and by the supply of gas to the boilers and hot-blast stoves falling short; and they would be recalled to their duty by the risk of all coming to a standstill from stoppage of the blowing engine.

This great sensitiveness of blast furnaces as to being kept charged full appears to suggest strongly the doubt whether the Staffordshire furnaces are now as high as the fuel would allow of their being worked profitably; and whether a gain of yield would not be found to result from raising the furnaces. The present height of from 40 to 50 ft. in Staffordshire has been increased in the Cleveland district up to 80 ft. and more, and it is considered the limit of height has not yet been arrived at in that district. There is no doubt that the Staffordshire fuel would not stand any very considerable increase of height, on account of its friability; but it would at least be desirable to ascertain by trial whether some increase of height would not be beneficial.

The result of the previous inquiries in which Addenbrooke was engaged as to the best mode of utilizing the waste gas from blast furnaces was the adoption of the open-topped plan, with Darby's bell inserted in the neck of the furnace to a depth of 4 ft. 6 in. below the surface of the materials. This plan was applied to two furnaces at Darlaston; and in carrying it out the special points attended to were to provide a large gas-main, a large chimney, and large flues to the chimney. For this purpose the gas-mains were made 5 ft. diameter, the chimney 10 ft. diameter inside throughout, with a height of 160 ft., and the main flues to the chimney very large, 5 ft. high to the crown of the arch and 4 ft. 6 in. wide; and experience has shown that these dimensions are none too large.

The two furnaces continued at work on this system till September, 1864, when the heavy cost of repairs and renewals, with the consequent stoppages and loss in wages, led G. Addenbrooke to design the plan which is shown in Figs. 725 to 731. Fig. 725 is an outside elevation of the furnace to which this system is applied. Fig. 726 is a vertical section of the same furnace, showing the gas-openings A A from the furnace into the neck-flue B, and the gas branch-pipe with stop-valve C for connecting or disconnecting this furnace from the range of gas-main. Fig. 727 is a sectional plan of the furnace taken through the gas-openings A A and neck-flue B. Fig. 728 is an enlarged section of the furnace-top; and Figs. 729, 730, 731, show a vertical section, outside elevation, and sectional plan of one of the segmental boxes or gas-openings.

These gas-openings or boxes are made of cast iron, and allows of the openings being made so wide that their combined area of passage is amply sufficient for the passage of the gas, without the depth of the opening being more than 15 in.; in consequence of which they do not require to be inserted lower than about 5 ft. down, and still leave a fair height of 4 ft. above them for variation in the level of the top of the materials charged. The boxes are cast very strong, as shown in Figs. 729 to 731, and the openings through them are made at such a slope that nothing except very light dust can be carried through them by the gas in regular working, unless it be a bit of material thrown over by one of the slight explosions that occasionally take place where any raw minerals are used in the furnace. The boxes are placed close together side by side, so as to form

725.

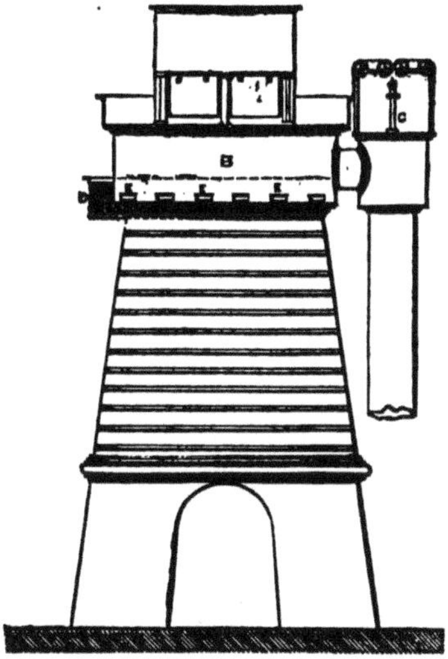

726.

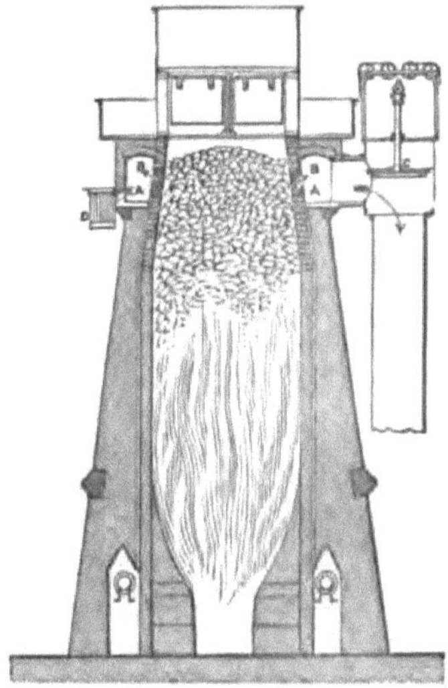

a continuous ring of openings round the furnace, as shown in the plan, Fig. 727, having the lower end of the slopes opening into the furnace and the upper end opening into the large external gas-flue B, Fig. 728, which surrounds the neck of the furnace. These castings take the place of so many courses of lining bricks, and after they have been fixed, the lining fire-bricks are continued above them to the top of the furnace. Considering their strength and situation, the castings appear likely to be almost permanent. As they stand flush with the face of the lining, the whole area of the throat of the furnace is left free for charging; and when the furnace is full, and any portion of the gas passing off at the surface of the materials, no damage can be done to the openings or any part of the gas apparatus. In case of the top of the materials sinking below the gas-openings, any damage is prevented by shutting the gas-valve C at once, when the whole of the gas will be burnt at the mouth of the furnace, but without injury occurring to any part of the apparatus, as is unavoidably the case with the wrought-iron gas-main proceeding from a bell inserted in the top of the furnace.

727.

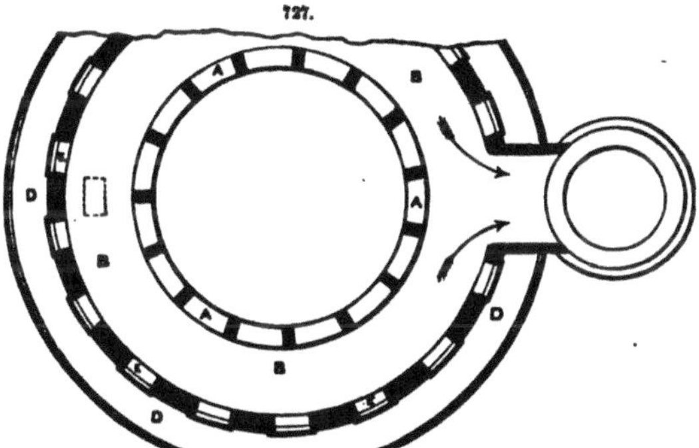

The large gas-flue B surrounding the neck of the furnace is lined with fire-brick, and is 4 ft. 3 in. high to the crown of the arch, by 3 ft. mean width. The outside of the furnace from a little below the bottom of the flue upwards is cased with wrought-iron plates, to which is fixed a light iron gallery D for the convenience of cleaning out the flue B. A series of openings E E are made in the outer side of the flue all round, as shown in the plan, Fig. 727, which are closed by pieces of boiler-plate, daubed with moistened fire-clay, and held in their places by crossbars and wedges; by means of these the whole of the neck-flue can be cleaned out in a few minutes any time that the blast is off the furnace. The bottom of the flue is placed at a lower level than the bottom edge of the gas-openings A A, Fig. 728, in order that the dust carried over with the gas may be allowed to accumulate in the flue, so long as it does not interfere with the gas-openings, and it can be easily cleaned out when required. Experience of the working of this plan of furnace-top proves that, from the increased area of the gas-openings as compared with other plans, the gas does not pass nearly so rapidly out of the furnace, and, consequently, has not the power to carry nearly so much dust into the flue. The sectional plan, Fig. 727, shows that there are fifteen gas-openings A A round the neck of the furnace, 23½ in. wide and 11½ in. high on the square, each giving 270 sq. in. clear opening, making a total area of 4050 sq. in. for drawing off the gas; whereas the single central opening of the bell of 4 ft. 6 in. diameter previously worked in the same furnace, which was as large a size as could be conveniently used, gave an area of only 2290 sq. in. for drawing off the gas,

728.

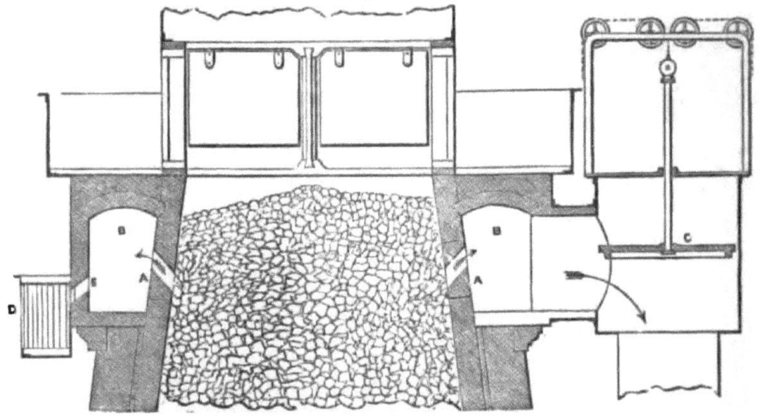

or only 56 per cent. of the area now obtained with the present neck-openings. As the gas-openings give a total area of 4050 sq. in. for the passage of the gas, while the descending gas-main supplied by them being 4 ft. 6 in. diameter has an area of only 2290 sq. in., the velocity of the current of gas through the openings is necessarily only half what it would be where a bell or centre opening is used for drawing off the gas, as in the latter case the gas-opening to the furnace cannot be made of larger area than the descending gas-main of 4 ft. 6 in. diameter.

Notwithstanding the temporary kind of construction that was adopted for trial in the first furnace, the gas-openings made with only 2-in. cast-iron plates lasted more than a year, and stood some of the severest treatment that a furnace-top can be exposed to, in consequence of a sort of fuel being tried at one time which proved a total failure, and with which the furnace was unable to drive at all; and consequently for two days the whole throat was at a red heat. It was expected that when the furnace did drive below where the openings had been, they would be found to have given way. They had not, however, completely given way in any instance; and, though the plates were very much bent, they remained at work a great part of a year afterwards. Their standing so well is to be attributed to their position, in the outside of the furnace instead of in the centre; and also to the effect caused by closing the valve on the top of the descending gas-main, so that

729.

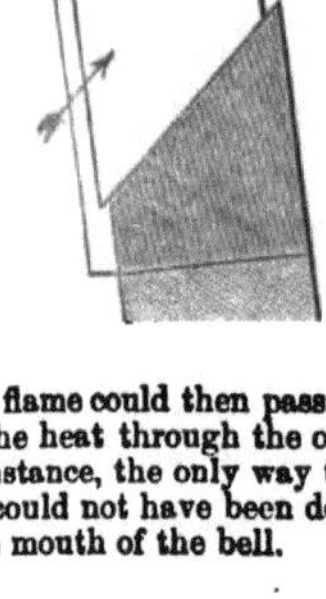

730.

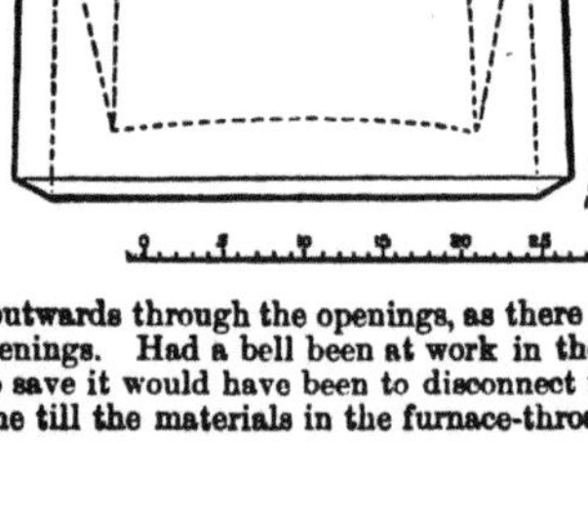

731.

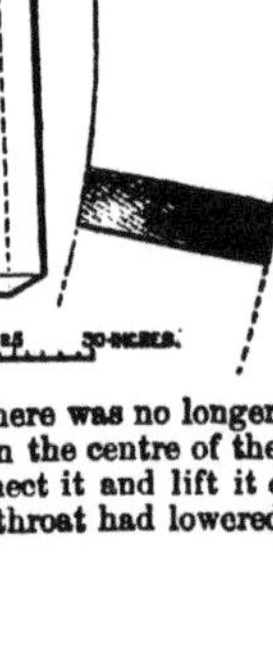

no gas or flame could then pass outwards through the openings, as there was no longer any current to draw the heat through the openings. Had a bell been at work in the centre of the furnace-top in this instance, the only way to save it would have been to disconnect it and lift it out entirely; but this could not have been done till the materials in the furnace-throat had lowered themselves below the mouth of the bell.

As the gas-openings are now cast, as shown in Figs. 729, 730, 731, it is anticipated they will stand for many years.

Details of Blast Furnaces.—Fig. 732 shows plan of hearth of blast furnace at Plymouth Iron Works. Fig. 733, plan of hearth of blast furnace at Rhymney Iron Works. Fig. 734, horizontal section through hearth at level of tuyeres, showing tuyeres and pipes in position, of large 18 ft. blast furnace, Dowlais Iron Works. Fig. 735 represents sectional plan of hearth of cupola furnace

732.

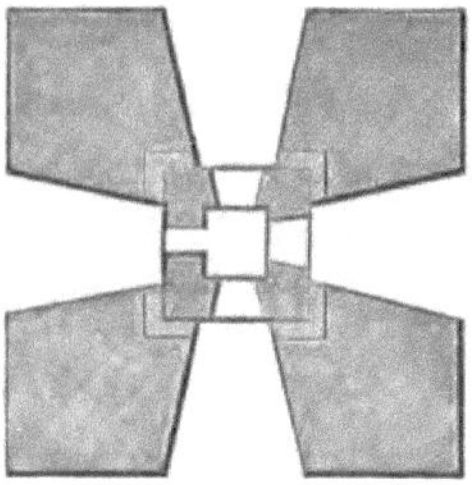

734.

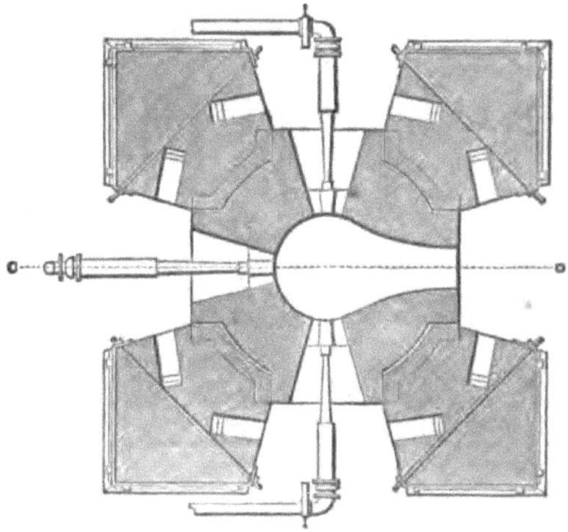

733.

735.

736.

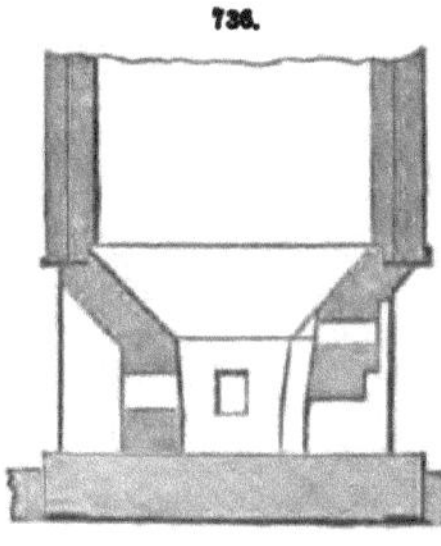

737.

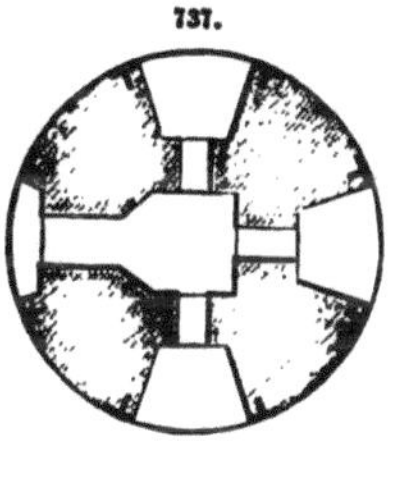

738.

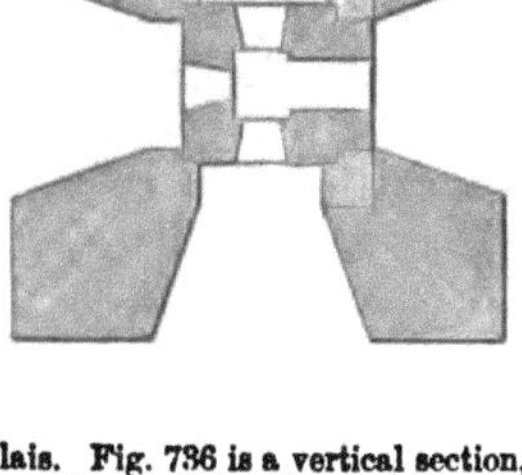

739.

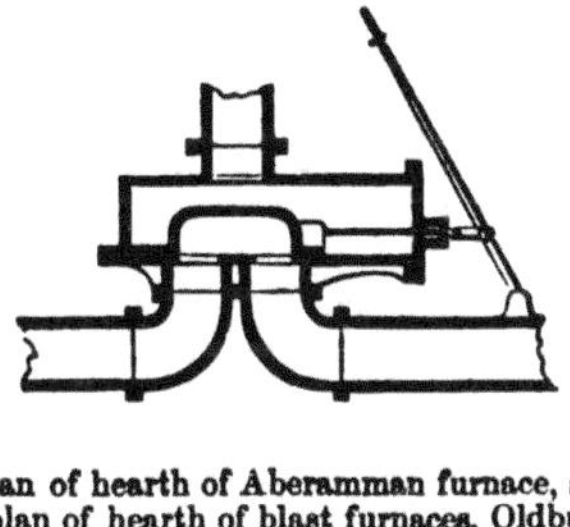

at Dowlais. Fig. 736 is a vertical section, and 737 plan of hearth of Aberamman furnace, showing tuyere-openings both in sides and breast; Fig. 738, plan of hearth of blast furnaces, Oldbury Iron Works; Fig. 739, section of valve-chest, with three passages, placed on blast-pipes of hot-blast furnaces, to direct the blast either through the stoves or at once into the furnaces at pleasure.

Figs. 740, 741, show a section and plan of a cinder-fall, in which is fixed a wrought-iron crane C; a cast-iron plate having its upper edge serrated to facilitate the removal of large masses of cinder; a cast-iron trough A, about 6 ft. long, to convey the fluid metal from the tapping-hole H, to the casting-bed or refinery B, B; two troughs *a a* are fixed on the cinder-bed, for guiding the fluid cinder into the tubs *t*.

Various plans for collecting furnace-gases are shown in Figs. 742 to 746. In Fig. 743, the cup, which is a funnel-shaped casting equal in its largest diameter to the throat of the furnace and 4 or 5 ft. deep, rests upon the top of the furnace by a flange round its outer edge. The orifice at the bottom measures from 3 to 5 ft. in diameter, and is closed by a conical casting, with the apex upwards. This casting is suspended by a chain from a lever which is counterbalanced at the other end. The materials are filled into the cup, and the workmen, by suitable gearing affixed to the lever, lower the cone, and the materials fall into the furnace, and the stopper is restored to its place by the counterpoise on the opposite end of the lever.

740.

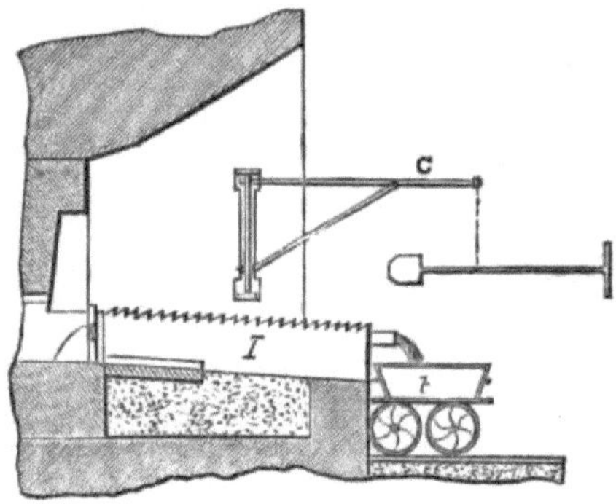

741.

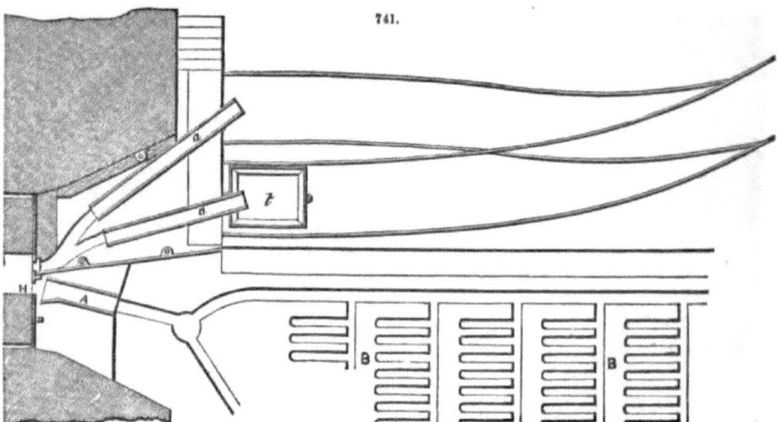

743.

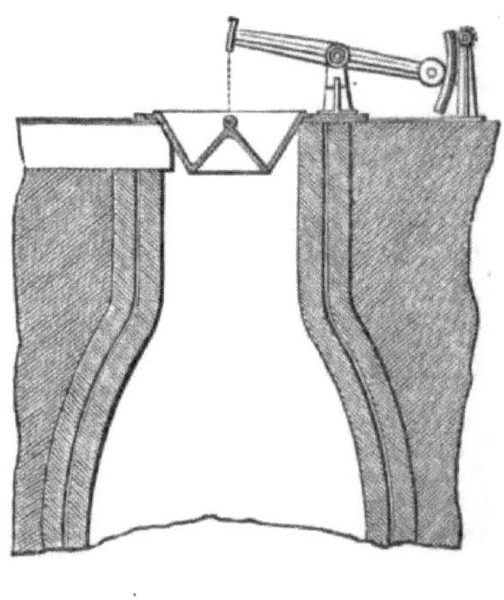

742.

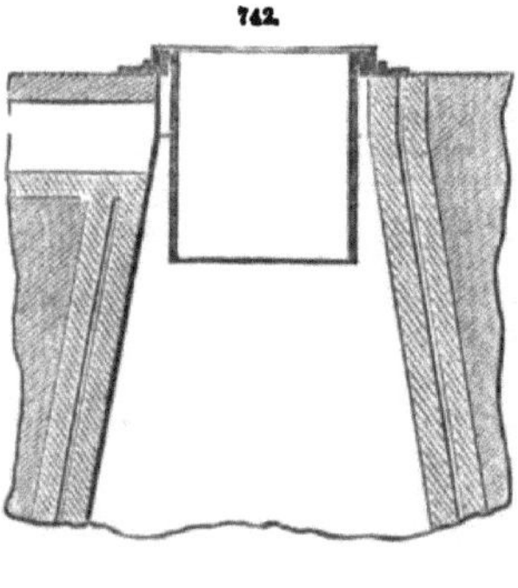

Another plan, Fig. 744, which has also been extensively adopted, consists of a lid fitting closely to the furnace · this lid is lifted by means of a counterbalance weight W, on each occasion of charging materials. No reduction in the working height of the furnace is caused by this arrangement, but the time during which the throat is open while the cover is being lifted, the materials filled in, and until it is again shut close, is very prejudicial to the quality of the gas. It is commonly stated, says Truran, that no gas passes while the cover is up; but this is an error: the same quantity of gas is evolved from the furnace whether the cover be open or closed; but if it is open, a large quantity of atmospheric air also passes into the pipes, increasing the bulk of the unprofitable gases, and thereby reducing the heating power of those that are combustible, as well as endangering the apparatus by causing a great increase of temperature in the pipes.

744.

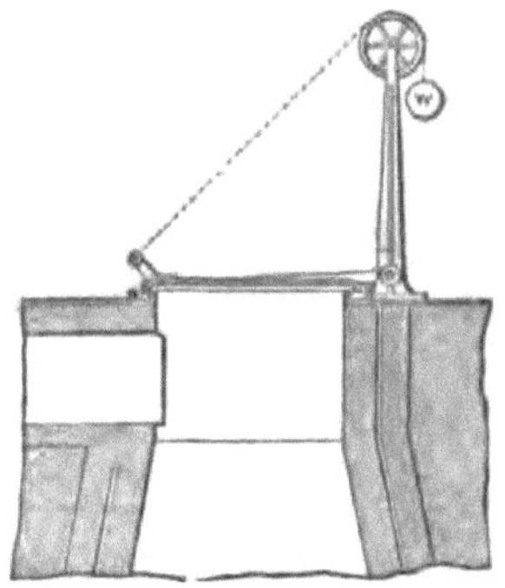

A third plan, Fig. 742, in use at some works is very readily applicable to existing furnaces (1862). An iron cylinder of 6 or 7 ft. in depth, and 6 or 8 in. smaller in diameter than the throat, is sunk into the furnace; a flange on the top, which rests upon the brickwork inside the tunnel-head, forms a joint and sustains the cylinder. The annular space between the cylinder and furnace underneath the flange and above the materials forms a chamber for the ascent of the gases, which are conveyed away through a suitable pipe or tunnel.

This plan also has its disadvantages. The duration of the cylinder is subject to great variation. In some cases it is burnt down in two or three weeks, and, under more favourable circumstances, seldom lasts longer than a few months. The cost of the cylinder, and the delay and expense attending its so frequent renewal, are formidable items in the working cost of this plan for collecting gas.

This method is also subject to the disadvantage of reducing the working height of the furnace. The cylinder, says Truran, is kept full of materials, it is true; but they receive very little heat while they are in it, as the hot gases are drawn into the outside flues, and do not enter the cylinders. It must be conceded that the capacity, and consequently the smelting power, of the furnace is diminished by the space occupied by the cylinder and chamber for collecting the gas. If the cylinder is immersed 7 ft., one-tenth of the capacity of the furnace will be useless so far as the reduction of metal is concerned. The deficiency of smelting power is still greater with the plan first described.

745.

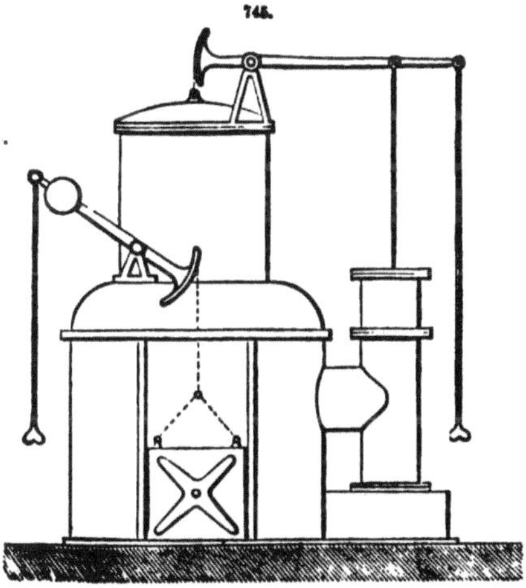

746.

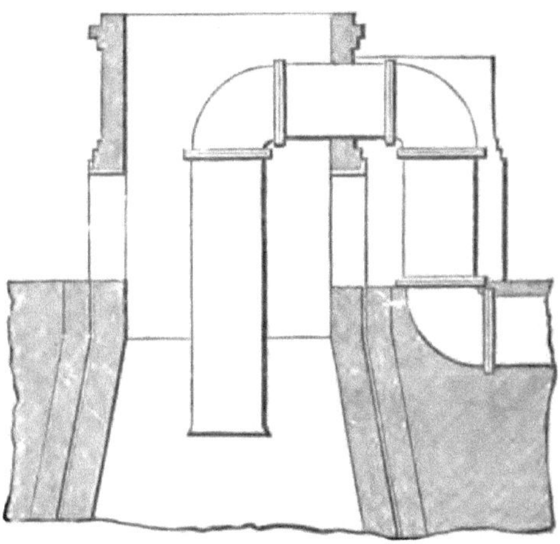

A fourth plan, Figs. 745, 746, of collecting the gases is considered by some engineers as the least objectionable. At a convenient depth, generally 8 or 10 ft. from the top of the furnace, an annular flue is constructed around the brick lining, with a number of orifices opening downwards into the body of the furnace. This plan leaves t e form of the throat and the arrangements for filling unaltered. From the descending direction taken by the orifices communicating with the furnace, they are not liable to obstruction from the materials, and the supply of gas is probably more regular than with either of the other plans.

Fig. 747 shows a front view of a cinder-fall. A, B, C, Fig. 748, are moulders' tools. Figs. 749 to 753 show ordinary furnace-keepers' tools.

747.

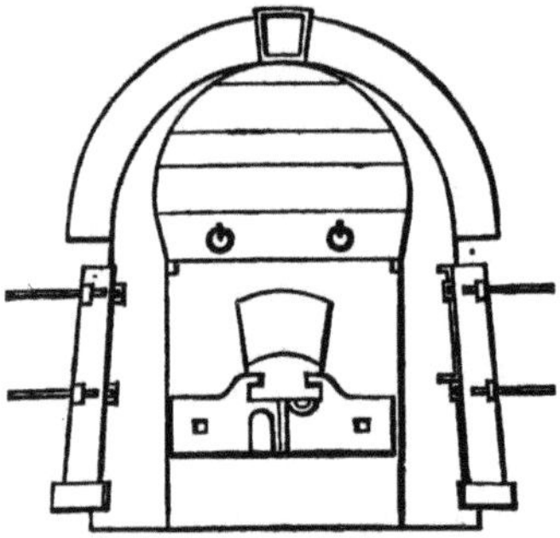

749.

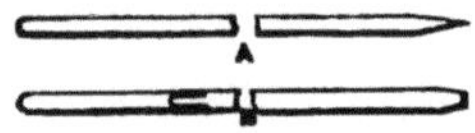

750.

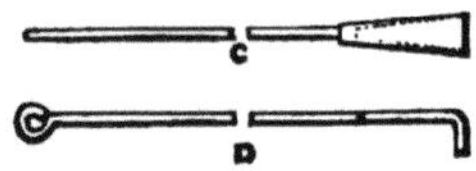

751.

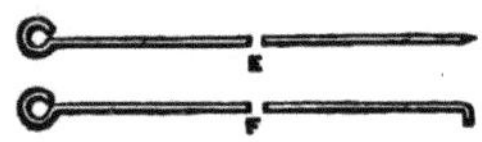

748.

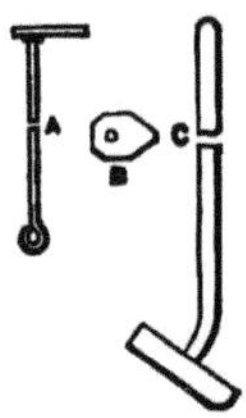

752.

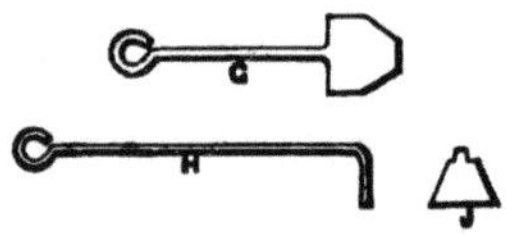

753.

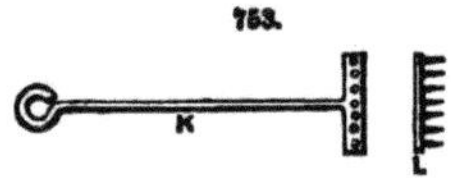

754.

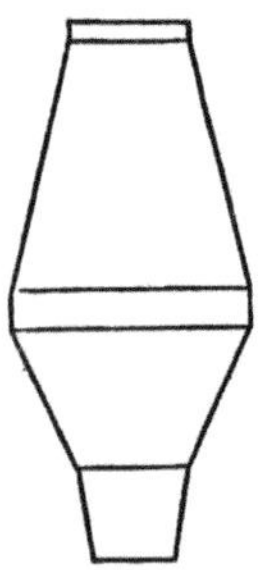

755.

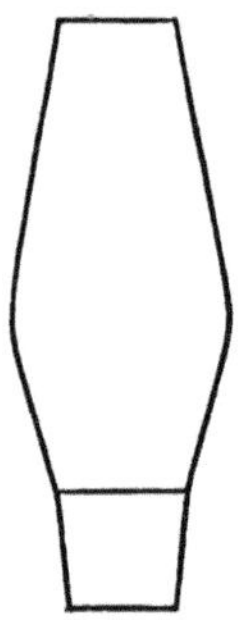

756.

757.

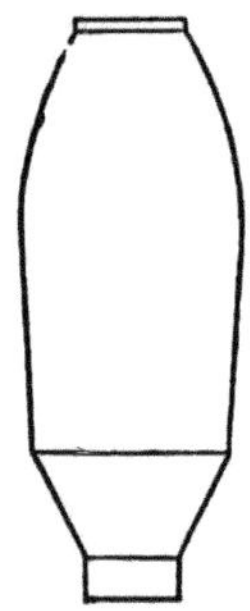

758.

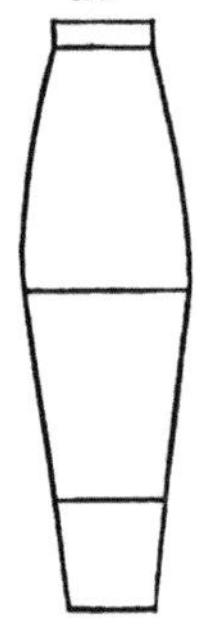

759.

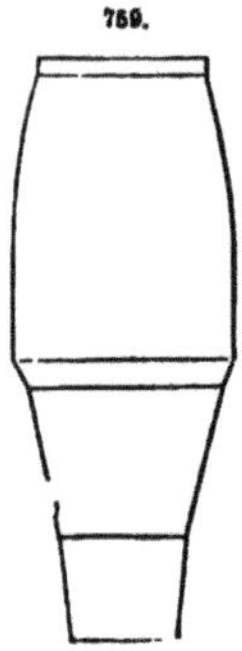

760.

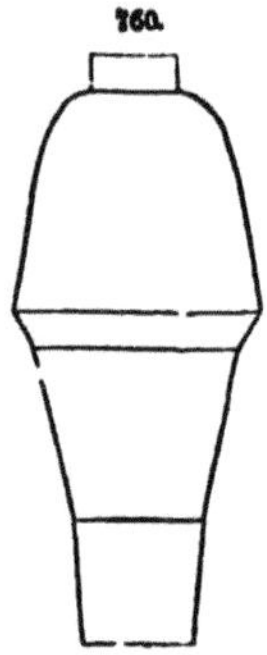

761.

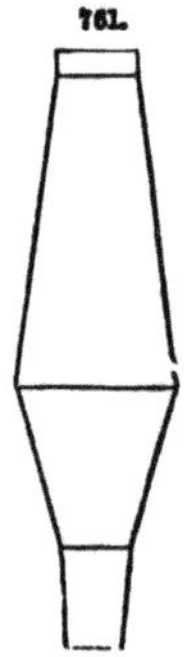

762.

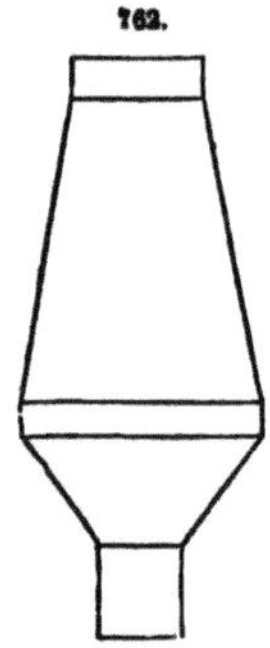

763.

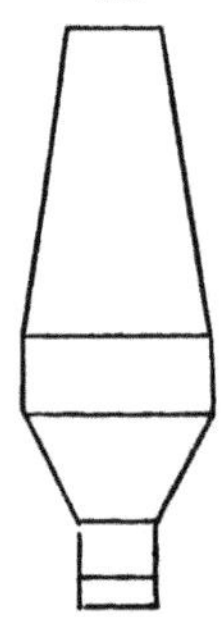

Outlines of blast-furnace interiors used at Rhymney are shown in Figs. 754, 755.

Fig. 756 shows the type adopted at Sirhowy; 757, Ebbw Vale; 758, 759, 760, 761, Dowlais Iron Works; 762, Hirwain; 763, Aberamman; 764, Pentyrch; 765, Landore; 766, Staffordshire; 767, Stafford; 768, Tipton; 769, 770, Shropshire; 771, 772, Corbyn's Hall; 773, Alfreton; 774, Wilkie; 775, Stockton; 776, Kinniel; 777, Dandyvan, 778, Muirkirk; 779, Yniscedwyn; 780 Ystalyfera New Furnace; 781, Ystalyfera Old Furnace, 782, Abernant; 786, 787, French; 783, 784, 785, American; 788, Silesian; 789, Norwegian; 790, Belgian; 791, Prussian; 792, Baermn; 793, 794, Hartz.

See Blowing Engines. Furnaces. Iron. Kilns. Ovens. Puddling and Puddling Machines. Rolling Mills. Squeezers. Steam-hammer. Steel. Tuyere.

764.

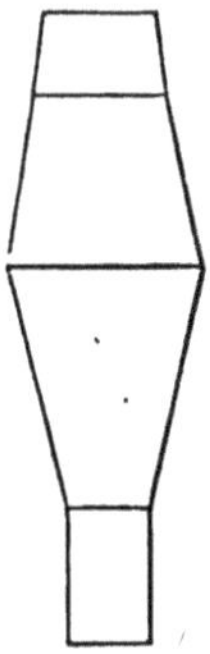

765.

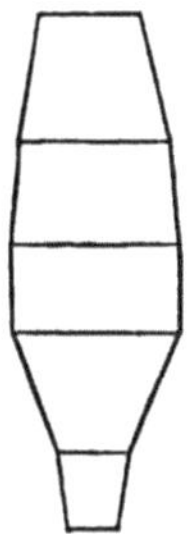

766.

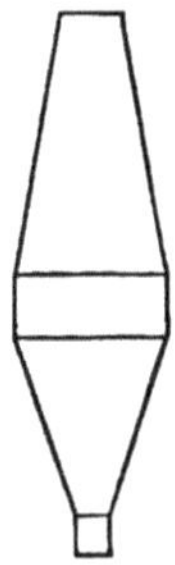

767.

768.

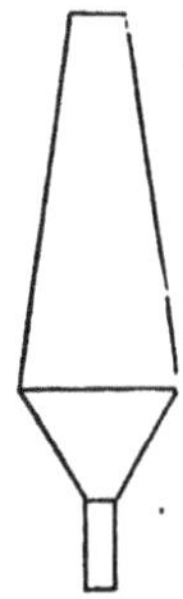

769. 770.

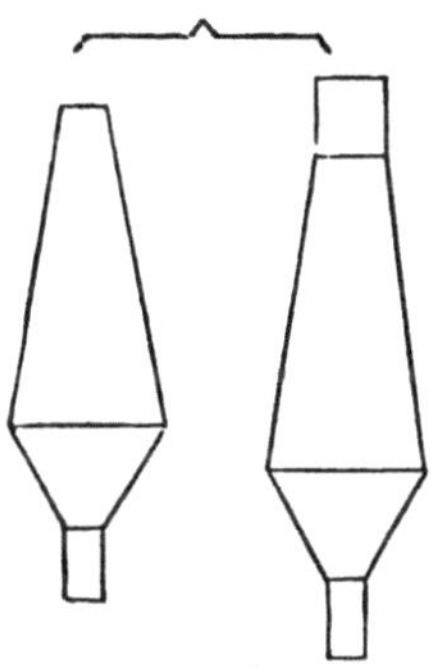

771. 772.

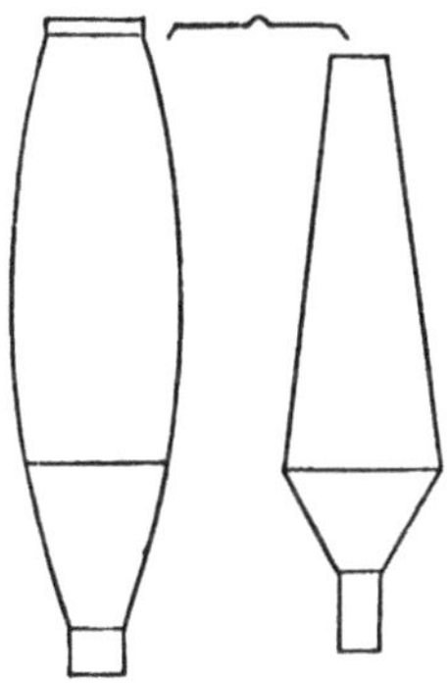

773.

774.

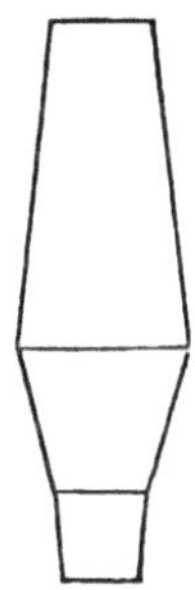

775.

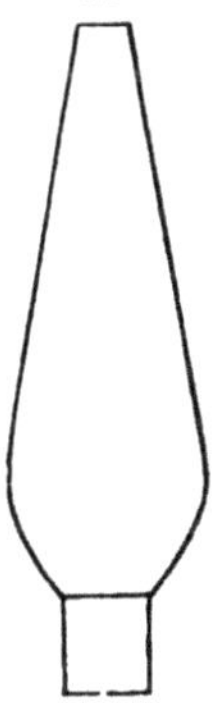

776.

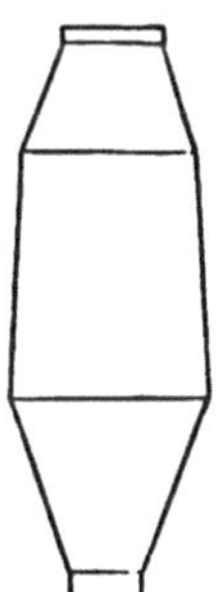

777.

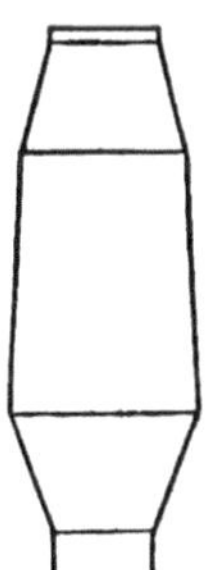

778.

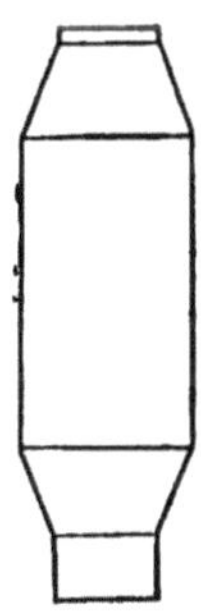

779.

788. 789 790. 91. 792.

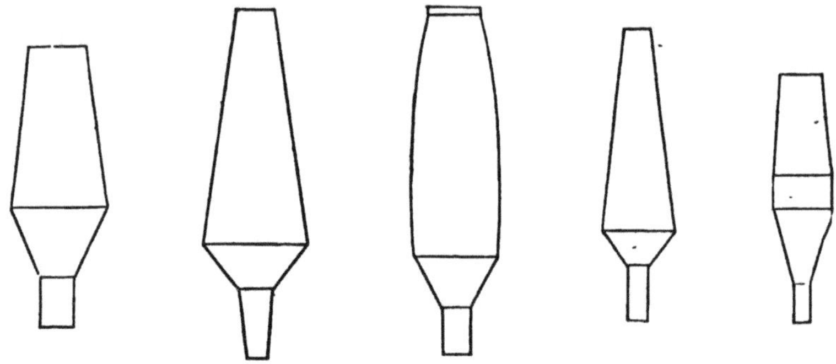

784. 785. 788.

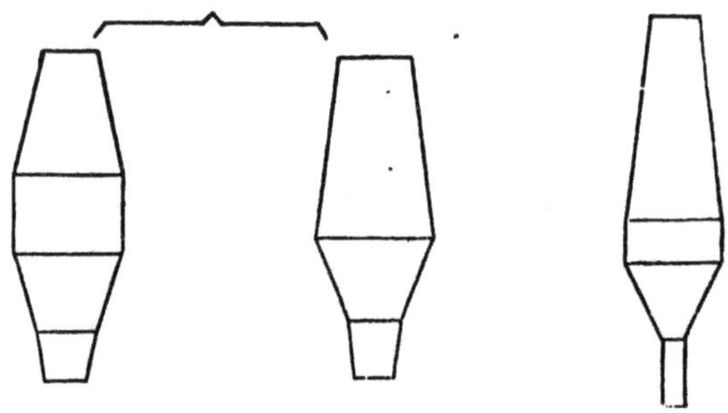

793. 794. 787. 786. 781. 782. 783.

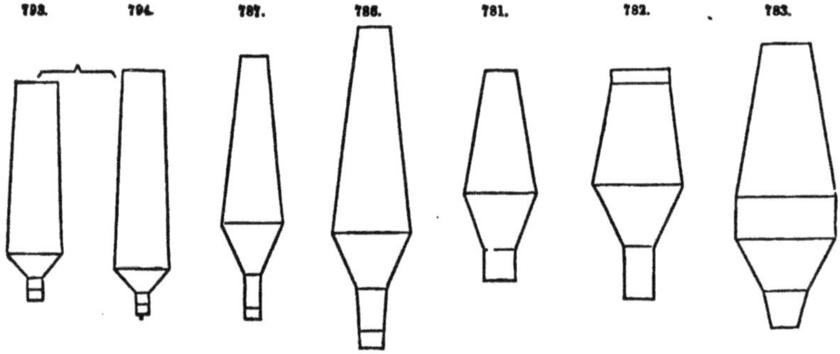

BLAST-PIPE. Fr., *Tuyau de raréfaction;* Ger., *Zugrohr;* Ital., *Tubo di scarico.*

See Details of Engines.

BLIND AREA.—The same as an *Area-drain*, with the addition of cross-walls at intervals, usually built to assist the dwarf-wall in supporting the earth at its back. A blind area is inferior to an area-drain, as the cross-walls interfere with the circulation of the air.

795. 796. 797.

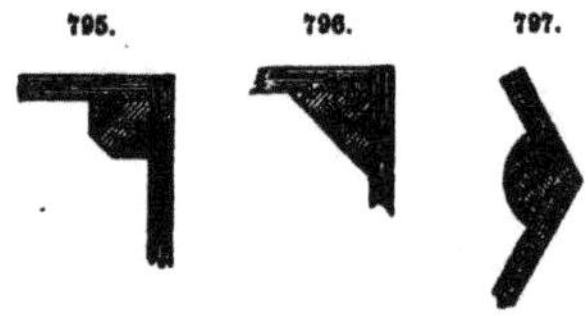

BLINDING. Fr., *Gravier;* Ger., *Grobe Sand;* Ital., *Intasatura Insabbiamento.*

See Roads.

BLOCK. Fr., *Poulie;* Ger., *Block;* Ital., *Girella;* Span., *Poléa.*

See Pulleys. Tackle.

BLOCK or BLOCKING. Fr., *Taquet;* Ger., *Eckleiste;* Ital., *Rinforzi.*

Small pieces of wood fitted and glued into the interior angles of two pieces of stuff to strengthen the joint. See Figs. 795 to 797.

BLOCKHOUSE. Fr., *Blockhaus;* Ger., *Blockhaus* · Ital., *Ridotto Coperto;* Span., *Palanque.* See Fortification.

BLOCKING-COURSE. Fr., *Le dessus d'un corniche;* Ger., *Oberer Theil eines Karniess;* Ital., *Ultimi corsi.*

Blocking-Course.—In masonry, a course of stone placed on the top of a cornice.

BLOOM. Fr., *Loupe;* Ger., *Frischluppe;* Ital., *Massello;* Span., *Changote.*

A mass of crude iron from the puddling furnace, while undergoing the first hammering previous to being rolled, is termed bloom.

BLOOMING MACHINE. Fr., *Machine à cingler;* Ger., *Presswerk Presse;* Ital., *Macchina da far i masselli.*

The blooming machine, invented by Jeremiah Brown, is shown in Figs. 798, 799, 800. It consists of three large eccentric rolls A B C, placed horizontally in the strong holsters D D, the centres of the rolls being arranged in a triangular position, and the bottom roll C being nearly central between the two top rolls A B. These rolls all rotate in the same direction, and are driven by a centre pinion E, working into three pinions of equal size F F F, fixed on the roll-spindles. In the present machine the driving power is applied direct to the bottom roll by means of the large wheel G, for the convenience of carrying the main shaft under the floor. The rolls are cast solid, with their journals like ordinary rolls, and are driven by coupling-boxes and spindles H H.

The roll-faces are 16 in. long, and the bottom roll has,

798.

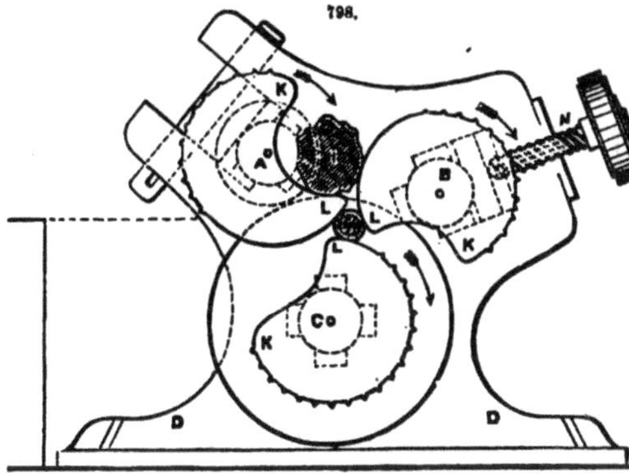

799.

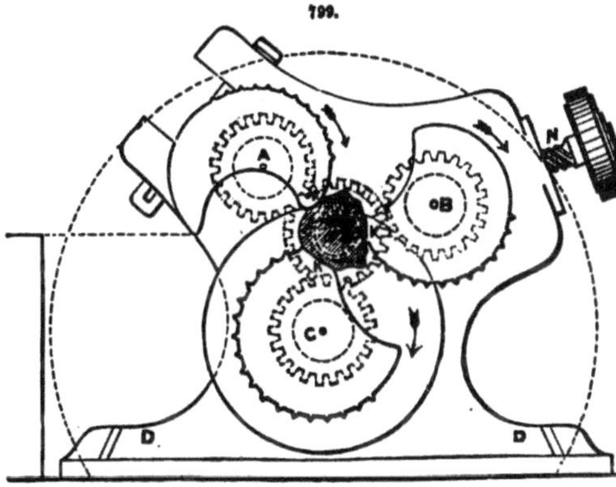

800.

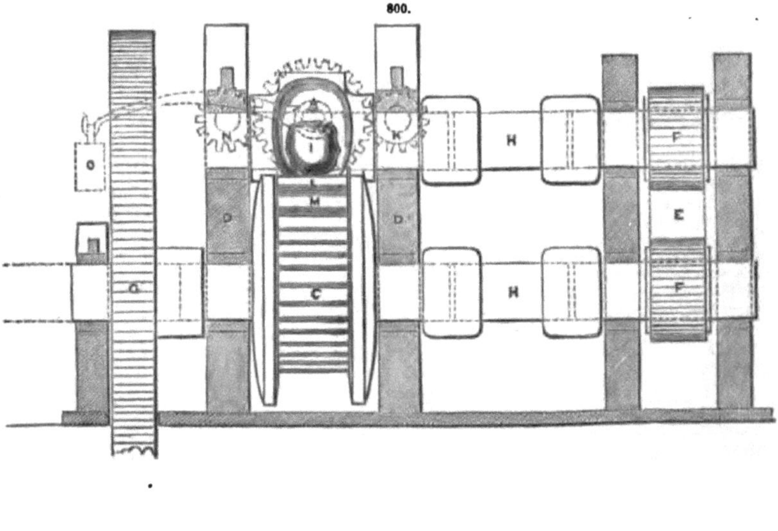

at each end, strong flanges, 8 in. deep, between which the two upper rolls work. The object of these flanges is to upset or compress the ends of the bloom, as the iron is elongated during the operation and the ends are forced against the flanges, which makes them square and sound. The top roll A has a large hollow in which the puddled ball I is placed by the puddler; this roll carries the ball round, and drops it into the space between the three rolls, as shown in Fig. 798, this space being at that moment at its largest capacity.

The three projecting points K K K of the rolls immediately impinge upon the ball, and compress it forcibly on three sides; and, giving a rotating motion to the ball at the same time, they have a powerful kneading action upon the iron, squeezing out the cinder, which falls down each side of the bottom roll. The space between the rolls gradually contracts, from the spiral or eccentric form of the rolls, and the iron is subjected to an increasing pressure, until it is liberated by the points L L L, simultaneously passing the bloom M, which falls in the direction of the arrow at the same moment that another ball is dropped in at the top of the machine. The projecting teeth on the surface of the rolls assist this action, by seizing the iron and kneading into it as it rotates; these teeth gradually diminish in projection, the last portion of each roll being plain, consequently the bloom is turned out in a smooth, compact form.

The space between the flanges of the bottom roll is widened for a short distance beyond the point L, for the purpose of allowing the bloom to fall out readily and admitting the fresh ball.

In order to prevent the rolls from being broken by any unusual size of ball, the machine is provided with two large triple-threaded screws N N, which bear upon the journals of one of the top rolls B; a small pinion on the head of each of these screws works into a large pinion fixed between them, which has a horizontal lever fixed to it, carrying a balance-weight O at the end. This weight causes a constant equal pressure of the roll, and, if a ball of extra size be put into the machine, the screws yield by turning back and lifting the weight to the extent required, so that a large ball may be worked in the same manner as those of smaller sizes.

A continual stream of water runs on to all the journals in the machine, thus preventing them from heating while the machine is at work.

BLOWING ENGINE. Fr., *Soufflerie*, Ger., *Gebläse*; Ital., *Macchina soffiante*; Span., *Bofeton*. See Engines, *Varieties of*.

BLOWING MACHINE. Fr., *Machine soufflante*; Ger., *Gebläse Maschine*; Ital., *Macchina soffiante*; Span., *Bofeton*.

A blowing machine is a machine or engine for forcing a strong and continuous blast of air into a furnace.

The first records show blowing cylinders to have been single-acting, that is, having the power of propelling the blast when the piston was moving in one direction only. Two or more of these blowing cylinders appear to have been attached to one crank-shaft, worked by a water-wheel, and thus a tolerably steady pressure of air was obtained. When the gradual improvements of the steam-engine and the demand for increased means of manufacture caused it almost entirely to supersede all other power, blowing apparatus appears to have been accommodated as much as possible to the steam-engine, so as to afford the character of engine for the time being the fullest development of its power.

In pursuance of this object, the single-acting atmospheric engine of Newcomen was attached to a blowing cylinder, which propelled the air from the upper side of the piston only; and in addition

801.

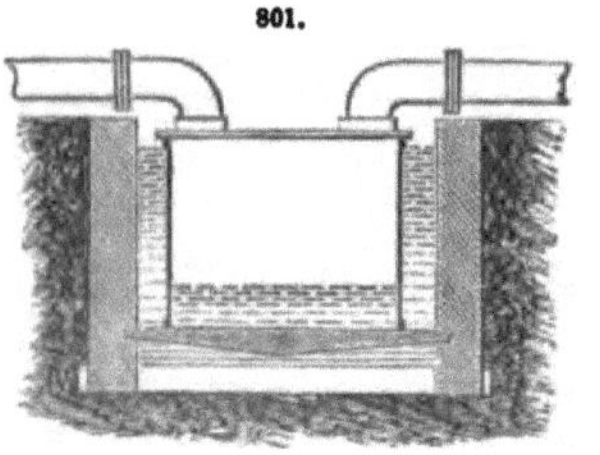

802.

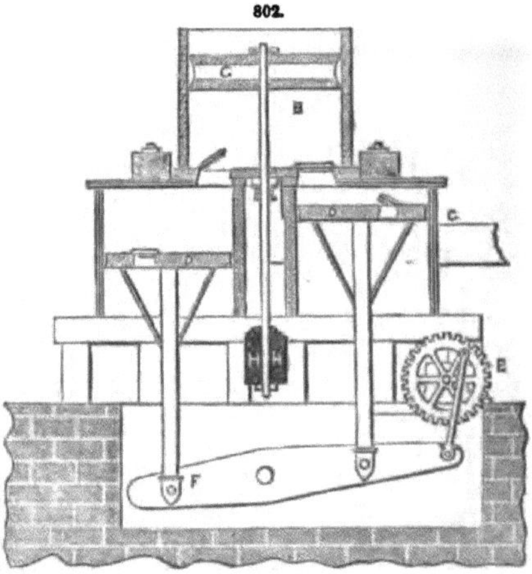

to the water-regulator, Fig. 801, which appears to have been known at an earlier date, there was attached a cylinder, B, Fig. 802, now known as the regulating tub, which was equal to or larger in diameter than the blowing cylinder. In this was fitted a piston G, with a rod moving in a guide fixed on the open top of the regulating tub, the bottom of the latter being close, and having an open connection to the main from the blowing cylinder. The piston in the tub was loaded at H to the pressure of blast required, and in the intervals between the discharges of the blowing cylinder, the descent of the piston in the tub kept up the discharge of air into the water-regulator, which intervened between it and the furnace; thus in effect, as far as possible, making the engine double-acting. To prevent the piston being blown out of the regulating tub, a large safety-valve was attached to the top of the rod by a strap, long enough to allow the desired play of the piston, and short enough to lift the safety-valve, or snorter, as it is usually termed, if the piston at any time exceeded its limits; and the number of strokes of the engine was also regulated by the tub-piston, as to it the cataracts were attached.

When the double-acting engines of Watt were introduced, the regulating tub was still retained, though not nearly so essential a part of the machine as in the former instance.

The next change that took place was the general abandonment of the water-regulator, though some of these are still at work, or have been within a few years. The reason for this change was the discovery that the air in summer, already surcharged with moisture, took up an additional quantity from passing over the surface of the water in the regulator, and that this was prejudicial to the working of the furnaces.

When the large area of the water-regulator was shut off, it was then found that the tub was by no means such a perfect regulator as it was supposed to be, as the momentum of the engine passed too suddenly into the heavy piston of the tub, and, throwing it up much beyond the height due to the pressure of the air, caused an irregularity that was even more aggravated by its descent. To counteract this, a spring-beam was placed on the top of the tub, so as gradually to check the momentum of the piston; and this had some effect, but not at all a satisfactory one.

The next alteration which appears to have suggested itself, was the application of large *air-chambers*, from twelve times to thirty times the area of the blowing cylinder, in which the elasticity of the compressed air acted as the regulator of the discharge; the tub, with its piston, being in some cases retained to work the cataracts, and as a tell-tale against the engine-men in case of their allowing the steam to slacken and the piston to descend. In other cases the tub was dispensed with altogether.

We now enter upon the last change which took place some thirty years ago, namely, the coupling of two double-acting engines and double-acting blowing cylinders upon the same crank-shaft at right angles, so as to keep up a regular discharge. This effect was in some measure obtained; but an air-chamber, or what is equivalent to it, very large mains, was still required to obtain a satisfactory result.

At this point the realized improvements of the blowing engine stop short, leaving it still a large, cumbrous, and expensive machine, and not capable of moving through its valves the HIGHLY ELASTIC MEDIUM AIR at a greater rate than the absolutely NON-ELASTIC FLUID WATER is moved through an ordinary pump. Under these circumstances it must be obvious that, after all the engineering talent that has been spent on this description of engine, there is still (if the expression may be applied) a wide range of discovery open.

The immediate cause of my attention being attracted to the improvement of the blowing engine, says Archibald Slate (to whose paper, read before the Inst. of Mechanical Engineers, we are indebted for the present article), was the difficulty experienced in regulating one of the old construction of blowing engine in the latter part of 1848, having at the same time occasion to employ some small 9-in. cylinders driven by the air of the large blowing engine. These small cylinders, when driving the shafting only, sometimes attained a velocity of upwards of 200 revolutions a minute, suggesting the idea of the possibility of reversing their motion, and taking in the air in place of blowing it out through them; there was, however, a difficulty in the slide-valve, which did not open and shut fast enough. After some consideration, it was agreed that another cylinder should be prepared, the centre-port made much larger, and the slide over-travelled nearly half its stroke in excess, which had the desired effect; a cylinder of 9 in. diameter, and 1 ft. stroke, having been driven 320 revolutions or 640 ft. a minute, discharging the air, at a pressure of 3½ lbs. the square inch, through a tuyere of 1⅛ in. diameter, or $\frac{1}{64}$th of the area of the blowing piston. This performance was in 1850 more than double that of any ordinary engine, the total area of the tuyeres with a 90-in. blowing cylinder being, at a pressure of 3½ lbs., about 52 circular in., or $\frac{1}{156}$th of the area of blowing piston.

We are all acquainted with the tremor which is felt even in the best form of the large-sized engines; but in the experiments at a high velocity with the small-sized cylinders, not the slightest jar was felt or noise heard; it was therefore proposed to increase the speed of the piston in actual practice, from 640 to 750 ft. a minute, the length of stroke being 2 ft. in place of 1 ft.; this is somewhat under the speed of a locomotive piston at 40 miles an hour, which is about 800 ft. a minute, so that it was conceived no difficulty could present itself to this. The proposed speed of 750 ft. a minute was three times the usual speed of the blowing engines then in use (250 ft. a minute).

The construction of the engine proposed by A. Slate is shown in the accompanying drawings. Fig. 804 is a plan, and Fig. 803 an elevation of the engine, showing the pair of steam-cylinders and

803.

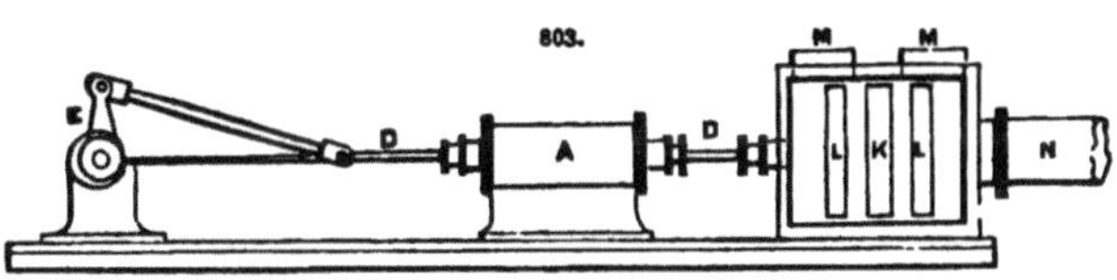

blowing cylinders; A A are the steam-cylinders, 10 in. diameter and 2 ft. stroke; B B blowing cylinders, 30 in. diameter and 2 ft. stroke, with their pistons C, fixed on the same piston-rods D, which are connected to two cranks E, fixed at right angles to each other on the same shaft. The slide-valves F of the steam-cylinders are worked by the eccentrics G on the cranked shaft, and the cranks H, at the outer ends of the same shaft, work the slide-valves I of the blowing cylinders. The centre-port K passes downwards to an external opening for the admission of the air, and the discharge-ports L L deliver into the passages M on the top of the cylinder, which communicate with the air-main N by the chest O formed between the cylinders. The piston of the blowing cylinder is intended to be made without any packing, being a light hollow cast-iron piston turned to an easy fit; and the slide-valve of the blowing cylinder to have a packing-plate at the back, working against the cover of the valve-box, with a ring of india-rubber inserted between this plate and the back of the valve, to give a little elasticity.

804.

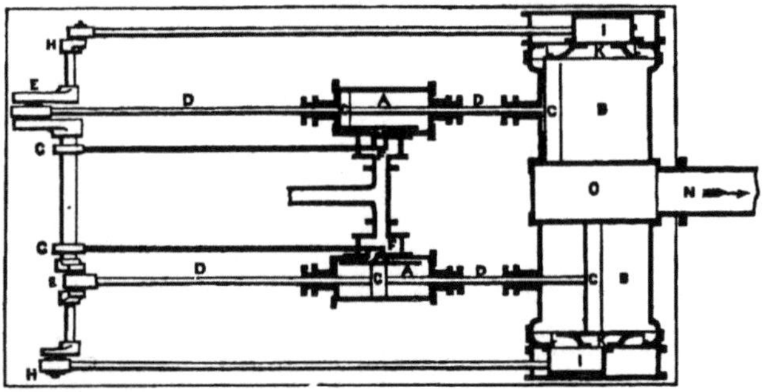

It appears that 30 in. diameter is about the most convenient size for a stroke of 2 ft.; and as it is considered an advantage to have the stroke as short as possible, to increase the regularity of the blast, the comparative cost of the different engines which follow has been taken upon this basis, $\frac{2}{10}$-in. steam-cylinders and $\frac{2}{30}$-in. blowing cylinders being reckoned equal to blow one of our largest furnaces, making 160 tons of iron a week, and having a surplus equal to blowing a cupola or refinery, as it is generally allowed that such an engine would give at 640 ft. a minute the same speed of piston as in the experiments, very nearly 30 circular in. of tuyere, at a pressure of $3\frac{1}{4}$ lbs. to the sq. in. The circular inch is used in speaking of the area of tuyere, as the blast that any furnace is taking is usually reckoned by simply squaring the diameter of the tuyere, but the pressure is taken on the square inch.

The experiments on which these calculations were founded, having been made in 1849, were repeated in 1850, and the results were found to be, as nearly as they could be measured, the same; the blowing cylinder had in the interval been driving the lathes in the pattern-shop, and the slide was found perfect. An indicator was applied with a view to test the amount of friction of the air in entering the cylinder at the high velocity, and a simple method was adopted of ascertaining this. A tuyere was made as large as the inlet-port, and the engine was driven to nearly or quite 700 ft. a minute, when the gauge showed a pressure of $\frac{1}{4}$ of a lb. to the square in., and as the friction would be the same through the same sized openings at other pressures, it follows that the loss by friction, on a pressure of blast of $3\frac{1}{4}$ lbs. the inch, would be $\frac{1}{13}$th or $6\frac{3}{4}$ per cent. loss; as the port in this case was $\frac{1}{11}$th of the area, and the port proposed is $\frac{1}{8}$th, it is assumed that the loss would not exceed 5 per cent. from this cause, or indeed from any other cause, as the friction from propelling the air through a given sized tuyere, at a given pressure, must be the same in both cases.

We extract a description of a set of six blast-engines, made for the East Indian Iron Company, from a paper by Edward A. Cowper, read before the Institute of Mechanical Engineers in 1855.

These engines were made to the plans and under the superintendence of Charles May, the consulting engineer to the East Indian Iron Company, by James Watt and Co., to the drawings prepared by E. A. Cowper.

The engines are six in number, two pairs of them being intended to blow air at 2 lbs. the square in. as a maximum pressure, and the other pair to blow air at 4 lbs. the square in. as a maximum pressure.

Fig. 805 is a side elevation of the engine complete, with crank-shaft, wheels, and other fittings.

Fig. 806 is a vertical section through the steam and air cylinders, and their valves and passages, and the branch air-pipes.

Fig. 807 shows a sectional plan taken through the air-valve, and the air-passages and branch air-pipes.

The general form and construction of the engine is that of a Pedestal or Table Engine; the air-cylinder A stands on a short pedestal, and itself forms the pedestal or table on which the steam-cylinder B stands. The foundation-plate is 6 ft. square, and carries a wrought-iron crank-shaft C in four plummer-blocks, having two light fly-wheels D D, one on each end of the shaft, and the two eccentrics E E for driving the air-valve F, one on each side of the air-cylinder, and the eccentric G for driving the steam-valve H, in the centre. The steam-piston has one piston-rod fixed in a short cross-head I at the top, and this cross-head has two other piston-rods, for driving the air-piston, which pass down outside the steam-cylinder through stuffing-boxes in the cover of the air-cylinder, and are attached to the air-piston. The long cross-head K, taking the connecting-rods to the cranks, is attached to the short cross-head by a pin, so as to allow a little freedom in case of unequal wear; the guides L L are attached to the steam-cylinder cover. V is section of air-valve.

The air-valve F is made under Archibald Slate's patent. It consists of a ring or crown valve entirely enclosing the air-cylinder, and is not self-acting by the pressure of the air in any way, but is moved by the pair of eccentrics E E at the proper times, so as to give ample passage for the air

805.

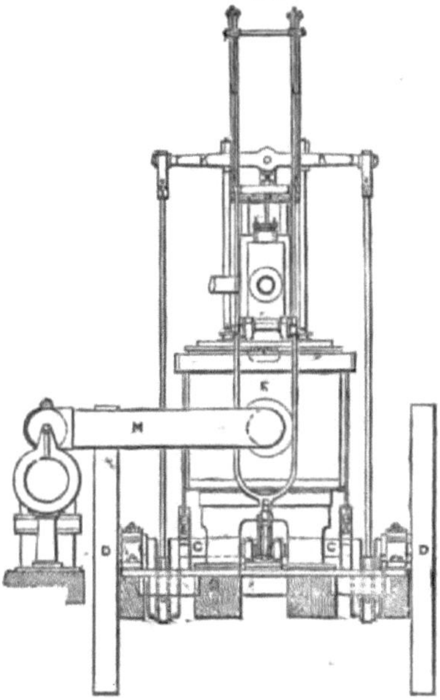

806.

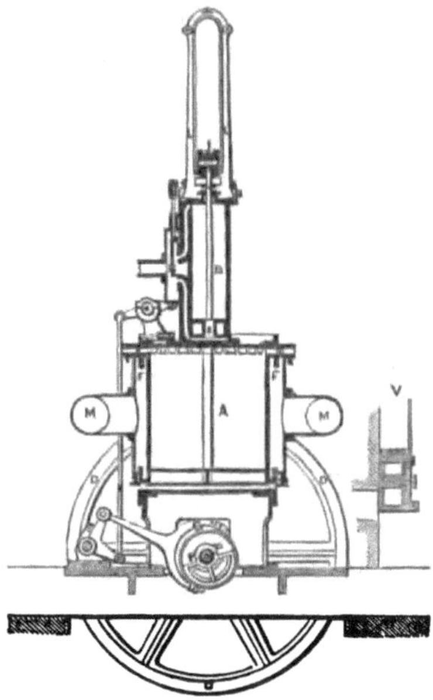

to move with the greatest freedom, and the valve has such a proportion of lap as to cause the air to be compressed up to the working pressure before it is delivered, thus giving the engine no more work to do than is necessary.

807.

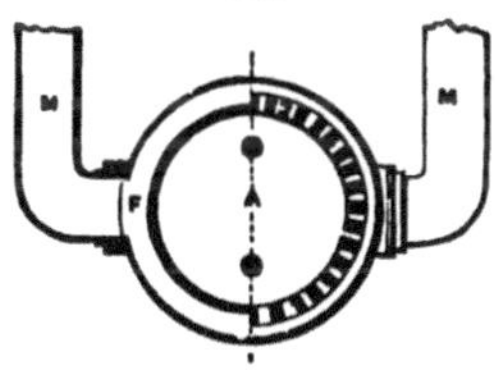

The openings or passages for the air from the air-cylinder to the valve are extremely short, and the bars between the openings are made inclined, so as to cause a regular wear on the brass packing-rings which form the rubbing-face of the valve. The body of the air-valve is made of thin sheet iron, neatly curved to two turned cast-iron rings, to which it is well secured by a great number of small bolts. These rings are bored out inside to receive the brass packing-rings before mentioned, which are secured in their places by bolts. There are no springs to the brass packing-rings, but they are bored out to a perfect fit to the outside of the air-cylinder, and are then cut into eight pieces, and, should any wear take place, they can be at once adjusted by introducing a thin sheet of paper behind them and screwing them fast in their places again. It should, however, be remarked that this valve is under totally different circumstances from any that have hitherto been made, as it is perfectly in balance, or rather it is suspended freely, slides up and down a turned cylindrical surface, and therefore there is no tendency or power to cause wear under any variation in the pressure of the air. The mode in which the two eccentrics drive the air-valve is by means of a Gymbal Ring; that is to say, there is a wrought-iron ring encircling the air-valve and attached to it by two pins opposite each other, and the eccentric rods are attached to the ring at two other points at right angles with the first: thus the air-valve is perfectly free.

The air-cylinder A is 30 in. diameter and 2 ft. 6 in. stroke, and the piston makes 30 strokes a minute. The air-piston is packed with hemp-packing, and has a ring to screw it down; the screws are so arranged that they can be got at by simply unscrewing small plugs in the cylinder-cover, when a socket-spanner can be introduced to screw the ring down. The air passes into the air-cylinder beyond the end of the valve, first at one end and then at the other, and is delivered into the hollow part of the valve, from which it escapes through two light copper branch-pipes M M, placed opposite each other, and having turned joints fitting turned collars fixed on the valve. The other ends of the pipes rest on a small surface or shelf prepared for them, and on which they slide backwards and forwards about $\frac{1}{4}$th in. These ends of the pipes are curved in the same manner as the other ends, so that the faces are in one plane, and the air-main, Fig. 805, has the faces of its branches surfaced to receive them; thus the air is taken equally from each side of the air-valve.

The steam-valve H has considerable lap, and is so proportioned as to cut off the steam just after the half-stroke and have a very free exhaust.

The boilers are on the Cornish plan, and will be chiefly used with wood as fuel, and the furnaces are made proportionately large for this purpose. The boilers are fed by a donkey engine entirely independent of the blast engine, so that they are complete in themselves, and there is no fear of getting short of water whilst the blast engines stand for tapping, at which time indeed the boiler should always be fed, if only to keep the steam down a little.

The engines having to be transported some distance up the country, a limit of weight was given, namely, 1 ton for any one part of the engine; and in accordance with this limitation the total weight of a pair of these engines is only 11 tons as compared with 25 tons, the weight of an ordinary blast engine of equal power; and the weight of the heaviest single piece of an ordinary engine is 4½ tons as compared with 1 ton, the weight of the heaviest piece in the new engines. It is, therefore, evident that the engine can be moved with the greatest facility; and the first pair put to work here for trial simply stood on some *balks* of timber, and a few small bolts through the bed-plates were sufficient to hold them and cause them to work quite steadily; whereas for the ordinary engine a strong building with massive foundations has to be erected.

The method by which a high speed for blast engines has been attained is simply that of moving the air-valves for the air, having of course very large valves and passages, instead of letting the air itself move the valves. This arrangement at once prevents all blow and jar in the working, provided that the lap and lead of the valve are properly proportioned, and allows of the piston being driven at a high velocity, and consequently its diameter may be reduced and its stroke shortened. This mode of working, combined with the fact of two engines working together as a pair with their cranks at right angles, causes such uniformity in the flow of the blast that no regulator of any kind is needed; indeed, the variation is hardly perceptible in a mercury gauge placed on a very short length of main, whereas the variation on the ordinary plan is very considerable. The pair of engines are arranged to blow 3600 cub. ft. a minute, and are speeded to 80 revolutions a minute, which with 2 ft. 6 in. stroke makes 400 ft. a minute, and this they do with the greatest ease and efficiency, owing to the exact manner in which the lap, lead, and area of passages, &c., are proportioned.

808.

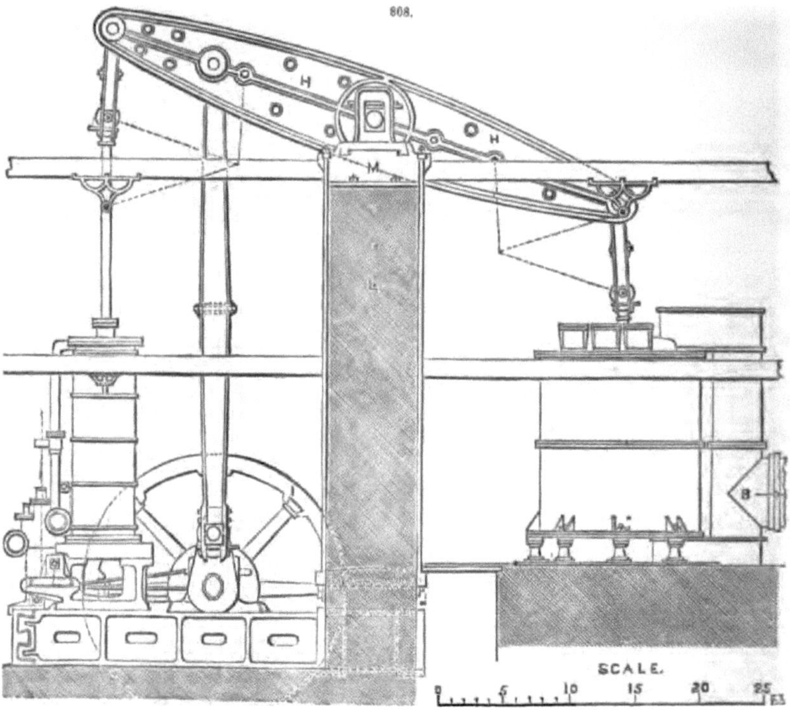

The following account of the blowing engine in use at the Dowlais Iron Works, is taken from a paper in the 'Transactions of the Institute of Mechanical Engineers,' read by Wm. Menelaus.

The blowing engine was erected in 1851, and is shown in Figs. 808 to 811. Fig. 808 is a side elevation of the engine, and Fig. 809 an end elevation. Fig. 810 is an enlarged vertical section of the blowing cylinder. Fig. 811 is a vertical section of steam-valves. The blowing cylinder is 144 in. in diameter, with a stroke of 12 ft., making 20 double strokes a minute, the pressure of the blast being $3\frac{1}{4}$ lbs. the square in. The discharge-pipe B is 5 ft. diameter, and about 140 yds. long, thus answering the purpose of a regulator. The area of the entrance air-valves is 56 sq. ft., and of the delivery air-valves 16 sq. ft. The quantity of air discharged at the above pressure is about 44,000 cub. ft. a minute.

The steam-cylinder C is 55 in. diameter, and has a stroke of 13 ft., with a steam-pressure of 60 lbs. the square in., and working up to 650 horse-power. The steam is cut off when the piston has made about one-third of its stroke, by means of a common gridiron-valve A near the back of the slide-valve E, as shown enlarged in Fig. 811; there is also on one side of the nozzle a small separate slide-valve B, for moving the engine by hand when starting. The cylinder-ports are 24 in. wide by 5 in. long, and the slide-valve E has a stroke of 11 in. with $\frac{1}{8}$-in. lap. The engine

809.

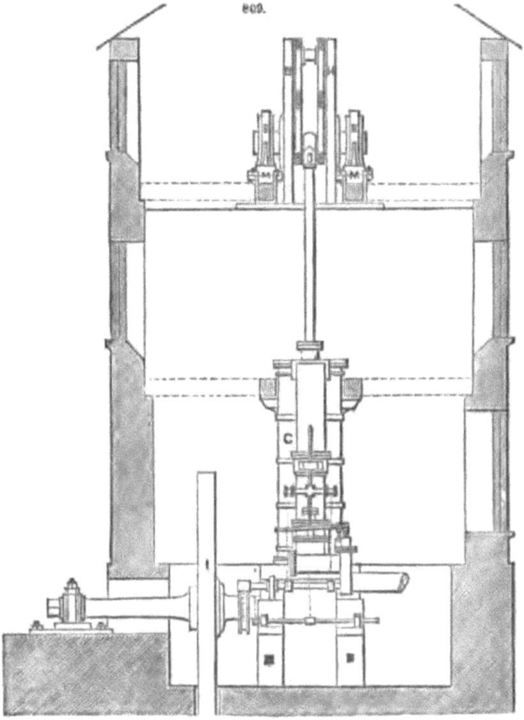

810.

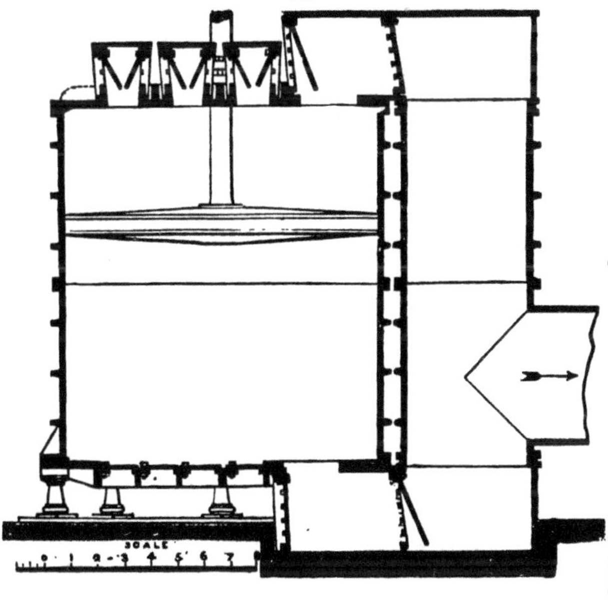

811.

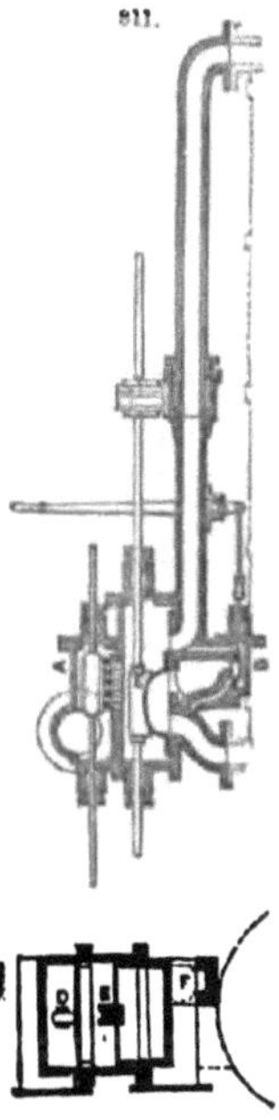

is non-condensing, and the steam is discharged into a cylindrical heating tank 7 ft. diameter and 36 ft. long, containing the feed-water from which the boilers are supplied. Under the steam-cylinder C there are about 75 tons of cast-iron framing, and 10,000 cub. ft. of limestone walling in large blocks, some of them weighing several tons each.

The beam H is cast in two parts, of about 16½ tons each, the total weight upon the beam-gudgeons being 44 tons; it is 40 ft. 1 in. long. from outside centre to outside centre, and is connected to the crank on the fly-wheel shaft I by an oak connecting-rod, strengthened from end to end by wrought-iron straps. The beam is supported by a wall L across the house, 7 ft. thick, built of dressed limestone blocks, to which the pedestals M are fastened down by twelve screw-bolts of 3 in. diameter. The fly-wheel I is 22 ft. diameter, and weighs about 35 tons.

Eight Cornish boilers are employed to supply the steam, each 42 ft. long and 7 ft. diameter, made of $\frac{3}{16}$-in. best Staffordshire plates, and having from end to end a single 4-ft. tube, in which is the fire-grate, 9 ft. long.

For some time this engine supplied blast to eight *furnaces* of large size, varying from 16 to 18 ft. across the *boshes*; it is now blowing, with three other engines of small dimensions, twelve furnaces, some of which make upwards of 235 tons of good forge pig-iron a week, the weekly make of the twelve furnaces being about 2000 tons of forge pig-iron. With the exception of the cylinders, made and fitted at the Perran Foundry, Truro, this engine and boilers were made at the Dowlais Iron Works, and erected according to the design and under the superintendence of Samuel Truran, the Company's engineer.

Blowing Engines at Creusot. — Schneider and Co's. works at Creusot include amongst their plant seven blowing engines, three of these being horizontal engines of an old type, and the other four direct-acting vertical engines; one of these latter is shown in Fig. 812. It will be seen that the blowing cylinder A, which is 108¼ in. in diameter, is placed below the floor of the engine-house, the steam-cylinder, which is 47¼ in. diameter, being overhead. The two pistons B B are fixed on one rod, and their stroke is 6 ft. 6⅝ in. The piston-rod C passes through the top of the steam-cylinder, and is attached at its upper end to a cross-head working in suitable guides. From the cross-head a *connecting-rod* extends to the *crank-shaft*, the centre of the latter being 25 ft. 10¼ in. above the floor of the engine-room, and no less than 44 ft. 3½ in. above the base of the engine.

812.

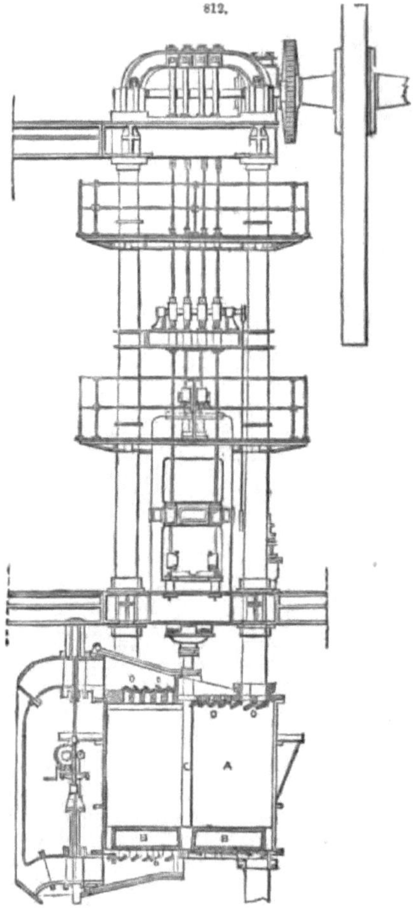

813.

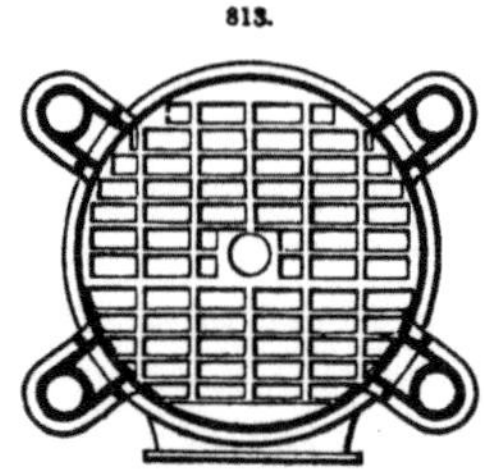

The admission of the steam to, and its release from, the cylinder are effected by equilibrium-*valves* worked by *cams* placed on a counter-shaft, which derives its motion from the crank-shaft through spur-*gearing*. The engine is of the non-condensing class, and the steam, which is supplied at a pressure of 60 lbs. the sq. in., is cut off at one-fourth of the stroke. The speed is usually 15 revolutions a minute.

814.

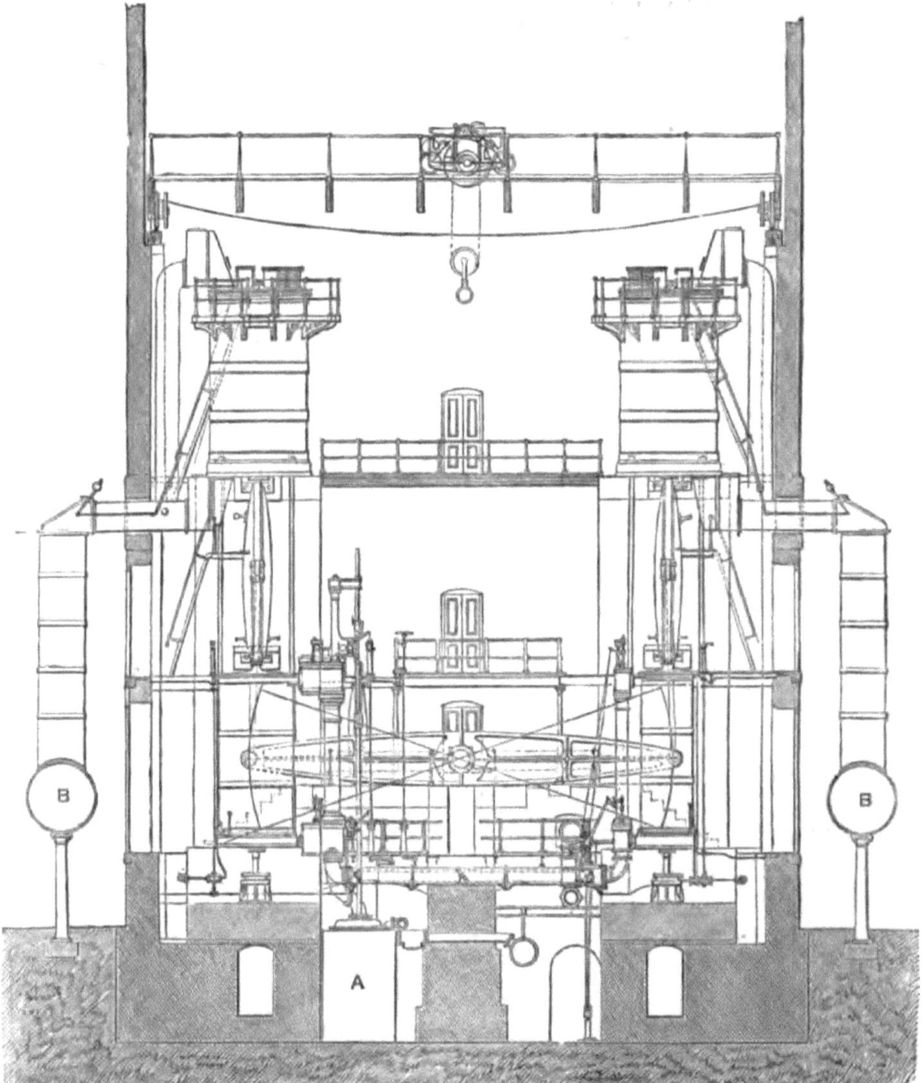

The Kirkless Hall Blowing Engine, designed by Robert Wilson.

The blowing cylinder is fitted with a number of small flap-valves, O, O, arranged on each cover, as shown by Figs. 812, 813. One-half of each cover, it will be noticed, is devoted to the inlet, and the other half to the delivery valves. The blast is delivered at a pressure of 6¼ in. of mercury, or rather more than 3 lbs the sq. in., and the quantity delivered is 90 per cent. of that due to the capacity of the blowing cylinder. The four engines, which are appropriately named the Simoun, Sirocco, Mistral, and Ouragan, are placed in one engine-house, and they serve to supply blast to twelve blast furnaces.

The Kirkless Hall Blowing Engines.—These fine engines were constructed by Naysmith, Wilson Company, from the designs of Robert Wilson, for the Wigan Iron and Coal Company. Their whole arrangement is shown in Fig. 814. The engine-house is a handsome detached structure about 25 yds. long., 60 ft. wide, and 72 ft. high, and is entered by means of a large square tower, having an internal spiral staircase, with doors communicating with the several galleries in the engine-house. On the top of this tower is placed a balcony, from which an extensive view of the surrounding country is obtained. The interior of this tower is entered by a flight of stone steps. The first impression on entering the engine-house is one of complete astonishment, there being little of the ordinary appearance of an engine-house to be seen. There are two handsome iron galleries extending round the whole of the interior; some feet below the first are seen the immense engine-beams, each beam being about 38 ft. long, and weighing upwards of 20 tons. These beams, notwithstanding their enormous size and weight, move as easily and gently as if they were the merest toys, and this without the usual control of a fly-wheel, which, in this instance, is entirely dispensed with. Here we see the beautiful and novel valve arrangement by means of which these monster engines are worked. When the engines were first designed, the valves were intended to be worked by the ordinary tappit-motion, because, having no rotary motion in any of their parts, the application of any other but this old arrangement was considered impracticable; but, on starting the engines, it was discovered that this motion did not allow sufficient latitude to admit of their being worked at the different speeds required to suit the varying number of *blast furnaces* that might from time to time be in operation. Robert Wilson seeing that the engines could not be worked satisfactorily with the old motion, at once applied himself to the analyzation of the difficulty, with a view of producing a more efficient arrangement for working the engines, and the result of his investigations was the invention of a modification of the Cornish valve-gear, which, on being applied, was immediately found to answer most admirably every requirement.

The engine-house contains three pairs of engines, each pair consisting of one high-pressure steam-cylinder, 45 in. diameter; one low-pressure steam-cylinder, 66 in. diameter; and two blowing cylinders, each 100 in. diameter—one of the latter being placed about 17 ft. above, and directly over each steam-cylinder; the stroke of all these cylinders being 12 ft.: the steam-cylinders are worked together by means of the large beams before described, of which there are two to each pair of engines. The high-pressure cylinders are placed on the left-hand side of the engine-house, and the low-pressure ones on the right-hand side, and are connected by beams working with *connecting-rods* from the cross-heads. The motion which works the valve-gear is on the low-pressure side, and is carried across the engine-house beneath the floor, so that by means of one set of hand-gear the eight valves of the two cylinders are easily controlled by one man, or worked by the engine itself, as may be desired. This, in itself, is considered to be a triumph of mechanical skill, and the smoothness of action, the perfect accuracy of every part, and the superior style of workmanship and finish of this motion, is of such a character as to attract the attention of even the most unprofessional spectator. The six air-cylinders, each weighing upwards of 25 tons, are placed upon stone piers, and on a level with the bottom of these cylinders is fixed the second gallery, and round the tops of them are fixed airy-like but substantial balconies, which are reached by means of an iron staircase at each end. These cylinders are fine specimens of English workmanship; in fact, there are few engineering establishments besides the Bridgewater Foundry where such cylinders could have been cast, bored, and finished, in the style these are.

815.

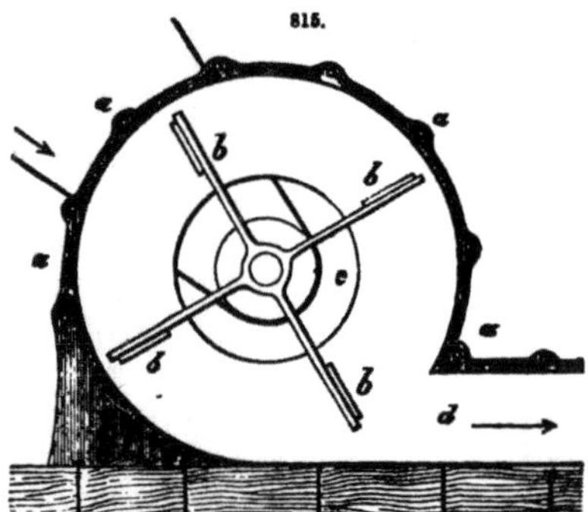

Common Fan.

Fan Blast Machines.—These machines are very common; they are used to urge the fire of steam *boilers*, and at *puddling*, *re-heating*, and *cupola furnaces*, where anthracite is burned; and at cupola furnaces, where coke is used for re-melting pig iron in foundries. Fig. 815 shows a section of a common fan. The two sides of the case are, in most instances, made of cast iron, and held together by the screw bolts *a*, *a*, *a*, *a*. These bolts reach through both sides, and their length is therefore equal to the width of the machine, which varies from 6 to 20 in. The space between the sides is occupied by a strip of sheet iron; this strip determines the width of the machine, and reaches all round the fan, forming the circular part of the case. The wings of the fan, marked *b*, *b*, *b*, *b*, are sometimes of sheet iron; they are fastened to iron arms set upon the axis, and rotate with it, and they occupy a different position in different fans. Some are set radially, others inclined more or less tangentially. Some are straight; others have a slight curvature. On the whole, no marked difference between the one form of wings and the other results, so far as effect is concerned, if no blunders against the laws of mechanics are made. The fans with curved and short wings do not make so much noise as those with straight, radial, and long wings. The opening *c*, which receives the air, to be pressed out at *d*, must be of greater or less diameter, according to the size of the fan or width of the wings. Broad fans require such an opening on each side. Small fans, of but 6 or 8 in. in width, work sufficiently well with one inlet. The diameter of a fan is seldom more than 3 ft.; and from various reasons it can be shown that a larger diameter is of no advantage. The number of revolutions of the axis, or the speed of the wings, is very seldom less than 700 a minute; this speed may be considered sufficient for the blast of a blacksmith's forge and small furnaces. At large furnaces or cupolas, we frequently find the number of revolutions as many as 1800 a minute. The motion of the axis is generally produced by means of a leather or india-rubber belt, and a pulley of from 4 to 6 in. in diameter.

Among the great variety of forms in which these fans have made their appearance, one, shown in Fig. 816, is worthy of notice. The wings of this fan are encased in a separate box; a wheel is thus formed, which rotates in the outer box. Fig. 816 shows a horizontal section through the axis. The wings are thus connected, and form a closed wheel, in which the air is whirled round, and thrown out at the periphery. The inner case, which revolves with the wings, is to be fitted as closely as possible to the outer case, at the centre near *a, a, a, a*; for no packing can, in this case, be applied, and there is a liability of losing blast if the two circles do not fit well.

816.

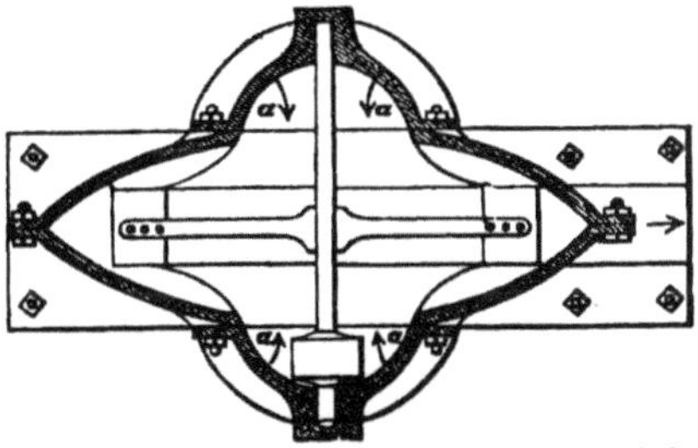

As the building of this apparatus receives much attention in machine shops, and as the leading principles involved in its construction are very little known, we shall designate such points as may be deemed of great importance by those who manufacture fans, which is frequently the lot of the iron manufacturer himself. The outward case should be strong and heavy; and the interior machinery, which revolves, as light as possible. For this reason it should be made of the best wrought iron, or, what is preferable, of steel. Four wings produce quite as much effect as a greater number. It is, therefore, useless to exceed that number. The greatest attention must be paid to the gudgeons and pans; it is advisable to make both of steel, or, better still, to run the two ends of the shaft in steel points. The wings are to be exactly at equal distances, and of equal weight; otherwise the strongest case will be shaken. The surface of each of the wings should be at least twice as large as the opening of the nozzle at the blowpipe.

The pressure of the blast from a fan is proportional to the square of the speed of the wings, with a given diameter of the fan. The pressure gains simply in the ratio of the diameter, or speed, provided there is the same number of revolutions. The increase of speed is in the ratio of the increase of the radius. The pressure in the blast is produced by centrifugal force. The atoms of air, after being whirled round by the wings, are thrown out at their periphery by a force equal to the centrifugal force resulting from the speed of the wings. This centrifugal force may be simply expressed by $\frac{c^2}{2gr}$; c is the speed in feet per second; g the speed of gravitation in the first second; and r the radius of the fan. According to this, the effects of a fan ought to be far greater than they actually are; therefore a remarkable loss of power must take place in these machines. It is thus very clear that the increase of diameter augments the effect of the machine in a numerical proportion, while an increase of revolutions adds to the effect in the proportion of the square. It is also very clear that an increased diameter greatly increases the friction, while the increase of speed does not augment it in the least. The friction, in these machines, is the greatest objection to their use; therefore the movable parts should be as light as possible. Friction increases in the ratio of the weight, where the materials are the same, but not with an augmentation of speed, at least, not in the same ratio. From practical observation, the following formula has been deduced, in which a is the speed of the fan, that is to say, it represents the number of feet which the wings make in a second; b, the surface of the nozzle; c, the surface of a wing; and d, the velocity of the escaping blast. This formula we conceive to be the proper dimensions of a fan: $d = (\cdot 73) \times \frac{a\sqrt{c}}{\sqrt{b}}$. See Centrifugal Pump.

Gwynne and Co.'s improved combined steam gas-exhauster and air-blower, shown in Figs. 817 to 821, is a design of one constructed to pass from 10,000 to 12,000 cub. ft. an hour; but these exhausters can be constructed to any capacity that may be required. On Figs. 817 to 821, B is the case, C the spindle, D the bed-plate, E suction-pipe, F discharge-pipe, G standard for engine, H H slides to piston, I steam-cylinder, K slide-jacket, L disc, M piston-rod, N connecting-rod, O eccentric-rod, P steam exhaust-pipe.

This blower is composed of an outer casing or cylinder B, fitted with top and bottom plates, one of which plates is constructed with a stuffing-box and gland, through which passes the spindle C, on which is firmly fixed an inner cylinder, dotted in the figures. This inner cylinder is slotted through the centre, which slot is fitted with two sliding plates or pistons made air-tight to the slots and working air-tight against the inner periphery of the outer cylinder.

The two cylinders, as shown in the figures, are set eccentrically to each other, so that the bottom of the inner cylinder touches the bottom of the outer cylinder.

When the spindle C is set in motion by the steam-engine or other method, as may be arranged, the slides continue to pass in and out, and thereby take up the air or gas which enters through the pipe E and is expelled through the pipe F.

The machine is fixed to a firm cast-iron bed-plate, so as to be perfectly portable; and combining in itself its own motive-power, it requires little or no foundation.

These blowers have been applied most successfully to gas-works and mining purposes, and may be constructed to work up to considerable pressure, if required. One of these exhausters is being very successfully worked by the Llanelly Gas Company.

Figs. 817, 818, 819, show the above-specified arrangement.

Figs. 820, 821, illustrate an arrangement to be driven by a strap or by hand-power.

817.

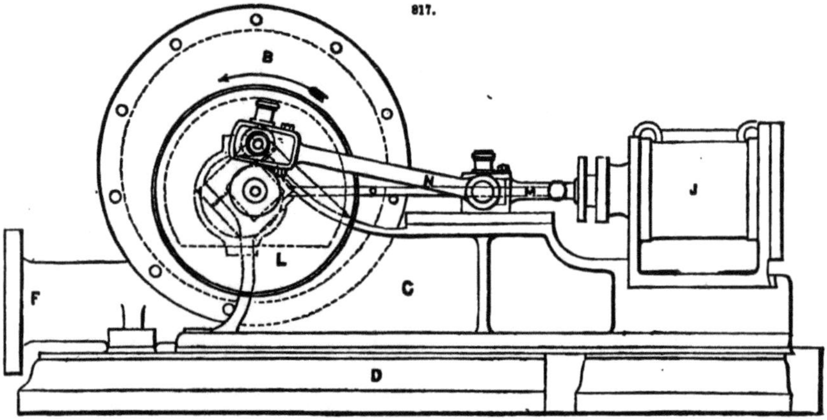

818.

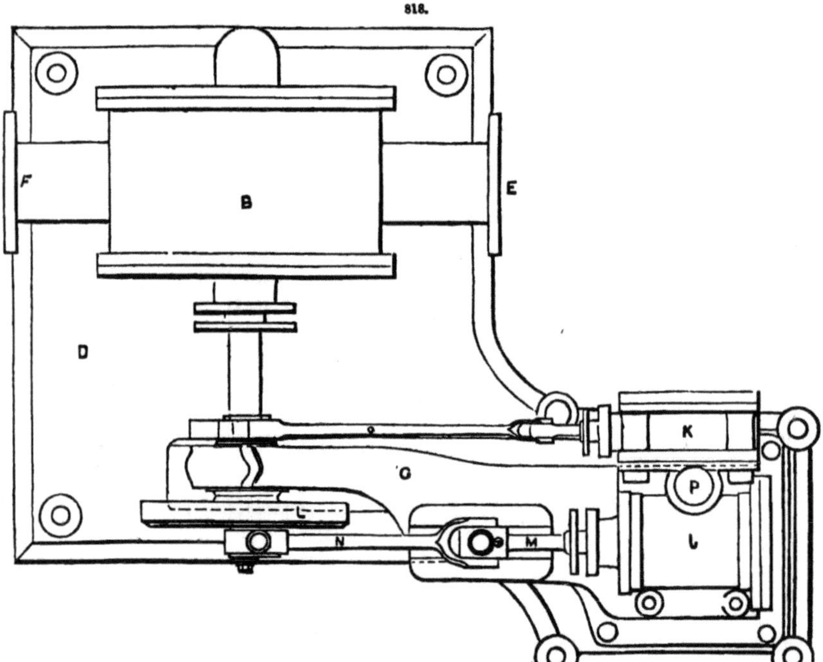

819.

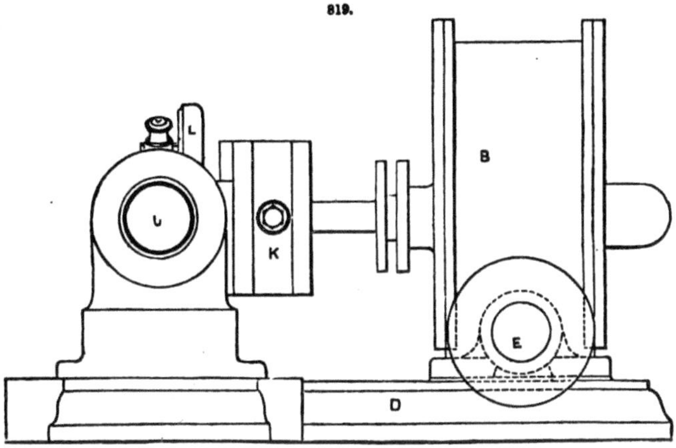

820.

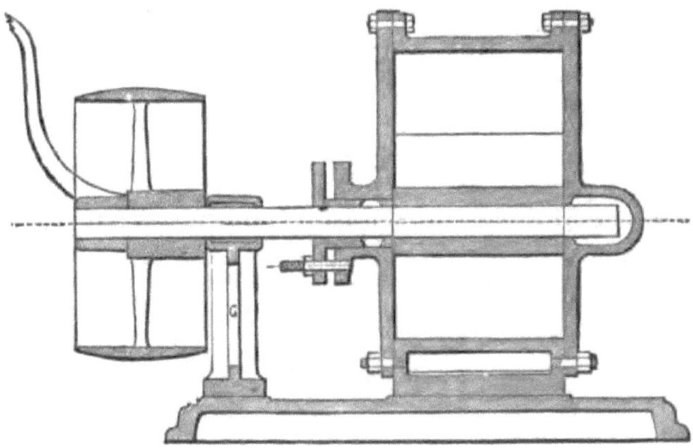

821.

BLOW-OFF COCK. Fr., *Robinet de vidange*; Ger., *Ausblashahn*; Ital., *Chiave di sfogo.*

See Details of Engines.

BLOWPIPE. Fr., *Chalumeau*; Ger., *Löthrohr*; Ital., *Tubo ferruminatorio*; Span., *Canutillo.*

See Assaying. Oxy-hydrogen Blowpipe.

BOARD. Fr., *Planche*; Ger., *Brett*; Ital., *Tavola*; Span., *Tabla.*

Board, or *piece.*—Timber cut into thin slabs less than 2½ in. in thickness and more than 4 in. wide are called boards. The term is usually applied to fir and elm, while the same thickness of oak, mahogany, and so on, is generally called *plank.*

822.

Boards cut from 7-in. stuff are called batten-boards, from 9-in. stuff deal-boards, and from 11-in. stuff plank-boards.

Fir boards 1¼ in. thick are called whole deal, and those which are full half-an-inch thick are called slit deal. Boards, Fig. 822, which are thinner on one edge than the other are called feather-edge boards.

BOARD and BRACE WORK. Fr., *Lier l'empanon avec l'arêtier*; Ger., *Verbindung der Halbsparren mit den Walmsparren*; Ital., *Commettitura a canale e tavola sottile.*

This work consists, Fig. 823, of boards with grooved edges, into which thinner boards are inserted.

823.

BOARDING. Fr., *Planchéiage*; Ger., *Bretterverschlag, Bretterwand*; Ital., *Tavolato.*

Boarding.—This is a general term for various kinds of work to which boards are applied, as *gutter-boarding, slate-boarding, weather-boarding, sound-boarding*, and so on.

BOARDING-JOISTS. Fr., *Soliveaux*; Ger., *Dielenlager, Polsterholz*; Ital., *Travi del palco.*

See Joists.

BOASTING. Fr., *Ebaucher*; Ger., *Roh-Behauen*; Ital., *Sbozzare.*

Boasting, in stone-cutting, is paring the stone with a broad chisel and mallet, so as to leave regular marks like ribbands or small chequers. It is also applied to a margin-draught round the edges of the stone in hammer-dressed work.

BODY-PLAN. Fr., *Section verticale*; Ger., *Spantenrisz*; Ital., *Proiezione verticale.*

In ship-building, the body-plan, Fig. 824, is descriptive of the largest vertical and athwart-ship section of a ship. This plan fixes by *orthographic projection* the heights and widths of the principal lines of a ship.

The orthographic projection is that projection which is made by drawing lines from every point to be projected perpendicular to the plane of projection.

A horizontal line supposed to be drawn about a ship's bottom at the surface of the water is called the *water-line*, which is higher or lower according to the depth of water necessary to float the vessel; light water-line, the lowest water-line, or that of a vessel when unloaded; load water-line, the highest water-line, or that of a loaded vessel.

The *sheer-plan* is an orthographic projection of the lines of a ship on a vertical longitudinal plane passing through the middle line of the vessel.

Body-plan, sheer-plan, and *half-breadth plan*, are on sectional planes supposed to pass through, at right angles to each other, the largest portions of the principal dimensions of the ship. The *half-breadth plan* is descriptive of half the widest and longest level section in the ship. This plane is a horizontal one passing through the length of the ship at the height of the greatest breadth. On this plane the position of any point in the vessel may be fixed by orthographic projection, as to width and length. The three planes on which the projections are established are rectangular co-ordinate planes passing through the centre of the vessel; the three perpendicular lines which transfer any given point to each of these planes are termed the co-ordinates of that point. Naval architects, John Scott Russell excepted, lay down the lines of a ship on the three co-ordinate planes by a good old rule, called the rule-of-thumb, which has not as yet been submitted to mathematical investigation.

In passing, it may be necessary to remark that one naval architect, J. W. Griffiths, works from a model which he whittles out of a piece of soft wood, by the good old rule before named. Upon this system Griffiths has written a large work.

Scott Russell lays down the lines of a ship by well-defined laws, which he establishes by abstract reasoning and experiment.

Russell is a great performer as well as a great thinker; he stands out in bold relief from among the great men of the extraordinary time in which we live. We give from his great work on Naval Architecture his method of laying down the lines of a ship, on what he terms the *compound wave-principle.* Scott Russell says:—

I will begin by taking the easiest problem which can ever be submitted to the constructor of a ship. I will suppose a case, which very frequently occurs in practice, that a certain length of ship is to be built—a certain breadth is given—a certain deepest draught of water and a certain lightest draught of water, and that these are about the ordinary proportions of a ship;—that no particular weight is to be carried, or work to be done, beyond sailing well, or steaming at a moderate speed, and that the purpose to be served is a fair, common mercantile trade, such as ordinary vessels will moderately well perform;—and I will take for granted that the owner expects from the naval architect, what he may reasonably expect from a man of science and skill, that his vessel will be somewhat faster, easier, safer, and more economical, and therefore somewhat more valuable, than a vessel built, without design or calculation, by an unskilled man. This is a task of the most ordinary kind to the naval architect.

There are two ways in which he may set about building his vessel: he may either take the model of the vessel, which is already the best that has been applied to the trade in question, and improve upon her; or he may at once throw all precedent overboard, and give his employer an entirely new design. The undertaking then will speedily shape itself as follows:—His extreme length and extreme breadth being given, he may determine a midship section, such as will give him the requisite carrying-power, with good sea-going qualities. Next, he will determine a water-line, which will give the highest speed and least resistance of which that length admits; or he may decide to fit her for a given speed only, and adopt a water-line of greater capacity fit for that slower speed. Thirdly, he will adopt a convenient form of deck, for the use and navigation of the ship; and on these principal points he will fill in, what I will call "a skeleton design," and frame an approximate calculation of the qualities of the ship, which we will also call "the skeleton calculation."

To Construct the Midship Section.—It is in the choice of midship section that the naval architect is left free to exercise, with the greatest liberty, his own absolute judgment. In the water-line he has little or no choice; Nature has fixed that for him. If he meddle with it, he shows his ignorance or presumption, and Nature sends her due punishment, by refusing to deal kindly with the spoilt water-line; but the midship section he may vary to his heart's content. He may give the ship every sort of quality by choosing it ill or well; and with a given water-line he may produce all sorts of ships.

To illustrate this latitude of choice, and to follow out the consequences which arise from each kind of choice, I will take three midship sections, and carry them through all the stages of design to their ultimate consequence; and I will further suppose it necessary that they should all have the greatest speed the length will allow.

The first of these sections is to carry extremely little cargo, to have little room, but to go as fast as she can be made to go with all the sail and steam-power she can carry. These are the practical conditions of the yacht or the cruiser, the opium clipper or the privateer. What such a vessel requires can readily be contrived, for the conditions given *make* the midship section, and leave very little to the choice of the constructor. Such a ship must be all shoulder and keel, and nothing else;—she will be like a racehorse, lanky and leggy: by being all shoulder, with very little under-water body to carry, she will possess the maximum of power with the minimum of weight:—her fault will be, that she must have an enormous keel, to prevent her from going to leeward; and this great mass of dead wood, or of solid iron keel, exposes a large surface to the adhesion and friction of the water. Nevertheless, it is the form of greatest power with least weight. The bottom of this midship section may be formed in two ways,—it may either be made elliptical, to have a minimum of skin for adhesion, and be reconciled to this deep keel by two hollow curves; or it may be reconciled to the keel by a long wedge-bottom. I prefer the elliptical bottom for iron ships; but the other, or peg-top shape, has been much used in wooden ones. See Figs. 824, 825, 826.

I will next suppose that the capacity, thus got, is too small for carrying remunerative cargo, and that a cargo hold, of a capacity more usual with mercantile vessels, is required; in that case I keep the same shoulders, and give a larger under-water body. See Figs. 827, 828, 829.

I will now take a third design. The ship is to carry as much as is not inconsistent with good sea-going qualities; and she is to have room, also, for boilers and machinery of considerable power. This requires her sides to be nearly upright, her bottom dead flat amidships, with only so much off her bilges as will not be inconsistent with what she is to receive inside. This the form and arrangement of her boilers and machinery will generally determine, and the boilers and machinery in such a vessel should be treated as ballast, and kept low. See Figs. 830, 831, 832.

In regard to these three midship sections, it is to be noticed, that they are prescribed in some measure by the uses of the ship; but the forms I have mentioned come entirely from the judgment of the constructor, and whether they have been wisely or injudiciously selected, must be judged of after the calculations have been made of the various qualities to which they give rise.

But there are one or two points which occur to a constructor, at the first glance at these forms of under-water body. It is plain, the first is easiest, and the last hardest, to drive. It would require much more sail to drive the two last, than the first; and it is equally plain, that the first is much better able to carry sail than the last. The area of midship section of under-water body is the thing to be driven; the area of the sail is the driving-power: but the power of the shoulders to carry the sail upright limits the quantity of sail the ship can carry. The bulk of the under-water body brings with it two evils,—resistance to being driven through the water, and under-water buoyancy tending to upset the ship.

It is plain, from these considerations, that the first shape is suited for a fast ship under sail alone, the last is suited for a fast ship under steam alone, and the middle form may do for a moderate quantity of both—for what is called "the mixed system."

Of these three vessels we may also form the following augury, before proceeding to precise calculation. The first ship may be powerful, weatherly, lively, and fast. The last vessel may be tender, easy, sluggish, and roomy. By a proper mixture, we may obtain, in the intermediate vessel, any compromise among these qualities for which we have a fancy. In this choice there is ample room for the display of skill and the exercise of judgment; but it will be first necessary to complete our skeleton design and our trial calculations.

We have said nothing as yet about the parts of midship section above water; but it will be noticed, that these grow naturally out of the form adopted under water; and it will be observed, that we have proportioned the above-water body to the under-water body. The object of this is to give adequate lifting-power in a sea-way, in proportion to the heavier under-water body: but of this we shall have more to say.

In three designs, the midship section (so-called) is far from being actually amidships, being placed at the point of greatest breadth, or nearer the stern than the bow, in the proportion of 4 to 6.

To Construct the Chief Water-line.—For the side-view of a ship, or vertical section, draw a horizontal line, representing her whole length at the main water-line, and erect at the ends of it two perpendiculars. This line I shall call henceforth *the length of the ship, the length between the perpendiculars, or the construction length.* These perpendiculars, also, are to be ruling elements of construction, and will be called *the perpendiculars.* Divide the length between the perpendiculars into ten equal parts; take four of these abaft for the length of the run, and six of these forward for the length of the entrance. Describe a semicircle on each half of the chief breadth. Divide the length of entrance and this semicircle into any, the same number, of equal parts; the distance of the water-line from the centre-line, opposite each division in length of the entrance, will be the distance of each corresponding division of the semicircle from the same centre-line, and a line through all the points thus found will be the true wave water-line of the bow. The water-line of the run will be different from this in one respect only,—it will be necessary to draw parallel lines to the centre-line through each of these divisions of the semicircle, which it may be convenient to call *the semicircle of construction*, as shown in Figs. 826, 829, 832. On each of these lines points may be found, just as if it were a bow-line. These lines must be prolonged aft beyond the points thus found to a distance equal to that part of the line intercepted between the semicircle of construction and the main breadth. The last-found points in the parallel lines give the line of main breadth.

Thus we have formed the chief water-line of the bow and of the stern, or the lines of entrance and of run for the greatest speed which the given length will admit. The lines thus given are absolute, and will admit of no deviation without some loss. Nevertheless, some modification in the application of these lines may be admitted as expedient, and one of them is obvious. It will be seen that the point of the bow is so extremely sharp, that it would be in continual danger of cutting everything which it touched, and, being as fine as a razor, would run risk of being crushed by rough usage. The stem also, for the rough work of a ship, must be of a considerable thickness; and the practical question at once arises: How shall we alter the line at the bow to get this thickness? Shall we cut the fine part off, and shorten the vessel? To this the answer is, that if you do, the vessel will have become too short for the length settled. The way I use, is to draw the bow to a few feet longer than I intend to retain, and then to cut off the excessive length of this water-line to the exact length that I want to keep. I thus find a thickness of a few inches remaining between the two sides of the water-line, which is just thick enough for the materials of the stem: by this means I get the extreme length I require, and I also get the strength of stem, which is necessary for durability and bad usage; and it will be noticed that the bow, which I thus gain, is of slightly greater capacity than the first attenuated line. This I call, therefore, *the corrected water-line of the bow.* No such adjustment is necessary at the stern: it is enough there to insert the stern-post, simply by increasing the breadth between the lines to admit the thickness of the stern-post,—a deviation insufficient to cause a sensible difference in the performance of the ship.

There is another modification of these water-lines of the fore-and-after bodies, which may require further consideration. We have put both entrance and run in the same plane, with the intention of leaving them so in the construction of the ship; but this is only our first intention, and we must be prepared afterwards, on good reason shown, to change it. The main water-line may require to go lower or higher in the vessel than we now propose, and the after-part of it may have to go higher or lower than the fore-part, in order to gain some advantage; but we shall effect this change, if necessary, without in any way altering the character of the line already drawn: we shall only alter the height at which it should be placed, and that we shall not do unless required.

The Sheer-plan.—This gives the entire outline of the ship, as we look at her sideways, Figs. 825, 828, 831. The top line, or upper boundary, is the line of her deck or bulwark, or, in short, the top of the ship, which must be laid down in order to construct the chief buttock-line in section (4); the bottom is the line of her keel; the front is the line of her stem, or cutwater; and the after-part is the line of her stern-post.

I begin with the stem. I make that line follow the form of the chief buttock-line, and gradually grow out of it. This ought to be so, because buttock-lines bound equal thicknesses of the ship, and a stem is merely a thin slice of the ship, and therefore follows one of the buttock-lines. As a matter of beauty and of reason, therefore, it must be made a buttock-line, in order that its outline may harmonize with the general form.

The form of the stem, therefore, will depend on the decision taken with regard to the deck-line and the buttock-line. If the deck-line be kept well aft, so that the main buttock-line tumbles home, the stem above the water will be curved backwards, and so be in unison with the tumble-home bow; and it will be the contrary, if the fashionable clipper-bow be adopted. Above the water, however, the mere form of the stem itself is a matter, to some considerable extent, of taste or fancy. If the general character of the bow give good buttock-lines, it will not much matter whether the stem to which they are joined curve out or in, except that it will be always more comely that the stem should harmonize with the character of the bow, of which it forms a conspicuous outline. Men who hesitate to give it a decided character, express their imbecility by leaving it perpendicular.

Below the water, on the contrary, the form of the bow is of the greatest practical value. It has long been usual in this country to carry the stem down to an angle with the keel, to continue the keel far forward to meet the stem, and so form what is called *the fore-foot* of the ship, giving it a great gripe, or hold, of the water. This gripe and fore-foot have every bad quality,—being weak in structure, and making the vessel hard to steer. I systematically cut it all off, following the shape of the other buttock-lines; and, further, I carry this rounding a great way back,—perhaps to one-sixth of the whole length of the ship. By this I not only diminish the fore-gripe and ease the steerage, but I keep the stem and the fore-foot out of harm's way; and I have often known this both to save repairs and to contribute to the safety of the ship. When turning round in narrow channels, when steering in intricate or shallow waters, or performing evolutions under difficult circumstances, the fact that there is no thin, protruding part near the bottom, to touch or strike the ground, to be broken off, or to impede or alter the movement or direction of the ship, is often of the very highest consequence. By a gentle curve at the stem, therefore, I keep the fore-keel and the fore-foot out of harm's way; and I generally do the same at the stern, where dead wood can be spared. By curving the after-part of the keel upwards, like the stem, we may often save both keel and rudder, and we certainly facilitate the ship's coming round. Of course it is done to a much less extent at the stern, as several feet of gripe at the bow will correspond with a few inches of trim by the stern.

But, while I thus curve the keel and fore-foot, I keep the whole central part of the keel perfectly straight, but for no other purpose than to be able to support the middle of the ship on blocks in the dock. This is an obvious necessity, for otherwise the keel would rest only on points, instead of being uniformly supported; and in the dock it is rather an advantage than otherwise, that the two ends of the ship should not be borne by the blocks, unless they should require special repairs, for which purpose they must be propped up when needful.

We have as yet said nothing about the inclination of the keel. It is not necessary that it should be parallel to the water-line, except where there is a narrow limit to the extreme draught of water, in which case the keel should be parallel to the load water-line: in most other cases it should incline downwards at the stern, so as to draw more water abaft than forward. In the case of screw-steamers of large power, it is imperative to have this draught, in order to get the screw sufficiently large and sufficiently under water for effective power; and further, it is convenient, in sailing vessels, to be able to carry a large sail area on the after-part of the ship, for which greater depth of keel aft than forward affords the necessary facility. It is convenient frequently, for the same purposes, that a vessel, when light, should draw very much more water aft than forward, and that her lading should bring her down gradually to even keel. It is usual to call this greater draught aft than forward, technically, *the difference of the ship;* and it is reckoned a main element in her trim.

Another element in the sheer-plan is the rake of the stern-post, and this also is a matter in which great licence can be allowed to the constructor, as it affords him great resources, out of which to further the uses of his ship. He may either plant his stern-post straight up and down, so as to make the rudder pivot fairly; or he may throw the head of the rudder back behind the perpendicular, at an angle; or he may throw the heel of the rudder forward of the perpendicular: indeed, he may make the line of the rudder cross the perpendicular at any angle he may choose. In the first case, he will maintain the balance of his original draught. In the second case, he will extend the dead wood and increase the lateral resistance of the ship to leewardliness. In the third case, he will diminish the lateral resistance, but increase handiness. In every intermediate degree between these two, he will gain one of these qualities at the sacrifice of the other.

The effect of this inclination on the rudder itself must not, however, be forgotten. The inclination of a rudder increases its power to turn the ship, but it also increases the resistance which the application of rudder offers in every degree to the progress of the ship through the water. The action of the rudder, as has been already stated, is of the nature of a hindrance to one side of the ship, so as to allow the other side to go forward with greater speed, and thus turning the ship; but the inclination of the rudder-post has a double effect, by which, when the rudder is held over, not only is one side of the ship hindered, but a certain quantity of the water which strikes the rudder is diverted upwards as well as to one side. Nevertheless I have no hesitation in recommending a certain amount of rake of stern-post to be given where very great power of rudder is wanted.

Next after rake of stern-post comes the question of rake of counter and rake or stern. I am in the habit of allowing the buttock-lines to decide for themselves the rake of the counter, so that when the stern is deep in the water the counter may be a continuation of the true form of the ship and of her lines. A good counter of this sort will, therefore, help the ship, instead of hindering it, especially when the stern happens to be buried in the waves. As to rake of stern, I am not sure that that is more than matter of fancy: excepting for convenience, it is better that the stern should rake outwards, than that it should tumble home; and even the Americans, who have extremely small sterns, always give them some rake.

We have not yet done with the sheer-plan. The upper bounding-plan, or that which appears to the eye to finish the ship above, is called the *sheer-line* This also has been called matter of taste merely; but some points of it have more or less reason. It is a matter of fact that, in looking at a ship the upper bounding-line of which is perfectly level, she is apt to have the appearance, contrary to the truth, of being rounded down,—that is to say, with a hump in the middle and a droop at the ends; and this is universally agreed to be so ugly, that considerable spring at the bow, and somewhat less spring at the stern, are necessary to counteract it. From experience one may say, that in a vessel 200 ft. long it requires about 2 ft. of rise at the bow and 8 in. at the stern to make her seem straight, and as matter of beauty and custom this is generally exceeded.

So much for the quantity of sheer, or spring, to be given fore and aft. The quality of it depends on the exact curve which may be adopted for the sheer. I adopt a parabola as the sheer-curve, and proceed as follows. Dividing the vessel into ten equal parts, six forward and four abaft, I rise forward successively 1, 4, 9, 16, 25, and 36 in.; and abaft, ½, 2, 4½, and 8 in. This gives a total spring of 3 ft. forward and of 8 in. aft, and makes the bow 28 in. higher out of the water than the stern. This proportion will serve very well from 200 ft. long up to 700 ft., as the larger vessel does not require an increasing spring; but for smaller vessels than 200 ft. these quantities would be in excess. Nevertheless it is to be observed that, even in very small vessels, especially when low on the water, a considerable spring forward is useful to keep them dry. I am, therefore, of opinion that there would not be much harm in extending my proportions, given above, to a considerably smaller size of vessel than I have stated. However, it is not a matter on which there is anything dogmatic to be said.

The sheer-line is important in its structure thus far, that it is usual to make the planking of the upper part of the ship, the line of ports, and the lines of the decks, follow the line of sheer: I say, *usual;* but I have found it very convenient and useful to deviate from this practice, and to make the decks follow any line that may be convenient for the internal arrangements. For example:—where I have wanted to obtain a large and roomy forecastle without deforming the vessel by raising this forecastle above the bulwarks, I have obtained it by running the line of the deck straight forward on the level, and so following the level of the water-line, instead of the sheer of the bulwark. In this way, the height of the top of the bulwark above the deck, which amidships might be 5 ft., might become 8 ft. at the stem; and I am not aware of any practical inconvenience arising from this, which is not much more than compensated by strength and usefulness.

To Construct the Chief Vertical Longitudinal Section, or Buttock-line.—In the construction of this line there is much room for judgment; for, although it can hardly be said to possess such remarkable properties of its own as the water-line and midship section, it has the power of either increasing the good qualities or aggravating the evils which the ship will derive from those two primary lines. It is only secondary in importance to these; because by its means all the possible good springing from the others may be favourably developed, marred, or neutralized. It happens also that this has not heretofore received the attention and study it deserves; in most designs it is not even to be found. In my belief, its good qualities tend materially to the ease, dryness, comfort, and safety of sea-going ships. Inland and fresh-water ships may perhaps neglect it with impunity; but a practised eye can detect in the faults of this line almost instantaneously the bad sea-going qualities of a defective design.

I place the chief buttock-line in a vertical plane parallel to the plane of the keel and the perpendiculars, or central plane of the ship, and at one-fourth of her breadth from this plane on both sides.

In ordinary ships this line will be found to be of a most variable, vague, and nondescript character. I have adopted for it the vertical line of a sea-wave, and I believe that its conformity to that shape has everything to do with the ease of the vessel at sea. The vertical section of the common sea-wave is the common cycloid. This must be elongated for a long, low vessel, and compressed for a short one. Three points through which it must pass have already been determined by the midship section, and by the water-line; because, as this line is distant from the centre one-fourth part of the breadth, it must cross those three lines where they cross this vertical plane.

These three are, however, the only points which do not admit of free choice; and it remains a part of the skill of the constructor to adopt such a cycloid as may consist with his general design and with the use of the ship. Each of the three midship sections, Figs. 824, 827, 830, which I have given, places the bottom of the buttock-line at a different depth under the water, and each of the three requires a different cycloidal line to fit it. The nature of this cycloidal line has been for a few centuries known to mathematicians and philosophers as the only line in which a pendulum can so swing that its vibrations, whatever their extent, shall be equal-timed. It happens that there is a remarkable analogy, as we have already seen, between the swing of a pendulum and the roll of a ship; there is an equally strong resemblance between the forces which exist in a wave and the forces which act on a pendulum: the mathematics of a wave and the mathematics of a cycloidal pendulum are nearly identical.

When therefore, says Russell, I discovered that the forces which replace the water in the run of a ship are of the same nature as the forces actuating a wind-wave at sea in the vertical position, I naturally found in this discovery a key to the vertical lines of the after-body of a ship; and I contrived the vertical lines of the fore-body, in the belief that wind-waves coming into collision with a body already perfectly fitted to the form which they themselves take in undulating, unresisted, free motion, would not be broken, but would have free way, and that they glide as smoothly over the face of a solid cycloid as the layers of the same wave glide over one another. On putting the question to the waves themselves, they decided that it was so, and a vertical cycloid thus became the buttock-line of the bow of an easy and dry ship above the water, just as it had already become the easy run of the wave of replacement in the stern of the ship.

The chief buttock-line, therefore, is described in the following manner:—The after-part is formed from a semicircle, the bottom of which is at the intersection of the midship section with the vertical plane, and of which the uppermost point is as high out of the water as we choose to carry the bulwark. From this we describe a cycloid, as shown in Figs. 825, 828, 831, and we cut off as much of this cycloid as we choose, to adapt the portion of the stern beyond the perpendicular, a point which is a matter of room and comfort merely. For the bow we have the choice whether it shall much overhang the water, or rise up pretty square, or tumble home. For a vessel low in the water I sometimes adopt the first; for a vessel high out of the water, never. I believe that, on the whole, a tumble-home form of bow is the dryest and easiest at sea; and there is also the vertical cycloid between the two. Each proportion and kind of vessel has its corresponding cycloid. In the case we are now considering, the vessel with a small under-water body has a flare-out bow. The fuller under-water body has a vertical buttock-line, and the fullest has a tumble-home bow. The first, Fig. 825, of these gives what is called *the clipper-bow.* The second and third, Figs. 828, 831, give what I may claim, says Russell, to call my bow. This bow has the quality of giving roundness, fulness, and capacity, in combination with the fine, hollow, unresisting entrance of the wave water-line.

On the Main-deck Line.—I call the main, or chief deck-line of a ship, the outline of that deck which is meant to be kept in all circumstances well out of the water, and in ordinary circumstances not to be swept by the waves. It is this which constitutes the chief gun-deck of a war-vessel, on which it is necessary, in all ordinary weather, that the ports should be open without the sea entering. There have been war-vessels in which this deck was generally under water; but they, after long experience, were named *coffins;* so I shall keep my definition.